Fundamental Constants

Avogadro's number (N_A)	6.0221413×10^{23}
Electron charge (e)	1.6022×10^{-19} C
Electron mass	9.109387×10^{-28} g
Faraday constant (F)	96,485.3 C/mol e^-
Gas constant (R)	0.08206 L · atm/K · mol
	8.314 J/K · mol
	62.36 L · torr/K · mol
	1.987 cal/K · mol
Planck's constant (h)	6.6256×10^{-34} J · s
Proton mass	1.672623×10^{-24} g
Neutron mass	1.674928×10^{-24} g
Speed of light in a vacuum (c)	2.99792458×10^8 m/s

Some Prefixes Used with SI Units

tera (T)	10^{12}	centi (c)	10^{-2}
giga (G)	10^9	milli (m)	10^{-3}
mega (M)	10^6	micro (μ)	10^{-6}
kilo (k)	10^3	nano (n)	10^{-9}
deci (d)	10^{-1}	pico (p)	10^{-12}

Useful Conversion Factors and Relationships

$$1 \text{ lb} = 453.6 \text{ g}$$

$$1 \text{ in} = 2.54 \text{ cm (exactly)}$$

$$1 \text{ mi} = 1.609 \text{ km}$$

$$1 \text{ km} = 0.6215 \text{ mi}$$

$$1 \text{ pm} = 1 \times 10^{-12} \text{ m} = 1 \times 10^{-10} \text{ cm}$$

$$1 \text{ atm} = 760 \text{ mmHg} = 760 \text{ torr} = 101{,}325 \text{ N/m}^2 = 101{,}325 \text{ Pa}$$

$$1 \text{ cal} = 4.184 \text{ J (exactly)}$$

$$1 \text{ L} \cdot \text{atm} = 101.325 \text{ J}$$

$$1 \text{ J} = 1 \text{ C} \times 1 \text{ V}$$

$$?°C = (°F - 32°F) \times \frac{5°C}{9°F}$$

$$?°F = \frac{9°F}{5°C} \times (°C) + 32°F$$

$$?K = (°C + 273.15°C)\left(\frac{1K}{1°C}\right)$$

Periodic Table of the Elements

Main group

Key

6
C
Carbon
12.01

Symbol

Average atomic mass

Atomic number

Name

An element

Transition metals

Period number

Group number

Period	1	2	3	4	5	6	7	8	9	10	11	12	13	14	15	16	17	18
1	1 H Hydrogen 1.008																	2 He Helium 4.003
2	3 Li Lithium 6.941	4 Be Beryllium 9.012											5 B Boron 10.81	6 C Carbon 12.01	7 N Nitrogen 14.01	8 O Oxygen 16.00	9 F Fluorine 19.00	10 Ne Neon 20.18
3	11 Na Sodium 22.99	12 Mg Magnesium 24.31											13 Al Aluminum 26.98	14 Si Silicon 28.09	15 P Phosphorus 30.97	16 S Sulfur 32.07	17 Cl Chlorine 35.45	18 Ar Argon 39.95
4	19 K Potassium 39.10	20 Ca Calcium 40.08	21 Sc Scandium 44.96	22 Ti Titanium 47.87	23 V Vanadium 50.94	24 Cr Chromium 52.00	25 Mn Manganese 54.94	26 Fe Iron 55.85	27 Co Cobalt 58.93	28 Ni Nickel 58.69	29 Cu Copper 63.55	30 Zn Zinc 65.41	31 Ga Gallium 69.72	32 Ge Germanium 72.64	33 As Arsenic 74.92	34 Se Selenium 78.96	35 Br Bromine 79.90	36 Kr Krypton 83.80
5	37 Rb Rubidium 85.47	38 Sr Strontium 87.62	39 Y Yttrium 88.91	40 Zr Zirconium 91.22	41 Nb Niobium 92.91	42 Mo Molybdenum 95.94	43 Tc Technetium (98)	44 Ru Ruthenium 101.1	45 Rh Rhodium 102.9	46 Pd Palladium 106.4	47 Ag Silver 107.9	48 Cd Cadmium 112.4	49 In Indium 114.8	50 Sn Tin 118.7	51 Sb Antimony 121.8	52 Te Tellurium 127.6	53 I Iodine 126.9	54 Xe Xenon 131.3
6	55 Cs Cesium 132.9	56 Ba Barium 137.3	57 La Lanthanum 138.9	72 Hf Hafnium 178.5	73 Ta Tantalum 180.9	74 W Tungsten 183.8	75 Re Rhenium 186.2	76 Os Osmium 190.2	77 Ir Iridium 192.2	78 Pt Platinum 195.1	79 Au Gold 197.0	80 Hg Mercury 200.6	81 Tl Thallium 204.4	82 Pb Lead 207.2	83 Bi Bismuth 209.0	84 Po Polonium (209)	85 At Astatine (210)	86 Rn Radon (222)
7	87 Fr Francium (223)	88 Ra Radium (226)	89 Ac Actinium (227)	104 Rf Rutherfordium (261)	105 Db Dubnium (262)	106 Sg Seaborgium (266)	107 Bh Bohrium (272)	108 Hs Hassium (277)	109 Mt Meitnerium (276)	110 Ds Darmstadtium (281)	111 Rg Roentgenium (280)	112 Cn Copernicium (285)	113 Nh Nihonium (285)	114 Fl Flerovium (287)	115 Mc Moscovium (289)	116 Lv Livermorium (291)	117 Ts Tennessine (293)	118 Og Oganesson (294)

Lanthanoids 6

58 Ce Cerium 140.1	59 Pr Praseodymium 140.9	60 Nd Neodymium 144.2	61 Pm Promethium (145)	62 Sm Samarium 150.4	63 Eu Europium 152.0	64 Gd Gadolinium 157.3	65 Tb Terbium 158.9	66 Dy Dysprosium 162.5	67 Ho Holmium 164.9	68 Er Erbium 167.3	69 Tm Thulium 168.9	70 Yb Ytterbium 173.0	71 Lu Lutetium 175.0

Actinoids 7

90 Th Thorium 232.0	91 Pa Protactinium 231.0	92 U Uranium 238.0	93 Np Neptunium (237)	94 Pu Plutonium (239)	95 Am Americium (243)	96 Cm Curium (247)	97 Bk Berkelium (249)	98 Cf Californium (252)	99 Es Einsteinium (252)	100 Fm Fermium (257)	101 Md Mendelevium (258)	102 No Nobelium (259)	103 Lr Lawrencium (262)

Metals

Nonmetals

Metalloids

List of the Elements with Their Symbols and Atomic Masses*

Element	Symbol	Atomic Number	Atomic Mass†	Element	Symbol	Atomic Number	Atomic Mass†
Actinium	Ac	89	(227)	Mendelevium	Md	101	(258)
Aluminum	Al	13	26.9815386	Mercury	Hg	80	200.59
Americium	Am	95	(243)	Molybdenum	Mo	42	95.94
Antimony	Sb	51	121.760	Moscovium	Mc	115	(289)
Argon	Ar	18	39.948	Neodymium	Nd	60	144.242
Arsenic	As	33	74.92160	Neon	Ne	10	20.1797
Astatine	At	85	(210)	Neptunium	Np	93	(237)
Barium	Ba	56	137.327	Nickel	Ni	28	58.6934
Berkelium	Bk	97	(247)	Niobium	Nb	41	92.90638
Beryllium	Be	4	9.012182	Nihonium	Nh	113	(285)
Bismuth	Bi	83	208.98040	Nitrogen	N	7	14.0067
Bohrium	Bh	107	(272)	Nobelium	No	102	(259)
Boron	B	5	10.811	Oganesson	Og	118	(294)
Bromine	Br	35	79.904	Osmium	Os	76	190.23
Cadmium	Cd	48	112.411	Oxygen	O	8	15.9994
Calcium	Ca	20	40.078	Palladium	Pd	46	106.42
Californium	Cf	98	(251)	Phosphorus	P	15	30.973762
Carbon	C	6	12.0107	Platinum	Pt	78	195.084
Cerium	Ce	58	140.116	Plutonium	Pu	94	(244)
Cesium	Cs	55	132.9054519	Polonium	Po	84	(209)
Chlorine	Cl	17	35.453	Potassium	K	19	39.0983
Chromium	Cr	24	51.9961	Praseodymium	Pr	59	140.90765
Cobalt	Co	27	58.933195	Promethium	Pm	61	(145)
Copernicium	Cn	112	(285)	Protactinium	Pa	91	231.03588
Copper	Cu	29	63.546	Radium	Ra	88	(226)
Curium	Cm	96	(247)	Radon	Rn	86	(222)
Darmstadtium	Ds	110	(281)	Rhenium	Re	75	186.207
Dubnium	Db	105	(268)	Rhodium	Rh	45	102.90550
Dysprosium	Dy	66	162.500	Roentgenium	Rg	111	(280)
Einsteinium	Es	99	(252)	Rubidium	Rb	37	85.4678
Erbium	Er	68	167.259	Ruthenium	Ru	44	101.07
Europium	Eu	63	151.964	Rutherfordium	Rf	104	(267)
Fermium	Fm	100	(257)	Samarium	Sm	62	150.36
Flerovium	Fl	114	(287)	Scandium	Sc	21	44.955912
Fluorine	F	9	18.9984032	Seaborgium	Sg	106	(271)
Francium	Fr	87	(223)	Selenium	Se	34	78.96
Gadolinium	Gd	64	157.25	Silicon	Si	14	28.0855
Gallium	Ga	31	69.723	Silver	Ag	47	107.8682
Germanium	Ge	32	72.64	Sodium	Na	11	22.98976928
Gold	Au	79	196.966569	Strontium	Sr	38	87.62
Hafnium	Hf	72	178.49	Sulfur	S	16	32.065
Hassium	Hs	108	(270)	Tantalum	Ta	73	180.94788
Helium	He	2	4.002602	Technetium	Tc	43	(98)
Holmium	Ho	67	164.93032	Tellurium	Te	52	127.60
Hydrogen	H	1	1.00794	Tennessine	Ts	117	(293)
Indium	In	49	114.818	Terbium	Tb	65	158.92535
Iodine	I	53	126.90447	Thallium	Tl	81	204.3833
Iridium	Ir	77	192.217	Thorium	Th	90	232.03806
Iron	Fe	26	55.845	Thulium	Tm	69	168.93421
Krypton	Kr	36	83.798	Tin	Sn	50	118.710
Lanthanum	La	57	138.90547	Titanium	Ti	22	47.867
Lawrencium	Lr	103	(262)	Tungsten	W	74	183.84
Lead	Pb	82	207.2	Uranium	U	92	238.02891
Lithium	Li	3	6.941	Vanadium	V	23	50.9415
Livermorium	Lv	116	(291)	Xenon	Xe	54	131.293
Lutetium	Lu	71	174.967	Ytterbium	Yb	70	173.04
Magnesium	Mg	12	24.3050	Yttrium	Y	39	88.90585
Manganese	Mn	25	54.938045	Zinc	Zn	30	65.409
Meitnerium	Mt	109	(276)	Zirconium	Zr	40	91.224

*These atomic masses show as many significant figures as are known for each element. The atomic masses in the periodic table are shown to four significant figures, which is sufficient for solving the problems in this book.

†Approximate values of atomic masses for radioactive elements are given in parentheses.

Introductory Chemistry

An Atoms First Approach

THIRD EDITION

Julia Burdge

COLLEGE OF WESTERN IDAHO

Michelle Driessen

UNIVERSITY OF MINNESOTA

INTRODUCTORY CHEMISTRY

Published by McGraw Hill LLC, 1325 Avenue of the Americas, New York, NY 10019. Copyright ©2024 by McGraw Hill LLC. All rights reserved. Printed in the United States of America. No part of this publication may be reproduced or distributed in any form or by any means, or stored in a database or retrieval system, without the prior written consent of McGraw Hill LLC, including, but not limited to, in any network or other electronic storage or transmission, or broadcast for distance learning.

Some ancillaries, including electronic and print components, may not be available to customers outside the United States.

This book is printed on acid-free paper.

1 2 3 4 5 6 7 8 9 LWI 28 27 26 25 24 23

ISBN 978-1-266-13707-5
MHID 1-266-13707-6

Cover Image: *Klaus Balzano/500px/Getty Images*

mheducation.com/highered

About the Authors

David Spurgeon/McGraw Hill

Julia Burdge holds a Ph.D. (1994) from The University of Idaho in Moscow, Idaho;
and a Master's Degree from The University of South Florida. Her research interests have included synthesis and characterization of cisplatin analogues, and development of new analytical techniques and instrumentation for measuring ultra-trace levels of atmospheric sulfur compounds.

She currently holds an adjunct faculty position at The College of Western Idaho in Nampa, Idaho, where she teaches general chemistry using an atoms first approach; but spent the lion's share of her academic career at The University of Akron in Akron, Ohio, as director of the Introductory Chemistry program. In addition to directing the general chemistry program and supervising the teaching activities of graduate students, Julia established a future-faculty development program and served as a mentor for graduate students and postdoctoral associates.

Julia relocated back to the Northwest to be near family. In her free time, she enjoys precious time with her three children, and with Erik Nelson, her husband and best friend.

Michelle Driessen/McGraw Hill

Michelle Driessen earned a Ph.D. in 1997 from the University of
Iowa in Iowa City, Iowa. Her research and dissertation focused on the thermal and photochemical reactions of small molecules at the surfaces of metal nanoparticles and high surface area oxides.

Following graduation, she held a tenure-track teaching and research position at Southwest Missouri State University for several years. A family move took her back to her home state of Minnesota where she held positions as adjunct faculty at both St. Cloud State University and the University of Minnesota. It was during these adjunct appointments that she became very interested in chemical education. Over the past several years she has transitioned the general chemistry laboratories at the University of Minnesota from verification to problem-based, and has developed both online and hybrid sections of general chemistry lecture courses. She is currently the Director of General Chemistry at the University of Minnesota where she runs the general chemistry laboratories, trains and supervises teaching assistants, and continues to experiment with active learning methods in her classroom.

Michelle and her husband love the outdoors and their rural roots. They take every opportunity to visit their family, farm, and horses in rural Minnesota.

To the people who will always matter the most: Katie, Beau, and Sam.
—Julia Burdge

To my family, the center of my universe and happiness, with special thanks
to my husband for his support and making me the person I am today.
—Michelle Driessen

Brief Contents

1 Atoms and Elements 2

2 Electrons and the Periodic Table 32

3 Compounds and Chemical Bonds 76

4 How Chemists Use Numbers 124

5 The Mole and Chemical Formulas 168

6 Molecular Shape 200

7 Solids, Liquids, and Phase Changes 242

8 Gases 276

9 Physical Properties of Solutions 316

10 Chemical Reactions and Chemical Equations 354

11 Using Balanced Chemical Equations 392

12 Acids and Bases 426

13 Equilibrium 464

14 Organic Chemistry 490

15 Biochemistry 516

16 Nuclear Chemistry 532

17 Electrochemistry 548

Appendix   Mathematical Operations   A-1

Glossary   G-1

Answers to Odd-Numbered Problems   AP-1

Index   I-1

Contents

Preface   xviii

1   ATOMS AND ELEMENTS 2

1.1   The Study of Chemistry 3
• Why Learn Chemistry? 3
• The Scientific Method 3

1.2   Atoms First 5

1.3   Subatomic Particles and the Nuclear Model of the Atom 6
■ Marie Skłodowska Curie 10

1.4   Elements and the Periodic Table 10
■ Elements in the Human Body 12
■ Helium 14

1.5   Organization of the Periodic Table 14
■ Elements in Earth's Crust 15

1.6   Isotopes 17
■ Mass Spectrometry 18

1.7   Atomic Mass 19
■ Iron-Fortified Cereal 21
■ Helen Mackay 21

rozbyshaka/iStock/Getty Images

2   ELECTRONS AND THE PERIODIC TABLE 32

2.1   The Nature of Light 33
■ Laser Pointers 35
■ Patricia Bath 36

2.2   The Bohr Atom 36

Visualizing Chemistry – Bohr Atom 38

■ Fireworks 40
■ The Photoelectric Effect 41

2.3   Atomic Orbitals 42
• *s* orbitals 45 • *p* orbitals 45
• *d* and *f* orbitals 46

2.4   Electron Configurations 48

2.5   Electron Configurations and the Periodic Table 53

2.6   Periodic Trends 57

2.7   Ions: The Loss and Gain of Electrons 63
• Electron Configuration of Ions 63
• Lewis Dot Symbols of Ions 65

David A. Tietz/McGraw Hill

3 COMPOUNDS AND CHEMICAL BONDS 76

3.1 Matter: Classification and Properties 77
- States of Matter 77 • Mixtures 78
- Properties of Matter 80

3.2 Ionic Bonding and Binary Ionic Compounds 83

3.3 Naming Ions and Binary Ionic Compounds 87
- Naming Atomic Cations 88
- Naming Atomic Anions 89
- Naming Binary Ionic Compounds 89

3.4 Covalent Bonding and Molecules 91
- Covalent Bonding 92 • Molecules 92
- Molecular Formulas 95
- ■ Fixed Nitrogen in Fertilizers 98
- ■ George Washington Carver 99

3.5 Naming Binary Molecular Compounds 99

3.6 Covalent Bonding in Ionic Species: Polyatomic Ions 101
- ■ Product Labels 102
- ■ Product Labels 103
- ■ Hydrates 106

3.7 Acids 107

3.8 Substances in Review 109

Visualizing Chemistry – Properties of Atoms 110

- Distinguishing Elements and Compounds 112
- Determining Whether a Compound Is Ionic or Molecular 113
- Naming Compounds 113

EpicStockMedia/Shutterstock

4 HOW CHEMISTS USE NUMBERS 124

4.1 Units of Measurement 125
- Base Units 125 • Mass, Length, and Time 126
- Metric Multipliers 126
- ■ Henrietta Swan Leavitt 127
- Temperature 130
- ■ The Fahrenheit Temperature Scale 131

4.2 Scientific Notation 134
- Very Large Numbers 135 • Very Small Numbers 136 • Using the Scientific Notation Function on Your Calculator 137

4.3 Significant Figures 139
- Exact Numbers 139 • Measured Numbers 139
- ■ Arthur Rosenfeld 143
- Calculations with Measured Numbers 144

David Clapp/Stone/Getty Images

4.4 Unit Conversion 148
- Conversion Factors 148
- ■ The Importance of Units 151
- Derived Units 152
- ■ The International Unit 154
- Dimensional Analysis 155

4.5 Success in Introductory Chemistry Class 157

5 THE MOLE AND CHEMICAL FORMULAS 168

5.1 Counting Atoms by Weighing 169
- The Mole (The "Chemist's Dozen") 169
- Molar Mass 171
- Interconverting Mass, Moles, and Numbers of Atoms 173

5.2 Counting Molecules by Weighing 175
- Calculating the Molar Mass of a Compound 175 • Interconverting Mass, Moles, and Numbers of Molecules (or Formula Units) 177 • Combining Multiple Conversions in a Single Calculation 179
- ■ Redefining the Kilogram 181
- ■ Derek Muller 182

5.3 Mass Percent Composition 182
- ■ Iodized Salt 184

5.4 Using Mass Percent Composition to Determine Empirical Formula 185
- ■ Fertilizer & Mass Percents 187

5.5 Using Empirical Formula and Molar Mass to Determine Molecular Formula 188

JULIAN STRATENSCHULTE/epa european pressphoto agency b.v./Alamy Stock Photo

6 MOLECULAR SHAPE 200

6.1 Drawing Simple Lewis Structures 201
- Lewis Structures of Simple Molecules 201
- Lewis Structures of Molecules with a Central Atom 203 • Lewis Structures of Simple Polyatomic Ions 203

6.2 Lewis Structures Continued 206
- Lewis Structures with Less Obvious Skeletal Structures 206 • Lewis Structures with Multiple Bonds 207 • Exceptions to the Octet Rule 208
- ■ Bleaching, Disinfecting, and Decontamination 209

Robin Treadwell/Science Source

6.3 Resonance Structures 209

6.4 Molecular Shape 211

■ Flavor, Molecular Shape, and Bond-Line Structures 213

• Bond Angles 216

■ Molecular Shapes Resulting from Expanded Octets 217

6.5 Electronegativity and Polarity 219

• Electronegativity 219 • Bond Polarity 221

• Molecular Polarity 223

■ How Bond Dipoles Sum to Determine Molecular Polarity 225

6.6 Intermolecular Forces 226

• Dipole-Dipole Forces 226 • Hydrogen Bonding 227

• Dispersion Forces 228

■ Linus Pauling 231

• Intermolecular Forces in Review 232

7 SOLIDS, LIQUIDS, AND PHASE CHANGES 242

7.1 General Properties of the Condensed Phases 243

7.2 Types of Solids 244

• Ionic Solids 244 • Molecular Solids 244

• Atomic Solids 246 • Network Solids 247

■ A Network Solid as Hard as Diamond 248

■ Carol V. Robinson 250

7.3 Physical Properties of Solids 251

• Vapor Pressure 251 • Melting Point 252

7.4 Physical Properties of Liquids 255

• Viscosity 255 • Surface Tension 255

■ Surface Tension and the Shape of Water Drops 256

• Vapor Pressure 257 • Boiling Point 258

■ High Altitude and High-Pressure Cooking 260

7.5 Energy and Physical Changes 261

• Temperature Changes 261

• Solid-Liquid Phase Changes: Melting and Freezing 263

• Liquid-Gas Phase Changes: Vaporization and Condensation 264

• Solid-Gas Phase Changes: Sublimation 265

Tony Campbell/Shutterstock

8 GASES 276

8.1 Properties of Gases 277
- Gaseous Substances 278
- Kinetic Molecular Theory of Gases 279

8.2 Pressure 280
- Definition and Units of Pressure 280
- Measurement of Pressure 283
- ■ Fritz Haber 284

8.3 The Gas Equations 285
- The Ideal Gas Equation 285
- ■ Pressure Exerted by a Column of Fluid 289
- The Combined Gas Equation 289
- The Molar Mass Gas Equation 290

8.4 The Gas Laws 293
- Boyle's Law: The Pressure-Volume Relationship 293
- Charles's Law: The Temperature-Volume Relationship 295
- ■ Automobile Airbags and Charles's Law 298
- Avogadro's Law: The Moles-Volume Relationship 298
- ■ Amanda Theodosia Jones 298

8.5 Gas Mixtures 301
- Dalton's Law of Partial Pressures 301
- Mole Fractions 302
- ■ Natural Gas 304

Eric Delmar/Vetta/Getty Images

9 PHYSICAL PROPERTIES OF SOLUTIONS 316

9.1 General Properties of Solutions 317
- ■ Honey – A Supersaturated Solution 318
- ■ Instant Hot Packs 319

9.2 Aqueous Solubility 319
- ■ Alice Ball 321

9.3 Solution Concentration 322
- Percent by Mass 322
- ■ Trace Concentrations 323
- Molarity 325 • Molality 327
- Comparison of Concentration Units 327

9.4 Solution Composition 330
- ■ Robert Cade, M.D. 332

9.5 Solution Preparation 334
- Preparation of a Solution from a Solid 334 • Preparation of a More Dilute Solution from a Concentrated Solution 335

Brian Rayburn/McGraw Hill

Visualizing Chemistry – Preparing a Solution from a Solid 336

■ Serial Dilution 338

9.6 Colligative Properties 340

• Freezing-Point Depression 340 • Boiling-Point Elevation 341

■ Ice Melters 342

• Osmotic Pressure 342

10 CHEMICAL REACTIONS AND CHEMICAL EQUATIONS 354

10.1 Recognizing Chemical Reactions 355

10.2 Representing Chemical Reactions with Chemical Equations 358

• Metals 359 • Nonmetals 359

• Noble Gases 359 • Metalloids 359

10.3 Balancing Chemical Equations 360

■ The Stoichiometry of Metabolism 364

10.4 Types of Chemical Reactions 365

• Precipitation Reactions 365

• Acid-Base Reactions 370

■ Oxygen Generators 371

• Oxidation-Reduction Reactions 373

■ Antoine Lavoisier 378

■ Dental Pain and Redox 380

10.5 Chemical Reactions and Energy 382

10.6 Chemical Reactions in Review 382

Lindsay Upson/Image Source/Getty Images

11 USING BALANCED CHEMICAL EQUATIONS 392

11.1 Mole to Mole Conversions 393

11.2 Mass to Mass Conversions 395

■ Marie-Anne Paulze Lavoisier 397

11.3 Limitations on Reaction Yield 398

• Limiting Reactant 398 • Percent Yield 401

■ Combustion Analysis 403

■ Alka-Seltzer 404

11.4 Aqueous Reactions 406

11.5 Gases in Chemical Reactions 411

• Predicting the Volume of a Gaseous Product 411 • Calculating the Required Volume of a Gaseous Reactant 412

■ Joseph Louis Gay-Lussac 414

11.6 Chemical Reactions and Heat 415

Piotr290/iStock/Getty Images

12 ACIDS AND BASES 426

12.1 Properties of Acids and Bases 427
- ■ James Lind 428

12.2 Definitions of Acids and Bases 429
- • Arrhenius Acids and Bases 429
- • Brønsted Acids and Bases 429
- • Conjugate Acid-Base Pairs 430

12.3 Water as an Acid; Water as a Base 432

12.4 Strong Acids and Bases 434

12.5 pH and pOH Scales 437
- ■ Commonly Encountered Acids and Bases 444
- ■ Lake Natron 445

12.6 Weak Acids and Bases 445
- ■ St. Elmo Brady 447

12.7 Acid-Base Titrations 450
- ■ Using Millimoles to Simplify Titration Calculations 452

12.8 Buffers 452

Aflo Co., Ltd./Alamy Stock Photo

13 EQUILIBRIUM 464

13.1 Reaction Rates 465

Visualizing Chemistry – Collision Theory 468

13.2 Chemical Equilibrium 470
- ■ How Do We Know That the Forward and Reverse Processes Are Ongoing in a System at Equilibrium? 472

13.3 Equilibrium Constants 472
- ■ Sweet Tea 473
- • Calculating Equilibrium Constants 473
- • Magnitude of the Equilibrium Constant 476

13.4 Factors That Affect Equilibrium 477
- ■ Hemoglobin Production at High Altitude 477
- • Addition or Removal of a Substance 478
- • Changes in Volume 480 • Changes in Temperature 481
- ■ Ruth Erica and Reinhold Benesch 484

SDI Productions/E+/Getty Images

14 ORGANIC CHEMISTRY 490

14.1 Why Carbon Is Different 491
14.2 Hydrocarbons 492
• Alkanes 493 • Alkenes and Alkynes 493
• Reactions of Hydrocarbons 495
14.3 Isomers 496
  ■ Partially Hydrogenated Vegetable
Oils 497
  ■ Representing Organic Molecules with
  Bond-Line Structures 499
14.4 Functional Groups 500
14.5 Alcohols and Ethers 501
14.6 Aldehydes and Ketones 503
  ■ Percy Lavon Julian 504
14.7 Carboxylic Acids and Esters 505
14.8 Amines and Amides 506
14.9 Polymers 508

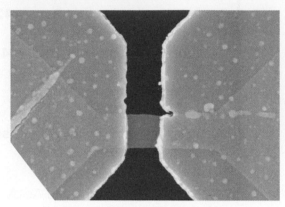

Andre Geim & Kostya Novoselov/Science Source

15 BIOCHEMISTRY 516

15.1 Biologically Important Molecules 517
• Glycerol 517 • Fatty Acids 517
• Amino Acids 517
  ■ Marie Maynard Daly 518
• Sugars 519
• Phosphates 519
• Organic Bases 519
15.2 Lipids 520
• Fats 520
• Phospholipids 521
• Steroids 522
15.3 Proteins 522
• Primary Structure 525
• Secondary Structure 525
• Tertiary Structure 525
• Quaternary Structure 526
15.4 Carbohydrates 526
• Monosaccharides 526
• Disaccharides 526
• Polysaccharides 527
15.5 Nucleic Acids 528
  ■ Rosalind Franklin 529

hlansdown/iStock/Getty Images

16 NUCLEAR CHEMISTRY 532

16.1 Radioactive Decay 533

16.2 Detection of Radiation and Its Biological Effects 536
- ■ Radioactivity in Tobacco 538

16.3 Dating Using Radioactive Decay 538

16.4 Medical Applications of Radioactivity 540
- ■ How Nuclear Chemistry Is Used to Treat Cancer 541

16.5 Nuclear Fission and Nuclear Fusion 541

Visualizing Chemistry – Nuclear Fission and Fusion 542

- ■ Lise Meitner 544

Andrey Gorulko/iStockphoto/Getty Images

17 ELECTROCHEMISTRY 548

17.1 Balancing Oxidation-Reduction Reactions Using the Half-Reaction Method 549

17.2 Batteries 553

Visualizing Chemistry – Construction of a Galvanic Cell 554

- • Dry Cells and Alkaline Batteries 557
- • Lead Storage Batteries 558
- • Lithium-Ion Batteries 559
- • Fuel Cells 559

17.3 Corrosion 560

17.4 Electrolysis 562
- • Electrolysis of Molten Sodium Chloride 562
- • Electrolysis of Water 562

Science Photo Library/Alamy Stock Photo

Appendix: Mathematical Operations A-1

Glossary G-1

Answers to Odd-Numbered Problems AP-1

Index I-1

Preface

Introductory Chemistry: An Atoms First Approach by Julia Burdge and Michelle Driessen has been developed and written using an atoms first approach *specific* to introductory chemistry. It is a carefully crafted text, designed and written with the introductory-chemistry student in mind.

The arrangement of topics facilitates the conceptual development of chemistry for the novice, rather than the historical development that has been used traditionally. Its language and style are student friendly and conversational; and the importance and wonder of chemistry in everyday life are emphasized at every opportunity. Continuing in the Burdge tradition, this text employs an outstanding art program, a consistent problem-solving approach, interesting applications woven throughout the chapters, and a wide range of end-of-chapter problems.

Features

- **Logical atoms first approach,** building first an understanding of atomic structure, followed by a logical progression of atomic properties, periodic trends, and how compounds arise as a consequence of atomic properties. Following that, physical and chemical properties of compounds and chemical reactions are covered—built upon a solid foundation of how all such properties and processes are the consequence of the nature and behavior of atoms.
- **Engaging real-life examples and applications.** Each chapter contains relevant, interesting stories in Familiar Chemistry segments that illustrate the importance of chemistry to other fields of study, and how the current material applies to everyday life. Many chapters also contain brief historical profiles of a diverse group of important people in chemistry and other fields of scientific endeavor.
- **Consistent problem-solving skill development.** Fostering a consistent approach to problem solving helps students learn how to approach, analyze, and solve problems. Each worked example (Sample Problem) is divided into logical steps: Strategy, Setup, Solution, and Think About It; and each is followed by three practice problems. Practice Problem A allows the student to solve a problem similar to the Sample Problem, using the same strategy and steps. Wherever possible, Practice Problem B probes understanding of the same concept(s) as the Sample Problem and Practice Problem A, but is sufficiently different that it requires a slightly different approach. Practice Problem C often uses concept art or molecular models, and probes comprehension of underlying concepts. The consistent use of this approach gives students the best chance for developing a robust set of problem-solving skills.
- **Outstanding pedagogy for student learning.** The Checkpoints and Student Notes throughout each chapter are designed to foster frequent self-assessment and to provide timely information regarding common pitfalls, reminders of important information, and alternative approaches. Rewind and Fast Forward links help to illustrate and reinforce

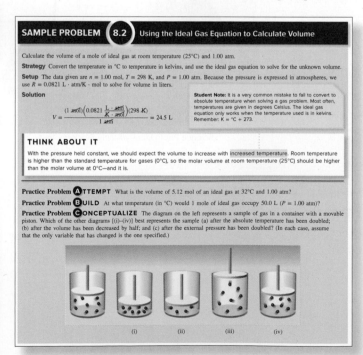

SAMPLE PROBLEM **8.2** Using the Ideal Gas Equation to Calculate Volume

Calculate the volume of a mole of ideal gas at room temperature (25°C) and 1.00 atm.

Strategy Convert the temperature in °C to temperature in kelvins, and use the ideal gas equation to solve for the unknown volume.

Setup The data given are $n = 1.00$ mol, $T = 298$ K, and $P = 1.00$ atm. Because the pressure is expressed in atmospheres, we use $R = 0.0821$ L · atm/K · mol to solve for volume in liters.

Solution

> **Student Note:** It is a very common mistake to fail to convert to absolute temperature when solving a gas problem. Most often, temperatures are given in degrees Celsius. The ideal gas equation only works when the temperature used is in kelvins. Remember: K = °C + 273.

$$V = \frac{(1 \text{ mol})\left(0.0821 \frac{L \cdot atm}{K \cdot mol}\right)(298 \text{ K})}{1 \text{ atm}} = 24.5 \text{ L}$$

THINK ABOUT IT

With the pressure held constant, we should expect the volume to increase with increased temperature. Room temperature is higher than the standard temperature for gases (0°C), so the molar volume at room temperature (25°C) should be higher than the molar volume at 0°C—and it is.

Practice Problem **A**TTEMPT What is the volume of 5.12 mol of an ideal gas at 32°C and 1.00 atm?

Practice Problem **B**UILD At what temperature (in °C) would 1 mole of ideal gas occupy 50.0 L ($P = 1.00$ atm)?

Practice Problem **C**ONCEPTUALIZE The diagram on the left represents a sample of gas in a container with a movable piston. Which of the other diagrams [(i)–(iv)] best represents the sample (a) after the absolute temperature has been doubled; (b) after the volume has been decreased by half; and (c) after the external pressure has been doubled? (In each case, assume that the only variable that has changed is the one specified.)

(i) (ii) (iii) (iv)

connections between material in different chapters, and enable students to find pertinent review material easily, when necessary.

- **Key Skills pages** are reviews of specific skills that the authors know will be important to students' understanding of later chapters. These go beyond simple reviews and actually preview the importance of the skills in later chapters. They are additional opportunities for self-assessment and are meant to be revisited when the specific skills are required later in the book.

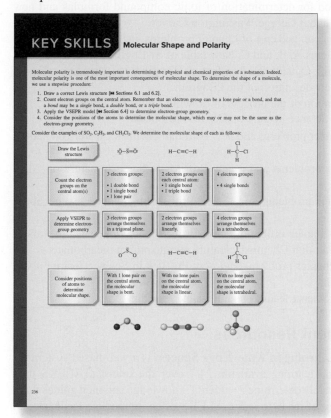

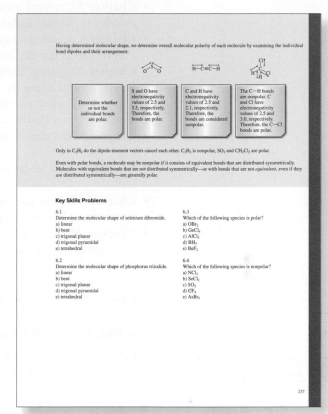

A Student-Focused Revision

For the third edition, real student data points and input were used to guide the revision. The following "New to This Edition" summary lists the more major additions and refinements.

Reflecting the Diverse World Around

McGraw Hill believes in unlocking the potential of every learner at every stage of life. To accomplish that, we are dedicated to creating products that reflect, and are accessible to, all the diverse, global customers we serve. Within McGraw Hill, we foster a culture of belonging, and we work with partners who share our commitment to equity, inclusion, and diversity in all forms. In McGraw Hill Higher Education, this includes, but is not limited to, the following:

- Refreshing and implementing inclusive content guidelines around topics including generalizations and stereotypes, gender, abilities/disabilities, race/ethnicity, sexual orientation, diversity of names, and age.
- Enhancing best practices in assessment creation to eliminate cultural, cognitive, and affective bias.
- Maintaining and continually updating a robust photo library of diverse images that reflect our student populations.
- Including more diverse voices in the development and review of our content.
- Strengthening art guidelines to improve accessibility by ensuring meaningful text and images are distinguishable and perceivable by users with limited color vision and moderately low vision.

New to This Edition

All periodic tables have been updated and specific links to Key Skills pages have been added to the "Before You Begin" section in many chapters.

- **Chapter 1** New and updated end-of-chapter problems and two new Profiles in Science features about Helen Mackay and Marie Curie.
- **Chapter 2** New Profiles in Science feature about Patricia Bath.
- **Chapter 3** Clarified method for determination of ionic compound formulas and new Profiles in Science feature about George Washington Carver.
- **Chapter 4** Additional in-chapter and end-of-chapter problems, including a new Sample Problem and Practice Problems A, B, and C. New table of SI-English unit conversions.
- **Chapter 5** Updated Thinking Outside the Box feature describes the changes in fundamental constants and the new definition of the kilogram.
- **Chapter 6** Further clarification of the process of drawing Lewis structures.
- **Chapter 7** New Profiles in Science Feature on Carol V. Robinson.
- **Chapter 8** New end-of-chapter problems and new Familiar Chemistry feature on natural gas.
- **Chapter 9** New in-chapter problems and a new Profiles in Science feature on Alice Ball.
- **Chapter 11** New in-chapter problems and a new Profiles in Science feature about Marie-Anne Paulze Lavoisier.
- **Chapter 12** New end-of-chapter problems, new Profiles in Science feature about St. Elmo Brady, and new Familiar Chemistry feature about sodium bicarbonate.
- **Chapter 13** New end-of-chapter problems and new Profiles in Science feature about Ruth and Reinhold Benesch.
- **Chapter 15** New Profiles in Science feature about Rosalind Franklin.

Instructor and Student Resources

ALEKS (Assessment and Learning in Knowledge Spaces) is a web-based system for individualized assessment and learning available 24/7 over the Internet. ALEKS uses artificial intelligence to accurately determine a students' knowledge and then guides them to the material that they are most ready to learn. ALEKS offers immediate feedback and access to ALEKSPedia—an interactive text that contains concise entries on chemistry topics. ALEKS is also a full-featured course management system with rich reporting features that allow instructors to monitor individual and class performance, set student goals, assign/grade online quizzes, and more. ALEKS allows instructors to spend more time on concepts while ALEKS teaches students practical problem-solving skills. And with ALEKS 360, your student also has access to this text's eBook. Learn more at **www.aleks.com/highered/science**

McGraw Hill Virtual Labs is a must-see, outcomes-based lab simulation. It assesses a student's knowledge and adaptively corrects deficiencies, allowing the student to learn faster and retain more knowledge with greater success. First, a student's knowledge is adaptively leveled on core learning outcomes: Questioning reveals knowledge deficiencies that are corrected by the delivery of content that is conditional on a student's response. Then, a simulated lab experience requires the student to think and act like a scientist: recording, interpreting, and analyzing data using simulated equipment found in labs and clinics. The student is allowed to make mistakes—a powerful part of the learning experience! A virtual coach provides subtle hints when needed, asks questions about the student's choices, and allows the student to reflect on and correct those mistakes. Whether your need is to overcome the logistical challenges of a traditional lab, provide better lab prep, improve student performance, or make your online experience one that rivals the real world, McGraw Hill Virtual Labs accomplishes it all.

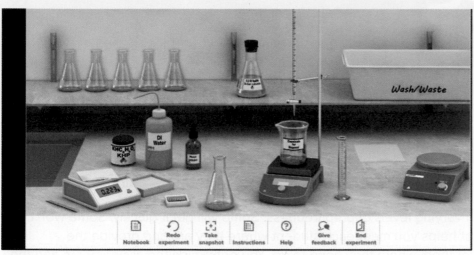

McGraw Hill

Instructors have access to the following instructor resources:

- A complete Instructor's Solutions Manual that includes solutions to all of the end-of-chapter problems
- Accessible lecture PowerPoint slides that facilitate classroom discussion of the concepts in the text
- Textbook images for repurposing in your personalized classroom materials
- Clicker questions for each chapter
- A comprehensive bank of assignable test questions. The testing software TestGen® can be used to create customized exams quickly. Hundreds of text-specific, open-ended, and multiple-choice questions are included in the question bank.

Students can purchase a Student Solutions Manual that contains detailed solutions and explanations for the odd-numbered problems in the text.

Create
Your Book, Your Way

McGraw Hill's Content Collections Powered by Create® is a self-service website that enables instructors to create custom course materials—print and eBooks—by drawing upon McGraw Hill's comprehensive, cross-disciplinary content. Choose what you want from our high-quality textbooks, articles, and cases. Combine it with your own content quickly and easily, and tap into other rights-secured, third-party content such as readings, cases, and articles. Content can be arranged in a way that makes the most sense for your course and you can include the course name and information as well. Choose the best format for your course: color print, black-and-white print, or eBook. The eBook can be included in your Connect course and is available on the free ReadAnywhere app for smartphone or tablet access as well. When you are finished customizing, you will receive a free digital copy to review in just minutes! Visit McGraw Hill Create®—www.mcgrawhillcreate.com—today and begin building!

Students start your course with varying levels of preparedness. Some will get it quickly. Some won't. ALEKS is a course assistant that helps you meet each student where they are and provide the necessary building blocks to get them where they need to go. You determine the assignments and the content, and ALEKS will deliver customized practice until they truly get it.

Experience The ALEKS Difference

Easily Identify Knowledge Gaps

Gain greater visibility into student performance so you know immediately if your lessons clicked.

- **ALEKS's "Initial Knowledge Check"** helps accurately evaluate student levels and gaps on day one, so you know precisely where students are at and where they need to go when they start your course.

- **You know when students are at risk of falling behind** through ALEKS Insights so, you can remediate – be it through prep modules, practice questions, or written explanations of video tutorials.

- **Students always know where they are**, how they are doing, and can track their own progress easily.

Gain More Flexibility and Engagement

Teach your course your way, with best-in-class content and tools to immerse students and keep them on track.

- **ALEKS gives you flexibility** to assign homework, share a vast library of curated content including videos, review progress and provide student support, anytime anywhere.

- **You save time** otherwise spent performing tedious tasks while having more control over and impact on your students' learning process.

- **Students gain a deeper level of understanding** through interactive and hands-on assignments that go beyond multiple-choice questions.

with ALEKS® Constructive Learning Paths.

Narrow the Equity Gap

Efficiently and effectively create individual pathways for students without leaving anyone behind.

- **ALEKS creates an equitable experience for all** students, making sure they get the support they need to successfully finish the courses they start.

- **You help reduce attrition,** falling enrollment, and further widening of the learning gap.

- **Student success rates improve** – not just better grades, but better learning.

Count on Hands-on Support

A dedicated Implementation Manager will work with you to build your course exactly the way you want it and your students need it.

- **An ALEKS Implementation Manager** is with you every step of the way through the life of your course.

- **You never have to figure it out on your own** or be your student's customer service. We believe in a consultative approach and take care of all of that for you, so you can focus on your class.

- **Your students benefit** from more meaningful in moments with you, while ALEKS—directed by you—does the rest.

Already benefitting from ALEKS?

Check out our New Enhancements:
mheducation.com/highered/aleks/new-releases.html

Acknowledgments

We wish to thank the many people who have contributed to the development of this new text. The following individuals reviewed the text and provided invaluable feedback.

Angela Bartholomay, Dakota College at Bottineau

Michele Berkey, San Juan College

Sridhar Budhi, Indiana Institute of Technology

Kristin Clark, Ventura College

Janina Eads, Northeast Oklahoma A&M College

Michael Fuertes, Monroe County Community College

Tammy Gummersheimer, SUNY Schenectady County Community College

John Andrew Hitron, Midway University

Janet Johannessen, County College of Morris

Saheba Khurana, Merced College

Bipin Pandey, Pensacola State College

Gloria Pimienta, Big Sandy Community and Technical College

Emily Rowland, The University of Mississippi

Elizabeth Sutton, Campbellsville University

Deborah Williams, Lone Star College-University Park

Additionally, we wish to thank our incredible team: Managing Director Kathleen McMahon, Senior Marketing Manager Cassie Cloutier, Senior Product Developer Mary Hurley, Senior Content Project Manager Melissa Leick, and Accuracy Checker John Murdzek.

Julia Burdge and Michelle Driessen

Introductory Chemistry

An Atoms First Approach

CHAPTER 1

Atoms and Elements

1.1 **The Study of Chemistry**
- Why Learn Chemistry?
- The Scientific Method

1.2 **Atoms First**

1.3 **Subatomic Particles and the Nuclear Model of the Atom**

1.4 **Elements and the Periodic Table**

1.5 **Organization of the Periodic Table**

1.6 **Isotopes**

1.7 **Atomic Mass**

The brilliant colors of a fireworks display result from the properties of the atoms they contain. These atoms give off specific colors when they are burned.
rozbyshaka/iStock/Getty Images

In This Chapter, You Will Learn

Some of what *chemistry* is and how it is studied using the *scientific method*. You will learn about *atomic structure* and you will become acquainted with the *periodic table,* how it is organized, and some of the information it embodies.

Things To Review Before You Begin

- Basic algebra

Have you ever wondered how an automobile airbag works? Or why iron rusts when exposed to water and air, but gold does not? Or why cookies "rise" as they bake? Or what causes the brilliant colors of fireworks displays? These phenomena, and countless others, can be explained by an understanding of the fundamental principles of *chemistry.* Whether or not we realize it, chemistry is important in every aspect of our lives. In the course of this book, you will come to understand the chemical principles responsible for many familiar observations and experiences.

1.1 The Study of Chemistry

Chemistry is the study of *matter* and the changes that matter undergoes. ***Matter,*** in turn, is anything that has mass and occupies space. ***Mass*** is one of the ways that scientists measure the *amount* of matter.

You may already be familiar with some of the terms used in chemistry—even if you have never taken a chemistry class. You have probably heard of *molecules;* and even if you don't know exactly what a *chemical formula* is, you undoubtedly know that "H_2O" is water. You may have used or at least heard the term *chemical reaction;* and you are certainly familiar with many processes that *are* chemical reactions.

Why Learn Chemistry?

Chances are good that you are using this book for a chemistry class you are required to take—even though you may not be a chemistry major. Chemistry is a required part of many degree programs because of its importance in a wide variety of scientific disciplines. It sometimes is called the "central science" because knowledge of chemistry supports the understanding of other scientific fields—including physics, biology, geology, ecology, oceanography, climatology, and medicine. Whether this is the first in a series of chemistry classes you will take or the only chemistry class you will ever take, we hope that it will help you to appreciate the beauty of chemistry—and to understand its importance in our daily lives.

The Scientific Method

Scientific experiments are the key to advancing our understanding of chemistry or any science. Although different scientists may take different approaches to experimentation, we all follow a set of guidelines known as the ***scientific method.*** This helps ensure the quality and integrity of new findings that are added to the body of knowledge within a given field.

The scientific method starts with the collection of data from careful observations and/or experiments. Scientists study the data and try to identify patterns. When a pattern is found, an attempt is made to describe it with a scientific **law.** In this context, a *law* is simply a concise statement of the observed pattern. Scientists may then formulate a **hypothesis,** an attempt to explain their observations. Experiments are then designed to *test* the hypothesis. If the experiments reveal that the hypothesis is incorrect, the scientists must go back to the drawing board and come up with a different interpretation of their data, and formulate a *new* hypothesis. The new hypothesis will then be tested by experiment. When a hypothesis stands the test of extensive experimentation, it may evolve into a *scientific* **theory** or **model.** A *theory* or *model* is a unifying principle that explains a body of experimental observations and the law or laws that are based on them. Theories are used both to explain past observations and to *predict* future observations. When a theory fails to predict correctly, it must be discarded or modified to become consistent with experimental observations. Thus, by their very nature, scientific theories must be subject to change in the face of new data that do not support them.

One of the most compelling examples of the scientific method is the development of the vaccine for *smallpox,* a viral disease responsible for an estimated half a *billion* deaths during the twentieth century alone. Late in the eighteenth century, English physician Edward Jenner observed that even during smallpox outbreaks in Europe, a particular group of people, *milkmaids,* seemed not to contract it.

Law: Milkmaids are not vulnerable to the virus that causes smallpox.

Based on his observations, Jenner proposed that perhaps milkmaids, who often contracted *cowpox,* a similar but far less deadly virus, from the cows they worked with, had developed a natural immunity to smallpox.

Hypothesis: Exposure to the cowpox virus causes the development of immunity to the smallpox virus.

Jenner tested his hypothesis by injecting a healthy child with the cowpox virus—and later with the smallpox virus. If his hypothesis were correct, the child would not contract smallpox—and in fact the child did *not* contract smallpox.

Theory: Because the child did not develop smallpox, immunity seemed to have resulted from exposure to cowpox.

Further experiments on many more people (mostly children and prisoners) confirmed that exposure to the cowpox virus imparted immunity to the smallpox virus.

The flowchart in Figure 1.1 illustrates the scientific method and how it guided the development of the smallpox vaccine.

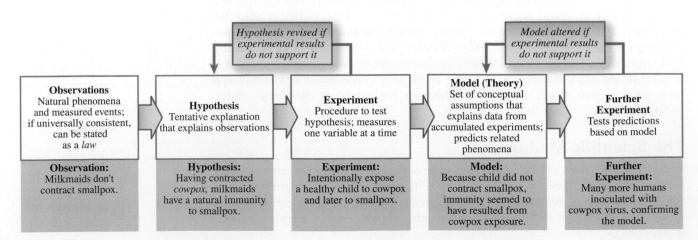

Figure 1.1 Flowchart of the scientific method and its importance to Edward Jenner's development of the smallpox vaccine.

1.2 Atoms First

Even if you have never studied chemistry before, you probably know already that atoms are the extraordinarily small building blocks that make up all matter. Specifically, an **atom** is the smallest quantity of matter that still retains the properties of matter. Further, an **element** is a substance that cannot be broken down into *simpler* substances by any means. Common examples of elements include aluminum, which we all have in our kitchens in the form of foil; carbon, which exists in several different familiar forms— including diamond and graphite (pencil "lead"); and helium, which can be used to fill balloons. The element aluminum consists entirely of *aluminum* atoms; the element carbon consists entirely of *carbon* atoms; and the element helium consists entirely of *helium* atoms. Although we can separate a sample of any element into smaller *samples* of that element, we cannot separate it into other substances.

Student Note: By contrast, consider a sample of salt water. We could divide it into smaller samples of salt water; but given the necessary equipment, we could also separate it into two different substances: water and salt. An element is different in that it is not made up of other substances. Elements are the *simplest* substances.

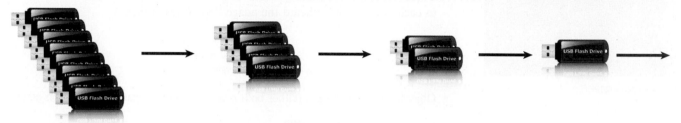

Figure 1.2 Repeatedly dividing this collection of flash drives into smaller and smaller collections eventually leaves us with a flash drive, which we cannot divide further without destroying it.
Sergii Korolko/RealVector/Shutterstock

Let's consider the example of helium. If we were to divide the helium in a balloon in half, and then divide one of the halves in half, and so on, we would eventually (after a very large number of these hypothetical divisions) be left with a sample of helium consisting of just one helium atom. This atom could not be further divided to give two smaller samples of helium. If this is difficult to imagine, think of a collection of eight identical flash drives. We could divide the collection in half three times before we were left with a single flash drive. Although we *could* divide the last flash drive in half, neither of the resulting pieces would be a flash drive (Figure 1.2).

The notion that matter consists of tiny, indivisible pieces has been around for a very long time, first having been proposed by the philosopher Democritus in the fifth century B.C. But it was first formalized early in the nineteenth century by John Dalton (Figure 1.3). Dalton devised a theory to explain some of the most important observations made by scientists in the eighteenth century. His theory included three statements, the first of which is:

- Matter is composed of tiny, indivisible particles called atoms; all atoms of a given element are identical; and atoms of one element are different from atoms of any other element.

We will revisit this statement later in this chapter and introduce the second and third statements to complete our understanding of Dalton's theory in Chapters 3 and 10.

We know now that atoms, although very small, are not indivisible. Rather, they are made up of still smaller *subatomic* particles. The type, number, and arrangement of subatomic particles determine the properties of atoms, which in turn determine the properties of everything we see, touch, smell, and taste.

Our goal in this book will be to understand how the nature of atoms gives rise to the properties of everything material. To accomplish this, we will take a somewhat unconventional approach. Rather than beginning with observations on the macroscopic scale and working our way backward to the atomic level of matter to explain these observations, we start by examining the structure of atoms, and the nature and arrangement of the tiny subatomic particles that atoms contain.

Figure 1.3 John Dalton (1766–1844) was an English chemist, mathematician, and philosopher. In addition to his atomic theory, Dalton formulated several laws governing the behavior of gases, and gave the first detailed description of a particular type of color blindness, from which he suffered. This form of color blindness, where red and green cannot be distinguished, is known as Daltonism.

Georgios Kollidas/GeorgiosArt/iStock/Getty Images

ErikaMitchell/iStock/Getty Images

believeinme33/123RF

Before we begin our study of atoms, it is important for you to understand a bit about the behavior of electrically charged objects. We are all at least casually familiar with the concept of electric charge. You may have brushed your hair in very low humidity and had it stand on end; and you have certainly experienced static shocks and seen lightning. All of these phenomena result from the interactions of electric charges. The following list illustrates some of the important aspects of electric charge:

- An object that is electrically charged may have a positive (+) charge or a negative (−) charge.

positive   negative

- Objects with *opposite* charges (one negative and one positive) are attracted to each other. (You've heard the adage "opposites attract.")

attraction

- Objects with *like* charges (either both positive or both negative) repel each other.

repulsion   repulsion

- Objects with larger charges interact more strongly than those with smaller charges.

repulsion   stronger repulsion

- Charged objects interact more strongly when they are closer together.

repulsion   stronger repulsion

- Opposite charges cancel each other.

positive   negative   no net charge

Keeping in mind how charged objects interact will greatly facilitate your understanding of chemistry.

1.3   Subatomic Particles and the Nuclear Model of the Atom

Experiments conducted late in the nineteenth century indicated that atoms, which had been considered the smallest possible pieces of matter, contained even *smaller* particles. The first of these experiments were done by J. J. Thomson, an English physicist. The experiments revealed that a wide variety of different materials could all be made to emit a stream of tiny, negatively charged particles—that we now know as **electrons.** Thomson reasoned that because all atoms appeared to contain these negative particles but were themselves electrically *neutral,* they must also contain something *positively* charged. This gave rise to a model of the atom as a sphere of positive charge, throughout which negatively charged electrons were uniformly distributed (Figure 1.4). This model was known as the "plum-pudding" model—named after a then-popular English dessert. Thomson's plum-pudding model was an early attempt to describe the internal structure of atoms. Although it was generally accepted for a number of years, this model ultimately was proven wrong by subsequent experiments.

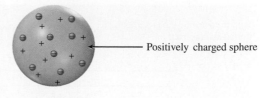

Positively charged sphere

Figure 1.4 Thomson's experiments indicated that atoms contained negatively charged particles, which he envisioned as uniformly distributed in a sphere of positive charge.

Working with Thomson, New Zealand physicist Ernest Rutherford (one of Thomson's own students) devised an experiment to test the plum-pudding model of atomic structure. By that time, Rutherford had already established the existence of another subatomic particle known as an *alpha (α) particle,* which is emitted by some *radioactive* substances. Alpha particles are positively charged, and are thousands of times more massive than electrons. In his most famous experiment, Rutherford directed a stream of alpha particles at a thin gold foil. A schematic of the experimental setup is shown in Figure 1.5. If Thomson's model of the atom were correct, nearly all of the alpha particles would pass directly through the foil—although a small number would be deflected slightly by virtue of passing very close to electrons. Remember that a positively charged object and a negatively charged object are attracted to each other. A positively charged alpha particle could be pulled slightly off course if it passed very close to one of the negatively charged electrons. Rutherford surrounded the gold foil target with a detector that produced a tiny flash of light each time an alpha particle collided with it. This allowed Rutherford to determine the paths taken by the alpha particles. Figure 1.6(a) illustrates the expected experimental result.

The actual experimental result, shown in Figure 1.6(b), was very different from what had been expected. Although most of the alpha particles did pass directly through the gold foil, some were deflected at much larger angles than had been anticipated. Some even bounced off the foil back toward the source—a result that Rutherford found absolutely shocking. He knew that alpha particles could only be deflected at such large angles, and occasionally bounce back in the direction of their source, if they encountered something within the gold atoms that was (1) positively charged, and (2) much more massive than themselves.

This experimental result gave rise to a new model of the internal structure of atoms. Rutherford proposed that atoms are mostly empty space, but that each has a tiny, dense core that contains *all* of its positive charge and *nearly* all of its mass. This core is called the atomic *nucleus.*

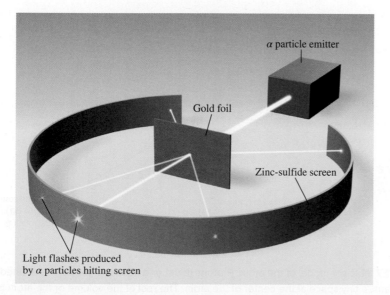

Figure 1.5 Rutherford's experiment directed a stream of positively charged alpha particles at a gold foil. The nearly circular detector emitted a flash of light when struck by an alpha particle.

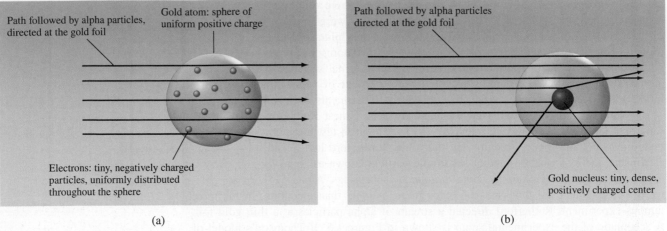

Path followed by alpha particles, directed at the gold foil

Gold atom: sphere of uniform positive charge

Electrons: tiny, negatively charged particles, uniformly distributed throughout the sphere

Path followed by alpha particles directed at the gold foil

Gold nucleus: tiny, dense, positively charged center

(a)   (b)

Figure 1.6 (a) If Thomson's plum-pudding model had been correct, the alpha particles would have passed directly through the foil, with a few being deflected slightly by interaction with electrons. (b) The actual result of Rutherford's gold foil experiment. Most of the alpha particles passed through the gold foil undeflected, but a few were deflected at angles much greater than expected—some even bounced back toward the source. This indicated that as they passed through the gold atoms, they encountered something positively charged and significantly more massive than themselves.

Subsequent experiments supported Rutherford's nuclear model of the atom; and we now know that all atomic *nuclei* (the plural of *nucleus*) contain positively charged particles called **protons.** And with the exception of *hydrogen,* the lightest element, atomic nuclei also contain electrically *neutral* particles called **neutrons.** Together, the protons and neutrons in an atom account for nearly all of its mass, but only a tiny fraction of its volume. The **mass number (A)** is the total number of neutrons and protons in an atom's nucleus. (Protons and neutrons are known, collectively, as **nucleons.**) The dense nucleus is surrounded by a "cloud" of electrons—and just as Rutherford proposed, atoms are mostly empty space. Figure 1.7 illustrates the nuclear model of the atom.

Student Note: An alpha particle is the combination of *two* protons and *two* neutrons.

Of the three subatomic particles in our model of the atom, the electron is the smallest and lightest. Protons and neutrons have very similar masses, and each is nearly

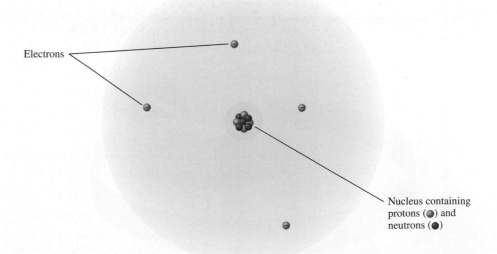

Electrons

Nucleus containing protons (⬤) and neutrons (⚫)

Figure 1.7 Nuclear model of the atom. Protons (blue) and neutrons (red) are contained within the nucleus, a tiny space at the center of the atom. The rest of the volume of the atom is nearly empty, but is occupied by the atom's electrons. This illustration exaggerates the size of the nucleus relative to the size of the atom. If the picture were actually done to scale, and the nucleus were the size shown here (1 centimeter), the atom would be on the order of 100 meters across—about the length of a football field.

2000 times as heavy as an electron. Further, because protons are positively charged and electrons are negatively charged, combination of equal numbers of each results in complete cancellation of the charges. The number of electrons is equal to the number of protons in a neutral atom. Because neutrons are electrically neutral, they do not contribute to an atom's overall charge.

Sample Problem 1.1 lets you practice identifying which combinations of subatomic particles constitute a neutral atom.

SAMPLE PROBLEM **1.1** **Identifying Neutral Atoms Using Numbers of Subatomic Particles**

The following table contains data sets that indicate numbers of subatomic particles. Which of the sets of data represent neutral atoms? For those that do not represent neutral atoms, determine what the charge is—based on the numbers of subatomic particles.

	neutrons	protons	electrons
(a)	11	10	10
(b)	13	12	10
(c)	10	9	9
(d)	18	17	18

Strategy You have learned that the charge on a proton is +1 and the charge on an electron is −1. Neutrons have no charge. The overall charge is the sum of charges of the protons and electrons, and a neutral atom has no charge. Therefore, a set of data in which the number of protons is equal to the number of electrons represents a neutral atom.

Setup Data sets (a) and (c) each contain equal numbers of protons and electrons. Data sets (b) and (d) do not.

Solution The data in sets (a) and (c) represent neutral atoms. Those in (b) and (d) represent charged species. The charge on the species represented by data set (b) is +2: 12 protons (+1 each) and 10 electrons (−1 each). The charge on the species represented by data set (d) is −1: 17 protons (+1 each) and 18 electrons (−1 each).

THINK ABOUT IT

By summing the charges of protons and electrons, we can determine the overall charge on a species. Note that the number of neutrons is not a factor in determining overall charge because neutrons have no charge.

Practice Problem **A**TTEMPT Which of the following data sets represent neutral atoms? For those that do not represent neutral atoms, determine the charge.

	neutrons	protons	electrons
(a)	38	31	28
(b)	26	22	20
(c)	12	11	11
(d)	6	5	5

Practice Problem **B**UILD Fill in the appropriate missing numbers in the following table:

	overall charge	protons	electrons
(a)	+2	23	
(b)	+3		30
(c)	0	53	
(d)		16	18

Practice Problem **C**ONCEPTUALIZE
Determine which of the following pictures represents a neutral atom. For any that does not represent a neutral atom, determine the overall charge. (Protons are blue, neutrons are red, and electrons are green.)

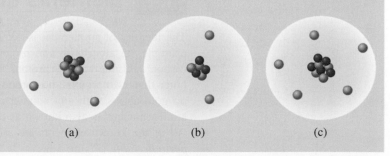

(a)   (b)   (c)

Profiles in Science

Marie Skłodowska Curie

One of the most accomplished and revered women in science, Marie Skłodowska Curie (1867–1934) coined the term "radioactivity" and did groundbreaking research with radioactive elements. She was the first woman to be awarded a Nobel Prize (the 1903 Nobel Prize in Physics), which she shared with her husband, Pierre Curie, and Henri Becquerel. After her husband's death in 1906, she continued the work they had done together and went on to win the 1911 Nobel Prize in Chemistry for her discovery of the elements polonium and radium. Madame Curie is the only person ever to have been awarded two Nobel Prizes in different scientific disciplines.

Marie Skłodowska Curie
IanDagnall Computing/Alamy Stock Photo

CHECKPOINT–SECTION 1.3   Subatomic Particles and the Nuclear Model of the Atom

1.3.1   Which of the following can change without changing the charge on an atom? (Select all that apply.)

a) Number of protons   c) Number of electrons

b) Number of neutrons

1.3.2   Which of the following can change without changing the elemental identity of an atom? (Select all that apply.)

a) Number of protons   c) Number of electrons

b) Number of neutrons

1.3.3   Which of the following must be equal for the combination to constitute a neutral atom?

a) Number of protons and number of neutrons

b) Number of protons and number of electrons

c) Number of neutrons and number of electrons

d) Number of protons, number of neutrons, and number of electrons

1.3.4   Which of the following could represent a neutral atom? (Select all that apply.)

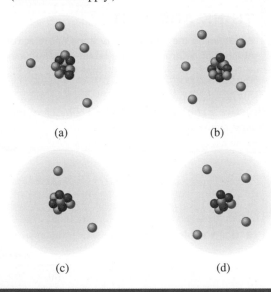

(a)   (b)

(c)   (d)

1.4   Elements and the Periodic Table

The identity of an element is determined by the number of protons in its nucleus. For example, an atom with two protons in its nucleus is helium; one with six protons is carbon; and one with 79 protons is gold. There are no helium atoms with any number other than two protons, no carbon atoms with any number other than six protons, and no gold atoms with any number other than 79 protons. The number of protons in an atom's nucleus is also known as the ***atomic number,*** for which we use the symbol Z. All of the known elements are arranged in order of increasing atomic number on the ***periodic table*** (Figure 1.8).

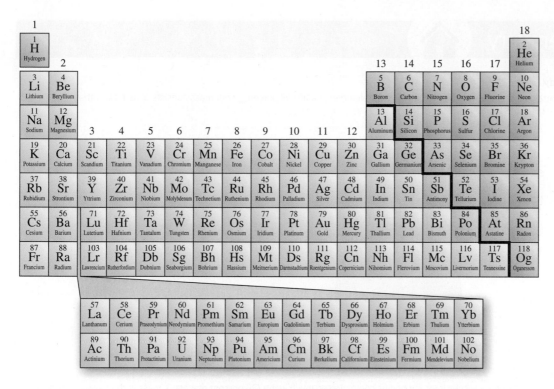

Figure 1.8 The modern periodic table. The elements are arranged in order of increasing atomic number, which is shown above each element's symbol.

Sample Problem 1.2 lets you practice identifying an element, given its atomic number.

SAMPLE PROBLEM 1.2 Identifying an Element by Its Atomic Number

Identify the element given the atomic number of 16.

Strategy You have learned that the atomic number of an element represents the number of protons the element contains. The atomic number is found above the element's symbol on the periodic table.

Setup The element contains 16 protons.

Solution The elements on the periodic table are arranged in order of increasing atomic number. The symbol on the periodic table that has a 16 above its symbol is S. This symbol represents the element sulfur.

THINK ABOUT IT

Remember that an element can be identified either by the number of protons in its nucleus (atomic number) or by its symbol. Every atom with 16 protons is a sulfur atom; and every sulfur atom has 16 protons.

Practice Problem ATTEMPT Identify the element with an atomic number of 35.

Practice Problem BUILD Determine the atomic number for iodine.

Practice Problem CONCEPTUALIZE
Identify the atomic number and identity of each atom shown. (Protons are blue, neutrons are red, and electrons are green.)

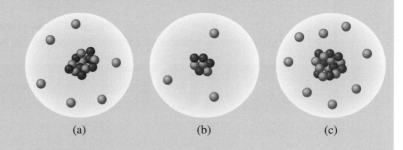

(a)   (b)   (c)

The periodic table also identifies each element with a *chemical symbol*. A chemical symbol consists of one capital letter, or a combination of two letters, one capital and one lowercase. The chemical symbol for helium, for example, is He, and that for carbon is C. Most chemical symbols, including He and C, are derived from the familiar English names of the elements.

Familiar Chemistry

Elements in the Human Body

Although the human body contains trace amounts of a large variety of elements, nearly 99 percent of our mass consists of just *six* of the 118 known elements:

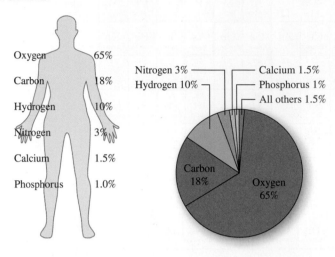

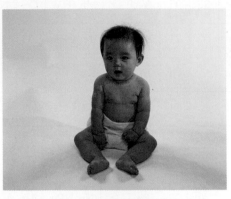

Kwame Zikomo/Purestock/SuperStock

The extraordinary abundance of oxygen results from our bodies containing so much water (89 percent of water's mass is the oxygen it contains). Depending on health and age, the water content of a human body can range from 50 percent in a dehydrated person to 75 percent in a healthy infant.

The second most abundant element in our bodies, carbon, actually has a relatively low natural abundance. Although it makes up only about 0.1 percent of Earth's crust, carbon is present in nearly all living systems.

Others are derived from an element's Greek or Latin name and may take some practice for you to recognize. Examples include Au (*aurium*) for gold, Sn (*stannum*) for tin, Na (*natrium*) for sodium, and K (*kalium*) for potassium. Many of the most recent additions to the periodic table (the highest atomic numbers) are named to honor the scientists involved in their discovery.

Spend some time looking at the periodic table shown in Figure 1.8, or at the beginning of this book. Note that each square on the table contains a chemical symbol and a number, along with the element's name. The number at the top of each square is the atomic number, which is always a *whole* number. (Remember that the atomic number, Z, is the number of protons.) Each element can be identified by its atomic number, its name, or its chemical symbol—and we need only one of these pieces of information to unambiguously specify the identity of an element. The periodic table squares for helium, carbon, and gold are:

Sample Problem 1.3 lets you practice using *atomic number, name,* and chemical *symbol* to identify an element.

SAMPLE PROBLEM 1.3   Relating an Element's Identity to Its Chemical Symbol and Atomic Number

Complete the following table:

	Element	Chemical Symbol	Atomic Number
(a)	calcium		
(b)		Cu	
(c)			13

Strategy You have learned that the atomic number of an element represents the number of protons the element contains and is found above the element's symbol on the periodic table.

Setup Using the one given piece of information in each part, the other two can be found. If the chemical name is given, the symbol should be determined and used to find the atomic number using the periodic table. If the chemical symbol is given, it should be used to determine the name of the element and the atomic number shown on the periodic table. If the atomic number is given, it should be found on the periodic table to determine the chemical symbol and element name.

Solution In part (a) the chemical symbol for calcium is Ca. Using the periodic table, locate Ca and find that its atomic number is 20. Part (b) gives the chemical symbol for copper. The chemical symbol Cu can be located on the periodic table to determine the atomic number is 29. Part (c) gives the atomic number, which can be located on the periodic table to find that the chemical symbol for the element is Al. This symbol represents the element aluminum.

THINK ABOUT IT

A strong grasp of the names and chemical symbols of the elements will allow you to use the periodic table to determine many properties of any element in question and vice versa.

Practice Problem **A**TTEMPT Complete the following table:

	Element	Chemical Symbol	Atomic Number
(a)	rubidium		
(b)			36

Practice Problem **B**UILD Identify the sets of data that are incorrect in the table:

	Element	Chemical Symbol	Atomic Number
(a)	iron	Ir	26
(b)	strontium	Sr	38
(c)	sodium	Na	23

Practice Problem **C**ONCEPTUALIZE Complete the following table:

	Element	Chemical Symbol	Atomic Number (Protons)	Neutrons	Electrons
(a)	potassium			20	
(b)		Be		5	
(c)			35	46	

CHECKPOINT–SECTION 1.4   Elements and the Periodic Table

1.4.1 For which sets of information do the atomic number and element symbol match? (Select all that apply.)

a) $Z = 6$, Cr

b) $Z = 8$, He

c) $Z = 38$, Sr

d) $Z = 16$, O

e) $Z = 82$, Pb

1.4.2 Which pieces of information are sufficient for you to identify an element? (Select all that apply.)

a) Atomic number only

b) Atomic number and element symbol

c) Elemental symbol only

d) Mass number only

e) Element name and element symbol

Familiar Chemistry

Helium

We have all seen helium balloons used as decorations and gifts; and most of us have been entertained by the silly-sounding high-pitched voice of a person who has breathed in the helium from a balloon. But as familiar as this may be, how much do you really know about helium? Where does it come from? How abundant is it? Why does a balloon filled with helium float in the air? And what other uses do we have for the element helium? Helium is actually the product of a radioactive decay process, and although you may not understand yet what that is, you are probably aware that uranium is "radioactive." As it turns out, part of what *makes* uranium radioactive is the process that produces helium. On Earth, helium is found in and around natural gas deposits, and although it is relatively rare here on Earth, it is the second most abundant element in the universe. The element helium was discovered late in the nineteenth century—and its value to society has been immense. It is indispensable as coolant for magnetic resonance imaging (MRI) machines; it is used in the manufacture of computer chips, in scuba diving gas mixtures, in arc welding operations, and in a host of military applications—including air-to-air missile guidance and surveillance operations. Helium balloons float because helium is "lighter" than air. (Technically, helium has a lower *density* [▶▶ Section 4.4] than air.) It is precisely because helium rises that we are facing a shortage here on Earth. Helium that is released into the air will rise until it leaves the atmosphere and floats out into space. Helium is considered a nonrenewable resource, prompting large-scale users of it (the military, the medical industry, scientific research facilities, and the silicon-chip industry) to develop methods for capturing and recycling the helium that they use.

Ireneusz Skorupa/Ericsphotography/iStock/ Getty Images

1.5   Organization of the Periodic Table

The periodic table (Figure 1.8) consists of 118 elements, arranged in vertical columns called *groups* and horizontal rows called *periods.* The groups are headed by numerical group designations 1 through 18 from left to right. The table is divided into *main group* elements and *transition* elements. The main group elements include Groups 1 and 2 on the left, and 13 through 18 on the right. (The transition elements are those in the sunken, middle section of the table.)

Although the periodic table is now arranged in order of increasing *atomic number* (left to right, starting at the top), it was arranged originally in groups of elements with similar properties—even before the concept of atomic number was known. Thus, the properties of elements within a group tend to be similar. Some of the groups have special names that refer to the shared properties of the elements they contain. Group 1, for example, is called the ***alkali metals;*** Group 2 is called the ***alkaline earth metals;*** Group 16 is called the ***chalcogens;*** Group 17 is called the ***halogens;*** and Group 18 is called the ***noble gases.***

In addition to groups (columns) and periods (rows), the periodic table is divided into *metals* and *nonmetals* by the diagonal zigzag line on the right side of the table. Most elements are ***metals*** (left of the zigzag line). ***Nonmetals*** are to the right of the zigzag line. A handful of elements have properties that are intermediate between metal and nonmetal and are referred to as ***metalloids.*** These are found adjacent to the zigzag line. By noting an element's position in the periodic table, you can determine whether it is a metal, a nonmetal, or a metalloid.

Sample Problem 1.4 gives you some practice classifying elements by their positions on the periodic table.

Student Note: The properties that distinguish metals and nonmetals are discussed in Chapter 2 [▶▶ Section 2.6], but you are undoubtedly familiar with the term *metal* and have a reasonably good sense of what metallic properties are. Metals conduct electricity and most are shiny solids.

SAMPLE PROBLEM **1.4** Identifying an Element as Metal, Nonmetal, or Metalloid by Its Position on the Periodic Table

Identify each of the following elements as a metal, nonmetal, or metalloid:

(a) N (b) Si (c) Ca (d) Cl (e) As

Strategy Find the given chemical symbol on the periodic table.

Setup The nonmetallic elements are found in the upper right corner of the periodic table, above the zigzag line. The metallic elements are found below the zigzag line. Note that the metalloids include the highlighted symbols next to the zigzag line. The metalloids are neither metals nor nonmetals.

Solution Part (a) gives the symbol for nitrogen which is found in the nonmetal portion of the periodic table. Part (b) describes silicon, which is a metalloid found along the zigzag line. Part (c) shows calcium, found in the metals area of the periodic table. Part (d) is chlorine, a nonmetal. Part (e) describes arsenic, a metalloid.

THINK ABOUT IT

Most of the periodic table is composed of metals, with the elements in the upper right corner being nonmetals. A small number of elements are shaded along the zigzag line and are considered metalloids.

Practice Problem **A**TTEMPT Identify each of the following elements as a metal, nonmetal, or metalloid:

(a) Se (b) Al (c) Na (d) Kr (e) Ge

Practice Problem **B**UILD Name an element that fits each of the following descriptions. Note that there may be more than one element that will work.

(a) A nonmetal found in Group 14. (c) A metal found in Group 15. (e) A metal found in Group 14.

(b) A metalloid found in Group 13. (d) A nonmetal found in Group 15.

Practice Problem **C**ONCEPTUALIZE Determine which categories each element (chemical symbol given) falls into. Rubidium is used as an example.

Symbol	Main Group Element	Transition Element	Metal	Nonmetal	Metalloid	Alkali Metal	Alkaline Earth Metal	Halogen	Noble Gas
Rb	✓		✓			✓			
B									
Zn									
K									

Familiar Chemistry

Elements in Earth's Crust

Earth's crust extends from the surface to an average depth of about 40 km (25 mi). Of the 118 known elements, just eight elements make up nearly 99 percent of our planet's crust. They are, in decreasing order of abundance, oxygen (O), silicon (Si), aluminum (Al), iron (Fe), calcium (Ca), sodium (Na), potassium (K), and magnesium (Mg). Beneath the crust is the *mantle,* a hot, fluid mixture of iron, carbon (C), silicon, and sulfur (S); and a solid core believed to consist mostly of iron.

Of the eight most abundant elements, oxygen and silicon alone constitute over 70 percent of the crust. These two elements combine (along with small amounts of other elements) to form a huge variety of silicate minerals, including the two most common minerals, feldspar and quartz. The feldspar and quartz families of minerals include many familiar rocks and gemstones.

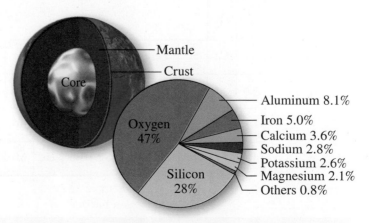

Feldspar minerals:

Amazonite

Marcopolo9442/iStock/Getty Images

Andesine

Doug Sherman/Geofile

Labradorite

Harry Taylor/Getty Images

Quartz minerals:

Milky quartz

Doug Sherman/Geofile

Smoky quartz

Scientifica/Getty Images

Rose quartz

Ron Evans/Photodisc/Getty Images

CHECKPOINT–SECTION 1.5   Organization of the Periodic Table

1.5.1 Which of the following series of elemental symbols lists a nonmetal, a metal, and a metalloid?

a) Ca, Cu, Si   c) Br, Ba, Ge   e) Ag, Cr, As

b) K, Mg, B   d) O, Na, S

1.5.2 Which of the following elements would you expect to have properties most similar to those of chlorine (Cl)?

a) Cu   c) Na   e) S

b) I   d) Cr

1.5.3 The periodic table shown here has four groups highlighted. Which of the highlighted groups contain(s) only one metal, which contain(s) only one nonmetal, and which contain(s) only one metalloid?

a) iv, iii, and i   d) iii & iv, ii, and i & iv

b) i, iv, and iii   e) i, ii, and iii & iv

c) iv, ii, and iii

1 H																	2 He
3 Li	4 Be											5 B	6 C	7 N	8 O	9 F	10 Ne
11 Na	12 Mg											13 Al	14 Si	15 P	16 S	17 Cl	18 Ar
19 K	20 Ca	21 Sc	22 Ti	23 V	24 Cr	25 Mn	26 Fe	27 Co	28 Ni	29 Cu	30 Zn	31 Ga	32 Ge	33 As	34 Se	35 Br	36 Kr
37 Rb	38 Sr	39 Y	40 Zr	41 Nb	42 Mo	43 Tc	44 Ru	45 Rh	46 Pd	47 Ag	48 Cd	49 In	50 Sn	51 Sb	52 Te	53 I	54 Xe
55 Cs	56 Ba	57 La	72 Hf	73 Ta	74 W	75 Re	76 Os	77 Ir	78 Pt	79 Au	80 Hg	81 Tl	82 Pb	83 Bi	84 Po	85 At	86 Rn
87 Fr	88 Ra	89 Ac	104 Rf	105 Db	106 Sg	107 Bh	108 Hs	109 Mt	110 Ds	111 Rg	112 Cn	113 Nh	114 Fl	115 Mc	116 Lv	117 Ts	118 Og
												(i)	(ii)		(iii)	(iv)	

1.6 Isotopes

We have learned that an atom can be identified by the number of protons contained in its nucleus—also known as its atomic number, Z. But remember that with just one exception (hydrogen), the nuclei of atoms also contain *neutrons*—and most elements consist of mixtures of atoms with *different* numbers of neutrons. For example, in a sample of pure chlorine, all of the atoms have 17 protons—but they do not all have the same number of neutrons. Roughly three-quarters of the chlorine atoms will have 18 neutrons and one-quarter will have 20 neutrons. An atom with 17 protons and 18 neutrons and an atom with 17 protons and 20 neutrons are both chlorine atoms. They are, however, different *isotopes* of chlorine. *Isotopes* are atoms of the same element, and therefore have the same number of protons, but have different numbers of neutrons.

Recall that an element's mass number (A) is the sum of the protons and neutrons in its nucleus. Returning to the example of chlorine, the mass number of a chlorine atom with 18 neutrons is 35 (17 protons + 18 neutrons), and the mass number of a chlorine atom with 20 neutrons is 37 (17 protons + 20 neutrons).

Isotope Symbol	Element Name	Protons	Neutrons
$^{35}_{17}\text{Cl}$	chlorine	17	18
		sum = 35	
$^{37}_{17}\text{Cl}$	chlorine	17	20
		sum = 37	

In general, the way to denote the identity of an atom is with its elemental symbol (shown here as X) with the superscript mass number (A) and the subscript atomic number (Z).

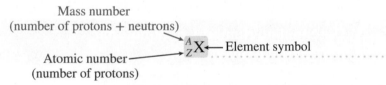

Mass number (number of protons + neutrons) — $^{A}_{Z}\text{X}$ — Element symbol
Atomic number (number of protons)

There are three isotopes of hydrogen, called *hydrogen, deuterium,* and *tritium*. Hydrogen has one proton and no neutrons in its nucleus, deuterium has one proton and one neutron, and tritium has one proton and two neutrons. Thus, to represent the isotopes of hydrogen, we write

$^{1}_{1}\text{H}$
protium

$^{2}_{1}\text{H}$
deuterium

$^{3}_{1}\text{H}$
tritium

Similarly, the two common isotopes of uranium ($Z = 92$), which have mass numbers of 235 and 238, respectively, can be represented as follows:

$$^{235}_{92}\text{U} \qquad ^{238}_{92}\text{U}$$

The first isotope, with $235 - 92 = 143$ neutrons in its nucleus, is used in nuclear reactors and atomic bombs, whereas the second isotope, with 146 neutrons, lacks the properties necessary for these applications. With the exception of hydrogen, which has different names for each of its isotopes, the isotopes of other elements are identified by their mass numbers. The two isotopes of uranium are called uranium-235 (pronounced "uranium two thirty-five") and uranium-238 (pronounced "uranium two thirty-eight"). Because the atomic number subscript can be deduced from the elemental symbol, it may be omitted from these representations without the loss of any information. The symbols ^{3}H and ^{235}U are sufficient to specify the isotopes tritium and uranium-235, respectively.

The chemical properties of an element are determined primarily by the number of protons, not by the number of neutrons. Therefore, isotopes of the same element typically exhibit very similar chemical properties.

Thinking Outside the Box

Mass Spectrometry

How do we know the mass of an atom? An instrument called a mass spectrometer is one very accurate method for determining the mass of an atom. A mass spectrometer works by bombarding a gaseous sample of a substance with a stream of electrons. When the electrons collide with the gaseous atoms, they dislodge an electron and create a positively charged ion (with a certain charge-to-mass, e/m ratio). The positive ions are accelerated between two plates of opposite charges. The beam of ions then passes through a magnetic field, which separates the ions on the basis of the e/m ratio. The smaller the e/m ratio, the more the magnetic field deflects the ion. The magnitude of deflection is used to determine the mass of the ion and therefore the mass of the parent atom.

The ability to identify different isotopes using mass spectrometry has facilitated some fascinating fields of study. Oxygen, for example, has three naturally occurring isotopes: ^{16}O, ^{17}O, and ^{18}O. ^{16}O is the most abundant, by far, making up more than 99.7 percent of the total. The ratio of oxygen-18 to oxygen-16 in the frozen water found at Earth's poles has proved invaluable

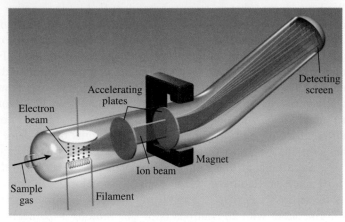

Diagram of one type of mass spectrometer.

for determining the long-term history of global temperatures. Oxygen-17 has been used in magnetic resonance imaging (MRI) experiments to elucidate details of physiology, such as oxygen usage in the brain.

Sample Problem 1.5 shows how to calculate the number of protons, neutrons, and electrons using atomic numbers and mass numbers.

SAMPLE PROBLEM 1.5 Determining Numbers of Subatomic Particles in a Given Atom

Determine the numbers of protons, neutrons, and electrons in each of the following species:

(a) $^{35}_{17}Cl$ (b) ^{37}Cl (c) ^{41}K (d) carbon-14

Strategy Recall that the superscript denotes the mass number, A, and the subscript denotes the atomic number, Z. In cases where no subscript is shown, as in parts (b), (c), and (d), the atomic number can be deduced from the elemental symbol or name. For the purpose of determining the number of electrons, remember that atoms are neutral, so the number of electrons is equal to the number of protons.

Setup Number of protons = Z, number of neutrons = $A - Z$, and number of electrons = number of protons. Recall that the 14 in carbon-14 is the *mass number*.

Solution (a) The atomic number is 17, so there are 17 protons. The mass number is 35, so the number of neutrons is $35 - 17 = 18$. The number of electrons equals the number of protons, so there are 17 electrons.

(b) Because the element is again Cl (chlorine), the atomic number is again 17, so there are 17 protons. The mass number is 37, so the number of neutrons is $37 - 17 = 20$. The number of electrons equals the number of protons, so there are 17 electrons, too.

(c) The atomic number of K (potassium) is 19, so there are 19 protons. The mass number is 41, so there are $41 - 19 = 22$ neutrons. There are 19 electrons.

(d) Carbon-14 can also be represented as ^{14}C. The atomic number of carbon is 6, so there are 6 protons and 6 electrons. There are $14 - 6 = 8$ neutrons.

THINK ABOUT IT

Verify that the number of protons and the number of neutrons for each example sum to the mass number that is given. In part (a), for example, there are 17 protons and 18 neutrons, which sum to give a mass number of 35, the value given in the problem. In part (b), 17 protons + 20 neutrons = 37. In part (c), 19 protons + 22 neutrons = 41. In part (d), 6 protons + 8 neutrons = 14.

Practice Problem **ATTEMPT** How many protons, neutrons, and electrons are there in an atom of (a) $^{10}_{5}B$, (b) ^{36}Ar, (c) $^{85}_{38}Sr$, and (d) carbon-11?

Practice Problem **BUILD** Give the correct symbols to identify an atom that contains (a) 4 protons, 4 electrons, and 5 neutrons; (b) 23 protons, 23 electrons, and 28 neutrons; (c) 54 protons, 54 electrons, and 70 neutrons; and (d) 31 protons, 31 electrons, and 38 neutrons.

Practice Problem **CONCEPTUALIZE** Fill in the missing information for neutral atoms:

Isotope Symbol	Element Name	Mass Number (A)	Neutrons ($n°$)	Protons (p^+)	Electrons (e^-)
N-15	nitrogen	15	8	7	7
	nitrogen	14			
^{23}Na		23		11	

CHECKPOINT–SECTION 1.6 Isotopes

1.6.1 How many neutrons are there in an atom of ^{60}Ni?

a) 60

b) 30

c) 28

d) 32

e) 29

1.6.2 What is the *mass number* of an oxygen atom with nine neutrons in its nucleus?

a) 8

b) 9

c) 17

d) 16

e) 18

1.7 Atomic Mass

As we have seen, there are two different isotopes of chlorine, ^{35}Cl and ^{37}Cl. However, if you examine the periodic table on the inside cover of the book, you will see a number under the element's symbol and name that is neither 35 nor 37. The number under the symbol and name of chlorine is 35.45. That number, 35.45, is the **atomic mass (M)** of chlorine, and it is what's known as a *weighted average.* To understand how weighted averages work, consider a collection of two different varieties of apples: Granny Smith and Pink Lady. A typical Granny Smith apple has a mass of 200 g, whereas a typical Pink Lady apple has a mass of 150 g. If we were to have one of each type of apple, the average "per-apple" mass would be $\frac{200\text{ g} + 150\text{ g}}{2\text{ apples}} = 175$ g per apple.

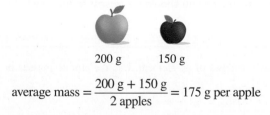

$$\text{average mass} = \frac{200\text{ g} + 150\text{ g}}{2\text{ apples}} = 175 \text{ g per apple}$$

However, if we were to have a collection containing different numbers of Granny Smith and Pink Lady apples, we would want to calculate a *weighted* average.

$$\text{average mass} = \frac{(2)200\text{ g} + (8)150\text{ g}}{10\text{ apples}} = 160 \text{ g per apple}$$

In this case, because there are more Pink Lady apples than Granny Smith apples, the average per-apple mass would be closer to the mass of a Pink Lady. Likewise, because there are more ^{35}Cl atoms than ^{37}Cl atoms, the average mass of a Cl atom is much closer to 35 than to 37.

In order to understand the concept of atomic mass, we need to become familiar with the units with which atomic masses are expressed—namely the *atomic mass unit* or *amu*. An atomic mass unit is defined as one-twelfth the mass of a ^{12}C atom. On this scale, the mass of a ^{35}Cl atom is 34.968852721 amu. (Note that the *mass* of an atom is not exactly equal to its mass *number*.) The mass of a ^{37}Cl atom is 36.96590262 amu.

The number on the periodic table, 35.45, is the *average atomic mass* of chlorine, and it can be calculated as follows:

Naturally occurring chlorine is 75.78 percent ^{35}Cl and 24.22 percent ^{37}Cl.

$$(0.7578)(34.968852721 \text{ amu}) + (0.2422)(36.96590262 \text{ amu}) = 35.45 \text{ amu}$$

Many elements have more than two naturally occurring isotopes. (Tin, Sn, holds the record with 10.) But in the case of elements with *two* isotopes, it is usually easy to tell which isotope is more abundant simply by looking at the atomic mass listed in the periodic table. Boron, for example, has two naturally occurring isotopes: ^{10}B and ^{11}B. Because the atomic mass on the periodic table is 10.81, we know that ^{11}B is the more abundant of the isotopes, because the average atomic mass is closer to 11 than it is to 10. In fact, the abundances of isotopes in natural boron are 80.1 percent ^{11}B and 19.9 percent ^{10}B.

Sample Problem 1.6 lets you practice determining which isotope is more abundant in elements with just two naturally occurring isotopes.

SAMPLE PROBLEM **1.6** Identifying the More Abundant Isotope Given Average Atomic Mass

Each set of data lists two isotopes of a particular element. Use the atomic masses on the periodic table to determine which one has a larger percent abundance.

(a) Ne-20 or Ne-22   (b) In-113 or In-115 (Z = 49)   (c) Cu-63 or Cu-65

Strategy The atomic masses shown on the periodic table are a weighted average of the naturally occurring isotopes of each element.

Setup The atomic mass shown on the periodic table will be closer to the isotope that is present in a higher abundance.

Solution (a) Ne-20 as the atomic mass of neon is listed as 20.18, which is closer to 20 than 22. (b) In-115 as the atomic mass of indium is listed as 114.82, which is closer to 115 than 113. (c) Cu-63 as the atomic mass of copper is listed as 63.55, which is closer to 63 than 65.

THINK ABOUT IT

The average atomic mass (shown on the periodic table for each element) should be closest to the most abundant isotope for that element.

Practice Problem ④TTEMPT Each set of data lists two isotopes of an element. Use the atomic masses on the periodic table to determine which one has a larger percent abundance.

(a) Mg-24 or Mg-25   (b) Li-6 or Li-7   (c) Ta-180 or Ta-181 (Z = 73)

Practice Problem ⑧UILD Which of the following statements could be true according to the atomic masses given on the periodic table?

(a) Silver is a roughly equal mix of Ag-107 and Ag-109.   (c) Vanadium is a roughly equal mix of V-50 and V-51.

(b) Rubidium is predominantly composed of Rb-87.

Practice Problem ⓒONCEPTUALIZE Determine the average per-apple mass of a collection of apples consisting of four Granny Smith apples and six Pink Lady apples.

Familiar Chemistry

Iron-Fortified Cereal

Iron deficiency is the single most common nutritional deficiency in the world. An estimated 25 percent of the world's population does not consume enough iron to maintain good health. Iron is necessary for the production of *hemoglobin*, the component in red blood cells responsible for the transport of oxygen, and insufficient iron causes anemia. People with anemia can suffer from a variety of symptoms, including fatigue, weakness, pale color, poor appetite, headache, and light-headedness. Although pharmacy shelves display a variety of over-the-counter iron supplements, one of the most popular ways to increase dietary intake of iron is by eating iron-fortified cereal. Such cereals are common and include many familiar brands. Have you ever thought about how the cereals become "fortified"? It may surprise you to learn that most cereals are fortified with iron simply by the addition of iron metal! Finely divided bits of iron are simply added to the grain and other ingredients during processing. The iron metal in fortified cereals is fairly simple to separate and observe—and this process is a popular chemistry demonstration. If you blend the cereal with water and apply a strong magnet, you can actually separate the iron filings.

David Moyer/McGraw Hill

David Moyer/McGraw Hill

Profiles in Science

Helen Mackay

Scottish-born pediatrician Helen Mackay (1891–1965) was a first in many achievements. Among other things, she was the first woman fellow of the Royal College of Physicians, the first female physician at the Queen Elizabeth Hospital for Children, and the first person to study the importance of iron in the diet of infants. While investigating rickets in infants, a disease caused by a vitamin D deficiency, she noted that all of the children in her study also suffered from anemia. Mackay's research revealed that the cause of anemia was dietary iron deficiency, and that iron supplementation reduced infections and other illnesses, and resulted in babies that were healthier overall. She was the first to establish quantitative criteria to define anemia, and today's World Health Organization's definition very closely resembles the one developed by Mackay.

Helen Mackay

Image courtesy of The Royal College of Physicians

Sample Problem 1.7 shows you how to calculate the average atomic mass for an element, knowing the relative abundance of its naturally occurring isotopes.

SAMPLE PROBLEM 1.7   Calculating Average Atomic Mass Given Isotope Abundance

The percent abundance of the two stable isotopes of copper, Cu-63 (62.929599 amu) and Cu-65 (64.927793 amu), are 69.17% and 30.83%, respectively. Determine the average atomic mass of copper.

Strategy Each isotope contributes to the average atomic mass based on its relative abundance. Multiplying the mass of each isotope by its fractional abundance (percent value divided by 100) will give its contribution to the average atomic mass.

Setup Each percent abundance must be converted to a fractional abundance: 69.17/100 or 0.6917 and 30.83/100 or 0.3083. Once we find the contribution to the average atomic mass for each isotope, we can then add the contributions together to obtain the average atomic mass. This is a weighted average.

Solution

$$0.6917 \times 62.929599 = 43.5284 \text{ amu}$$
$$+ \underline{0.3083 \times 64.927793 = 20.0172 \text{ amu}}$$
$$63.5456 \text{ amu}$$

THINK ABOUT IT

The average atomic mass should be closest to the atomic mass of the most abundant isotope (in this case the Cu-63 isotope) and, to two places past the decimal point, should be the same number that appears in the periodic table at the beginning of this book (in this case, 63.55 amu).

Practice Problem ⒶTTEMPT The percent abundance of the two stable isotopes of nitrogen, N-14 (14.003074002 amu) and N-15 (15.000108898 amu), are 99.63% and 0.37%, respectively. Determine the average atomic mass of nitrogen. (Report your answer to two places past the decimal point.)

Practice Problem ⒷUILD The percent abundance of the three stable isotopes of neon, Ne-20 (19.9924401754 amu), Ne-21 (20.99384668 amu), and Ne-22 (21.991385114 amu), are 90.5%, 0.3%, and 9.3%, respectively. Determine the average atomic mass of neon. (Report your answer to two places past the decimal point.)

Practice Problem ⒸONCEPTUALIZE The figure is a representation of 15 atoms of a fictitious element with the symbol Rr. Rr has two isotopes represented by the colors: Rr-285 (green) and Rr-294 (purple). Use the drawing (statistically representative of naturally occurring Rr) and the following masses of each isotope to calculate the atomic mass of Rr.

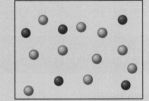

Rr-285 (green) = 284.9751 amu
Rr-294 (purple) = 293.9855 amu

CHECKPOINT–SECTION 1.7   Atomic Mass

1.7.1 Boron has two naturally occurring isotopes, ^{10}B and ^{11}B, which have masses of 10.0129 and 11.0093 amu, respectively. Given the average atomic mass of boron (10.81 amu), determine the percent abundance of each isotope.

   a) 50% ^{10}B, 50% ^{11}B

   b) 20% ^{10}B, 80% ^{11}B

   c) 98% ^{10}B, 2% ^{11}B

   d) 93% ^{10}B, 7% ^{11}B

   e) 22% ^{10}B, 78% ^{11}B

1.7.2 The two naturally occurring isotopes of antimony, ^{121}Sb (57.21 percent) and ^{123}Sb (42.79 percent), have masses of 120.904 and 122.904 amu, respectively. What is the average atomic mass of Sb?

   a) 121.90 amu

   b) 122.05 amu

   c) 121.76 amu

   d) 121.34 amu

   e) 122.18 amu

Chapter Summary

Section 1.1

- *Chemistry* is the study of matter and of the changes that matter undergoes. *Matter* is everything material—essentially everything we can see, smell, taste, and touch. *Mass* refers to the *amount* of matter in an object or group of objects.

- The *scientific method* is a systematic approach to scientific experimentation in which observations can be used to formulate a scientific *law,* the experimental data collected can be used to develop a scientific *hypothesis,* and a successful hypothesis can give rise to a scientific *theory* or *model.* Scientific theories are always being tested by further experiment and must be changed or replaced if newly discovered data are inconsistent with them.

Section 1.2

- *Atoms* are the tiny building blocks of all matter. *Elements* are substances that consist of just one type of atom and that cannot be broken down into simpler substances.

Section 1.3

- *Electrons* are negatively charged particles that are part of all atoms. The *nucleus* is the tiny, dense core that contains all of an atom's positive charge, and nearly all of its mass. The positively charged particles in the nucleus of an atom are *protons,* and the electrically neutral particles in the nucleus are *neutrons.* With the exception of hydrogen atoms, all atoms contain neutrons. An *alpha particle* is the positively charged combination of two protons and two neutrons. The sum of protons and neutrons in an atom is the atom's *mass number (A).* Collectively, protons and neutrons (the particles in the nucleus) are called *nucleons.*

Section 1.4

- The *atomic number (Z)* is the number of protons in the nucleus of an atom. Atomic number determines the elemental identity of an atom. The *periodic table* arranges the elements in order of increasing atomic number. Each element has a unique *chemical symbol,* consisting of one capital letter (e.g., C, H, F, O) or of one capital and one lowercase letter (e.g., He, Cl, Na, Al).

Section 1.5

- The periodic table is divided into vertical columns called *groups* and horizontal rows called *periods.* Elements in the same group exhibit similar properties. Further, the periodic table is separated into *metals* and *nonmetals* by a zigzag line running down the right side of the table. Most of the elements that lie adjacent to the zigzag line are neither metal nor nonmetal, but are referred to as *metalloids.* Some of the groups (vertical columns) have special names. Group 1 is referred to as the *alkali metals,* Group 2 is the *alkaline earth metals,* Group 16 is the *chalcogens,* Group 17 is the *halogens,* and Group 18 is the *noble gases.*

Section 1.6

- Atoms of a given element are not all identical. Those with the same atomic number (Z) but different numbers of neutrons are called *isotopes.*

Section 1.7

- *Atomic mass (M)* is the mass of an atom in *atomic mass units (amu).* One atomic mass unit is defined as one-twelfth the mass of a carbon-12 atom (a carbon atom with six protons and six neutrons in its nucleus). The *average atomic mass* of an element is the weighted average of the element's naturally occurring isotopes.

Key Terms

Alkali metal 14	Chemistry 3	Mass number (A) 8	Nucleus 7
Alkaline earth metal 14	Electron 6	Matter 3	Period 14
Alpha (α) particle 7	Element 5	Metal 14	Periodic table 10
Atom 5	Group 14	Metalloid 14	Proton 8
Atomic mass (M) 19	Halogen 14	Model 4	Scientific method 3
Atomic mass unit (amu) 20	Hypothesis 4	Neutron 8	Theory 4
Atomic number (Z) 10	Isotope 17	Noble gas 14	
Average atomic mass 20	Law 4	Nonmetal 14	
Chalcogens 14	Mass 3	Nucleon 8	

Organization of the Periodic Table

As you proceed through this course, you will find yourself using the periodic table a lot. It is a powerful tool that embodies much more information than may be immediately apparent. For you to make the best possible use of this tool, it is very important that you become familiar with the information it provides and comfortable using it to understand new concepts and solve problems.

There are several ways to view the organization of the periodic table. One way is to divide the table into metals, nonmetals, and metalloids. This is shown with color coding here. Metals are shown in green, nonmetals in blue, and metalloids in yellow.

H																	He
Li	Be											B	C	N	O	F	Ne
Na	Mg											Al	Si	P	S	Cl	Ar
K	Ca	Sc	Ti	V	Cr	Mn	Fe	Co	Ni	Cu	Zn	Ga	Ge	As	Se	Br	Kr
Rb	Sr	Y	Zr	Nb	Mo	Tc	Ru	Rh	Pd	Ag	Cd	In	Sn	Sb	Te	I	Xe
Cs	Ba	La	Hf	Ta	W	Re	Os	Ir	Pt	Au	Hg	Tl	Pb	Bi	Po	At	Rn
Fr	Ra	Ac	Rf	Db	Sg	Bh	Hs	Mt	Ds	Rg	Cn	Nh	Fl	Mc	Lv	Ts	Og

Another way to look at the organization is that elements are arranged in order of increasing atomic number. We read across the periods of the periodic table the same way we read a page of text: left to right in the first row, then left to right in the second row, and so on.

1 H																	2 He
3 Li	4 Be											5 B	6 C	7 N	8 O	9 F	10 Ne
11 Na	12 Mg											13 Al	14 Si	15 P	16 S	17 Cl	18 Ar
19 K	20 Ca	21 Sc	22 Ti	23 V	24 Cr	25 Mn	26 Fe	27 Co	28 Ni	29 Cu	30 Zn	31 Ga	32 Ge	33 As	34 Se	35 Br	36 Kr
37 Rb	38 Sr	39 Y	40 Zr	41 Nb	42 Mo	43 Tc	44 Ru	45 Rh	46 Pd	47 Ag	48 Cd	49 In	50 Sn	51 Sb	52 Te	53 I	54 Xe
55 Cs	56 Ba	57 La	72 Hf	73 Ta	74 W	75 Re	76 Os	77 Ir	78 Pt	79 Au	80 Hg	81 Tl	82 Pb	83 Bi	84 Po	85 At	86 Rn
87 Fr	88 Ra	89 Ac	104 Rf	105 Db	106 Sg	107 Bh	108 Hs	109 Mt	110 Ds	111 Rg	112 Cn	113 Nh	114 Fl	115 Mc	116 Lv	117 Ts	118 Og

The elements we will encounter most often will be those in the main group of the periodic table. Each element's tile gives its atomic number, its elemental symbol, its name, and its average atomic mass. Note that each column of elements has a group number. The main group includes Groups 1, 2, and 13–18. Very often, solving a problem in a later chapter will require you to use the periodic table to look up atomic masses, elemental symbols, group numbers, etc.—and it is very important that you practice this enough to do it quickly and accurately.

1 ⌐ Group number							18
1 H Hydrogen 1.008	2						2 He Helium 4.003
3 Li Lithium 6.941	4 Be Beryllium 9.012	5 B Boron 10.81	6 C Carbon 12.01	7 N Nitrogen 14.01	8 O Oxygen 16.00	9 F Fluorine 19.00	10 Ne Neon 20.18
11 Na Sodium 22.99	12 Mg Magnesium 24.31	13 Al Aluminum 26.98	14 Si Silicon 28.09	15 P Phosphorus 30.97	16 S Sulfur 32.07	17 Cl Chlorine 35.45	18 Ar Argon 39.95
19 K Potassium 39.10	20 Ca Calcium 40.08	31 Ga Gallium 69.72	32 Ge Germanium 72.64	33 As Arsenic 74.92	34 Se Selenium 78.96	35 Br Bromine 79.90	36 Kr Krypton 83.80
37 Rb Rubidium 85.47	38 Sr Strontium 87.62	49 In Indium 114.8	50 Sn Tin 118.7	51 Sb Antimony 121.8	52 Te Tellurium 127.6	53 I Iodine 126.9	54 Xe Xenon 131.3
55 Cs Cesium 132.9	56 Ba Barium 137.3	81 Tl Thallium 204.4	82 Pb Lead 207.2	83 Bi Bismuth 209.0	84 Po Polonium (209)	85 At Astatine (210)	86 Rn Radon (222)
87 Fr Francium (223)	88 Ra Radium (226)	113 Nh Nihonium (284)	114 Fl Flerovium (289)	115 Mc Moscovium (288)	116 Lv Livermorium (293)	117 Ts Tennessine (293)	118 Og Oganesson (294)

Key Skills Questions

1.1
Of the main group elements, which *group* contains the largest number of metals?
a) 1
b) 3
c) 4
d) 16
e) 17

1.2
Of the main group elements, which *period* (of periods 1–6) contains the largest number of nonmetals?
a) 2
b) 3
c) 4
d) 5
e) 6

1.3
Within the main group, how many instances are there where the order of elements does not proceed in order of increasing average atomic mass?
a) None
b) One
c) Two
d) Three
e) More than three

1.4
What causes isotopes of an element to have different atomic masses?
a) Different numbers of protons
b) Different numbers of electrons
c) Different numbers of neutrons
d) Different charges
e) Different numbers of protons and neutrons

Questions and Problems

SECTION 1.1: THE STUDY OF CHEMISTRY

1.1 Describe a scientific theory in your own words.

1.2 In your own words, describe a scientific law.

1.3 Describe a hypothesis in your own words.

1.4 What is the difference between a law and a theory?

SECTION 1.2: ATOMS FIRST

1.5 Can an atom be broken down into smaller parts? If this is possible, will the smaller parts behave like the atom?

1.6 We say that opposite charges "cancel." Which of the following sets of data show cancelled charges? If the charges do not cancel, what is the "leftover" charge?

(a) 10+ and 10−   (c) 15+ and 18−
(b) 20+ and 21−   (d) 16+ and 13−

SECTION 1.3: SUBATOMIC PARTICLES AND THE NUCLEAR MODEL OF THE ATOM

1.7 If you "disassembled" an atom of helium into its separate components, what would you have?

1.8 List the components (or subatomic particles) that make up atoms. Describe the charge on each of these subatomic particles.

1.9 Identify the following statements as true or false. If the statement is false, correct it to make it true.

(a) A neutral atom always contains the same number of protons and neutrons.

(b) The number of electrons is always equal to the number of protons in a neutral atom.

(c) The number of neutrons does not affect the charge on an atom.

(d) A proton is the smallest identifiable "piece" of an element and has the same properties as the element.

1.10 Which of the following sets of data represent a neutral atom?

	Neutrons	Protons	Electrons
(a)	7	6	6
(b)	16	16	18
(c)	20	18	20

1.11 Which of the following sets of data represent a neutral atom?

	n°	p^+	e^-
(a)	28	24	21
(b)	28	24	24
(c)	62	46	46

1.12 Which of the following diagrams represent neutral atoms? For any that does not represent a neutral atom, determine the charge.

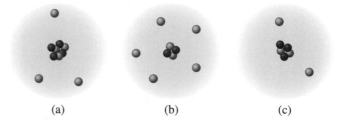

(a)   (b)   (c)

SECTION 1.4: ELEMENTS AND THE PERIODIC TABLE

1.13 Which of the following are correct element symbol/name combinations? Correct the error in any that are not correct combinations.

(a) Ca = carbon   (c) Na = sodium
(b) B = bromine   (d) Cu = copper

1.14 Which of the following are correct element symbol/name combinations? Correct the error in any that are not correct combinations.

(a) K = krypton   (c) Ne = neon
(b) S = silicon   (d) Se = selenium

1.15 Which of the following are correct element symbol/name combinations? Correct the error in any that are not correct combinations.

(a) Pt = plutonium   (c) Pb = lead
(b) Ni = nitrogen   (d) P = phosphorus

1.16 Which of the following are correct element symbol/name combinations? Correct the error in any that are not correct combinations.

(a) Mg = manganese   (c) Xe = xerxes
(b) Cu = copper   (d) F = flourine

1.17 Fill in the missing information for each of the neutral atoms indicated.

Element Symbol	Element Name	Atomic Number	Mass Number	Number of Protons	Number of Neutrons	Number of Electrons
	silicon		29			
Mg					14	12
		15	31			
				30	36	
			127			53

1.18   Fill in the missing information for each of the neutral atoms indicated.

Element Symbol	Element Name	Atomic Number	Mass Number	Number of Protons	Number of Neutrons	Number of Electrons
K					22	
			14			7
					18	16
	strontium		87			
				18	22	

SECTION 1.5: ORGANIZATION OF THE PERIODIC TABLE

1.19   Label each of the colored regions of the periodic table with the correct term. Draw outlines identifying the main group elements.

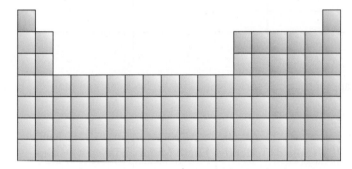

1.20   Label the groups/families on the periodic table with the correct name.

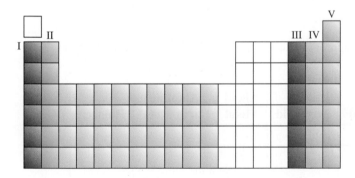

1.21   Which of the following elements are metals?

Li   Ba   C   Cl   Ar   Cu   V   I   Kr   O   F   S

1.22   Which of the following elements are nonmetals?

Li   Ba   C   Cl   Ar   Cu   V   I   Kr   O   F   S

1.23   Which of the following elements are metalloids?

Li   Ba   C   Cl   Ar   Cu   V   I   Kr   O   F   S

1.24   Which of the following elements are halogens?

Li   Ba   C   Cl   Ar   Cu   V   I   Kr   O   F   S

1.25   Which of the following elements are noble gases?

Li   Ba   C   Cl   Ar   Cu   V   I   Kr   O   F   S

1.26   Which of the following elements are alkali metals?

Li   Ba   C   Cl   Ar   Cu   V   I   Kr   O   F   S

1.27   Which of the following elements are alkaline earth metals?

Li   Ba   C   Cl   Ar   Cu   V   I   Kr   O   F   S

1.28   Which of the following elements are metals?

Fe   Br   Xe   Se   As   Si   Ni   K   Sr   Pb   P   N

1.29   Which of the following elements are nonmetals?

Fe   Br   Xe   Se   As   Si   Ni   K   Sr   Pb   P   N

1.30   Which of the following elements are metalloids?

Fe   Br   Xe   Se   As   Si   Ni   K   Sr   Pb   P   N

1.31   Which of the following elements are halogens?

Fe   Br   Xe   Se   As   Si   Ni   K   Sr   Pb   P   N

1.32   Which of the following elements are noble gases?

Fe   Br   Xe   Se   As   Si   Ni   K   Sr   Pb   P   N

1.33   Which of the following elements are alkali metals?

Fe   Br   Xe   Se   As   Si   Ni   K   Sr   Pb   P   N

1.34   Which of the following elements are alkaline earth metals?

Fe   Br   Xe   Se   As   Si   Ni   K   Sr   Pb   P   N

1.35   Classify the following elements as belonging to all appropriate categories by placing a check mark in the correct column(s). An example is shown.

Symbol	Main Group Element	Transition Element	Metal	Nonmetal	Metalloid	Alkali Metal	Alkaline Earth Metal	Halogen	Noble Gas
Rb	✓		✓			✓			
Be									
Ag									
Zn									

1.36   Classify the following elements as belonging to all appropriate categories by placing a check mark in the correct column(s).

Symbol	Main Group Element	Transition Element	Metal	Nonmetal	Metalloid	Alkali Metal	Alkaline Earth Metal	Halogen	Noble Gas
Fe									
Pb									
Se									

1.37   Classify the following elements as belonging to all appropriate categories by placing a check mark in the correct column(s).

Symbol	Main Group Element	Transition Element	Metal	Nonmetal	Metalloid	Alkali Metal	Alkaline Earth Metal	Halogen	Noble Gas
Cl									
P									
Mg									

1.38   Classify the following elements as belonging to all appropriate categories by placing a check mark in the correct column(s).

Symbol	Main Group Element	Transition Element	Metal	Nonmetal	Metalloid	Alkali Metal	Alkaline Earth Metal	Halogen	Noble Gas
S									
Ne									
Si									

1.39   Classify the following elements as belonging to all appropriate categories by placing a check mark in the correct column(s).

Symbol	Main Group Element	Transition Element	Metal	Nonmetal	Metalloid	Alkali Metal	Alkaline Earth Metal	Halogen	Noble Gas
I									
Ar									
K									

SECTION 1.6: ISOTOPES

1.40   Write the isotope symbol for the following atoms:
(a) An atom with 7 protons and 8 neutrons
(b) An atom with 9 protons and 10 neutrons
(c) An atom with 6 neutrons and 6 protons

1.41   Write the isotope symbol for the atoms that contain the following:
(a) Four protons and five neutrons
(b) Twelve protons and thirteen neutrons
(c) Twenty protons and twenty neutrons

1.42   Sketch an atom of each of the following isotopes:
(a) $^{22}_{10}Ne$ (b) $^{7}_{3}Li$ (c) $^{11}_{5}B$

1.43   Sketch an atom of each of the following isotopes:
(a) $^{9}_{4}Be$ (b) $^{4}_{2}He$ (c) $^{10}_{5}B$

1.44   Fill in the missing information for neutral atoms.

Isotope Symbol	Element Name	Mass Number (A)	Neutrons (n°)	Protons (p⁺)	Electrons (e⁻)
C-13	carbon	13	7	6	6
	carbon	12			
		66		30	
^{65}Cu					

1.45   Fill in the missing information for neutral atoms.

Isotope Symbol	Element Name	Mass Number (A)	Neutrons (n°)	Protons (p⁺)	Electrons (e⁻)
^{109}Ag					
Si-28					
		40			18

1.46   Which of the following sets of data indicate that the two atoms described are isotopes of one another?

	Atom #1			Atom #2		
	n°	p⁺	e⁻	n°	p⁺	e⁻
(a)	60	47	47	62	47	47
(b)	7	7	7	7	7	10
(c)	74	53	53	74	54	54

1.47   Which of the following sets of data indicate that the two atoms described are isotopes of one another?

	Atom #1			Atom #2		
	n°	p⁺	e⁻	n°	p⁺	e⁻
(a)	22	20	20	20	22	20
(b)	12	12	12	14	12	12
(c)	20	19	19	19	20	19

SECTION 1.7: ATOMIC MASS

1.48   Each set of data below lists two isotopes of a given element. Indicate which one has a higher abundance by referring to the atomic mass on the periodic table.
(a) Ti-48 or Ti-46
(b) Ca-40 or Ca-42
(c) Ba-134 or Ba-138

1.49   Each set of data below lists two isotopes of a given element. Indicate which one has a higher abundance by referring to the atomic mass on the periodic table.
(a) Ni-58 or Ni-60
(b) K-39 or K-41
(c) Fe-54 or Fe-56

1.50   Which of the following statements *could* be true according to the atomic masses on the periodic table?
(a) Si-28 is higher in abundance than Si-30.
(b) Argon consists of a roughly 50:50 mix of Ar-36 and Ar-40.
(c) Strontium is predominantly composed of Sr-86.
(d) Neon is composed of a roughly equal mix of Ne-20 and Ne-21.

1.51   Which of the following statements *could* be true according to the atomic masses on the periodic table?
(a) La-138 is higher in abundance than La-139.
(b) Nitrogen is composed of a roughly equal mix of N-14 and N-15.
(c) Magnesium is predominantly composed of Mg-26.
(d) Carbon consists of a roughly 50:50 mix of C-13 and C-12.

1.52   Calculate the atomic mass of bromine, given the following information about bromine's stable isotopes.

	Mass	Percent Abundance
Br-79	78.9183 amu	50.70%
Br-81	80.9163 amu	49.32%

1.53   Determine the atomic mass of potassium, given the following information about potassium's stable isotopes.

	Mass	Percent Abundance
K-39	38.96370 amu	93.258%
K-41	40.96183 amu	6.730%

1.54   Determine the atomic mass of magnesium, given the following information about magnesium's stable isotopes.

	Mass	Percent Abundance
Mg-24	23.9850 amu	78.99%
Mg-25	24.9858 amu	10.00%
Mg-26	25.9826 amu	11.01%

1.55   Determine the atomic mass of silicon, given the following information about silicon's stable isotopes.

	Mass	Percent Abundance
Si-28	27.9769265 amu	92.223%
Si-29	28.9764947 amu	4.685%
Si-30	29.9737702 amu	3.092%

1.56   Determine the atomic mass of strontium, given the following information about strontium's stable isotopes.

	Mass	Percent Abundance
Sr-86	85.9092607 amu	9.861%
Sr-87	86.9088775 amu	7.001%
Sr-88	87.9056122 amu	82.581%

1.57   Determine the atomic mass of iron, given the following information about iron's stable isotopes.

	Mass	Percent Abundance
Fe-54	53.9396 amu	5.845%
Fe-56	55.9349 amu	91.754%
Fe-57	56.9354 amu	2.119%

1.58   Determine the percent abundance of the two stable isotopes of silver, given the information below and the average atomic mass of silver on the periodic table.

	Mass	Percent Abundance
Ag-107	106.9051 amu	? %
Ag-109	108.9048 amu	? %

1.59   Determine the percent abundance of the two stable isotopes of boron, given the information below and the average atomic mass of boron on the periodic table.

	Mass	Percent Abundance
B-10	10.01294 amu	? %
B-11	11.00931 amu	? %

1.60   The following figure is a representation of 20 atoms of a fictitious element with the symbol Mx and atomic number 125. Mx has two isotopes represented by the colors: Mx-315 (white) and Mx-318 (black). Use the drawing (statistically representative of naturally occurring Mx) and the following masses of each isotope to calculate the atomic mass of Mx.

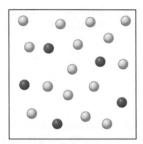

Mx-315 (white) = 314.965 amu
Mx-318 (black) = 317.985 amu

1.61   The following figure is a representation of twelve atoms of a fictitious element with the symbol Bu and atomic number 128. Bu has two isotopes represented by the colors: Bu-333 (green) and Bu-337 (orange). Use the drawing (statistically representative of naturally occurring Bu) and the following masses of each isotope to calculate the atomic mass of Bu.

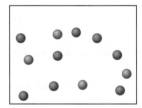

Bu-333 (green) = 332.979 amu
Bu-337 (orange) = 336.991 amu

Answers to In-Chapter Materials

Answers to Practice Problems

1.1A Only (c) and (d) are neutral atoms. The charges on the others are: (a) +3 and (b) +2. **1.1B** (a) 21, (b) 33, (c) 53, (d) −2. **1.2A** Bromine. **1.2B** 53. **1.3A** (a) Rb, 37; (b) krypton, Kr. **1.3B** (a) The chemical symbol of iron is Fe. (c) The atomic number of sodium is 11. **1.4A** (a) nonmetal, (b) metal, (c) metal, (d) nonmetal, (e) metalloid. **1.4B** (a) carbon, (b) boron, (c) bismuth or moscovium, (d) nitrogen or phosphorus, (e) tin, lead, or flerovium. **1.5A** (a) 5, 5, 5; (b) 18, 18, 18; (c) 38, 47, 38; (d) 6, 5, 6. **1.5B** (a) ^{9}Be, (b) ^{51}V, (c) ^{124}Xe, (d) ^{69}Ga. **1.6A** (a) Mg-24, (b) Li-7, (c) Ta-181. **1.6B** (a). **1.7A** 14.01 amu. **1.7B** 20.20 amu.

Answers to Checkpoints

1.3.1 b. **1.3.2** b and c. **1.3.3** b. **1.3.4** b. **1.4.1** (c) and (e). **1.4.2** a, b, c, e. **1.5.1** c. **1.5.2** b. **1.5.3** d. **1.6.1** d. **1.6.2** c. **1.7.1** b. **1.7.2** c.

CHAPTER 2

Electrons and the Periodic Table

2.1 **The Nature of Light**

2.2 **The Bohr Atom**

2.3 **Atomic Orbitals**
- *s* orbitals
- *p* orbitals
- *d* and *f* orbitals

2.4 **Electron Configurations**

2.5 **Electron Configurations and the Periodic Table**

2.6 **Periodic Trends**

2.7 **Ions: The Loss and Gain of Electrons**
- Electron Configuration of Ions
- Lewis Dot Symbols of Ions

Sodium chloride is a ubiquitous additive in processed foods, and is a staple on the dinner table for most people. The main ingredient in some salt substitutes is *potassium* chloride, which looks the same as sodium chloride and, according to some, tastes very similar. These two substances are so similar because sodium (Na) and potassium (K) have very similar properties. This is the reason they appear in the same column on the periodic table.

David A. Tietz/McGraw Hill

In This Chapter, You Will Learn

About some of the interesting properties of *light,* also known as *electromagnetic radiation;* and about how light has been used to further our understanding of *atomic structure.* You will also learn how to determine the arrangement of electrons in an atom, and how that arrangement impacts some of the atom's properties.

Things To Review Before You Begin

- Nuclear model of the atom [◄◄ Section 1.3]
- Chapter 1 Key Skills [◄◄ pages 24–25]

In Chapter 1, we learned about the nuclear atom, in which all of an atom's positive charge, and *nearly* all of its mass, are concentrated in a tiny, dense central core called the *nucleus.* We also learned that most atomic nuclei contain both *protons,* which are positively charged, and *neutrons,* which carry no charge. Although this nuclear model of the atom was a tremendous leap forward in our understanding of atomic structure, it did not provide much information about the location of the negatively charged particles in the atom, the *electrons.* In this chapter, we will learn more about the internal structure of the atom, including the arrangement of electrons, and about how the number, nature, and arrangement of the subatomic particles give rise to the properties of the atoms that contain them. Much of what we know about the nature of atoms is the result of experiments with *light.* Therefore, before delving into our current view of the atom and atomic structure, we will spend a little bit of time discussing light itself.

2.1 The Nature of Light

Light is one of the ways that energy can be transmitted in the form of *waves.* (Sound is another.) All waves have certain features in common, including wavelength and frequency, which are illustrated for two different waves in Figure 2.1. *Wavelength* (symbolized with the Greek letter lambda, λ) is the distance between identical points (usually *peaks*) on successive waves. *Frequency* (symbolized with the Greek letter nu, ν) is the number of waves that pass through a particular point in 1 second. Note that there is a reciprocal relationship between wavelength and frequency. As wavelength increases, frequency decreases—and vice versa. Further, the *energy* transmitted by light is *inversely* proportional to wavelength and *directly* proportional to frequency:

- As wavelength increases, energy decreases.
- As frequency increases, energy increases.

When we say "light," we almost always mean *visible* light. Many of us remember a project in grade school science in which a prism was used to separate sunlight into its colored components [Figure 2.2(a)]. Even if you have never done such a project, you have certainly seen a rainbow, which is separation of sunlight into colors by water droplets in the atmosphere—usually after a rainstorm [Figure 2.2(b)]. The

> **Student Note:** Visible light refers to light that humans can see. For some other species, the visible portion of the spectrum includes wavelengths not visible to us. For example, many insects, reptiles, and birds can see wavelengths in the ultraviolet range—although most that can cannot see as far into the red range as humans can. Interestingly, in some bird species, to us, males and females appear identical—but with ultraviolet vision, the birds themselves can see the difference.

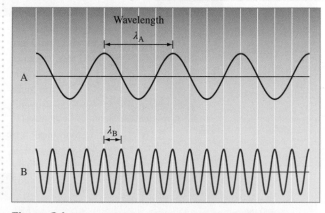

Figure 2.1 Two waves with different wavelengths and frequencies. Wave A has the greater wavelength, by a factor of four; and Wave B has the greater frequency, also by a factor of four. Wave A is lower energy light, and wave B is higher energy light.

Figure 2.2 (a) A prism is used to separate sunlight into its colored components. (b) Water droplets serve as tiny prisms to separate sunlight into a rainbow. (c) Both phenomena separate sunlight into its visible spectrum.

(a) Don Farrall/Photodisc/Getty Images
(b) Ron Worobec/500px Prime/Getty Images
(c) Dorling Kindersley/Getty Images

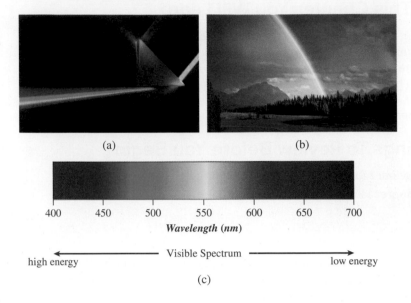

(a)   (b)

400   450   500   550   600   650   700
Wavelength (nm)

Visible Spectrum

high energy   low energy

(c)

array of colors that make up sunlight, also called "white light," is known as the Sun's *visible **emission spectrum*** [Figure 2.2(c)].

Although visible light is by far the most familiar form, we know that there are other kinds of light. Most of us wear sunscreen to protect our skin from *ultraviolet* light, which is not visible to humans. In fact, both visible light and ultraviolet light are parts of the ***electromagnetic spectrum*** (Figure 2.3). The electromagnetic spectrum arranges all types of *electromagnetic radiation* (a more technical term for *light*) in order of increasing wavelength. The shortest wavelengths are gamma rays, and the longest are radio waves, which can be many meters long. The wavelengths of the various colors in Figure 2.2 are listed in *nanometers* (nm). (There are 1,000,000,000 nanometers in a meter—nanometers are *very* small.) As you can see, our familiar visible spectrum occupies only a very small portion of the overall electromagnetic spectrum.

Student Note: We explain more about the units used in chemistry, including those used to express wavelengths, in Chapter 4.

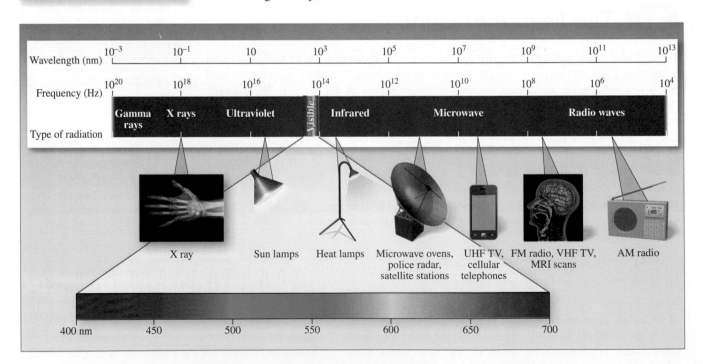

Figure 2.3 The electromagnetic spectrum.

(hand x-ray) itsmejust/Shutterstock; (head MRI) Don Farrall/Photodisc/Getty Images

Figure 2.4 The visible emission spectra of (a) a neon light, (b) white light (sunlight), (c) helium, and (d) hydrogen.
(a–d) H.S. Photos/Alamy Stock Photo

The Sun is not the only thing that gives off visible light. In fact, most anything that is visible to us is either *radiating* or *reflecting* visible light. Using a technique similar to that used to separate sunlight into its component colors, we can separate the red light given off by a neon sign as illustrated in Figure 2.4(a). You will note that the visible spectrum of light emitted by neon is different from that emitted by the Sun. The emission spectrum of sunlight [Figure 2.4(b)] is *continuous*—meaning that it contains *all* of the visible wavelengths. The emission spectrum of the neon light contains many fewer wavelengths, leaving much of the spectrum black. The emission spectra of helium [Figure 2.4(c)] and hydrogen [Figure 2.4(d)] contain even fewer lines, with hydrogen having only *four* distinct wavelengths in its visible emission spectrum. Emission spectra that are not continuous, but contain only discrete *lines* (such as those of helium and hydrogen), are called ***line spectra***. Line spectra turned out to be critical to our development of models of atomic structure.

Student Note: "Spectra" is the plural of "spectrum."

Thinking Outside the Box

Laser Pointers

Popular laser pointers usually emit light in the red region of the visible spectrum, with wavelengths ranging from 630 to 680 nm. Low cost and wide availability have made the devices popular not only with instructors and other speakers, but also with teenagers and even children—raising some significant concern over safety. Although the human blink reflex is generally sufficient to protect against serious or permanent injury to the eye by one of these devices, intentional prolonged exposure of the eye to the beam from a laser pointer can be dangerous. Of particular concern are the new green laser pointers that emit a wavelength of 532 nm. The lasers in these devices also produce radiation in the infrared region of the electromagnetic spectrum (1064 nm), but they are equipped with filters to prevent the infrared radiation from leaving the device. However, some of the inexpensive lasers do not have adequate safety labeling, and the filters are easily removed—potentially allowing the emission of dangerous radiation. Although the 1064-nm laser beam delivers less energy than the shorter wavelengths, it poses a greater danger to the eye because it is not visible and does

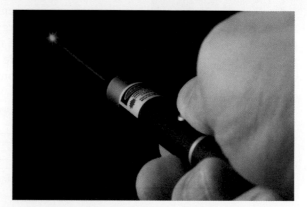

Common laser pointer. Robin Lund/Alamy Stock Photo

not trigger the blink response that visible wavelengths do. A 1064-nm beam can pass through the eye and damage the retina. Because the beam is not visible, the damage is not noticed immediately, but it can be permanent.

Profiles in Science

Patricia Bath

A first in many endeavors, Patricia Bath (1942–2019) earned her MD at Howard University and was the first African American to complete a residency in ophthalmology (UCLA). She was also the first female faculty member and first female chair of the Department of Ophthalmology at UCLA. She co-founded the American Institute for the Prevention of Blindness and, in 1988, was the first African American woman to receive a medical patent. The patent was for her invention of the Laserphaco Probe, which fundamentally transformed the treatment of cataracts and is now used internationally. This and other inventions facilitated the treatment of various causes of blindness. The prevention and treatment of eyesight disorders, especially in underserved communities, was of profound importance to Dr Bath. In 1993, she was named a Howard University Pioneer in Academic Medicine; and in 2018, she was awarded the New York Academy of Medicine John Stearns Medal for Distinguished Contributions in Clinical Practice.

Patricia Bath

Jemal Countess/Stringer/
Getty Images

2.2   The Bohr Atom

Niels Bohr

Library of Congress Prints & Photographs
Division [LC-DIG-ggbain-35303]

Early in the twentieth century, a radical notion emerged in the scientific community. Several brilliant scientists, including Max Planck and Albert Einstein, proposed that some of their experimental results could only be explained by considering light to be a stream of tiny particles, rather than simply energy being transmitted via waves. It's relatively easy for us to think of matter as **quantized,** meaning it consists of tiny, discrete *particles* (atoms). But it may be more difficult to think of energy this way. Energy had always been thought to be *continuous,* rather than quantized. To visualize the difference on a macroscopic scale, consider two possible routes between platforms at different levels (Figure 2.5). One route is a set of steps. The other is a ramp. If we were to place an object on the ramp, it could rest at any elevation, simply depending on where we put it. If we were to place an identical item on the steps, however, there would be only certain, specific elevations available. Now imagine the elevation of the rubber duck in the illustration to represent an amount of energy. On the ramp, the duck's elevation can represent any amount of energy. On the stairs, however, the duck's elevation can represent only five different, specific amounts of energy. This is the gist of quantization. Energy, specifically *light,* cannot exist in just *any* amount. Like matter, light must consist of tiny, discrete amounts—analogous to atoms. These tiny, discrete "packets" of light are known as **photons.** This radical notion, that light is *quantized,* was used in clever combination with line spectra to develop the next model of the atom.

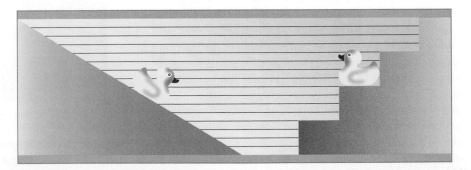

Figure 2.5 The ramp on the left represents a continuous path between the upper and lower platforms. The rubber duck can be placed at any level between the two platforms. The steps on the right represent a quantized path. Only five different levels (including the upper and lower platforms) are possible via this route.

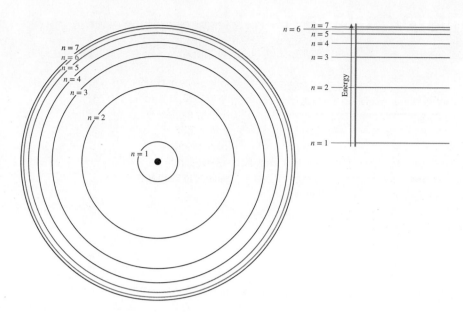

Figure 2.6 The Bohr model of the atom.

Niels Bohr, a Danish physicist, combined the line spectrum of hydrogen [Figure 2.4(d)] with the concept of quantization to conceive a model that we now refer to as the **Bohr atom** (Figure 2.6). Because the line spectrum of hydrogen always contained the same four lines—never a *different* number of lines, always four, and always at the *same* wavelengths—Bohr envisioned the hydrogen atom as having discrete energy levels, like steps, at which its electron could reside. (Remember that a hydrogen atom has just *one* electron.) He described these as a series of concentric, circular paths around the nucleus called **orbits.** Each orbit was at a specific energy level, which was designated by the use of a **quantum number, n.** He assumed that ordinarily, the electron in a hydrogen atom resided at the lowest possible energy level ($n = 1$), which he called the **ground state.** Further, he reasoned that when energy was added to a sample of hydrogen atoms, the atoms absorbed some of the energy. The absorption of energy caused electrons to be promoted to higher energy levels called **excited states,** where n was any integer greater than 1 (n could be 2, 3, 4, 5, etc.). Then, when the electrons in excited states subsequently *relaxed* (returned to lower energy levels), they were emitting, in the form of electromagnetic radiation (light), some of the energy that had been used to excite them. Because his model only contained certain, specific energy levels, the energy given off could only be certain, specific amounts—corresponding to certain, specific *wavelengths* of emitted light. This explained the four specific, discrete lines in the visible emission spectrum of hydrogen.

According to Bohr's model, excited electrons can relax to any orbit with a lower value of n. So an electron that has been excited to the $n = 9$ orbit may relax to the $n = 4$ orbit. Another may relax from $n = 4$ to $n = 1$, still another from $n = 5$ to $n = 2$. The number of possible downward transitions is endless. But when the electron transition is one of these:

$n = 6$ to $n = 2$

$n = 5$ to $n = 2$

$n = 4$ to $n = 2$

$n = 3$ to $n = 2$

the light emitted is in the visible region of the electromagnetic spectrum. Although energy is given off whenever an excited electron relaxes to a lower orbit, only these four transitions result in the emission of *visible* light—and each one corresponds to one of the lines in hydrogen's emission spectrum. The electronic transitions responsible for the line spectrum of hydrogen are illustrated in detail in Figure 2.7 (Visualizing Chemistry).

Animation
Hydrogen Emission Spectrum

Figure 2.7

Visualizing Chemistry – Bohr Atom

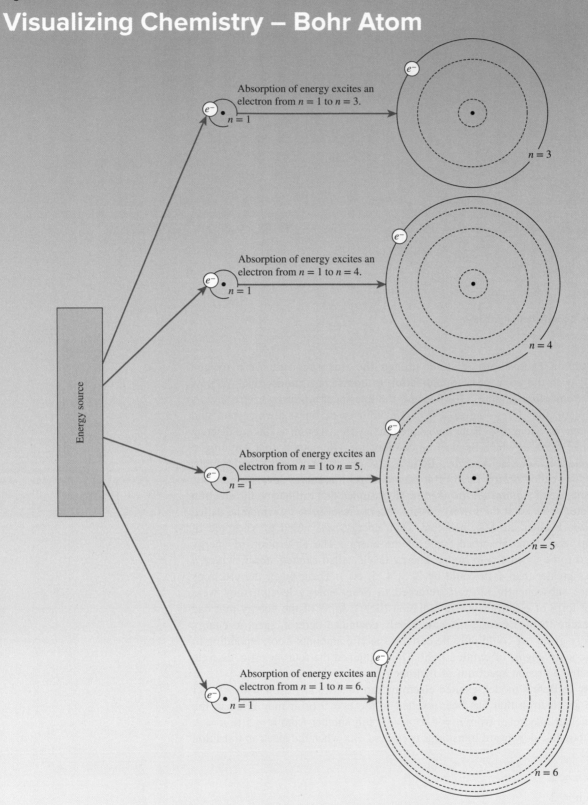

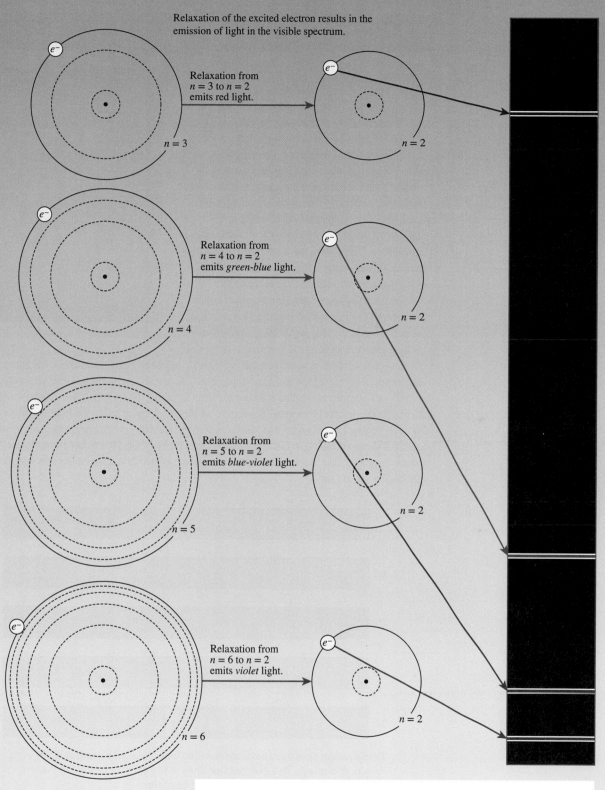

Relaxation of the excited electron results in the emission of light in the visible spectrum.

Relaxation from $n = 3$ to $n = 2$ emits red light.

$n = 3$

$n = 2$

Relaxation from $n = 4$ to $n = 2$ emits *green-blue* light.

$n = 4$

$n = 2$

Relaxation from $n = 5$ to $n = 2$ emits *blue-violet* light.

$n = 5$

$n = 2$

Relaxation from $n = 6$ to $n = 2$ emits *violet* light.

$n = 6$

$n = 2$

What's the point?

Each line in the visible emission spectrum of hydrogen is the result of an electronic transition from a higher excited state ($n = 3, 4, 5,$ or 6) to a lower excited state ($n = 2$). The energy gap between the initial and final states determines the wavelength of the light emitted.

(See Visualizing Chemistry questions VC 2.1–VC 2.4 on page 71.)

Familiar Chemistry

Fireworks

Most of us have watched colorful fireworks displays, but have you ever wondered what produces the different colors that light up the sky? Just as the hydrogen atom emits light of particular colors (wavelengths), so does every other element when electrons (excited by the addition of energy) relax to lower energy levels. A fireworks shell is loaded with gunpowder and a carefully selected combination of substances, including certain metals. When the gunpowder explodes, it provides energy to the metals and excites electrons within them to higher energy levels. When these electrons subsequently return to a lower energy level, visible light is emitted, and we see the resulting brilliant colors of fireworks.

rozbyshaka/iStock/Getty Images

The following figure shows the simplified spectra of five metals commonly used in fireworks. As you can see, each spectrum is unique. Although an emission spectrum may contain several different wavelengths of light, each metal has a dominant color that is emitted to a greater extent (indicated in the figure by the exaggerated width of that line in the spectrum). This gives each metal a characteristic emission color that we observe when they are ignited in a fireworks display. Every element has its own unique emission spectrum that is a sort of elemental "fingerprint." In addition to being useful for fireworks, these unique elemental fingerprints can be used to determine both the identities and the amounts of the elements that make up a sample of matter. In fact, emission spectra have been used to identify elements since the nineteenth century, long before scientists knew about the atomic structure and electron transitions that actually give rise to them.

Emission line spectra of several elements that are used as colorants in fireworks displays. The exaggerated lines in the spectra represent the emissions responsible for the colors we observe.

Familiar Chemistry

The Photoelectric Effect

One of the fascinating phenomena observed by scientists in the nineteenth century, and that helped us develop our current model of the atom, is the *photoelectric effect*. The photoelectric effect is the emission of electrons from a metal surface that occurs when light of appropriate wavelength shines on the metal. This may sound like a foreign and remote concept, but the photoelectric effect is something that you very likely encounter every day. One of its common uses is in the type of device that prevents a garage door from closing when something (or someone) is in the door's path. Another is in motion-detection systems used in museums and other high-security environments. Still another is the mechanism by which barcode scanners operate. These sorts of devices work simply by responding to an interruption in a beam of light. When the beam of light shining on a metal target is interrupted, electrons cease to be emitted by the metal, and the flow of electrons (electricity) stops. In the case of the garage-door safety device, when the flow of electrons from the metal stops, the movement of the door stops. In the case of motion-detection systems, an alarm may sound or a light may turn on. The mechanism by which a barcode scanner works is less straightforward, but it still relies on the interruption of a beam of light by a specific pattern of dark lines.

RTimages/Alamy Stock Photo

Light from
Barcode
Scanner

Bar code

Sample Problem 2.1 gives you some practice identifying electron transitions using the Bohr model of the atom.

SAMPLE PROBLEM 2.1 Using the Bohr Model to Identify Electron Transitions That Emit Visible Light

According to the Bohr model of the atom, indicate which of the following electron transitions would be expected to emit visible light: (a) $n = 7$ to $n = 4$, (b) $n = 2$ to $n = 3$, (c) $n = 5$ to $n = 2$, (d) $n = 6$ to $n = 1$, (e) $n = 4$ to $n = 2$, (f) $n = 7$ to $n = 2$.

Strategy You have learned that in the Bohr model of the atom, energy is required to excite electrons from the ground state to various excited states. Further, there are specific transitions that result in the emission of *visible* wavelengths.

Setup To emit energy, an electron transition must be from a higher value of n to a lower value of n. For the energy emitted to be a visible wavelength of light, the transition must have a final n value of 2, and an initial n value of 3, 4, 5, or 6.

Solution The transitions in (c) and (e) would result in emission of visible light.

THINK ABOUT IT

Although the transition in (f) is to $n = 2$, it would not result in a visible emission because it does not begin at $n = 3$, 4, 5, or 6. The light emitted by the transition from $n = 7$ to $n = 2$ would occur in the *ultraviolet* region of the electromagnetic spectrum.

Practice Problem Ⓐ**TTEMPT** Indicate which of the following electron transitions would be expected to emit visible light in the Bohr model of the atom:

(a) $n = 6$ to $n = 2$ (c) $n = 6$ to $n = 3$ (e) $n = 3$ to $n = 2$
(b) $n = 3$ to $n = 8$ (d) $n = 2$ to $n = 5$ (f) $n = 9$ to $n = 2$

Practice Problem Ⓑ**UILD** Indicate which of the following electron transitions would be expected to emit *any* wavelength of light (not just visible) in the Bohr model of the atom:

(a) $n = 6$ to $n = 1$ (c) $n = 7$ to $n = 3$ (e) $n = 6$ to $n = 3$
(b) $n = 4$ to $n = 5$ (d) $n = 1$ to $n = 6$ (f) $n = 2$ to $n = 8$

Practice Problem Ⓒ**ONCEPTUALIZE** By looking at Figure 2.3, you can see that the shortest wavelengths are associated with the highest-energy radiation. X rays, for example, are used to image bone precisely because they are sufficiently energetic to penetrate soft tissue. Considering this, indicate the region of the electromagnetic spectrum where you would expect to find the radiation given off by the following electron transitions:

(a) $n = 6$ to $n = 1$ (b) $n = 6$ to $n = 3$

(Recall that the *visible* transitions are those from $n = 6$ to $n = 2$, $n = 5$ to $n = 2$, $n = 4$ to $n = 2$, and $n = 3$ to $n = 2$.)

CHECKPOINT–SECTION 2.2   The Bohr Atom

2.2.1 Which of the following quantum numbers corresponds to the ground state in the Bohr model of the hydrogen atom? (Select all that apply.)

a) $n = 0$ d) $n = 3$
b) $n = 1$ e) $n = 4$
c) $n = 2$

2.2.2 According to the Bohr model, which of the following electron transitions would result in the emission of energy in the form of electromagnetic radiation? (Select all that apply.)

a) $n = 1$ to $n = 3$ d) $n = 4$ to $n = 7$
b) $n = 3$ to $n = 1$ e) $n = 6$ to $n = 3$
c) $n = 3$ to $n = 2$

2.3 Atomic Orbitals

Although Bohr's model elegantly described the electron transitions responsible for the line spectrum of hydrogen (his work earned him the Nobel Prize in Physics in 1922), it failed to explain or predict the emission spectra of any other elements. It turns out that when an atom has more than one electron, the model has to be somewhat more sophisticated. Enter the *quantum mechanical* model of the atom.

With the failure of Bohr's model to apply to anything having more than one electron, scientists knew they needed a new approach to advance their understanding of atomic structure. In 1924, French physicist Louis de Broglie provided that new approach. He postulated that if energy (light) can behave like a stream of particles (photons) under certain circumstances, then perhaps *particles* such as electrons can, under certain circumstances, exhibit *wavelike* properties. The upshot is that our current model of atomic structure, the *quantum mechanical* or **QM model,** does not try to describe the location of an electron, but rather defines a region of space called an *orbital,* in which the electron is most likely to be found. Unlike Bohr's *orbits,* which were all circular and varied only in size, *orbitals* also vary in shape. Although the mathematical development of the QM model of the atom is beyond the scope of this book, the result is important to your understanding of atomic structure. Consider the following:

When you light a sparkler, you get a roughly spherical cloud of very hot, glowing sparks. If you were able to assess, at any point in time, the locations of the sparks in the region of space around the burning flame of a sparkler, you would find the highest density of sparks near the center, as Figure 2.8(a) shows. The density of glowing sparks is higher near the center because the sparks cool as they move away from the center of the flame. So, although some sparks do get relatively distant from the center of the flame, the density of sparks is greatest near the center, and it diminishes with increasing distance from the center.

If we were to plot the coordinates of every spark in the flame pictured here, we could call it a *spark density map.* For this particular sparkler, we might estimate that 90 percent of the sparks are contained within a spherical region with a radius of 5 centimeters [Figure 2.8(b) and (c)].

Unlike the sparks that emanate from a burning sparkler, we cannot know or map the coordinates of electrons in an atom. However, the use of quantum mechanics enables us to calculate the *probability* of finding an electron in a region of space around the nucleus and to construct a *probability density map* of possible locations for an electron. Such a probability map of electron density within an atom is called an **atomic orbital.**

> **Student Note:** These calculations are quite complex and you will not be required to do them in this course.

To calculate the sizes and shapes of atomic orbitals, we need more than just the integer quantum number from Bohr's model, which simply indicated an orbit's distance from the nucleus. In the QM model, n is called the **principal quantum number,** and it can still have any integer value. It tells us the **principal energy level** (sometimes referred to as the *shell*) in which an electron resides. In addition to the principal quantum number, the QM model requires a letter designation: *s, p, d,* or *f,* which indicates the **energy sublevel** (sometimes referred to as the *subshell*).

Each *principal* energy level contains one or more energy *sublevels;* and each energy sublevel contains one or more *orbitals.* But before you get the impression that this is terribly complicated, consider this: The number of energy sublevels and the number of orbitals are *predictable,* and they depend on the value of the principal quantum number, which is an integer. Table 2.1 shows the possible designations for atomic orbitals for principal energy levels 1 through 4.

Note that the number of energy sublevels in each principal energy level is equal to the principal quantum number, n. There is one sublevel (*s*) in $n = 1$, there are two sublevels (*s* and *p*) in $n = 2$, and so on. Note, too, that the number of *orbitals* in energy sublevels with the same *letter* is always the same. That is, an energy sublevel designated *s* always contains just one orbital; an energy sublevel designated *p* always contains *three* orbitals; and so on. At this point, it will be useful to construct a clarifying picture.

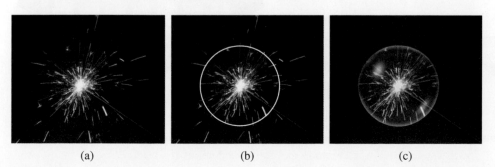

(a)   (b)   (c)

Figure 2.8 (a) A burning sparkler produces a roughly spherical cloud of sparks. (b) Most (90 percent) of the sparks from this sparkler are located within a sphere with a 5-centimeter radius. (c) The sphere that contains 90 percent of the sparks represented by a solid surface.

(a–c) Imagemore/Getty Images

TABLE 2.1		
When *n* is	the energy sublevel letter designation can be	and the number of orbitals in the energy sublevel is
1	*s*	1
2	*s*	1
	p	3
3	*s*	1
	p	3
	d	5
4	*s*	1
	p	3
	d	5
	f	7

Figure 2.9 shows a series of trendy V-shaped cabinets designed to display shoes in boxes at a new shoe store. These cabinets will help illustrate the concepts of principal energy levels, energy sublevels, and orbitals.

The smallest of the cabinets [Figure 2.9(a)] represents the lowest principal energy level, where $n = 1$. In this cabinet, there is just one shelf—at floor level, which represents the energy sublevel *s*. On this shelf, there is room for just one shoebox, which represents an orbital—specifically, the 1*s* orbital.

An orbital is specified by its principal quantum number and its letter designation.

1*s*

The second cabinet [Figure 2.9(b)] represents the second principal energy level, where $n = 2$. This cabinet has two shelves: one at floor level, corresponding to the *s* energy sublevel, and one above it, corresponding to the *p* energy sublevel. As with the first cabinet, the lower shelf has room for just one shoebox, which represents the 2*s* orbital. On the next shelf, though, there is room for *three* shoeboxes, which represent the three 2*p* orbitals.

The third cabinet [Figure 2.9(c)] represents the third principal energy level, where $n = 3$. This cabinet has three shelves: one at floor level, corresponding to the *s* energy sublevel, one directly above it, representing the *p* energy sublevel, and one directly above *that,* representing the *d* energy sublevel. As before, the lowest shelf (*s* sublevel) contains one shoebox (the 3*s* orbital) and the second shelf up (*p* sublevel) contains three shoeboxes (the three 3*p* orbitals). The third shelf up (*d* sublevel) contains five shoeboxes, representing the five 3*d* orbitals.

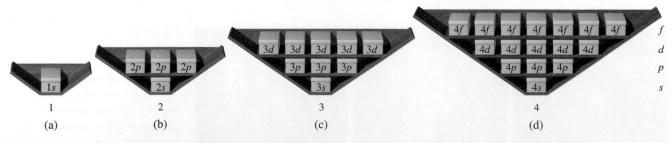

Figure 2.9 (a)–(d) V-shaped cabinets illustrate the concepts of principal energy levels, energy sublevels, and orbitals.

The fourth cabinet [Figure 2.9(d)] follows the same pattern, which by now should be familiar, but adds a *fourth* shelf corresponding to the *f* sublevel, which contains seven 4*f* orbitals.

Student Note: The orbitals represented by shoeboxes in the fourth cabinet, from bottom to top, are: one 4*s* orbital, three 4*p* orbitals, five 4*d* orbitals, and seven 4*f* orbitals.

So far, our discussion of atomic orbitals has been fairly abstract and analogy-driven. Now let's look at some representations of the orbitals—and learn what their designations tell us.

s orbitals

The 1*s* orbital [Figure 2.10(a)] looks a lot like the sparkler flame in Figure 2.8(a). It is spherical, the probability of finding an electron (the *electron density*) is highest near the nucleus, and the electron density decreases as we move away from the nucleus. For the sake of neat illustration, we generally picture atomic orbitals as solid shapes, rather than as electron-density maps. As with the sphere superimposed on the sparkler flame in Figure 2.8, the solid shape encompasses a volume within which 90 percent of the electron density is contained. The probability of finding an electron does not drop to zero at the surface of the solid shape; rather, orbitals by their very nature have *fuzzy* edges. Figure 2.10(b) shows a solid sphere superimposed on the electron-density map of the 1*s* orbital, and Figure 2.10(c) shows just the solid sphere generally used to represent the 1*s* orbital. In a hydrogen atom in the ground state, the lowest energy state possible, the electron resides in the 1*s* orbital.

Figure 2.10 (a) An electron-density map of the 1s orbital; (b) a solid sphere superimposed on the electron-density map; and (c) the solid sphere typically used to represent the 1s orbital.

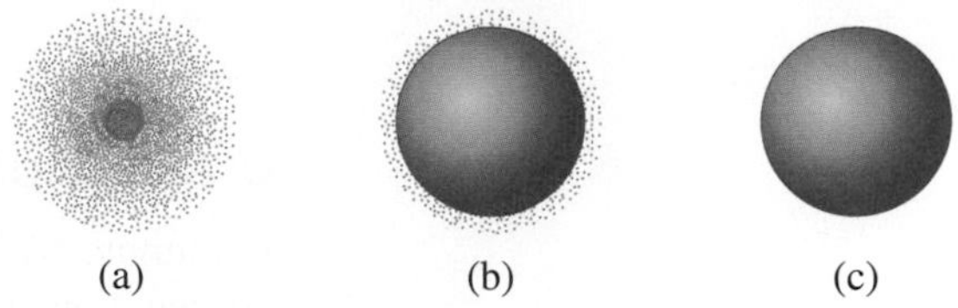

The 2*s* orbital looks very much like the 1*s* orbital, just bigger; and so on for the 3*s* and 4*s* orbitals (Figure 2.11). All of the *s* orbitals are spherical, with the electron density highest near the nucleus and decreasing with increasing distance from the nucleus.

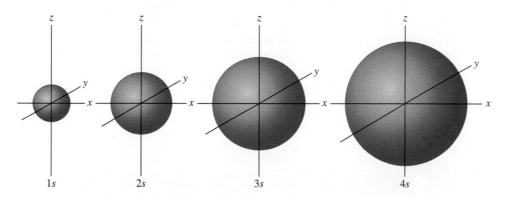

Figure 2.11 1*s*, 2*s*, 3*s*, and 4*s* orbitals represented with solid spheres.

p orbitals

Beginning with the second principal energy level ($n = 2$), in addition to the *s* sublevel, there is also a *p* sublevel. A *p* sublevel contains three *p* orbitals that are shaped like dumbbells. Each of the *p* orbitals lies along one of the axes, making them all perpendicular to one another. Figure 2.12(a) shows the electron-density map of a 2*p* orbital that lies along the *z* axis, and Figure 2.12(b) through (d) show the solid-surface representations of all three orbitals in the 2*p* sublevel. Note that the axis along which a *p* orbital lies is indicated with a subscript in the orbital designation.

Student Note: Remember each of these shapes is the result of a quantum mechanical calculation and represents the probability of finding an electron in a region of space around a nucleus. The principal energy level of an orbital determines its size. The letter designation determines its shape.

Figure 2.12 (a) Electron density map of a $2p_z$ orbital and solid-surface representations of (b) the $2p_x$ orbital; (c) the $2p_y$ orbital; and (d) the $2p_z$ orbital.

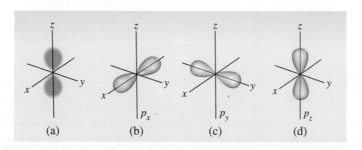

Figure 2.13 The *d* orbitals.

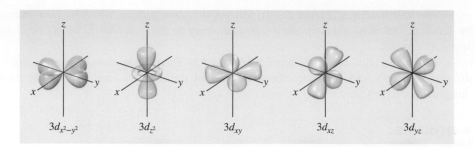

$3d_{x^2-y^2}$ $3d_{z^2}$ $3d_{xy}$ $3d_{xz}$ $3d_{yz}$

d and *f* orbitals

Beginning with the third principal energy level ($n = 3$), there is also a *d* sublevel containing five *d* orbitals, which are represented as solid surfaces in Figure 2.13. The shapes and designations are somewhat more complex than those of the *p* orbitals.

Finally, beginning in the fourth principal energy level ($n = 4$), there is an *f* sublevel, which contains seven *f* orbitals. The shapes of the *f* orbitals are beyond the scope of this book and are not shown.

Consider for a moment how much more complex these *orbitals* are than the orbits in Bohr's model of the atom—and how many *more* of them there are! Why then does the QM model of the atom predict the same lines in the emission spectrum of hydrogen as those correctly predicted by the Bohr model? According to the QM model, it appears that a hydrogen atom in the ground state, upon absorption of energy, might have its electron excited to a *multitude* of different orbitals. Why are there only four lines in the visible line spectrum of hydrogen? The answer to this question lies in the reason Bohr's model worked for the hydrogen in the first place: hydrogen has just *one* electron. In an atom with just one electron, the energy of an orbital depends only on the principal quantum number. Figure 2.14 shows energies of orbitals in the first four principal energy levels for a hydrogen atom. Note that regardless of the letter designations, orbitals in the same principal energy level have the same energy. Thus, if the electron in a hydrogen atom transitions from the *fourth* principal energy level to the *second* principal energy level, it doesn't matter if the transition is from 4*s* to 2*s*, or 4*d* to 2*p*, or 4*p* to 2*s*—each of those transitions

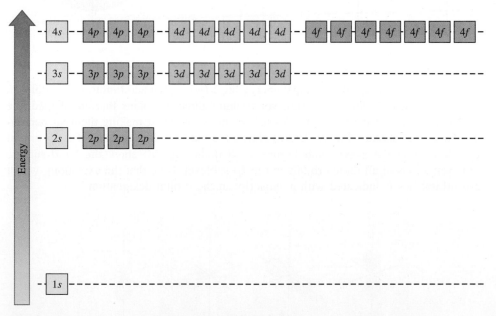

Figure 2.14 Orbital energy levels for a hydrogen atom. Each box represents one orbital. Orbitals with the same principal quantum number (*n*) all have the same energy.

corresponds to the same amount of energy being given up by the electron and so results in the same *wavelength* of light being emitted. The energy of an orbital in a hydrogen atom is determined only by the value of n, not by the letter designation. As we will see shortly, this is not true for atoms with more than one electron, which is precisely why the Bohr model did not work for other atoms.

All atoms other than hydrogen contain more than one electron. Because of the interaction between charged particles [◄◄ Section 1.2], the presence of multiple electrons in an atom changes the orbital energies. Figure 2.15 shows the energies of orbitals in the first four principal energy levels of a multielectron atom. In such cases, the energy of an orbital depends not only on its n value, which indicates its principal energy level, but also on its letter designation, s, p, d, or f. Within a given principal energy level, the relative orbital energies are:

$$s < p < d < f$$

One of the interesting results of the changes in orbital energies is that the $3d$ orbitals, which are in the *third* principal energy level, end up higher in energy than the $4s$ orbital, which is in the *fourth* principal energy level. The significance of this will become apparent when you have to write electron configurations for atoms in Section 2.4.

Sample Problem 2.2 lets you practice identifying orbital designations.

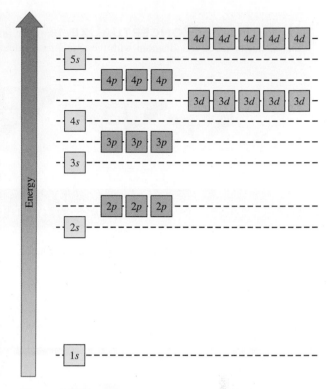

Figure 2.15 Orbital energy levels in a multielectron atom. The energy of an orbital depends both on its principal energy level, n, and on its letter designation—s, p, d, or f.

SAMPLE PROBLEM **2.2** Identifying Legitimate Combinations of Quantum Numbers to Designate Atomic Orbitals

Consider each of the following sublevel designations. Determine if each is a legitimate designation and if not, explain why: (a) $1p$, (b) $2s$, (c) $3f$, (d) $4p$.

Strategy You have learned that each energy level in an atom is divided into sublevels (Table 2.1). Further, the n value is equal to the number of sublevels present.

Setup The first level, $n = 1$, only contains an s sublevel, no p sublevel. The second level, $n = 2$, contains s and p sublevels. The $n = 3$ level contains s, p, and d sublevels. The $n = 4$ level contains s, p, d, and f sublevels.

Solution The sublevel designations in (b) and (d) are legitimate. As for the designations in (a) and (c), there is no p sublevel in the $n = 1$ level, so the $1p$ designation in (a) is not legitimate. And there is no f sublevel present in the third level, so the $3f$ sublevel designation in (c) is not legitimate.

THINK ABOUT IT

There is an s sublevel in the second level, so $2s$ is an appropriate sublevel designation. There is a p sublevel in all levels except $n = 1$, so $4p$ is a legitimate sublevel designation.

Practice Problem **A**TTEMPT Consider each of the following sublevel designations. Determine if each is a legitimate designation and if not, explain why: (a) $3f$, (b) $3s$, (c) $4d$, (d) $2d$.

Practice Problem **B**UILD Of principal energy levels 1–4, list all that can have each of the following sublevel letter designations: (a) p, (b) s, (c) f, (d) d.

(Continued on next page)

Practice Problem ⒸONCEPTUALIZE Using the analogy of the shoe cabinets from Figure 2.9, label each of the colored shoeboxes in this diagram with the appropriate value of *n* and the appropriate letter designation.

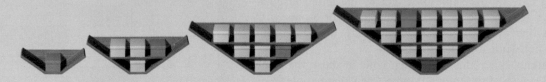

CHECKPOINT–SECTION 2.3   Atomic Orbitals

2.3.1 How many orbitals are there in the 3*p* subshell?

  a) 3   d) 7

  b) 5   e) 10

  c) 6

2.3.2 In a hydrogen atom, which orbitals are higher in energy than a 3*s* orbital? (Select all that apply.)

  a) 3*p*   d) 3*d*

  b) 4*s*   e) 4*p*

  c) 2*p*

2.4 Electron Configurations

The behavior of an atom, how it interacts with *other* atoms, depends on its **electron configuration,** the specific arrangement of electrons in the atom's orbitals. In this section, we will learn how to determine electron configurations using orbital diagrams—and how to write electron configurations correctly.

In the construction of orbital diagrams, we represent each orbital with a square box; we label each orbital with the appropriate *n* value and letter designation; and we represent each electron with an arrow. The simplest orbital diagram is that of hydrogen, which in the ground state has just one electron in the 1*s* orbital:

H   ⬛
     1*s*

We can use the orbital diagram to write the electron configuration. The electron configuration is a string of numbers and letters that indicates which energy sublevels are occupied, and by how many electrons. The energy sublevels are represented as they have been to this point, and the number of electrons is indicated with a superscript. (This will be made more clear by the examples that follow.) The only energy sublevel occupied in a ground-state hydrogen atom is 1*s,* and it contains just one electron. Therefore, the electron configuration of hydrogen is:

$1s^1$ ⟵ number of electrons
          in the energy sublevel

energy sublevel

> **Student Note:** Electrons don't literally spin like tops, but they have an intrinsic property with two possible orientations, which we call "spin."

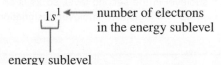

Before proceeding to the orbital diagram and electron configuration of helium, which has *two* electrons, it is necessary to introduce a property of electrons known as **spin.** Imagine an electron spinning on its vertical axis like a top. There are two possible directions for it to spin—and two electrons that are spinning in *opposite* directions are said to have *paired spins* (Figure 2.16).

Figure 2.16 Two electrons spinning in opposite directions.

When we represent two electrons with paired spins in the same box, we orient the arrows in opposite directions. The first arrow typically is pointed up, so the second arrow must point down.

To construct the orbital diagram of helium, we start with the diagram for hydrogen and simply add another arrow to represent the second electron:

Student Note: Remember that atoms are neutral because they have equal numbers of protons and electrons. So the number of electrons in an atom's orbital diagram is equal to its atomic number.

He ⟦↑↓⟧
     $1s$

From this, we can write the electron configuration for helium. $1s$ is still the only occupied energy sublevel, but in helium, the number of electrons in the sublevel is 2.

$1s^2$ ⟵ number of electrons
          in the energy sublevel

energy sublevel

We will now continue to construct orbital diagrams and write electron configurations for elements 3 (Li) through 18 (Ar). To do this, we will use our knowledge of the ordering of orbital energies (Figure 2.15), and apply a few simple rules:

1. Electrons will occupy the *lowest energy* orbital available.
2. Each orbital can accommodate a maximum of *two* electrons.
3. Any two electrons occupying the same orbital must have *paired* spins; that is, they must be spinning in opposite directions. (This is known as the **Pauli exclusion principle.**)
4. Orbitals of equal energy (those in the same energy sublevel) will each get *one* electron before any of them gets a *second* electron. (This is known as **Hund's rule.**)

After helium, the next element on the periodic table is lithium, which has three electrons. We begin with the orbital diagram of helium and add another electron. Because of rule 2: Each orbital can accommodate a maximum of *two* electrons, so we cannot put the third electron in with the first two. The $1s$ orbital, with two electrons in it, is full. The *third* electron must go into the next available orbital, with the lowest possible energy. According to Figure 2.15, that's the $2s$ orbital. The orbital diagram of lithium is:

Li ⟦↑↓⟧   ⟦↑⟧
     $1s$ $2s$

In lithium, there are two energy sublevels occupied by electrons: $1s$, which contains two electrons, and $2s$, which contains one electron. Therefore, the electron configuration of lithium is written as:

numbers of electrons in the energy sublevels

$1s^2\ 2s^1$

energy sublevels

Next is beryllium, with four electrons. The lowest energy orbital with room for another electron is the $2s$ orbital, so the orbital diagram and electron configuration of beryllium are:

numbers of electrons in the energy sublevels

$1s^2\ 2s^2$

Be ⟦↑↓⟧   ⟦↑↓⟧   and
     $1s$ $2s$

energy sublevels

Now the $1s$ and $2s$ orbitals are full. To construct a diagram for the next element, boron, with *five* electrons, we must identify the next available orbital. According to Figure 2.15, the next available orbital is one of *three* orbitals in the $2p$ energy sublevel. (By convention, when there are multiple orbitals within an energy sublevel, we put the first arrow in the leftmost box.) So the orbital diagram and electron configuration of boron are:

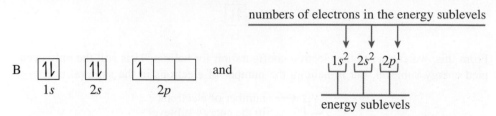

The next element, carbon, with six electrons, causes us to invoke rule 4: Orbitals of equal energy (those in the same energy sublevel) will each get one electron before any of them gets a second electron (Hund's rule). Although the three orbitals in the $2p$ sublevel have equal energies, the next electron we put in, to construct the diagram for carbon, will go into one of the *empty* orbitals rather than the one that is already occupied. Thus, the orbital diagram and electron configuration for carbon are:

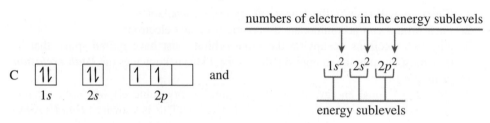

With the next element, nitrogen, we add a seventh electron—which also goes into an empty orbital, rather than into one that already contains an electron. The orbital diagram and electron configuration are:

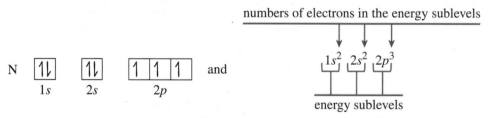

Now that all three of the orbitals in the $2p$ sublevel are *singly* occupied, the next electron to be added must go into one of the *occupied* orbitals and pair with the electron that is already there. Therefore, the orbital diagram and electron configuration of oxygen are:

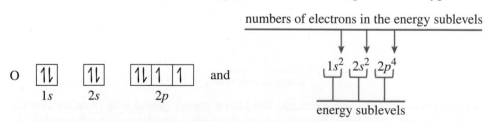

The addition of another electron gives the orbital diagram and electron configuration of fluorine:

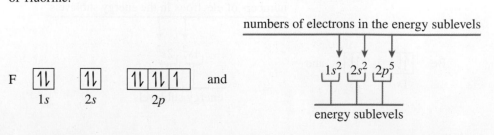

And finally, the addition of one more electron gives the orbital diagram and electron configuration of neon:

numbers of electrons in the energy sublevels

$$1s^2 \quad 2s^2 \quad 2p^6$$

energy sublevels

Ne $\boxed{\uparrow\downarrow}$ $\boxed{\uparrow\downarrow}$ $\boxed{\uparrow\downarrow}\,\boxed{\uparrow\downarrow}\,\boxed{\uparrow\downarrow}$ and
  1s   2s   2p

At this point, we have completely filled the *first* and *second* principal energy levels. We can continue with orbital diagrams and electron configurations for elements in the third period of the periodic table, but the electron configurations are beginning to get rather long. To write them a bit more compactly, we can abbreviate the part of an electron configuration that constitutes the *noble gas core*. For example, consider a nitrogen atom. We can write the electron configuration as

$$1s^2 2s^2 2p^3$$

However, because $1s^2$ is actually the configuration of *helium,* we could abbreviate nitrogen's configuration as

$$[\text{He}] \longrightarrow [\text{He}]2s^2 2p^3$$

This practice does not save much effort when the noble gas core is *helium,* which has just two electrons. But at higher atomic numbers, when the noble gas core is neon, or argon, or krypton, this abbreviation technique does make it considerably easier to write electron configurations.

As an example, consider the sodium atom. According to Figure 2.15, after neon, the 3s orbital will fill next. The orbital diagram of sodium is

$\boxed{\uparrow\downarrow}$ $\boxed{\uparrow\downarrow}$ $\boxed{\uparrow\downarrow}\,\boxed{\uparrow\downarrow}\,\boxed{\uparrow\downarrow}$ $\boxed{\uparrow}$
 1s   2s   2p   3s

And we can write its electron configuration the long way, or the shorter way:

$$1s^2 2s^2 2p^6 3s^1$$

$$\text{Ne} \longrightarrow [\text{Ne}]3s^1$$

Sample Problem 2.3 gives you some practice drawing orbital diagrams and writing electron configurations for elements in the third period of the periodic table.

SAMPLE PROBLEM 2.3 — Writing Electrons Configurations and Drawing Orbital Diagrams for Elements in the Third Period

Determine both the full and condensed ground-state electron configurations and orbital diagram for each of the following elements: (a) Mg, (b) Al, and (c) S.

Strategy You have seen the order in which orbitals fill in Figure 2.15. Start by determining the electron configurations for each element.

Setup Locate the element on the periodic table and determine how many electrons need to be filled in to represent each separate element. Magnesium contains 12 electrons, aluminum 13, and sulfur contains 16 electrons.

For the full electron configuration, begin filling electrons in the 1s sublevel and continue according to Figure 2.15.

In order to write the *condensed* electron configuration, we must locate the nearest noble gas *before* the element we are working with. The nearest noble gas is *neon* (Ne) for all three of these elements; it represents the first 10 electrons.

The orbital diagram for each element represents each sublevel with the appropriate number of boxes to match the number of orbitals present. For an *s* sublevel there is one orbital/box and for a *p* sublevel there are three orbitals/boxes.

(Continued on next page)

Solution The full electron configurations are:

(a) $1s^2 2s^2 2p^6 3s^2$ (b) $1s^2 2s^2 2p^6 3s^2 3p^1$ (c) $1s^2 2s^2 2p^6 3s^2 3p^4$

For the condensed electron configurations, we replace $1s^2 2s^2 2p^6$ with [Ne], and we arrive at (a) [Ne]$3s^2$, (b) [Ne]$3s^2 3p^1$, and (c) [Ne]$3s^2 3p^4$. The orbital diagrams are as follows:

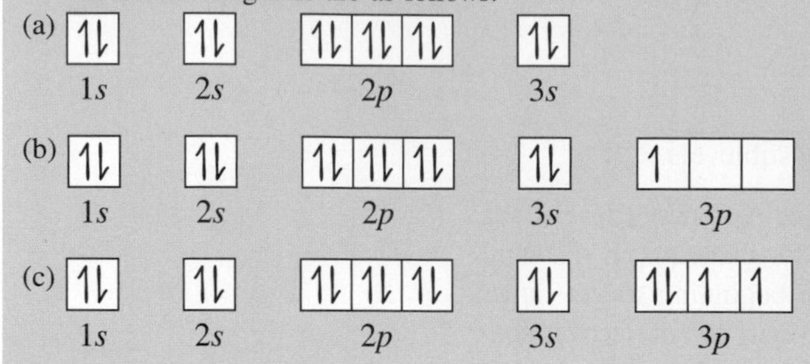

(a)

↑↓	↑↓	↑↓ ↑↓ ↑↓	↑↓
$1s$	$2s$	$2p$	$3s$

(b)

↑↓	↑↓	↑↓ ↑↓ ↑↓	↑↓	↑
$1s$	$2s$	$2p$	$3s$	$3p$

(c)

↑↓	↑↓	↑↓ ↑↓ ↑↓	↑↓	↑↓ ↑ ↑
$1s$	$2s$	$2p$	$3s$	$3p$

THINK ABOUT IT

Make sure you understand that the orbital diagram and full electron configuration represent the same information with one distinction. The orbital diagram also shows how electrons are paired in each orbital. Remember to only pair electrons in an orbital after each orbital within a sublevel has one electron.

Practice Problem **A**TTEMPT Determine both the full and condensed ground-state electron configurations and orbital diagram for each of the following elements.

(a) C   (b) Ne   (c) Cl

Practice Problem **B**UILD Given the following ground-state electron configurations, identify the element that each represents.

(a) [Ne]$3s^2$ (b) [Ne]$3s^2 3p^3$

Practice Problem **C**ONCEPTUALIZE An incorrect orbital diagram for the silicon atom is shown here. Make the necessary change(s) to correct it.

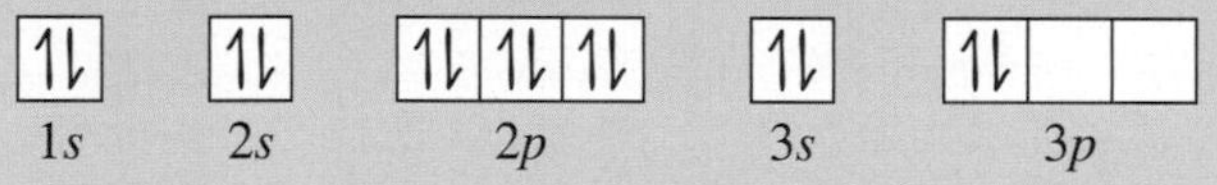

↑↓	↑↓	↑↓ ↑↓ ↑↓	↑↓	↑↓
$1s$	$2s$	$2p$	$3s$	$3p$

CHECKPOINT–SECTION 2.4   Electron Configurations

2.4.1 Which of the following electron configurations correctly represents the Cl atom?

a) [Ne]$3p^7$

b) [Ne]$3s^2 3p^5$

c) [Ne]$3s^6 3p^1$

d) [He]$3s^3 3p^4$

e) [He]$2s^2 2p^5$

2.4.2 What element is represented by the following electron configuration? [Ne]$3s^2 3p^1$

a) B

b) Ga

c) K

d) Al

e) Na

2.4.3 Which orbital diagram is correct for the ground-state S atom?

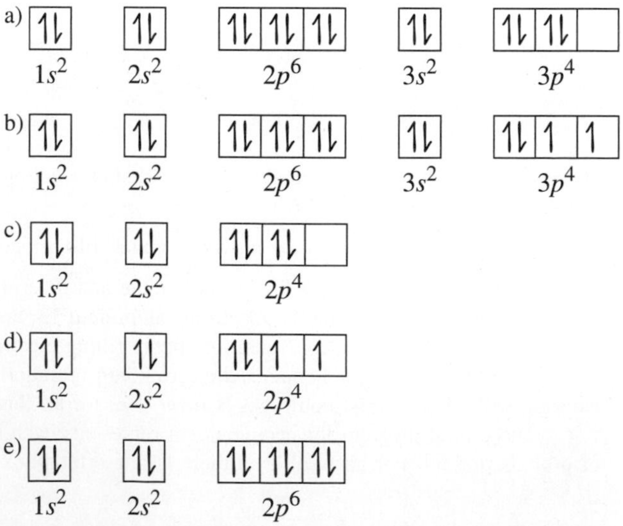

a)

↑↓	↑↓	↑↓ ↑↓ ↑↓	↑↓	↑↓ ↑↓
$1s^2$	$2s^2$	$2p^6$	$3s^2$	$3p^4$

b)

↑↓	↑↓	↑↓ ↑↓ ↑↓	↑↓	↑↓ ↑ ↑
$1s^2$	$2s^2$	$2p^6$	$3s^2$	$3p^4$

c)

↑↓	↑↓	↑↓ ↑↓
$1s^2$	$2s^2$	$2p^4$

d)

↑↓	↑↓	↑↓ ↑ ↑
$1s^2$	$2s^2$	$2p^4$

e)

↑↓	↑↓	↑↓ ↑↓ ↑↓
$1s^2$	$2s^2$	$2p^6$

2.5 Electron Configurations and the Periodic Table

Up until now, it has probably been easy to remember the order in which orbitals will fill—you should be able to construct orbital diagrams and write electron configurations for the first three periods (up through argon) without having to refer to Figure 2.15. But remember that in a multielectron atom, the orbital energies change such that the $4s$ sublevel is lower than the $3d$ sublevel—meaning that it will fill first. This can make it harder to remember the order in which orbitals should be filled—especially if you don't have access to Figure 2.15. A variety of visual aids have been developed to make it easier to remember the order of orbital filling, but none is better than the periodic table itself. You will generally have access to a periodic table, so it makes sense to learn how to use the information it provides.

Animation
Electron Configurations

Figure 2.17 shows a periodic table that has been color coded in "blocks." Elements with grey boxes are in the s block; those with green boxes are in the p block; and those in the blue boxes are in the d block. (There is also an f block, but because f orbitals and the atoms that contain them are beyond the scope of this book, they are not shown in the figure.) The block in which an element is found indicates the type of orbital that would be filled *last* when its electron configuration is written.

1	1s																1s
2	2s													2p			
3	3s													3p			
4	4s					3d								4p			
5	5s					4d								5p			
6	6s					5d								6p			
7	7s					6d								7p			

Figure 2.17 The simplest way to remember the order in which orbitals fill with electrons is to consult the periodic table.

To determine the ground-state electron configuration of any element, we start with the most recently completed noble gas core and count across the following period to determine the rest of the configuration. For example, for chlorine (atomic number 17), the most recently completed noble gas is neon, so we begin writing chlorine's configuration with the neon core:

Cl [Ne]…

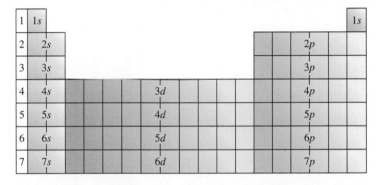

Then we count across period 3, adding an electron for each successive element until we reach Cl. The block (s, p, or d) gives us the type of orbital being filled with electrons. Across period 3, we count two in the *s* block, and five in the *p* block, so we add $3s^2$ and $3p^5$ to the configuration. This gives the electron configuration

$$\text{Cl } [\text{Ne}]3s^2 3p^5$$

1	1s																	1s
2	2s												2p					10 Ne
3	3s												3p					
4	4s					3d							4p					
5	5s					4d							5p					
6	6s					5d							6p					
7	7s					6d							7p					

Now let's use this technique to write an electron configuration for an element with a higher atomic number, and for which you might not easily remember the orbital filling order. Note that the *d* block in Figure 2.17 has a principal energy level one lower than the rest of the period—because of the relative energies of orbitals in multielectron atoms. (See Figure 2.15.) Arsenic, atomic number 33, has the noble gas core of argon. We begin writing arsenic's electron configuration as

$$\text{As } [\text{Ar}] \dots$$

1	1s																	1s
2	2s												2p					
3	3s												3p					18 Ar
4	4s					3d							33 As	4p				
5	5s					4d							5p					
6	6s					5d							6p					
7	7s					6d							7p					

Then we count across period 4 as follows:

two in the *s* block adds $4s^2$

ten in the *d* block adds $3d^{10}$

three in the *p* block adds $4p^3$

giving a final configuration of

$$\text{As } [\text{Ar}] \, 4s^2 3d^{10} 4p^3$$

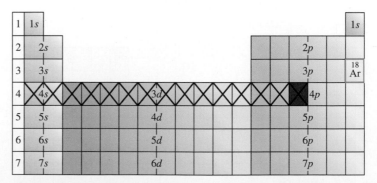

Sample Problem 2.4 lets you practice writing abbreviated electron configurations for elements beyond the third period of the periodic table.

SAMPLE PROBLEM **2.4** Writing Electron Configurations Beyond the Third Period

Write the ground-state electron configuration, using the noble gas core abbreviation, for each of the following elements: (a) Ca, (b) Te, and (c) Br.

Strategy Find the nearest noble gas core for each element and use the periodic table to determine the order in which the sublevels will fill.

Setup (a) The nearest noble gas core for Ca ($Z = 20$) is [Ar] ($Z = 18$). There are two electrons beyond the noble gas core and they are added to the $n = 4$ level in the s block.

(b) The nearest noble gas core for Te ($Z = 52$) is [Kr] ($Z = 36$). There are $52 - 36 = 16$ electrons beyond the noble gas core, and they are added beginning in the fifth row of the periodic table. The sublevels that fill (in order) are: $5s$, $4d$, $5p$. Remember the d block is always in the level that is equal to the row minus 1.

(c) The nearest noble gas core for Br ($Z = 35$) is [Ar] ($Z = 18$). There are $35 - 18 = 17$ electrons beyond the noble gas core that are added beginning in the fourth row of the periodic table. The sublevels that fill (in order) are: $4s$, $3d$, $4p$. Remember the d block is always in the level that is equal to the row minus 1.

Solution (a) $[\text{Ar}]4s^2$ (b) $[\text{Kr}]5s^24d^{10}5p^4$ (c) $[\text{Ar}]4s^23d^{10}4p^5$

THINK ABOUT IT

Make certain to find the most recently *completed* noble gas to the element in question. This means that it will be the noble gas in the period (row) *above* the element you are working with.

Practice Problem (A)TTEMPT Write the ground-state electron configuration, using the noble gas core abbreviation, for each of the following elements: (a) Rb, (b) Se, and (c) I.

Practice Problem (B)UILD Identify the element represented by each of the following ground-state electron configurations: (a) $[\text{Ar}]4s^23d^{10}4p^2$, (b) $[\text{Kr}]5s^24d^{10}5p^3$, and (c) $[\text{Xe}]6s^2$.

Practice Problem (C)ONCEPTUALIZE Identify the element represented by each of the following valence orbital diagrams and specify the noble gas core for each.

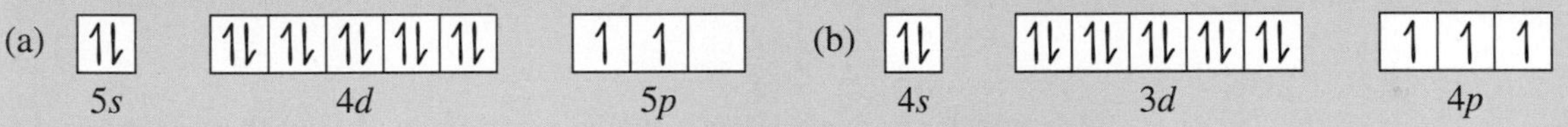

Electron configuration is extraordinarily important in the determination of an element's properties. The outermost electrons in an atom, those with the highest principal quantum number, are called the **valence electrons**. (The inner electrons in an atom are called the **core electrons**.) It is the valence electrons that give elements their "personalities." Figure 2.18 shows the main group elements of the periodic table and their valence electron configurations.

Note that within a group, the valence electron configurations are essentially the same—only the value of n changes as we move from the top of the periodic table to the bottom. Scientists knew for centuries prior to the development of the quantum mechanical (QM) model of the atom that certain groups of elements had similar properties. But the QM model shows us *why* those elements behave in similar ways—because they have the same number of *valence electrons*.

H $1s^1$							He $1s^2$
Li $2s^1$	Be $2s^2$	B $2s^22p^1$	C $2s^22p^2$	N $2s^22p^3$	O $2s^22p^4$	F $2s^22p^5$	Ne $2s^22p^6$
Na $3s^1$	Mg $3s^2$	Al $3s^23p^1$	Si $3s^23p^2$	P $3s^23p^3$	S $3s^23p^4$	Cl $3s^23p^5$	Ar $3s^23p^6$
K $4s^1$	Ca $4s^2$	Ga $4s^24p^1$	Ge $4s^24p^2$	As $4s^24p^3$	Se $4s^24p^4$	Br $4s^24p^5$	Kr $4s^24p^6$
Rb $5s^1$	Sr $5s^2$	In $5s^25p^1$	Sn $5s^25p^2$	Sb $5s^25p^3$	Te $5s^25p^4$	I $5s^25p^5$	Xe $5s^25p^6$
Cs $6s^1$	Ba $6s^2$	Tl $6s^26p^1$	Pb $6s^26p^2$	Bi $6s^26p^3$	Po $6s^26p^4$	At $6s^26p^5$	Rn $6s^26p^6$
Fr $7s^1$	Ra $7s^2$	Nh $7s^27p^1$	Fl $7s^27p^2$	Mc $7s^27p^3$	Lv $7s^27p^4$	Ts $7s^27p^5$	Og $7s^27p^6$

Figure 2.18 Periodic table showing valence electron configurations of the main group elements.

Sample Problem 2.5 gives you some practice identifying the valence electrons in electron configurations.

SAMPLE PROBLEM (2.5)   Identifying the Valence Electrons in Electron Configurations

Identify the valence electrons in the electron configurations that you wrote for each element in Sample Problem 2.4.

Strategy Valence electrons are the electrons in the highest energy level (n value) in the electron configuration or orbital diagram.

Setup For Ca (a), the highest n value is 4.
For Te (b), the highest n value is 5. (The $4d$ sublevel is not in the highest n value shell.)
For Br (c), the highest n value is 4. (The $3d$ sublevel is not in the highest n value shell.)

Solution (a) There are 2 valence electrons, both in the $4s$ sublevel: $4s^2$

(b) There are 6 valence electrons, 2 in the $5s$ and 4 in the $5p$ sublevels: $5s^2 5p^4$

(c) There are 7 valence electrons, 2 in the $4s$ and 5 in the $4p$ sublevels: $4s^2 4p^5$

THINK ABOUT IT

Note that electrons in the d sublevel are not counted as valence electrons as they are in a lower n-value level. They are not contained in the *outermost* level.

Practice Problem Ⓐ**TTEMPT** Identify the valence electrons in the electron configurations that you wrote for each element in Practice Problem 2.4A.

Practice Problem Ⓑ**UILD** Identify the valence electrons in the electron configurations given for each element in Practice Problem 2.4B.

Practice Problem Ⓒ**ONCEPTUALIZE** When you construct an orbital diagram, are the last electrons you add to the diagram necessarily valence electrons? Explain.

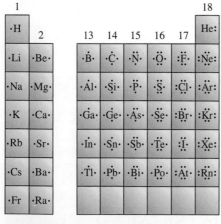

Figure 2.19 Main group elements represented with Lewis dot symbols.

Student Note: For elements in the second period, the maximum of eight dots would represent the combined total of two electrons in the 2s orbital and six electrons in the three 2p orbitals.

We will learn more about valence electrons and how they determine an element's properties in Section 2.6. But first, let's look at one more way to represent an atom and indicate its electron configuration. In **Lewis dot symbols,** valence electrons are shown as dots around the element's symbol. For example, sodium, with the electron configuration [Ne]$3s^1$, has one valence electron. Thus, we can represent a sodium atom as

·Na

Similarly, we can write Lewis dot symbols for the rest of the elements in the third period as

·Mg·   ·Äl·   ·S̈i·   ·P̈·   ·S̈·   :C̈l·   :Är:

Note that the maximum number of dots around an element's symbol is *eight*, representing the combined total of two electrons in the s orbital (in this case, the $3s$ orbital) and six electrons in the three p orbitals (in this case, the three $3p$ orbitals). Note also that when we write Lewis dot symbols, we don't put a second dot in any position (left, right, top, or bottom) until there are single dots in all the positions. (For example, the Lewis dot symbol for Mg would not be correct if both of its two dots were to the left of the symbol.) Figure 2.19 shows the main group elements of the periodic table, represented with Lewis dot symbols.

CHECKPOINT–SECTION 2.5 Electron Configurations and the Periodic Table

2.5.1 Which of the following electron configurations correctly represents the Sn atom?

a) $[Kr]5s^25p^2$

b) $[Kr]5s^24d^{10}5p^2$

c) $[Kr]4d^{10}5p^2$

d) $[Xe]5s^24d^{10}$

e) $[Xe]5s^24d^{10}5p^2$

2.5.2 What element is represented by the following electron configuration: $[Kr]5s^24d^{10}5p^5$?

a) Tc   b) Br   c) I   d) Xe   e) Te

2.5.3 Which of the following is a *d*-block element? (Select all that apply.)

a) Sb   b) Au   c) Ca   d) Zn   e) Fe

2.5.4 Which of the following is a *p*-block element? (Select all that apply.)

a) Pb   b) C   c) Sr   d) Xe   e) Na

2.5.5 How many valence electrons does a phosphorus atom have?

a) 2   b) 3   c) 4   d) 5   e) 8

2.5.6 Select the correct Lewis dot symbol for the phosphorus atom.

a) $:\!\overset{\cdot\cdot}{P}\!:$ b) $\cdot\!\overset{\cdot\cdot}{P}\!:$ c) $\cdot\overset{\cdot\cdot}{\underset{}{P}}\cdot$ d) $:\!\dot{P}\cdot$ e) $\cdot\dot{P}\cdot$

2.6 Periodic Trends

There are several important trends in the properties of the elements on the periodic table that can be explained in terms of our current model of the atom. One such property is *atomic size*. Figure 2.20 depicts the relative atomic sizes of the main group

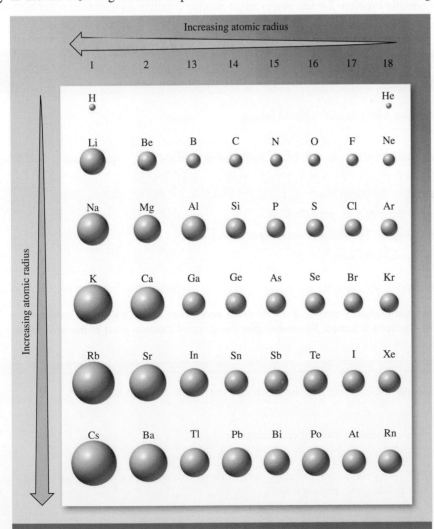

Figure 2.20 Atomic size increases from top to bottom within a group and decreases from left to right within a period.

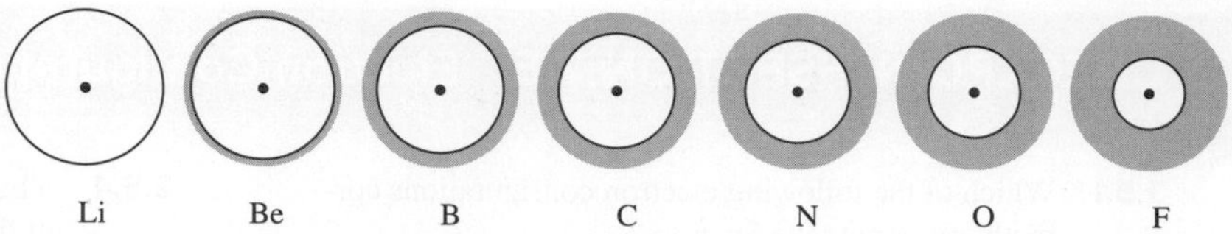

Figure 2.21 Attractive forces between the nucleus and valence electrons increase as the charges on both increase.

Charge on nucleus	Li +3	Be +4	B +5	C +6	N +7	O +8	F +9
Charge on valence shell	−1	−2	−3	−4	−5	−6	−7

Student Note: Remember that the *valence* electrons are the *outermost* electrons. [◄◄ Section 2.5]

elements. Note that as we move down the table within a group, atomic size increases. This should not be at all surprising, because the principal quantum number of the valence electrons increases down the table; and remember that the principal quantum number, *n,* tells us the *size* of an orbital.

From left to right, as atomic number and atomic mass increase, atomic size actually *decreases.* This trend is somewhat less intuitive, but it can be explained using the QM model of the atom and our knowledge of how charged particles interact [◄◄ Section 1.2]. Remember that opposite charges are attracted to each other and that all of an atom's positive charge is contained within its nucleus. As we move across a period, the numbers of protons and electrons both increase. However, the electrons that are added all go into the same principal energy level (see Figure 2.18). There is an attractive force between the nucleus and the valence electrons—and the larger the *charges* of the nucleus and the principal energy level that contains the valence electrons, the *greater* the attractive force between them. The larger attractive force pulls the valence electrons *closer* to the nucleus, causing atomic size to *decrease.* Figure 2.21 illustrates how this results in the downward trend in atomic size as we move from left to right on the periodic table.

Sample Problem 2.6 lets you practice making predictions about the relative atomic sizes of various main group elements using the trends you have just learned.

Student Hot Spot

Student data indicate that you may struggle with ranking elements by atomic size. Access the eBook to view additional Learning Resources on this topic.

SAMPLE PROBLEM **2.6** **Predicting Relative Atomic Sizes Based on Position on the Periodic Table**

For each pair of elements, indicate which one you would expect to be the largest:

(a) Al or P   (b) Se or O

Strategy Consider the periodic trend where atoms decrease in size, left to right, within a period (row) and *increase* in size from top to bottom within a group (column).

Setup (a) Al is in the same period as P and is to the left of P.

(b) Se and O are in the same group with Se further down the periodic table.

Solution (a) Al is larger than P because it lies to the left of P within the same period.

(a) Se is larger than O because it lies below O within the same group.

THINK ABOUT IT

When comparing the sizes of atoms in the same group (column), it should make sense that for each element encountered down the column, a new level containing electrons is added. Remember that the electron cloud is most of the volume of an atom and determines its overall size.

Practice Problem Ⓐ**TTEMPT** For each pair of elements, indicate which one you would expect to be larger.

(a) As or Kr   (b) Br or I

Practice Problem Ⓑ**UILD** Place the given elements in order of decreasing size.

(a) K, Cl, and Se   (b) Br, I, and Sb

Practice Problem Ⓒ**ONCEPTUALIZE** Using periodic trends alone, rank each set of elements in order of increasing size. If it is not possible to rank them using periodic trends alone, explain why.

(a) P, S, and Br   (b) Cs, Rb, and I

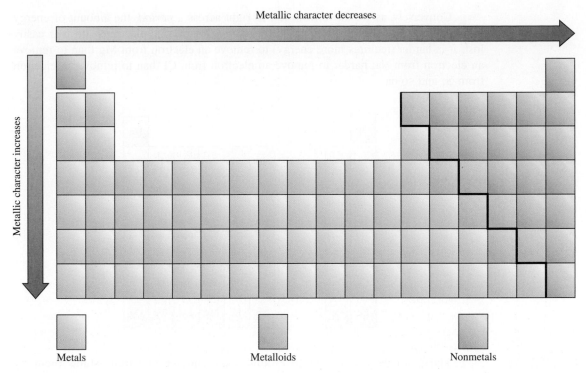

Figure 2.22 The periodic table, divided into metals, nonmetals, and metalloids. Metallic character increases from top to bottom, and from right to left.

Now let's look again at the periodic table and how it is organized (Figure 2.22). The zigzag line separates the metals (on the left) from the nonmetals (mostly in the top right corner). Most of the elements are metals, and they exhibit properties that we refer to collectively as ***metallic character.*** For example, metals tend to be shiny solids that conduct electricity. Further, one of the defining properties of metallic character is the relative ease with which electrons can be removed from metal atoms. The more an element exhibits properties such as these, the greater its *metallic character* is said to be. Metallic character tends to *increase* from top to bottom in a group, and tends to *decrease* from left to right across a period. Thus, the *most* metallic elements are those in the lower left of the periodic table—and the *least* metallic elements (the *nonmetals*) are those in the top right corner of the periodic table.

One of the properties shared by metals, the *ease* with which electrons can be removed, is itself an important property that exhibits periodic trends. It requires energy to remove an electron from *any* atom, but it typically requires significantly *less* energy to remove an electron from a *metal* atom than to remove an electron from a nonmetal atom. As we move *down* a group, less and less energy is required. Thus, it is easier to remove an electron from K than to remove an electron from Na; easier to remove an electron from Se than to remove an electron from S; and so on.

Student Note: Recall that there are elements bordering the zig-zag line, known as *metalloids*, that have properties intermediate between metals and nonmetals.

Student Note: The energy required to remove an electron from an atom is called the ***ionization energy.*** Although the term ionization energy has an explicit definition that is not included here, the important thing to understand is that some atoms lose electrons more easily than others—and that the trend in this property is related to an element's position on the periodic table.

Conversely, as we move from left to right across a *period,* the amount of energy required to remove an electron generally *increases*—that is, electrons are *less* easily lost. It is harder (requires more energy) to remove an electron from Mg than to remove an electron from Na; harder to remove an electron from Cl than to remove an electron from S; and so on.

These trends are related to atomic size, and we can understand them by again invoking our knowledge of how charged particles behave. As we move down a group, atomic size increases, and the valence electrons are farther and farther away from the nucleus—making the attractive force between them *weaker,* and making an electron *easier* to remove. As we move from left to right, not only does the magnitude of charge on the nucleus *increase,* but the distance between it and the electron to be removed *decreases* (because atomic size gets *smaller*). Both factors result in a *greater* attractive force between the nucleus and the valence electrons and contribute to it being *more* difficult to pull an electron away from the atom.

Figure 2.23 shows the periodic trends in how easily electrons are removed from atoms of the main group elements.

Sample Problems 2.7 and 2.8 let you practice comparing the metallic character of several elements and understand how easily atoms lose electrons based on the trends in those properties.

Figure 2.23 Trends in how easily electrons are removed from atoms of the main group elements. Ease of electron-removal increases from top to bottom within a group and decreases from left to right within a period.

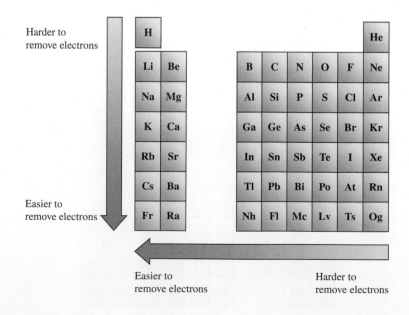

SAMPLE PROBLEM 2.7   Assessing Relative Metallic Character Based on Position on the Periodic Table

For each pair of elements, indicate which one you would expect to have the *greatest* metallic character: (a) Na or Cl, (b) I or F, (c) Li or Rb.

Strategy Consider the periodic trend where metallic character increases within a period from right to left and increases within a group from top to bottom.

Setup (a) Na is to the left of Cl in the same period.   (b) I is below F in the same group.   (c) Rb is below Li in the same group.

Solution (a) Na is more metallic than Cl.   (b) I is more metallic than F.   (c) Rb is more metallic than Li.

THINK ABOUT IT

The closer an element is to the lower left-hand corner of the periodic table, the more metallic it is. The metals are on the left side of the zigzag line and comprise most of the periodic table.

Practice Problem **TTEMPT** For each pair of elements, indicate which one you would expect to have the least metallic character.

(a) Cs or Mg   (b) N or As

Practice Problem B UILD Place the following elements in order of increasing metallic character:

(a) Li, C, and K   (b) O, S, and Si

Practice Problem C ONCEPTUALIZE Is it possible to rank the elements F, Ne, and Ar in order of decreasing metallic character using only the periodic trend? If not, explain why.

SAMPLE PROBLEM 2.8   Comparing the Ease with Which Atoms of Different Elements Can Lose Electrons

For each pair of elements, indicate which one you would expect to be *more* difficult from which to remove an electron.

(a) Rb or Li   (b) Ca or Br

Strategy Consider the periodic trend where difficulty of electron removal increases, from left to right, within a period (row) and decreases top to bottom within a group (column).

Setup (a) Li is in the same group as Rb and lies above it on the periodic table.

(b) Ca and Br are in the same period, with Br farther to the right.

Solution (a) It is more difficult to remove an electron from Li than Rb.

(b) It is more difficult to remove an electron from Br than Ca.

THINK ABOUT IT

The more metallic an element is, the easier it is to remove an electron from it, because metals lose electrons more easily than nonmetals.

Practice Problem A TTEMPT For each pair of elements, indicate which one you would expect to be *more* difficult from which to remove an electron.

(a) Mg or Sr   (b) Rb or Te

Practice Problem B UILD Place the following elements in order of decreasing difficulty of electron removal.

(a) Li, C, and K   (b) Ba, Ca, and As

Practice Problem C ONCEPTUALIZE Is it possible to rank the elements N, S, and Br in order of increasing difficulty of electron removal using only the periodic trend? If not, explain why.

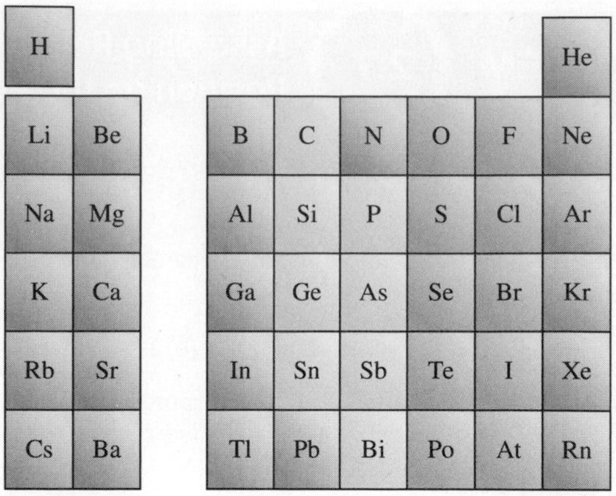

Figure 2.24 Atoms on the right of the periodic table, especially in Groups 16 and 17, gain electrons most readily. Elements shown in yellow can gain electrons. Those shown in darker yellow are those that gain electrons most readily.

H									He
Li	Be			B	C	N	O	F	Ne
Na	Mg			Al	Si	P	S	Cl	Ar
K	Ca			Ga	Ge	As	Se	Br	Kr
Rb	Sr			In	Sn	Sb	Te	I	Xe
Cs	Ba			Tl	Pb	Bi	Po	At	Rn

Student Note: The technical term for this is *electron affinity*. (It is easier to add an electron to an atom with a high *affinity* for electrons.) As with *ionization energy*, we will not be concerned with the technical definition of *electron affinity*. Nevertheless, it is important to understand that atoms of some elements gain electrons more easily than others.

Student Note: There is not a clear trend in atoms' ability to gain electrons from top to bottom within a group.

One more important trend in properties is the ability of an atom to *gain* an electron. In general, an atom that *loses* an electron easily will not *gain* an electron easily. And vice versa: an atom that does *not* lose an electron easily will readily *gain* an electron. (Most atoms, but not all, will do one or the other.) This leads to essentially opposite trends for the two processes from left to right within a period. Atoms at the far right of the periodic table (not including the noble gases) gain electrons most easily. Figure 2.24 shows the elements whose atoms readily gain electrons. Although the ability to *gain* an electron is less straightforward than the ability to *lose* an electron, and the trend is not as regular, it is important for you to know that just as some atoms can *lose* electrons, others can *gain* electrons.

Sample Problem 2.9 lets you practice comparing how easily elements gain electrons.

SAMPLE PROBLEM (**2.9**)   **Comparing the Ease with Which Atoms of Different Elements Can Gain Electrons**

In each of the following pairs of elements, identify the element to which it is easier to add an electron.

(a) Li or O   (b) Al or Cl   (c) S or Ca

Strategy Metals lose electrons most easily; nonmetals gain electrons most easily.

Setup Locate the two elements in each pair on the periodic table and identify each as a metal or nonmetal. In each pair, it will be easier to add an electron to the nonmetal atom.

Solution (a) Li is a metal and O is a nonmetal. Oxygen will gain an electron more easily than lithium.

(b) Al is a metal and Cl is a nonmetal. Chlorine will gain an electron more easily than aluminum.

(c) S is a nonmetal and Ca is a metal. Sulfur will gain an electron more easily than calcium.

THINK ABOUT IT

The nonmetals in the upper right-hand corner of the periodic table tend to gain electrons to become negatively charged ions. The metals in the lower left-hand corner of the periodic table tend to lose electrons to form positively charged ions.

Practice Problem Ⓐ**TTEMPT** In each of the following pairs of elements, identify the element to which it is easier to add an electron.

(a) Ga or O   (b) N or Pb   (c) I or Fe

Practice Problem Ⓑ**UILD** In each of the following pairs of elements, identify the element to which it is more difficult to add an electron.

(a) Ga or S   (b) As or Sn   (c) Br or Zn

Practice Problem Ⓒ**ONCEPTUALIZE** Why aren't the noble gases included in the periodic trend that ranks elements in order of ease of electron gain?

CHECKPOINT–SECTION 2.6 Periodic Trends

2.6.1 Arrange the elements Ca, Sr, and Ba in order of increasing amount of energy required to remove an electron.

a) Ca < Sr < Ba   d) Sr < Ba < Ca

b) Ba < Sr < Ca   e) Sr < Ca < Ba

c) Ba < Ca < Sr

2.6.2 For each of the following pairs of elements, indicate which will have the greater tendency to gain an electron: Rb or Sr, C or N, O or F.

a) Rb, C, O   d) Sr, N, O

b) Sr, N, F   e) Rb, C, F

c) Sr, C, F

2.6.3 Which element, K or Br, will have the greater tendency to *lose* an electron and which will have the greater tendency to *gain* an electron?

a) K, Br

b) Br, K

c) K, K

d) Br, Br

e) Both will have the same tendencies.

2.7 Ions: The Loss and Gain of Electrons

In the last section, we discussed trends in the ability to lose an electron or to gain an electron. In fact some atoms can lose more than *one* electron, and some can *gain* more. When an atom loses or gains one or more electrons, it becomes an *ion,* specifically, an **atomic ion,** which is simply an atom in which the number of electrons is not equal to the number of protons. Because an electron has a negative charge, the loss or gain of electrons causes an atom to become charged. An atom that *loses* one or more electrons becomes *positively* charged, and we refer to it as a **cation** (pronounced **cat**-ion). An atom that *gains* one or more electrons becomes *negatively* charged, and we refer to it as an **anion** (pronounced **an**-ion). Let's look at some specific examples of ions and how we represent them.

Electron Configuration of Ions

Sodium, with atomic number 11, is at the far left of the periodic table in Group 1. According to the trend in how easily atoms lose electrons, we expect sodium to lose an electron easily—and it does. In fact, sodium loses an electron so readily that it is never found in nature in its *elemental* (neutral) form. Rather, all of the sodium atoms found in nature are sodium *ions,* each having lost an electron. The sodium ion is represented as Na^+, with the charge indicated by a superscript. Recall that the orbital diagram of sodium is:

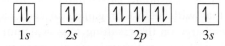

And we can write its electron configuration as:

$$[Ne]3s^1$$

When a sodium atom loses an electron to become a sodium ion, the electron lost is sodium's only *valence* electron—in the $3s$ orbital. With the loss of the $3s$ electron, the orbital diagram becomes:

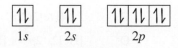

And the electron configuration becomes simply:

$$[Ne]$$

Chlorine, with atomic number 17, is at the far right of the periodic table in Group 17. According to Figure 2.24, we expect chlorine to *gain* an electron readily—and it does. The orbital diagram of the chlorine atom is:

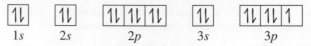

And its electron configuration can be written as:

$$1s^2 2s^2 2p^6 3s^2 3p^5 \text{ or } [Ne]3s^2 3p^5$$

When an atom gains an electron, the electron goes into the lowest energy orbital available. In the case of Cl, that's the $3p$ orbital that has just one electron in it. When a Cl atom has gained an electron, it becomes the ion Cl^-, which, like the sodium ion, is written with a superscript charge. The orbital diagram of the Cl^- ion is:

And its electron configuration is

$$[Ne]3s^2 3p^6$$

which is the same as the electron configuration for argon: [Ar].

Now let's take a moment to contemplate the significance of the electron configurations of these ions. In each case, the resulting electron configuration is identical to the electron configuration of a noble gas. The way chemists say this is that each ion is **isoelectronic** with a noble gas. Na^+ is isoelectronic with Ne, and Cl^- is isoelectronic with Ar. Recall that the trend in ease of electron loss (electrons being harder to remove as we go from left to right across a period) indicates that the element with the *smallest* tendency to gain an electron in each period is the noble gas. Recall also that the noble gases (Group 18) were not included in our discussion of how easily atoms *gain* electrons. This is because the noble gases neither *lose* nor *gain* electrons readily. The electron configurations of the noble gases are inherently *stable*— so there is no benefit to a noble gas atom from either losing or gaining electrons. But for most main group elements, achieving the electron configuration of a noble gas significantly increases their stability. Both sodium and chlorine are very *unstable* as neutral atoms. But when each has achieved a noble gas electron configuration (sodium by *losing* an electron and chlorine by *gaining* one), they become the highly *stable* ions, Na^+ and Cl^-, that make up the familiar, stable substance most of us sprinkle on our food every day: *salt*.

Knowing that atoms of the main group elements will lose or gain electrons to achieve a noble gas electron configuration—to become *isoelectronic* with a noble gas—enables us to predict the number of electrons that will be lost or gained in each case—and therefore to predict the charge on the resulting ions. Calcium, for example, atomic number 20, will *lose* its two valence electrons to become isoelectronic with argon:

$$Ca \qquad [Ar]4s^2$$
$$Ca^{2+} \qquad [Ar]$$

Oxygen will *gain* two electrons to become isoelectronic with neon:

$$O \qquad [He]2s^2 2p^4$$
$$O^{2-} \qquad [He]2s^2 2p^6 \text{ or } [Ne]$$

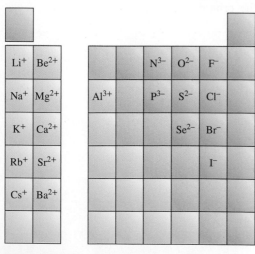

Figure 2.25 Common ions of main group elements.

Figure 2.25 shows the predictable charges on common ions of the main-group elements.

Sample Problem 2.10 lets you practice writing electron configurations for ions of main group elements.

SAMPLE PROBLEM **(2.10)** **Writing Electron Configurations for Common Ions of Main Group Elements**

Predict the charge on the common ion formed from each element listed and write the electron configuration for the ion.

(a) Na   (b) Ca   (c) O

Strategy Locate the element on the periodic table and note that main group elements gain or lose electrons to take on the electron configuration of the nearest noble gas.

Setup (a) Sodium has 11 protons and 11 electrons as a neutral atom. Sodium will take on the electron configuration of Ne when it becomes an ion by losing one electron.

(b) Calcium has 20 protons and 20 electrons as a neutral atom. It will take on the electron configuration of argon when it becomes an ion by losing 2 electrons.

(c) Oxygen has 8 protons and 8 electrons as a neutral atom. It will take on the electron configuration of neon when it becomes an ion by gaining 2 electrons.

Solution (a) The sodium ion contains 11 protons and 10 electrons, leaving one "extra" positive charge for a +1 charge. Its electron configuration is the same as that of neon, $[He]2s^2 2p^6$.

(b) The calcium ion contains 20 protons and 18 electrons (after the loss of two), leaving an "extra" two positive charges for a +2 charge. The electron configuration for the calcium ion is the same as that of argon, $[Ne]3s^2 3p^6$.

(c) The oxygen ion contains 8 protons and 10 electrons (after the gain of two), leaving an "extra" two negative charges for a −2 charge. The electron configuration for the oxygen ion is the same as that of neon, $[He]2s^2 2p^6$.

> **THINK ABOUT IT**
>
> Nonmetals gain electrons to take on the electron configuration of the noble gas to their right. Metals lose electrons to take on the electron configuration of the noble gas in the row *above* them.

Practice Problem ATTEMPT Predict the charge on the common ion formed from each element listed and write the electron configuration for the ion.

(a) Br   (b) K   (c) S

Practice Problem BUILD Given the atomic number (Z) and the electron configuration, identify the ion represented.

(a) Z = 34, $[Ar]4s^2 3d^{10} 4p^6$ (b) Z = 35, $[Ar]4s^2 3d^{10} 4p^6$ (c) Z = 19, $[Ne]3s^2 3p^6$

Practice Problem CONCEPTUALIZE Would it be possible to identify the ions listed in Practice Problem B if you were not given the atomic number? Why or why not?

Lewis Dot Symbols of Ions

Just as we can represent atoms of main group elements with Lewis dot symbols, showing each valence electron as a dot, we can represent *ions* of main group elements with Lewis dot symbols. To do this, we start with the Lewis dot symbol of an atom and remove a dot for each electron it loses—or *add* a dot for each electron it gains. Using the examples of sodium, which loses one electron to become the cation Na$^+$, and chlorine, which gains one electron to become the anion Cl$^-$, the Lewis dot symbols are

$$Na^+ \quad \text{and} \quad \left[:\ddot{Cl}: \right]^-$$

Note that there are no dots around the symbol for sodium. As a neutral atom, sodium has just one valence electron. It loses its only valence electron to become the ion Na$^+$, which has the same electron configuration as the noble gas neon (Ne).

There are eight dots around the symbol for chlorine. As a neutral atom, chlorine has *seven* valence electrons. To become the Cl$^-$ ion, it *gains* one electron and achieves the electron configuration of the noble gas argon (Ar). Note also that the Lewis symbol for the anion is enclosed in square brackets, with the charge shown *outside* the brackets.

Sample Problem 2.11 lets you practice writing Lewis dot symbols for common main group ions.

SAMPLE PROBLEM (**2.11**)   Writing Lewis Dot Symbols for Common Ions of Main Group Elements

Write the Lewis dot symbols for the atom and the ion commonly formed by each element.

(a) P   (b) Se   (c) Sr

Strategy Locate the element on the periodic table and determine how many valence electrons it has. Each valence electron is represented by a dot around the element symbol. To determine the Lewis dot symbol of the ion, decide if the atom will gain or lose electrons to become like its nearest noble gas.

Setup (a) P has 5 valence electrons as an atom, and will gain 3 more to become an ion with the electron configuration of argon.

(b) Se has 6 valence electrons as an atom and will gain 2 more to become an ion with the electron configuration of krypton.

(c) Sr has 2 valence electrons as an atom and will lose both to become an ion with the electron configuration of krypton.

Solution (a) $\cdot\ddot{P}\cdot$ and $\left[:\ddot{P}:\right]^{3-}$ (b) $\cdot\ddot{S}e:$ and $\left[:\ddot{S}e:\right]^{2-}$ (c) $\dot{S}r\cdot$ and Sr^{2+}

THINK ABOUT IT

You should notice that negatively charged monatomic ions have the same Lewis dot symbol containing 8 valence electrons. You can see a similar pattern for positively charged monatomic ions in Practice Problem A.

Practice Problem Ⓐ**TTEMPT** Write the Lewis dot symbols for the atom and the ion commonly formed by each element.

(a) K   (b) Mg   (c) Al

Practice Problem Ⓑ**UILD**

$$\left[:\ddot{X}:\right]^{n-}$$

Given this generic Lewis dot symbol, the value of n, and the period where the element is located, determine the identity of the ion described.

(a) $n = 2$, period 4   (b) $n = 3$, period 2   (c) $n = 2$, period 3

Practice Problem Ⓒ**ONCEPTUALIZE**

$$X^{n+}$$

Given this generic Lewis dot symbol and the group the element belongs to, identify the charge (value of n).

(a) Group 1   (b) Group 2

We have seen that atoms have periodic variation in their ability to lose or gain electrons. In fact, some atoms have no appreciable tendency to do either. In Chapter 3, we will examine the results of these tendencies and how they govern the interactions of atoms and give rise to most of the matter we encounter every day.

CHECKPOINT–SECTION 2.7   Ions: The Loss and Gain of Electrons

2.7.1 Which of the following ions are isoelectronic with a noble gas? (Select all that apply.)

a) K^{2+} c) Br^- e) F^-

b) Ca^{2+} d) O^{2+}

2.7.2 Which of the following pairs are isoelectronic with each other? (Select all that apply.)

a) Ca^{2+} and Sr^{2+} c) I^- and Kr   e) He and H^+

b) O^{2-} and Mg^{2+} d) S^{2-} and Cl^-

2.7.3 Select the correct ground-state electron configuration for S^{2-}.

a) $[Ne]3p^4$ d) $[Ne]3p^6$

b) $[Ne]3s^23p^6$ e) $[Ne]$

c) $[Ne]3s^23p^2$

Chapter Summary

Section 2.1

- Light has been very important in the experiments that led to the development of our current model of the atom. Light is a form of energy that has certain characteristics, including **wavelength (λ)** and **frequency (ν).**

- The familiar rainbow is actually the Sun's visible **emission spectrum,** meaning that those are the visible wavelengths of light emitted by the Sun. What we usually refer to as "light" is actually the visible portion of the **electromagnetic spectrum.**

- When atoms of an element such as hydrogen are excited by the addition of energy, they emit a characteristic set of discrete wavelengths known as a **line spectrum.**

Section 2.2

- Experiments with light led scientists to propose that energy (specifically light) is **quantized,** meaning that it consists of tiny, discrete "packages," rather than being continuous. The **Bohr atom** model explained the line spectrum of hydrogen with the concepts of electrons orbiting the nucleus like planets around the Sun. In the Bohr model, an orbit was defined by a **quantum number (n).** The lowest energy orbit ($n = 1$) was the **ground state,** and any orbit of higher energy ($n > 1$) was called an **excited state.** Each of the four lines in the hydrogen line spectrum was the result of an electron transition from a higher orbit ($n = 3, 4, 5,$ or 6) to a lower orbit ($n = 2$).

Section 2.3

- The most current model of the atom is the quantum mechanical (QM) model. The QM model does not specify the location of an electron within an atom, but rather allows calculation of the probability of finding an electron within a region of space called an **atomic orbital.** To define an atomic orbital requires a **principal quantum number (n)** (designated with an integer: 1, 2, 3, 4…), which specifies the **principal energy level,** and a letter designation (*s, p, d,* or *f*), which specifies the **energy sublevel.**

Section 2.4

- An **electron configuration** specifies the arrangement of an atom's electrons in its atomic orbitals. Each atomic orbital can accommodate a maximum of two electrons. Two electrons occupying the same atomic orbital must have opposite **spin,** meaning that they are spinning in opposite directions. This is the **Pauli exclusion principle. Hund's rule** states that electrons in an atom in the ground state will occupy the lowest energy orbitals possible.

Section 2.5

- The electron configuration of an atom can be determined using the periodic table. Atoms can be represented with **Lewis dot symbols,** which consist of the element's symbol surrounded by up to eight dots—each representing an electron in the atom's outermost *s* and *p* energy sublevels.

Section 2.6

- Elements exhibit periodic trends in certain properties including **atomic size, metallic character, ionization energy** (how easily an atom *loses* an electron), and **electron affinity** (how easily an atom *gains* an electron).

Section 2.7

- When an atom either loses or gains one or more electrons, it becomes an **atomic ion.** An atom that loses one or more electrons becomes positively charged and is called a **cation.** An atom that gains one or more electrons becomes negatively charged and is called an **anion.** Atoms of main group elements typically lose or gain enough electrons to become **isoelectronic** with a noble gas. Isoelectronic species have the same number of electrons. This enables us to predict the charges on the ions of main group elements.

Key Terms

Anion 63	Emission spectrum 34	Line spectra 35	Quantization 36
Atomic ion 63	Energy sublevel 43	Metallic character 59	Quantum number, *n* 37
Atomic orbital 43	Excited state 37	Orbit 37	Spin 48
Bohr atom 37	Frequency (ν) 33	Pauli exclusion principle 49	Valence electrons 55
Cation 63	Ground state 37	Photon 36	Wavelength (λ) 33
Core electrons 55	Hund's rule 49	Principal energy level 43	
Electromagnetic spectrum 34	Isoelectronic 64	Principal quantum number, *n* 43	
Electron configuration 48	Lewis dot symbol 56	QM model 43	

Determining Ground-State Valence Electron Configurations Using the Periodic Table

An easy way to determine the electron configuration of an element is by using the periodic table. Although the table is arranged by atomic number, it is also divided into blocks that indicate the type of orbital occupied by an element's outermost electrons. Outermost valence electrons of elements in the *s* block (shown below in yellow), reside in *s* orbitals; those of elements in the *p* block (blue) reside in *p* orbitals; and so on.

To determine the ground-state electron configuration of any element, we start with the most recently completed noble gas core and count across the following period to determine the valence electron configuration. Consider the example of Cl, which has atomic number 17. The noble gas that precedes Cl is Ne, with atomic number 10. Therefore, we begin by writing [Ne]. The noble gas symbol in square brackets represents the core electrons with a completed *p* subshell. To complete the electron configuration, we count from the left of period 3 as shown by the red arrow, adding the last (rightmost) configuration label from each block the arrow touches:

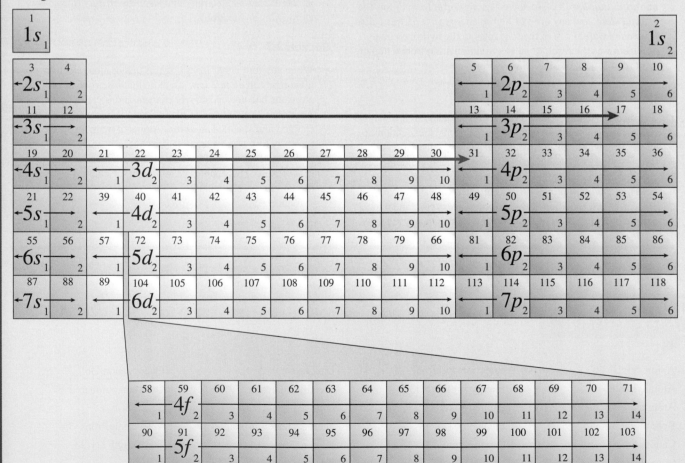

There are seven electrons in addition to the noble gas core. Two of them reside in an s subshell, and five of them reside in a p subshell. By simply counting across the third period, we can determine the specific subshells that contain the valence electrons and arrive at the correct ground-state electron configuration: $[Ne]3s^2 3p^5$.

For Ga, with atomic number 31, the preceding noble gas is Ar, with atomic number 18. Counting across the fourth period (green arrow) gives the ground-state electron configuration: $[Ar]4s^2 3d^{10} 4p^1$.

We can also use the periodic table to determine the identity of an element, given its ground-state electron configuration. For example, given the configuration $[Ne]3s^2 3p^4$, we focus on the last entry in the configuration: $3p^4$. This tells us that the element is in the *third* period (3), in the p block (p), and that it has *four* electrons in its p subshell (superscript 4). This corresponds to atomic number 16, which is the element sulfur (S).

Key Skills Problems

2.1
Determine the atomic number of tin (Sn) using the periodic table on the inside cover of your book. What is the noble gas core for Sn?
a) Ar
b) Kr
c) Xe
d) Ne
e) Rn

2.2
Determine the atomic number of vanadium (V) using the periodic table on the inside cover of your book. Which of the following electron configurations correctly represents the V atom?
a) $[Ar]3d^5$
b) $[Ar]4s^2 3d^2$
c) $[Ar]4s^2 3d^4$
d) $[Ar]4s^2 3d^3$
e) $[Kr]4s^2 3d^3$

2.3
What element is represented by the electron configuration $[Kr]5s^2 4d^{10} 5p^1$?
a) Sn
b) Ga
c) In
d) Tl
e) Zr

2.4
Determine the atomic number of iodine (I) using the periodic table on the inside cover of your book. What is the electron configuration of the I atom?
a) $[Xe]5s^2 4d^{10} 5p^5$
b) $[Kr]5s^2 5p^5$
c) $[Xe]5s^2 5p^5$
d) $[Kr]5s^2 4d^{10} 5p^5$
e) $[Kr]5s^2 5d^{10} 5p^5$

2.5
What is the electron configuration of the common ion that forms from the K atom?
a) $[Ar]4s^1$
b) $[Ar]$
c) $[Ar]4s^2$
d) $[Kr]$
e) $[Kr]4s^1$

2.6
What common ions of main group elements are isoelectronic with Ne?
a) S^{2-}, Cl^-, and K^+
b) O^{2-}, F^-, and Na^+
c) F^-, Cl^-, and Br^-
d) O^{2-}, S^{2-}, and Se^{2-}
e) Li^+, Na^+, and K^+

Questions and Problems

SECTION 2.1: THE NATURE OF LIGHT

2.1 How are the wavelength and frequency of light related to one another?

2.2 How is the energy of light related to its frequency?

2.3 How is the energy of light related to its wavelength?

2.4 Which of the following colors of light has the longest wavelength: yellow, green, or violet?

2.5 Which of the following colors of light has the longest wavelength: red, blue, or orange?

2.6 Which of the following colors of light has the largest frequency: yellow, green, or violet?

2.7 Which of the following colors of light has the largest frequency: red, blue, or orange?

2.8 Place the following wavelengths of light in order of increasing frequency: 350 nm, 450 nm, and 550 nm.

2.9 Place the following wavelengths of light in order of decreasing frequency: 400 nm, 550 nm, and 700 nm.

2.10 Place the following types of electromagnetic radiation in order of increasing energy: infrared, X rays, and radio waves.

2.11 Place the following types of electromagnetic radiation in order of increasing frequency: microwave, gamma rays, and visible.

2.12 Place the following colors of light in order of increasing energy: violet, yellow, and red.

2.13 Place the following colors of light in order of decreasing energy: blue, green, and orange.

SECTION 2.2: THE BOHR ATOM

2.14 Describe what the term *quantized* means in your own words.

2.15 Define the term *photon* in your own words.

2.16 What is meant by the term *ground state* when referring to an atom?

2.17 What is meant by the term *excited state* when referring to an atom?

2.18 Sketch the Bohr model of the hydrogen atom and include the following:
(a) Label $n = 1$ through $n = 6$.
(b) Label the absorption of light from $n = 1$ (ground state) to $n = 3$ (excited state).
(c) Label the emission of light from $n = 3$ to $n = 2$.

2.19 Indicate which of the following electron transitions would be expected to emit visible light in the Bohr model of the atom:
(a) $n = 6$ to $n = 1$
(b) $n = 5$ to $n = 2$
(c) $n = 8$ to $n = 3$

2.20 Indicate which of the following electron transitions would be expected to emit visible light in the Bohr model of the atom:
(a) $n = 7$ to $n = 8$
(b) $n = 2$ to $n = 6$
(c) $n = 4$ to $n = 2$

2.21 Indicate which of the following electron transitions would be expected to emit any light in the Bohr model of the atom:
(a) $n = 6$ to $n = 1$
(b) $n = 1$ to $n = 8$
(c) $n = 3$ to $n = 5$

2.22 Indicate which of the following electron transitions would be expected to emit any light in the Bohr model of the atom:
(a) $n = 2$ to $n = 8$
(b) $n = 7$ to $n = 5$
(c) $n = 6$ to $n = 3$

2.23 Indicate which of the following electron transitions would be expected to absorb any light in the Bohr model of the atom:
(a) $n = 1$ to $n = 8$
(b) $n = 2$ to $n = 3$
(c) $n = 5$ to $n = 3$

2.24 Indicate which of the following electron transitions would be expected to absorb any light in the Bohr model of the atom:
(a) $n = 3$ to $n = 4$
(b) $n = 6$ to $n = 5$
(c) $n = 4$ to $n = 7$

2.25 The hydrogen atom emission spectrum contains four wavelengths or colors of visible light. Match the wavelength/color to the transition that it comes from.

Transition	Wavelength/Color
$n = 6$ to $n = 2$	657 nm/red
$n = 5$ to $n = 2$	486 nm/green
$n = 4$ to $n = 2$	434 nm/blue
$n = 3$ to $n = 2$	410 nm/violet

2.26 How many photons of light are emitted during each of the following processes in a hydrogen atom?
(a) one electron undergoes a transition from $n = 9$ to $n = 4$.
(b) one electron undergoes a transition from $n = 7$ to $n = 5$.
(c) twelve electrons (in 12 separate hydrogen atoms) undergo transitions from $n = 6$ to $n = 3$.
(d) fifty electrons (in 50 separate hydrogen atoms) undergo transitions from $n = 8$ to $n = 2$.

Visualizing Chemistry
Figure 2.7 Visualizing Chemistry–Bohr Atom

VC 2.1   Which of the following best explains why we see only four lines in the emission spectrum of hydrogen?
 a) Hydrogen has only four different electronic transitions.
 b) Only four of hydrogen's electronic transitions correspond to visible wavelengths.
 c) The other lines in hydrogen's emission spectrum can't be seen easily against the black background.

VC 2.2   One way to see the emission spectrum of hydrogen is to view the hydrogen in an electric discharge tube through a *spectroscope*, a device that separates the wavelengths. Why can we not view the emission spectrum simply by pointing the spectroscope at a sample of hydrogen confined in a glass tube or flask?
 a) Without the electrons being in excited states, there would be no emission of light.
 b) The glass would make it impossible to see the emission spectrum.
 c) Hydrogen alone does not exhibit an emission spectrum—it must be combined with oxygen.

VC 2.3   How many lines would we see in the emission spectrum of hydrogen if the downward transitions from excited states all ended at $n = 1$ and no transitions ended at $n = 2$?
 a) We would still see four lines.
 b) We would see five lines.
 c) We would not see any lines.

VC 2.4   For a hydrogen atom in which the electron has been excited to $n = 4$, how many different transitions can occur as the electron eventually returns to the ground state?
 a) 1   b) 3   c) 6

SECTION 2.3: ATOMIC ORBITALS

2.27   What does an atomic orbital represent?

2.28   How is a sublevel different from an orbital?

2.29   Sketch the 3s, 3p, and 3d orbitals.

2.30   How is the 2s orbital different from the 3s orbital you sketched in Problem 2.29? How is it similar?

2.31   How is a 4p orbital different from the 3p orbital you sketched in Problem 2.29? How is it similar?

2.32   How is a 4d orbital different from the 3d orbital you sketched in Problem 2.29? How is it similar?

2.33   Choose the larger orbital of each given pair.
 (a) 2s or 4s
 (b) $2p_x$ or $2p_y$
 (c) 3p or 4p

2.34   Choose the larger orbital of each given pair.
 (a) 1s or 5s
 (b) $3p_z$ or $4p_y$
 (c) 3p or 4p

2.35   Choose the smaller orbital of each given pair.
 (a) 3d or 4d
 (b) 1s or 2s
 (c) $2p_x$ or $3p_x$

2.36   Choose the smaller orbital of each given pair.
 (a) 3s or 4s
 (b) 5p or 2p
 (c) $4d_y$ or $3d_y$

2.37   Consider an electron in each of the following orbitals. On average, which electron would be closer to the nucleus in each pair?
 (a) 1s or 3s
 (b) 2p or 4p
 (c) 3d or 4d

2.38   Consider an electron in each of the following orbitals. On average, which electron would be closer to the nucleus in each pair?
 (a) 2s or 2p
 (b) 3p or 5f
 (c) 4s or 4p

2.39   Consider each of the following sublevel designations. Determine if each is a legitimate designation and if not, explain why.
 (a) 5p   (c) 2f
 (b) 4s   (d) 1p

2.40   Consider each of the following sublevel designations. Determine if each is a legitimate designation and if not, explain why.
 (a) 1p   (c) 3f
 (b) 6s   (d) 4p

2.41   Determine whether each of the following designations represents a single orbital, a sublevel, or both.
 (a) 4d   (c) $3p_x$
 (b) 2s   (d) 2p

2.42   Determine whether each of the following designations represents a single orbital, a sublevel, or both.
 (a) 5p   (c) $2p_z$
 (b) 4s   (d) 4p

2.43   Match each of the following descriptions with the correct orbital designation.

Description	Orbital Designation
Spherical orbital in the fourth level	1s
Dumbbell-shaped orbital in the second level	2p
Cloverleaf-shaped orbital in the third level	4s
Spherical orbital in the first level	3d

2.44   Place the following sublevels in order of decreasing energy in a multielectron atom.
(a) 3*s*, 3*p*, 3*d*
(b) 1*s*, 2*s*, 3*s*
(c) 2*s*, 2*p*, 3*s*

2.45   Place the following sublevels in order of decreasing energy in a multielectron atom.
(a) 4*s*, 4*p*, 4*d*
(b) 2*p*, 3*p*, 4*p*
(c) 1*s*, 2*p*, 3*d*

SECTION 2.4: ELECTRON CONFIGURATIONS

2.46   What is meant by the term *ground state* when referring to electron configurations?

2.47   How many electrons can one orbital of any type hold?

2.48   How many electrons can be held in an orbital with the following designation?
(a) *s*   (c) *d*
(b) *p*   (d) *f*

2.49   How many orbitals does each subshell contain?
(a) 2*s*   (c) 2*p*
(b) 4*d*   (d) 4*f*

2.50   How many electrons can be held in a subshell with the following designation?
(a) *s*   (c) *d*
(b) *p*   (d) *f*

2.51   What is the maximum number of electrons that can be held in the following subshells?
(a) 2*p*   (c) 4*s*
(b) 3*d*   (d) 3*p*

2.52   Determine the ground-state electron configuration of the following atoms:
(a) chlorine
(b) calcium
(c) nitrogen

2.53   Determine the ground-state electron configuration of the following atoms:
(a) potassium
(b) arsenic
(c) selenium

2.54   Determine the ground-state electron configuration of the following atoms:
(a) oxygen
(b) sulfur
(c) sodium

2.55   Determine the ground-state electron configuration of the following atoms:
(a) lithium
(b) silicon
(c) magnesium

2.56   Draw the orbital diagram for each of the elements in Problem 2.52.

2.57   Draw the orbital diagram for each of the elements in Problem 2.53.

2.58   Draw the orbital diagram for each of the elements in Problem 2.54.

2.59   Draw the orbital diagram for each of the elements in Problem 2.55.

2.60   Write the condensed (using noble gas cores) ground-state electron configuration for the following atoms:
(a) bromine
(b) tellurium
(c) cesium

2.61   Write the condensed (using noble gas cores) ground-state electron configuration for the following transition metals:
(a) Fe   (b) Zn   (c) Ni

2.62   Write the condensed (using noble gas cores) ground-state electron configuration for the following transition metals:
(a) Zr   (b) Co   (c) Mn

2.63   Write the condensed (using noble gas cores) ground-state electron configuration for the following transition metals:
(a) Cd   (b) Pd   (c) V

SECTION 2.5: ELECTRON CONFIGURATIONS AND THE PERIODIC TABLE

2.64   Describe the difference between core and valence electrons.

2.65   Determine the number of core electrons and valence electrons in each of the following elements:
(a) phosphorus   (c) calcium
(b) iodine   (d) potassium

2.66   Determine the number of core electrons and valence electrons in each of the following elements:
(a) chlorine   (c) cesium
(b) nitrogen   (d) arsenic

2.67   Label the diagram with the appropriate term describing each shaded "block."

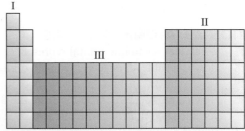

2.68   Determine the ground-state electron configuration of the following atoms. How many valence electrons does each have?
(a) strontium
(b) bromine
(c) xenon

2.69 Determine the ground-state electron configuration of the following atoms. How many valence electrons does each have?
(a) antimony
(b) barium
(c) tin

2.70 Determine the ground-state electron configuration of and draw the valence orbital diagram for each of the following atoms:
(a) aluminum   (c) phosphorus
(b) rubidium   (d) indium

2.71 Determine the ground-state electron configuration of and draw the valence orbital diagram for each of the following atoms:
(a) iodine   (c) krypton
(b) selenium   (d) strontium

2.72 How many unpaired electrons does each of the atoms in Problem 2.70 contain?

2.73 How many unpaired electrons does each of the atoms in Problem 2.71 contain?

2.74 Write out the ground-state electron configuration for the following atoms. How many valence electrons does each have? What prediction can you make about the ions of these three elements? Is this related to their location on the periodic table?
(a) lithium
(b) sodium
(c) potassium

2.75 Write out the ground-state electron configuration for the following atoms. How many valence electrons does each have? What prediction can you make about the ions of these three elements? Is this related to their location on the periodic table?
(a) fluorine
(b) chlorine
(c) bromine

2.76 Determine the element represented by each of the following ground-state electron configurations.
(a) $[Ar]4s^23d^8$
(b) $[Ne]3s^23p^3$
(c) $[Ar]4s^23d^{10}4p^4$

2.77 Determine the element represented by each of the following ground-state electron configurations.
(a) $[Kr]5s^24d^{10}5p^2$
(b) $[Xe]6s^1$
(c) $[Ar]4s^23d^{10}4p^5$

2.78 Determine the identity of each element represented by the following orbital diagrams:

(a) [↑↓] [↑↓] [↑ |↑ |]
   1s   2s   2p

(b) [↑↓] [↑↓] [↑↓|↑↓|↑↓] [↑]
   1s   2s   2p   3s

(c) [↑↓] [↑↓] [↑↓|↑↓|↑↓] [↑↓] [↑ |↑ |↑]
   1s   2s   2p   3s   3p

2.79 Determine the error(s) in the following ground-state electron configurations and identify the element represented.
(a) $1s^22s^12p^53s^23p^24s^2$
(b) $1s^32s^22p^1$
(c) $1s^22s^52p^23s^1$

2.80 Determine the error(s) in the following ground-state electron configurations and identify the element represented.
(a) $1s^22p^63s^23p^84s^2$
(b) $1s^22s^22p^63p^24s^23d^5$
(c) $1s^22s^22p^43s^23p^44s^23d^{10}$

2.81 Which of the following elements would you expect to have properties most similar to F? Why?
O   Ar   Ne   C   Br

2.82 Which of the following elements would you expect to have properties most similar to S? Why?
Br   Cl   P   Se   Na

2.83 Draw the Lewis dot symbol for each of the following atoms:
(a) magnesium   (c) fluorine
(b) phosphorus   (d) argon

2.84 Draw Lewis dot symbols for the following atoms:
(a) carbon   (c) neon
(b) chlorine   (d) potassium

2.85 Draw Lewis dot symbols for the following atoms:
(a) nitrogen   (c) calcium
(b) bromine   (d) lithium

2.86 Draw the Lewis dot symbol for each of the following atoms. What do they have in common? How is this related to their location on the periodic table?
(a) oxygen
(b) sulfur
(c) selenium

2.87 Draw the Lewis dot symbol for each of the following atoms. What do they have in common? How is this related to their location on the periodic table?
(a) nitrogen
(b) phosphorus
(c) arsenic

SECTION 2.6: PERIODIC TRENDS

2.88 What does it mean for an atom to lose an electron easily?

2.89 What types of elements gain electrons most easily?

2.90 Sketch an outline of the periodic table and draw a single arrow showing the overall trend of increasing atomic size.

2.91 Sketch an outline of the periodic table and draw a single arrow showing the overall trend of increasing metallic character.

2.92 Sketch an outline of the periodic table and draw a single arrow showing the overall trend of increasing difficulty of electron removal from an atom.

2.93   Sketch an outline of the periodic table and draw a single arrow showing the overall trend of increasing ease of electron gain.

2.94   Based on the periodic trend, arrange the following sets of elements in order of decreasing atomic size.
(a) P, S, Cl
(b) O, S, Se
(c) Ca, K, Kr

2.95   Based on the periodic trend, arrange the following sets of elements in order of increasing atomic size.
(a) Rb, S, Sr
(b) Ca, Li, Mg
(c) K, Ca, Br

2.96   Can you place N, S, and Br in order of increasing atomic size based only on the periodic trend? Why or why not?

2.97   Select the element in each group that will lose an electron most easily, based on the periodic trend.
(a) F, O, S
(b) Si, C, Ne
(c) Li, Na, K

2.98   Select the element in each group from which it will be most difficult to remove an electron, based on the periodic trend.
(a) Ca, Se, As
(b) S, Se, Ge
(c) Li, C, O

2.99   Select the least metallic element in each group, based on the periodic trend.
(a) Si, S, K
(b) Br, F, Ba
(c) Na, Mg, P

2.100   Select the most metallic element in each group, based on the periodic trend.
(a) Cs, Ba, Sr
(b) Li, Be, Na
(c) K, Br, As

SECTION 2.7: IONS: THE LOSS AND GAIN OF ELECTRONS

2.101   What is an ion? How is it different from an atom?

2.102   Describe a cation in your own words.

2.103   Describe an anion in your own words.

2.104   Determine the charge on each of the following ions:
(a) Bromine with 36 electrons
(b) Aluminum with 10 electrons
(c) Zinc with 28 electrons
(d) Arsenic with 36 electrons

2.105   Predict the common ion, including the charge, formed by each of the following elements:
(a) magnesium   (d) oxygen
(b) potassium   (e) iodine
(c) phosphorus

2.106   Predict the common ion, including the charge, formed by each of the following elements:
(a) barium   (d) aluminum
(b) nitrogen   (e) tellurium
(c) sulfur

2.107   Write the ground-state electron configuration for the common *ion* formed by each of the following elements:
(a) nitrogen
(b) strontium
(c) bromine

2.108   Write the ground-state electron configuration for the common *ion* formed by each of the following elements:
(a) chlorine
(b) sulfur
(c) calcium

2.109   Write the ground-state electron configuration for the common *ion* formed by each of the following elements:
(a) rubidium
(b) aluminum
(c) phosphorus

2.110   Write the ground-state electron configuration for the common *ion* formed by each of the following elements. What do the ions have in common? How are they different from one another?
(a) O   (b) F   (c) Na

2.111   Write the ground-state electron configuration for the common *ion* formed by each of the following elements. What do they have in common? How are they different from one another?
(a) S   (b) K   (c) Ca

2.112   Write the ground-state electron configuration for the common *ion* formed by each of the following elements:
(a) lithium
(b) selenium
(c) iodine

2.113   Draw the Lewis dot symbol for the common *ion* formed by each of the following elements:
(a) Ca   (d) O
(b) K   (e) N
(c) F

2.114   Draw the Lewis dot symbol for the common *ion* formed by each of the following elements:
(a) As   (d) Li
(b) S   (e) Cs
(c) Br

ADDITIONAL PROBLEMS

2.115   A common salt substitute contains potassium chloride. Draw the Lewis dot symbol for the potassium ion and the ion formed by chlorine.

2.116   Sodium metal is highly reactive, but the sodium ion is stable. Explain.

FAMILIAR CHEMISTRY FIREWORKS

2.117 Copper is responsible for the blue color that is observed in fireworks. Write the ground-state electron configuration and valence orbital diagram of copper. The transition (movement of the electron between energy levels) in copper that is responsible for the blue light emitted is $4p$ to $4s$. Represent this transition using the orbital diagram you've drawn.

2.118 Barium is responsible for the green color that is observed in fireworks. Write the ground-state electron configuration and draw the valence orbital diagram of barium. The transition (movement of the electron between energy levels) in barium that is responsible for this observed color is $6p$ to $6s$. Represent this transition using the orbital diagram you've drawn.

2.119 Write the electron configuration and orbital diagram of the lithium atom. If the $2s$ electron is excited to the $2p$ sublevel, and then subsequently returns to the ground state, it emits light with a wavelength of 670 nm, or 6.70×10^{-7} m.

Determine the energy gap between the $2s$ and $2p$ sublevels in lithium. (The relationship between energy and wavelength is expressed by the equation $E = hc/\lambda$, where h and c are constants with values of 6.626×10^{-34} J · s and 3.00×10^8 m/s, respectively; and λ is the wavelength of light expressed in meters.)

2.120 Write the electron configuration for the *excited* barium atom using information from Problem 2.118.

2.121 Write the electron configuration for the *excited* lithium atom using information from Problem 2.119.

2.122 A certain electron transition in an atom results in the emission of visible light. If the difference in energy between the initial and final sublevels is 4.32×10^{-19} J, what is the wavelength of the emitted light? In what region of the visible spectrum does this wavelength fall? See the relationship between energy and wavelength in Problem 2.119.

Answers to In-Chapter Materials

Answers to Practice Problems

2.1A a, e. **2.1B** a, c, e. **2.2A** (a) No, the third principal energy level has no f sublevel; (b) yes; (c) yes; (d) no, the second principal energy level has no d sublevel. **2.2B** (a) 2, 3, and 4; (b) 1, 2, 3, and 4; (c) 4; (d) 3 and 4. **2.3A** (a) $1s^2 2s^2 2p^2$, [He]$2s^2 2p^2$; (b) $1s^2 2s^2 2p^6$, [He]$2s^2 2p^6$; $1s^2 2s^2 2p^6 3s^2 3p^5$, [Ne]$3s^2 3p^5$. **2.3B** (a) Mg, (b) P. **2.4A** (a) [Kr]$5s^1$, (b) [Ar]$4s^2 3d^{10} 4p^4$, (c) [Kr]$5s^2 4d^{10} 5p^5$. **2.4B** (a) Ge, (b) Sb, (c) Ba. **2.5A** (a) $5s^1$, (b) $4s^2 4p^4$, (c) $5s^2 5p^5$. **2.5B** (a) $4s^2 4p^2$, (b) $5s^2 5p^3$, (c) $6s^2$. **2.6A** (a) As, (b) I. **2.6B** (a) K > Se > Cl, (b) Sb > I > Br. **2.7A** (a) Mg, (b) N. **2.7B** (a) C < Li < K, (b) O < S < Si.

2.8A (a) Mg, (b) Te. **2.8B** (a) C > Li > K, (b) As > Ca > Ba. **2.9A** (a) O, (b) N, (c) I. **2.9B** (a) Ga, (b) Sn, (c) Zn. **2.10A** (a) Br⁻, [Ar]$4s^2 3d^{10} 4p^6$; (b) K⁺, [Ar]; (c) S²⁻, [Ne]$3s^2 3p^6$. **2.10B** (a) Se²⁻, (b) Br⁻, (c) K⁺. **2.11A** (a) ·K, K⁺; (b) ·Mg·, Mg²⁺; (c) ·Àl·, Al³⁺. **2.11B** (a) Se²⁻, (b) N³⁻, (c) S²⁻.

Answers to Checkpoints

2.2.1 b. **2.2.2** b, c, e. **2.3.1** a. **2.3.2** b, e. **2.4.1** b. **2.4.2** d. **2.4.3** b. **2.5.1** b. **2.5.2** c. **2.5.3** b, d, e. **2.5.4** a, b, d. **2.5.5** d. **2.5.6** c. **2.6.1** b. **2.6.2** b. **2.6.3** a. **2.7.1** b, c, e. **2.7.2** b, d. **2.7.3** b.

CHAPTER 3

Compounds and Chemical Bonds

3.1 Matter: Classification and Properties
- States of Matter
- Mixtures
- Properties of Matter

3.2 Ionic Bonding and Binary Ionic Compounds

3.3 Naming Ions and Binary Ionic Compounds
- Naming Atomic Cations
- Naming Atomic Anions
- Naming Binary Ionic Compounds

3.4 Covalent Bonding and Molecules
- Covalent Bonding
- Molecules
- Molecular Formulas

3.5 Naming Binary Molecular Compounds

3.6 Covalent Bonding in Ionic Species: Polyatomic Ions

3.7 Acids

3.8 Substances in Review
- Distinguishing Elements and Compounds
- Determining Whether a Compound Is Ionic or Molecular
- Naming Compounds

Much of the matter we encounter exists in the form of mixtures. Seawater is a mixture consisting predominantly of two familiar substances: water and salt.

EpicStockMedia/Shutterstock

In This Chapter, You Will Learn

How some atoms lose or gain electrons to become *ions;* and how some atoms combine to form *molecules.* You will also learn about *chemical bonding*—the forces that hold atoms together in *compounds;* and about *nomenclature*—how to associate a compound's formula with its name.

Things To Review Before You Begin

- The charges on common main group ions [◀◀ Figure 2.25]
- Chapter 2 Key Skills [◀◀ pages 68–69]

Most of the substances we encounter every day are not elements. Rather, they are **compounds**—substances that consist of more than one element. Two examples of compounds are *water* (H_2O), which is a combination of the elements *hydrogen* and *oxygen;* and *sodium chloride* (NaCl), commonly known as *salt,* which is a combination of the elements *sodium* and *chlorine.* The relationship between electron configuration and periodic properties of the elements that we learned about in Chapter 2 gives us a way to understand the existence and the formation of compounds. In this chapter, we explain what compounds are and how to identify and name them. To do this, we must begin by discussing *matter* in general.

3.1 Matter: Classification and Properties

Chemists classify matter as either a *substance* or a *mixture* of substances. A **substance** is a form of matter that has a specific, universally *constant* composition and distinct properties, such as color, smell, and taste. Familiar examples include water, salt, iron, mercury, carbon dioxide, and oxygen. These examples differ from one another in composition, and we can tell them apart because they have different properties. For example, salt can be dissolved in water; iron cannot. Mercury is a liquid with a silvery appearance; carbon dioxide is a colorless gas. Carbon dioxide can be used to extinguish a flame; oxygen, which is also a colorless gas, actually *feeds* a flame.

States of Matter

All substances can, in principle, exist as a solid, a liquid, and a gas. These three physical states are depicted in Figure 3.1. In a solid, particles are held close together in an orderly fashion with very little freedom of motion. As a result, a solid does not conform to the shape of its container. Particles in a liquid are also close together, but are not held rigidly in position; they can move around within the liquid. This freedom of motion allows a liquid to conform to the shape of a container. In a gas, the particles are separated by distances that are very large compared to the size of the particles. A sample of gas conforms not only to the *shape* but also to the *volume* of its container.

We can convert a substance from one physical state to another without changing the identity of the substance. For example, if we start with an ice cube and heat it, it will melt to form liquid water. If we continue to heat the resulting liquid, it will boil, and vaporize to form a gas (water vapor). If we were to cool that water vapor, it would condense into liquid water. Further cooling will cause it to freeze back into ice. Despite the changes the water has undergone, it has remained *water.* The *identity* of the substance has not changed despite our having converted it several times from one *state* to another. Figure 3.2 illustrates the three physical states of water.

Figure 3.1 Molecular-level illustrations of a substance as a solid, a liquid, and a gas.

Mixtures

A *mixture* is a combination of two or more substances in which each substance retains its distinct identity. Like pure substances, mixtures can be solids, liquids, or gases. Some familiar examples of mixtures are 14 carat gold, seawater, and air. Mixtures do *not* have universally constant composition; 14 carat gold, for example, is a mixture of gold and various other metals to augment durability and/or color. Some of the supplemental metals used for this purpose are silver, platinum, zinc, nickel, and copper, and the relative amounts of the supplemental metals will vary depending on the manufacturer. Seawater, which we think of as principally water and salt, actually contains a variety of other substances as well—the amounts of which will be different at different locations. The composition of air varies as well. We would expect a sample of air taken downwind of a coal-burning power plant to be different from a sample taken over the open ocean.

Mixtures are either *homogeneous* or *heterogeneous*. When we dissolve a teaspoon of sugar in a glass of water, the result is a **homogeneous mixture** because the composition of the mixture is uniform throughout. If we mix sand with iron filings, however, the sand and the iron filings remain distinct and discernible from each other [Figure 3.3(a)]. This type of mixture is called a **heterogeneous mixture** because the composition is *not* uniform.

A mixture, whether homogeneous or heterogeneous, can be separated into the substances it contains without changing the identities of the individual substances. Thus, sugar can be recovered from sugar water by evaporating the mixture to dryness. The solid sugar will be left behind, and the water component can be recovered by condensing the water vapor that evaporates. To separate the sand-iron mixture, we can use a magnet to remove the iron filings from the sand because sand is not attracted to the magnet [Figure 3.3(b)]. After being separated, the individual components of each mixture will have the same composition and properties that they did prior to being mixed.

Figure 3.4 illustrates several of the physical processes that are commonly used to separate mixtures.

Figure 3.2 Water as a solid (ice), liquid, and gas. (We can't actually see water vapor, any more than we can see the nitrogen and oxygen that make up most of the air we breathe. When we see steam, what we are actually seeing is water vapor that has condensed to liquid water upon encountering cold air.)

Charles D. Winters/Timeframe Photography/McGraw Hill

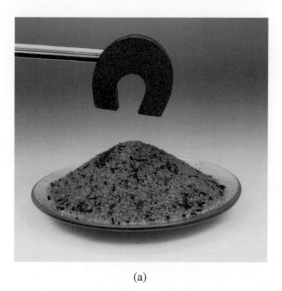

(a)

(b)

Figure 3.3 (a) A heterogeneous mixture of iron filings and sand. (b) A magnet can be used to separate the mixture by removing the iron filings. The sand is left behind because it is not attracted to the magnet.
(both) Charles D. Winters/Timeframe Photography/McGraw Hill

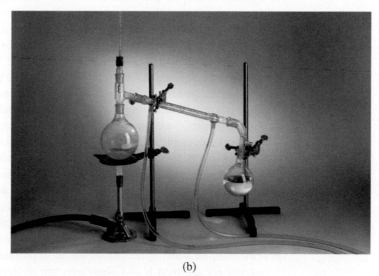

(a)

(b)

Figure 3.4 (a) Filtration can be used to separate a heterogeneous mixture of a liquid and a solid, such as coffee and coffee grounds. The filter, in this case a coffee filter, allows only the liquid coffee to pass through. (b) Distillation can be used to separate components with different boiling points. The vapor can be condensed and recovered by cooling. (c) A variety of chromatographic techniques, including paper chromatography, can be used to separate mixtures. Here, paper chromatography is used to separate the dye in candy coatings into its individual colored components.
(a) Bloomimage/Collage/Corbis; (b & c) ©Richard Megna/Fundamental Photographs, NYC

(c)

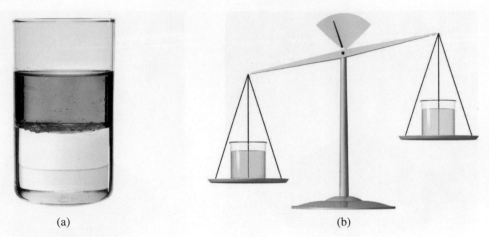

(a)   (b)

Figure 3.5 (a) Vegetable oil and water. (b) Equal volumes of vegetable oil and water weigh different amounts because they have different densities.
David A. Tietz/McGraw Hill

Properties of Matter

Substances are identified by their properties as well as by their composition. Properties of a substance may be **quantitative** (measured and expressed with a *number*) or **qualitative** (not requiring an explicit measurement and expressed without the use of a number). You are undoubtedly familiar with the fact that oil and water do not mix and that oil floats on water [Figure 3.5(a)]. The reason oil floats on water is that it has a lower *density*—meaning that if we were to weigh equal volumes of the two liquids, the oil would weigh less [Figure 3.5(b)]. To determine the densities of these two liquids would require measurements, and expression of the results would require the use of numbers. Density is a *quantitative* property. On the other hand, we can easily distinguish between these two liquids simply by observing that they are different colors. Water is colorless, whereas oil (vegetable oil, in this case) is yellowish. The determination of color requires only observation, no measurement. Color is a *qualitative* property.

Color, melting point, boiling point, density, and physical state are all physical properties. A **physical property** of a substance is a property that can be observed and measured without changing the *identity* of the substance. For example, we can determine the melting point of ice by heating a block of ice and measuring the temperature at which the ice is converted to liquid water. Melting is a **physical change** or **physical process.** Liquid water differs from ice in appearance but not in composition. (Liquid water and ice are both H_2O.) Therefore, the melting point of a substance is a physical property. Similarly, the color, boiling point, density, and physical state of a substance are physical properties.

Substances are also identified by their *chemical properties*. The statement "iron rusts when exposed to water and air," describes a **chemical property** of iron, because for us to observe this property, a **chemical change** or **chemical process** must occur. In this case, the chemical change is *corrosion* or *oxidation* of iron. After a chemical change, the original substance (iron metal, in this case) is no longer there. What remains is a different substance (*rust,* in this case). There is no *physical process* by which we can recover the iron from the rust.

Often when we cook, we bring about chemical changes. Baking a cake, for example, involves heating batter that contains a leavening agent, typically baking powder. One of the ingredients in baking powder undergoes a chemical change that produces a gas, which forms numerous tiny bubbles during the baking process, causing the cake to "rise." After the cake is baked, we cannot recover the baking powder by cooling the cake—or by any *physical* process. When we eat the cake, we initiate further chemical changes that occur during digestion and metabolism.

Sample Problems 3.1 and 3.2 give you some practice classifying matter and distinguishing physical and chemical properties and changes.

SAMPLE PROBLEM (**3.1**) Distinguishing Pure Substances and Mixtures

Classify each of the following as a pure substance or mixture: (a) the aluminum in a soda can, (b) a sports drink (like Powerade®), (c) salt, (d) carbonated water.

Strategy You know that pure substances are either elements or compounds and that mixtures are combinations of two or more elements or compounds.

Setup If there is only one substance present, it is either an element or a compound. If there is more than one substance present, it is a mixture.

Solution (a) The aluminum of a soda can consists only of the element aluminum, so it is a pure substance.

(b) A sports drink contains water and several other ingredients, including sugar and salt. It is a mixture.

(c) Salt is a compound composed of the elements sodium and chlorine. It is a pure substance.

(d) Carbonated water contains two substances: water and carbon dioxide. It is a mixture.

THINK ABOUT IT

Remember that pure substances have constant composition (e.g., aluminum and salt) and that mixtures contain two or more pure substances and may vary in composition (e.g., sports drinks and carbonated water).

Practice Problem **A**TTEMPT Classify each of the following as a pure substance or mixture: (a) rocky road ice cream, (b) helium gas, (c) air, (d) ice (solid water).

Practice Problem **B**UILD Classify each of the pure substances in Sample Problem 3.1 (aluminum and salt) as an element or a compound. Classify each of the mixtures in Sample Problem 3.1 (sports drink and carbonated water) as heterogeneous or homogeneous.

Practice Problem **C**ONCEPTUALIZE Classify each of the following diagrams as representing a pure substance or a mixture. For any pure substances, indicate whether it is an element or a compound. For any mixtures, identify the components of the mixture as elements or compounds.

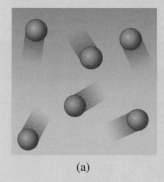
(a)

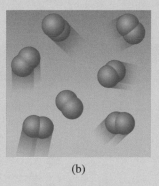
(b)

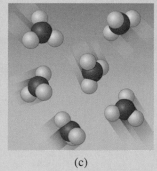
(c)

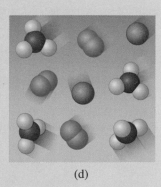
(d)

SAMPLE PROBLEM (**3.2**) Distinguishing Physical and Chemical Properties

Classify each of the following as a physical property or a chemical property: (a) Water has a density of 1.00 g/mL at room temperature; (b) iron rusts in moist air; (c) copper turns green over time; (d) aluminum melts at 660°C.

Strategy You know that a physical property is one that can be observed or measured *without* changing the identity of the substance. Observing a chemical property *does* require a change in the identity of a substance.

Setup Determine which of the properties can be observed only by the original substance changing to another substance—these are the *chemical* properties. Those that can be observed without the original substance changing to another substance are *physical* properties.

Solution (a) The density of a substance can be measured without changing it, so this is a physical property.

(b) Iron metal is shiny and silvery in appearance. Rust is a flaky reddish-brown substance. The process of rusting changes iron metal into rust—a different substance. Therefore, iron's tendency to rust is a chemical property.

(Continued on next page)

(c) Copper changing color indicates a change in its composition. This is a chemical property.

(d) The melting point of a substance is measured when the substance changes from the solid state to the liquid state. Liquid aluminum is still *aluminum,* though, so melting point is a physical property.

THINK ABOUT IT

A change in the identity of a substance is evidence of a *chemical* property of the substance. Physical properties of a substance are those that can be determined *without* changing its identity.

Practice Problem **A**TTEMPT Classify each of the following as a physical property or a chemical property: (a) Water boils at 100°C; (b) sugar dissolves in water; (c) sodium reacts violently with water; (d) magnesium is a shiny, silvery metal.

Practice Problem **B**UILD Classify each of the following as a physical change or a chemical change: (a) paper burns; (b) a piece of paper is torn into smaller pieces; (c) a silver fork tarnishes in air; (d) helium gas is buoyant in air.

Practice Problem **C**ONCEPTUALIZE The diagram on the left represents the result of a change. Which of the diagrams [(a)–(c)] could represent the starting material if the process were physical, and which could represent the starting material if the change were chemical?

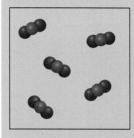 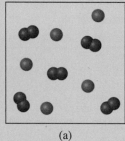  

	(a)	(b)	(c)

For the remainder of this chapter, we look in some detail at how the periodic properties of atoms result in the formation of two different *types* of compounds; and we describe how chemists name and identify these compounds using a specific system of nomenclature.

Student Note:
Nomenclature refers to the *naming* of compounds.

CHECKPOINT–SECTION 3.1   Matter: Classification and Properties

3.1.1 Which of the following illustrations [(a)–(f)] represent a physical change? (Select all that apply.)

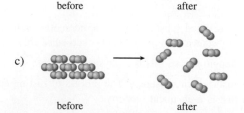

3.1.2 Which of the following illustrations [(a)–(f)] represent a chemical change? (Select all that apply.)

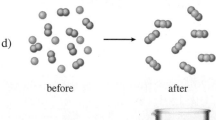

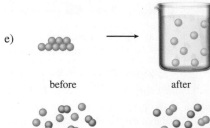

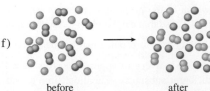

3.2 Ionic Bonding and Binary Ionic Compounds

Recall from Chapter 2 that atoms of elements that lose electrons easily (metals) form positively charged ions, known as *cations;* and those that *gain* electrons easily (nonmetals) form negatively charged ions, known as *anions.* It is important to recognize that in the context of chemical interactions, these two things do not happen independently. The electron lost by one atom is gained by another; and the electrostatic attraction between the resulting oppositely charged species [◀◀ Section 1.2], also known as Coulombic attraction, draws them together to form an *ionic compound.*

Consider the example of sodium and chlorine. The electron configuration of sodium is $[Ne]3s^1$ and that of chlorine is $[Ne]3s^2 3p^5$. When a sodium atom and a chlorine atom come into contact with each other, the valence electron of the sodium atom is transferred to the chlorine atom, giving *both* atoms noble gas electron configurations. We can imagine these processes taking place separately and represent each process using Lewis dot symbols.

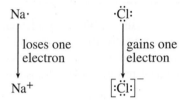

(same electron configuration as Ne) (same electron configuration as Ar)

The positively charged Na^+ ion and the negatively charged Cl^- ion are attracted to each other,

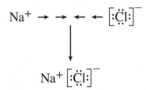

and they are drawn together to form the electrically neutral compound, sodium chloride, which we represent with the chemical formula NaCl. The **chemical formula,** or simply *formula,* of an ionic compound identifies the elements in the compound and indicates the ratio in which they combine. Sodium chloride, for example, consists of equal numbers of Na^+ ions and Cl^- ions.

Despite what the formula NaCl may seem to imply, solid sodium chloride does not consist of a collection of discrete NaCl units. Rather, it consists of a vast three-dimensional array of alternating Na^+ and Cl^- ions as illustrated in Figure 3.6. Note that although an ionic compound consists of oppositely charged ions, the compound overall is electrically neutral, and we do not include the charges of the ions in the chemical formula. The electrostatic attractions between oppositely charged ions are what we refer to as **ionic bonding.** Ionic bonds form most often between metals, which tend to *lose* electrons easily, and nonmetals, which tend to *gain* electrons easily.

The ratio of combination of Na^+ and Cl^- ions in sodium chloride is 1:1 because although their charges are opposite in *sign,* they are *equal* in magnitude. The same is true of combinations of any ions with opposite charges of equal magnitude, such as:

Ca^{2+} and O^{2-}, which combine in a 1:1 ratio to form the compound CaO (calcium oxide)

Al^{3+} and N^{3-}, which combine in a 1:1 ratio to form the compound AlN (aluminum nitride)

When ions with charges that are *not* equal in magnitude combine to form an ionic compound, we can determine the formula by remembering that the compound overall

Animation
Formation of Ionic Compounds

Student Hot Spot

Student data indicate that you may struggle to understand the formation of ionic bonds when electrons are transferred between atoms. Access the eBook to view additional Learning Resources on this topic.

Student Note: Although the name "sodium chloride" is probably familiar, we explain in the next section how this name and the names of other ionic compounds are determined.

Student Note: When we write the formula of this type of ionic compound, we write the metal (Na) first and the nonmetal (Cl) second.

Student Note: Now is a good time to take another look at Figure 2.25 in Section 2.7. It shows the charges on the common ions of main group elements. Being able to deduce these charges using a periodic table is very important.

Figure 3.6 A Na atom loses one electron to become a Na$^+$ ion. A Cl atom gains one electron to become a Cl$^-$ ion. The oppositely charged ions are pulled together by the electrostatic attraction between them. Solid sodium chloride (salt) consists of a three-dimensional array of alternating Na$^+$ and Cl$^-$ ions.

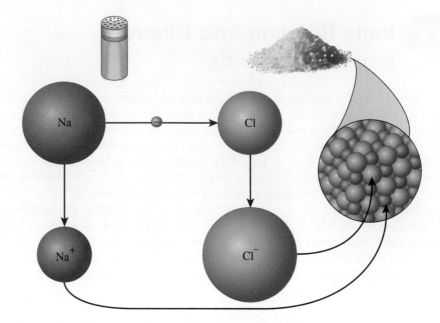

must be electrically neutral. For example, when the elements barium (Ba) and iodine (I) combine to form an ionic compound, the ions are Ba^{2+} and I$^-$ (see Figure 2.25). For the resulting compound to be neutral, there must be twice as many I$^-$ ions as Ba^{2+} ions. We denote this in a chemical formula with a subscript 2:

$$BaI_2$$

A simple method for determining the ratio of combination in an ionic compound when the charges are not numerically equal, is to simply write a subscript for the *cation* that is numerically equal to the charge on the *anion,* and write a subscript for the *anion* that is numerically equal to the charge on the *cation.* The following examples illustrate this method. Note that when a subscript would be 1, it is not shown in the formula.

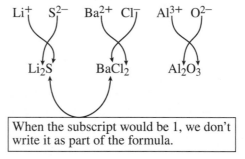

When the subscript would be 1, we don't write it as part of the formula.

Occasionally, this method will give a formula with subscripts that are *not* the smallest possible whole numbers. One example is *barium oxide.* The numerical value of the charges on both ions is 2, so this method would give the formula Ba$_2$O$_2$. In a case such as this, we must reduce the subscripts to the smallest possible whole numbers, giving us the formula BaO.

Sample Problem 3.3 shows how to deduce formulas for ionic compounds, given the elements they contain.

SAMPLE PROBLEM   **3.3**   Determining Formulas for Binary Ionic Compounds

Write the formula for the ionic compound that forms from the following pairs of elements: (a) Li and F, (b) Mg and F, (c) Mg and O.

Strategy You have learned how to predict the charges on ions formed from main group elements. Knowing the charges on the ions and knowing that compounds must be neutral (no extra positive or negative charges), you can determine formulas for ionic compounds.

Setup Determine the charges on each ion in the pair and then determine how many of each are necessary to arrive at a neutral compound, keeping in mind that the smallest whole numbers are to be used in the formula.

Solution (a) Lithium forms a +1 ion and fluorine forms a −1 ion. One of each ion provides the same number of positive and negative charges, meaning a neutral compound with the formula LiF.

The positive charge on lithium becomes the subscript on fluorine, and the negative charge on fluorine becomes the subscript on lithium in the formula. In this case, the magnitudes of the charges are both 1, making the subscripts both 1 as well. Remember that we do not write the subscript 1 in a formula.

(b) Magnesium forms a +2 ion and fluorine forms a −1 ion. To obtain a neutral compound, we need *two* of the negative fluorine ions for each positive magnesium ion to balance the charges. The formula is MgF_2.

In this case, we place a subscript 2 on fluorine in the compound—to indicate that two fluorine ions, each with a charge of −1, are necessary to balance the +2 charge on magnesium.

(c) Magnesium forms a +2 ion and oxygen forms a −2 ion. One of each ion is needed to form a neutral compound. The formula of the compound is MgO.

$$Mg^{2+} \quad O^{2-}$$

$$Mg_2O_2 \rightarrow MgO$$

Using the charge magnitudes (in this case, both 2s) as subscripts in the formula may result in something other than the smallest possible whole numbers. When this happens, reduce the subscripts to the smallest possible whole numbers—in this case by dividing both subscripts by 2.

THINK ABOUT IT

The shortcut of "swapping" the charges to determine the subscripts as shown in the preceding illustrations is an easy way to predict the formula of an ionic compound. However, you must keep in mind that the *smallest* whole-number ratio gives the correct formula.

Practice Problem ATTEMPT Write the formula for the ionic compound that can form from each of the following pairs of elements: (a) Ca and N, (b) K and Br, (c) Ca and Br.

Practice Problem BUILD Determine the most probable charges on each of the ions present in the following hypothetical compounds: (a) AX_3, (b) D_3E_2, (c) LM_2.

Practice Problem CONCEPTUALIZE Why is it impossible to determine the charges of the ions in the hypothetical compound YZ?

The ionic compounds for which we have determined formulas thus far are known as **binary ionic compounds,** meaning that they are ionic substances consisting of just *two* elements. Further, they have all contained main group metal cations whose charges are *predictable*. The charge on a sodium ion is always +1, that on a barium ion is always +2, that on an aluminum ion is always +3, and so on. This type of binary ionic compound, in which the cation has only one possible charge, is known specifically as a *type I compound.*

Unlike most main group metals, many transition metals can form cations with more than one possible charge. Figure 3.7 shows the common ions of some transition metals, along with some main group metals that can form cations with more than one

Student Note: There are some main group metals that form cations of variable charge (e.g., lead); and there are transition metals that form only one cation with a predictable charge (e.g., silver).

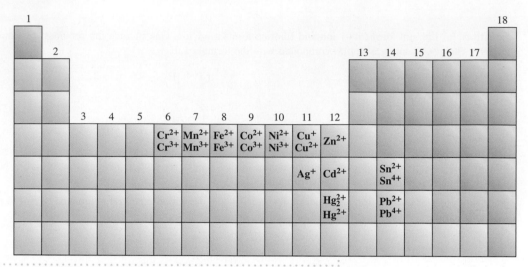

Figure 3.7 Common transition-metal ions and main group ions with more than one possible charge.

possible charge. A binary ionic compound in which the metal cation's charge is *not* always the same is known as a ***type II compound.*** Chromium, for example, forms two different cations: Cr^{2+} and Cr^{3+}. Therefore, it can form two different binary ionic compounds with the element chlorine.

$$Cr^{2+} \quad Cl^- \qquad Cr^{3+} \quad Cl^-$$
$$CrCl_2 \qquad\qquad CrCl_3$$

Sample Problem 3.4 lets you practice deducing formulas for type II binary ionic compounds.

SAMPLE PROBLEM   (3.4)   Determining Formulas for Type II Binary Ionic Compounds

Write the formula for the compound that would form from each of the following pairs of ions: (a) Fe^{3+} and Cl^-, (b) Fe^{2+} and Cl^-, (c) Pb^{4+} and O^{2-}.

Strategy Knowing the charges on the ions and the fact that compounds must be neutral (no extra positive or negative charges), the formula can be determined.

Setup Determine how many of each ion are necessary to arrive at a neutral compound, keeping in mind that the smallest whole numbers are to be used in the formula.

Solution (a) One Fe^{3+} ion requires three Cl^- ions (three negative charges) to form a neutral compound. This gives the formula $FeCl_3$.

$$Fe^{3+} \quad Cl^-$$
$$FeCl_3$$

(b) One Fe^{2+} ion requires two Cl^- ions to form a neutral compound. The formula is $FeCl_2$.

$$Fe^{2+} \quad Cl^-$$
$$FeCl_2$$

(c) One Pb^{4+} ion requires two O^{2-} ions to form a neutral compound. The formula of the compound is PbO_2.

$$Pb^{4+} \quad O^{2-}$$
$$Pb_2O_4 \rightarrow PbO_2$$

THINK ABOUT IT

The shortcut of "swapping" the charges to determine the subscripts as shown in the preceding illustrations is an easy way to predict the formula of an ionic compound. However, you must keep in mind that the smallest whole-number ratio must be used as shown in part (c).

Practice Problem **A**TTEMPT Determine the formula for the compound formed by each of the following pairs of ions: (a) Sn^{4+} and N^{3-}, (b) Ni^{2+} and F^-, (c) Mn^{4+} and O^{2-}.

Practice Problem **B**UILD Determine the charge on the metal ion present in each of the following binary ionic compounds: (a) $HgCl_2$, (b) CuO, (c) Fe_2O_3.

Practice Problem **C**ONCEPTUALIZE Write formulas for all of the possible binary ionic compounds that can form from the following elements: (a) iron and phosphorus, (b) lead and iodine, and (c) copper and oxygen.

CHECKPOINT–SECTION 3.2 Ionic Bonding and Binary Ionic Compounds

3.2.1 How many different binary ionic compounds can form by combining the elements Fe and F?

a) one   d) four

b) two   e) five

c) three

3.2.2 What is the charge on the metal ion in the compound Cr_2O_3?

a) 2+   d) 5+

b) 3+   e) 0

c) 4+

3.2.3 How many different binary ionic compounds can be made using combinations of the ions shown here?

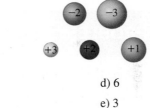

a) 12   d) 6

b) 9   e) 3

c) 8

3.3 Naming Ions and Binary Ionic Compounds

In centuries past, when chemistry was a *new* science, it was possible for chemists to memorize the names of the relatively small number of known compounds. Often a compound's name was related to its appearance, its properties, its origin, or its common uses. For example, "milk of magnesia" resembles milk; "laughing gas" can make dental patients feel silly; formic acid (*formica* is the Latin word for *ant*) is the compound responsible for the sting of an ant bite.

Today, there are many *millions* of known compounds—with many more being produced every year. It would be impossible for anyone to memorize all of their names. Fortunately, such memorization is unnecessary, because over the years, chemists have devised a system of *nomenclature* for naming chemical substances. The rules of chemical nomenclature are universal, making it possible for scientists the world over to understand one another, and providing a practical way to identify a seemingly overwhelming number of different substances. Mastering these relatively simple rules now will benefit you tremendously as you progress through your chemistry course. You must be able to *name* a compound, given its chemical formula; and you must be able to write the correct *formula* for a compound, given its name.

Naming Atomic Cations

Student Note: Remember that an *atomic* ion (also called a *monatomic* ion) consists of just one atom that has either lost or gained one or more electrons. [◀◀ Section 2.7]

We begin our discussion of chemical nomenclature with the naming of atomic cations. Recall that charges on the common ions of main group elements are predictable, based on their position in the periodic table. (See Figure 2.23.) For common ions of main group metals, the charge is equal to the group number for metals on the left side of the periodic table, and equal to the second *digit* of the group number for metals on the right side of the periodic table. Metals in Group 1 form cations with a charge of +1; metals in Group 2 form cations with a charge of +2; and the metal in Group 13 that forms a common ion (Al) forms a cation with a charge of +3. An atomic cation is named simply by adding the word *ion* to the name of the element. Thus, the ion of potassium, K^+, is called *potassium ion;* the ion of magnesium, Mg^{2+}, is called *magnesium ion;* and the ion of aluminum, Al^{3+}, is called *aluminum ion.*

For metals that form ions with more than one possible charge, the ion's name must also *include* the charge. This is done with a Roman numeral in parentheses, which immediately follows the element's name. Thus, the +2 cation of chromium, Cr^{2+}, is called *chromium(II) ion;* and the +3 cation of chromium, Cr^{3+}, is called *chromium(III) ion.* Note that we do not put a space between the element name and the parenthetical Roman numeral that indicates the ion's charge.

Sample Problem 3.5 lets you practice naming atomic cations.

SAMPLE PROBLEM   3.5   Naming Metal Cations

Name each of the following ions: (a) Ca^{2+}, (b) Pb^{4+}, (c) Ag^+

Strategy Use Figure 3.7 to determine if the metal can form ions of more than one charge.

Setup If the metal can form ions of different charges, a parenthetical Roman numeral must be included in the name to specify the charge.

Solution (a) Calcium is a main group metal that does not form more than one ion. Its charge is always +2 and it is simply called the *calcium ion.*

(b) Lead is one of the few main group metals that *does* form more than one ion (+2 and +4). The name of this ion must include a Roman numeral denoting the +4 charge. It is called the *lead(IV) ion.*

(c) Silver is a transition metal that does not form more than one ion. Its charge is always +1 and its name is simply *silver ion.*

THINK ABOUT IT

Most main group metals do not form more than one ion and will not need Roman numerals in their names. Only some of the transition metals and a few main group metals form more than one ion and will need Roman numerals in their names.

Practice Problem Ⓐ**TTEMPT** Name each of the following ions: (a) Fe^{3+}, (b) Mn^{2+}, (c) Rb^+.

Practice Problem Ⓑ**UILD** Write the symbol and charge for each of the following ions: (a) lead(II) ion, (b) sodium ion, (c) zinc ion.

Practice Problem Ⓒ**ONCEPTUALIZE** Why do you think the Hg_2^{2+} ion is called the mercury(I) ion, even though both the charge and the superscript charge are 2?

TABLE 3.1	Names and Formulas of Some Common Atomic Ions				
Name	**Formula**	**Name**	**Formula**	**Name**	**Formula**
Cations		**Cations**		**Anions**	
aluminum	Al^{3+}	lead(II)	Pb^{2+}	bromide	Br^-
barium	Ba^{2+}	lithium	Li^+	chloride	Cl^-
cadmium	Cd^{2+}	magnesium	Mg^{2+}	fluoride	F^-
calcium	Ca^{2+}	manganese(II)	Mn^{2+}	hydride	H^-
cesium	Cs^+	mercury(II)	Hg^{2+}	iodide	I^-
chromium(III)	Cr^{3+}	potassium	K^+	nitride	N^{3-}
cobalt(II)	Co^{2+}	silver	Ag^+	oxide	O^{2-}
copper(I)	Cu^+	sodium	Na^+	sulfide	S^{2-}
copper(II)	Cu^{2+}	strontium	Sr^{2+}		
hydrogen	H^+	tin(II)	Sn^{2+}		
iron(II)	Fe^{2+}	zinc	Zn^{2+}		
iron(III)	Fe^{3+}				

Naming Atomic Anions

An atomic anion is named by changing the ending of the element's name to *–ide* and adding the word *ion*. Thus, the anion of chlorine, Cl^-, is called *chloride ion*. The anions of nitrogen and oxygen (N^{3-} and O^{2-}) are called *nitride ion* and *oxide ion,* respectively. Because nonmetals form anions with predictable charges (in general, equal to the element's group number −18), it is not necessary for the ion's name to specify its charge. Table 3.1 gives alphabetical lists of some common atomic cations and anions.

Naming Binary Ionic Compounds

A binary ionic compound is named using the name of the cation followed by the name of the anion—eliminating the word *ion* from each. We have already encountered several examples of compounds named in this way: sodium chloride, calcium oxide, and aluminum nitride. Let's apply this approach to the compounds for which we have determined formulas: Li_2S, $BaCl_2$, and Al_2O_3.

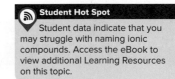

Student Hot Spot

Student data indicate that you may struggle with naming ionic compounds. Access the eBook to view additional Learning Resources on this topic.

	Cation	**Anion**	**Name**
Li_2S	lithium ion	sulfide ion	lithium sulfide
$BaCl_2$	barium ion	chloride ion	barium chloride
Al_2O_3	aluminum ion	oxide ion	aluminum oxide

Note that the compounds' names make no reference to the subscript numbers in the formulas. This is because the charges on each of these ions is predictable, making the names unambiguous. In the case of lithium sulfide, the lithium ion is Li^+. Its charge is always +1. The sulfide ion is S^{2-}. Its charge is always −2. The only combination of lithium and sulfide ions that results in a neutral formula is two lithium ions for every one sulfide ion: Li_2S. In the case of barium chloride, the barium ion is Ba^{2+} (always +2) and the chloride ion is Cl^- (always −1). There is only one possible combination ratio that results in a neutral formula: $BaCl_2$. The charge on the aluminum ion is always +3, and that on the oxide ion is always −2; so again, there is only one possible ratio of combination: Al_2O_3. It is simply unnecessary to indicate the numbers of each ion in an ionic compound's name.

Now let's consider some type II compounds, in which the charge on the metal cation is not always the same. You will find that the nomenclature is essentially the

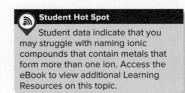

Student Hot Spot

Student data indicate that you may struggle with naming ionic compounds that contain metals that form more than one ion. Access the eBook to view additional Learning Resources on this topic.

Student Note: It is important to remember that the *number* in an ionic compound's name refers to the *charge* on the metal cation—not to its subscript in the formula.

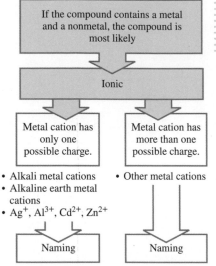

Figure 3.8 Flowchart for naming type I and type II binary ionic compounds.

same, except that the charge on the cation, in the form of a parenthetical Roman numeral, is part of the name.

	Cation	**Anion**	**Name**
$CoCl_2$	cobalt(II) ion	chloride ion	cobalt(II) chloride
Cr_2O_3	chromium(III) ion	oxide ion	chromium(III) oxide
Cu_3N_2	copper(II) ion	nitride ion	copper(II) nitride

When we *say* the names of these three compounds, we say "cobalt *two* chloride," "chromium *three* oxide," and "copper *two* nitride." You may be wondering how we know from the formula what the charge *is* on a variable-charge cation. This is a case where we use what we know, the charge on the *anion,* to figure out what we need to know, the charge on the *cation.* The anion in $CoCl_2$ is the chloride ion, which we know has a charge of −1, because the chloride ion is predictable and its charge is always the same. With *two* chloride ions, in order for the formula to be neutral, the charge on cobalt must be +2. Therefore, we know that $CoCl_2$ contains the cobalt(II) ion. The cancellation of charges can be calculated as 2(−1) + 1(+2) = 0. The same strategy can be applied to the other two compounds:

Cr_2O_3 contains three oxide ions, each with a predictable charge of −2. For the formula to be neutral, with *two* chromium ions, the charge on each chromium ion must be +3. Cancellation of charges: 3(−2) + 2(+3) = 0.

Cu_3N_2 contains two nitride ions, each with a predictable charge of −3. For the formula to be neutral, with three copper ions, the charge on each copper ion must be +2. Cancellation of charges: 2(−3) + 3(+2) = 0.

Figure 3.8 summarizes the process of naming type I and type II binary ionic compounds.

Sample Problem 3.6 lets you practice naming ionic compounds.

SAMPLE PROBLEM (3.6) Naming Binary Ionic Compounds

Determine the name of each of the ionic compounds: (a) Ca_3N_2, (b) $MgCl_2$, (c) Cu_2O, (d) $ZnCl_2$, (e) Mn_2O_3, (f) Li_3P.

Strategy Remember that the name of an ionic compound is simply the name of the cation (positive ion), followed by the anion (negative ion). Only metals that form more than one ion should include a Roman numeral as part of their name.

Setup Determine which of the metals have ions with different charges and plan to use Roman numerals in naming them. Where necessary, use the known charge on the anion and the ratio expressed in the formula to determine the charge on the metal ion.

Solution

(a) The ions present are the calcium ion and the nitride ion. Putting them together gives the name calcium nitride.

(b) The ions present are the magnesium ion and the chloride ion. Putting them together gives the name magnesium chloride.

(c) Copper can form several ions, so we must use the charge on the oxygen and the formula to determine the charge on the copper ion. In this case, because two copper ions are combined with one oxide ion, which always has a charge of −2, we see that the charge on copper is +1. Therefore, the ions present are the copper(I) ion and the oxide ion. Putting them together gives the name copper(I) oxide.

(d) The ions present are the zinc ion and the chloride ion. Putting them together gives the name zinc chloride.

(e) The ions present are the manganese(III) ion and the chloride ion. Putting them together gives the name manganese(III) chloride.

(f) The ions present are the lithium ion and the phosphide ion. Putting them together gives the name lithium phosphide.

THINK ABOUT IT

Remember that the subscript following an atom's symbol in a chemical formula is not its charge. The charge must be determined using the periodic table and the chemical formula.

Practice Problem **A**TTEMPT   Write the name of each of the following ionic compounds: (a) FeN, (b) AgF, (c) Na_2O, (d) Co_2O_3, (e) Rb_2S, (f) PbO.

Practice Problem **B**UILD   Write the formula for each of the following ionic compounds: (a) iron(II) oxide, (b) zinc bromide, (c) strontium sulfide, (d) potassium oxide, (e) nickel(IV) nitride, (f) lithium phosphide.

Practice Problem **C**ONCEPTUALIZE   Determine the charge on the copper ion in the compound shown.

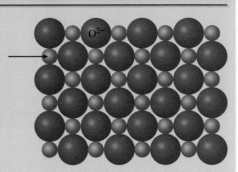

O^{2-}

CHECKPOINT–SECTION 3.3   Naming Ions and Binary Ionic Compounds

3.3.1   What is the correct name of the compound PbS?

　　a) lead sulfide

　　b) lead(I) sulfide

　　c) lead(II) sulfide

　　d) lead(III) sulfide

　　e) lead(IV) sulfide

3.3.2   What is the correct formula for the compound iron(III) nitride?

　　a) Fe_3N

　　b) FeN_3

　　c) FeN

　　d) Fe_2N_3

　　e) Fe_3N_2

3.3.3   What is the correct formula for the compound sodium nitride?

　　a) NaN

　　b) Na_3N

　　c) NaN_3

　　d) Na_2N_3

　　e) Na_3N_2

3.3.4   What are the charges on the cation (smaller) and the anion (larger), respectively, in the binary ionic compound shown here?

　　a) +2, −1

　　b) +1, −2

　　c) +2, −2

　　d) +3, −2

　　e) +2, −3

3.4   Covalent Bonding and Molecules

We learned in Section 3.2 that ionic bonds form when metal atoms lose electrons, nonmetal atoms gain electrons, and the resulting oppositely charged particles are drawn together by Coulombic attraction. The driving force for the formation of ionic compounds is the tendency of both main group metals and nonmetals to achieve stability by acquiring a noble gas electron configuration. Main group metals do this by *losing* their valence electrons, and nonmetals do this by gaining enough electrons to result in their having *eight* valence electrons. However, there are elements on the periodic table, such as carbon (C), silicon (Si), and phosphorus (P), that neither lose nor gain electrons readily. Moreover, there are nonmetals that *can* gain electrons, but will only do so when they are combined with elements that lose electrons *very* easily, such as the metals in Groups 1 and 2. In cases where the nature of the atoms involved precludes

Student Note: Remember that metals tend to *lose* electrons easily and nonmetals tend to *gain* electrons easily. [◄◄ Section 2.7]

there being a transfer of electrons from one to the other, another type of chemical bond can form: a *covalent* bond.

Covalent Bonding

We begin by examining the element chlorine. We have seen that if Cl and Na atoms are combined, the Cl atom gains the electron lost by the Na atom; and the binary ionic compound NaCl (sodium chloride) forms. But what if we have only Cl atoms and no Na atoms? Each Cl atom still has the tendency to achieve stability by acquiring a noble gas electron configuration; but without metal atoms to provide the necessary electrons, Cl atoms cannot accomplish this by *gaining* electrons. Instead, the Cl atoms will acquire the noble gas configuration by *sharing* electrons. Recall that we can represent atoms with Lewis dot symbols [◄◄ Section 2.5]. The Lewis dot symbols for two Cl atoms are shown here:

$$:\ddot{Cl}\cdot \qquad \cdot\ddot{Cl}:$$

> **Student Note:** We can draw the Lewis symbols for Cl atoms with the single dot on either side of the element's symbol. These are drawn with the single dots facing each other to illustrate the formation of a chemical bond.

With no ready supply of electrons for the atoms to gain, they move close enough together to *share* a pair of electrons.

> **Student Note:** Even though two of the electrons around each atom are shared by the other atom, both atoms "think" they own both of the shared electrons.

In this way, both Cl atoms are surrounded by eight valence electrons, giving them both the noble gas configuration they need for stability. In this type of arrangement, where two atoms share a pair of electrons, the shared pair of electrons constitutes a chemical bond known as a ***covalent bond.*** The chemical formula for the resulting species is Cl_2, which is also known as *chlorine* and represents the most stable form of this element.

We can illustrate covalent bond formation equally well with hydrogen atoms. Lewis dot symbols for two H atoms are shown here:

$$H\cdot \qquad \cdot H$$

Each H atom has just one electron. Hydrogen is the only nonmetal that achieves a noble gas configuration with just *two* electrons, giving it the electron configuration of helium. When the two H atoms move close enough to share their combined pair of electrons, a covalent bond forms between them to form H_2, also known as *hydrogen.*

Molecules

The formation of a covalent bond between two Cl atoms or between two H atoms produces a species known as a *molecule*. A ***molecule*** is a neutral combination of at least two atoms that is held together by covalent bonds. A molecule may contain two or more atoms of the same element, or it may contain atoms of different elements. For example, H and Cl atoms can form the molecule HCl by moving close enough to share a pair of electrons:

> **Student Note:** Substances held together by only covalent bonds (no ionic bonds) are called *molecular substances.*

> **Student Note:** These are actually molecular representations known as *Lewis structures,* which we discuss in more detail in Chapter 6.

The resulting species is the molecule HCl. In each of these examples, the shared pair of electrons can be represented with a dash in place of the two dots:

$$:\ddot{Cl}-\ddot{Cl}: \qquad H-H \qquad H-\ddot{Cl}:$$

> **Student Note:** Unlike ionic substances, which can only be *compounds,* a *molecular* substance can be an element or a compound.

Thus, a molecule can be of an *element,* as in the case of Cl_2 and H_2; or it can be of a compound, as in the case of HCl. Recall that a *compound,* by definition, is a substance made up of two or more elements. This definition actually stems from Dalton's atomic theory, which we first encountered in Chapter 1. Dalton's theory consists of three hypotheses, the first of which is that *matter is composed of atoms* [◀◀ Section 1.2]. The second hypothesis is that *compounds are made of atoms of more than one element; and in any given compound, the same types of atoms are always present in the same relative numbers.* HCl is a compound that consists of molecules, each of which always contains one H atom and one Cl atom. We revisit Dalton's theory and learn about his third hypothesis when we discuss chemical reactions [▶▶ Chapter 10].

Again using HCl as an example, Dalton's second hypothesis suggests that in order for this compound to form, we need not only atoms of the right kind (H and Cl), but also specific *numbers* of each kind of atom. This idea is an extension of a law published at the end of the eighteenth century by French chemist Joseph Proust (1754–1826). According to Proust's **law of definite proportions,** also known as the **law of constant composition,** different samples of any given compound always contain the *same* elements in the *exact* same mass ratio. Any sample of HCl from anywhere in the world, regardless of the source, is 2.765 percent hydrogen and 97.235 percent chlorine.

To use another example, all carbon dioxide gas, regardless of where we get it, has the same mass ratio of oxygen to carbon. Consider the following results of the analysis of three samples of carbon dioxide, each from a different source:

Sample	Mass of O (grams)	Mass of C (grams)	Mass ratio (grams O : grams C)
123 g carbon dioxide	89.4	33.6	2.66:1
50.5 g carbon dioxide	36.7	13.8	2.66:1
88.6 g carbon dioxide	64.4	24.2	2.66:1

In any sample of pure carbon dioxide, there are 2.66 grams of oxygen for every gram of carbon present. This constant mass ratio can only be explained by assuming that the elements exist in tiny particles of specific mass (atoms), and that compounds are formed by the combination of specific *numbers* of each type of particle.

Dalton's second hypothesis also supports the **law of multiple proportions,** which he himself published early in the nineteenth century. According to this law, if two elements (in this case, C and O) can combine with each other to form two or more *different* compounds, the ratio of masses of one element that combine with a fixed mass of the other element can be expressed in small whole numbers. To unravel and make sense of this statement, we'll use carbon dioxide again. Carbon can combine with oxygen to form two different compounds: carbon *dioxide* (CO_2) and carbon *monoxide* (CO). We have just seen that in any sample of pure CO_2, there are 2.66 grams of oxygen for every gram of carbon. In any sample of pure CO, there are 1.33 grams of oxygen for every gram of carbon.

Student Note: We learn how to name such compounds in Section 3.5, but these compounds' names should be familiar to you.

Sample	Mass of O (grams)	Mass of C (grams)	Mass ratio (grams O : grams C)
16.3 g carbon monoxide	9.31	6.99	1.33:1
25.9 g carbon monoxide	14.8	11.1	1.33:1
88.4 g carbon monoxide	50.5	37.9	1.33:1

Thus, the mass ratio of oxygen to carbon in carbon *di*oxide is 2.66:1; and the mass ratio of oxygen to carbon in carbon *mon*oxide is 1.33:1. According to the law of multiple proportions, the *ratio* of two such *mass ratios* can be expressed as small whole numbers.

$$\frac{\text{Mass ratio of O to C in carbon dioxide}}{\text{Mass ratio of O to C in carbon monoxide}} = \frac{2.66}{1.33} = 2:1$$

For samples containing equal masses of carbon, the mass ratio of oxygen in carbon *di*oxide to oxygen in carbon *mon*oxide is 2:1. Our conclusion, which has been proven with modern measurement techniques, is that carbon *di*oxide molecules each consist of *one* C atom and *two* O atoms; and carbon *mon*oxide molecules each consist of one C atom and *one* O atom. This result is illustrated in Figure 3.9.

Figure 3.9 Molecular illustration of the law of multiple proportions.

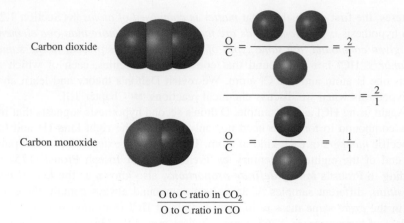

Carbon dioxide

$$\frac{O}{C} = \frac{}{} = \frac{2}{1}$$

$$\frac{}{} = \frac{2}{1}$$

Carbon monoxide

$$\frac{O}{C} = \frac{}{} = \frac{1}{1}$$

$$\frac{\text{O to C ratio in CO}_2}{\text{O to C ratio in CO}}$$

Student Note: Figure 3.9 contains molecular art representations of carbon dioxide and carbon monoxide.

Student Note: Molecules that contain two atoms of the *same* element, such as Cl_2, H_2, and F_2, are called *homonuclear* diatomic molecules. Those that contain two *different* elements, such as HCl and CO, are called *heteronuclear* diatomic molecules.

Molecules are far too small for us to observe them directly, but it is important that you learn to visualize them. Throughout this book, we will represent matter at the molecular level using two-dimensional models called molecular art, in which atoms are depicted as spheres, and different colors are used to represent different elements. Table 3.2 lists some of the elements that you will encounter most often and the colors we will use to represent them in this book.

Two different styles of molecular art are used: *ball-and-stick* and *space-filling*. Ball-and-stick models show the bonds between atoms as sticks [Figure 3.10(a)], and space-filling models show atoms that are connected by a bond as overlapping each other [Figure 3.10(b)]. Ball-and-stick and space-filling models are used to illustrate the specific, three-dimensional arrangements of atoms that are connected by chemical bonds. The ball-and-stick models generally do a good job of illustrating the *arrangement* of atoms, but they exaggerate the distances between atoms, relative to their sizes. By contrast, the space-filling models give a more accurate picture of the distances *between* atoms, but they can sometimes make it harder to see all of the details of the three-dimensional arrangement.

Many of the molecules we have seen so far (Cl_2, H_2, HCl, and CO) belong to a specific category. These are called ***diatomic molecules*** because each contains just *two atoms*. Several elements normally exist as diatomic molecules: nitrogen (N_2), oxygen (O_2), and the Group 17 elements—fluorine (F_2), chlorine (Cl_2), bromine (Br_2), and iodine (I_2). However, most molecules contain *more* than two atoms. They can all be atoms of the same element, as in ozone (O_3), or they can be combinations of two or more different elements, as in water (H_2O). Molecules containing *more* than two atoms are called ***polyatomic molecules.***

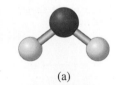

(a)

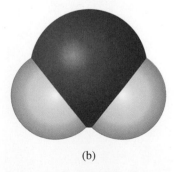

(b)

Figure 3.10 Molecular art representing water with (a) a ball-and-stick model, and (b) a space-filling model.

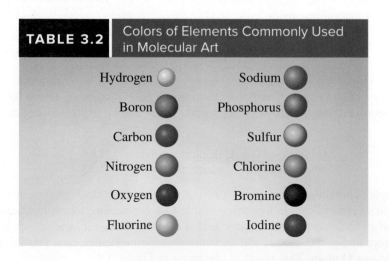

TABLE 3.2	Colors of Elements Commonly Used in Molecular Art
Hydrogen	Sodium
Boron	Phosphorus
Carbon	Sulfur
Nitrogen	Chlorine
Oxygen	Bromine
Fluorine	Iodine

Molecular Formulas

A *chemical formula* [◂◂ Section 3.2] can be used to denote the composition of any substance—ionic or molecular. A ***molecular formula*** shows the exact number of atoms of each element in a *molecule*. In our discussion of molecules, each example was given with its molecular formula in parentheses. Thus, H_2 is the molecular formula for hydrogen, O_2 is that for oxygen, O_3 is that for ozone, and H_2O is that for water. A subscript number indicates the number of atoms of an element present in the molecule. There is no subscript for O in H_2O because there is only one oxygen atom in a molecule of water, and the number 1 is never used as a subscript in a chemical formula. Oxygen (O_2) and ozone (O_3) are two of the element oxygen.

Sample Problem 3.7 shows how to write a molecular formula from the corresponding molecular art.

> **Student Note:** Two or more distinct forms of an element are known as allotropes. Two of the *allotropes* (there are actually many more) of the element carbon—*diamond* and *graphite*—have dramatically different properties. (They also have dramatically different *prices*.)

SAMPLE PROBLEM (**3.7**)   Determining Molecular Formula Given a Molecular Model

Determine the molecular formula for each of the following models.

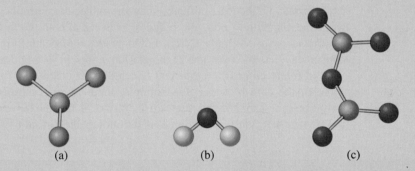

(a)   (b)   (c)

Strategy See Table 3.2 to determine the element that corresponds to each colored sphere in the models.

Setup Each type of colored sphere present is counted to determine the formula from the model.

Solution (a) There is one boron atom and three chlorine atoms represented by the model. The chemical formula is BCl_3.

(b) There is one oxygen atom and two fluorine atoms represented by the model. The chemical formula is OF_2.

(c) There are two nitrogen atoms and five oxygen atoms represented by the model. The chemical formula is N_2O_5.

THINK ABOUT IT

You may be wondering about the bonding in some of these molecules as they are a bit more complex than the examples shown. We go into more depth on the arrangement of atoms and the bonding between them in Chapter 6.

Practice Problem Ⓐ**TTEMPT** Determine the molecular formula for each of the following models.

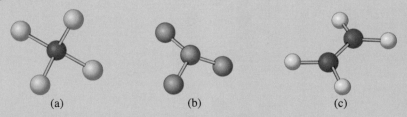

(a)   (b)   (c)

Practice Problem Ⓑ**UILD** Draw a model of each of the following compounds: (a) PF_3, (b) CS_2, (c) C_2H_6.

(Continued on next page)

Practice Problem CONCEPTUALIZE Determine if each of the molecules shown is an element or a compound.

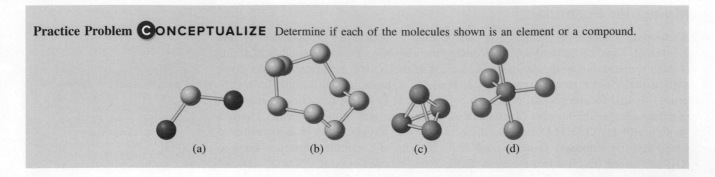

(a)   (b)   (c)   (d)

One more way that we can represent a molecular substance is by using an *empirical formula.* The word *empirical* means "from experience" or, in the context of chemical formulas, "from experiment." The empirical formula tells what elements are present in a molecule and in what whole-number ratio they are combined. For example, the molecular formula of hydrogen peroxide is H_2O_2, but because the ratio of hydrogen atoms to oxygen atoms is 1:1, its empirical formula is simply HO. Hydrazine, which has been used as a rocket fuel, has the molecular formula N_2H_4, so its empirical formula is NH_2. Although empirical formulas are useful, and we use them extensively in Chapter 5, they provide less information about a substance than the molecular formula. Using the example of hydrazine, the ratio of nitrogen to hydrogen is 1:2 in both the molecular formula (N_2H_4) and the empirical formula (NH_2), but only the molecular formula tells us the actual *number* of N atoms (two) and H atoms (four) present in a hydrazine molecule.

In many cases, the empirical and molecular formulas for a compound are the same. In the case of water, for example, there is no combination of smaller whole numbers that can convey the ratio of two H atoms for every one O atom, so the empirical formula is the same as the molecular formula: H_2O. Table 3.3 lists the

TABLE 3.3	Molecular and Empirical Formulas			
Compound	Molecular Formula	Model	Empirical Formula	Molecular Model
Water	H_2O		H_2O	
Hydrogen peroxide	H_2O_2		HO	
Ethane	C_2H_6		CH_3	
Propane	C_3H_8		C_3H_8	
Acetylene	C_2H_2		CH	
Benzene	C_6H_6		CH	

molecular and empirical formulas for several compounds and shows molecular models for each.

Empirical formulas are the *simplest* chemical formulas; they are written by reducing the subscripts in molecular formulas to the smallest possible whole numbers (without altering the relative numbers of atoms). Molecular formulas are the *true* formulas of molecules.

Sample Problem 3.8 lets you practice determining empirical formulas from molecular formulas.

SAMPLE PROBLEM (**3.8**)   Determining Empirical Formula from Molecular Formula

Write the empirical formulas for the following molecules: (a) glucose ($C_6H_{12}O_6$), a substance known as blood sugar; (b) adenine ($C_5H_5N_5$), also known as vitamin B_4; and (c) nitrous oxide (N_2O), a gas that is used as an anesthetic ("laughing gas") and as an aerosol propellant for whipped cream.

Strategy To write the empirical formula, the subscripts in the molecular formula must be reduced to the smallest possible whole numbers (without altering the relative numbers of atoms).

Setup The molecular formulas in parts (a) and (b) each contain subscripts that are divisible by common numbers. Therefore, we will be able to express the formulas with smaller whole numbers than those in the molecular formulas. In part (c), the molecule has only one O atom, so it is impossible to simplify this formula further.

Solution (a) Dividing each of the subscripts in the molecular formula for glucose by 6, we obtain the empirical formula, CH_2O. If we had divided the subscripts by 2 or 3, we would have obtained the formulas $C_3H_6O_3$ and $C_2H_4O_2$, respectively. Although the ratio of carbon to hydrogen to oxygen atoms in each of these formulas is correct (1:2:1), neither is the simplest formula because the subscripts are not in the smallest possible whole-number ratio. (b) Dividing each subscript in the molecular formula of adenine by 5, we get the empirical formula, CHN. (c) Because the subscripts in the formula for nitrous oxide are already the smallest possible whole numbers, its empirical formula is the same as its molecular formula, N_2O.

THINK ABOUT IT

Make sure that the *ratio* in each empirical formula is the same as that in the corresponding molecular formula and that the subscripts are the smallest possible whole numbers. In part (a), for example, the ratio of C:H:O in the molecular formula is 6:12:6, which is equal to 1:2:1, the ratio expressed in the empirical formula.

Practice Problem Ⓐ**TTEMPT**   Write empirical formulas for the following molecules: (a) caffeine ($C_8H_{10}N_4O_2$), a stimulant found in tea and coffee; (b) butane (C_4H_{10}), which is used in cigarette lighters; and (c) glycine ($C_2H_5NO_2$), an amino acid.

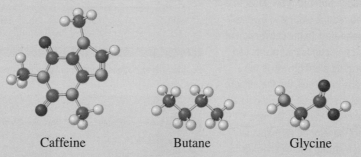

| Caffeine | Butane | Glycine |

Practice Problem Ⓑ**UILD**   For which of the following molecular formulas is the formula shown in parentheses the correct empirical formula? (a) $C_{12}H_{22}O_{11}$ ($C_{12}H_{22}O_{11}$), (b) $C_8H_{12}O_4$ ($C_4H_6O_2$), (c) H_2O_2 (H_2O)

Practice Problem Ⓒ**ONCEPTUALIZE**   Which of the following molecules has/have the same empirical formula as acetic acid ($HC_2O_2H_3$)?

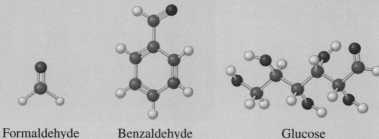

Formaldehyde   Benzaldehyde   Glucose

CHECKPOINT–SECTION 3.4   Covalent Bonding and Molecules

3.4.1   What is the empirical formula of the compound shown?

a) C_6H_6

b) C_4H_4

c) C_2H_2

d) CH_2

e) CH

3.4.2   What is the empirical formula of the compound shown?

a) HO

b) HO_2

c) H_2O_2

d) H_2O

e) H_4O_2

Familiar Chemistry

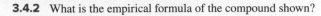

Fixed Nitrogen in Fertilizers

In the fall of 2011, the United Nations estimated that the world's population had reached 7 billion people. Although it is impossible to know the *exact* number of humans, it is a certainty that Earth could not sustain its current population without the intervention of chemistry. You may know that most of the air we breathe is nitrogen gas (N_2). (The atmosphere is about 80 percent N_2.) You may also know that plant fertilizers *contain* nitrogen atoms. Yet in spite of such an abundance of atmospheric nitrogen, plants cannot use the nitrogen in the air directly. The chemical bond that holds nitrogen molecules together is simply too strong to be broken easily, which would be necessary for plants to make use of the nitrogen atoms they need in order to grow. Most of the plants that we cultivate for food require what is known as "fixed nitrogen." Fixation of nitrogen refers to the conversion of nitrogen gas in the atmosphere to compounds that plants can incorporate. Nature's way of supporting plant growth is the fixation of nitrogen by certain bacteria in the soil. But the amount of usable nitrogen produced by this mechanism is far too small to meet our needs. In the early twentieth century, German chemists Fritz Haber and Carl Bosch developed an industrial process that enables us to break the strong chemical bonds in atmospheric N_2 on a large enough scale to produce a practically unlimited supply of usable nitrogen for use as plant fertilizers. Since the development of this process, commonly known as the *Haber process*, Earth's population has increased far more rapidly than at any other time in history—increasing from about 2 billion in 1927 to over 7 billion today.

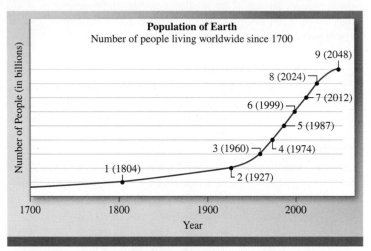

Source: United Nations World Population Prospects

Jeff Vanuga/USDA Natural Resources Conservation Service

Profiles in Science

George Washington Carver

Perhaps best known for his work developing hundreds of uses for peanuts, George Washington Carver (1864–1943) was actually an early luminary of *sustainability*—a term not coined until long after his death. Although Carver was born into slavery in Missouri (one year prior to slavery being outlawed in the United States), he would ultimately become a world-renowned chemist whose work revolutionized agriculture. Early in life, he developed a passion for horticulture, often advising local farmers as to how they might improve the health of the plants they grew. In his first attempt at higher education, he was initially accepted to study at Highland College in Kansas, but administrators rescinded his acceptance when they learned that he was Black. Encouraged by those who knew him, he eventually earned a bachelor's degree in science (the first African American ever to do so) at Iowa State Agricultural School—now Iowa State University, where a prominent academic building bears his name. Carver's professors recognized his brilliance and persuaded him to stay on and earn a master's degree in agricultural science. He spent his career advocating for farmers and

encouraging them to *rotate* crops. Farms where cotton was planted year after year saw a steady depletion of nitrogen in the soil, making it difficult or impossible to grow enough cotton to make a profit. Carver taught farmers to plant crops that *return* nitrogen to the soil, including sweet potatoes, peanuts, and other legumes, making farming practices *sustainable*. Soon after Carver's death, his childhood home was made a national monument by then President Franklin D. Roosevelt. Carver was not only the first African

George Washington Carver

Johnston (Frances Benjamin) Collection/Library of Congress/ LC-J601-302

American to whom a national monument was dedicated, he was the first individual not elected president to be honored in this way. In 1977, Carter was posthumously inducted into the National Inventors Hall of Fame.

3.5 Naming Binary Molecular Compounds

All of the molecular compounds we have encountered so far have been binary compounds. The nomenclature of binary *molecular* compounds is very similar to the nomenclature of binary *ionic* compounds. Most binary molecular compounds are composed of two non-metals (see Figure 2.22). In order to name such a compound, we first name the element that appears first in the formula. For HCl that would be hydrogen. We then name the second element, changing the ending of its name to *–ide,* just as we did when naming anions of nonmetals in Section 3.3. For HCl, the second element is chlorine, so we would change chlor*ine* to chlor*ide*. Thus, the systematic name of HCl is *hydrogen chloride*. Similarly, HI is hydrogen iodide (iod*ine* → iod*ide*) and SiC is silicon carbide (carb*on* → carb*ide*).

It is very common for one pair of nonmetals to form two or more different binary molecular compounds, as we have seen with carbon and oxygen. In these cases, confusion in naming the compounds is avoided by the use of Greek prefixes to denote the number of atoms of each element present. Some of the Greek prefixes are listed in Table 3.4, and several compounds named using these prefixes are listed in Table 3.5.

> **Student Note:** Recall that a *binary* compound is one that consists of just two different elements. [◀◀ Section 3.2]

TABLE 3.4	Greek Prefixes			
Prefix	**Meaning**		**Prefix**	**Meaning**
Mono–	1		Hexa–	6
Di–	2		Hepta–	7
Tri–	3		Octa–	8
Tetra–	4		Nona–	9
Penta–	5		Deca–	10

TABLE 3.5	Some Compounds Named Using Greek Prefixes		
Compound	**Name**	**Compound**	**Name**
CO	Carbon monoxide	SO_3	Sulfur trioxide
CO_2	Carbon dioxide	NO_2	Nitrogen dioxide
SO_2	Sulfur dioxide	N_2O_5	Dinitrogen pentoxide

The prefix *mono–* is generally omitted for the first element. For example, as you know, CO_2 is named *carbon dioxide,* not *monocarbon dioxide.* Thus, the absence of a prefix for the first element usually means there is only one atom of that element present in the molecule. In addition, for ease of pronunciation, we usually eliminate the last letter of a prefix that ends in "o" or "a" when naming an oxide. Thus, N_2O_5 is *dinitrogen pentoxide,* rather than *dinitrogen pentaoxide.*

Sample Problems 3.9 and 3.10 give you some practice naming binary molecular compounds from their formulas, and determining formulas based on compound names.

SAMPLE PROBLEM **3.9**   Naming Binary Molecular Compounds

Name the following binary molecular compounds: (a) NF_3 and (b) N_2O_4.

Strategy Each compound will be named using the systematic nomenclature including, where necessary, appropriate Greek prefixes.

Setup With binary compounds, we start with the name of the element that appears *first* in the formula, and we change the ending of the *second* element's name to *–ide.* We use prefixes, where appropriate, to indicate the number of atoms of each element. In part (a), the molecule contains one nitrogen atom and three fluorine atoms. We will omit the prefix *mono–* for nitrogen because it is the first element listed in the formula, and we will use the prefix *tri–* to denote the number of fluorine atoms. In part (b), the molecule contains two nitrogen atoms and four oxygen atoms, so we will use the prefixes *di–* and *tetra–* in naming the compound. Recall that in naming an oxide, the last letter of a prefix that ends in "a" or "o" is omitted.

Solution (a) nitrogen trifluoride and (b) dinitrogen tetroxide

THINK ABOUT IT

Make sure that the prefixes match the subscripts in the molecular formulas and that the word *oxide* is not preceded immediately by an "a" or an "o."

Practice Problem **ATTEMPT** Name the following binary molecular compounds: (a) Cl_2O and (b) $SiCl_4$.

Practice Problem **BUILD** Name the following binary molecular compounds: (a) ClO_2 and (b) CBr_4.

Practice Problem **CONCEPTUALIZE** Name the binary molecular compound shown.

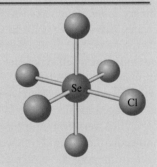

SAMPLE PROBLEM **3.10**   Determining Molecular Formula Given a Compound's Name

Write the chemical formulas for the following binary molecular compounds: (a) sulfur tetrafluoride and (b) tetraphosphorus decasulfide.

Strategy The formula for each compound will be deduced using the systematic nomenclature guidelines.

Setup In part (a) there is no prefix for sulfur, so there is only one sulfur atom in a molecule of the compound. Therefore, we will use no prefix for the S in the formula. The prefix *tetra–* means that there are four fluorine atoms. In part (b), the prefixes *tetra–* and *deca–* denote four and ten, respectively.

Solution (a) SF_4 and (b) P_4S_{10}

THINK ABOUT IT

Double-check that the subscripts in the formulas match the prefixes in the compound names: (a) 4 = tetra and (b) 4 = tetra and 10 = deca.

Practice Problem **A**TTEMPT Give the molecular formula for each of the following compounds: (a) carbon disulfide and (b) dinitrogen trioxide.

Practice Problem **B**UILD Give the molecular formula for each of the following compounds: (a) sulfur hexafluoride and (b) disulfur decafluoride.

Practice Problem **C**ONCEPTUALIZE Draw a molecular model of sulfur trioxide.

CHECKPOINT–SECTION 3.5 Naming Binary Molecular Compounds

3.5.1 What is the correct systematic name of PCl_5?

a) Phosphorus chloride

b) Phosphorus pentachloride

c) Monophosphorus chloride

d) Pentachlorophosphorus

e) Pentaphosphorus chloride

3.5.2 What is the correct formula for the compound carbon tetrachloride?

a) C_4Cl_4

b) C_4Cl

c) CCl_4

d) CCl_2

e) CCl

3.6 Covalent Bonding in Ionic Species: Polyatomic Ions

So far, the compounds we have encountered have been either *ionic,* held together by ionic bonding; or *molecular,* held together by covalent bonding. Further, all of the ionic compounds we have encountered so far have contained only *atomic* ions. However, many common ionic substances contain *poly*atomic ions, which themselves are held together by *covalent* bonding. In this section, we examine substances that are held together by a combination of ionic and covalent bonding.

An ion that consists of a combination of two or more atoms is called a ***polyatomic ion.*** Because we will encounter the same polyatomic ions over and over in this course, you must know the names, formulas, and charges of those that are listed in Table 3.6. Although most of the common polyatomic ions are anions, a few are cations—including the mercury(I) ion, which we first saw in Figure 3.7. For compounds containing polyatomic ions, formulas are determined following the same rule as for ionic compounds containing only atomic ions: ions must be combined in a ratio that gives a neutral formula overall. The following examples illustrate how this is done.

Ammonium Chloride

The cation is NH_4^+ and the anion is Cl^-. The sum of the charges is $1 + (-1) = 0$, so the ions combine in a 1:1 ratio and the resulting formula is NH_4Cl.

Calcium Phosphate

The cation is Ca^{2+} and the anion is PO_4^{3-}. The following diagram can be used to determine the subscripts:

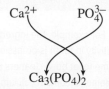

The sum of the charges is $3(+2) + 2(-3) = 0$. Thus, the formula for calcium phosphate is $Ca_3(PO_4)_2$. When we add a subscript to a polyatomic ion, we must first put parentheses

Student Note: Some oxoanions occur in series of ions that contain the same central atom and have the same charge, but contain different numbers of oxygen atoms.

perchlorate	ClO_4^-
chlorate	ClO_3^-
chlorite	ClO_2^-
hypochlorite	ClO^-
nitrate	NO_3^-
nitrite	NO_2^-
phosphate	PO_4^{3-}
phosphite	PO_3^{3-}
sulfate	SO_4^{2-}
sulfite	SO_3^{2-}

TABLE 3.6	Common Polyatomic Ions
Name	**Formula/Charge**
Cations	
ammonium	NH_4^+
hydronium	H_3O^+
mercury(I)	Hg_2^{2+}
Anions	
acetate	$C_2H_3O_2^-$
azide	N_3^-
carbonate	CO_3^{2-}
chlorate	ClO_3^-
chlorite	ClO_2^-
chromate	CrO_4^{2-}
cyanide	CN^-
dichromate	$Cr_2O_7^{2-}$
dihydrogen phosphate	$H_2PO_4^-$
hydrogen carbonate or bicarbonate	HCO_3^-
hydrogen phosphate	HPO_4^{2-}
hydrogen sulfate or bisulfate	HSO_4^-
hydroxide	OH^-
hypochlorite	ClO^-
nitrate	NO_3^-
nitrite	NO_2^-
oxalate	$C_2O_4^{2-}$
perchlorate	ClO_4^-
permanganate	MnO_4^-
peroxide	O_2^{2-}
phosphate	PO_4^{3-}
phosphite	PO_3^{3-}
sulfate	SO_4^{2-}
sulfite	SO_3^{2-}
thiocyanate	SCN^-

around the ion's formula to indicate that the subscript applies to *all* the atoms in the polyatomic ion. Other examples are sodium cyanide (NaCN), potassium permanganate ($KMnO_4$), and ammonium sulfate [$(NH_4)_2SO_4$].

Familiar Chemistry

Product Labels

Have you ever looked at the label of a product at the drugstore only to find that you don't recognize most of the ingredients? Many of the ingredients are compounds named using the systematic nomenclature described in this chapter. The active ingredients on the labels shown here are compounds that you now have the ability to identify.

For example, the ingredient primarily responsible for diminishing sensitivity in teeth is potassium nitrate or KNO_3. The active ingredient in some toilet bowl cleaners is hydrochloric acid, HCl. Many calcium supplements contain calcium carbonate, $CaCO_3$, as the major active ingredient.

Drug Facts

Active ingredients	Purpose
Potassium nitrate 5%	Antisensitivity
Sodium fluoride 0.24% (0.14% w/v fluoride ion)......................	Anticavity

Uses • builds increasing protection against painful sensitivity of the teeth to cold, heat, acids, sweets or contact
• helps protect against cavities

Warnings
When using this product, if pain/sensitivity still persists after 4 weeks of use, please visit your dentist.

Stop use and ask a dentist if the problem persists or worsens. Sensitive teeth may indicate a serious problem that may need prompt care by a dentist.

Keep out of reach of children. If more than used for brushing is accidentally swallowed, get medical help or contact a Poison Control Center right away.

Toothpaste for sensitive teeth

Mark Dierker/McGraw Hill

Toilet bowl cleaner

David A. Tietz/McGraw Hill

Supplement Facts

Serving Size: 2 caplets
Servings Per Container: 40

	Amount Per Serving	% Daily Value
Vitamin D	1000 IU	250%
Calcium (elemental)	1200 mg	120%
Magnesium	80 mg	20%
Sodium	5 mg	< 1%

INGREDIENTS: Calcium Carbonate, Calcium Citrate, Magnesium Hydroxide, Acacia, Hydroxypropyl Methylcellulose, Croscarmellose Sodium, Magnesium Silicate, Titanium Dioxide (color), Propylene Glycol Dicaprylate/Dicaprate, Magnesium Stearate, Inulin (Oligofructose Enriched), Vitamin D$_3$ (Cholecalciferol).

If pregnant, breast-feeding, taking medication, or have any underlying medical condition ask a health professional before use.

Calcium supplement

Mark Dierker/McGraw Hill

Thinking Outside the Box

Product Labels

Although there are many systematic compound names found on product labels, you will see names for some ingredients that do not follow the systematic nomenclature presented in this chapter—see the purple highlighted compounds in the baby formula label. These compounds have been named using an older system that is not used much any longer and is not used in this text. However, it may be useful for you to be able to translate these older names into systematic ones if your curiosity gets the better of you.

This older naming system was designed to differentiate between the multiple ions that are formed by many of the transition metals, but in a less convenient fashion than our systematic nomenclature. The following table shows examples of the systematic names that you have learned, the older names they correspond to, and the ions that they represent.

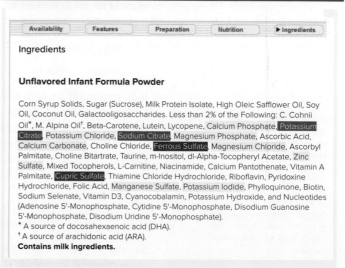

Label from powdered baby formula.

Systematic and older* names of metal ions.

Metal	Ion Symbol	Systematic Name	Older* Name
chromium	Cr^{2+}	chromium(II)	chromous
	Cr^{3+}	chromium(III)	chromic
copper	Cu^+	copper(I)	cuprous
	Cu^{2+}	copper(II)	cupric
iron	Fe^{2+}	iron(II)	ferrous
	Fe^{3+}	iron(III)	ferric
tin	Sn^{2+}	tin(II)	stannous
	Sn^{4+}	tin(IV)	stannic
lead	Pb^{2+}	lead(II)	plumbous
	Pb^{4+}	lead(IV)	plumbic

Using the table to translate, ferrous sulfate from the baby formula label would be called iron(II) sulfate using systematic nomenclature. The other compound on the label that uses this older system is cupric sulfate. Cupric sulfate would be named copper(II) sulfate using the systematic nomenclature.

As with the nomenclature of binary ionic compounds [◀◀ Section 3.3], the names of ionic compounds containing polyatomic ions give the name of the cation first and that of the anion second; and they do *not* require the use of Greek prefixes. For example, Li_2CO_3 is *lithium carbonate,* not *di*lithium carbonate, even though there are *two* lithium ions for every *one* carbonate ion. Prefixes are unnecessary because each of the ions has a specific, *known* charge. Lithium ion always has a charge of +1, and carbonate ion always has a charge of −2. The only ratio in which they *can* combine to form a neutral compound is two Li^+ ions for every one CO_3^{2-} ion. Therefore, the name *lithium carbonate* is sufficient to convey the compound's empirical formula.

Oxoanions are polyatomic anions that contain one or more oxygen atoms and one atom (the "central atom") of another element. Examples include the chlorate (ClO_3^-), nitrate (NO_3^-), and sulfate (SO_4^{2-}) ions. Often, oxoanions occur in series of two ions that have the same central atom but different numbers of O atoms (e.g., NO_3^- and NO_2^-), and we have a systematic way to name the ions within such a series. When a series of oxoanions consists of two ions, as in the case of NO_3^- and NO_2^-, the ion with *more* O atoms is named with the *–ate* ending, and the one with *fewer* O atoms is named with the *–ite* ending.

NO_3^-	NO_2^-	SO_4^{2-}	SO_3^{2-}	PO_4^{3-}	PO_3^{3-}
nitrate	nitrite	sulfate	sulfite	phosphate	phosphite

It is important to note that the endings *–ate* and *–ite* do not indicate the *number* of O atoms in a polyatomic ion's formula. These endings simply indicate the *relative* numbers of O atoms: greater number (*–ate*), or smaller number (*–ite*).

One series of oxoanions in Table 3.6 consists of *four* ions with the same central atom (Cl), each with a different number of O atoms. To name this series of ions, we use prefixes *per–* and *hypo–* in addition to the *–ate* and *–ite* endings:

ClO_4^-	ClO_3^-	ClO_2^-	ClO^-
perchlorate	chlorate	chlorite	hypochlorite

Sample Problems 3.11 and 3.12 let you practice associating names and formulas of compounds that contain polyatomic ions.

Student Note: In these names, you can still think of the *–ate* and *–ite* endings as loosely meaning "more oxygen" and "less oxygen," respectively. The combination of the prefix *per–* and the ending *–ate* can be interpreted as "even *more* oxygen"; and the combination of *hypo–* and *–ite* can be interpreted as "even *less* oxygen."

SAMPLE PROBLEM (3.11) Naming Ionic Compounds with Polyatomic Ions

Name the following ionic compounds: (a) NH_4F, (b) $Al(OH)_3$, and (c) $Fe_2(SO_4)_3$.

Strategy Begin by identifying the cation and the anion in each compound, and then combine the names for each, eliminating the word *ion*.

Setup NH_4F contains NH_4^+ and F^-, the ammonium ion and the fluoride ion; $Al(OH)_3$ contains Al^{3+} and OH^-, the aluminum ion and the hydroxide ion; and $Fe_2(SO_4)_3$ contains Fe^{3+} and SO_4^{2-}, the iron(III) ion and the sulfate ion. We know that the iron in $Fe_2(SO_4)_3$ is iron(III), Fe^{3+}, because it is combined with the sulfate ion in a 2:3 ratio.

Solution (a) Combining the cation and anion names, and eliminating the word *ion* from each of the individual ions' names, we get *ammonium fluoride* as the name of NH_4F; (b) $Al(OH)_3$ is *aluminum hydroxide;* and (c) $Fe_2(SO_4)_3$ is *iron(III) sulfate.*

THINK ABOUT IT

Be careful not to confuse the subscript in a formula with the charge on the metal ion. In part (c), for example, the subscript on Fe is 2, but this is an iron(III) compound.

Practice Problem **A**TTEMPT Name the following ionic compounds: (a) Na_2SO_4, (b) $Cu(NO_3)_2$, (c) $Fe_2(CO_3)_3$.

Practice Problem **B**UILD Name the following ionic compounds: (a) $K_2Cr_2O_7$, (b) $Li_2C_2O_4$, (c) $CuNO_3$.

Practice Problem **C**ONCEPTUALIZE The diagram represents a small sample of an ionic compound, where red spheres represent nitrate ions and grey spheres represent iron ions. Deduce the correct formula and name of the compound.

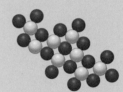

SAMPLE PROBLEM (3.12) Determining Formulas from Compound Names

Deduce the formulas of the following ionic compounds: (a) mercury(I) chloride, (b) lead(II) chromate, and (c) potassium hydrogen phosphate.

Strategy Identify the ions in each compound, and determine their ratios of combination using the charges on the cation and anion in each.

Setup (a) Mercury(I) chloride is a combination of Hg_2^{2+} and Cl^-. [Mercury(I) is one of the few cations listed in Table 3.6.] In order to produce a neutral compound, these two ions must combine in a 1:2 ratio. (b) Lead(II) chromate is a combination of Pb^{2+} and CrO_4^{2-}. These ions combine in a 1:1 ratio. (c) Potassium hydrogen phosphate is a combination of K^+ and HPO_4^{2-}. These ions combine in a 2:1 ratio.

Solution The formulas are (a) Hg_2Cl_2, (b) $PbCrO_4$, and (c) K_2HPO_4.

THINK ABOUT IT

Make sure that the charges sum to zero in each compound formula. In part (a), for example, $Hg_2^{2+} + 2Cl^- = (2+) + 2(-1) = 0$; in part (b), $(+2) + (-2) = 0$; and in part (c), $2(+1) + (-2) = 0$.

Practice Problem **A**TTEMPT Deduce the formulas of the following ionic compounds: (a) lead(II) chlorate, (b) magnesium carbonate, and (c) ammonium phosphate.

Practice Problem **B**UILD Deduce the formulas of the following ionic compounds: (a) iron(III) phosphite, (b) mercury(II) nitrate, and (c) potassium sulfite.

Practice Problem **C**ONCEPTUALIZE The diagram represents a small sample of an ionic compound, where yellow spheres represent sulfite ions and blue spheres represent copper ions. Deduce the correct formula and name of the compound.

Familiar Chemistry 🏠

Hydrates

A hydrate is a compound that has a specific number of water molecules in its chemical formula. For example, under ordinary conditions, copper(II) sulfate ($CuSO_4$) has five water molecules associated with it. The systematic name for this compound is copper(II) sulfate pentahydrate, and its formula is written as $CuSO_4 \cdot 5H_2O$. The water molecules can be removed from the hydrate by heating. When this occurs, the resulting compound is simply $CuSO_4$, and it is sometimes referred to as *anhydrous* copper(II) sulfate. (*Anhydrous* means "without water.") Hydrates and their anhydrous counterparts have distinctly different physical and chemical properties. Shown here are samples of copper(II) sulfate pentahydrate (blue) and anhydrous copper(II) sulfate (white). Note that the anhydrous compound must be kept in a sealed container or it will absorb water from the atmosphere and revert to the hydrate form.

(both) Charles D. Winters/McGraw Hill

One of the common hydrates that may be familiar to you is $MgSO_4 \cdot 7H_2O$ (magnesium sulfate heptahydrate), which is the active ingredient in milk of magnesia—a popular antacid and laxative.

Jill Braaten/McGraw Hill

CHECKPOINT–SECTION 3.6   Covalent Bonding in Ionic Species: Polyatomic Ions

3.6.1 What is the correct name of the compound $PbSO_4$?

    a) Lead sulfoxide   d) Monolead sulfate

    b) Lead(I) sulfate   e) Lead monosulfate

    c) Lead(II) sulfate

3.6.2 What is the correct formula for the compound iron(III) carbonate?

    a) $FeCO_3$ d) $Fe_2(CO_3)_3$

    b) Fe_3CO_3 e) $Fe_3(CO_3)_2$

    c) Fe_2CO_3

3.6.3 What is the correct formula for sodium nitride?

a) NaN

b) NaN$_3$

c) Na$_3$N

d) NaNO$_3$

e) NaNO$_2$

3.6.4 What is the correct name of the compound Hg$_2$CrO$_4$?

a) Mercury(I) chromate

b) Mercury(II) chromate

c) Mercury dichromate

d) Dimercury chromate

e) Monomercury chromate

3.7 Acids

Acids constitute an important class of molecular compounds. One way we can define the term **acid** is as any substance that produces hydrogen ions (H$^+$) when dissolved in water. Although acids are *not* ionic compounds, they are *molecular* compounds, we can envision an acid as one or more hydrogen ions attached to an anion. The anion may be a *simple* anion or it may be an oxoanion. Some examples of compounds that are acids are:

Student Note: Here, the term *simple anion* refers to an anion that is *not* an oxoanion. This includes any atomic anion, CN$^-$ and SCN$^-$.

HCl   HF   HCN   HClO$_4$ HNO$_3$ H$_2$SO$_4$ H$_3$PO$_4$

Note that in these acid formulas, the H atoms are written first and the anions' formulas are written last. The first two examples, HCl and HF, are binary molecular compounds, which you learned how to name in Section 3.5. Their systematic names are *hydrogen chloride* and *hydrogen fluoride*. However, when compounds such as these are dissolved in water, we use a different system of nomenclature that identifies them specifically as acids. The rules for naming an acid depend on whether the anion it contains is a *simple* anion or an *oxoanion*.

The rules for naming simple acids, in which the anion is not an oxoanion, are as follows: remove the −*gen* ending from hydrogen (leaving *hydro*−), change the −*ide* ending on the anion to −*ic,* combine the two words, and add the word *acid.* To name the acid HCl:

hydrogen (−*gen*) becomes hydro

+

chloride (−*ide,* +*ic*) becomes chloric

+

acid ⟶ *hydrochloric acid*

To name HF:

hydrogen (−*gen*) becomes hydro

+

fluoride (−*ide,* +*ic*) becomes fluoric

+

acid ⟶ *hydrofluoric acid*

To name HCN:

hydrogen (−*gen*) becomes hydro

+

cyanide (−*ide,* +*ic*) becomes cyanic

+

acid ⟶ *hydrocyanic acid*

TABLE 3.7	Some Common Simple Acids	
Formula	Binary Compound Name	Acid Name
HF	Hydrogen fluoride	Hydrofluoric acid
HCl	Hydrogen chloride	Hydrochloric acid
HBr	Hydrogen bromide	Hydrobromic acid
HI	Hydrogen iodide	Hydroiodic acid
HCN	Hydrogen cyanide	Hydrocyanic acid
H_2S	Hydrogen sulfide	Hydrosulfuric acid

Note that the number of hydrogen ions in an acid's formula depends on the anion's charge. For the anions of Group 17, the corresponding acid formulas contain just one hydrogen ion. The acid containing the sulfide ion (S^{2-}) has *two* hydrogen ions. As with any compounds, the formulas of acids must be neutral overall. Table 3.7 lists several examples of simple acid formulas and names.

An acid in which the anion is an oxoanion is called an **oxoacid.** The names of oxoacids are derived from the oxoanions they contain, using the following guidelines:

1. When the oxoanion's name ends in *–ate,* the ending changes from *–ate* to *–ic* and we add the word *acid.* Thus, $HClO_3$, in which the anion is the *chlorate* ion (ClO_3^-), is called *chloric acid.*

$$\text{chlorate } (-ate, +ic) \text{ becomes chloric}$$
$$+$$
$$\text{acid} \longrightarrow chloric\ acid$$

2. When the oxoanion's name ends in *–ite,* the ending changes from *–ite* to *–ous* and we add the word *acid.* Thus, $HClO_2$, based on the *chlorite* ion (ClO_2^-), is called *chlorous acid.*

$$\text{Chlorite } (-ite, +ous) \text{ becomes chlorous}$$
$$+$$
$$\text{acid} \longrightarrow chlorous\ acid$$

3. When the oxoanion's name contains a prefix, the prefix is retained in the name of the acid. Thus, $HClO_4$ and $HClO$ are called *perchloric acid* and *hypochlorous acid,* respectively.

$$\text{perchlorate } (-ate, +ic) \text{ becomes perchloric}$$
$$+$$
$$\text{acid} \longrightarrow perchloric\ acid$$

$$\text{hypochlorite } (-ite, +ous) \text{ becomes hypochlorous}$$
$$+$$
$$\text{acid} \longrightarrow hypochlorous\ acid$$

As with simple acids, the number of hydrogen ions in an oxoacid's formula depends on the charge on the corresponding anion. For example, the formulas of oxoacids containing the nitrate (NO_3^-) and sulfate (SO_4^{2-}) ions are HNO_3 and H_2SO_4, respectively.

Practice Problem 3.13 gives you some practice naming acids.

SAMPLE PROBLEM 3.13 Naming Acids

Name each of the following acids: (a) $HClO_2$, (b) H_2CO_3, (c) H_2S.

Strategy Determine if the anion contained in the acid is an oxoanion. Recall the name of the oxoanion and apply the rules to determine the acid name. If the anion is not an oxoanion, determine the name of the ion and apply the rules to determine the acid name.

Setup The anions present are: (a) chlorite, (b) carbonate, and (c) sulfide.

Solution The acids in parts (a) and (b) both contain oxoanions, and therefore are named by simply changing the ending of the ion name to either –*ic* (if it ends in –*ate*) or –*ous* (if it ends in –*ite*) and adding the word *acid* to the end.

(a) chlorite becomes *chlorous acid*.

(b) carbonate becomes *carbonic acid*.

(c) H_2S contains the sulfide ion. The base of the acid name becomes *sulfuric acid*. Since it is a binary acid, the prefix *hydro–* is added to give *hydrosulfuric acid*.

THINK ABOUT IT

None of the oxoacids' names begin with the prefix *hydro–*. The prefix *hydro–* is only used to name a binary acid like H_2S.

Practice Problem **A** **TTEMPT** Name each of the following acids: (a) HCN, (b) HNO_2, (c) H_3PO_4.

Practice Problem **B** **UILD** Determine the formula for each of the following acids: (a) sulfurous acid, (b) chromic acid, (c) chloric acid.

Practice Problem **C** **ONCEPTUALIZE** The diagrams show models of a series of oxoanions (whose identity isn't important). Which of these models, when combined with H^+ to form an acid, would have an –*ic* ending in the acid name?

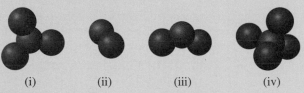

(i)   (ii)   (iii)   (iv)

CHECKPOINT–SECTION 3.7 Acids

3.7.1 Which of the following is the correct formula for nitrous acid?

a) HNO

b) HN_2O

c) N_2O

d) HNO_2

e) HNO_3

3.7.2 What is the systematic name of the acid H_2S?

a) hydrosulfuric acid

b) sulfuric acid

c) sulfurous acid

d) hydrosulfurous acid

e) hyposulfuric acid

3.8 Substances in Review

In this chapter, we have seen how the nature of atoms can lead to the formation of chemical bonds. Figure 3.11 illustrates how the properties of atoms determine what kinds of bonds (if any) they can form—and the types of substances that result. In this section, we review how to classify substances as elements or compounds and as consisting of atoms, ions, or molecules. We also see how a compound's formula can tell us whether it is ionic or molecular, and we review the procedures for naming compounds.

Figure 3.11
Properties of Atoms

Na

$1s^2 2s^2 2p^6 3s^1$

Metals, such as sodium, easily lose one or more electrons to become cations.

Cl

$1s^2 2s^2 2p^6 3s^2 3p^5$

Nonmetals, such as chlorine, easily gain one or more electrons to become anions.

Nonmetals can also achieve a noble gas electron configuration by sharing electrons to form covalent bonds.

C

$1s^2 2s^2 2p^2$

Although it is a nonmetal, carbon neither loses nor gains electrons easily. It achieves a noble gas electron configuration by sharing electrons and forming covalent bonds.

H

$1s^1$

Also a nonmetal, hydrogen typically achieves a noble gas electron configuration by sharing electrons.

He

$1s^2$

As a noble gas, helium has no tendency to gain, lose, or share electrons. It exists as individual atoms.

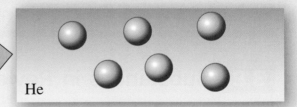

He

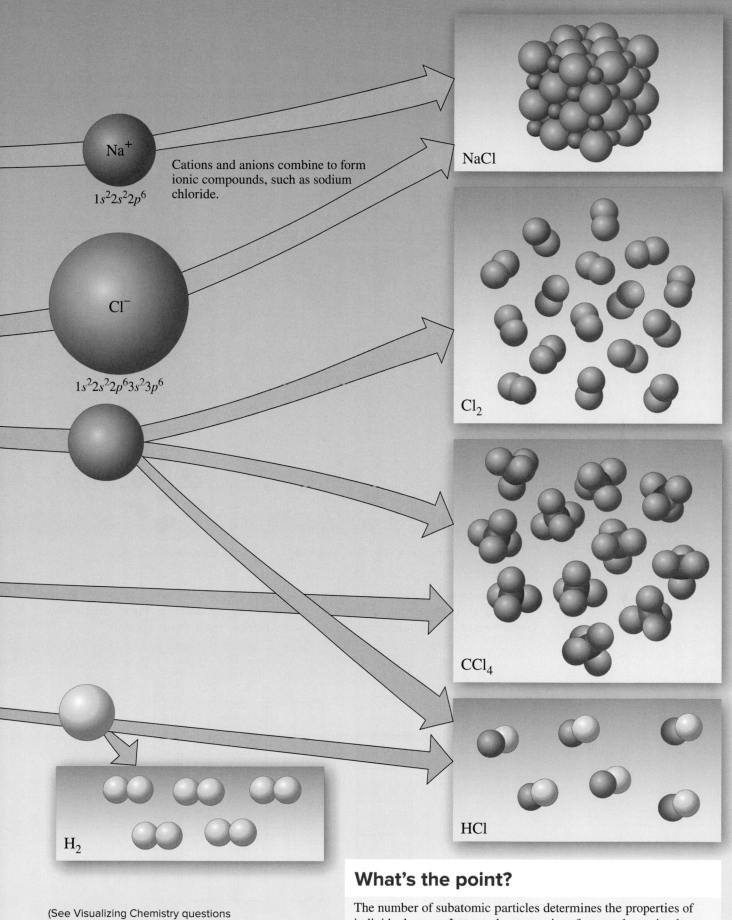

Na⁺

$1s^22s^22p^6$

Cations and anions combine to form ionic compounds, such as sodium chloride.

Cl⁻

$1s^22s^22p^63s^23p^6$

NaCl

Cl₂

CCl₄

HCl

H₂

(See Visualizing Chemistry questions VC 3.1–VC 3.4 on page 122.)

What's the point?

The number of subatomic particles determines the properties of individual atoms. In turn, the properties of atoms determine how they interact with other atoms and what compounds, if any, they form.

Distinguishing Elements and Compounds

An element is a substance that contains only one type of atom. Elements may exist as independent atoms, as in the case of helium (He), or they may exist as molecules, as in the case of oxygen (O_2).

Douglas Pulsipher/Alamy Stock Photo   B.A.E. Inc./Alamy Stock Photo

Most elements, including the metals and the noble gases, exist as isolated atoms. Nonmetals generally exist as molecules, many of which are diatomic. Figure 3.12 shows the formulas of the elemental forms of some familiar main group elements.

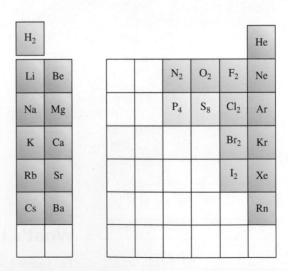

Figure 3.12 Elemental forms of some main group elements. Elements shown in purple exist as independent atoms; those shown in green exist as diatomic molecules; and those shown in yellow exist as polyatomic molecules.

A compound is a substance consisting of more than one type of atom. Compounds may be ionic, as in the case of sodium chloride (NaCl), or molecular, as in the case of water (H_2O).

Determining Whether a Compound Is Ionic or Molecular

A compound can be classified as ionic if its formula meets any of the following three criteria:

- The formula consists of just a <u>metal</u> and a nonmetal.

    Examples: NaCl <u>Li</u>$_2$S <u>Fe</u>$_2O_3$ <u>Al</u>Cl$_3$ <u>Zn</u>O

- The formula consists of a <u>metal</u> and a polyatomic anion.

    Examples: <u>K</u>NO$_3$ <u>Cr</u>$_2$(SO$_4$)$_3$ <u>Mn</u>CO$_3$ <u>Sr</u>ClO$_3$ <u>Hg</u>$_2$CrO$_4$

- The formula consists of the ammonium ion (NH_4^+) and an anion (atomic or polyatomic).

    Examples: NH$_4$Cl (NH$_4$)$_2$S (NH$_4$)$_2$CO$_3$ (NH$_4$)$_2$SO$_4$ (NH$_4$)$_3$PO$_4$

A compound can be classified as molecular if its formula consists of only nonmetals.

    Examples: HI CS$_2$ N$_2$O ClF SF$_6$

Naming Compounds

The rules of nomenclature are different for ionic compounds and molecular compounds. First, determine the compound type:

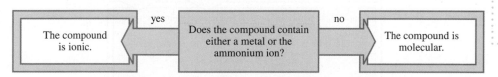

When you have determined the type of compound, you can name it using the guidelines in Sections 3.3, 3.5, and 3.7. The flowcharts in Figures 3.13 and 3.14 summarize these processes.

NaCl H_2O

Richard Hutchings/Digital Light/McGraw Hill

Student Note: You must be able to recognize the common polyatomic ions and know their charges when you see them in a chemical formula—even though their *charges* do not *appear* in the formula.

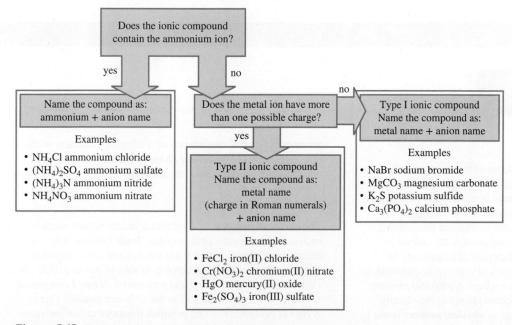

Figure 3.13 Flowchart for naming ionic compounds.

Figure 3.14 Flowchart for naming binary molecular compounds and acids.

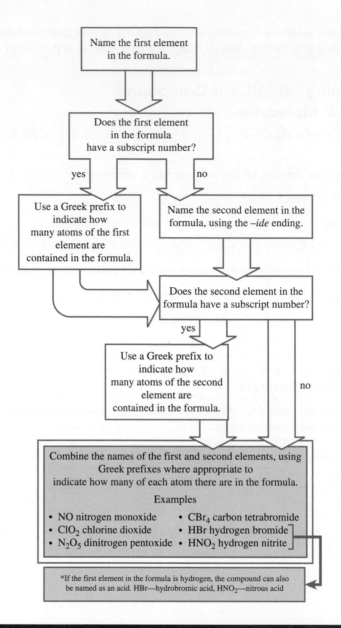

Name the first element in the formula.

Does the first element in the formula have a subscript number?

yes   no

Use a Greek prefix to indicate how many atoms of the first element are contained in the formula.

Name the second element in the formula, using the *–ide* ending.

Does the second element in the formula have a subscript number?

yes

Use a Greek prefix to indicate how many atoms of the second element are contained in the formula.

no

Combine the names of the first and second elements, using Greek prefixes where appropriate to indicate how many of each atom there are in the formula.

Examples

- NO nitrogen monoxide
- ClO_2 chlorine dioxide
- N_2O_5 dinitrogen pentoxide
- CBr_4 carbon tetrabromide
- HBr hydrogen bromide
- HNO_2 hydrogen nitrite

*If the first element in the formula is hydrogen, the compound can also be named as an acid. HBr—hydrobromic acid, HNO_2—nitrous acid

Chapter Summary

Section 3.1

- A *compound* is a substance made up of two or more different elements. A *substance* is matter with a composition that is universally constant. Substances may be elements or compounds. A *mixture* is a combination of substances. A *homogeneous mixture* has a constant composition throughout the mixture. A *heterogeneous mixture* has a composition that varies in different parts of the mixture. Properties of matter may be *quantitative,* which require numbers to express; or *qualitative,* which do not require numbers to express. A *physical property* is one that can be determined without changing the identity of a substance. A *physical change* or *physical process* is one that does not change the identity of a substance. A *chemical*

property is one that describes how a substance interacts with other substances. A *chemical change* or *chemical process* is one that changes the identity of a substance.

Section 3.2

- The *chemical formula* of a substance indicates how much of each element the substance contains. *Ionic bonding* refers to electrostatic attraction that holds cations and anions together. A *binary ionic compound* consists of ions of just two different elements—typically a metal and a nonmetal. A *type I compound* is one in which the metal cation has only one possible charge. A *type II compound* is one in which the metal cation has more than one possible charge.

Section 3.3

- *Nomenclature* refers to a system used to name ionic and molecular compounds. Cations are named using the name of the element from which they are derived. Anions are named similarly, but the element's ending is changed to *–ide*. The names of ionic compounds do not make reference to the relative numbers of cations and anions in the formula because the charges on each determine the ratio of combination. The overall formula of an ionic compound must be electrically neutral. For type II ionic compounds, the charge on the metal cation is included in the name as a Roman numeral in parentheses immediately following the name of the metal.

Section 3.4

- *Covalent bonds* form between atoms that neither lose nor gain electrons, but that achieve noble gas electron configurations by *sharing* electrons. The resulting species is a *molecule,* which is a neutral combination of two or more atoms. According to the *law of definite proportions* (also called the *law of constant composition*), any sample of a given compound will always contain the same elements in the same mass ratio. The *law of multiple proportions* states that if two elements can form more than one compound with one another, the mass ratio of one will be related to the mass ratio of the other by a small whole number. A *diatomic molecule* is one that contains just two atoms. A *polyatomic molecule* contains three or more atoms. A *molecular formula* specifies with subscript numbers the exact number of each type of atom in a molecule. An *empirical formula* indicates with subscript numbers the smallest whole number ratio of combination, and is sometimes but not always the same as the molecular formula.

Section 3.5

- Binary molecular compounds are composed of two different nonmetals. The nomenclature of binary molecular compounds is similar to that of binary ionic compounds but uses Greek prefixes to specify the numbers of atoms of each element. When the first element's prefix would be *mono,* it is left out of the name.

Section 3.6

- Many chemical species contain both ionic and covalent bonds. A *polyatomic ion* is one that contains more than one atom held together by covalent bonds. An *oxoanion* is a polyatomic ion that contains one or more oxygen atoms. Most common polyatomic ions are oxoanions. The names and formulas of the common polyatomic ions (Table 3.6) should be committed to memory and will be used often throughout this course.

Section 3.7

- An *acid* is a compound that produces H^+ ion when dissolved in water. Although acids are molecular compounds, we can picture them as compounds containing H^+ ions and anions, either atomic or polyatomic. Simple acids are those whose anions are not oxoanions. *Oxoacids* are those whose anions are oxoanions. Simple acids can be named using the rules of nomenclature for binary molecular compounds, or using acid nomenclature. The acid nomenclature of simple acids is different from that of oxoacids.

Section 3.8

- Compounds are named using a system of nomenclature specific to the compound type. Binary ionic compounds are named as cation first, anion second—without Greek prefixes to specify numbers (because they are unnecessary). If the cation has more than one possible charge, the charge is given in the name as a Roman numeral in parentheses immediately following the name of the metal. Ionic compounds containing polyatomic ions are named using the same system, cation name followed by anion name—with the cation's charge specified only if necessary. Binary molecular compounds are named using the same nomenclature, except that numbers of atoms must be specified in the name using Greek prefixes—because unlike with ionic compounds, there is generally more than one possible ratio of combination.

Key Terms

Acid 107	Diatomic molecule 94	Mixture 78	Physical property 80
Binary ionic compound 85	Empirical formula 96	Molecular formula 95	Polyatomic ion 101
Chemical change 80	Heterogeneous mixture 78	Molecule 92	Polyatomic molecule 94
Chemical formula 83	Homogeneous mixture 78	Nomenclature 88	Qualitative 80
Chemical process 80	Ionic bonding 83	Oxoacid 108	Quantitative 80
Chemical property 80	Law of constant composition 93	Oxoanion 104	Substance 77
Compound 77	Law of definite proportions 93	Physical change 80	Type I compound 85
Covalent bond 92	Law of multiple proportions 93	Physical process 80	Type II compound 86

The process of naming binary molecular compounds follows the procedure outlined in Section 2.6. The element that appears first in the formula is named first, followed by the name of the second element—with its ending changed to –ide. Greek prefixes are used to indicate numbers of atoms, but the prefix mono– is not used when there is only one atom of the *first* element in the formula.

Examples:

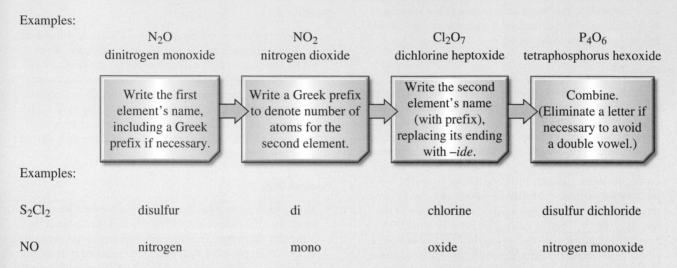

N_2O	NO_2	Cl_2O_7	P_4O_6
dinitrogen monoxide	nitrogen dioxide	dichlorine heptoxide	tetraphosphorus hexoxide

Write the first element's name, including a Greek prefix if necessary. → Write a Greek prefix to denote number of atoms for the second element. → Write the second element's name (with prefix), replacing its ending with –ide. → Combine. (Eliminate a letter if necessary to avoid a double vowel.)

Examples:

S_2Cl_2	disulfur	di	chlorine	disulfur dichloride
NO	nitrogen	mono	oxide	nitrogen monoxide

The process of naming binary ionic compounds follows the simple procedure outlined in Section 3.3. Naming compounds that contain polyatomic ions follows essentially the same procedure, but it does require you to recognize the common polyatomic ions [◀◀ Table 3.6]. Because many ionic compounds contain polyatomic ions, it is important that you know their names, formulas, and charges—well enough that you can identify them readily. In ionic compounds with ratios of combination other than 1:1, subscript numbers are used to denote the number of each ion in the formula.

Examples: $CaBr_2$, Na_2S, $AlCl_3$, Al_2O_3, FeO, Fe_2O_3

Recall that because the common ions of main group elements have predictable charges, it is unnecessary to use prefixes to denote their numbers when naming compounds that contain them. Thus, the names of the first four examples above are calcium bromide, sodium sulfide, aluminum chloride, and aluminum oxide. The last two contain transition metal ions, many of which have more than one possible charge. In these cases, in order to avoid ambiguity, the charge on the metal ion is designated with a Roman numeral in parentheses. The names of these two compounds are iron(II) oxide and iron(III) oxide, respectively.

When a subscript number is required for a polyatomic ion, the ion's formula must first be enclosed in parentheses.

Examples: $Ca(NO_3)_2$, $(NH_4)_2S$, $Ba(C_2H_3O_2)_2$, $(NH_4)_2SO_4$, $Fe_3(PO_4)_2$, $Co_2(CO_3)_3$

Names: calcium nitrate, ammonium sulfide, barium acetate, ammonium sulfate, iron(II) phosphate, cobalt(III) carbonate
The process of naming ionic compounds given their formulas can be summarized with the following flowchart:

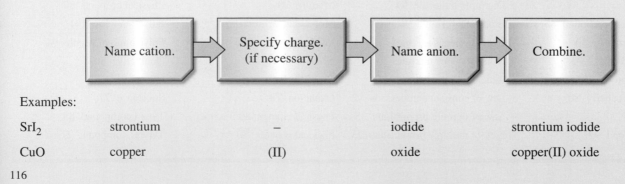

Name cation. → Specify charge. (if necessary) → Name anion. → Combine.

Examples:

SrI_2	strontium	–	iodide	strontium iodide
CuO	copper	(II)	oxide	copper(II) oxide

It is equally important that you be able to write the formula of an ionic compound given its name. Again, knowledge of the common polyatomic ions is critical. The process of writing an ionic compound's formula given its name is summarized as follows:

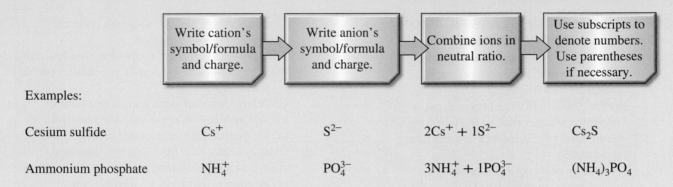

Examples:

Cesium sulfide	Cs^+	S^{2-}	$2Cs^+ + 1S^{2-}$	Cs_2S
Ammonium phosphate	NH_4^+	PO_4^{3-}	$3NH_4^+ + 1PO_4^{3-}$	$(NH_4)_3PO_4$

The process of writing a molecular compound's formula given its name is summarized as follows:

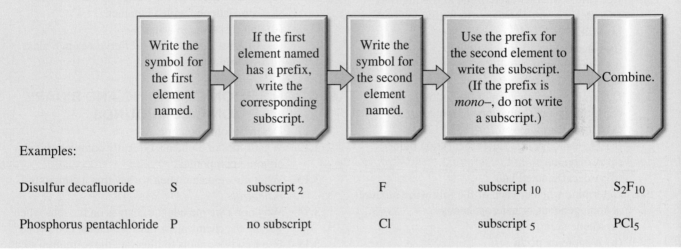

Examples:

Disulfur decafluoride	S	subscript 2	F	subscript 10	S_2F_{10}
Phosphorus pentachloride	P	no subscript	Cl	subscript 5	PCl_5

Key Skills Problems

3.1
Determine the atomic number of calcium (Ca) using the periodic table on the inside cover of your book. What is the electron configuration of the Ca atom?
a) $[Ne]4s^2$ b) $[Ar]4s^2$ c) $[Kr]4s^2$ d) $[Ar]$ e) $[Kr]$

3.2
What is the charge on the common ion that forms from the Ca atom?
a) −1   b) −2   c) 0   d) +1   e) +2

3.3
What is the electron configuration of the common ion that forms from the Ca atom?
a) $[Ar]4s^1$ b) $[Ar]$ c) $[Ar]4s^2$ d) $[Kr]$ e) $[Kr]4s^1$

3.4
With what noble gas is the calcium ion isoelectronic?
a) He   b) Ne   c) Ar   d) Kr   e) None of these.

3.5
What is the correct name for $CaSO_4$?
a) calcium sulfoxide   b) calcium sulfite   c) calcium sulfur oxide   d) calcium sulfate   e) calcium sulfide tetroxide

3.6
What is the correct formula for nickel(II) perchlorate?
a) $NiClO_4$ b) Ni_2ClO_4 c) $Ni(ClO_4)_2$ d) $NiClO_3$ e) $Ni(ClO_3)_2$

3.7
What is the correct name for NCl_3?
a) trinitrogen chloride   b) mononitrogen chloride   c) nitrogen trichloride   d) nitride trichloride   e) mononitride chloride

3.8
What is the correct formula for phosphorus pentachloride?
a) PCl_5 b) P_5Cl c) $P(ClO)_5$ d) PO_4Cl e) $PClO$

Questions and Problems

SECTION 3.1: MATTER: CLASSIFICATION AND PROPERTIES

3.1 Determine whether each of the following constitutes a pure substance or a mixture:
(a) Soil   (c) Sugar
(b) Tea   (d) The iron in a nail

3.2 Determine whether each of the following constitutes a pure substance or a mixture:
(a) Seawater
(b) A shiny new copper bracelet
(c) A piece of aluminum foil
(d) A fruit smoothie

3.3 Determine whether each of the following constitutes a pure substance or a mixture:
(a) Milk
(b) Chocolate chip cookie dough
(c) The helium in a balloon
(d) Distilled water

3.4 Determine whether each of the following constitutes a pure substance or a mixture:
(a) Chicken noodle soup
(b) The mercury in a fluorescent light bulb
(c) A diamond
(d) A decaffeinated latte

3.5 Determine whether each of the following mixtures is homogeneous or heterogeneous:
(a) Maple syrup
(b) A supreme pizza
(c) Italian dressing
(d) Concrete (such as that used in a driveway)

3.6 Determine whether each of the following mixtures is homogeneous or heterogeneous:
(a) Orange juice with pulp
(b) Iced tea
(c) A carbonated soft drink
(d) A "flat" soft drink

3.7 Describe each property of sodium metal as a physical property or a chemical property:
(a) It has a silvery-white metallic appearance.
(b) It reacts violently with water.
(c) Its density is 0.968 g/cm³.
(d) It melts at 98°C.
(e) It reacts with chlorine gas to form sodium chloride (salt).

3.8 Describe each property of chlorine gas as a physical or chemical property:
(a) It is a pale yellow-green gas at room temperature.
(b) It boils at −34°C.
(c) It reacts with sodium metal to form sodium chloride (salt).
(d) It smells like bleach.
(e) It reacts with organic compounds.

3.9 Classify each of the following as a chemical change or a physical change:
(a) A penny is stamped flat when run over by a train.
(b) A large wooden log is split into smaller pieces for kindling.
(c) A large wooden log is burned in a bonfire.
(d) A marshmallow is toasted over a bonfire.
(e) Ice melts inside an insulated cooler.

3.10 Classify each of the following as a chemical change or a physical change:
(a) Propane is burned in a barbeque grill.
(b) An iron nail rusts.
(c) An iron horseshoe is heated to red-hot, and then flattened with a hammer.
(d) Salt dissolves in water.
(e) An Alka Seltzer® tablet effervesces in a glass of warm water.

SECTION 3.2: IONIC BONDING AND BINARY IONIC COMPOUNDS

3.11 Describe ionic bonding.

3.12 What types of elements typically combine to form an ionic compound?

3.13 What do we mean when we say that a compound is *neutral?*

3.14 Why don't we include the charges on the ions when writing the chemical formula of an ionic compound?

3.15 Use Lewis symbols to illustrate the formation of a unit of calcium oxide (CaO) from one calcium atom and one oxygen atom.

3.16 Use Lewis symbols to illustrate the formation of aluminum fluoride (AlF_3) from one aluminum atom and three fluorine atoms.

3.17 Determine the chemical formula of the compound shown here.

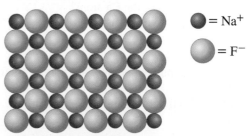

3.18 Determine the chemical formula of the compound shown here.

3.19 Determine the formula of the compound formed from each of the following element pairs:
(a) Potassium and oxygen
(b) Lithium and oxygen
(c) Magnesium and fluorine
(d) Strontium and nitrogen

3.20 Determine the formula of the compound formed from each of the following element pairs:
(a) Rubidium and iodine
(b) Calcium and chlorine
(c) Sodium and sulfur
(d) Aluminum and bromine

3.21 Complete the following table with the formulas of the compounds that form from each pair of ions indicated.

Ions	N^{3-}	Cl^-	O^{2-}
Fe^{2+}			
Fe^{3+}			
Zn^{2+}			
Al^{3+}			
Sr^{2+}			
NH_4^+			

3.22 Complete the following table with the formulas of the compounds that form from each pair of ions indicated.

Ions	I^-	P^{3-}	S^{2-}
Ni^{2+}			
Ti^{4+}			
Ca^{2+}			
Cr^{3+}			
Ag^+			
Li^+			

3.23 Determine the charge on the unknown ion, X, in each of the following compounds:
(a) X_2O (b) SrX (c) K_3X (d) AlX_3 (e) X_2O_3

3.24 Determine the charge on copper in each of the following ionic compounds:
(a) $CuCl_2$ (c) CuO
(b) Cu_3N (d) Cu_2O

3.25 Determine the charge on iron in each of the following ionic compounds:
(a) Fe_2O_3 (c) FeO
(b) $FeCl_2$ (d) FeN

SECTION 3.3: NAMING IONS AND BINARY IONIC COMPOUNDS

3.26 Why do we not use Greek prefixes to specify the number of ions of each type when we write the name of an ionic compound?

3.27 Name the following ions:
(a) Na^+ (b) Mg^{2+} (c) Al^{3+} (d) S^{2-} (e) F^-

3.28 Name the following ions:
(a) Ba^{2+} (b) Cl^- (c) O^{2-} (d) Li^+ (e) N^{3-}

3.29 Name the following ions:
(a) Ti^{2+} (b) Ag^+ (c) Ni^{4+} (d) Pb^{2+} (e) Zn^{2+}

3.30 Name the following ions:
(a) Cu^+ (b) Cr^{3+} (c) Fe^{2+} (d) Ti^{4+} (e) Co^{2+}

3.31 Indicate the number of protons and electrons in each of the ions in Problem 3.27.

3.32 Indicate the number of protons and electrons in each of the ions in Problem 3.28.

3.33 Indicate the number of protons and electrons in each of the ions in Problem 3.29.

3.34 Indicate the number of protons and electrons in each of the ions in Problem 3.30.

3.35 Write the Lewis dot symbol for each ion listed in Problem 3.27.

3.36 Write the Lewis dot symbol for each ion listed in Problem 3.28.

3.37 Write the systematic name of each compound:
(a) $RbCl_2$ (b) Na_2O (c) $CuCl_2$ (d) $NiCl_4$

3.38 Write the systematic name of each compound:
(a) CaS (b) $ZnBr_2$ (c) Li_3N (d) $CuCl$

3.39 Write the systematic name of each compound:
(a) CrF_3 (b) AgI (c) Li_2S (d) CoO

3.40 Write the systematic name of each compound:
(a) SnO (b) SnO_2 (c) Hg_2Cl_2 (d) MnN

3.41 Write the systematic name of each compound:
(a) Cs_3N (b) Sr_3P_2 (c) FeP (d) Pb_3N_4

3.42 Write the chemical formula of each compound:
(a) Potassium sulfide   (c) Magnesium chloride
(b) Aluminum oxide   (d) Sodium selenide

3.43 Write the chemical formula of each compound:
(a) Strontium nitride   (c) Aluminum sulfide
(b) Lithium phosphide   (d) Barium oxide

3.44 Write the chemical formula of each compound:
(a) Silver oxide   (c) Strontium chloride
(b) Zinc bromide   (d) Cadmium sulfide

3.45 Write the chemical formula of each compound:
(a) Titanium(IV) fluoride   (c) Copper(II) oxide
(b) Iron(III) oxide   (d) Nickel(IV) sulfide

SECTION 3.4: COVALENT BONDING AND MOLECULES

3.46 Describe covalent bonding.
3.47 What is an empirical formula?
3.48 Determine the molecular and empirical formulas of the compounds shown here. (Black spheres are carbon, and white spheres are hydrogen.)

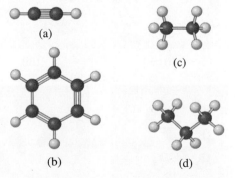

(a)

(c)

(b)

(d)

3.49 Write the empirical formula for each of the following molecular formulas:
(a) C_2N_2 (b) C_6H_6 (c) C_6H_{12} (d) P_4O_{10}

3.50 Write the empirical formula for each of the following molecular formulas:
(a) B_2H_6 (b) N_2O_4 (c) N_2O_5 (d) Al_2Br_6

3.51 What type(s) of elements typically combine to form molecular substances?

3.52 Use Lewis symbols to illustrate the formation of a chlorine molecule (Cl_2) from two individual chlorine atoms.

3.53 Use Lewis symbols to illustrate the formation of a water molecule from its individual constituent atoms.

3.54 Identify each diagram as a pure substance or a mixture. For each pure substance, identify it as an element or a compound. For each mixture, identify the components as elements or compounds.

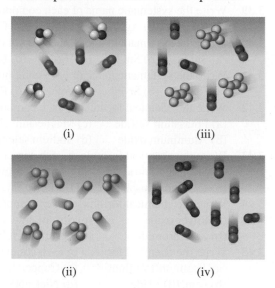

(i)   (iii)

(ii)   (iv)

3.55 Identify each diagram as a pure substance or a mixture. For each pure substance, identify it as an element or a compound. For each mixture, identify the components as elements or compounds.

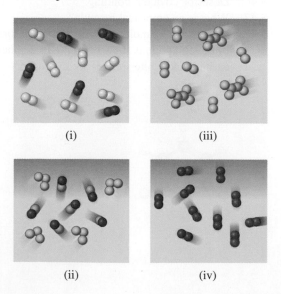

(i)   (iii)

(ii)   (iv)

SECTION 3.5: NAMING BINARY MOLECULAR COMPOUNDS

3.56 Determine the formula for each of the following compounds:
(a) Dinitrogen tetroxide   (c) Carbon tetrachloride
(b) Phosphorus trichloride   (d) Nitrogen trifluoride

3.57 Determine the formula for each of the following compounds:
(a) Silicon disulfide   (c) Selenium hexabromide
(b) Sulfur tetrafluoride   (d) Phosphorus trihydride

3.58 Determine the name for each of the following compounds:
(a) P_2O_5 (b) N_2O (c) PF_3 (d) OCl_2

3.59 Determine the name for each of the following compounds:
(a) CS_2 (b) SF_6 (c) SO_2 (d) ICl_5

3.60 Determine the name for each of the following compounds:
(a) SeO (b) As_2O_5 (c) SO_3 (d) N_2F_4

3.61 Write the molecular formulas and names of the following compounds:

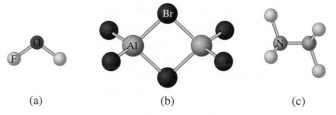

(a)   (b)   (c)

3.62 Write the molecular formulas and names of the following compounds:

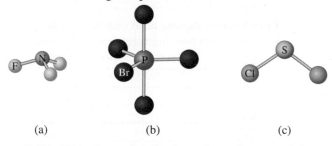

(a)   (b)   (c)

SECTION 3.6: COVALENT BONDING IN IONIC SPECIES: POLYATOMIC IONS

3.63 Is it possible to have an ionic compound that does not contain a metal? Explain.

3.64 Name each of the following polyatomic ions:
(a) PO_4^{3-} (b) NO_3^- (c) SO_3^{2-} (d) CO_3^{2-}

3.65 Name each of the following polyatomic ions:
(a) PO_3^{3-} (b) NO_2^- (c) CN^- (d) OH^-

3.66 Determine the charge on the unknown ion, A, in each of the following compounds:
(a) NH_4A (c) A_3PO_4 (e) $A(CO_3)_2$
(b) $(NH_4)_3A$ (d) A_2SO_3 (f) $A(CN)_3$

3.67 Determine the formula for the compound formed between each pair of ions listed:
(a) Li^+ and ClO_3^- (c) Ca^{2+} and $C_2H_3O_2^-$
(b) Ba^{2+} and SO_3^{2-} (d) Al^{3+} and ClO_4^-

3.68 Determine the formula for the compound formed between each pair of ions listed:
(a) Pb^{2+} and HCO_3^- (c) Ti^{4+} and ClO_3^-
(b) Zn^{2+} and NO_3^- (d) Ti^{4+} and SO_4^{2-}

3.69 Determine the formula for the compound formed between each pair of ions listed:
(a) NH_4^+ and HCO_3^- (c) Al^{3+} and NO_2^-
(b) Ca^{2+} and PO_4^{3-} (d) K^+ and $Cr_2O_7^{2-}$

3.70 Name the ionic compounds in Problem 3.67.

3.71 Name the ionic compounds in Problem 3.68.

3.72 Name the ionic compounds in Problem 3.69.

3.73 For each of the following ionic compounds, determine the identity and number of ions present. Note that the first one is answered for you.
(a) Li_3N <u>3 Li^+ and 1 N^{3-}</u>   (d) $Sr(ClO_3)_2$
(b) $Ca(CN)_2$ (e) $(NH_4)_3PO_3$
(c) $Fe_2(SO_4)_3$

3.74 For each of the following ionic compounds, determine the identity and number of ions present. Note that the first one is answered for you.
(a) BaO <u>1 Ba^{2+} and 1 O^{2-}</u>   (d) K_2CrO_4
(b) Mg_3P_2 (e) $NaHCO_3$
(c) $Mg(C_2H_3O_2)_2$

3.75 Determine the formula for the compound formed between each pair of ions listed:
(a) Na^+ and PO_4^{3-} (c) Mg^{2+} and CN^-
(b) Al^{3+} and SO_4^{2-} (d) Ca^{2+} and CO_3^{2-}

3.76 Complete the table with the formula of the compound that forms from each pair of ions.

Ions	OH^-	$C_2O_4^{2-}$	PO_3^{3-}	SO_3^{2-}
Fe^{2+}				
Fe^{3+}				
Zn^{2+}				
Al^{3+}				
Sr^{2+}				
NH_4^+				

3.77 Complete the table with the formula of the compound that forms from each pair of ions.

Ions	$Cr_2O_7^{2-}$	HCO_3^-	$C_2H_3O_2^-$	CO_3^{2-}
Ni^{2+}				
Ti^{4+}				
Ca^{2+}				
Cr^{3+}				
Ag^+				
Li^+				

3.78 Which of the following compounds could be represented by the image?

(a) $MgCl_2$ (c) $NaBr$ (e) $Ca(NO_3)_2$
(b) AlF_3 (d) CaO

3.79 Which of the following compounds could be represented by the image?

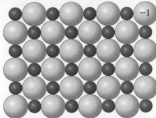

(a) AlN (c) KBr (e) $LiC_2H_3O_2$
(b) BaO (d) $Sr(ClO_4)_2$

3.80 Which of the following compounds could be represented by the image?

(a) $Zn(NO_3)_2$ (c) Na_2S
(b) $(NH_4)_2SO_4$ (d) TiO_2

3.81 Determine the formula that matches each of the compound names:
(a) Ammonium bromide   (c) Calcium chlorate
(b) Aluminum nitrate   (d) Lithium carbonate

3.82 Determine the chemical formula for each of the eight yellow-highlighted compounds listed on the baby formula label on page 103.

3.83 The citrate ion is a polyatomic ion with the formula $C_6H_5O_7^{3-}$. Determine the formula for each of the two blue-highlighted compounds on the baby formula label on page 103.

SECTION 3.7: ACIDS

3.84 Name the following acids:
(a) HCN (b) HBr (c) H_2S

3.85 Name the following acids:
(a) H_2Se (b) HF (c) HI

3.86 Name the following acids:
(a) H_3PO_3 (b) HNO_2 (c) H_2SO_4

3.87 Name the following acids:
(a) $HClO_3$ (b) $HSCN$ (c) H_2CO_3

3.88 Write the formula of each of the following acids:
(a) Phosphoric acid   (c) Hydrosulfuric acid
(b) Oxalic acid

3.89 Write the formula of each of the following acids:
(a) Acetic acid   (c) Hydrothiocyanic acid
(b) Chromic acid

SECTION 3.8: SUBSTANCES IN REVIEW

3.90 List the seven elements that exist as diatomic molecules.

3.91 What is the difference between an ionic compound and a molecular compound?

3.92   Give an example of each of the following:
(a) Monatomic cation   (c) Polyatomic cation
(b) Monatomic anion   (d) Polyatomic anion

3.93   Which of the following exist as molecules in their elemental form?
(a) Mg   (b) N   (c) K   (d) S   (e) Ba   (f) Ar

3.94   Which of the following exist as molecules in their elemental form?
(a) Cl   (b) P   (c) Kr   (d) Rb   (e) Ba   (f) O

3.95   Which of the following compounds are ionic? Explain how you know.
(a) KCl   (c) CF_4 (e) SrF_2
(b) CaO   (d) NO_2 (f) $AlBr_3$

3.96   Which of the following compounds are molecular? Explain how you know.
(a) P_2O_5 (c) K_2O (e) PCl_3
(b) MgS   (d) SO_2 (f) NF_3

3.97   Classify each of the following substances as an element or a compound. For each substance that you identify as an element, indicate whether or not it is a molecular substance.
(a) $AlPO_4$ (b) CF_4 (c) NH_3 (d) O_2

3.98   Classify each of the following substances as an element or a compound. For each substance that you identify as an element, indicate whether or not it is a molecular substance.
(a) Al   (b) $MgCl_2$ (c) NH_4Cl (d) I_2

3.99   Classify each of the following substances as an element or a compound. For each substance that you identify as an element, indicate whether or not it is a molecular substance.

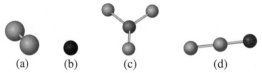

(a)   (b)   (c)   (d)

3.100   Classify each of the following substances as an element or a compound. For each substance that you identify as an element, indicate whether or not it is a molecular substance.

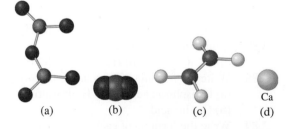

Ca

(a)   (b)   (c)   (d)

3.101   Draw a model, similar to those shown in Problems 3.99 and 3.100, of each of the following substances:
(a) K   (b) SO_2 (c) NO   (d) SCl_2

3.102   Draw a model, similar to those shown in Problems 3.99 and 3.100, of each of the following substances:
(a) Mg   (b) MgO   (c) CO   (d) H_2O

3.103   Give an example of each of the following:
(a) An atomic element
(b) A molecular element
(c) A molecular compound
(d) An ionic compound
(e) A homogeneous mixture
(f) A heterogeneous mixture

3.104   Identify each of the following as an element or a compound:
(a) NH_3 (c) S_8 (e) H_2 (g) P_4
(b) N_2 (d) NO   (f) CO   (h) H_2O

▶▶▶ **Visualizing Chemistry Questions**
Figure 3.11

VC 3.1   Which pair of elements would most likely form an ionic compound?
(a) S and Cl   (b) Rb and Na   (c) K and F

VC 3.2   Which pair of elements would most likely form a molecular compound?
(a) N and O   (b) Li and Ti   (c) Mg and Br

VC 3.3   Which element can likely form either ionic or covalent bonds?
(a) Ba   (b) B   (c) F

VC 3.4   Which element will likely form neither ionic nor covalent bonds?
(a) C   (b) Ar   (c) Cs

Answers to In-Chapter Materials

Answers to Practice Problems

3.1A (a) mixture, (b) pure substance, (c) mixture, (d) pure substance.
3.1B Aluminum is an element, salt is a compound, a sports drink is a homogeneous mixture, and carbonated water is a heterogeneous mixture. **3.2A** (a) physical, (b) physical, (c) chemical, (d) physical.
3.2B (a) chemical, (b) physical, (c) chemical, (d) physical.
3.3A (a) Ca_3N_2, (b) KBr, (c) $CaBr_2$. **3.3B** (a) A^{3+}, X^-; (b) D^{2+}, E^{3-};
(c) L^{2+}, M^-. **3.4A** Sn_3N_4, (b) NiF_2, (c) MnO_2. **3.4B** (a) 2+, (b) 2+,
(c) 3+. **3.5A** (a) iron(III) ion, (b) manganese(II) ion, (c) rubidium ion.
3.5B (a) Pb^{2+}, (b) Na^+, (c) Zn^{2+}. **3.6A** (a) iron(III) nitride, (b) silver fluoride, (c) sodium oxide, (d) cobalt(III) oxide, (e) rubidium sulfide, (f) lead(II) oxide. **3.6B** (a) FeO, (b) $ZnBr_2$, (c) SrS,
(d) K_2O, (e) Ni_3N_4, (f) Li_3P. **3.7A** (a) CF_4, (b) NCl_3, (c) C_2H_4.

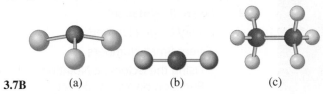

3.7B   (a)   (b)   (c)

3.8A (a) $C_4H_5N_2O$, (b) C_2H_5, (c) $C_2H_5NO_2$. **3.8B** (a).
3.9A (a) dichlorine monoxide, (b) silicon tetrachloride.
3.9B (a) chlorine dioxide, (b) carbon tetrabromide. **3.10A** (a) CS_2,
(b) N_2O_3. **3.10B** (a) SF_6, (b) S_2F_{10}. **3.11A** (a) sodium sulfate,
(b) copper(II) nitrate, (c) iron(III) carbonate. **3.11B** (a) potassium dichromate, (b) lithium oxalate, (c) copper(I) nitrate.
3.12A (a) $Pb(ClO_3)_2$, (b) $MgCO_3$, (c) $(NH_4)_3PO_4$.
3.12B (a) $FePO_3$, (b) $Hg(NO_3)_2$, (c) K_2SO_3. **3.13A** (a) hydrocyanic acid, (b) nitrous acid, (c) phosphoric acid. **3.13B** (a) H_2SO_3,
(b) H_2CrO_4, (c) $HClO_3$.

Answers to Checkpoints

3.1.1 b, c, e. **3.1.2** a, d, f. **3.2.1** b. **3.2.2** b. **3.2.3** d. **3.3.1** c. **3.3.2** c.
3.3.3 b. **3.3.4** e. **3.4.1** e. **3.4.2** d. **3.5.1** b. **3.5.2** c. **3.6.1** c. **3.6.2** d.
3.6.3 c. **3.6.4** a. **3.7.1** d. **3.7.2** a.

CHAPTER 4

How Chemists Use Numbers

4.1   Units of Measurement
- Base Units
- Mass, Length, and Time
- Metric Multipliers
- Temperature

4.2   Scientific Notation
- Very Large Numbers
- Very Small Numbers
- Using the Scientific Notation Function on Your Calculator

4.3   Significant Figures
- Exact Numbers
- Measured Numbers
- Calculations with Measured Numbers

4.4   Unit Conversion
- Conversion Factors
- Derived Units
- Dimensional Analysis

4.5   Success in Introductory Chemistry Class

California's Giant Sequoias are some of the world's largest and longest-lived living organisms. Tree heights can exceed 300 feet and sequoias can live for thousands of years. Scientific measurements enable us to quantify such natural wonders—and to get a sense of their extraordinary magnitude.

David Clapp/Stone/Getty Images

In This Chapter, You Will Learn

About the system of units used by scientists and about the proper way to record scientific measurements. You will also learn how to perform calculations with measured numbers, and how to report the results of your calculations.

Things To Review Before You Begin

• How to calculate percentage

So far, our discussion of chemistry has been mostly qualitative—classification of matter, descriptions of atoms and subatomic particles, how the nature of atoms can lead to the formation of ions and molecules. The only explicitly quantitative detail we have encountered to this point is the expression of atomic mass, for which we use the *atomic mass unit* [◄◄ Section 1.7]. At this point, it is important that you learn more about the quantitative aspects of science. In this chapter, we explain how chemists use measurements and a specific system of units to record observations, solve problems, and report their results.

4.1 Units of Measurement

The measured quantities that you will see often in this course include mass, length, time, and temperature. You are already familiar with some of the units used to express length (inches, feet, yards, miles), time (seconds, minutes, hours), and temperature (degrees Fahrenheit)—you have probably used them since you were a child. In science, however, we use a particular set of units that have been agreed upon by scientists worldwide to facilitate the sharing and reporting of scientific data. This system is the **International System of Units (SI units),** and most of the units with which you are probably familiar are *not* SI units. It is important that you become comfortable with SI units now, because they are used extensively throughout this book.

Student Note: The *atomic mass unit,* although very commonly used throughout science, is not actually one of the official SI units.

Base Units

Table 4.1 gives the SI base units for mass, length, time, and temperature. The *base* units are essentially the starting points for expression of these four quantities. In many cases, when it is convenient, we use something other than the base unit to describe a measurement. For example, reporting the period of time required for a sequoia tree to grow to maturity using the base unit *seconds* would not be altogether convenient.

TABLE 4.1	SI Base Units for Mass, Length, Time, and Temperature	
Base Quantity	**Name of Unit**	**Symbol**
Mass	kilogram	kg
Length	meter	m
Time	second	s
Temperature	kelvin	K

To express such a length of time in seconds would require an enormous number. In this case, we would most likely use the unit *years*. Likewise, if we wanted to report the mass of a single leaf from such a tree, we would probably report it in *grams,* rather than in the base unit *kilograms,* because this would give us a number of more reasonable magnitude.

Mass, Length, and Time

Although the terms *mass* and *weight* often are used interchangeably, they do not mean the same thing. Strictly speaking, weight is the force exerted by an object due to gravity. **Mass** is a measure of the *amount of matter* in an object or a sample. Because gravity varies from location to location (gravity on the Moon is only about one-sixth that on Earth), the weight of an object varies depending on where it is measured. The mass of an object remains the same regardless of where it is measured. The SI base unit of mass is the kilogram (kg), but in chemistry the smaller gram (g) often is more convenient for the amounts of matter used in typical laboratory experiments. One kilogram is equal to one thousand grams:

$$1 \text{ kg} = 1000 \text{ g}$$

The SI unit for length is the meter (m), which is probably familiar to you even if you don't use it in everyday conversation. If you are an athlete or a fan of certain sports, you may recognize the use of the meter to describe the depth or length of a swimming pool, or the kilometer (km) to describe the length of a marathon. One kilometer is equal to one thousand meters:

$$1 \text{ km} = 1000 \text{ m}$$

The SI unit for time is the second (s), which is also the unit commonly used outside of scientific circles. (For longer periods of time, when it is convenient, we also use minutes, hours, days, and years—the same as the units of everyday usage.)

Metric Multipliers

As we have seen, SI base units are not always convenient for expressing very large or very small masses or lengths; or for expressing especially long or short periods of time. In fact, it is very often the case that *base* units are not ideal for expression of scientific measurements. When a base unit such as the kilogram or the meter is not well suited to the magnitude of a measurement, we use a Greek prefix to tailor the unit to a particular purpose. For example, the amounts of active ingredients in drugs are typically expressed in *milligrams*. The dimensions of an envelope might be given in *centimeters*. And the duration of an extremely fast process may be given in *nanoseconds*. In each case, we have used a Greek prefix as a metric multiplier—to make the size of the *unit* appropriate for the size of the *measurement*.

> **Student Note:** Some confusion can arise regarding the terms *mass* and *weight,* partly because we determine an object's mass by *weighing* it.

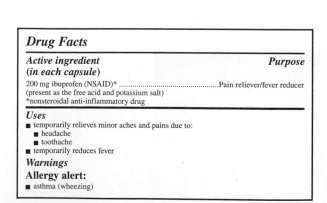

Drug Facts

Active ingredient   *Purpose*
(in each capsule)

200 mg ibuprofen (NSAID)* ....Pain reliever/fever reducer
(present as the free acid and potassium salt)
*nonsteroidal anti-inflammatory drug

Uses
■ temporarily relieves minor aches and pains due to:
  ■ headache
  ■ toothache
■ temporarily reduces fever

Warnings

Allergy alert:
■ asthma (wheezing)

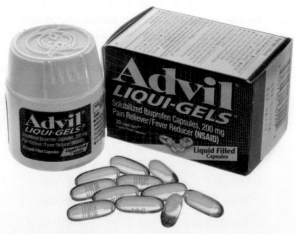

David A. Tietz/McGraw Hill

Profiles in Science

Henrietta Swan Leavitt

Early in the twentieth century, Henrietta Swan Leavitt (1868–1921) made a groundbreaking discovery in the field of astronomy. Although Leavitt never completed a degree in astronomy, she was hired as a "computer" at the Harvard College Observatory. At the time, the term *computer* referred to a woman employed by the observatory to gather and catalogue astronomical data from photographic plates. The work involved painstaking and tedious measurement of the size and brightness of thousands of stars on plates that had been generated over many years.

Leavitt was assigned by the observatory's director to focus on a particular type of pulsating star known as a *Cepheid variable*. These stars regularly change in both size and temperature. By analyzing her measurements of more than a thousand stars from photographic plates of the Magellanic Cloud galaxies, she discovered that Cepheid variables exhibit a distinct relationship between the time it takes for one complete cycle of pulsation (the *period*)—and the total amount of light produced by the star (the *luminosity*).

Although it did not earn her a great deal of notoriety during her lifetime, Leavitt's discovery would ultimately enable astronomers to measure distances on an intergalactic scale never before possible. Edwin Hubble (for whom the *Hubble Space Telescope* is named) used her work to further much of his own. One of his most noteworthy accomplishments, establishing that the universe is expanding, would not have been possible without Leavitt's work. Hubble, who is regarded as one of the most important astronomers of all time, is said to have opined often that Leavitt should have received a Nobel Prize. Although there was an effort made to nominate her, it came too late as she died prematurely from cancer. (Nobel Prizes are never awarded posthumously.)

An asteroid, a crater on the Moon, and a telescope in the All Sky Automated Survey for SuperNovae (ASAS-SN) have been named in Leavitt's honor.

(Left): Henrietta Swan Leavitt. (Middle): This pulsating star is one of the brightest Cepheid variables in the Milky Way. (Right): Night sky in the southern hemisphere. The Magellanic Clouds are two of only a very few galaxies, outside of our own, that are visible to the naked eye.

(Left): WorldPhotos/Alamy Stock Photo; (Middle): GSFC/NASA; (Right): Jon G. Fuller/VW PICS/Universal Images Group/Getty Images

Table 4.2 gives the Greek prefixes that are most commonly used in chemistry for this purpose and shows how each affects the magnitude of a unit. (Note that not all of the units in Table 4.2 are SI base units, but they should all be relatively familiar units.)

Selecting the right prefix to tailor a unit to a measurement usually results in a number between 1 and 1000. This range of numbers is, in a sense, *comfortable* for the human brain. Consider the following example:

One of the wavelengths of visible light is 0.000000487 meter. But the magnitude of that number, expressed in the SI base unit *meters,* is hard to grasp. If you were to read the number and then try to reproduce it without looking at the original, you may write too many, or too few, zeros. By changing the unit to *nanometers,* we change the number to 487. The new number is one that you could easily reproduce, having looked at it just once.

Look again at Table 4.2 and consider the last four rows. The relationship between the number given in the *Meaning* column and the number 1 tells us how far to the right we move the decimal point when we add the corresponding prefix. For example, the number in the Meaning column for nano– (bottom row) is 0.000000001. To change this number to 1, we would have to move its decimal point *nine* places to the right.

$$0.\underset{\smile}{000000001} \text{ second} \longrightarrow 1 \text{ nanosecond}$$

When we add the prefix nano–, we move the decimal point nine places to the right.

TABLE 4.2	Common Greek Prefixes		
Prefix	Symbol	Meaning	Example
tera–	T	1,000,000,000,000	1 terabyte (TB) = 1,000,000,000,000 bytes* (B)
			1 B = 0.000000000001 TB
giga–	G	1,000,000,000	1 gigawatt (GW) = 1,000,000,000 watts* (W)
			1 W = 0.000000001 GW
mega–	M	1,000,000	1 megahertz (MHz) = 1,000,000 hertz* (Hz)
			1 Hz = 0.000001 MHz
kilo–	k	1,000	1 kilometer (km) = 1,000 meters (m)
			1 m = 0.001 km
deci–	d	0.1	1 deciliter (dL) = 0.1 liter (L)
			1 L = 10 dL
centi–	c	0.01	1 centimeter (cm) = 0.01 meter (m)
			1 m = 100 cm
milli–	m	0.001	1 millisecond (ms) = 0.001 second (s)
			1 s = 1000 ms
micro–	μ	0.000001	1 microgram (μg) = 0.000001 gram (g)
			1 g = 1,000,000 μg
nano–	n	0.000000001	1 nanosecond (ns) = 0.000000001 second (s)
			1 s = 1,000,000,000 ns

*You are familiar with a *byte* as an amount of computer memory and with *watts* as a measure of lightbulb power. You may also be familiar with *hertz* as a measure of computer processing speed—or as a measure of sound frequency; and the unit *liter* is a familiar measure of volume, as in 2-liter bottles of soda.

The number in the Meaning column for micro– is 0.000001. To change this number to 1, we would have to move its decimal point *six* places to the right.

$$0.\underset{\smile}{000001} \text{ gram} \longrightarrow 1 \text{ microgram}$$

When we add the prefix micro–, we move the decimal point six places to the right.

For the prefixes milli– and centi–, we move the decimal point *three* and *two* places to the right, respectively.

$$0.\underset{\smile}{001} \text{ second} \longrightarrow 1 \text{ millisecond}$$

When we add the prefix milli–, we move the decimal point three places to the right.

$$0.\underset{\smile}{01} \text{ meter} \longrightarrow 1 \text{ centimeter}$$

When we add the prefix centi–, we move the decimal point two places to the right.

Returning to our visible-wavelength example, to transform the number 0.000000487 into a more manageable number (a number between 1 and 1000), we could move its decimal point to the right by *seven* places (to give 4.87), by *eight* places (to give 48.7), or by *nine* places (to give 487). Each of these numbers lies between 1 and 1000. However, the only prefix available to us in this case is nano–, which moves the decimal point to the right by *nine* places to give 487. (There is no prefix corresponding to either of the other options.)

$$0.\underset{\smile}{000000487} \text{ meter} \longrightarrow 487 \text{ nanometers}$$

We add the prefix nano–, and we move the decimal point nine places to the right.

This tells us that *nanometer* is the best unit to express this wavelength.

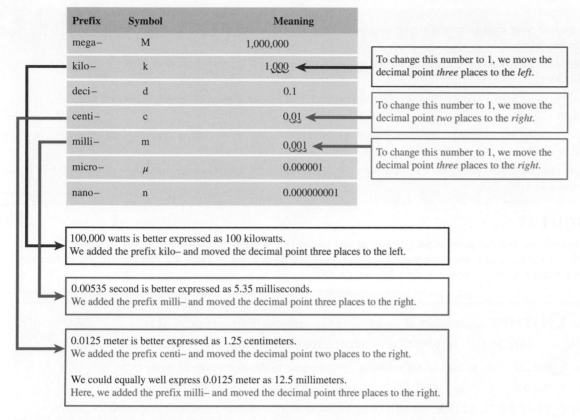

Prefix	Symbol	Meaning
mega–	M	1,000,000
kilo–	k	1,000
deci–	d	0.1
centi–	c	0.01
milli–	m	0.001
micro–	μ	0.000001
nano–	n	0.000000001

To change this number to 1, we move the decimal point *three* places to the *left*.

To change this number to 1, we move the decimal point *two* places to the *right*.

To change this number to 1, we move the decimal point *three* places to the *right*.

100,000 watts is better expressed as 100 kilowatts.
We added the prefix kilo– and moved the decimal point three places to the left.

0.00535 second is better expressed as 5.35 milliseconds.
We added the prefix milli– and moved the decimal point three places to the right.

0.0125 meter is better expressed as 1.25 centimeters.
We added the prefix centi– and moved the decimal point two places to the right.

We could equally well express 0.0125 meter as 12.5 millimeters.
Here, we added the prefix milli– and moved the decimal point three places to the right.

Figure 4.1 Choosing the best prefix to tailor a unit to the corresponding measurement.

The foregoing example starts with an unmanageably *small* number (0.000000487). But we can apply a similar approach to numbers that are unmanageably *big*. In these cases, we consult Table 4.2 to determine how many places to the *left* a decimal point is moved when a Greek prefix is added. Figure 4.1 shows several examples of choosing the best unit for a particular measurement.

Sample Problem 4.1 gives you some practice choosing the appropriate prefix to convert measured numbers into units of reasonable magnitude.

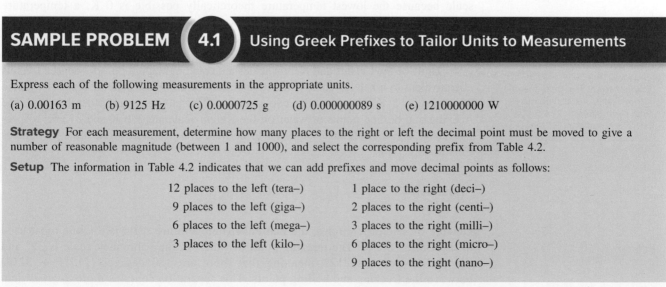

SAMPLE PROBLEM **4.1** Using Greek Prefixes to Tailor Units to Measurements

Express each of the following measurements in the appropriate units.

(a) 0.00163 m   (b) 9125 Hz   (c) 0.0000725 g   (d) 0.000000089 s   (e) 1210000000 W

Strategy For each measurement, determine how many places to the right or left the decimal point must be moved to give a number of reasonable magnitude (between 1 and 1000), and select the corresponding prefix from Table 4.2.

Setup The information in Table 4.2 indicates that we can add prefixes and move decimal points as follows:

12 places to the left (tera–)	1 place to the right (deci–)
9 places to the left (giga–)	2 places to the right (centi–)
6 places to the left (mega–)	3 places to the right (milli–)
3 places to the left (kilo–)	6 places to the right (micro–)
	9 places to the right (nano–)

(Continued on next page)

Solution (a) The decimal point must be moved three places to the right to give 1.63, making the appropriate prefix *milli–*. 0.00163 m should be expressed as 1.63 mm.

(b) The decimal point must be moved three places to the left to give 9.125, making the appropriate prefix *kilo–*. 9125 Hz should be expressed as 9.125 kHz.

(c) The decimal point must be moved six places to the right to give 72.5, making the appropriate prefix *micro–*. 0.0000725 g should be expressed as 72.5 µg.

(d) The decimal point must be moved nine places to the right to give 89, making the appropriate prefix *nano–*. 0.000000089 m should be expressed as 89 nm.

(e) The decimal point must be moved nine places to the left to give 1.21, making the appropriate prefix *giga–*. 1210000000 W should be expressed as 1.21 GW.

THINK ABOUT IT

We could have moved the decimal point in part (a) *four* or *five* places to the right—to give the numbers 16.3 and 163, respectively. Both are numbers of reasonable magnitude. But there are no prefixes in Table 4.2 that correspond to these options. The same applies to the numbers in parts (b) through (e).

Practice Problem ATTEMPT Express each of the following measurements in the appropriate SI units.

(a) 0.00000000825 L   (b) 68,250 g   (c) 0.00552 m   (d) 41,800,000,000,000 B   (e) 0.000375 s

Practice Problem BUILD Express each of the following measurements in the appropriate SI units.

(a) 9520 mL   (b) 23,800,000 µm   (c) 12,000 ms   (d) 3,650,000,000 nm   (e) 0.00082 MHz

Practice Problem CONCEPTUALIZE For each of the following objects, identify the SI unit best suited to describe it:

(a) The mass of a 2-lb block of cheese   (b) the volume of a gallon of milk

(c) the length of a football field   (d) the length of a pencil.

Temperature

There are two temperature scales used in chemistry: the **Celsius scale** and the **Kelvin scale** (also known as the **absolute temperature scale**). Their units are the degree Celsius (°C) and the **kelvin (K),** respectively. The Celsius scale was originally defined using the freezing point (0°C) and the boiling point (100°C) of pure water at sea level. Although both scales are used in science, the official SI base unit of temperature is the kelvin. The Kelvin temperature scale is also known as the *absolute* temperature scale because the lowest temperature theoretically possible is 0 K, a temperature referred to as "absolute zero." When we express a temperature on the Kelvin scale, we do not use a degree sign. The theoretical temperature of absolute zero is written as 0 K (*not* as 0°K).

Units of the Celsius and Kelvin scales are equal in magnitude, so a degree Celsius is equivalent to a kelvin. Thus, if the temperature of an object increases by 5°C, it also increases by 5 K. However, the two scales are offset from each other by 273. The freezing and boiling points of water on the Kelvin scale are 273 K and 373 K, respectively. We use the following equation to convert a temperature from units of degrees Celsius to kelvins:

Student Note: The Celsius and Kelvin scales actually are offset from each other by 273.15, but 273 is sufficient for the calculations you will encounter in this book.

Equation 4.1	$K = °C + 273$

Outside of scientific circles, the Fahrenheit temperature scale is the one most used in the United States. The freezing point of water on the Fahrenheit scale is 32° and the boiling point is 212°, meaning that there are 180 degrees (212°F − 32°F) between the freezing and boiling points. This separation contains more *degrees* than the 100 between the freezing point and boiling point of water on the Celsius scale. This is because the size of a degree on the Fahrenheit scale is smaller than a degree on the Celsius scale. In fact, a *degree* on the Fahrenheit scale is only five-ninths the size of a degree on the Celsius scale. The following equations show how we convert a temperature on the Celsius scale to a temperature on the Fahrenheit scale, and vice versa.

$$\text{temp in }°F = \left[\frac{9°F}{5°C} \times (\text{temp in }°C)\right] + 32°F \qquad \textbf{Equation 4.2}$$

$$\text{temp in }°C = (\text{temp in }°F - 32°F) \times \frac{5°C}{9°F} \qquad \textbf{Equation 4.3}$$

Familiar Chemistry

The Fahrenheit Temperature Scale

Outside of scientific circles, the Fahrenheit temperature scale is the one most used in the United States. Before the work of Daniel Gabriel Fahrenheit (German physicist, 1686–1736), there were many different, arbitrarily defined temperature scales, none of which gave consistent measurements. Accounts of exactly how Fahrenheit devised his temperature scale vary from source to source. In one account, in 1724, Fahrenheit labeled as 0° was the lowest artificially attainable temperature at the time (the temperature of a mixture of ice, water, and ammonium chloride). Using a traditional scale consisting of 12 degrees, he labeled the temperature of a healthy human body as the twelfth degree. On this scale, the freezing point of water occurred at the fourth degree. For better precision, he divided each degree into eight smaller degrees. This convention makes the freezing point of water 32° and normal body temperature 96°. (Today we consider normal body temperature to be somewhat higher than 96°F.) The boiling point of water on the Fahrenheit scale is 212°, meaning that there are 180 degrees (212 − 32) between the freezing and boiling points. This separation is considerably more *degrees* than the 100 between the freezing and boiling points of water on the Celsius scale. The size of a degree on the Fahrenheit scale is only 100/180 or five-ninths the size of a degree on the Celsius scale. Equations 4.2 and 4.3 are used to convert between these two temperature scales. The picture shows a baby's temperature being taken with a digital tympanic thermometer—a common practice in doctors' offices.

Ruth Jenkinson/Science Source

Figure 4.2 Thermometers showing various temperatures on the Celsius, Kelvin, and Fahrenheit scales.

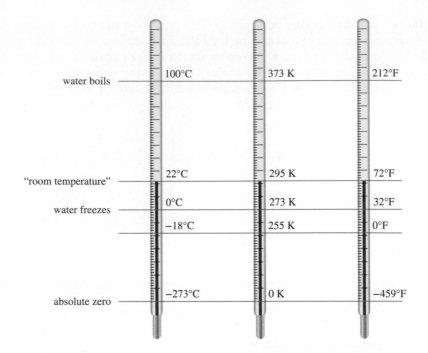

water boils	100°C	373 K	212°F
"room temperature"	22°C	295 K	72°F
water freezes	0°C	273 K	32°F
	−18°C	255 K	0°F
absolute zero	−273°C	0 K	−459°F

Figure 4.2 shows a thermometer for each temperature scale, and Sample Problems 4.2 and 4.3 let you practice converting from one temperature scale to another.

SAMPLE PROBLEM **4.2** **Conversion Between Celsius and Kelvin Temperature Scales**

Normal human body temperature can range over the course of the day from about 36°C in the early morning to about 37°C in the afternoon. Express these two temperatures and the range that they span (difference) using the Kelvin scale.

Strategy Use Equation 4.1 to convert from the Celsius scale to the Kelvin scale. Then convert the range of temperatures from degrees Celsius to kelvins, keeping in mind that 1°C is equivalent to 1 K.

Setup Equation 4.1 is already set up to convert the two temperatures from degrees Celsius to kelvins. No further manipulation of the equation is needed. The range in kelvins will be the same as the range in degrees Celsius.

Solution 36°C + 273 = 309 K, 37°C + 273 = 310 K, and the range of 1°C is equal to a range of 1 K.

THINK ABOUT IT

Check your math and remember that converting a temperature from degrees Celsius to kelvins is different from converting a *difference* in temperature from degrees Celsius to kelvins.

Practice Problem **A**TTEMPT Express the freezing point of water (0°C), the boiling point of water (100°C), and the range spanned by the two temperatures using the Kelvin scale.

Practice Problem **B**UILD According to the website of the National Aeronautics and Space Administration (NASA), the average temperature of the universe is 2.7 K. Convert this temperature to degrees Celsius.

Practice Problem **C**ONCEPTUALIZE If a single degree on the Celsius scale is represented by the first rectangle, which of the rectangles (i) through (iv) best represents a single kelvin?

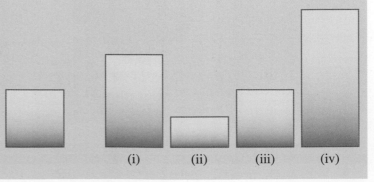

(i)   (ii)   (iii)   (iv)

SAMPLE PROBLEM **4.3**   Conversion of Temperatures to the Fahrenheit Scale

A body temperature above 39°C constitutes a high fever. Convert this temperature to the Fahrenheit scale.

Strategy We are given a temperature in Celsius and are asked to convert it to Fahrenheit.

Setup Equation 4.3 is correctly arranged to use for this conversion.

$$\text{temp in }°F = \left[\frac{9°F}{5°C} \times (\text{temp in }°C)\right] + 32°F$$

Solution

$$\text{temp in Fahrenheit} = \frac{9°F}{5°\cancel{C}} \times 39°\cancel{C} + 32°F = 102.2°F$$

THINK ABOUT IT

Knowing that "normal" body temperature on the Fahrenheit scale is approximately 99°F (98.6°F is the number most often cited), 102.2°F seems like a reasonable answer.

Practice Problem Ⓐ**TTEMPT** Convert the temperatures 45.0°C and 90.0°C, and the difference between them, to degrees Fahrenheit.

Practice Problem Ⓑ**UILD** In Ray Bradbury's 1953 novel *Fahrenheit 451,* 451°F is said to be the temperature at which books, which have been banned in the story, ignite. Convert 451°F to the Celsius scale.

Practice Problem Ⓒ**ONCEPTUALIZE** If a single degree on the Fahrenheit scale is represented by the first rectangle, which of the rectangles (i) through (iv) best represents a single degree on the Celsius scale? Which best represents a single kelvin?

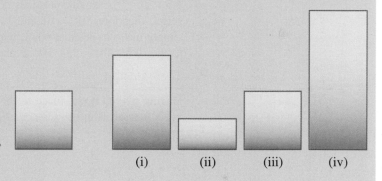

(i)   (ii)   (iii)   (iv)

CHECKPOINT–SECTION 4.1   Units of Measurement

4.1.1 The coldest temperature ever recorded on Earth was −129°F (recorded at Vostok Station, Antarctica, on July 21, 1983). Express this temperature in degrees Celsius and in kelvins.

a) −89°C, −89 K

b) −289°C, −16 K

c) −89°C, 184 K

d) −174°C, 99 K

e) −7.0°C, 266 K

4.1.2 A sample of water is heated from room temperature to just below the boiling point. The overall change in temperature is 72°C. Express this temperature change in kelvins.

a) 345 K

b) 72 K

c) 0 K

d) 201 K

e) 273 K

4.1.3 The highest recorded temperature on Earth was 134°F in Mitribah, Kuwait, on July 21, 2016. Express this temperature in °C and K.

a) 184°C, 354 K

b) 102°C, 330 K

c) 56.7°C, 330 K

d) 134°C, 354 K

e) 70.0°C, 330 K

4.1.4 Determine the *difference* between the boiling point of water (212°F) and the freezing point of water (32°F) in °C and in kelvins.

a) 100°C, 273 kelvins

b) 100°C, 373 kelvins

c) 100°C, 100 kelvins

d) 180°C, 453 kelvins

e) 180°C, 180 kelvins

4.2 Scientific Notation

Scientists often have to work with numbers that are either very large or very small. Consider the following examples:

> In a single teaspoon of water (about 5 grams), there are approximately 167,100,000,000,000,000,000,000 water molecules—an *unimaginably* big number.

> Each of the water molecules has a mass of approximately 0.0000000000000000000000002994 kilograms—an unimaginably *small* number.

David A. Tietz/McGraw Hill

Numbers such as these are difficult to manage because it is all too easy to transcribe them or enter them into a calculator with too many or too few zeros. Because scientists need to perform calculations with such numbers, we need a way to express them reliably and unambiguously. The solution to this problem is the use of *scientific notation*. **Scientific notation** is the expression of a number as $N \times 10^n$, where N is a number between 1 and 10, and n is either a positive integer—for the purpose of expressing a very large number; or a negative integer, for the purpose of expressing a very small number.

To get a feel for how scientific notation works, let's look at some examples with numbers that are neither extraordinarily small nor extraordinarily large. Consider the number 15. Although 15 is a manageable number, and is not one that would require scientific notation to express unambiguously, we could express it in scientific notation as 1.5×10^1. The exponent 1 does not change the value of the number to which it is super-script; therefore, $10^1 = 10$, and $1.5 \times 10 = 15$. What we have done to the number 15 is simply move its decimal point (which is not shown ordinarily, but which could be written to the right of the digit 5) one place to the left—to get a number between 1 and 10.

$$15 \longrightarrow 1.5$$

Having moved it one place to the left, we must indicate this with the integer to which 10 is raised (n) in the scientific-notation expression.

15 is equal to 1.5×10, or 1.5×10^1

one 10   10 to the 1

Now consider the number 150. To determine the value of N, we must move the decimal point (originally to the right of the zero) *two* places to the left—again, to get a number between 1 and 10.

$$150 \longrightarrow 1.50$$

This makes the value of n (the power to which 10 is raised) 2. Therefore, we can express 150 as 1.50×10^2.

150 is equal to $1.5 \times 10 \times 10$, or 1.5×10^2

two 10s   10 to the 2

Student Note: We could also express 150 as 1.5×10^2. Whether or not we include the zero depends on the number of digits that are appropriate—a subject we discuss in Section 4.3.

We can also consider some examples of small numbers to illustrate how scientific notation works. To express the number 0.5 in scientific notation, we determine the value of N by moving the decimal point one place to the right. This gives $N = 5$. Because we have moved the decimal point to the *right,* the power to which we raise 10 will be *negative.* This may seem slightly less intuitive than the determination of positive powers of 10. Think of it this way: $10^{-1} = 0.1$. So

$$0.5 = 5 \times 0.1 = 5 \times 10^{-1}$$

one 0.1   10 to the −1

Student Note: If this is not immediately obvious, enter it into your calculator as 10 ^ (−) 1 and hit enter. If you have entered it properly, you will get 0.1.

$$10^{-1} = 0.1$$

For another example, 0.025 can be expressed in scientific notation as 2.5×10^{-2}. We had to move the decimal point *two* places to the right to get 2.5 (N)

$$0.025 \longrightarrow 2.5$$

and therefore must multiply 2.5 by 10^{-2}.

$$0.025 = 2.5 \times 0.1 \times 0.1 = 2.5 \times 10^{-2}$$

two 0.1s   10 to the -2

Very Large Numbers

Now let's look at the expression of *very* large numbers. Using the example of the number of water molecules in a teaspoon of water, we have the number 167,100,000,000,000,000,000,000. To express this number using scientific notation, we move the decimal point 23 places to the left

original position of the decimal point

167,100,000,000,000,000,000,000

final position of the decimal point   moved 23 places to the left

and we multiply the result by 10^{23}, to indicate where the decimal point was in the original number.

167,100,000,000,000,000,000,000 is equal to

1.671×10

Twenty-three 10s   10 to the 23

The resulting number in scientific notation is 1.671×10^{23}. Therefore, expressed in scientific notation, there are approximately 1.671×10^{23} water molecules in a teaspoon of water.

Sample Problem 4.4 lets you practice expressing very large numbers using scientific notation.

SAMPLE PROBLEM (4.4)   Using Scientific Notation to Express Large Numbers

Write the following numbers in scientific notation:

(a) 277,000,000   (b) 93,800,000,000   (c) 5,500,000

Strategy These are all large numbers, so they should all have large positive exponents when written in scientific notation.

Setup Move the decimal point from its current location to the left until there is only one digit to the left of the decimal point. Count the number of places the decimal point was moved and this becomes the exponent ($\times 10^?$).

Solution (a) 277,000,000 must have the decimal point (to the right of the last zero, but not shown in the original number) moved eight places to the left to give us a number between 1 and 10: 2.77. This gives an answer of 2.77×10^8.

(b) 93,800,000,000 must have the decimal point moved 10 places to the left to give 9.38×10^{10}.

(c) 5,500,000 must have the decimal point moved six places to the left to give 5.5×10^6.

THINK ABOUT IT

We only included nonzero values in these answers. We explain significant figures, and when to include them in scientific notation, in Section 4.3.

Practice Problem **A**TTEMPT   Write the following numbers in scientific notation:

(a) 48,000,000,000,000   (b) 299,000,000   (c) 8,000,000,000,000,000

(Continued on next page)

Practice Problem ⓑ**UILD** Write the following values in decimal format:

(a) 3.55×10^8 (b) 1.899×10^5 (c) 5.7×10^9

Practice Problem ⓒ**ONCEPTUALIZE** Express the following examples from Table 4.2 in scientific notation:

(a) number of grams in a Tg   (b) number of W in a GW   (c) number of Hz in a MHz   (d) number of m in a km

Very Small Numbers

Finally, let's look at the expression of very *small* numbers using the example of the mass of a water molecule in kilograms: 0.00000000000000000000000002994 kg. In this case, to determine N, the number between 1 and 10, we must move the decimal point 26 places to the right:

original position of the decimal point

0.00000000000000000000000002994

final position of the decimal point

This gives $N = 2.994$. Because we have moved the decimal point to the *right,* our exponent must be *negative*—to convey the actual value of the number we are expressing.

0.00000000000000000000000002994 is equal to

2.994×0.1

Twenty-six 0.1s   10 to the −26

The resulting number in scientific notation is 2.994×10^{-26}. Therefore, in scientific notation, the mass of a water molecule is 2.994×10^{-26} kg.

Sample Problem 4.5 lets you practice expressing very small numbers using scientific notation.

SAMPLE PROBLEM (**4.5**)   Using Scientific Notation to Express Small Numbers

Write the following values in scientific notation:

(a) 0.0000000338   (b) 0.000021   (c) 0.00000000000244

Strategy These are all very small values (less than 1), so they will have negative exponents when written in scientific notation.

Setup Move the decimal point from its current location to the right until there is only one nonzero digit to the left of the decimal point. Count the number of places the decimal point was moved and this becomes the exponent ($\times 10^?$).

Solution (a) 0.0000000338 must have the decimal point moved eight places to the right to give us a number between 1 and 10, giving us an answer of 3.38×10^{-8}.

(b) 0.000021 must have the decimal point moved five places to the right, giving 2.1×10^{-5}.

(c) 0.00000000000244 must have the decimal point moved 12 places to the right, giving 2.44×10^{-12}.

THINK ABOUT IT

These values all look reasonable compared to one another. The *smallest* value (c) has the *largest* negative exponent.

Practice Problem Ⓐ**TTEMPT** Write the following values in scientific notation:

(a) 0.0006   (b) 0.0000000057   (c) 0.000000000000399

Practice Problem Ⓑ**UILD** Write the following values in decimal format:

(a) 4.27×10^{-5} (b) 2.77×10^{-8} (c) 7.33×10^{-11}

Practice Problem Ⓒ**ONCEPTUALIZE** Express the following examples from Table 4.2 in scientific notation:

(a) number of m in a cm   (b) number of s in a ms   (c) number of m in a μm   (d) number of s in a ns

Using the Scientific Notation Function on Your Calculator

A scientific calculator is a necessity for chemistry class, and it is important that you familiarize yourself with the proper use of its functions. A mistake that students commonly make with their scientific calculators is the improper entry of scientific notation. Examine your calculator and determine what sequence of buttons you must push to enter a number in scientific notation. Figure 4.3 shows several popular brands of scientific calculator and highlights the scientific notation function of each.

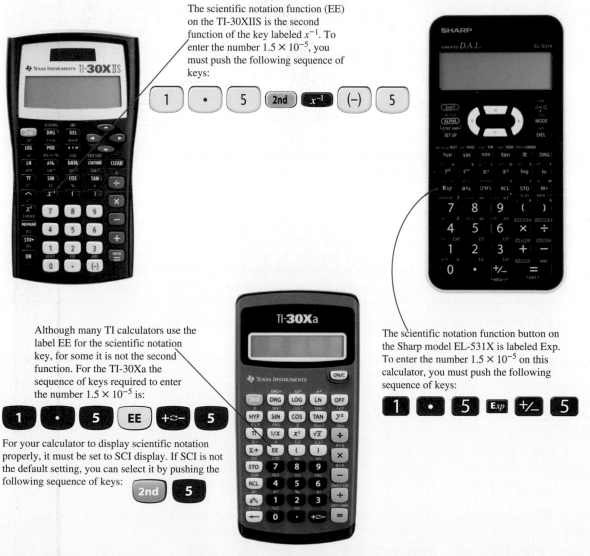

The scientific notation function (EE) on the TI-30XIIS is the second function of the key labeled x^{-1}. To enter the number 1.5×10^{-5}, you must push the following sequence of keys:

| 1 | · | 5 | 2nd | x^{-1} | (−) | 5 |

Although many TI calculators use the label EE for the scientific notation key, for some it is not the second function. For the TI-30Xa the sequence of keys required to enter the number 1.5×10^{-5} is:

| 1 | · | 5 | EE | +◦− | 5 |

For your calculator to display scientific notation properly, it must be set to SCI display. If SCI is not the default setting, you can select it by pushing the following sequence of keys:

| 2nd | 5 |

The scientific notation function button on the Sharp model EL-531X is labeled Exp. To enter the number 1.5×10^{-5} on this calculator, you must push the following sequence of keys:

| 1 | · | 5 | Exp | +/− | 5 |

Figure 4.3 Scientific notation shown on several popular brands of scientific calculator.

(all): David A. Tietz/McGraw Hill

If your calculator is not one of the brands shown in Figure 4.3, or if you do not recognize the scientific notation function key sequence, consult your user's manual or an Internet search to be sure you know how to enter scientific notation properly on your specific model of calculator. Without this knowledge, and without regular practice, it is very easy to make simple entry errors that result in wrong answers.

Sample Problem 4.6 gives you some practice entering scientific notation on your calculator and making sure you are doing it correctly.

SAMPLE PROBLEM 4.6 — Using the Scientific Notation Function on Your Scientific Calculator

Use your calculator to determine the result of each of the following calculations:

(a) $55.0 \times (6.20 \times 10^{-9})$

(b) $(3.67 \times 10^4) \times 231$

(c) $6.88 \times 10^{-8} \div 4922$

Strategy Plan to enter the values in your calculator using the correct key sequences.

Setup You must enter any values written in scientific notation using the protocol for your specific calculator.

Solution Push the following sequence of buttons in your calculator for each calculation:

(a) 3.41×10^{-7}

(b) 8.48×10^6

(c) 1.40×10^{-11}

> **Student Note:** When you calculate this result, the number your calculator actually displays is probably 8477700. (If your calculator is set to display results in scientific notation, it may display as 8.4777E6.) The reason we report the answer as 8.48×10^6 rather than as 8.477700×10^6, has to do with *rounding* to a reasonable number of digits. What constitutes a reasonable number of digits in a particular circumstance, namely *significant figures*, and how to round numbers properly, is the subject of the next section. For the remainder of this section of Practice Problems, you should round your answers to no more than two places to the right of the decimal point.

THINK ABOUT IT

If you do not get the answers shown here, you are entering the values into your calculator incorrectly. Refer to your user's manual or seek help from your instructor. A very common error is the manual entry of × 10 prior to use of the calculator's exponential function. Note that *none* of the illustrated key sequences in the solutions to this Sample Problem includes pushing × 10. It's important that you become comfortable with the appropriate key sequences by practicing.

Practice Problem ATTEMPT Use your calculator to determine the result of each of the following calculations:

(a) $4.77 \times 10^6 \div 323$

(b) $4925 \times (1.55 \times 10^4)$

(c) $55.99 + 6.55 \times 10^2$

Practice Problem BUILD Use your calculator to perform the following calculations. Your answers should be given in scientific notation.

(a) $(4.88 \times 10^{-4}) \times (3.99 \times 10^{-5})$

(b) $357 \times 1,569,000$

(c) $7.88 \times 10^{12} \div 6.56 \times 10^3$

Practice Problem CONCEPTUALIZE How far off would the answer to $(1.3 \times 10^7) \times 5.1$ be if the calculation were entered by pushing the following sequence of calculator keys rather than the proper sequence?

CHECKPOINT–SECTION 4.2   Scientific Notation

4.2.1 Express the number 0.00000000000585 in scientific notation.

 a) 5.85×10^{12} 　　　 d) 5.85×10^{-12}

 b) 5.85×10^{-14} 　　 e) 5.85×10^{-11}

 c) 5.85×10^{14}

4.2.2 Express the number 1,257,000,000 in scientific notation.

 a) 1.257×10^{6} 　　 d) 1.257×10^{-9}

 b) 1.257×10^{-6} 　 e) 1257×10^{6}

 c) 1.257×10^{9}

4.2.3 Express the number 3.772×10^{7} in decimal format.

 a) 37,720,000,000 　 d) 0.0003772

 b) 37,720,000 　　　 e) 3,772,000

 c) 0.0000003772

4.2.4 Perform the calculation and report the correct answer.
$4.55 \times 10^{12} \div 2.00$

 a) 2.28×10^{13} 　　 d) 2.28×10^{24}

 b) 22.8 　　　　　　 e) 2.28×10^{12}

 c) 2.28×10^{6}

4.2.5 Perform the calculation and report the correct answer.
$(3.76 \times 10^{13}) \times (4.91 \times 10^{10})$

 a) 1.85×10^{3} 　　 d) 1.85×10^{23}

 b) 1.85×10^{-3} 　 e) 1.85×10^{26}

 c) 1.85×10^{24}

4.2.6 Perform the calculation and report the correct answer.
$1.6 \times 10^{-4} \div 6.0 \times 10^{5}$

 a) 2.7×10^{-2} 　　 d) 2.7×10^{-10}

 b) 2.7×10^{9} 　　 e) 3×10^{-10}

 c) 9.6×10^{9}

4.3 Significant Figures

Chemists use two types of numbers: *exact* numbers, which have no uncertainty associated with them, and *measured* numbers, which by their nature are *inexact,* meaning that they always have some uncertainty associated with them. In this section, we explain which numbers are exact and which are measured, and the inherent uncertainty in measured numbers. We also describe how to write measured numbers with the appropriate amount of uncertainty—and the importance of doing so.

Student Note: The triple equal sign indicates that two quantities are equal to each other by definition.

Exact Numbers

An *exact number* is one that is counted or that results from a definition. If you count a group of objects, such as eggs, or kittens, or people, you get an exact number. Figure 4.4 illustrates collections of these objects and indicates the number of each.

 When a number is determined by counting, the number has no uncertainty. There are *exactly* 14 eggs, *exactly* three kittens, and *exactly* 5 people. Likewise, when a number is the result of a definition, such as 1 dozen ≡ 12, both the numbers 1 and 12 are *exact* numbers. Unlike numbers obtained by counting or from a definition, *measured* numbers always have some uncertainty associated with them. Have you ever noticed that a group of 2-L bottles of soda don't all appear to contain exactly the same amount of soda? How can that be—despite the fact that all of the bottles are labeled 2 L? The fact is, the amount of a packaged product such as soda is a *measured* quantity. All measured quantities are subject to error, and the numbers used to express them contain unavoidable uncertainty.

Student Note: Other exact numbers that we have already encountered include the metric multipliers indicated by the Greek prefixes in Table 4.2; and the fractions 5/9 and 9/5 in the equations used to convert between Celsius and Fahrenheit temperatures.

Measured Numbers

In Section 4.2, we encountered an example of a number that could be expressed two different ways using scientific notation: 1.50×10^{2} and 1.5×10^{2}. Although these two ways of expressing the number 150 are nearly equivalent, there is an important difference between them. The latter, 1.5×10^{2}, implies greater *uncertainty*.

2-L soda bottles do not necessarily all contain exactly the same volume.

David A. Tietz/McGraw Hill

Figure 4.4 Discrete collections of objects. (a) The number of eggs, 14, is an exact number; (b) the number of kittens, 3, is an exact number; and (c) the number of people, 5, is an exact number.

(a) Ken Welsh/Pixtal/age fotostock; (b) DAJ/Getty Images; (c) Prostock-Studio/ iStock/Getty Images

(a)

(b)

(c)

To understand why, let's look in some detail at the measurement of an everyday object: a cell phone. Figure 4.5 shows a cell phone beside a ruler, which we might use to measure the phone's length. The ruler is marked in centimeters, and we can tell that the phone's length is slightly greater than 16 cm. Because of the ruler's markings (called *gradations*), we can only *estimate* how much greater than 16 cm the phone is. Depending on the person doing the measurement, the length of the phone might be recorded as 16.3 cm, 16.4 cm, or maybe 16.5 cm. In any event, the third digit, in this case a 3, a 4, or a 5, is one about which we are not certain. When we report the length of the phone using this ruler, we should report it with three digits. Let's say that our estimate of the last digit is 4. We therefore report the length of the phone as 16.4 cm. The measured number 16.4 consists of three digits. Two of the digits, the 1 and the 6, are certain. The last digit, the 4, is uncertain—because it is an *estimate*.

two certain

∧

16.4 cm

↖

one uncertain

All of the digits in the number 16.4 are significant figures. **Significant figures** are digits that convey the precision with which a number is known. In this number, we are absolutely sure that the 1 is a 1, and that the 6 is a 6. We are less sure about the 4. It might actually be a 3—or it might actually be a 5. Given the limitations of the ruler we have used to measure the phone, we cannot be entirely sure. In this case, with this ruler, our best estimate of that last digit is 4. Another way for us to state explicitly our uncertainty about the last digit is to write the measured number as 16.4 cm ± 0.1 cm. This indicates

Figure 4.5 The length of a cell phone measured with a ruler marked only in gradations of 1 cm.

jacek lasa/leonardo255/123RF

Figure 4.6 The length of a cell phone measured with a ruler marked in gradations of 1 mm.

jacek lasa/leonardo255/123RF

that the uncertainty is in the last digit, and implies that the measured number may be as low as 16.3 cm, or as high as 16.5 cm. If we were to have a ruler marked with finer gradations, say every millimeter instead of just every centimeter, we would be able to measure the length of the phone with greater precision, and we could report the measurement using a number with more significant figures. Figure 4.6 illustrates this.

When we measure the phone using a ruler with finer gradations, in this case marked every millimeter, we can be certain of more of the numbers in our measurement. Now we can be certain of the third digit in the measurement, which appears to be a 4 after all. However, the measurement is not *exactly* 16.4 cm—in fact it now appears to be between 16.4 cm and 16.5 cm. Nevertheless, our third digit is definitely a 4, because the measurement falls between the fourth and fifth millimeter gradations (after the 16-centimeter gradation). Because we are certain of three digits, the 1, the 6, and the 4, we now must estimate the fourth digit. Although the measurement is between the fourth and fifth millimeter marks, it does not appear to be halfway between them—it is considerably closer to the fifth mark. So our estimate of the fourth digit in the measurement must be greater than 5. In fact, it may be an 8 or a 9. It is impossible to tell for sure, so we will estimate that it is an 8. Now we can report the length of the cell phone as 16.48 cm, where we are sure of the 1, the 6, and the 4, and we have made our best estimate of the 8. Now we are reporting the measurement with *four* significant figures: three certain digits and one uncertain digit.

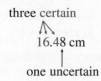

As before, we can show the uncertainty explicitly by writing 16.48 cm ± 0.01 cm. Note that now our uncertainty is much smaller than it was before. Now we know the dimension within a hundredth of a centimeter, whereas before we knew it only within a tenth of a centimeter.

Consider for a moment what these reported measurements tell us about the precision of the measurements. In the first case, when we reported the measurement with three significant figures, we were implying that the uncertainty was 0.1 cm on a length of 16.4 cm. In the second case, when we reported the measurement with four significant figures, we were implying that the uncertainty was 0.01 cm on a length of 16.48 cm. In each case, the implied uncertainty can be expressed as a percentage of the total measurement:

$$\frac{0.1 \text{ cm}}{16.4 \text{ cm}} \times 100\% = 0.6\% \text{ uncertainty} \quad \text{vs.} \quad \frac{0.01 \text{ cm}}{16.48 \text{ cm}} \times 100\% = 0.06\% \text{ uncertainty}$$

In the second case, the implied uncertainty in our measurement is only one-tenth of that in the first case. The second measurement has greater precision.

It's important to be able to determine the number of significant figures in a number, which is not always as straightforward as the examples of 16.4 (three significant figures) and 16.48 (four significant figures). Although the importance of this will become more apparent in Section 4.4, now is a good time to learn how to assess the number of significant figures in a number, and to know what uncertainty is implied by the way a number is written.

The following set of guidelines is useful for determining the number of significant figures in a measured number.

1. Any digit that is not zero is significant.

18,911 (five significant figures)	1, 8, 9, 1, and 1
4.1 (two significant figures)	4 and 1
58.63 (four significant figures)	5, 8, 6, and 3

2. Zeros located *between* nonzero digits are significant.

101 (three significant figures)	1, 0, and 1
5002.1 (five significant figures)	5, 0, 0, 2, and 1
8.05 (three significant figures)	8, 0, and 5

3. Zeros to the *left* of the first nonzero digit are *not* significant.

0.11 (two significant figures)	1 and 1
0.00006 (one significant figure)	6
0.0575 (three significant figures)	5, 7, and 5

4. Zeros to the right of the last nonzero digit are significant if the number contains a decimal point.

9.10 (three significant figures)	9, 1, and 0
0.1500 (four significant figures)	1, 5, 0, and 0
0.00030100 (five significant figures)	3, 0, 1, 0, and 0

5. Zeros to the right of the last nonzero digit in a number that does *not* contain a decimal point may or may not be significant—depending on the circumstances. The number 150 may have two or three significant figures. Without additional information, it is not possible to know for sure. To avoid ambiguity in such cases, it is best to express the number in scientific notation. If the intended number of significant figures is two, meaning that the zero is not significant, the number should be written as 1.5×10^2. If the intended number of significant figures is three, meaning that the zero *is* significant, the number should be written as 1.50×10^2. [Another way to express this number with three significant figures is to simply add a decimal point to the right of the zero (writing it as 150.), and although technically correct, this is not a common practice in chemistry.]

Sample Problem 4.7 lets you practice determining how many significant figures a number has.

SAMPLE PROBLEM (4.7) Determining the Number of Significant Figures in a Number

Determine the number of significant figures in the following measurements: (a) 443 cm, (b) 15.03 g, (c) 0.0356 kg, (d) 3.000×10^{-7} L, (e) 50 mL, (f) 0.9550 m

Strategy All nonzero digits are significant, so the goal will be to determine which of the zeros are significant.

Setup Zeros are significant if they appear between nonzero digits or if they appear after a nonzero digit in a number that contains a decimal point. They may or may not be significant if they appear to the right of the last nonzero digit in a number that does not contain a decimal point.

Solution (a) 443, all of the digits are significant: *three* significant figures; (b) 15.03, again, all digits are significant (the zero is captive): *four* significant figures; (c) 0.0356 contains two zeros that are not significant because they are *leading* zeros (serving only to identify the placement of the decimal point): *three* significant figures; (d) 3.000 contains three zeros that are significant because they are to the *right* of the decimal point: *four* significant figures; (e) the zero in 50 may or may not be significant—we would need more information to be sure; *one* or *two* significant figures (an ambiguous case); (f) 0.9550 contains two zeros, one of which is significant (the one to the right of the decimal point), one of which is not significant (the leading zero): *four* significant figures.

THINK ABOUT IT

Be sure that you have identified each zero correctly as either significant or not significant. Zeros are significant in (b) and (d); they are not significant in (c); it is not possible to tell in (e); and the number in (f) contains one zero that is significant, and one that is not.

Practice Problem ATTEMPT Determine the number of significant figures in the following measurements:

(a) 1129 m   (c) 1.094 cm   (e) 150 mL

(b) 0.0003 kg   (d) 3.5×10^{12} atoms   (f) 9.550 km

Practice Problem BUILD Determine the number of significant figures in each of the following numbers. Then write each number in decimal format and determine the number of significant figures.

(a) 1.080×10^{-4} (c) 2.910×10^{3}

(b) 5.5×10^{6} (d) 8.100×10^{-5}

Practice Problem CONCEPTUALIZE Report the number of colored objects contained within each square and, in each case, indicate the number of significant figures in the number you report.

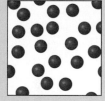

Profiles in Science

Arthur Rosenfeld

You may have heard your parents or grandparents tell of having lived through the "energy crisis." Originally, this term was used to describe events that occurred in the 1970s, during which the lives of average Americans were affected significantly by the global result of political instability in the Middle East. In 1973, in response to military involvement in the Yom Kippur war, a group of oil-producing Arab states declared an oil embargo against the United States, the United Kingdom, Canada, Japan, and the Netherlands. In 1979, following the Iranian revolution, the global supply of oil was again reduced, triggering widespread anxiety over the reliability of our supply of oil. These events resulted in, among other significant geopolitical effects, a drastic increase in the price of crude oil—and ultimately in the price of gasoline.

Arthur Rosenfeld
U.S. Department of Energy

    Although it may seem less than catastrophic now, when the price of regular gasoline first rose above $1.00/gal in the United States (in 1980), it constituted a significant burden for American consumers. "Sunday drives," once a common and inexpensive family activity, became largely a thing of the past.

Gasoline was rationed, and long lines could be seen at every filling station—from open to close. Locking gas caps were invented to prevent the theft of gasoline by siphoning the contents of gas tanks of unattended vehicles—a crime that had become common.

In 1974, in the wake of the first oil crisis, renowned Berkeley physicist Arthur Rosenfeld (1926–2017) turned his attention to the study of energy efficiency. In 1975, he founded the Center for Building Science at the Lawrence Berkeley National Laboratory, where he developed widely used computer programs for the analysis of energy efficiency in buildings. He also oversaw the development of many of the technologies in use today worldwide, including key components of energy-efficient fluorescent lighting and transparent window coatings that keep heat in or out of a building, as circumstance requires.

You have probably seen the Energy Star stickers on new appliances. They indicate the estimated annual cost of operating an appliance. The label shown here is for a small, energy-efficient refrigerator. A similar appliance manufactured a few decades ago might have consumed four to five times as much energy. The technological advancements resulting from Rosenfeld's work have amounted to energy-cost savings to Americans on the order of a *trillion* dollars.

Roger Loewenberg/McGraw Hill

Rosenfeld served as an energy advisor on both the state and national levels, and he received numerous medals and awards, nationally and internationally, in recognition of his considerable contributions to the science of energy efficiency. In 2013, President Obama presented Rosenfeld with a National Medal of Technology and Innovation for his role in the development of energy-efficient technologies and his spearheading of consequent changes to building standards and policies that support energy efficiency. It has been suggested in a publication coauthored by dozens of scientists that a new unit, the *Rosenfeld*, be adopted as a standard measure of electricity savings. As proposed in the 2010 article in *Environmental Research Letters*, one Rosenfeld would equal 3 billion kilowatt-hours per year—the amount of energy savings necessary to replace the energy output of one 500-megawatt coal-fired power plant.

Calculations with Measured Numbers

Much of what we do in chemistry requires calculations, which is why a scientific calculator is necessary for this class. Suppose you need to know the total weight of a group of 575 objects, each of which weighs about 2.5 lb. You may punch these numbers into your calculator as:

And your calculator will display the result as 1437.5. But is that really the correct answer? Is the total weight of 575 of these 2.5-lb objects *really* 1437.5 lb? Remember, the objects were described as weighing *about* 2.5 lb each. The weight of the objects is given by a number with two significant figures. What is the implied uncertainty in that number? Recall that a number with two significant figures has one certain digit and one uncertain digit. The last digit, the 5, might really be a 5—or it might be a 4, or a 6. The uncertainty in the last digit is generally presumed to be ± 1. So if the weight of these objects is given as 2.5 lb, the implied uncertainty is $\pm$ 0.1 lb. We can express this uncertainty as a percentage of the weight of an object:

$$\frac{0.1 \text{ lb}}{2.5 \text{ lb}} \times 100\% = 4\% \text{ uncertainty}$$

But what is the implied uncertainty if we report the total weight as 1437.5 lb? Again, with the uncertainty in the last digit being $\pm$ 0.1 lb:

$$\frac{0.1 \text{ lb}}{1437.5 \text{ lb}} \times 100\% = 0.007\% \text{ uncertainty}$$

How can the uncertainty in the *total* weight be so much smaller than the uncertainty in the weight of a single object? The answer is, it can't. Simply multiplying one number by another number cannot make us *more* certain of the result than we were of the original numbers. Therefore, the correct answer to the question is *not* 1437.5 lb—even though that's what the calculator displays. The correct answer to this question should be reported with *two* significant figures as 1.4×10^2 lb.

$$\frac{0.1 \times 10^2 \text{ lb}}{1.4 \times 10^2 \text{ lb}} \times 100\% = 7\% \text{ uncertainty}$$

This uncertainty, although not identical, is far closer to the uncertainty in the weight of a single object. Because so many calculations that we do involve one or more measured numbers, we have a set of guidelines to use that help us determine the appropriate number of significant figures to report in calculated results. Note that the guidelines for addition and subtraction are different from those for multiplication and division. For the final result of any calculation to have the appropriate number of significant figures, it is important that you apply the correct guidelines to each mathematical operation.

1. In addition and subtraction, the answer cannot have more digits after the decimal point than the original number with the smallest number of digits after the decimal point.

    102.50 ⟵ two digits after the decimal point
    $+ \ \ 0.231$ ⟵ three digits after the decimal point
    102.731 ⟵ must be rounded so the result has only two digits after the decimal point

    Correct result: 102.73

    143.29 ⟵ two digits after the decimal point
    $- \ \ 20.1$ ⟵ one digit after the decimal point
    123.19 ⟵ must be rounded so the result has only one digit after the decimal point

    Correct result: 123.2

    To round a number to the appropriate number of significant figures, we can draw a line after the last digit we are able to keep. In the case of 102.731, where we are only able to keep two digits after the decimal point, we get

    $$102.73|1$$

    Because the digit immediately following the line is less than 5, we simply drop everything after the line, in this case just the last digit, to get

    $$102.73$$

    This is known as "rounding down."

    In the case of 123.19, where we are only able to keep one digit after the decimal point, again we draw a line to show how many digits we can keep.

    $$123.1|9$$

    In this case, though, where the digit immediately following the line is 5 or greater (in this case it is a 9), we increase (by 1) the last digit to be kept. This is known as "rounding up," and the result is

    $$123.2$$

2. In multiplication and division, the number of significant figures in the final product or quotient cannot have more significant figures than the *original* number with the *smallest* number of significant figures. (This is the rule we applied in the calculation of the total weight of a group of 2.5-lb objects.)

    Multiplication

    $\ \ \ 8.011$ ⟵ four significant figures
    $\times \ \ 1.4$ ⟵ two significant figures
    11.2154 ⟵ must be rounded to *two* significant figures (limited by the number 1.4)

    11 The first digit to be dropped is a 2, so we round down.

Division

four significant figures $\longrightarrow$ $\dfrac{11.57}{305.88}$ = 0.03782529096 must be rounded to *four*
five significant figures $\longrightarrow$ significant figures (limited
by the number 11.57)

0.03783 The first digit to be dropped is a 5, so we
round up.

3. Exact numbers do not limit the number of significant figures in a calculated
result. For example, a penny minted after 1982 has a mass of 2.5 g. If we have
three such pennies, the total mass is

$$3 \times 2.5 \text{ g} = 7.5 \text{ g}$$

We do *not* round the answer to just one significant figure because of the 3. In
this case, 3 is an *exact* number determined by counting.

4. In calculations with multiple steps, rounding, to the extent possible, should be
done at the end of the *final* step. Rounding the result of each step can result in
"rounding error." Consider the following two-step calculation:

$$(13.597 + 101.45) \times 7.9891 = ?$$

This calculation requires us to use the rules for both addition and multiplication.
If we were to perform the addition first, and apply the rule of keeping only the
minimum number of digits after the decimal point, the addition would yield:

13.597 $\longleftarrow$ three digits after the decimal point
+ 101.45 $\longleftarrow$ two digits after the decimal point
115.047 $\longleftarrow$ according to guideline #1 should be rounded to 115.05

Then we perform the multiplication and get

115.05 × 7.9891 = 919.145955, which rounded to five significant figures gives 919.15

However, had we not rounded the result of the addition prior to performing the
multiplication step, our final answer would have been slightly different:

13.597 $\longleftarrow$ three digits after the decimal point
+ 101.45 $\longleftarrow$ two digits after the decimal point
115.047

We recognize that the result of the addition step should have just two places after
the decimal point, giving it five significant figures. However, because it is not the
final step in the calculation, we retain all of the digits for the multiplication step:

115.047 × 7.9891 = 919.1219877, which rounded to five significant figures gives 919.12

Although the difference between 919.15 and 919.12 may seem very small, over the
course of a multistep calculation, rounding error can add up to enough difference
to make an answer wrong. When performing a multistep calculation, keep track of
how many significant figures are appropriate at each step, but do not round to the
appropriate number of significant figures until the very end of the calculation.

Sample Problem 4.8 gives you the opportunity to practice reporting calculated results
with the proper number of significant figures.

SAMPLE PROBLEM   **4.8**   **Expressing Calculated Results with the
Correct Number of Significant Figures**

Perform the following arithmetic operations and report the result to the proper number of significant figures: (a) 317.5 mL +
0.675 mL, (b) 47.80 L − 2.075 L, (c) 13.5 g ÷ 45.18 L, (d) 6.25 cm × 1.175 cm, (e) 5.46×10^2 g + 4.991×10^3 g

Strategy Apply the rules for significant figures in calculations, and round each answer to the appropriate number of digits.

Setup (a) The answer will contain one digit to the right of the decimal point to match 317.5, which has the fewest digits to the right of the decimal point. (b) The answer will contain two digits to the right of the decimal point to match 47.80. (c) The answer will contain three significant figures to match 13.5, which has the fewest number of significant figures in the calculation. (d) The answer will contain three significant figures to match 6.25. (e) To add numbers expressed in scientific notation, first write both numbers to the same power of 10. That is, $4.991 \times 10^3 = 49.91 \times 10^2$, so the answer will contain two digits to the right of the decimal point (when multiplied by 10^2) to match both 5.46 and 49.91.

Solution

(a)   317.7 mL
    +   0.675 mL
    ─────────────────
    318.175 mL ⟵ round to 318.2 mL

(b)   47.80 L
    −   2.075 L
    ─────────────────
    45.725 L ⟵ round to 45.73 L

(c) $\dfrac{13.5 \text{ g}}{45.18 \text{ L}} = 0.298804781$ g/L ⟵ round to 0.299 g/L

(d) 6.25 cm × 1.175 cm = 7.34375 cm² ⟵ round to 7.34 cm²

(e) 5.46×10^2 g
    + 49.91×10^2 g
    ─────────────────────────
    55.37×10^2 g = 5.537×10^3 g

THINK ABOUT IT

It may look as though the rule of addition has been violated in part (e) because the final answer (5.537×10^3 g) has three places past the decimal point, not two. However, the rule was applied to get the answer 55.37×10^2 g, which has *four* significant figures. Changing the answer to correct scientific notation doesn't change the number of significant figures, but in this case it changes the number of places past the decimal point.

Practice Problem Ⓐ**TTEMPT** Perform the following arithmetic operations, and report the result to the proper number of significant figures: (a) 105.5 L + 10.65 L, (b) 81.058 m − 0.35 m, (c) 3.801×10^{21} atoms + 1.228×10^{19} atoms, (d) 1.255 dm × 25 dm, (e) 139 g ÷ 275.55 mL

Practice Problem Ⓑ**UILD** Perform the following arithmetic operations, and report the result to the proper number of significant figures: (a) 1.0267 cm × 2.508 cm × 12.599 cm, (b) 15.0 kg ÷ 0.036 m³, (c) 1.113×10^{10} kg − 1.050×10^9 kg, (d) 25.75 mL + 15.00 mL, (e) 46 cm³ + 180.5 cm³

Practice Problem Ⓒ**ONCEPTUALIZE** A citrus dealer in Florida sells boxes of 100 oranges at a roadside stand. The boxes routinely are packed with one to three extra oranges to help ensure that customers are happy with their purchases. The average weight of an orange is 7.2 ounces, and the average weight of the boxes in which the oranges are packed is 3.2 pounds. Determine the total weight of five of these 100-orange boxes.

Ken Welsh/Pixtal/age fotostock

CHECKPOINT–SECTION 4.3 Significant Figures

4.3.1 What volume of water does the graduated cylinder contain (to the proper number of significant figures)?

a) 32.2 mL

b) 30.25 mL

c) 32.5 mL

d) 32.50 mL

e) 32.500 mL

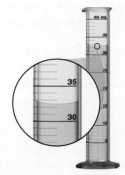

4.3.2 Which of the following is the sum of the following numbers to the correct number of significant figures?

3.115 + 0.2281 + 712.5 + 45 =

a) 760.8431   d) 760.8

b) 760.843   e) 761

c) 760.84

4.3.3 What is the result of the following calculation to the correct number of significant figures?

(6.266 − 6.261) ÷ 522.0 =

a) 9.5785×10^{-6} d) 9.6×10^{-6}

b) 9.579×10^{-6} e) 1×10^{-5}

c) 9.58×10^{-6}

4.3.4 The two volumes of water are poured together into a beaker for an experiment. What total volume should be reported for the water?

a) 1.2×10^2 mL

d) 119.00 mL

b) 119 mL

e) 119.000 mL

c) 119.0 mL

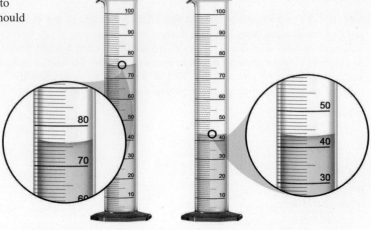

4.4 Unit Conversion

Much of what you do in this course will involve problem solving. Solving chemistry problems correctly requires careful manipulation of both numbers and units. Paying close attention to *units* will benefit you greatly as you proceed through this or any other science course.

Conversion Factors

A **conversion factor** is a fraction in which the same quantity is expressed one way in the numerator and another way in the denominator. By definition, for example, 1 inch is equal to 2.54 centimeters:

$$1 \text{ in} \equiv 2.54 \text{ cm}$$

We can derive a conversion factor from this equality by writing it as the following fraction:

$$\frac{1 \text{ in}}{2.54 \text{ cm}}$$

Because the numerator and denominator express the same length, this fraction is equal to 1. As a result, we can equally well write the conversion factor as

$$\frac{2.54 \text{ cm}}{1 \text{ in}}$$

Because both forms of this conversion factor are equal to 1, we can multiply a quantity by either form without changing the value of that quantity. This is useful for changing the units in which a given quantity is expressed—something you will do often throughout this text. For instance, if we need to convert a length of 12.00 inches to centimeters, we multiply the length in inches by the appropriate conversion factor.

$$12.00 \text{ in} \times \frac{2.54 \text{ cm}}{1 \text{ in}} = 30.48 \text{ cm}$$

Student Note: Remember that conversion factors such as those used in temperature conversions (9/5, 5/9, 32) are *exact* numbers. They do not limit the number of significant figures in the results they are used to calculate.

The units of inches cancel, and we are left with units of centimeters. (We specifically chose the form of the conversion factor that would cancel inches and give us the desired unit, centimeters.) Note that the result contains *four* significant figures because exact numbers, such as those obtained from definitions, do not limit the number of significant figures in the result of a calculation. Thus, the number of significant figures in the answer to this calculation is based on the number 12.00, not the number 2.54.

Table 4.3 revisits the metric multipliers introduced in Section 4.1 and gives the conversion factors necessary to convert between SI units that differ by orders of magnitude.

TABLE 4.3	Conversion Factors Based on Metric Multipliers			
Metric Multiplier	**Meaning**		**Conversion Factors**	
tera (T)	1×10^{12}	$\dfrac{1\ \text{Tg}}{1 \times 10^{12}\ \text{g}}$	or	$\dfrac{1 \times 10^{12}\ \text{g}}{1\ \text{Tg}}$
	for conversion from	g → Tg	or from	Tg → g
giga (G)	1×10^{9}	$\dfrac{1\ \text{GW}}{1 \times 10^{9}\ \text{W}}$	or	$\dfrac{1 \times 10^{9}\ \text{W}}{1\ \text{GW}}$
	for conversion from	W → GW	or from	GW → W
mega (M)	1×10^{6}	$\dfrac{1\ \text{MHz}}{1 \times 10^{6}\ \text{Hz}}$	or	$\dfrac{1 \times 10^{6}\ \text{Hz}}{1\ \text{MHz}}$
	for conversion from	Hz → MHz	or from	MHz → Hz
kilo (k)	1×10^{3}	$\dfrac{1\ \text{km}}{1 \times 10^{3}\ \text{m}}$	or	$\dfrac{1 \times 10^{3}\ \text{m}}{1\ \text{km}}$
	for conversion from	m → km	or from	km → m
deci (d)	1×10^{-1}	$\dfrac{1 \times 10^{1}\ \text{dL}}{1\ \text{L}}$	or	$\dfrac{1\ \text{L}}{1 \times 10^{1}\ \text{dL}}$
	for conversion from	L → dL	or from	dL → L
centi (c)	1×10^{-2}	$\dfrac{1 \times 10^{2}\ \text{cm}}{1\ \text{m}}$	or	$\dfrac{1\ \text{m}}{1 \times 10^{2}\ \text{cm}}$
	for conversion from	m → cm	or from	cm → m
milli (m)	1×10^{-3}	$\dfrac{1 \times 10^{3}\ \text{ms}}{1\ \text{s}}$	or	$\dfrac{1\ \text{s}}{1 \times 10^{3}\ \text{ms}}$
	for conversion from	s → ms	or from	ms → s
micro (μ)	1×10^{-6}	$\dfrac{1 \times 10^{6}\ \mu\text{m}}{1\ \text{m}}$	or	$\dfrac{1\ \text{m}}{1 \times 10^{6}\ \mu\text{m}}$
	for conversion from	m → μm	or from	μm → m
nano (n)	1×10^{-9}	$\dfrac{1 \times 10^{9}\ \text{ns}}{1\ \text{s}}$	or	$\dfrac{1\ \text{s}}{1 \times 10^{9}\ \text{ns}}$
	for conversion from	s → ns	or from	ns → s

Sample Problem 4.9 lets you practice using conversion factors to convert from one unit to another.

SAMPLE PROBLEM (**4.9**) Using Conversion Factors to Convert from One Unit to Another

Perform each of the following conversions using the conversion factors shown in Table 4.3:

(a) Convert 4.5×10^{5} m to km.   (b) Convert 3.78×10^{6} mg to g.   (c) Convert 8.22×10^{8} μL to L.

Strategy Find the relevant metric multiplier prefixes and the corresponding conversion factors in Table 4.3.

Setup Write the two versions of each metric multiplier for each conversion and determine which is the correct one to use in the given conversion.

Solution

(a) $\dfrac{1\ \text{km}}{1000\ \text{m}}$ or $\dfrac{1000\ \text{m}}{1\ \text{km}}$ $\qquad 4.5 \times 10^{5}\ \cancel{\text{m}} \times \dfrac{1\ \text{km}}{1000\ \cancel{\text{m}}} = 4.5 \times 10^{2}\ \text{km}$

(b) $\dfrac{1 \times 10^{3}\ \text{mg}}{1\ \text{g}}$ or $\dfrac{1\ \text{g}}{1 \times 10^{3}\ \text{mg}}$ $\qquad 3.78 \times 10^{6}\ \cancel{\text{mg}} \times \dfrac{1\ \text{g}}{1000\ \cancel{\text{mg}}} = 3.78 \times 10^{3}\ \text{g}$

(c) $\dfrac{1\ \text{L}}{1 \times 10^{6}\ \mu\text{L}}$ or $\dfrac{1 \times 10^{6}\ \mu\text{L}}{1\ \text{L}}$ $\qquad 8.22 \times 10^{8}\ \cancel{\mu\text{L}} \times \dfrac{1\ \text{L}}{1 \times 10^{6}\ \cancel{\mu\text{L}}} = 8.22 \times 10^{2}\ \text{L}$

(Continued on next page)

THINK ABOUT IT

The selected form of the conversion factor must have the new unit on the top (numerator) and the unit that you are converting from on the bottom (denominator). You should also note that the conversion factors are exact and do not limit the significant figures in the reported answers.

Practice Problem **A**TTEMPT   Perform each of the following conversions using the conversion factors shown in Table 4.3:

(a) Convert 2.10×10^2 Gg to g.   (b) Convert 9.31×10^9 nL to L.   (c) Convert 5.88×10^7 m to Mm.

Practice Problem **B**UILD   Given each relationship, write the two versions of a conversion factor. Select the correct conversion factor and perform the conversion.

(a) Convert 135 lb to kg. (1 kg = 2.205 lb)   (c) Convert 8.1 quarts to L. (1 L = 1.057 quarts)

(b) Convert 15.9 yards to m. (1.094 yards = 1 m)

Practice Problem **C**ONCEPTUALIZE   For each of the following conversion factors, write three more that are equivalent:

(a) $\dfrac{1 \times 10^9 \text{ nm}}{1 \text{ m}}$ (b) $\dfrac{1 \text{ cm}}{0.394 \text{ in}}$ (c) $\dfrac{1 \text{ } \mu g}{1 \times 10^{-6} \text{ g}}$

In the United States, most consumer products that list a volume or a weight (such as beverages and food) display them in both English and SI units. A 12-oz can of soda, for example, also gives the volume as 355 mL. And a 1-lb box of macaroni also gives the weight as 454 g. Table 4.4 lists some of the most commonly used conversions.

TABLE 4.4	Conversions between SI Units and English Units		
Kilograms (kg) and pounds (lb)	1 kg = 2.20462 lb	1 lb = 0.453592 kg
Grams (g) and pounds (lb)	1 g = 0.002205 lb	1 lb = 453.592 g
Grams (g) and ounces (oz)	1 g = 0.035274 oz	1 oz = 28.3495 g
Kilometers (km) and miles (mi)	1 km = 0.621371 mi	1 mi = 1.60934 km
Meters (m) and feet (ft)	1 m = 3.28084 ft	1 ft = 0.3048 m
Centimeters (cm) and inches (in)	1 cm = 0.393701 in	1 in ≡ 2.54 cm
Liters (L) and gallons (gal)	1 L = 0.264172 gal	1 gal = 3.78541 L
Milliliters (mL) and fluid ounces (oz)	1 mL = 0.033814 oz	1 oz = 29.5735 mL

Sample Problem 4.10 lets you practice converting traditional English units to SI units.

SAMPLE PROBLEM 4.10

Convert each quantity to the equivalent quantity in SI units.

(a) 1.25 lb   (b) 68 ft   (c) 3.25 gal   (d) 315 mi   (e) 32.5 fluid oz

Strategy We select the appropriate conversion for each from Table 4.4.

Setup For each part, we must multiply by a conversion factor that cancels the English unit and introduces the SI unit. Part (a) requires the conversion from pounds to grams or pounds to kilograms. The necessary conversion factors are:

$$\frac{453.592 \text{ g}}{1 \text{ lb}} \quad \text{and} \quad \frac{0.453592 \text{ kg}}{1 \text{ lb}}$$

Part (b) requires conversion from feet to meters using the conversion factor:

$$\frac{0.3048 \text{ m}}{1 \text{ ft}}$$

Part (c) requires conversion from gallons to liters using the conversion factor:

$$\frac{3.78541 \text{ L}}{1 \text{ gal}}$$

Part (d) requires conversion from miles to kilometers using the conversion factor:

$$\frac{1.60934 \text{ km}}{1 \text{ mi}}$$

Part (e) requires conversion from fluid ounces to milliliters using the conversion factor:

$$\frac{29.5735 \text{ mL}}{1 \text{ oz}}$$

Solution

(a) $1.25 \text{ lb} \times \dfrac{453.592 \text{ g}}{1 \text{ lb}} = 567 \text{ g}$ and $1.25 \text{ lb} \times \dfrac{0.453592 \text{ kg}}{1 \text{ lb}} = 0.567 \text{ kg}$ (b) $68 \text{ ft} \times \dfrac{0.3048 \text{ m}}{1 \text{ ft}} = 21 \text{ m}$

(c) $3.25 \text{ gal} \times \dfrac{3.78541 \text{ L}}{1 \text{ gal}} = 12.3 \text{ L}$ (d) $315 \text{ mi} \times \dfrac{1.60934 \text{ km}}{1 \text{ mi}} = 507 \text{ km}$ (e) $32.5 \text{ oz} \times \dfrac{29.5735 \text{ mL}}{1 \text{ oz}} = 961 \text{ mL}$

THINK ABOUT IT

Although you may be more familiar and comfortable with English units, especially if you grew up in the United States, it is a good idea to try and develop a sense of magnitude with regard to SI units. Learn your height and weight in the appropriate SI units, and familiarize yourself with the common weights and volumes of foods and beverages.

Practice Problem **TTEMPT** Convert each quantity to the indicated SI unit.

(a) 16.0 oz to g   (b) 11.55 in to cm   (c) 575 mi to km   (d) 172 in to m   (e) 375 fluid oz to L

Practice Problem **B**UILD Convert each quantity to the indicated English unit.

(a) 10.75 kg to lb   (b) 10.0 km to mi   (c) 750.0 L to gal   (d) 15.5 L to fluid oz   (e) 0.675 m to in

Practice Problem **C**ONCEPTUALIZE How many significant figures should there be in the answer to Practice Problem A part (b)? Explain.

Thinking Outside the Box

The Importance of Units

On December 11, 1998, NASA launched the *Mars Climate Orbiter,* which was intended to be the Red Planet's first weather satellite. After a 416-million-mile journey, the spacecraft was supposed to go into orbit around Mars on September 23, 1999. This never happened, though, and the *Climate Orbiter* was lost. Mission controllers later determined the cause to have been failure to convert English units to metric units in the navigation software.

As the *Mars Climate Orbiter* approached Mars, thrust engines were supposed to alter its course so that it entered the atmosphere more than 200 km above the planet's surface as shown in the planned trajectory. Engineers at Lockheed Martin Corporation, where the spacecraft was built, had specified its thrust in *pounds* (lb), which is a familiar English unit of force. Scientists at NASA's Jet Propulsion Laboratory, who were responsible for deployment, thought the data they were given expressed the thrust in *newtons* (N), a derived metric unit. The conversion between these two units is:

$$1 \text{ lb} = 4.45 \text{ N}$$

Failure to convert the thrust from pounds to newtons resulted in the thrust being less than one-fourth what it should have been, resulting in the actual trajectory shown in the picture. The 125-million-dollar *Orbiter*

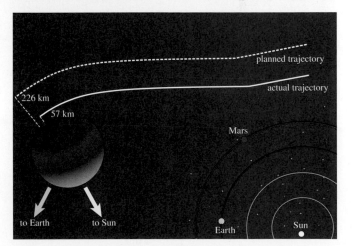

entered Mars's atmosphere far closer to the planet than had been planned and was destroyed by heat.

According to the U.S. Metric Association (USMA), the United States is "the only significant holdout" with regard to adoption of the metric system. The other countries that continue to use traditional units are Myanmar (formerly Burma) and Liberia.

Derived Units

There are many quantities that cannot be expressed using SI base units. One such quantity that we have encountered already is *volume*. Another is *density*. In each of these cases, we can combine base units to derive appropriate units for the quantity.

The derived SI unit for volume, the meter cubed (m^3), is a much larger volume than is practical in most laboratory settings. The more commonly used metric unit, the liter (L), is derived by cubing the decimeter (dm). Figure 4.7 illustrates the relationship between the liter and the *milliliter,* which is also known as the *cubic centimeter* (cm^3).

Density is the ratio of mass to volume. Oil floats on water, for example, because it has a lower density than water. That is, equal volumes of oil and water have different masses. For a given volume, the water has a greater mass. Additionally, because oil and water do not mix, they form distinct layers with water at the bottom and oil at the top.

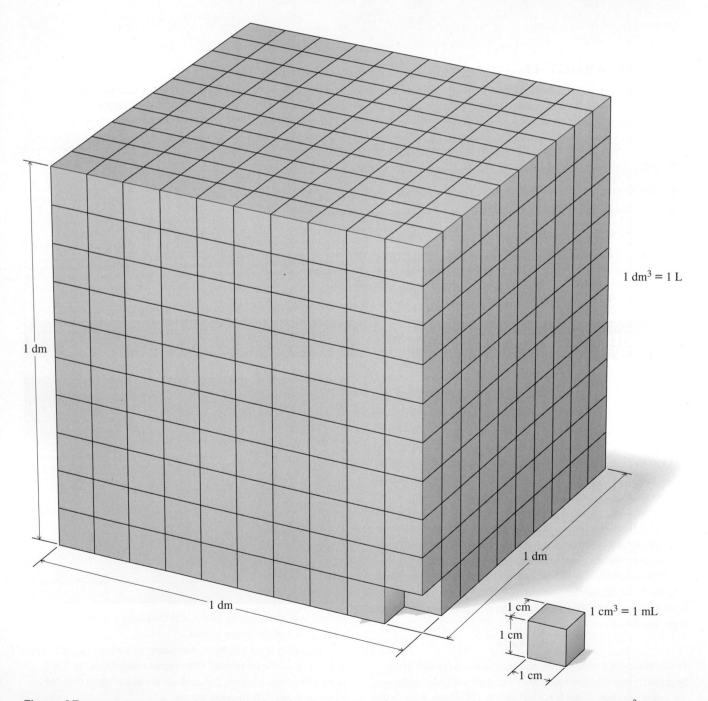

Figure 4.7 The larger cube has 1-dm (10-cm) sides and a volume of 1 L. The smaller cube has 1-cm sides and a volume of 1 cm^3 or 1 mL.

Density can be calculated as

$$d = \frac{m}{V}$$

Equation 4.4

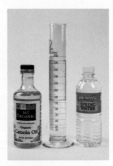

David A. Tietz/Editorial Image, LLC

where d is density, m is mass, and V is volume. The unit derived from SI base units for density is kilogram per cubic meter (kg/m^3), but like the derived unit m^3 for volume, this is not a very practical unit for everyday use. Instead, we use g/cm^3 and its equivalent g/mL to express the densities of most solids and liquids. Water, for example, has a density of 1.00 g/cm^3 at 4°C. Because gas densities tend to be much lower than those of liquids and solids, we typically express them in units of grams per liter (g/L).

Sample Problem 4.11 illustrates the calculation of density of a substance from mass and volume.

SAMPLE PROBLEM (4.11) Calculating Density from Mass and Volume

Ice cubes float in a glass of water because solid water is less dense than liquid water. (a) Calculate the density of ice given that, at 0°C, a cube that is 2.0 cm on each side has a mass of 7.36 g, and (b) determine the volume occupied by 23 g of ice at 0°C.

Strategy (a) Determine density by dividing mass by volume (Equation 4.4), and (b) use the calculated density to determine the volume occupied by the given mass.

Setup (a) We are given the mass of the ice cube, but we must calculate its volume from the dimensions given. The volume of the ice cube is $(2.0 \text{ cm})^3$, or 8.0 cm^3. (b) Rearranging Equation 4.4 to solve for volume gives $V = \frac{m}{d}$.

Solution

$$d = \frac{7.36 \text{ g}}{8.0 \text{ cm}^3} = 0.92 \text{ g/cm}^3 \quad \text{or} \quad 0.92 \text{ g/mL}$$

$$V = 23 \text{ g} \times \frac{1 \text{ cm}^3}{0.92 \text{ g}} = 25 \text{ cm}^3 \quad \text{or} \quad 25 \text{ mL}$$

THINK ABOUT IT

For a sample with a density less than 1 g/cm^3, the number of cubic centimeters should be greater than the number of grams. In this case, 25 (cm^3) > **23 (g).**

Practice Problem (A)TTEMPT Given that 25.0 mL of mercury has a mass of 340.0 g, calculate (a) the density of mercury and (b) the mass of 120.0 mL of mercury.

Practice Problem (B)UILD Calculate (a) the density of a solid substance if a cube measuring 2.33 cm on one side has a mass of 117 g and (b) the mass of a cube of the same substance measuring 7.41 cm on one side.

Practice Problem (C)ONCEPTUALIZE Using the picture of the graduated cylinder and its contents, arrange the following in order of increasing density: blue liquid, pink liquid, yellow liquid, grey solid, blue solid, green solid.

Student Note: A related term, *specific gravity,* refers to the *ratio* of the density of a substance to the density of water. Specific gravity is used in urinalysis to detect evidence of urinary-tract infection, dehydration, and other medical conditions.

Familiar Chemistry

The International Unit

The units and conversion factors discussed in this chapter ultimately are used to share data that can be interpreted easily by others. One set of units that you have probably encountered are *International Units,* or IUs—typically found on vitamin supplement labels. Unlike the metric and English units we have seen in this chapter, the "International Unit" does not have a standard definition. Instead, it is defined differently for each supplement, and refers to the amount of a substance believed to provide a certain level of desired biological effect. For example, one IU of vitamin D_3 is equivalent to 0.025 μg of cholecalciferol, whereas one IU of vitamin A is equivalent to 0.3 μg of retinol (the active form of vitamin A) or to 0.6 μg of beta-carotene (a "pro-vitamin" from which the body synthesizes the active form).

Supplement Facts
Serving Size 1 Tablet

Each Tablet Contains	% DV	Each Tablet Contains	% DV
Vitamin A 3500 IU	70%	Biotin 30 mcg	10%
(14% as beta-carotene)		Pantothenic Acid 5 mg	50%
Vitamin C 60 mg	100%	Calcium 210 mg	21%
Vitamin D₃ 700 IU	175%	Magnesium 120 mg	30%
Vitamin E 22.5 IU	75%	Zinc 15 mg	100%
Vitamin K 20 mcg	25%	Selenium 110 mcg	157%
Thiamin 1.2 mg	80%	Copper 2 mg	100%
Riboflavin 1.7 mg	100%	Manganese 2 mg	100%
Niacin 16 mg	80%	Chromium 120 mcg	100%
Vitamin B₆ 3 mg	150%	Lycopene 300 mcg	
Folate 400 mcg	100%		
Vitamin B₁₂ 18 mcg	300%	*Daily Value (DV) not established.	

Suggested Use: Adults - Take one tablet daily with food as a dietary supplement.

SAFETY SEALED. DO NOT USE IF PRINTED SEAL UNDER CAP IS CUT, TORN, OR MISSING.

Keep out of reach of children. Store at 15° - 30° C (59° - 86° F).

Warning: Individuals taking medication(s) or persons who have a health condition should consult their physician before using this product.

†One A Day® Men's Health Formula is distributed by Bayer HealthCare LLC.

David A. Tietz/McGraw Hill

To determine the *mass* of vitamin D_3 per serving, according to the label shown here, we would construct a conversion factor using the definition of the IU for vitamin D_3:

$$1\ IU \equiv 0.025\ \mu g\ cholecalciferol$$

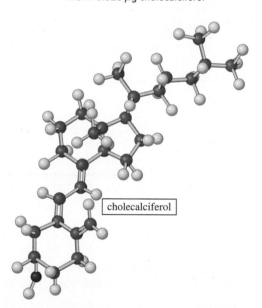

cholecalciferol

Our conversion factor can be written either as

$$\frac{1\ IU\ vitamin\ D_3}{0.025\ \mu g\ cholecalciferol} \quad or \quad \frac{0.025\ \mu g\ cholecalciferol}{1\ IU\ vitamin\ D_3}$$

We choose the version that enables us to cancel units appropriately, and calculate the mass of cholecalciferol per serving of this supplement as:

$$400\ \cancel{IU}\ vitamin\ D_3\ per\ serving \times \frac{0.025\ \mu g\ cholecalciferol}{1\ \cancel{IU}\ vitamin\ D_3} = 10\ \mu g\ cholecalciferol\ per\ serving$$

Dimensional Analysis

The use of conversion factors in problem solving is called *dimensional analysis.* Many problems require the use of more than one conversion factor. The conversion of 12.00 inches to meters, for example, takes two steps: one to convert inches to centimeters, which we have already seen, and one to convert centimeters to meters. The additional conversion factor required is derived from the equality

$$1 \text{ m} = 100 \text{ cm}$$

and, as shown in Table 4.3, can be expressed as either

$$\frac{1 \times 10^2 \text{ cm}}{1 \text{ m}} \quad \text{or} \quad \frac{1 \text{ m}}{1 \times 10^2 \text{ cm}}$$

We must choose the conversion factor that will introduce the unit *meter* and cancel the unit *centimeter*. (That would be the one on the right.) We can set up a problem of this type as the following series of unit conversions so that it is unnecessary to calculate an intermediate answer at each step:

$$12.00 \text{ in} \times \frac{2.54 \text{ cm}}{1 \text{ in}} \times \frac{1 \text{ m}}{1 \times 10^2 \text{ cm}} = 0.3048 \text{ m}$$

Careful tracking of units and their cancellation can be a valuable tool in checking your work. If we had accidentally used the reciprocal of one of the conversion factors, the resulting units would have been something other than meters—and would have made no sense. For example, if we had accidentally used the reciprocal of the conversion from cm to m, the result would have been 3048 cm^2/m, which would make no sense—both because the units are nonsensical, and because the magnitude of the numerical result is not at all reasonable. You know intuitively that 12 inches is a foot and that a foot is not equal to thousands of meters.

Sample Problem 4.12 shows how to do multistep dimensional analysis with careful tracking of units.

> **Student Note:** Unexpected or nonsensical units can reveal an error in your problem-solving strategy, and can help you find your mistakes before they cost you points on an exam.

SAMPLE PROBLEM (4.12) Multistep Dimensional Analysis

The Food and Drug Administration (FDA) recommends that dietary sodium intake be no more than 2400 mg per day. What is this mass in pounds (lb), if 1 lb = 453.6 g?

Strategy This problem requires a two-step dimensional analysis because we must convert milligrams to grams and then grams to pounds. Assume the number 2400 has four significant figures.

Setup The necessary conversion factors are derived from the equalities 1 g = 1000 mg and 1 lb = 453.6 g.

$$\frac{1 \text{ g}}{1000 \text{ mg}} \quad \text{or} \quad \frac{1000 \text{ mg}}{1 \text{ g}} \quad \text{and} \quad \frac{1 \text{ lb}}{453.6 \text{ g}} \quad \text{or} \quad \frac{453.6 \text{ g}}{1 \text{ lb}}$$

From each pair of conversion factors, we select the one that will result in the proper unit cancellation.

Solution

$$2400 \text{ mg} \times \frac{1 \text{ g}}{1000 \text{ mg}} \times \frac{1 \text{ lb}}{453.6 \text{ g}} = 0.005291 \text{ lb}$$

THINK ABOUT IT

Make sure that the magnitude of the result is reasonable and that the units have canceled properly. Because pounds are much larger than milligrams, a given mass will be a much smaller number of pounds than of milligrams. If we had mistakenly multiplied by 1000 and 453.6 instead of dividing by them, the result (2400 mg × 1000 mg/g × 453.6 g/lb = 1.089 × 10^9 mg^2/lb) would be unreasonably large—and the units would not have canceled properly.

(Continued on next page)

Practice Problem **ATTEMPT** The American Heart Association recommends that healthy adults limit dietary cholesterol to no more than 300 mg per day. Convert this mass of cholesterol to ounces (1 oz = 28.3459 g). Assume 300 mg has just one significant figure.

Practice Problem **BUILD** An object has a mass of 24.98 oz. What is its mass in grams?

Practice Problem **CONCEPTUALIZE** The diagram contains several objects that are constructed using colored blocks and grey connectors. Note that each of the objects is essentially identical, consisting of the same number and arrangement of blocks and connectors. Give the appropriate conversion factor for each of the specified operations.

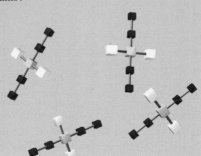

(a) We know the number of objects and wish to determine the number of red blocks.

(b) We know the number of yellow blocks and wish to determine the number of objects.

(c) We know the number of yellow blocks and wish to determine the number of white blocks.

(d) We know the number of grey connectors and wish to determine the number of yellow blocks.

Many familiar quantities require units raised to specific powers. For example, an area is expressed in units of length squared. Examples include square inches (in^2) and square meters (m^2). Volumes can be expressed in units of length *cubed*. Examples include cubic feet (ft^3) and cubic centimeters (cm^3). More often, though, volumes are expressed in liters (L) or milliliters (mL). It is important to realize that *milliliter* and *liter* are the common names given to specific units of length cubed. The liter is defined as a decimeter (dm) cubed: 1 L ≡ 1 dm^3; and the milliliter is defined as a centimeter cubed: 1 mL ≡ 1 cm^3. (See Figure 4.7.) Special care must be taken when dimensional analysis involves units raised to powers. For example, converting from cubic meters to cubic centimeters requires the following operation:

$$1 \, \cancel{m^3} \times \frac{100 \text{ cm}}{1 \, \cancel{m}} \times \frac{100 \text{ cm}}{1 \, \cancel{m}} \times \frac{100 \text{ cm}}{1 \, \cancel{m}} = 1.00 \times 10^6 \text{ cm}^3$$

or

$$1 \, \cancel{m^3} \times \left(\frac{100 \text{ cm}}{1 \, \cancel{m}}\right)^3 = 1.00 \times 10^6 \text{ cm}^3$$

Forgetting to raise the conversion factor to the same power as the unit itself is a common error.

Sample Problem 4.13 lets you practice dimensional analysis with units raised to powers.

Student Hot Spot

Student data indicate that you may struggle to convert units that contain squared or cubed units. Access the eBook to view additional Learning Resources on this topic.

SAMPLE PROBLEM (4.13) Dimensional Analysis with Units Raised to Powers

An average adult has 5.2 L of blood. What is the volume of blood in cubic meters?

Strategy There are several ways to solve a problem such as this. One way is to convert liters to cubic centimeters and then cubic centimeters to cubic meters.

Setup 1 L = 1000 cm^3 and 1 cm = 1 × 10^{-2} m. When a unit is raised to a power, the corresponding conversion factor must also be raised to that power in order for the units to cancel appropriately.

Solution

$$5.2 \, \cancel{L} \times \frac{1000 \, \cancel{cm^3}}{1 \, \cancel{L}} \times \left(\frac{1 \times 10^{-2} \text{ m}}{1 \, \cancel{cm}}\right)^3 = 5.2 \times 10^{-3} \text{ m}^3$$

THINK ABOUT IT

Based on the preceding conversion factors, 1 L = 1 × 10^{-3} m^3. Therefore, 5 L of blood would be equal to 5 × 10^{-3} m^3, which is close to the calculated answer.

Practice Problem **A**TTEMPT The density of silver is 10.5 g/cm³. What is its density in kg/m³?

Practice Problem **B**UILD The density of mercury is 13.6 g/cm³. What is its density in mg/mm³?

Practice Problem **C**ONCEPTUALIZE Each diagram [(i) or (ii)] shows the objects contained within a cubical space. In each case, determine to the appropriate number of significant figures the number of objects that would be contained within a cubical space in which the length of the cube's edge is exactly five times that of the cube shown in the diagram.

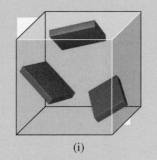

(i)

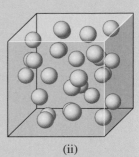

(ii)

CHECKPOINT–SECTION 4.4 Unit Conversion

4.4.1 The density of lithium metal is 535 kg/m³. What is this density in g/cm³?

a) 0.000535 g/cm³

b) 0.535 g/cm³

c) 0.0535 g/cm³

d) 0.54 g/cm³

e) 53.5 g/cm³

4.4.2 Convert 43.1 cm³ to liters.

a) 43.1 L

b) 43,100 L

c) 0.0431 L

d) 4310 L

e) 0.431 L

4.4.3 What is the volume of a 5.75-g object that has a density of 3.97 g/cm³?

a) 1.45 cm³

b) 0.690 cm³

c) 22.8 cm³

d) 0.0438 cm³

e) 5.75 cm³

4.4.4 How many cubic centimeters are there in a cubic meter?

a) 10

b) 100

c) 1000

d) 1×10^4

e) 1×10^6

4.4.5 Convert 1.75 lb to grams.

a) 0.794 g

b) 0.00386 g

c) 794 g

d) 1750 g

e) 3.86 g

4.4.6 Convert 0.255 g/mL to pounds per gallon.

a) 0.149 lb/gal

b) 1490 lb/gal

c) 255 lb/gal

d) 2.13 lb/gal

e) 47.2 lb/gal

4.5 Success in Introductory Chemistry Class

Success in a chemistry class depends largely on the development of problem-solving skills. The Sample Problems throughout this text are designed specifically to help you develop those skills. As you have seen, each Sample Problem is divided into four steps: *Strategy, Setup, Solution,* and *Think About It.*

Strategy

Read the problem carefully and determine what is being asked and what information is provided. The Strategy step is where you should think about what skills are required and lay out a plan for solving the problem. Give some thought to what you expect the result to be. If you are asked to determine the number of atoms in a sample of matter, for example, you should expect the answer to be a whole number. Determine what, if any, units should be associated with the result. When possible, make a ballpark estimate of the magnitude of the correct result, and make a note of your estimate. If your calculated answer is "outside the ballpark" (wrong units, wrong magnitude, wrong sign), check your work carefully and try to figure out where you might have gone wrong.

Setup

Next, gather the information necessary to solve the problem. Some of the information will have been given in the problem itself. Other information, such as equations, constants, and tabulated data (including atomic masses), should also be brought together in this step. Write down and label clearly all of the information you will use to solve the problem. Be sure to write the appropriate unit with each piece of information.

Solution

Using the necessary equations, constants, and other information, calculate the answer to the problem. Pay particular attention to the units associated with each number, tracking and canceling units carefully throughout the calculation. In the event that multiple calculations are required, label any intermediate results, but don't round to the necessary number of significant figures until the final calculation. Always carry at least one extra digit in intermediate calculations to avoid rounding error—and make sure that the final answer has the correct number of significant figures.

Think About It

Consider your calculated result and ask yourself whether or not it makes sense. Compare the units and the magnitude of your result with your ballpark estimate from the Strategy step. If your result does not have the appropriate units, or if its magnitude or sign is not reasonable, check your solution for possible errors. A very important part of problem solving is being able to judge whether the answer is reasonable. It is relatively easy to spot a wrong sign or incorrect units, but you should also develop a sense of magnitude and be able to tell when an answer is either way too big or way too small. For example, if a problem asks how many molecules are in a sample and you calculate a number that is less than 1, you should know that it cannot be correct.

Each Sample Problem is followed by three Practice Problems. The first, "Attempt," typically is a very similar problem that can be solved using the same strategy as that shown in the Sample Problem. The second and third, "Build" and "Conceptualize," generally test the same skills but require approaches slightly different from the one used to solve the preceding Sample and Practice Problems. Regular use of the Sample Problems and Practice Problems in this text can help you develop an effective set of problem-solving skills. They can also help you assess whether you are ready to move on to the next new concepts. If you struggle with the Practice Problems, then you probably need to review the corresponding Sample Problem and the concepts that led up to it.

Finally, each section that contains Sample and Practice Problems ends with a set of multiple-choice *Checkpoint* problems. These problems give you frequent opportunities for self-assessment and can help you gauge your readiness to move on to the next section.

Chemistry is a required part of many college degree programs precisely because of the critical-thinking and problem-solving skills that it teaches and requires. Regular problem-solving practice on your part will greatly enhance your chances of success in this class and any degree program of which it is a part.

Chapter Summary

Section 4.1

- Scientists use a system of units referred to as the ***International System of Units*** or ***SI units***.

- ***Mass*** is a measure of the amount of matter in an object or a sample. The SI unit for mass is the kilogram (kg). SI units for length and time are the meter (m) and the second (s). Greek prefixes (metric multipliers) are used to make the magnitude of the unit appropriate for a measurement.

- Two temperature scales are used in science: the ***Celsius scale*** and the ***Kelvin scale*** or ***absolute temperature scale.*** The units of temperature for these two scales are the *degree Celsius (°C)* and the ***kelvin (K),*** and the two units are equal in magnitude. Temperatures on the Kelvin scale are written *without* a degree sign.

Section 4.2

- *Scientific notation* is used to express very small or very large numbers as $N \times 10^n$, where N is a number between 1 and 10, and n is an exponent that indicates the position of the decimal point.

- It is important to learn and practice using the scientific notation function on your particular model of calculator.

Section 4.3

- Chemists use two types of numbers: *exact numbers* and measured numbers. An exact number results from counting objects, or from a definition such as 1 ft ≡ 12 in, or 1 in ≡ 2.54 cm.

- A measured number contains uncertainty. *Significant figures* are used to specify the uncertainty in a measured number or a number calculated using measured numbers.

- Significant figures must be carried through calculations so that the final result conveys the appropriate amount of uncertainty.

Rounding to the appropriate number of significant figures should be done at the end of a multistep calculation to avoid introducing *rounding error*.

Section 4.4

- A *conversion factor* is a fraction in which the numerator and denominator are the same quantity expressed in different units. Multiplying by a conversion factor is called *unit conversion*.

- *Density* is the ratio of mass to volume of an object or a substance. Densities of most liquids and solids are expressed as grams per milliliter (g/mL) or the equivalent grams per cubic centimeter (g/cm^3). Densities of gases are usually expressed in units of grams per liter (g/L).

- *Dimensional analysis* refers to the use of conversion factors in the solution of a multistep problem.

Key Terms

Absolute temperature scale 130	Dimensional analysis 155	Kelvin (K) 130	Scientific notation 134
Celsius scale 130	Exact number 139	Kelvin scale 130	Significant figures 140
Conversion factor 148	International System of Units (SI units) 125	Mass 126	
Density 152			

Key Equations

4.1 $K = {}^\circ C + 273$	Temperature in kelvins is determined by adding 273 to the temperature in degrees Celsius.
4.2 temp in $^\circ F = \left[\dfrac{9^\circ F}{5^\circ C} \times (\text{temp in } ^\circ C)\right] + 32^\circ F$	Temperature in degrees Fahrenheit can be calculated from a temperature in degrees Celsius.
4.3 temp in $^\circ C = (\text{temp in } ^\circ F - 32^\circ F) \times \dfrac{5^\circ C}{9^\circ F}$	Temperature in degrees Celsius can be calculated from a temperature in degrees Fahrenheit.
4.4 $d = \dfrac{m}{V}$	Density is the ratio of mass to volume. The most common units used to express density of solids and liquids are g/mL or the equivalent g/cm^3. Substances with very low densities, such as gases, may be expressed in units of g/L.

Student Note: Remember that the numbers 5, 9, and 32 are *exact* numbers in these conversions. They do not affect the number of significant figures in the results.

KEY SKILLS   Dimensional Analysis

Solving problems in chemistry often involves mathematical combinations of measured values and constants. A conversion factor is a fraction (equal to 1) derived from an equality. For example, 1 inch is, by definition, equal to 2.54 centimeters:

$$\boxed{1 \text{ in}} = \boxed{2.54 \text{ cm}}$$

We can derive two different conversion factors from this equality:

$$\boxed{\dfrac{1 \text{ in}}{2.54 \text{ cm}}} \quad \text{or} \quad \boxed{\dfrac{2.54 \text{ cm}}{1 \text{ in}}}$$

Which fraction we use depends on what units we start with, and what units we expect our result to have. If we are converting a distance given in centimeters to inches, we multiply by the first fraction:

$$\boxed{5.23 \text{ in}} \times \boxed{\dfrac{2.54 \text{ cm}}{1 \text{ in}}} = \boxed{13.3 \text{ cm}}$$

In each case, the units cancel to give the desired units in the result.

When a unit is raised to a power to express, for example, an area (cm^2) or a volume (cm^3), the conversion factor must be raised to the same power. For example, converting an area expressed in square centimeters to square inches requires that we square the conversion factor; converting a volume expressed in cubic centimeters to cubic meters requires that we cube the conversion factor. The following individual flowcharts converting an area in cm^2 to m^2 show why this is so:

$$\boxed{48.5 \text{ cm}^2} = \boxed{48.5 \text{ cm}} \times \boxed{\text{cm}} \qquad \boxed{\left(\dfrac{1 \text{ in}}{2.54 \text{ cm}}\right)^2} = \boxed{\dfrac{1 \text{ in}}{2.54 \text{ cm}}} \times \boxed{\dfrac{1 \text{ in}}{2.54 \text{ cm}}}$$

$$\boxed{48.5 \text{ cm}} \times \boxed{\text{cm}} \times \boxed{\dfrac{1 \text{ in}}{2.54 \text{ cm}}} \times \boxed{\dfrac{1 \text{ in}}{2.54 \text{ cm}}} = \boxed{7.52 \text{ in}^2}$$

$$\boxed{380.75 \text{ cm}^3} = \boxed{380.75 \text{ cm}} \times \boxed{\text{cm}} \times \boxed{\text{cm}}$$

$$\boxed{\left(\dfrac{1 \text{ m}}{100 \text{ cm}}\right)^3} = \boxed{\dfrac{1 \text{ m}}{100 \text{ cm}}} \times \boxed{\dfrac{1 \text{ m}}{100 \text{ cm}}} \times \boxed{\dfrac{1 \text{ m}}{100 \text{ cm}}}$$

$$\boxed{380.75 \text{ cm}} \times \boxed{\text{cm}} \times \boxed{\text{cm}} \times \boxed{\dfrac{1 \text{ m}}{100 \text{ cm}}} \times \boxed{\dfrac{1 \text{ m}}{100 \text{ cm}}} \times \boxed{\dfrac{1 \text{ m}}{100 \text{ cm}}} = 3.8075 \times 10^{-4} \text{ m}^3$$

Failure to raise the conversion factor to the appropriate power would result in units not canceling properly.

Often the solution to a problem requires several different conversions, which can be combined on a single line. For example: If we know that a 157-lb athlete running at 7.09 miles per hour consumes 55.8 cm³ of oxygen per kilogram of body weight for every minute spent running, we can calculate how many liters of oxygen this athlete consumes by running 10.5 miles $(1 \text{ kg} = 2.2046 \text{ lb}, 1 \text{ L} = 1 \text{ dm}^3)$:

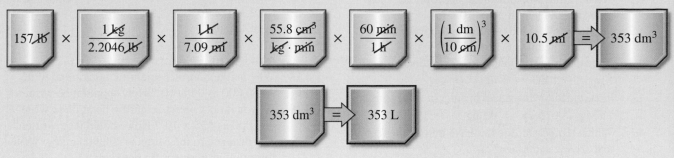

Key Skills Problems

4.1
Given that the density of gold is 19.3 g/cm³, calculate the volume (in cm³) of a gold nugget with a mass of 5.98 g.
a) 3.23 cm³
b) 5.98 cm³
c) 115 cm³
d) 0.310 cm³
e) 13.3 cm³

4.2
The SI unit for energy is the joule (J), which is equal to the kinetic energy possessed by a 2.00-kg mass moving at 1.00 m/s. Convert this velocity to mph (1 mi = 1.609 km).
a) 4.47×10^{-7} mph
b) 5.79×10^6 mph
c) 5.79 mph
d) 0.0373 mph
e) 2.24 mph

4.3
Determine the density of the following object in g/cm³:
A cube with edge length = 0.750 m and mass = 14.56 kg.
a) 0.0345 g/cm³
b) 1.74 g/cm³
c) 670 g/cm³
d) 53.8 g/cm³
e) 14.6 g/cm³

4.4
A 28-kg child can consume a maximum of 23 children's acetaminophen tablets in an 8-h period without exceeding the safety-limit maximum allowable dose. Given that each children's tablet contains 80 mg of acetaminophen, determine the maximum allowable dose in mg per pound of body weight for one day.
a) 80 mg/lb
b) 90 mg/lb
c) 430 mg/lb
d) 720 mg/lb
e) 3.7 mg/lb

Questions and Problems

SECTION 4.1: UNITS OF MEASUREMENT

4.1 What SI unit would be most appropriate when reporting the length 3.9×10^{-9} m?

4.2 What SI unit would be most appropriate when reporting the mass 6.12×10^{3} g?

4.3 Place the following values in order of increasing size:
1 mL   1 kL   1 nL

4.4 Place the following values in order of decreasing size:
15 Mg   15 μg   15 cg

4.5 Which temperature scales have units ("degrees") that are the same size?

4.6 Write the equation to convert from °F to °C.

4.7 Write the equation to convert from °C to K.

4.8 Write the equation to convert from °C to °F.

4.9 Convert the following temperatures to °C:
(a) 212°F (boiling point of water)
(b) 32°F (freezing point of water)
(c) 0 K (absolute zero)
(d) 273 K
(e) 350°F (oven temperature at which many items are baked)

4.10 Convert the following temperatures to °F:
(a) 191°C   (c) 417°C   (e) 45 K
(b) 325 K   (d) −22°C

4.11 Convert the following temperatures to K:
(a) 475°F   (c) 935°C   (e) 139°C
(b) 25°C   (d) 1299°F

4.12 Place the following in order of *increasing* temperature:
(a) 25°C   (c) 355°F   (e) 455°F
(b) 355 K   (d) 175°C

4.13 Place the following in order of *decreasing* temperature:
(a) 489 K   (c) 288°F   (e) 452°C
(b) 193°C   (d) 212 K

SECTION 4.2: SCIENTIFIC NOTATION

4.14 Write the following numbers in scientific notation:
(a) 4955   (d) 4,955,000
(b) 3,100,000   (e) 0.0004
(c) 0.000000000276

4.15 Write the following numbers in scientific notation:
(a) 1,900,000
(b) 3,450,000,000
(c) 0.0000005568
(d) 0.00028
(e) 21,000,000,000

4.16 Write the following numbers in decimal format:
(a) 3.89×10^{6} (d) 1.14×10^{-3}
(b) 1.94×10^{4} (e) 6.093×10^{5}
(c) 7.594×10^{-6}

4.17 Write the following numbers in decimal format:
(a) 9.4×10^{9}
(b) 2.751×10^{6}
(c) 9.4×10^{-9}
(d) 2.751×10^{-6}
(e) 4.8×10^{4}

4.18 Why is it not correct scientific notation to write 2,400,000 as 24×10^{5}? How should it be written?

4.19 Why is it not correct scientific notation to write 455,000 as 455×10^{3}? How should it be written?

4.20 The distance to the Moon is 238,900 miles. Write this number in scientific notation.

4.21 There are roughly 100,000 hairs on the human head. Express this value in scientific notation.

4.22 There are 9.2×10^{7} acres of land planted with corn in the United States each year. Write this number in decimal format.

SECTION 4.3: SIGNIFICANT FIGURES

4.23 What measurement should you record from the following pieces of glassware?

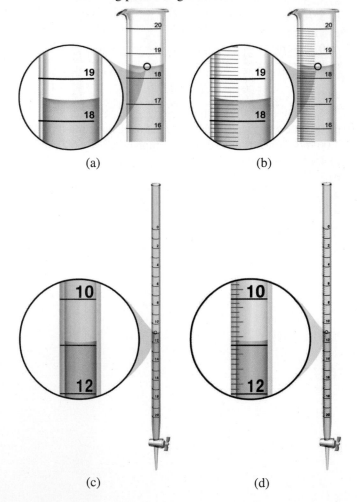

(a)   (b)

(c)   (d)

4.24 What measurement should you record for each object?

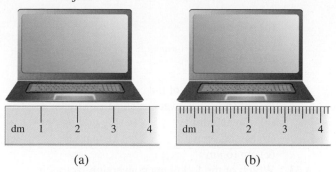

(a)   (b)

(c) Use the ruler above the object shown below.

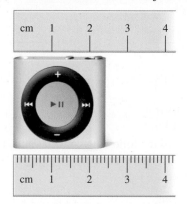

(d) Use the ruler below the object shown above.

4.25 What measurement should you record from the following balances?

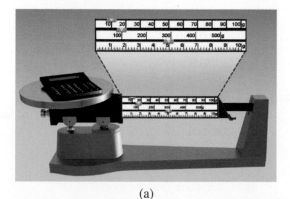

(a)

(b)

(c)

(d)

4.26 The following measurements were taken using the balances shown in Problem 4.25. Match each of the values with the balance (a–d) that was used to take the measurement.
   (a) 112.2 g
   (b) 112 g
   (c) 112.23 g
   (d) 112.233 g

4.27 How many significant figures does each of the following numbers contain?
   (a) 135   (c) 4000   (e) 8001
   (b) 833.0   (d) 4000.0

4.28 How many significant figures does each of the following numbers contain?
   (a) 0.000045
   (b) 0.0000450
   (c) 1.000045
   (d) 1.0000450
   (e) 10,000,450

4.29 How many significant figures does each of the following numbers contain?
   (a) 10.093
   (b) 10,093,000
   (c) 1.00930×10^7
   (d) 1.0093×10^7
   (e) 1.93×10^4

4.30 How many significant figures does each of the following numbers contain?
   (a) 976,001
   (b) 9.8×10^5
   (c) 0.002040
   (d) 10.02040
   (e) 9.766×10^8

4.31 Round each of the following numbers to three significant figures:
   (a) 2983   (d) 34,785.0
   (b) 21.655   (e) 0.00011136
   (c) 585.311

4.32 Round each of the following numbers to two significant figures:
   (a) 1.399×10^6 (d) 0.00000044211
   (b) 1.06333   (e) 1.00000044211
   (c) 39,569,321

4.33 Perform the following calculations and report each answer with the correct number of significant figures:
(a) $25.11 + 0.22$
(b) $493.750 - 1.00$
(c) $595.101 + 0.2325$
(d) $3696 + 0.4228$
(e) $3696 - 0.4228$

4.34 Perform the following calculations and report each answer with the correct number of significant figures:
(a) $102.1 + 0.0000051$
(b) $654 - 0.000699$
(c) $125.1 + 945.2$
(d) $1004.1 - 6.0099$
(e) $31.5 - 0.599$

4.35 Perform the following calculations and report each answer with the correct number of significant figures:
(a) 45.8×3 (d) $5.1 \div 89.3$
(b) 798.61×25.8 (e) 4005×941
(c) $3589 \div 1.5$

4.36 Perform the following calculations and report each answer with the correct number of significant figures:
(a) $1 \times 10^6 \div 61.7$
(b) 89.377×0.1000
(c) 89.377×0.0001
(d) $4099 \div 5001$
(e) $7.5 \div 0.0985$

4.37 Perform the following calculations and report each answer with the correct number of significant figures:
(a) $(4162.95 - 4161.95) \div 2253.8$
(b) $(233.995 + 1.11) \times 2.44747$
(c) $(2.75 \times 3994) - 0.114$
(d) $175.6 + (1.00 \times 10^6 \div 1.00 \times 10^{-3})$
(e) $1.756 \times 10^9 + (1.00 \times 10^6 \div 1.00 \times 10^{-3})$

4.38 Perform the following calculations and report each answer with the correct number of significant figures:
(a) $(925 \times 1.2) + (1.559 \times 10^3 - 1.559 \times 10^2)$
(b) $(925 \times 1.200) - (1.559 \times 10^3 - 1.559 \times 10^2)$
(c) $(43.6 - 0.1) \div 23.22$
(d) $(43.6 - 0.1) \times 23.22$
(e) $(23.995 + 1.1) \div 33.29$

SECTION 4.4: UNIT CONVERSION

4.39 Which of the following units are derived units? Explain.
g/mL   cm^3 ng   m^2 mm

4.40 Which of the following units are derived units? Explain.
kL   Mg   Gm   kg/L   mg/cm^3

4.41 Write both forms of the conversion factor needed for conversion between each pair of units:
(a) ng and g
(b) g and mg
(c) mL and L
(d) L and kL
(e) Mm and m

4.42 Write both forms of the conversion factor needed for conversion between each pair of units:
(a) mm and dm   (d) Gg and ng
(b) Mg and mg   (e) kg and Gg
(c) cm to μm

4.43 Which of the following conversion factors is incorrect? Correct any incorrect conversion factors.

(a) $\dfrac{1000 \text{ kg}}{1 \text{ g}}$ (d) $\dfrac{1 \times 10^2 \text{ cg}}{1 \text{ g}}$

(b) $\dfrac{1 \times 10^{-9} \text{ nL}}{1 \text{ L}}$ (e) $\dfrac{1 \times 10^6 \text{ } \mu\text{g}}{1 \text{ g}}$

(c) $\dfrac{1 \times 10^3 \text{ m}}{1 \text{ km}}$

4.44 Which of the following conversion factors is incorrect? Make corrections to any that are incorrect.

(a) $\dfrac{1 \text{ ng}}{1 \times 10^9 \text{ g}}$ (e) $\dfrac{1 \text{ } \mu\text{m}}{1 \times 10^{-9} \text{ m}}$

(b) $\dfrac{1 \text{ cL}}{1 \times 10^2 \text{ L}}$ (f) $\dfrac{1 \text{ mm}}{1 \times 10^{-3} \text{ m}}$

(c) $\dfrac{1 \text{ } \mu\text{m}}{1 \times 10^{-9} \text{ m}}$ (g) $\dfrac{1 \text{ Gg}}{1 \times 10^9 \text{ g}}$

(d) $\dfrac{1 \text{ mm}}{1 \times 10^{-3} \text{ m}}$

4.45 Which of the following conversions is set up correctly? Make corrections to any that are incorrect.
(a) Converting from mL to L.

$$2.8 \text{ mL} \times \frac{1 \text{ mL}}{1 \times 10^3 \text{ L}} = 2.8 \times 10^{-3} \text{ L}$$

(b) Converting from kg to g.

$$56 \text{ kg} \times \frac{1 \text{ g}}{1 \times 10^3 \text{ kg}} = 5.6 \times 10^{-2} \text{ g}$$

(c) Converting from μm to m.

$$7.65 \times 10^6 \text{ } \mu\text{m} \times \frac{1 \text{ m}}{1 \times 10^6 \text{ } \mu\text{m}} = 7.65 \text{ m}$$

(d) Converting from g to ng.

$$1.35 \times 10^3 \text{ g} \times \frac{1 \text{ g}}{1 \times 10^9 \text{ ng}} = 1.35 \times 10^{-6} \text{ ng}$$

(e) Converting from L to dL.

$$9.84 \times 10^3 \text{ L} \times \frac{10 \text{ dL}}{1 \text{ L}} = 9.84 \times 10^4 \text{ dL}$$

4.46 Convert the following length measurements to meters:
(a) 256 km
(b) 9.6 Mm
(c) 1.22×10^{12} nm
(d) 14,339 cm
(e) $3.55 \times 10^5 \ \mu m$

4.47 Convert the following mass measurements to grams:
(a) 9651 ng
(b) 2.33×10^{-3} Gg
(c) 499 mg
(d) 6.77×10^6 cg
(e) 62 μg

4.48 Convert the following volume measurements to liters:
(a) 5.87×10^{-4} kL
(b) 2.091×10^{-7} cL
(c) 138 ML
(d) 99.83 TL
(e) 7.44×10^6 nL

4.49 Determine the density (in g/mL) of a liquid with a mass of 366 g and a volume of 26.9 mL.

4.50 Determine the density (in g/mL) of a liquid sample with a mass of 1.73×10^9 mg and a volume of 1.99 m³.

4.51 The density of ethanol, the alcohol in alcoholic beverages, is 0.789 g/mL.
(a) Determine the mass (in g) of ethanol that occupies 250.0 mL.
(b) Determine the volume (in mL) of a 250.0-g sample of ethanol.

4.52 The Group 13 metal aluminum has been melted and poured into anthills to create strange and striking sculptures. The density of molten aluminum is 2.70 g/cm³.
(a) Determine the mass (in g) of molten aluminum that occupies 685 cm³.
(b) Determine the volume (in L) of a 52.3-kg sample of molten aluminum.

Courtesy of Anthillart.com

4.53 Which of the following conversions is set up correctly? Make corrections to any that are incorrect.
(a) Converting from kg to mg.

$$25.6 \ kg \times \frac{1 \ g}{1 \times 10^{-3} \ kg} \times \frac{1000 \ mg}{1 \ g} = 2.56 \times 10^7 \ mg$$

(b) Converting from μL to nL.

$$659 \ \mu L \times \frac{1 \ \mu L}{1 \times 10^{-6} \ L} \times \frac{1 \times 10^{-9} \ nL}{1 \ L} = 6.59 \times 10^{-1} \ nL$$

(c) Converting from Mm to cm.

$$12 \ Mm \times \frac{1 \times 10^6 \ m}{1 \ Mm} \times \frac{1 \times 10^2 \ cm}{1 \ m} = 1.2 \times 10^9 \ cm$$

(d) Converting from ng to kg.

$$9.42 \ ng \times \frac{1 \times 10^9 \ g}{1 \ ng} \times \frac{1 \times 10^3 \ kg}{1 \ g} = 9.42 \times 10^{12} \ kg$$

(e) Converting from km to nm.

$$8.8 \ km \times \frac{1 \ km}{1 \times 10^3 \ m} \times \frac{1 \ m}{1 \times 10^9 \ nm} = 8.8 \times 10^{-12} \ nm$$

4.54 The distance to the Moon is 3.84×10^5 km. What is this distance in meters?

4.55 A bee collects and carries about 20 mg of pollen per trip from a flower to its hive. What is this mass in kilograms? If a hive of bees collects 22 kg of pollen over the course of a summer, how many total collection trips have the bees made?

4.56 Scientists have recently recorded very high levels of particulate pollution (700 μg per cubic meter) in Beijing. What is this value in milligram per cubic centimeter?

4.57 In 2010, wind energy in the United States accounted for 3.91×10^3 MW (W = watts). How many watts is this equivalent to?

4.58 The concentration of iron in human blood is around 110 μg/dL. What is this concentration in g/L?

4.59 Due to the helium shortage, helium prices are predicted to rise as high as $84.00 per thousand cubic feet in the near future. At this price, how much would it cost to fill a typical party balloon with a volume of 4.00 liters?

4.60 Carbonate ion concentration in seawater has a value near 155 μmol/kg. What is this value in mol/g?

4.61 Perform the following conversions. Before doing so, predict whether the numerical value will get larger or smaller after conversion.
(a) 49.2 kg to mg
(b) $7.542 \times 10^9 \ \mu L$ to mL
(c) 299 Gm to nm
(d) 1.75×10^{-6} Mg to cg
(e) $3.22 \times 10^4 \ \mu m$ to dm

4.62 Perform the following conversions. Before doing so, predict whether the numerical value will get larger or smaller after conversion.
(a) 2.119×10^5 mg to ng
(b) $6.59 \times 10^8 \ \mu g$ to kg
(c) 3.24×10^{12} nL to ML
(d) 19 Gm to mm
(e) 5088 dg to μg

4.63 Perform the following conversions:
(a) 493 ft to m
(b) 648 mi to km
(c) 247 lb to g
(d) 311 gal to L

4.64   Perform the following conversions:
(a) 557 cm to in
(b) 122 kg to oz
(c) 941 fluid oz to mL
(d) 18.6 oz to g

4.65   Perform the following conversions:
(a) 23.6 kg to oz
(b) 319 L to fluid oz
(c) 48.6 mi to m
(d) 3.89 gal to mL

4.66   Perform the following conversions:
(a) 547 gal/min to L/s
(b) 12.9 ft/s to km/min
(c) 174 in^3 to mL
(d) 32.6 g/cm^2 to oz/in^2

4.67   Perform the following conversions:
(a) 9.78 g/mL to mg/L
(b) 25.8 m/s to km/hour
(c) 3.56×10^3 cm/min to m/s
(d) 1.76×10^5 ng/dL to g/L
(e) 1.4 g/L to μg/cL

4.68   Perform the following conversions:
(a) 14.1 m^2 to km^2
(b) 9.45×10^3 cm^2 to m^2
(c) 66.82 km^2 to mm^2
(d) 596 μm^2 to cm^2
(e) 32.7 nm^2 to km^2

4.69   Perform the following conversions:
(a) 33.8 cm^3 to mm^3
(b) 2.89×10^4 nm^3 to m^3
(c) 73.6 km^3 to μm^3
(d) 249 dm^3 to cm^3
(e) 5.75×10^{-5} nm^3 to μm^3

4.70   Perform the following conversions:
(a) 39.6 mg/cm^3 to g/m^3
(b) 89.0 ng/mm^3 to g/cm^3
(c) 45.6 m/s^2 to cm/ms^2
(d) 9.8 mm/ns^2 to m/s^2
(e) 2.83×10^4 kg/m^3 to g/mm^3

4.71   In the most productive areas of the Minnesota Soudan iron mine, iron ore was found to contain nearly 69% iron by mass. How many kilograms of iron ore are needed to provide enough iron to build a large aircraft carrier, if about 95,000 tons of iron, or 1.9×10^8 lb (1 lb = 453.6 g), are needed? If the density of the ore is 5.15 g/cm^3, how big a hole would be left after mining was completed? Assume a 100-foot-deep square hole and report your answer as the length of the square's side in feet.

Answers to In-Chapter Materials

Answers to Practice Problems

4.1A (a) 8.25 nL, (b) 68.25 kg, (c) 5.52 mm, (d) 41.8 TB, (e) 375 μs. **4.1B** (a) 9.52 L, (b) 23.8 m, (c) 12 s, (d) 3.65 m, (e) 820 Hz. **4.2A** 273 K, 373 K, range = 100 kelvins. **4.2B** −270.5°C. **4.3A** 113°F, 194°F, difference = 81°F. **4.3B** 233°C. **4.4A** (a) 4.8 × 10^{13}, (b) 2.99 × 10^8, (c) 8 × 10^{15}. **4.4B** (a) 355,000,000, (b) 189,900, (c) 5,700,000,000. **4.5A** (a) 6 × 10^{-4}, (b) 5.7 × 10^{-9}, (c) 3.99 × 10^{-13}. **4.5B** (a) 0.0000427, (b) 0.0000000277, (c) 0.0000000000733. **4.6A** (a) 1.48 × 10^4, (b) 7.63 × 10^7, (c) 7.11 × 10^2. **4.6B** (a) 1.95 × 10^{-8}, (b) 5.60 × 10^8, (c) 1.20 × 10^9. **4.7A** (a) 4, (b) 1, (c) 4, (d) 2, (e) 2 or 3—ambiguous, (f) 4. **4.7B** (a) 4, 0.0001080, 4; (b) 2, 5,500,000, 2; (c) 4, 2910, 3 or 4—ambiguous; (d) 4, 0.00008100, 4. **4.8A** (a) 116.2 L, (b) 80.71 m, (c) 3.813 × 10^{21} atoms, (d) 31 dm², (e) 0.504 g/mL. **4.8B** (a) 32.44 cm³, (b) 4.2 × 10^2 kg/m³, (c) 1.008 × 10^{10} kg, (d) 40.75 mL, (e) 227 cm³. **4.9A** (a) 2.10 × 10^{11} g, (b) 9.31 L, (c) 58.8 Mm. **4.9B** (a) 61.2 kg, (b) 14.5 m, (c) 7.7 L. **4.10A** (a) 454 g, (b) 29.34 cm, (c) 925 km, (d) 4.37 m, (e) 11.1 L. **4.10B** (a) 23.70 lb, (b) 6.21 mi, (c) 198.1 gal, (d) 524 oz, (e) 26.6 in. **4.11A** (a) 13.6 g/mL, (b) 1.63 × 10^3 g. **4.11B** (a) 9.25 g/cm³, (b) 3.76 × 10^3 g. **4.12A** 0.01 oz. **4.12B** 708.1 g. **4.13A** 1.05 × 10^4 kg/m³. **4.13B** 13.6 mg/mm³.

Answers to Checkpoints

4.1.1 c. **4.1.2** b. **4.1.3** c. **4.1.4** c. **4.2.1** d. **4.2.2** c. **4.2.3** b. **4.2.4** e. **4.2.5** c. **4.2.6** d. **4.3.1** c. **4.3.2** e. **4.3.3** e. **4.3.4** c. **4.4.1** b. **4.4.2** c. **4.4.3** a. **4.4.4** e. **4.4.5** c. **4.4.6** d.

CHAPTER 5

The Mole and Chemical Formulas

5.1 **Counting Atoms by Weighing**
- The Mole (The "Chemist's Dozen")
- Molar Mass
- Interconverting Mass, Moles, and Numbers of Atoms

5.2 **Counting Molecules by Weighing**
- Calculating the Molar Mass of a Compound
- Interconverting Mass, Moles, and Numbers of Molecules (or Formula Units)
- Combining Multiple Conversions in a Single Calculation

5.3 **Mass Percent Composition**

5.4 **Using Mass Percent Composition to Determine Empirical Formula**

5.5 **Using Empirical Formula and Molar Mass to Determine Molecular Formula**

A nearly perfect sphere of silicon, known as "the world's roundest object," is part of the Avogadro Project—an ongoing effort to redefine one of the base SI units: the kilogram.

JULIAN STRATENSCHULTE/epa european pressphoto agency b.v./Alamy Stock Photo

In This Chapter, You Will Learn

How chemists count atoms and molecules by weighing macroscopic samples of matter—and how knowing the numbers of atoms or molecules facilitates our understanding of substances and chemical processes.

Things To Review Before You Begin

- Average atomic mass [◀◀ Section 1.7]
- Empirical formula [◀◀ Section 3.4]
- Chapter 4 Key Skills [◀◀ pages 160–161]

Now that we have discussed some of the qualitative properties of matter, and we have seen some of the numerical and mathematical practices of scientists, we can begin to learn more about some of the quantitative aspects of chemistry. In this chapter, we discuss the importance of knowing how many atoms, molecules, or ions a sample of matter contains; how chemists *determine* such numbers; and how they *use* those numbers to solve problems.

5.1 Counting Atoms by Weighing

Atoms are so tiny that even the *smallest* macroscopic quantity of matter contains an *enormous* number of them. Often it is important, however, for chemists to know how many atoms there are in a sample. It would not be convenient to express the quantities used in the laboratory in terms of the actual number of individual atoms involved. Instead, for convenience, chemists use a unit of measurement called the *mole*.

The Mole (The "Chemist's Dozen")

If you go to a doughnut shop to get doughnuts for yourself, you probably buy them individually. Most people eat just one or two at a time. But if you were to buy doughnuts for everyone in your chemistry class, you would almost certainly order them by the dozen. One dozen doughnuts contains 12 doughnuts. In fact, a dozen of anything contains exactly 12 of that thing. 1 dozen ≡ 12. Pencils often come 12 to a box, and 12 such boxes typically are packaged together for shipment to bookstores. The larger package, containing 12 boxes of 12 pencils each, contains one *gross* of pencils. 1 gross ≡ 144. Whether a dozen or a gross, each is a quantity convenient for expression of a group of particular items—in these cases, doughnuts and pencils. Further, the dozen and the gross each constitutes a *reasonable, specific, exact* number of items.

Chemists, too, have adopted such a number to make it easy to express the number of atoms in a typical macroscopic sample of matter. Atoms are *so* much smaller than doughnuts or pencils, though, that the number used by chemists is *significantly* bigger than a dozen or a gross. The quantity used by chemists is the *mole (mol)*. Just as the dozen is equal to 12, and the gross is equal to 144, the mole is equal to 6.022×10^{23}. This unimaginably big number is known as *Avogadro's number (N_A)* in honor of Italian scientist Amedeo Avogadro (1776–1856) and is defined as the number of carbon atoms in exactly 12 grams of carbon-12. (Recall that a carbon-12 atom is one that has six protons and six neutrons in its nucleus. [◀◀ Section 1.6])

So a dozen of something is 12 of those things; a gross of something is 144 of those things; and a mole of something is 6.022×10^{23} of those things. A dozen doughnuts is 12 doughnuts [Figure 5.1(a)]; a gross of pencils is 144 pencils [Figure 5.1(b)];

Student Note: The value of this number is known to many more digits past the decimal point, but for our purposes, the use of four significant figures generally is sufficient.

Rewind: Remember that all carbon atoms have six protons. Most also have six neutrons. A carbon atom with six protons and six neutrons is a carbon-12 atom, where 12 is the mass number (sum of protons and neutrons). There are other isotopes of carbon containing different numbers of neutrons. For example: carbon-13 has seven neutrons; and carbon-14 has eight neutrons.

Figure 5.1 (a) A dozen doughnuts, (b) a gross of pencils, and (c) a mole of helium.

(a) lovethephoto/Alamy Stock Photo;
(b) David A. Tietz/McGraw Hill;
(c) Dorling Kindersley/Getty Images

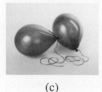

(a)   (b)   (c)

and a mole of an element, such as helium, is 6.022×10^{23} helium atoms—an amount that would fill a couple of average size helium balloons [Figure 5.1(c)].

Note that Avogadro's number (6.022×10^{23}) has no units—like the dozen (12) and the gross (144) it is just a number. When we do calculations with moles, what we use is actually called Avogadro's *constant*, which is expressed as 6.022×10^{23} items *per* mole:

$$\frac{6.022 \times 10^{23} \text{ items}}{1 \text{ mol of items}}$$

Student Note: This can also be written as:
6.022×10^{23} items · mol⁻¹

Sample Problem 5.1 shows how to convert between moles of an element and the number of atoms.

SAMPLE PROBLEM   5.1   Converting Between Moles of an Element and Number of Atoms

Calcium is the most abundant metal in the human body. A typical human body contains roughly 30 moles of calcium. Determine (a) the number of Ca atoms in 30.00 moles of calcium and (b) the number of moles of calcium in a sample containing 1.00×10^{20} Ca atoms.

Strategy Use Avogadro's constant to convert from moles to atoms and from atoms to moles.

Setup When the number of moles is known, we multiply by Avogadro's constant to convert to atoms. When the number of atoms is known, we divide by Avogadro's constant to convert to moles.

Solution

(a) $30.00 \text{ mol Ca} \times \dfrac{6.022 \times 10^{23} \text{ Ca atoms}}{1 \text{ mol Ca}} = 1.807 \times 10^{25}$ Ca atoms

(b) $1.00 \times 10^{20} \text{ Ca atoms} \times \dfrac{1 \text{ mol Ca}}{6.022 \times 10^{23} \text{ Ca atoms}} = 1.66 \times 10^{-4}$ mol Ca

THINK ABOUT IT

Make sure that units cancel properly in each solution and that the result makes sense. In part (a), for example, the number of moles (30) is greater than one, so the number of atoms is greater than Avogadro's number. In part (b), the number of atoms (1×10^{20}) is less than Avogadro's number, so there is less than a mole of substance.

Practice Problem **A**TTEMPT Potassium is the second most abundant metal in the human body. Calculate (a) the number of atoms in 7.31 moles of potassium and (b) the number of moles of potassium that contains 8.91×10^{25} atoms.

Practice Problem **B**UILD Calculate (a) the number of atoms in 1.05×10^{26} moles of helium and (b) the number of moles of helium that contain 2.33×10^{21} atoms.

Practice Problem **C**ONCEPTUALIZE These diagrams show collections of objects. For each diagram, express the number of objects using units of *dozen* and using units of *gross*. (Report each answer to four significant figures, but explain why the answers to this problem actually have more than four significant figures.)

(i)   (ii)   (iii)

Molar Mass

The number of atoms in a macroscopic sample is simply too big to be counted. Instead, the number of atoms is determined by weighing the sample—like nails in a hardware store (Figure 5.2). The number of nails needed to build a house, for example, is a very large number. When nails are purchased for such a project, they are not counted, but rather they are weighed to determine their number. How many nails there are to a pound depends on the size and type of nail.

For example, if we want to buy 1000 1.5-in 4d common nails, we would not count out a thousand nails. Instead, we would weigh out an amount just over three pounds. According to the table in Figure 5.2, there are 316 4d common nails per pound. We use this equality,

$$316 \text{ 4d common nails} = 1 \text{ lb 4d common nails}$$

to derive our conversion factor, which can be written two different ways:

$$\frac{1 \text{ lb 4d common nails}}{316 \text{ 4d common nails}} \quad \text{and} \quad \frac{316 \text{ 4d common nails}}{1 \text{ lb 4d common nails}}$$

Because we know how many nails we need, but want to know how many pounds to weigh out, we use the first version to convert from number of nails to weight:

$$1000 \ \cancel{\text{4d common nails}} \times \frac{1 \text{ lb 4d common nails}}{316 \ \cancel{\text{4d common nails}}} = 3.16 \text{ lb 4d common nails}$$

Therefore, we need 3.16 pounds of 4d common nails to have 1000. However, just as we would not count every nail, we also would not spend an inordinate amount of time adding and removing nails to get *exactly* 3.16 lb. We would weigh a quantity *close* to this, and then determine how many nails were in the batch using the other version of our derived conversion factor. Let's say that we loaded the scale with what we thought would be a little over 3 pounds, and the actual weight was 3.55 lb.

$$3.55 \text{ lb} \ \cancel{\text{4d common nails}} \times \frac{316 \text{ 4d common nails}}{1 \text{ lb} \ \cancel{\text{4d common nails}}} = 1121.8 \text{ 4d common nails}$$

We cannot have something other than a whole number of nails, so we might decide that our 3.55 lb consisted of 1122 4d common nails. But the weight of a batch of nails is a *measured* number and so contains uncertainty. Because both the number of nails

Type of nail	Size (in)	Nails/lb
3d box	1.5	635
6d box	2	236
10d box	3	94
4d casing	1.5	473
8d casing	2.5	145
2d common	1	876
4d common	1.5	316
6d common	2	181
8d common	2.5	106

Figure 5.2 Bulk nails are sold by the pound. How many nails there are in a pound depends on the size and type of nail.

Ryan McVay/Photodisc/Getty Images

Student Note: Remember that in a number such as this, the zero may or may not be significant. In this case, it is not.

Student Note: Like any conversion factor, this one can be written two ways:

$$\frac{145 \text{ 8d casing nails}}{1 \text{ lb 8d casing nails}} \text{ or}$$

$$\frac{1 \text{ lb 8d casing nails}}{145 \text{ 8d casing nails}}$$

We always choose the form that enables us to cancel units appropriately. In this case, the necessary conversion is:

$$\frac{145 \text{ 8d casing nails}}{1 \text{ lb 8d casing nails}}$$

from pounds of nails to number of nails.

per pound and the total weight of our nails are expressed with three significant figures, the number of nails in the batch should also be expressed with three significant figures as 1.12×10^3 or 1120 4d common nails.

Note that the same weight of a different *type* of nail would contain a different *number* of nails. For example, we might weigh out 3.55 lb of 8d casing nails:

$$3.55 \text{ lb 8d casing nails} \times \frac{145 \text{ 8d casing nails}}{1 \text{ lb 8d casing nails}} = 514.75 \text{ 8d casing nails}$$

We would likely round this number to 515 8d casing nails.

The number of nails in each case is determined by weighing (determining the *mass* of) a sample of nails; and the number of nails in a given mass depends on the type of nail. Numbers of atoms, too, are determined by weighing a sample; and the number of atoms per unit mass also depends on the *types* of atoms in the sample.

The **_molar mass_** (**_M_**) of a substance is the mass in grams of one mole of the substance. Using helium as an example, one mole of helium consists of 6.022×10^{23} He atoms. The average atomic mass of helium, from the periodic table, is 4.003 amu. The mass of a mole of helium, then, is

$$\frac{6.022 \times 10^{23} \text{ He atoms}}{1 \text{ mol}} \times \frac{4.003 \text{ amu}}{\text{He atom}} = \frac{2.4106066 \times 10^{24} \text{ amu}}{1 \text{ mol}}$$

(Note that we have not rounded to four significant figures because this is not the final answer.) To convert this mass to grams, we need a conversion factor from atomic mass units to grams. The necessary equality is $1 \text{ amu} = 1.661 \times 10^{-24} \text{ g}$. We write both forms of the conversion factor:

$$\frac{1.661 \times 10^{-24} \text{ g}}{1 \text{ amu}} \quad \text{and} \quad \frac{1 \text{ amu}}{1.661 \times 10^{-24} \text{ g}}$$

And multiplying by the form that will give us the desired cancellation of units (amu → g) gives

$$\frac{2.4106066 \times 10^{24} \text{ amu}}{1 \text{ mol}} \times \frac{1.661 \times 10^{-24} \text{ g}}{1 \text{ amu}} = \frac{4.003 \text{ g}}{1 \text{ mol}}$$

Student Note: Although the term *molar mass* specifies the mass of one mole of a substance, making the appropriate units simply *grams* (g), we usually express molar masses in units of *grams per mole* (g/mol) to facilitate cancellation of units in our calculations. Examples from the periodic table:

Element	Molar mass
Na	22.99 g/mol
S	32.07 g/mol
K	39.10 g/mol
Br	79.90 g/mol

It is no coincidence that the *molar* mass of He, 4.003 g, is numerically the same as the *atomic* mass, 4.003 amu. This is because the conversion factor between amu and g is the reciprocal of Avogadro's number. In other words, we could also express the pertinent equality as:

$$6.022 \times 10^{23} \text{ amu} = 1 \text{ g}$$

In fact, the molar mass of any element is numerically equal to its atomic mass, with molar masses being expressed in *grams* and atomic masses being expressed in *atomic mass units*. (This is also why there are no units after the numbers on the periodic table.)

One of the important conversions you will need to make often is that of mass of an element to moles. This is done by dividing the mass in grams by the molar mass in grams per mole, as shown in Equation 5.1:

Equation 5.1 $$\frac{\text{mass of element (g)}}{\text{molar mass of element (g/mol)}} = \text{moles of element}$$

Sample Problem 5.2 shows how molar mass is used to convert between mass and moles of an element.

SAMPLE PROBLEM (5.2) Converting Between Mass of an Element and Number of Moles

Determine (a) the number of moles of C in 25.00 g of carbon, (b) the number of moles of He in 10.50 g of helium, and (c) the number of moles of Na in 15.75 g of sodium.

Strategy Molar mass of an element is numerically equal to its average atomic mass. Use the molar mass for each element to convert from mass to moles.

Setup (a) The molar mass of carbon is 12.01 g/mol. (b) The molar mass of helium is 4.003 g/mol. (c) The molar mass of sodium is 22.99 g/mol.

Solution

(a) $25.00 \text{ g C} \times \dfrac{1 \text{ mol C}}{12.01 \text{ g C}} = 2.082 \text{ mol C}$ (c) $15.75 \text{ g Na} \times \dfrac{1 \text{ mol Na}}{22.99 \text{ g Na}} = 0.6851 \text{ mol Na}$

(b) $10.50 \text{ g He} \times \dfrac{1 \text{ mol He}}{4.003 \text{ g He}} = 2.623 \text{ mol He}$

THINK ABOUT IT

Always double-check unit cancellations in problems such as these—errors are common when molar mass is used as a conversion factor. Also make sure that the results make sense. For example, in the case of part (c), a mass smaller than the molar mass corresponds to less than a mole.

Practice Problem ATTEMPT Determine the number of moles in (a) 12.25 g of argon (Ar), (b) 0.338 g of gold (Au), and (c) 59.8 g of mercury (Hg).

Practice Problem BUILD Determine the mass in grams of (a) 2.75 moles of calcium, (b) 0.075 mole of helium, and (c) 1.055×10^{-4} mole of potassium.

Practice Problem CONCEPTUALIZE Plain doughnuts from a particular bakery have an average mass of 32.6 g, whereas jam-filled doughnuts from the same bakery have an average mass of 40.0 g. (a) Determine the mass of a dozen plain doughnuts and the mass of a dozen jam-filled doughnuts. (b) Determine the number of doughnuts in a kilogram of plain and the number in a kilogram of jam-filled. (c) Determine the mass of plain doughnuts that contains the same number of doughnuts as a kilogram of jam-filled. (d) Determine the total mass of a dozen doughnuts consisting of three times as many plain as jam-filled.

Interconverting Mass, Moles, and Numbers of Atoms

Molar mass is the conversion factor that we use to convert from mass to moles, and vice versa. We use Avogadro's constant (N_A) to convert from number of moles to number of atoms, and vice versa. The flowchart in Figure 5.3 summarizes the operations involved in these conversions.

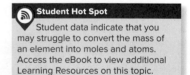

Student Hot Spot

Student data indicate that you may struggle to convert the mass of an element into moles and atoms. Access the eBook to view additional Learning Resources on this topic.

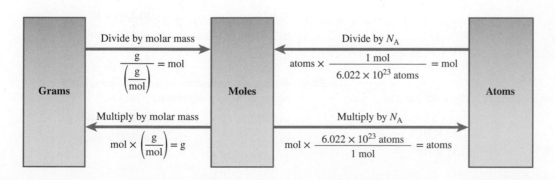

Figure 5.3 Flowchart for conversions among mass, moles, and number of atoms.

Sample Problem 5.3 lets you practice these conversions.

SAMPLE PROBLEM 5.3   Converting Between Mass of an Element, Moles, and Number of Atoms

Determine (a) the number of C atoms in 0.515 g of carbon, and (b) the mass of helium that contains 6.89×10^{18} He atoms.

Strategy Use the conversions depicted in Figure 5.3 to convert (a) from grams to moles to atoms, and (b) from atoms to moles to grams.

Setup (a) The molar mass of carbon is 12.01 g/mol. (b) The molar mass of helium is 4.003 g/mol. $N_A = 6.022 \times 10^{23}$.

Solution

(a) $0.515 \text{ g C} \times \dfrac{1 \text{ mol C}}{12.01 \text{ g C}} \times \dfrac{6.022 \times 10^{23} \text{ C atoms}}{1 \text{ mol C}} = 2.58 \times 10^{22} \text{ C atoms}$

(b) $6.89 \times 10^{18} \text{ He atoms} \times \dfrac{1 \text{ mol He}}{6.022 \times 10^{23} \text{ He atoms}} \times \dfrac{4.003 \text{ g He}}{1 \text{ mol He}} = 4.58 \times 10^{-5} \text{ g He}$

THINK ABOUT IT

A ballpark estimate of your result can help you prevent common errors. For example, the mass in part (a) is smaller than the molar mass of carbon. Therefore, you should expect a number of atoms *smaller* than Avogadro's number. Likewise, the number of atoms in part (b) is smaller than Avogadro's number. Therefore, you should expect a mass of helium *smaller* than the molar mass of helium.

Practice Problem ATTEMPT Determine (a) the number of atoms in 105.5 g of gold, and (b) the mass of calcium that contains 8.075×10^{12} Ca atoms.

Practice Problem BUILD Determine the mass of calcium that contains the same number of atoms as 81.06 g of helium, and (b) the number of gold atoms that has the same mass as 7.095×10^{31} argon atoms.

Practice Problem CONCEPTUALIZE Several types of coins were minted to commemorate the anniversary of an important historical event. Among them were silver coins with a mass of 28.35 g each and gold coins with a mass of 9.45 g each. Determine the mass of 50 silver coins, and the mass of 50 gold coins. How many gold coins would have a total mass of 425.25 g? How many silver coins would have a total mass of 425.25 g? What number of silver coins would have the same mass as 135 gold coins?

CHECKPOINT–SECTION 5.1   Counting Atoms by Weighing

5.1.1 How many moles of arsenic are there in 6.50 g of arsenic?

  a) 0.0868   d) 0.115

  b) 0.0487   e) 0.0193

  c) 0.00205

5.1.2 How many atoms are there in 3.559×10^{-6} mol of krypton?

  a) 2.982×10^{-4} d) 2.143×10^{18}

  b) 5.910×10^{-30} e) 1.652×10^{23}

  c) 3.353×10^{3}

5.1.3 How many atoms are there in 30.1 g of magnesium?

  a) 1.24   d) 2.06×10^{-24}

  b) 7.46×10^{23} e) 4.55×10^{23}

  c) 1.81×10^{25}

5.1.4 What mass of mercury contains the same number of atoms as 90.15 g of helium?

  a) 22.53 g   d) 1.357×10^{25} g

  b) 90.15 g   e) 3.22×10^{2} g

  c) 4.518×10^{3} g

5.2 Counting Molecules by Weighing

We learned in Chapter 3 how to write chemical formulas for compounds. We can use the chemical formula and atomic masses from the periodic table to determine a compound's *molecular* mass, in the case of a molecular compound; or its *formula* mass, in the case of an ionic compound. **Molecular mass** is the sum of atomic masses for all of the atoms a molecule contains. **Formula mass** is the sum of atomic masses of a **formula unit**, which is represented by the empirical formula of an ionic compound. [◄◄ Section 3.4]

> **Rewind:** Recall that an empirical formula denotes the smallest whole-number ratio of combination of elements in a compound. We can write both molecular formulas, which denote actual numbers of atoms in a molecule, and empirical formulas for molecular compounds—and in some cases they are the same. We generally write just empirical formulas for ionic compounds.

Thus, the molecular mass of water, H_2O, is

$$\underbrace{2(1.008 \text{ amu})}_{\text{two H atoms}} + \underbrace{16.00 \text{ amu}}_{\text{one O atom}} = 18.016 \text{ amu}$$

and the formula mass of sodium chloride, NaCl, is

$$\underbrace{22.99 \text{ amu}}_{\text{one Na atom}} + \underbrace{35.45 \text{ amu}}_{\text{one Cl atom}} = 58.44 \text{ amu}$$

> **Student Note:** If we are asked simply to calculate the molecular mass of water, we would round this to 18.02 amu. However, if the problem asks a question that requires two or more calculation steps, we would not round this number before using it in subsequent calculations.

Calculating the Molar Mass of a Compound

In Section 5.1, we learned that molar mass ($\mathcal{M}$) is the mass in grams of one mole of a substance. The example we used was that of helium, an element in which the elementary particles are atoms. But what about substances that do not consist of isolated atoms? Most do not. In fact, many *elements* do not consist of atoms, but rather of *molecules,* as in the cases of nitrogen and oxygen, which exist under ordinary conditions as the diatomic gases N_2 and O_2. In the case of a molecular substance (such as water, nitrogen, or oxygen), a mole consists of Avogadro's number of molecules.

Consider again the compound water, H_2O. Recall that the subscript 2 following the H and the lack of any subscript following the O indicate that each water molecule consists of two hydrogen atoms and one oxygen atom. Now imagine that we have a dozen water molecules. How many of each atom would there be? The chemical formula tells us the ratio of combination of the different elements that make up a compound. Water will always consist of twice as many H atoms as O atoms—no matter how much water we have. So in a dozen water molecules, there will be two dozen H atoms and one dozen O atoms. Likewise, in a *gross* of water molecules, there will be two gross H atoms and one gross O atoms. In a *thousand* water molecules, there will be two thousand H atoms and one thousand O atoms. The ratio is always the same. Therefore, if we have a *mole* of water (6.022×10^{23} water molecules), we will have *two* moles of H atoms (1.2044×10^{24}) and *one* mole of O atoms (6.022×10^{23}). To determine the molar mass of a *compound,* then, all we need to do is sum the molar masses of the *elements* it contains. In the case of water:

$$\underbrace{2(1.008 \text{ g})}_{\text{two mol H}} + \underbrace{16.00 \text{ g}}_{\text{one mol O}} = \underbrace{18.016 \text{ g}}_{\text{one mol } H_2O}$$

which we round to 18.02 g (unless there are additional calculations in which the molar mass is to be used). The same applies to nitrogen, N_2, a mole of which consists of *two* moles of N. The molar mass of N_2 is 2(14.01 g) = 28.02 g. Similarly, the molar mass of O_2, a mole of which consists of two moles of O, is 2(16.00 g) = 32.00 g.

In the case of an ionic compound, a mole consists of Avogadro's number of *formula units;* and to determine the molar mass of an ionic compound, we do essentially the same thing, sum the molar masses of the elements indicated by the compound's formula. In the case of sodium chloride:

$$\underbrace{22.99 \text{ g}}_{\text{one mol Na}} + \underbrace{35.45 \text{ g}}_{\text{one mol Cl}} = \underbrace{58.44 \text{ g}}_{\text{one mol NaCl}}$$

When it comes to expressing the molar mass of an element such as oxygen, we must be careful to specify what *form* of the element we mean. For instance, the element oxygen exists predominantly as diatomic molecules (O_2). Thus, if we say *one mole of oxygen* and by *oxygen* we mean O_2, the molar mass is 32.00 g. If, on the other hand, we mean a mole of oxygen *atoms* (O), then the molar mass is only 16.00 g. As you progress through your chemistry course, it should become easier for you to tell from context which form of the element is intended, as the following examples illustrate:

Context	The underlined word means	and the molar mass is
How many moles of **oxygen** react with 2 moles of hydrogen to produce water?	O_2	32.00 g
How many moles of **oxygen** are there in 1 mole of water?	O	16.00 g
Air is approximately 21% **oxygen**.	O_2	32.00 g
Many organic compounds contain **oxygen**.	O	16.00 g
Air is approximately 79% **nitrogen**.	N_2	28.02 g
Protein molecules contain **nitrogen**.	N	14.01 g
The airship Hindenburg was filled with flammable **hydrogen**.	H_2	2.016 g
Most acids contain **hydrogen**.	H	1.008 g

Sample Problem 5.4 lets you practice determining molar mass using a chemical formula.

SAMPLE PROBLEM (5.4) Determining Molar Mass Using a Chemical Formula

Determine the molar mass of each of the following compounds:

(a) N_2O_5 (b) Na_2CO_3 (c) $Mg(CN)_2$ (d) $Sr(ClO_3)_2$

Strategy Use the ratios represented in the chemical formulas and molar masses of the elements from the periodic table. [Remember that the subscript 2 in $Mg(CN)_2$ indicates that there are two of everything inside the parentheses.]

Setup The molar masses required are N (14.01 g), O (16.00 g), Na (22.99 g), C (12.01 g), Mg (24.31 g), Sr (87.62 g), and Cl (35.45 g).

Solution

(a) $\dfrac{28.02 \text{ g}}{2 \text{ mol N}}$ + $\dfrac{80.00 \text{ g}}{5 \text{ mol O}}$ = $\dfrac{108.02 \text{ g}}{1 \text{ mol } N_2O_5}$

(b) $\dfrac{45.98 \text{ g}}{2 \text{ mol Na}}$ + $\dfrac{12.01 \text{ g}}{1 \text{ mol C}}$ + $\dfrac{48.00 \text{ g}}{3 \text{ mol O}}$ = $\dfrac{108.02 \text{ g}}{1 \text{ mol } Na_2CO_3}$

(c) $\dfrac{24.31 \text{ g}}{1 \text{ mol Mg}}$ + $\dfrac{24.02 \text{ g}}{2 \text{ mol C}}$ + $\dfrac{28.02 \text{ g}}{2 \text{ mol N}}$ = $\dfrac{76.35 \text{ g}}{1 \text{ mol } Mg(CN)_2}$

(d) $\dfrac{87.62 \text{ g}}{1 \text{ mol Sr}}$ + $\dfrac{70.90 \text{ g}}{2 \text{ mol Cl}}$ + $\dfrac{96.00 \text{ g}}{6 \text{ mol O}}$ = $\dfrac{254.52 \text{ g}}{1 \text{ mol } Sr(ClO_3)_2}$

THINK ABOUT IT

Be extra careful counting when formulas include parentheses—especially if there are subscripts within the parentheses. It's an easy place to make a mistake.

Practice Problem (A)TTEMPT Determine the molar mass of each of the following compounds:

(a) PCl_3 (b) H_2SO_4 (c) $(NH_4)_2S$ (d) SF_6

Practice Problem (B)UILD Determine the value of the missing subscript in each chemical formula, given the molar mass.

(a) $N_2O_?$ ($\mathscr{M} = 108.02$ g/mol)   (c) $H_2C_?O_4$ ($\mathscr{M} = 90.04$ g/mol)
(b) $HClO_?$ ($\mathscr{M} = 68.46$ g/mol)   (d) $CCl_?$ ($\mathscr{M} = 153.81$ g/mol)

Practice Problem (C)ONCEPTUALIZE Determine the identity of element X in each chemical formula, given the molar mass.

(a) $X(NO_3)_2$ ($\mathscr{M} = 187.57$ g/mol)   (c) H_3XO_4 ($\mathscr{M} = 97.99$ g/mol)
(b) $SnX_{4?}$ ($\mathscr{M} = 197.7$ g/mol)   (d) SiX_4 ($\mathscr{M} = 347.69$ g/mol)

Interconverting Mass, Moles, and Numbers of Molecules (or Formula Units)

One of the most common and important conversions that you will have to do frequently is that of mass of a compound to moles of that compound. This is done using Equation 5.2, which is very similar to Equation 5.1:

$$\frac{\text{mass of compound (g)}}{\text{molar mass of compound (g/mol)}} = \text{moles of compound} \qquad \textbf{Equation 5.2}$$

Consider a typical 500-mL bottle of water, and for the sake of this calculation, let's assume that it contains exactly 500 grams of water. To determine the number of moles of water in this bottle, we use the molar mass of water, which we have previously determined to be 18.016 g/mol, to convert the mass of water in grams to moles of water:

Student Note: Remember that the density of water, the conversion factor between volume and mass, at 4°C, is 1 g/mL. [◄◄ Section 4.4]

$$500 \text{ g } H_2O \times \frac{1 \text{ mol } H_2O}{18.016 \text{ g } H_2O} = 27.753 \text{ mol } H_2O$$

We would round this to 27.75 mol H_2O if this were our final result—remember that the molar mass we determined for H_2O can have only four significant figures. If, however, we wish to determine the number of *molecules* in the bottle of water, we would retain all of the digits to continue the calculation:

$$27.753 \text{ mol } H_2O \times \frac{6.022 \times 10^{23} \text{ } H_2O \text{ molecules}}{1 \text{ mol } H_2O} = 1.6713 \times 10^{25} \text{ } H_2O \text{ molecules}$$

(If this were our final answer, we would round to 1.671×10^{25} H_2O molecules.) What if we want to know the numbers of H and O atoms? As is often the case, there is more than one way to solve such a problem. What follows are two different approaches to answer the same question.

Approach 1: We can use the chemical formula to determine how many atoms of a specific element a given number of *molecules* of that compound contains. In the case of our 500-g H_2O sample, which we have determined to contain 1.6713×10^{25} H_2O molecules, we can use the molecular formula to determine the number of atoms of each H and O:

Student Note: Remember that if we use a calculated number in a subsequent calculation, we use the unrounded number—retaining extra digits until the very end of the calculation.

$$1.6713 \times 10^{25} \text{ } H_2O \text{ molecules} \times \frac{2 \text{ H atoms}}{1 \text{ molecule } H_2O} = 3.3426 \times 10^{25} \text{ H atoms}$$

$$1.6713 \times 10^{25} \text{ } H_2O \text{ molecules} \times \frac{1 \text{ O atom}}{1 \text{ molecule } H_2O} = 1.6713 \times 10^{25} \text{ O atoms}$$

Rounding to the appropriate number of significant figures (*four*) for both results gives:

$$3.343 \times 10^{25} \text{ H atoms} \quad \text{and} \quad 1.671 \times 10^{25} \text{ O atoms}$$

Approach 2: We also have seen how a chemical formula can be used to determine how many moles of a specific element a mole of compound contains. Again, using the case of our 500-g H_2O sample, which we have determined to contain 27.753 mol H_2O, we can use the molecular formula to determine the number of moles of each H and O, and then use Avogadro's constant to convert to numbers of atoms:

$$27.753 \text{ mol } H_2O \times \frac{2 \text{ mol H}}{1 \text{ mol } H_2O} = 55.506 \text{ mol H}$$

and

$$55.506 \text{ mol H} \times \frac{6.022 \times 10^{23} \text{ H atoms}}{1 \text{ mol H}} = 3.343 \times 10^{25} \text{ H atoms}$$

$$27.753 \text{ mol } H_2O \times \frac{1 \text{ mol O}}{1 \text{ mol } H_2O} = 27.753 \text{ mol O}$$

and

$$27.753 \text{ mol O} \times \frac{6.022 \times 10^{23} \text{ O atoms}}{1 \text{ mol O}} = 1.671 \times 10^{25} \text{ O atoms}$$

The results are the same as those we got using Approach 1.

Now let's consider a similar problem with an ionic compound, sodium chloride (NaCl). Suppose we want to know the number of moles of NaCl in a box of salt containing 735 grams. We follow the same procedure as the one we used for water, first dividing the mass in grams by the molar mass of NaCl:

$$735 \text{ g NaCl} \times \frac{1 \text{ mol NaCl}}{58.44 \text{ g NaCl}} = 12.577 \text{ mol NaCl}$$

If this were our final result, because only three significant figures are appropriate (determined by the mass 735 g) we would round to 12.6 mol NaCl. If we further wished to determine the number of formula units of NaCl in this sample, we would use the unrounded result of our first calculation and Avogadro's constant:

$$12.577 \text{ mol NaCl} \times \frac{6.022 \times 10^{23} \text{ NaCl formula units}}{1 \text{ mol NaCl}} = 7.574 \times 10^{24} \text{ NaCl formula units}$$

And if this were our final answer, we would round to 7.57×10^{24} NaCl formula units. If we further want to determine the numbers of sodium and chloride ions in the sample, we would use the formula of the compound, just as we did with water:

$$7.574 \times 10^{24} \text{ NaCl formula units} \times \frac{1 \text{ Na}^+ \text{ ion}}{1 \text{ NaCl formula unit}} = 7.574 \times 10^{24} \text{ Na}^+ \text{ ions}$$

$$7.574 \times 10^{24} \text{ NaCl formula units} \times \frac{1 \text{ Cl}^- \text{ ion}}{1 \text{ NaCl formula unit}} = 7.574 \times 10^{24} \text{ Cl}^- \text{ ions}$$

And for both, we would round to three significant figures to get 7.57×10^{24} each Na^+ ions and 7.57×10^{24} Cl^- ions. Note that this analysis was analogous to the first approach we used to determine atoms of H and O in our 500-g sample of water. We could equally well have done this analysis using the second approach, converting moles of NaCl to moles of Na^+ and Cl^- ions using the ratio given by the chemical formula; and then to number of ions using Avogadro's constant.

Figure 5.4 summarizes the operations involved in these conversions.

Figure 5.4 Flowchart for conversions among mass, moles, and number of elementary particles.

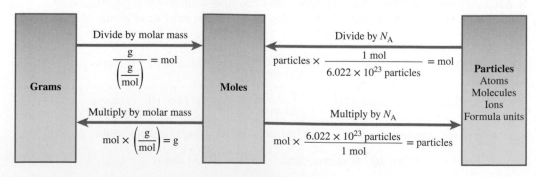

Combining Multiple Conversions in a Single Calculation

Many of the problems we have solved in this section have involved multiple calculations, with repeated admonitions not to round the intermediate results—to minimize rounding error in the final answers. In practice, scientists carry out multistep calculations without ever explicitly determining the intermediate results. For example, in our determination of the number of H atoms in the 500-g sample of water, we carried out three separate calculations:

> **Student Note:** Remember that we have assumed our sample to be *exactly* 500 g H_2O; therefore, the number 500 does not limit the number of significant figures in the calculated answer.

1. mass H_2O to moles H_2O
2. moles H_2O to H_2O molecules
3. H_2O molecules to H atoms

Instead, we could combine the entire problem into a single calculation as follows:

$$500 \text{ g } H_2O \times \frac{1 \text{ mol } H_2O}{18.016 \text{ g } H_2O} \times \frac{6.022 \times 10^{23} \text{ } H_2O \text{ molecules}}{1 \text{ mol } H_2O} \times \frac{2 \text{ H atoms}}{1 \text{ } H_2O \text{ molecule}} = 3.343 \times 10^{25} \text{ H atoms}$$

When combining the steps of a multistep calculation such as this, it is especially important to track and cancel units carefully to minimize the chance of common errors, such as using the wrong form of a conversion factor.

Sample Problems 5.5 through 5.7 let you practice these conversions.

> **Student Hot Spot**
>
> Student data indicate that you may struggle to convert between mass, moles, and molecules of a compound. Access the eBook to view additional Learning Resources on this topic.

SAMPLE PROBLEM (5.5) Converting Mass, Moles, and Molecules (or Formula Units) by Combining Multiple Conversions in a Single Calculation

Determine (a) the number of moles of CO_2 in a 10.00-g sample of CO_2 and (b) the mass of 0.905 mole of sodium chloride.

Strategy Use molar mass to convert from mass to moles and to convert from moles to mass.

Setup The molar mass of CO_2 is 44.01 g/mol. The molar mass of a compound is numerically equal to its formula mass. The molar mass of sodium chloride (NaCl) is 58.44 g/mol.

Solution (a) $10.00 \text{ g } CO_2 \times \frac{1 \text{ mol } CO_2}{44.01 \text{ g } CO_2} = 0.2272 \text{ mol } CO_2$

(b) $0.905 \text{ mol NaCl} \times \frac{58.44 \text{ g NaCl}}{1 \text{ mol NaCl}} = 52.9 \text{ g NaCl}$

THINK ABOUT IT

Always double-check unit cancellations in problems such as these—errors are common when molar mass is used as a conversion factor. Also make sure that the results make sense. In both cases, a mass smaller than the molar mass corresponds to less than a mole of substance.

Practice Problem A TTEMPT (a) Determine the mass in grams of 2.75 moles of glucose ($C_6H_{12}O_6$). (b) Determine the number of moles in 59.8 g of sodium nitrate ($NaNO_3$).

Practice Problem B UILD Determine the number of moles of SCl_4 that has the same mass as 3.212 moles of CF_4.

Practice Problem C ONCEPTUALIZE A combination of plastic snap-together blocks is used here to represent a molecule. Each "molecule" consists of three large red blocks and two small blue blocks. The masses of the individual blocks are 4.75 g and 2.80 g, respectively. Determine the mass of a dozen of these molecules. How many dozen molecules are there in 714.6 g of molecules? How many *molecules* are there in that mass?

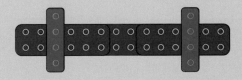

SAMPLE PROBLEM (5.6) Converting Between Mass and Numbers of Atoms/Molecules Using a Combined Calculation

(a) Determine the number of water molecules and the numbers of H and O atoms in 3.26 g of water. (b) Determine the mass of 7.92×10^{19} carbon dioxide molecules.

Strategy Use molar mass and Avogadro's constant to convert from mass to molecules, and vice versa. Use the molecular formula of water to determine the numbers of H and O atoms.

Setup (a) Starting with mass (3.26 g of water), we use molar mass (18.02 g/mol) to convert to moles of water. From moles, we use Avogadro's constant to convert to number of water molecules. In part (b), we reverse the process in part (a) to go from number of molecules to mass of carbon dioxide.

Solution (a) $3.26 \text{ g } H_2O \times \dfrac{1 \text{ mol } H_2O}{18.02 \text{ g } H_2O} \times \dfrac{6.022 \times 10^{23} \text{ } H_2O \text{ molecules}}{1 \text{ mol } H_2O} = 1.09 \times 10^{23} \text{ } H_2O \text{ molecules}$

Using the molecular formula, we can determine the number of H and O atoms in 3.26 g of H_2O as follows:

$$1.09 \times 10^{23} \text{ } H_2O \text{ molecules} \times \dfrac{2 \text{ H atoms}}{1 \text{ } H_2O \text{ molecule}} = 2.18 \times 10^{23} \text{ H atoms}$$

$$1.09 \times 10^{23} \text{ } H_2O \text{ molecules} \times \dfrac{1 \text{ O atom}}{1 \text{ } H_2O \text{ molecule}} = 1.09 \times 10^{23} \text{ O atoms}$$

(b) $7.92 \times 10^{19} \text{ } CO_2 \text{ molecules} \times \dfrac{1 \text{ mol } CO_2}{6.022 \times 10^{23} \text{ } CO_2 \text{ molecules}} \times \dfrac{44.01 \text{ g } CO_2}{1 \text{ mol } CO_2} = 5.79 \times 10^{-3} \text{ g } CO_2$

THINK ABOUT IT

Again, check the cancellation of units carefully and make sure that the magnitudes of your results are reasonable.

Practice Problem ATTEMPT (a) Calculate the number of oxygen molecules and the number of oxygen atoms in 35.5 g of O_2. (b) Calculate the mass of 9.95×10^{14} SO_3 molecules.

Practice Problem BUILD (a) Determine the number of nitrogen and oxygen atoms in 1.00 kg of N_2O_5. (b) Calculate the mass of calcium carbonate that contains 1.00 mol of oxygen atoms.

Practice Problem CONCEPTUALIZE Another combination of plastic snap-together blocks is used here to represent a molecule. It consists of two large white blocks and three small blue blocks with masses of 5.20 g and 2.80 g, respectively. Determine the mass of a dozen of these "molecules." How many *dozen* of these molecules and how many *molecules* are there in a collection with a total mass of 789.6 g? Determine the number of white blocks and the number of blue blocks in this collection and express them both in dozens, and in individual blocks.

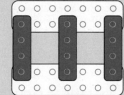

SAMPLE PROBLEM (5.7) Converting Between Mass and Numbers of Ions Using a Combined Calculation

Using approach number 1, determine the number of magnesium and cyanide ions present in 73.8 grams of $Mg(CN)_2$. Try to string together all steps of the process in one long calculation.

Strategy Use the molar mass of $Mg(CN)_2$ and Avogadro's constant to convert from mass to moles to formula units of $Mg(CN)_2$. Use the chemical formula of $Mg(CN)_2$ to convert to number of Mg^{2+} and CN^- ions.

Setup Starting with the mass 73.8 g $Mg(CN)_2$, we use molar mass (76.35 g/mol) to convert to moles of $Mg(CN)_2$. From moles, we use Avogadro's constant to convert to number of formula units of $Mg(CN)_2$. We then use conversion factors from the chemical formula $\left(\dfrac{1 \text{ } Mg^{2+} \text{ ion}}{1 \text{ } Mg(CN)_2 \text{ formula unit}} \text{ and } \dfrac{2 \text{ } CN^- \text{ ions}}{1 \text{ } Mg(CN)_2 \text{ formula unit}} \right)$ to convert to ions of magnesium and cyanide.

Solution

$$73.8 \text{ g } Mg(CN)_2 \times \dfrac{1 \text{ mol } Mg(CN)_2}{76.34 \text{ g } Mg(CN)_2} \times \dfrac{6.022 \times 10^{23} \text{ } Mg(CN)_2}{1 \text{ mol } Mg(CN)_2} \times \dfrac{1 \text{ } Mg^{2+} \text{ ion}}{1 \text{ } Mg(CN)_2 \text{ formula unit}} = 5.82 \times 10^{23} \text{ } Mg^{2+} \text{ ions}$$

$$73.8 \text{ g } Mg(CN)_2 \times \dfrac{1 \text{ mol } Mg(CN)_2}{76.34 \text{ g } Mg(CN)_2} \times \dfrac{6.022 \times 10^{23} \text{ } Mg(CN)_2}{1 \text{ mol } Mg(CN)_2} \times \dfrac{2 \text{ } CN^- \text{ ions}}{1 \text{ } Mg(CN)_2 \text{ formula unit}} = 1.16 \times 10^{24} \text{ } CN^- \text{ ions}$$

THINK ABOUT IT

Check the cancellation of your units and the size of your answers carefully. When determining numbers of atoms or ions present in a sample, the answer should be very large.

Practice Problem 5.6A asks you to answer this same question using a different approach. Your answers should be the same.

Practice Problem Ⓐ**TTEMPT** Using approach number 2, determine the number of magnesium and cyanide ions present in 73.8 grams of $Mg(CN)_2$.

Practice Problem Ⓑ**UILD** Determine the mass (in grams) of $LiNO_3$ that contains 3.55×10^{24} nitrate (NO_3^-) ions.

Practice Problem Ⓒ**ONCEPTUALIZE** What mass of the molecules in Practice Problem 5.5C (left) would contain the same number of blue blocks as 1191 g of the molecules in Practice Problem 5.6C (right)?

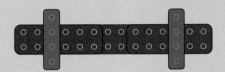
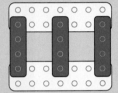

Thinking Outside the Box

Redefining the Kilogram

For over a century, scientific balances worldwide were calibrated using the *international prototype kilogram* (IPK), a 1-kilogram platinum-iridium alloy cylinder that was forged in London in the late nineteenth century. This cylinder, one of 40 such cylinders forged at the same time, was deemed by international treaty in 1889 to be the official definition of the kilogram. The remainder of the original cylinders, the masses of which were measured against the IPK in 1889, were distributed internationally to provide other countries, including the United States, with their own official standard kilograms. The IPK is carefully stored, along with six of the other original cylinders, in a high-security vault at the International Bureau of Weights and Measures near Paris, France, under scrupulously controlled conditions. It has been removed from its vault only *three* times since becoming international standard—for the purpose of comparing the masses of all the official cylinders. Unfortunately, the mass of this 1-kg standard changed over the decades, relative to the other copies forged at the same time. But it is impossible to know whether the IPK has *lost* some of its mass—or the other official copies have *gained* mass. Although the differences in mass were extremely small, on the order of micrograms, the definition of one of the SI base units being tied inextricably to the mass of an antique object that appeared not to be absolutely constant over time raised significant concerns in the scientific community. In recent decades, scientists have agreed that a new standard for the kilogram must be devised. One method that was considered would have defined the kilogram using a specific number of atoms. This method essentially would have used Avogadro's constant to define the kilogram—hence the name of the endeavor: *The Avogadro Project*.

In this attempt to redefine the kilogram, a pure silicon-28 crystal was grown and machined into the world's most perfect sphere at the Australia Center for Precision Optics (ACPO).

Silicon was chosen because it is possible to make an isotopically pure sample of Si-28 and to grow very large, nearly perfect crystals from it. The purity and exact volume of the sphere were carefully measured. Knowing the volume of the sphere and the volume of one Si-28 atom, it was possible to determine the number of Si-28 atoms contained in the sphere. The intent was to determine a more accurate value for Avogadro's number, and define the kilogram based on a precise number of Si-28 atoms.

Ultimately, The Avogadro Project was not used to redefine the kg. That was done in 2019 by fixing the values of Avogadro's number and the Planck constant, a constant that relates the energy and frequency of light. Although the details of how this was achieved are beyond the scope of this text, with these two constants established as *exact* numbers, scientists were able to use the equations relating mass to energy to establish a definition of the kg that does not depend on a physical object.

A perfect sphere of solid Si-28; and one of the original 40 1-kg cylinders forged in 1889.
(left) Julian Stratenschulte/dpa picture alliance/Alamy Stock Photo; (right) Jacques Brinon/AP Images

Profiles in Science

Derek Muller

Derek Muller, Australian physicist and world-renowned science educator, is the creator of Vertasium, an English-language educational YouTube science channel. Muller has made videos of a vast range of scientific topics, including radioactivity, climate change, and The Avogadro Project. His in-depth analyses of complicated subjects—coupled with his exceptional ability to communicate complex ideas to a general audience—have made his YouTube channel enormously popular. His Vertasium channel has hundreds of uploads and millions of subscribers.

Muller has created two additional YouTube channels: 2Vertasium and Sciencium, both of which are growing rapidly in popularity.

Derek Muller

Courtesy of Veritasium

Derek Muller

Courtesy of Veritasium

CHECKPOINT—SECTION 5.2   Counting Molecules by Weighing

5.2.1 Determine the molar mass of $Ca(ClO_4)_2$.

   a) 91.53 g/mol   d) 179.61 g/mol

   b) 174.98 g/mol   e) 238.98 g/mol

   c) 139.53 g/mol

5.2.2 How many formula units are contained in 29.4 g of $Ca(ClO_4)_2$?

   a) 0.123   d) 7.41×10^{22}

   b) 1.77×10^{25} e) 8.15×10^{23}

   c) 2.22×10^{23}

5.2.3 How many molecules are in 30.1 g of SO_2?

   a) 1.81×10^{25} d) 1.02×10^{24}

   b) 2.83×10^{23} e) 5.00×10^{-23}

   c) 6.02×10^{23}

5.2.4 How many moles of hydrogen are there in 6.55 g of ammonia (NH_3)?

   a) 0.382   d) 1.15

   b) 1.39   e) 2.66

   c) 0.215

5.3   Mass Percent Composition

As we have seen, the formula of a compound indicates the numbers of atoms of each element in a unit (molecule or formula unit) of a compound. From a molecular formula or an empirical formula, we can calculate what percentage of the total mass is contributed by each element in a compound. A list of the percent by mass of each element in a compound is known as the compound's *mass percent composition* often called simply the *percent composition*. The mass percent composition of a compound can be calculated using Equation 5.3:

Equation 5.3 $\text{percent by mass of an element} = \dfrac{n \times \text{molar mass of element}}{\text{molar mass of compound}} \times 100\,\%$

where n is the number of moles of an element in a mole of the compound. For example, in a mole of hydrogen peroxide (H_2O_2), there are two moles of hydrogen (H) and two moles of oxygen (O). The molar masses of H and O are 1.008 g and 16.00 g, respectively; and the molar mass of H_2O_2 is:

$$\underbrace{2(1.008 \text{ g})}_{\text{two mol H}} + \underbrace{2(16.00) \text{ g}}_{\text{two mol O}} = 34.016 \text{ g}$$

The mass percent composition of H_2O_2 is:

$$\%H = \frac{2 \times 1.008 \text{ g H}}{34.016 \text{ g } H_2O_2} \times 100\% = 5.927\% \quad \text{and} \quad \%O = \frac{2 \times 16.00 \text{ g O}}{34.016 \text{ g } H_2O_2} \times 100\% = 94.073\%$$

Rounding both to the appropriate number of significant figures gives the mass percent composition of hydrogen peroxide as 5.927 percent hydrogen and 94.07 percent oxygen. The sum of percentages is $5.927\% + 94.07\% = 99.997\%$, which rounds to 100%. (The very small discrepancy between the result before rounding and 100% is the result of our using only four significant figures for the molar masses of elements.)

We could equally well have used the empirical formula of hydrogen peroxide (HO) for the calculation of percent composition. In this case, we would have used the molar mass of the empirical formula in place of the molar mass of the compound:

$$\underbrace{1.008 \text{ g}}_{\text{one mol H}} + \underbrace{16.00 \text{ g}}_{\text{one mol O}} = 17.008 \text{ g}$$

$$\%H = \frac{1.008 \text{ g H}}{17.008 \text{ g HO}} \times 100\% = 5.927\% \quad \text{and} \quad \%O = \frac{16.00 \text{ g O}}{17.008 \text{ g HO}} \times 100\% = 94.073\%$$

Because both the molecular formula and the empirical formula tell us the composition of the compound, they both give the same mass percent composition.

Sample Problem 5.8 lets you practice determining the mass percent composition of a compound.

Student Note: If you determine the mass percent composition of a compound, the resulting percentages must add to 100 (within rounding error). If they do not, go back and check your work for errors.

SAMPLE PROBLEM **5.8** Determining Mass Percent Composition Using Chemical Formula

Lithium carbonate (Li_2CO_3) was the first "mood-stabilizing" drug approved by the FDA for the treatment of mania and manic-depressive illness, also known as bipolar disorder. Calculate the percent composition by mass of lithium carbonate.

Strategy Use Equation 5.3 to determine the percent by mass contributed by each element in the compound.

Setup Lithium carbonate is an ionic compound that contains Li, C, and O. In a formula unit, there are two Li atoms, one C atom, and three O atoms with atomic masses 6.941, 12.01, and 16.00 amu, respectively. The formula mass of Li_2CO_3 is $2(6.941 \text{ amu}) + 12.01 \text{ amu} + 3(16.00 \text{ amu}) = 73.89 \text{ amu}$.

Solution For each element, multiply the number of atoms by the atomic mass, divide by the formula mass, and multiply by 100 percent.

$$\%Li = \frac{2 \times 6.941 \text{ amu Li}}{73.89 \text{ amu } Li_2CO_3} \times 100\% = 18.79\%$$

$$\%C = \frac{12.01 \text{ amu C}}{73.89 \text{ amu } Li_2CO_3} \times 100\% = 16.25\%$$

$$\%O = \frac{3 \times 16.00 \text{ amu O}}{73.89 \text{ amu } Li_2CO_3} \times 100\% = 64.96\%$$

THINK ABOUT IT

Make sure that the percent composition results for a compound sum to approximately 100. (In this case, the results sum to exactly 100 percent—$18.79\% + 16.25\% + 64.96\% = 100.00\%$—but remember that because of rounding, the percentages may sum to very slightly more or very slightly less.)

Practice Problem **A**TTEMPT Determine the percent composition by mass of the artificial sweetener aspartame ($C_{14}H_{18}N_2O_5$).

Practice Problem **B**UILD Determine the percent composition by mass of Atorvastatin, a cholesterol-lowering drug, if its chemical formula is $C_{33}H_{35}FN_2O_5$.

Practice Problem **C**ONCEPTUALIZE Determine the percent composition by mass of acetaminophen, the active ingredient in over-the-counter pain relievers such as Tylenol. A molecular model of the acetaminophen molecule is shown here.

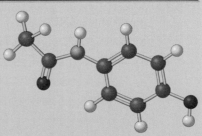

Acetaminophen

Familiar Chemistry

Iodized Salt

Dietary iodine deficiency is a worldwide problem and, according to the United Nations International Children's Emergency Fund (UNICEF), is the leading preventable cause of mental retardation. This deficiency can be addressed at minimal cost by the addition of a small amount of iodine, usually in the form of *sodium iodide* (NaI) or *potassium iodide* (KI), to the salt used by consumers every day.

David A. Tietz/McGraw Hill

Nutrition Facts		
Serving Size 1/4 tsp (1.5g)		
Servings Per Container 491		
Amount Per Serving		
Calories 0		
		%Daily Value*
Total Fat 0g		0%
Sodium 590mg		25%
Total Carbohydrate 0g		0%
Protein 0g		
Iodine		45%

Not a significant source of calories from fat, saturated fat, *trans* fat, cholesterol, dietary fiber, sugars, vitamin A, vitamin C, calcium and iron.

* Percent Daily Values are based on a 2,000 calorie diet.

Most of us grew up in homes where the familiar round blue box of iodized salt was a pantry staple. What many people don't realize is how recent the routine use of iodized salt is. Less than a century ago, a significant fraction of the population of the United States suffered from a condition called *goiter,* which causes bulbous, sometimes extreme, swelling of the neck. Scientists of the nineteenth and early twentieth centuries knew that the incidence of goiter was highest in an area including the Great Lakes region and the Pacific Northwest—collectively known as the "goiter belt."

Between the years 1916 and 1920, two Cleveland, Ohio, doctors (David Marine and Oliver P. Kimball) conducted a study in which thousands of schoolgirls in Akron, Ohio, were given iodine supplements. The incidence of goiter, which had been quite high in this population, decreased significantly—establishing a rationale for the public health initiative that followed. By 1924, the Morton Salt Company had begun distributing iodized salt nationwide. Although goiter caused by insufficient dietary iodine has been essentially eradicated in affluent nations, it persists as a serious health problem in less economically developed parts of the world.

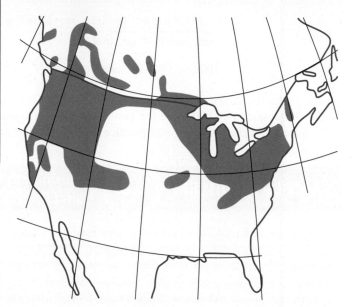

CHECKPOINT—SECTION 5.3   Mass Percent Composition

5.3.1 What is the percent composition by mass of aspirin ($C_9H_8O_4$)?

   a) 44.26% C, 3.28% H, 52.46% O

   b) 60.00% C, 4.47% H, 35.53% O

   c) 41.39% C, 3.47% H, 55.14% O

   d) 42.86% C, 6.35% H, 50.79% O

   e) 42.86% C, 38.09% H, 19.05% O

5.3.2 What is the percent composition by mass of sodium bicarbonate ($NaHCO_3$)?

   a) 20.89% Na, 2.75% H, 32.74% C, 43.62% O

   b) 44.20% Na, 1.94% H, 23.09% C, 30.76% O

   c) 21.28% Na, 0.93% H, 33.35% C, 44.43% O

   d) 44.20% Na, 1.94% H, 23.09% C, 30.76% O

   e) 27.37% Na, 1.20% H, 14.30% C, 57.14% O

5.3.3 What is the mass of oxygen in a 20.0-g sample of H_2SO_4?

   a) 0.0206 g

   b) 16.31 g

   c) 66.02 g

   d) 13.05 g

   e) 3.26 g

5.3.4 An extra-strength tablet contains 500 mg of acetaminophen. What is the mass of nitrogen in two extra-strength acetaminophen tablets? (See Practice Problem 5.8C.)

   a) 0.093 g

   b) 0.21 g

   c) 0.046 g

   d) 0.18 g

   e) 9.3 g

5.4 Using Mass Percent Composition to Determine Empirical Formula

In Section 5.3, we learned how to use a compound's chemical formula (either molecular or empirical) to determine the mass percent composition. Using the concepts of the mole and molar mass, we can also use the *mass percent composition* to determine *empirical formula*.

Consider a compound with a mass percent composition of 92.26 percent carbon and 7.743 percent hydrogen. The percent composition is the same regardless of how much of the compound we have. To keep the math simple, we can assume that we have exactly 100 grams of the compound. We first determine what mass of each element is present in the 100-g sample:

Student Note: It isn't necessary to use a mass of 100 grams; this just simplifies the math involved in empirical-formula determination.

$$\text{g C} = \frac{92.26 \text{ g C}}{100 \text{ g compound}} \times 100 \text{ g compound} = 92.26 \text{ g C}$$

and

$$\text{g H} = \frac{7.743 \text{ g H}}{100 \text{ g compound}} \times 100 \text{ g compound} = 7.743 \text{ g H}$$

We then use the molar mass of each element (C and H) to determine the number of moles of each in our 100-g sample:

$$92.26 \text{ g C} \times \frac{1 \text{ mol C}}{12.01 \text{ g C}} = 7.682 \text{ mol C} \quad \text{and} \quad 7.743 \text{ g H} \times \frac{1 \text{ mol H}}{1.008 \text{ g H}} = 7.682 \text{ mol H}$$

Thus, our 100-g sample of compound contains equal numbers of moles of H and C. We can initially write the formula using the actual numbers of moles as subscripts:

$$C_{7.682}H_{7.682}$$

But an *empirical* formula by definition is the smallest possible whole-number ratio, making the empirical formula of this compound (with a 1:1 mole ratio of C to H) simply CH. The molar mass of this empirical formula is 12.01 g + 1.008 g = 13.018 g (or 13.02 g, rounded to two places past the decimal point).

Now let's consider a compound with a mass percent composition of 53.31 percent carbon, 11.19 percent hydrogen, and 35.51 percent oxygen. Again, we assume a sample size of exactly 100 g, and we determine the mass of each element in that sample:

$$\text{g C} = \frac{53.31 \text{ g C}}{100 \text{ g compound}} \times 100 \text{ g compound} = 53.31 \text{ g C}$$

$$\text{g H} = \frac{11.19 \text{ g H}}{100 \text{ g compound}} \times 100 \text{ g compound} = 11.19 \text{ g H}$$

$$\text{g O} = \frac{35.51 \text{ g O}}{100 \text{ g compound}} \times 100 \text{ g compound} = 35.51 \text{ g O}$$

We then use the mass of each element to determine the corresponding number of moles:

$$53.31 \text{ g C} \times \frac{1 \text{ mol C}}{12.01 \text{ g C}} = 4.439 \text{ mol C}$$

$$11.19 \text{ g H} \times \frac{1 \text{ mol H}}{1.008 \text{ g H}} = 11.10 \text{ mol H}$$

$$35.51 \text{ g O} \times \frac{1 \text{ mol O}}{16.00 \text{ g O}} = 2.219 \text{ mol O}$$

This gives us a preliminary result of $C_{4.439}H_{11.10}O_{2.219}$, which we must reduce to a formula containing only whole numbers. When an analysis such as this results in different numbers of moles for each element, we reduce the subscripts to whole numbers as follows: we first identify the *smallest* number of moles, in this case 2.219 (moles of oxygen), and divide each of the numerical results by it:

$$C_{(4.439/2.219)} \quad H_{(11.10/2.219)} \, O_{(2.219/2.219)}$$

$$\frac{4.439}{2.219} \approx 2 \quad \frac{11.10}{2.219} \approx 5 \quad \frac{2.219}{2.219} = 1$$

We then use the results to write the correct empirical formula, which in this case is C_2H_5O.

Sample Problem 5.9 illustrates the determination of empirical formula from mass percent composition for a compound.

Student Note: This type of determination is especially sensitive to rounding error. Be sure to carry extra digits through the calculation until the end, so as not to end up with the wrong empirical formula.

SAMPLE PROBLEM   5.9   Using Mass Percent Composition to Determine Empirical Formula

Determine the empirical formula of a compound with mass percent composition: 37.51 percent carbon, 2.52 percent hydrogen, and 59.97 percent oxygen.

Strategy Assume a 100-g sample to simplify the math. Using the appropriate molar masses, convert the grams of each element to moles. Use the resulting numbers as subscripts in the empirical formula, reducing them to the lowest possible whole numbers for the final answer.

Setup The necessary molar masses (from the periodic table) are:

$$C\ 12.01\ g \quad H\ 1.008\ g \quad O\ 16.00\ g$$

A 100-g sample of the compound consists of

$$\frac{37.51\ g\ C}{100\ g\ compound} \times 100\ g\ compound = 37.51\ g\ C$$

$$\frac{2.52\ g\ H}{100\ g\ compound} \times 100\ g\ compound = 2.52\ g\ H$$

$$\frac{59.97\ g\ O}{100\ g\ compound} \times 100\ g\ compound = 59.97\ g\ O$$

Solution Converting the mass of each compound to moles:

$$37.51\ g\ C \times \frac{1\ mol\ C}{12.01\ g\ C} = 3.123\ mol\ C$$

$$2.52\ g\ H \times \frac{1\ mol\ H}{1.008\ g\ H} = 2.500\ mol\ H$$

$$59.97\ g\ O \times \frac{1\ mol\ O}{16.00\ g\ O} = 3.747\ mol\ O$$

This gives the empirical formula $C_{3.123}H_{2.500}O_{3.747}$. To reduce the subscripts to whole numbers, we divide each by the smallest subscript, in this case 2.500.

$$C_{(3.123/2.500)}\ H_{(2.500/2.500)}\ O_{(3.747/2.500)}$$

$$\frac{3.123}{2.500} \approx 1.25 \quad \frac{2.500}{2.500} \approx 1 \quad \frac{3.747}{2.500} = 1.5$$

Note that after dividing each subscript by 2.500, we still have an empirical formula with two noninteger subscripts: $C_{1.25}HO_{1.5}$. Empirical formulas must have integer subscripts, so this necessitates one additional step. We must now multiply all of the subscripts by an integer that will result in each subscript being an integer. In this case, multiplying by 4

$$C_{(1.25\times4)}\ H_{(1\times4)}\ O_{(1.5\times4)}$$
$$5 \quad 4 \quad 6$$

giving us the empirical formula $C_5H_4O_6$.

THINK ABOUT IT

It takes practice to recognize when a subscript can be *rounded* to a whole number, and when all of the subscripts must be *multiplied by an integer* to arrive at a correct empirical formula. There are certain decimal values that will crop up in this type of problem that should alert you to the need to multiply rather than round. The following table shows these common decimal values and the integer by which each should be multiplied.

Decimal	should be multiplied by
#.10	10
#.20	5
#.25	4
#.33	3
#.50	2
#.66	3
#.75	4

Practice Problem **A**TTEMPT Determine the empirical formula of a compound with mass percent composition 52.15 percent carbon, 13.13 percent hydrogen, and 34.73 percent oxygen.

Practice Problem **B**UILD Determine the empirical formula of a compound with mass percent composition 60.00 percent carbon, 4.48 percent hydrogen, and 35.53 percent oxygen.

Practice Problem **C**ONCEPTUALIZE Imagine another "molecule" made from the white and blue blocks from Practice Problem 5.6C, but with a different ratio of combination. A collection of these molecules is known to be 57.38 percent blue blocks by mass, and 42.62 percent white blocks by mass. Determine how many of each color block a molecule contains.

Familiar Chemistry

Fertilizer & Mass Percents

You may have looked at a lawn fertilizer label and been overwhelmed by the notation and wealth of information that it provides. These labels simply give the amount, in a mass percent, of the fertilizer that comes from each of the critical elements needed for lawn health. These are sometimes referred to as the NPK values, standing for nitrogen, phosphorus (as P_2O_5), and potassium (as K_2O) content. They are always listed in this order. The fertilizer label below shows 26 – 4 – 12 content, or 26% N, 4% P_2O_5, and 12% K_2O by mass.

Let's say you have a 100.0-g sample of this particular fertilizer. You could determine the mass of each substance using the NPK values. From that information, you could also determine the amount of the elements phosphorus and potassium that are present using the conversion factors from the compound formulas of P_2O_5 and K_2O, as follows:

V J Matthew/Shutterstock

$$100.0 \text{ g fertilizer} \times \frac{26 \text{ g N}}{100.0 \text{ g fertilizer}} = 26 \text{ g N}$$

$$100.0 \text{ g fertilizer} \times \frac{4 \text{ g P}_2\text{O}_5}{100.0 \text{ g fertilizer}} = 4 \text{ g P}_2\text{O}_5$$

$$4 \text{ g P}_2\text{O}_5 \times \frac{1 \text{ mol P}_2\text{O}_5}{141.88 \text{ g P}_2\text{O}_5} \times \frac{2 \text{ mol P}}{1 \text{ mol P}_2\text{O}_5} \times \frac{30.97 \text{ g P}}{1 \text{ mol P}} = 1.75 \text{ or } \sim 2 \text{ g P}$$

$$100.0 \text{ g fertilizer} \times \frac{12 \text{ g K}_2\text{O}}{100.0 \text{ g fertilizer}} = 12 \text{ g K}_2\text{O}$$

$$12 \text{ g K}_2\text{O} \times \frac{1 \text{ mol K}_2\text{O}}{94.20 \text{ g K}_2\text{O}} \times \frac{2 \text{ mol K}}{1 \text{ mol K}_2\text{O}} \times \frac{39.10 \text{ g K}}{1 \text{ mol K}} = 9.96 \text{ or } 10. \text{ g K}$$

These conversion factors are a convenient method for representing a percentage in a way that facilitates the cancellation of units.

Practice Problem **A**TTEMPT Determine the empirical formula of a compound that is 53.3 percent C, 11.2 percent H, and 35.5 percent O by mass. What is the molecular formula if the molar mass is approximately 90 g/mol?

Practice Problem **B**UILD Determine the empirical formula of a compound that is 89.9 percent C and 10.1 percent H by mass. What is the molecular formula if the molar mass is approximately 120 g/mol?

Practice Problem **C**ONCEPTUALIZE The percent composition by mass was determined for a compound containing only C and H; and the percent C was found to be four times that of H. Further, the compound's molecular mass is twice its empirical formula mass. Determine the molecular formula of the compound.

CHECKPOINT–SECTION 5.5   Using Empirical Formula and Molar Mass to Determine Molecular Formula

5.5.1 Determine the molecular formula of a compound that has the following composition: 48.6% C, 8.2% H, and 43.2% O, and the molar mass is approximately 148 g/mol.

    a) $C_3H_6O_2$

    b) $C_4H_4O_3$

    c) $C_4H_4O_2$

    d) C_2H_3O

    e) $C_6H_{12}O_4$

5.5.2 Determine the molecular formula of a compound that has the following composition: 43.64% P and 56.36% O, and the molar mass is approximately 284 g/mol.

    a) P_4O_{12}

    b) P_2O_6

    c) PO_3

    d) P_4O_{10}

    e) P_2O_5

5.5.3 The diagram shows a molecule composed of carbon (black) and hydrogen (light grey). Determine the empirical formula and the percent composition by mass of this compound.

    a) C_2H_4, 85.6% C, 14.4% H

    b) C_6H_{14}, 30.0% C, 70.0% H

    c) C_6H_{14}, 83.6% C, 16.4% H

    d) C_3H_7, 83.6% C, 16.4% H

    e) CH_2, 85.6% C, 14.4% H

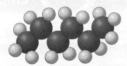

Chapter Summary

Section 5.1

- The unit of measure used to specify numbers of atoms is the *mole (mol),* which is similar to the *dozen* and the *gross* in that it specifies a number of items. One mole of items, such as atoms, is equal to 6.022×10^{23} of those items. The number 6.022×10^{23} is known as *Avogadro's number (N_A).* In most calculations, we use Avogadro's *constant,* which is expressed as 6.022×10^{23} items per mol:

$$\frac{6.022 \times 10^{23} \text{ items}}{1 \text{ mol of items}}$$

- The *molar mass ($\mathscr{M}$)* of a substance is the mass in grams of one mole of the substance. Chemists determine the number of atoms in a macroscopic sample of matter by weighing the sample and dividing the mass in grams by the molar mass in grams per mole.

- The molar mass of an element is numerically equal to its atomic mass—with molar masses generally expressed in g/mol and atomic masses expressed as amu. Molar masses are used as conversion factors to interconvert mass and moles; and Avogadro's constant is used to interconvert moles and numbers of atoms in an element such as helium.

Section 5.2

- A chemical formula can be used to calculate the *molecular mass* of a molecular compound, or the *formula mass* of an ionic compound. This is done by summing the atomic masses of the atoms indicated by the chemical formula. Just as molecular compounds consist of *molecules,* ionic compounds consist of *formula units.*

THINK ABOUT IT

It takes practice to recognize when a subscript can be *rounded* to a whole number, and when all of the subscripts must be *multiplied by an integer* to arrive at a correct empirical formula. There are certain decimal values that will crop up in this type of problem that should alert you to the need to multiply rather than round. The following table shows these common decimal values and the integer by which each should be multiplied.

Decimal	should be multiplied by
#.10	10
#.20	5
#.25	4
#.33	3
#.50	2
#.66	3
#.75	4

Practice Problem Ⓐ**TTEMPT**   Determine the empirical formula of a compound with mass percent composition 52.15 percent carbon, 13.13 percent hydrogen, and 34.73 percent oxygen.

Practice Problem Ⓑ**UILD**   Determine the empirical formula of a compound with mass percent composition 60.00 percent carbon, 4.48 percent hydrogen, and 35.53 percent oxygen.

Practice Problem Ⓒ**ONCEPTUALIZE**   Imagine another "molecule" made from the white and blue blocks from Practice Problem 5.6C, but with a different ratio of combination. A collection of these molecules is known to be 57.38 percent blue blocks by mass, and 42.62 percent white blocks by mass. Determine how many of each color block a molecule contains.

Familiar Chemistry

Fertilizer & Mass Percents

You may have looked at a lawn fertilizer label and been overwhelmed by the notation and wealth of information that it provides. These labels simply give the amount, in a mass percent, of the fertilizer that comes from each of the critical elements needed for lawn health. These are sometimes referred to as the NPK values, standing for nitrogen, phosphorus (as P_2O_5), and potassium (as K_2O) content. They are always listed in this order. The fertilizer label below shows 26 – 4 – 12 content, or 26% N, 4% P_2O_5, and 12% K_2O by mass.

Let's say you have a 100.0-g sample of this particular fertilizer. You could determine the mass of each substance using the NPK values. From that information, you could also determine the amount of the elements phosphorus and potassium that are present using the conversion factors from the compound formulas of P_2O_5 and K_2O, as follows:

V J Matthew/Shutterstock

$$100.0 \text{ g fertilizer} \times \frac{26 \text{ g N}}{100.0 \text{ g fertilizer}} = 26 \text{ g N}$$

These conversion factors are a convenient method for representing a percentage in a way that facilitates the cancellation of units.

$$100.0 \text{ g fertilizer} \times \frac{4 \text{ g } P_2O_5}{100.0 \text{ g fertilizer}} = 4 \text{ g } P_2O_5$$

$$4 \text{ g } P_2O_5 \times \frac{1 \text{ mol } P_2O_5}{141.88 \text{ g } P_2O_5} \times \frac{2 \text{ mol } P}{1 \text{ mol } P_2O_5} \times \frac{30.97 \text{ g } P}{1 \text{ mol } P} = 1.75 \text{ or } \sim 2 \text{ g } P$$

$$100.0 \text{ g fertilizer} \times \frac{12 \text{ g } K_2O}{100.0 \text{ g fertilizer}} = 12 \text{ g } K_2O$$

$$12 \text{ g } K_2O \times \frac{1 \text{ mol } K_2O}{94.20 \text{ g } K_2O} \times \frac{2 \text{ mol } K}{1 \text{ mol } K_2O} \times \frac{39.10 \text{ g } K}{1 \text{ mol } K} = 9.96 \text{ or } 10. \text{ g } K$$

Guaranteed Analysis   26–4–12

Total Nitrogen	26%
3.2% Ammoniacal Nitrogen	
9.7% Water Insoluble Nitrogen*	
3.4% Urea Nitrogen	
9.7% Other Water Soluble Nitrogen*	
Available Phosphate (P_2O_5)	4%
Soluble Potash (K_2O)	12%
Total Sulfur (S)	1.5%
1.5% Combined Sulfur (S)	

Nutrient Sources: Ammonium Phosphate, Ammonium Sulfate, Isobutylidene Diurea, Urea, Methylene Urea, Muriate of Potash.

Chlorine (Cl) not more than	**10.0%**

***19.4% Slowly Available Nitrogen from Methylene Ureas and IBDU.**   **F699**

Information regarding the contents and levels of metals in this product is available on the Internet at http://www.regulatory-info-lebsea.com

Label from a common lawn fertilizer.

CHECKPOINT–SECTION 5.4   Using Mass Percent Composition to Determine Empirical Formula

5.4.1 Determine the empirical formula of a compound that has the following composition: 92.3% C and 7.7% H.

  a) CH

  b) C_2H_3

  c) C_4H_6

  d) C_6H_7

  e) C_4H_3

5.4.2 Determine the empirical formula of a compound that has the following composition: 48.6% C, 8.2% H, and 43.2% O.

  a) C_3H_8O

  b) C_3H_6O

  c) $C_2H_5O_2$

  d) C_2H_6O

  e) $C_3H_6O_2$

5.5 Using Empirical Formula and Molar Mass to Determine Molecular Formula

As useful as an empirical formula can be, we must keep in mind that the empirical formula gives only the *ratio* of combination of the atoms in a molecule, not the exact numbers—so there are many cases where multiple compounds have the *same* empirical formula. Therefore, to determine the *molecular* formula of a compound, we need more than just its percent composition—we also need at least an estimate of its molar mass. Knowing both the empirical formula and the approximate molar mass, we can use Equation 5.4 to determine a compound's molecular formula:

Equation 5.4 $$\frac{\text{molar mass}}{\text{molar mass of empirical formula}} = n$$

where n is the number of empirical formula units contained within a molecular formula. Using the two compounds for which we determined empirical formulas in Section 5.4, we can see how this is done. In the case of the first compound, with empirical formula CH, if we know that its molar mass is approximately 78 g, we can determine its molecular formula as follows:

$$\frac{\text{molar mass}}{\text{molar mass of empirical formula}} = \frac{78 \text{ g}}{13.018 \text{ g}} = 5.99$$

The ratio of a compound's molar mass to the molar mass of its empirical formula, *n*, is typically a whole number—or something very *close* to a whole number, in this case 6.

What this tells us is that there are six empirical formulas in each molecular formula. To get the molecular formula, we simply multiply the subscripts in the empirical formula by 6:

$$C_{(1\times6)}H_{(1\times6)}$$

which gives us a molecular formula of C_6H_6.

In the second example, we determined the empirical formula to be C_2H_5O, for which the molar mass of the empirical formula is:

$$\underbrace{2(12.01 \text{ g})}_{\text{two mol C}} + \underbrace{5(1.008 \text{ g})}_{\text{five mol H}} + \underbrace{16.00 \text{ g}}_{\text{one mol O}} = 45.06 \text{ g}$$

If we know that the molar mass is approximately 90 g, we determine the molecular formula the same way as before:

$$\frac{\text{molar mass}}{\text{molar mass of empirical formula}} = \frac{90 \text{ g}}{45.06 \text{ g}} \approx 2$$

Multiplying the subscripts in our empirical formula by 2

$$C_{(2\times2)}H_{(5\times2)}O_{(1\times2)}$$

gives a molecular formula of $C_4H_{10}O_2$.

Sample Problem 5.10 lets you practice determining empirical and molecular formulas of a compound given percent composition and approximate molar mass.

SAMPLE PROBLEM **5.10** Using Percent Composition and Molar Mass to Determine Empirical and Molecular Formulas

Determine the empirical formula of a compound that is 30.45 percent nitrogen and 69.55 percent oxygen by mass. Given that the molar mass of the compound is approximately 92 g/mol, determine the molecular formula of the compound.

Strategy Assume a 100-g sample so that the mass percentages of nitrogen and oxygen given in the problem statement correspond to the masses of N and O in the compound. Then, using the appropriate molar masses, convert the grams of each element to moles. Use the resulting numbers as subscripts in the empirical formula, reducing them to the lowest possible whole numbers for the final answer. To calculate the molecular formula, first divide the molar mass given in the problem statement by the empirical formula mass. Then, multiply the subscripts in the empirical formula by the resulting number to obtain the subscripts in the molecular formula.

Setup The empirical formula of a compound consisting of N and O is N_xO_y. The molar masses of N and O are 14.01 and 16.00 g/mol, respectively. One hundred grams of a compound that is 30.45 percent nitrogen and 69.55 percent oxygen by mass contains 30.45 g N and 69.55 g O.

Solution

$$30.45 \ \cancel{\text{g N}} \times \frac{1 \text{ mol N}}{14.01 \ \cancel{\text{g N}}} = 2.173 \text{ mol N}$$

$$69.55 \ \cancel{\text{g O}} \times \frac{1 \text{ mol O}}{16.00 \ \cancel{\text{g O}}} = 4.347 \text{ mol O}$$

This gives a formula of $N_{2.173}O_{4.347}$. Dividing both subscripts by the smaller of the two to get the smallest possible whole numbers (2.173/2.173 = 1, 4.347/2.173 ≈ 2) gives an empirical formula of NO_2.

Finally, dividing the approximate molar mass (92 g/mol) by the empirical formula mass [14.01 g/mol + 2(16.00 g/mol) = 46.01 g/mol] gives 92/46.01 ≈ 2. Then, multiplying both subscripts in the empirical formula by 2 gives the molecular formula, N_2O_4.

THINK ABOUT IT

Use the method described in Sample Problem 5.8 to calculate the percent composition of the molecular formula N_2O_4 and verify that it is the same as that given in this problem.

Practice Problem ATTEMPT Determine the empirical formula of a compound that is 53.3 percent C, 11.2 percent H, and 35.5 percent O by mass. What is the molecular formula if the molar mass is approximately 90 g/mol?

Practice Problem BUILD Determine the empirical formula of a compound that is 89.9 percent C and 10.1 percent H by mass. What is the molecular formula if the molar mass is approximately 120 g/mol?

Practice Problem CONCEPTUALIZE The percent composition by mass was determined for a compound containing only C and H; and the percent C was found to be four times that of H. Further, the compound's molecular mass is twice its empirical formula mass. Determine the molecular formula of the compound.

CHECKPOINT–SECTION 5.5   Using Empirical Formula and Molar Mass to Determine Molecular Formula

5.5.1 Determine the molecular formula of a compound that has the following composition: 48.6% C, 8.2% H, and 43.2% O, and the molar mass is approximately 148 g/mol.

a) $C_3H_6O_2$

b) $C_4H_4O_3$

c) $C_4H_4O_2$

d) C_2H_3O

e) $C_6H_{12}O_4$

5.5.2 Determine the molecular formula of a compound that has the following composition: 43.64% P and 56.36% O, and the molar mass is approximately 284 g/mol.

a) P_4O_{12}

b) P_2O_6

c) PO_3

d) P_4O_{10}

e) P_2O_5

5.5.3 The diagram shows a molecule composed of carbon (black) and hydrogen (light grey). Determine the empirical formula and the percent composition by mass of this compound.

a) C_2H_4, 85.6% C, 14.4% H

b) C_6H_{14}, 30.0% C, 70.0% H

c) C_6H_{14}, 83.6% C, 16.4% H

d) C_3H_7, 83.6% C, 16.4% H

e) CH_2, 85.6% C, 14.4% H

Chapter Summary

Section 5.1

- The unit of measure used to specify numbers of atoms is the *mole (mol)*, which is similar to the *dozen* and the *gross* in that it specifies a number of items. One mole of items, such as atoms, is equal to 6.022×10^{23} of those items. The number 6.022×10^{23} is known as *Avogadro's number (N_A)*. In most calculations, we use Avogadro's *constant,* which is expressed as 6.022×10^{23} items per mol:

$$\frac{6.022 \times 10^{23} \text{ items}}{1 \text{ mol of items}}$$

- The *molar mass ($\mathcal{M}$)* of a substance is the mass in grams of one mole of the substance. Chemists determine the number of atoms in a macroscopic sample of matter by weighing the sample and dividing the mass in grams by the molar mass in grams per mole.

- The molar mass of an element is numerically equal to its atomic mass—with molar masses generally expressed in g/mol and atomic masses expressed as amu. Molar masses are used as conversion factors to interconvert mass and moles; and Avogadro's constant is used to interconvert moles and numbers of atoms in an element such as helium.

Section 5.2

- A chemical formula can be used to calculate the *molecular mass* of a molecular compound, or the *formula mass* of an ionic compound. This is done by summing the atomic masses of the atoms indicated by the chemical formula. Just as molecular compounds consist of *molecules,* ionic compounds consist of *formula units.*

- The molar mass of a *compound* is determined by summing the molar masses of the elements specified by the compound's chemical formula. Molar masses are also used to convert from mass to moles of *compounds;* and Avogadro's constant is used to interconvert moles of compounds and numbers of molecules (for molecular compounds) or numbers of formula units (for ionic compounds).

Section 5.3

- A chemical formula can be used to calculate the ***mass percent composition*** or simply the ***percent composition*** of a compound.

- For a molecular compound, either the molecular formula or the empirical formula can be used to calculate percent composition.

Section 5.4

- Mass percent composition of a compound can be used to determine the empirical formula of a compound. By assuming a mass (typically 100 g), the mass of each element in a sample of the compound can be determined and converted to moles. The numbers of moles of each element are used as preliminary subscripts in the empirical formula—and must then be reduced to the smallest possible whole numbers.

Section 5.5

- If both the empirical formula and the molar mass of a compound are known, the molecular formula can be determined. Dividing the molar mass by the molar mass of the empirical formula gives a number (n) by which all of the subscripts in the empirical formula must be multiplied to get the molecular formula.

Key Terms

Avogadro's number (N_A) 169
Formula mass 175

Formula unit 175
Mass percent composition 182

Molar mass ($\mathcal{M}$) 172
Mole (mol) 169

Molecular mass 175
Percent composition 182

Key Equations

5.1	$\dfrac{\text{mass of element (g)}}{\text{molar mass of element (g/mol)}} = \text{moles of element}$	Moles of an element can be determined by dividing the mass of the element in grams by the molar mass of the element in grams per mole.
5.2	$\dfrac{\text{mass of compound (g)}}{\text{molar mass of compound (g/mol)}} = \text{moles of compound}$	Moles of a compound can be determined by dividing the mass of the compound in grams by the molar mass of the compound in grams per mole.
5.3	percent by mass of an element = $\dfrac{n \times \text{molar mass of element}}{\text{molar mass of compound}} \times 100\%$	Mass percent composition of a compound is determined by dividing the mass of each element in a mole of the compound (determined by number of moles of the element, n, times molar mass of the element) by the molar mass of the compound—and multiplying by 100%.
5.4	$\dfrac{\text{molar mass}}{\text{molar mass of empirical formula}} = n$	To determine the molecular formula from empirical formula and molar mass, we divide the molar mass of the compound by the molar mass of the empirical formula. We then multiply each subscript in the empirical formula by the result, n, to get the molecular formula. (In cases where the empirical and molecular formulas of a compound are the same, the result of Equation 5.4 will be $n = 1$.)

Molar Mass Determination and Conversion from Mass to Moles

Knowing the chemical formula of a compound enables us to calculate the compound's molar mass. For molecular compounds, this process is fairly straightforward. We simply multiply each atom's molar mass (from the periodic table) by its subscript in the molecular formula. For example, we calculate the molar mass of the compound sucrose ($C_{12}H_{22}O_{11}$) as:

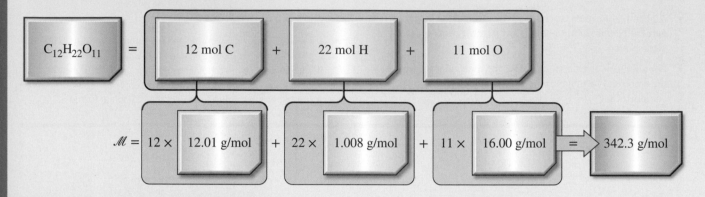

Determining the molar mass of some ionic compounds can be slightly more complicated, and is the source of some common mistakes. When determining the molar mass of an ionic compound, be sure to look carefully at the formula to determine the correct numbers of atoms. For example, the formula for barium acetate is $Ba(C_2H_3O_2)_2$. This formula contains one barium, *four* carbons, *six* hydrogens, and *four* oxygens. Its molar mass is calculated as:

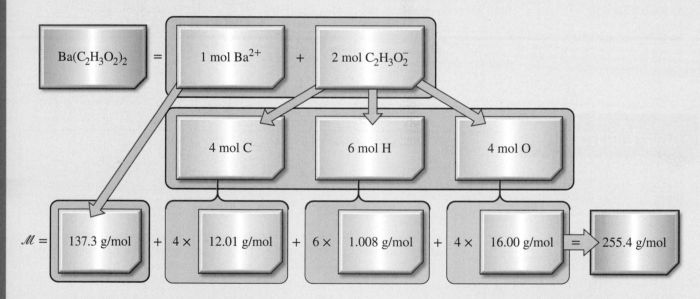

Having a correct molar mass allows us to convert from the mass of a compound to the number of moles—a skill that will be required throughout your chemistry course. Remember, to convert from mass of a compound in *grams* to *moles* of compound, divide the mass by the molar mass of the compound. Consider two 10.0-g samples: one of sucrose, and one of barium acetate. Because these two substances have *different* molar masses, samples of equal mass will contain *different* numbers of moles.

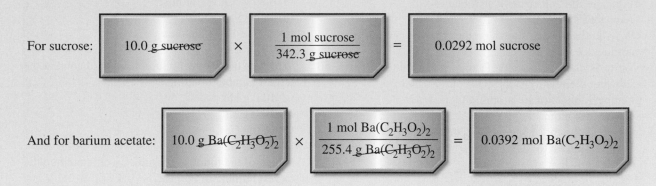

For sucrose: 10.0 g sucrose   × $\dfrac{1 \text{ mol sucrose}}{342.3 \text{ g sucrose}}$ =   0.0292 mol sucrose

And for barium acetate: 10.0 g Ba(C$_2$H$_3$O$_2$)$_2$ × $\dfrac{1 \text{ mol Ba(C}_2\text{H}_3\text{O}_2)_2}{255.4 \text{ g Ba(C}_2\text{H}_3\text{O}_2)_2}$ =   0.0392 mol Ba(C$_2$H$_3$O$_2$)$_2$

Key Skills Problems

5.1
Determine the molar mass of Co(NO$_2$)$_2$.
a) 151.0 g/mol
b) 209.9 g/mol
c) 163.9 g/mol
d) 119.0 g/mol
e) 104.0 g/mol

5.2
Determine the molar mass of iron(III) sulfate.
a) 151.9 g/mol
b) 344.1 g/mol
c) 399.9 g/mol
d) 271.9 g/mol
e) 359.7 g/mol

5.3
Determine the molar mass of nickel(II) perchlorate.
a) 110.14 g/mol
b) 158.14 g/mol
c) 257.59 g/mol
d) 222.14 g/mol
e) 316.28 g/mol

5.4
Determine the molar mass of XeF$_4$.
a) 171.29 g/mol
b) 207.29 g/mol
c) 150.29 g/mol
d) 544.16 g/mol
e) 525.16 g/mol

5.5
Determine the moles of Co(NO$_2$)$_2$ present in a 455-g sample of the compound.
a) 3.01 mol
b) 6.86 mol
c) 0.332 mol
d) 2.18 mol
e) 1.68 mol

5.6
Determine the moles of XeF$_4$ present in a 455-g sample of the compound.
a) 4.56 mol
b) 3.84 mol
c) 1.69 mol
d) 2.20 mol
e) 0.943 mol

5.7
Determine the mass of nickel(II) perchlorate that contains the same number of moles as 244 g of iron(III) sulfate.
a) 379 g
b) 94.7 g
c) 237 g
d) 61.0 g
e) 157 g

Questions and Problems

SECTION 5.1: COUNTING ATOMS BY WEIGHING

5.1 What are the units usually associated with Avogadro's constant?

5.2 Why do the masses shown on the periodic table not include units?

5.3 Write as many conversion factors as possible showing the relationships between Avogadro's number, a mole of He atoms, and the mass of a mole of He atoms.

5.4 Which of the following are not correct versions of a conversion factor using Avogadro's constant? Correct any that are incorrect.

(a) $\dfrac{6.022 \times 10^{23} \text{ atoms}}{1 \text{ gram}}$

(b) $\dfrac{6.022 \times 10^{23} \text{ Br atoms}}{1 \text{ mole Br}}$

(c) $\dfrac{6.022 \times 10^{23} \text{ g O atoms}}{1 \text{ mole O}}$

(d) $\dfrac{16.00 \text{ grams O atoms}}{6.022 \times 10^{23} \text{ mole O}}$

(e) $\dfrac{1 \text{ mol F}_2 \text{ molecules}}{6.022 \times 10^{23} \text{ g F}_2}$

5.5 Determine the number of atoms contained in
(a) 2.6 moles of atomic bromine.
(b) 8.1 moles of magnesium.
(c) 4.9 moles of argon.

5.6 Determine the number of moles of each element that contains
(a) 3.65×10^{24} P atoms.
(b) 9.21×10^{23} Al atoms.
(c) 5.38×10^{23} Ca atoms.

5.7 Determine the mass of neon that contains
(a) 4.38 moles of Ne.
(b) 3.44×10^{23} Ne atoms.

5.8 Determine the mass of carbon that contains
(a) 4.92×10^{24} C atoms.
(b) 5.88 moles of C.

5.9 Determine the mass of potassium that contains
(a) 8.11×10^{23} K atoms.
(b) 2.99 moles of K.

5.10 Determine the mass of xenon that contains
(a) 7.63×10^{24} Xe atoms.
(b) 9.55 moles of Xe.

5.11 Determine the moles of each substance:
(a) 84.4 g Si
(b) 5.09×10^{23} Si atoms
(c) 27.11 g Ag
(d) 3.82×10^{24} Ag atoms

5.12 Determine the moles of each substance:
(a) 6.89×10^{24} Se atoms
(b) 121 g Se
(c) 2.77×10^{23} Cs atoms
(d) 489 g Cs

5.13 Which of the following samples would contain the largest number of atoms?
(a) 10.0 g He
(b) 10.0 g Ar
(c) 10.0 g Ca
(d) 10.0 g Ba

5.14 Which of the following samples would contain the largest number of atoms?
(a) 1.00 g O
(b) 1.00 g Xe
(c) 1.00 g Be
(d) 1.00 g Li

5.15 Determine the mass in grams of each of the following:
(a) 15 Fe atoms (c) 15 K atoms
(b) 15 Ne atoms (d) 15 Sr atoms

5.16 Determine the mass in grams of each of the following:
(a) 12 N atoms (c) 12 Pb atoms
(b) 12 S atoms (d) 12 Au atoms

5.17 Fill in the following table.

Moles of Sample	Mass of Sample	Atoms in Sample
3.75 moles Ag		
	90.3 g Fe	
		2.38×10^{23} N atoms

5.18 Fill in the following table.

Moles of Sample	Mass of Sample	Atoms in Sample
0.118 mole K		
	644 g Cr	
		9.13×10^{24} Ne atoms

5.19 Determine the mass of Mg that contains the same number of atoms as 3.48 grams of Xe.

5.20 Determine the mass of Zn that is required to contain the same number of atoms as 92.6 g of Pb.

SECTION 5.2: COUNTING MOLECULES BY WEIGHING

5.21 Describe how to determine the molar mass of a compound from its chemical formula.

5.22 Explain how a 20.0-g sample of H_2 can contain more molecules than a 20.0-g sample of CO_2.

5.23 How many molecules are contained in each of the following samples?
(a) 25.7 g CO_2
(b) 98.3 g C_2Cl_4
(c) 38.7 g SO_2
(d) 55.4 g SF_4

5.24 How many molecules are contained in each of the following samples?
(a) 1.9 kg OCl_2
(b) 275 g $AsBr_3$
(c) 933 g P_2S_5
(d) 345 mg Br_2

5.25 Determine the number of molecules and the number of oxygen atoms in a 0.983-g sample of SO_3.

5.26 Determine the number of molecules and the number of hydrogen atoms in a 12.05-g sample of H_2SO_4.

5.27 Determine the mass of SO_2 that contains 1.525×10^{26} sulfur atoms.

5.28 Determine the mass of CO_2 that contains 4.32×10^{18} oxygen atoms.

5.29 Determine the mass in grams of each of the following:
(a) 25 molecules H_2O
(b) 25 molecules PCl_3
(c) 25 formula units of $LiNO_3$
(d) 25 formula units of $Mg(ClO_4)_2$

5.30 Determine the mass in grams of each of the following:
(a) 1 molecule CCl_4
(b) 1 molecule SO_2
(c) 1 formula unit Rb_2SO_4
(d) 1 formula unit $Al(C_2H_3O_2)_3$

5.31 What mass (in grams) of hydrogen is contained in each of the following samples?
(a) 3.76×10^{24} molecules H_2O
(b) 4.74×10^{23} formula units NH_4Cl
(c) 1.09×10^{24} molecules C_3H_8
(d) 6.22×10^{23} molecules NH_3

5.32 What mass (in grams) of chlorine is contained in each of the following samples?
(a) 7.81×10^{23} formula units $Ca(ClO_2)_2$
(b) 2.65×10^{24} molecules SCl_6
(c) 3.50×10^{23} formula units $CrCl_3$
(d) 1.47×10^{24} molecules $SiCl_4$

5.33 Explain how the moles of hydrogen can outnumber the moles of CH_4 in a sample.

5.34 Write a conversion factor that shows the relationship between the moles of Cl and moles of each compound.
(a) CCl_4 (c) NCl_3
(b) C_2Cl_4 (d) PCl_5

5.35 Write a conversion factor that shows the relationship between the moles of oxygen and moles of each compound.
(a) $Al_2(SO_4)_3$ (c) $Mg_3(PO_4)_2$
(b) N_2O_5 (d) OCl_2

5.36 Write a conversion factor that shows the relationship between moles of oxygen and moles of nitrogen in each compound.
(a) N_2O
(b) $Mg(NO_3)_2$
(c) $LiNO_2$
(d) N_2O_5

5.37 Write a conversion factor that shows the relationship between moles of hydrogen and moles of carbon in each compound.
(a) C_2H_4
(b) C_2H_6
(c) $Al(C_2H_3O_2)_3$
(d) H_2CO_3

5.38 Determine the number of molecules and the number of moles of F in each sample.
(a) 3.55 moles C_2F_4
(b) 1.78 moles NF_3
(c) 5.77 moles OF_2
(d) 2.11 moles SF_6

5.39 Determine the number of molecules and the number of moles of C in each sample.
(a) 3.55 moles C_2H_6
(b) 1.78 moles C_3H_8
(c) 5.77 moles H_2CO_3
(d) 2.11 moles $C_6H_{12}O_6$

5.40 Determine the number of formula units and the number of moles of O in each sample.
(a) 8.44 moles Na_2SO_4
(b) 6.38 moles $LiClO_4$
(c) 4.99 moles $Sr(ClO)_2$
(d) 2.33 moles $Cr(ClO_2)_3$

5.41 Determine the number of formula units and the number of moles of N in each sample.
(a) 6.03 moles NaCN
(b) 1.05 moles $Ca(NO_3)_2$
(c) 10.3 moles $(NH_4)_2SO_4$
(d) 8.55 moles $Cr(CN)_3$

5.42 Which of the following samples has the largest number of moles of O?
(a) 1.0 mole NO
(b) 1.0 mole N_2O_5
(c) 1.0 mole $CaCO_3$
(d) 1.0 mole TiO_2

5.43 Which of the following samples has the largest number of moles of Br?
(a) 2.5 moles $SrBr_2$
(b) 2.5 moles PBr_3
(c) 2.5 moles C_2Br_2
(d) 2.5 moles CBr_4

5.44 Determine the molar mass for each of the following compounds:
(a) $CuCl_2$
(b) CF_4
(c) Na_2O
(d) SCl_4

5.45   Determine the molar mass for each of the following compounds:
(a) $Sr(ClO)_2$
(b) Rb_2CO_3
(c) $(NH_4)_2O$
(d) $Al_2(SO_3)_3$

5.46   Determine the number of molecules of CO_2 and the number of O atoms in the following samples:
(a) 0.223 mole of CO_2
(b) 12.4 grams of CO_2

5.47   Determine the number of molecules of $AsCl_3$ and the number of Cl atoms in the following samples:
(a) 1.88 moles of $AsCl_3$
(b) 43.7 grams of $AsCl_3$

5.48   Find the mass of potassium (in grams) contained in a 5.00-g sample of each of the following compounds:
(a) K_2SO_4
(b) $KClO_4$
(c) K_3N
(d) K_2O

5.49   Find the mass of sulfur (in grams) contained in a 7.88-g sample of each of the following compounds:
(a) $Al_2(SO_3)_3$
(b) SF_6
(c) Al_2S_3
(d) $MgSO_4$

5.50   For each compound, determine the number of moles that contains 1.55×10^{23} O atoms.
(a) Al_2O_3
(b) CO_2
(c) OF_2
(d) $Mg(ClO_2)_2$

5.51   For each compound, determine the number of moles that contains 9.67×10^{22} Mg atoms.
(a) Mg_3P_2
(b) $Mg_3(PO_3)_2$
(c) $Mg(CN)_2$
(d) MgS

5.52   For each compound, determine the mass in grams that contains 3.44×10^{21} N atoms.
(a) NO_2 (c) Mg_3N_2
(b) N_2O_5 (d) $Mg(NO_2)_2$

5.53   For each compound, determine the mass in grams that contains 7.32×10^{19} C atoms.
(a) C_2H_4 (c) Li_2CO_3
(b) $Ca(C_2H_3O_2)_2$ (d) MgC_2O_4

5.54   Which of the following are not correct conversion factors for molecular bromine? Correct any that are incorrect.

(a) $\dfrac{6.022 \times 10^{23}\ Br_2}{1\ mol\ Br_2}$

(b) $\dfrac{6.022 \times 10^{23}\ g\ Br_2}{1\ mol\ Br_2}$

(c) $\dfrac{79.90\ g\ Br_2}{1\ mol\ Br_2}$

(d) $\dfrac{159.80\ g\ Br_2}{2\ mol\ Br_2}$

(e) $\dfrac{79.90\ g\ Br}{1\ mol\ Br}$

5.55   Which of the following are not correct conversion factors for molecular oxygen? Correct any that are incorrect.

(a) $\dfrac{16.00\ g\ O}{1\ mol\ O_2}$

(b) $\dfrac{6.022 \times 10^{23}\ O_2}{1\ mol\ O_2}$

(c) $\dfrac{1\ mol\ O_2}{32.00\ g\ O_2}$

(d) $\dfrac{6.022 \times 10^{23}\ O}{1\ mol\ O}$

(e) $\dfrac{2\ mol\ O_2}{32.00\ g\ O_2}$

5.56   Which of the following are not correct conversion factors for SCl_4? Correct any that are incorrect.

(a) $\dfrac{1\ mol\ Cl}{1\ mol\ SCl_4}$

(b) $\dfrac{173.87\ g\ SCl_4}{1\ mol\ SCl_4}$

(c) $\dfrac{6.022 \times 10^{23}\ molecules\ SCl_4}{1\ mol\ SCl_4}$

(d) $\dfrac{2\ mol\ Cl_2}{1\ mol\ SCl_4}$

(e) $\dfrac{6.022 \times 10^{23}\ molecules\ SCl_4}{1\ mol\ SCl_4}$

5.57   Which of the following are not correct conversion factors for $(NH_4)_2O$? Correct any that are incorrect.

(a) $\dfrac{1\ mol\ O}{1\ mol\ (NH_4)_2O}$

(b) $\dfrac{4\ mol\ H}{1\ mol\ (NH_4)_2O}$

(c) $\dfrac{34.04\ g\ (NH_4)_2O}{1\ mol\ (NH_4)_2O}$

(d) $\dfrac{6.022 \times 10^{23}\ formula\ units\ (NH_4)_2O}{1\ mol\ (NH_4)_2O}$

(e) $\dfrac{1\ mol\ NH_4^+}{1\ mol\ (NH_4)_2O}$

5.58 Fill in the following table.

Moles of Sample	Mass of Sample	O Atoms in Sample
1.32 moles $Ca(ClO_2)_2$		
	90.3 g Fe_2O_3	
_____ moles $Na_2C_2O_4$	_____ g $Na_2C_2O_4$	5.69×10^{22} atoms O

5.59 Fill in the following table.

Moles of Sample	Mass of Sample	N Atoms in Sample
6.44 moles $Al(NO_2)_3$		
	4.31 g Mg_3N_2	
_____ moles $(NH_4)_2CO_3$	_____ g $(NH_4)_2CO_3$	2.38×10^{23} atoms N

SECTION 5.3: MASS PERCENT COMPOSITION

5.60 Describe the steps necessary to determine the mass percent of one element in a compound, given the chemical formula.

5.61 To what number must the mass percent values of all elements sum for any compound?

5.62 Determine the mass percent of carbon in each of the following compounds:
(a) $Rb_2C_2O_4$
(b) K_2CO_3
(c) $Fe(C_2H_3O_2)_3$
(d) CF_4

5.63 Determine the mass percent of nitrogen in each of the following compounds:
(a) $Mg(NO_3)_2$
(b) N_2Cl_4
(c) N_2O_5
(d) NBr_3

5.64 Determine the mass percent of fluorine in each of the following compounds:
(a) AlF_3 (c) PF_5
(b) CF_4 (d) SF_6

5.65 Determine the mass percent of oxygen in each of the following compounds:
(a) $Ba(ClO_3)_2$ (c) Li_2O
(b) $Fe(OH)_3$ (d) H_3PO_4

5.66 Determine the mass percent of the *anion* in each compound.
(a) Calcium nitrate   (c) Sodium chlorate
(b) Aluminum sulfite   (d) Strontium acetate

5.67 Determine the mass percent of *cation* in each compound.
(a) Ammonium chloride
(b) Copper(II) oxide
(c) Magnesium phosphate
(d) Aluminum oxide

SECTION 5.4: USING MASS PERCENT COMPOSITION TO DETERMINE EMPIRICAL FORMULA

5.68 Azithromycin is a popular antibiotic used to treat bacterial infections. Its chemical formula is $C_{38}H_{72}N_2O_{12}$. What is its empirical formula?

5.69 Nicotine is an addictive substance found in cigarettes. Its chemical formula is $C_{10}H_{14}N_2$. What is its empirical formula?

5.70 Fructose is an ingredient found in many prepared foods. Its chemical formula is $C_6H_{12}O_6$. What is its empirical formula?

5.71 Caffeine is found in coffee and many soft drinks. Its chemical formula is $C_8H_{10}N_4O_2$. What is its empirical formula?

5.72 A compound is found to be 64.19% Cu and 35.81% Cl by mass. Determine its empirical formula.

5.73 A compound is found to be 63.65% N and 36.35% O by mass. Determine its empirical formula.

5.74 A compound is found to be 43.64% P and 56.36% O by mass. Determine its empirical formula.

5.75 A compound is found to be 51.95% Cr and 48.05% S by mass. Determine its empirical formula.

5.76 A 20.00-mg sample of a compound is decomposed and found to contain 5.05 mg Ti and 14.95 mg Cl. Determine its empirical formula.

5.77 A 79.22-g sample of a compound is decomposed and found to contain 30.71 g Ca, 15.82 g P, and 32.69 g O. Determine its empirical formula.

5.78 An 8.819-mg sample of P reacts with Cl_2 to form 39.11 mg of product. Determine the empirical formula of the product.

5.79 A 29.3-g sample of Ti reacts with O_2 to form 48.9 grams of product. Determine the empirical formula of the product.

SECTION 5.5: USING EMPIRICAL FORMULA AND MOLAR MASS TO DETERMINE MOLECULAR FORMULA

5.80 What's the difference between *empirical formula* and *molecular formula*? Is it possible for the empirical and molecular formulas of a compound to be the same? Explain.

5.81 A compound composed of nitrogen and hydrogen is found to have an empirical formula of NH_2. Determine the molecular formula of the compound if its molar mass is 32.05 g/mol.

5.82   A compound composed of carbon, hydrogen, and oxygen is found to have an empirical formula of C_2H_4O. Determine the molecular formula of the compound if its molar mass is 88.10 g/mol.

5.83   A compound containing only carbon and hydrogen was found to be 85.62% carbon. Determine the molecular formula of the compound if its molar mass is 84.16 g/mol.

5.84   A compound containing carbon, hydrogen, and oxygen was found to be 55.80% C and 37.18% O by mass. Determine the molecular formula of the compound if its molar mass is found to be 86.08 g/mol.

CUMULATIVE PROBLEMS

5.85   Calculate the number of chlorine atoms in each of the following samples:
(a) 24.3 g Cl_2
(b) 83.5 g Cl
(c) 4.22 moles $CHCl_3$
(d) 67.5 g CCl_4
(e) 3.45×10^{22} molecules $CHCl_3$

5.86   Calculate the number of oxygen atoms in each of the following samples:
(a) 2.98×10^{23} molecules P_2O_5
(b) 1.55×10^{21} molecules O_2
(c) 3.77 moles CO_2
(d) 27.6 g $Al_2(CO_3)_3$
(e) 9.14 g $Mg(ClO_3)_2$

5.87   Calculate the number of hydrogen atoms contained in each of the following samples:
(a) 198 mg $C_2H_3O_2$
(b) 3.82 kg H_3PO_4
(c) 7.44×10^{10} ng C_2H_6

5.88   Determine the mass (in the units indicated) of each compound that contains 2.19×10^{23} Na atoms.
(a) mg of Na_2O
(b) kg of Na_3PO_3
(c) ng of Na_2SO_3

5.89   Determine the mass (in g) of each compound that contains 2.97×10^{23} N atoms and convert each mass to moles of compound.
(a) N_2O_5
(b) $(NH_4)_2O$
(c) $Al(NO_3)_3$

5.90   Determine the mass (in g) of each compound that contains 7.98×10^{23} S atoms and convert each mass to moles of compound.
(a) $Ti(SO_3)_2$
(b) S_2F_6
(c) Fe_2S_3

5.91   Determine the number of sodium ions in each sample.
(a) 7.95 moles sodium phosphate
(b) 4.36×10^{24} formula units sodium oxide
(c) 213 g sodium sulfide

5.92   Determine the number of sulfate ions in each sample.
(a) 682 g aluminum sulfate
(b) 2.11 moles potassium sulfate
(c) 1.79×10^{23} formula units magnesium sulfate

5.93   A particular commemorative set of coins contains two 1.00-oz silver coins and three 0.500-oz gold coins. (a) How many gold coins are there in 49.0 lb of coin sets? (b) How many silver coins are there in a collection of sets that has a total mass of 63.0 lb? (c) What is the total mass (in lb) of a collection of sets that contains 93 gold coins? (d) What is the mass of silver coins (in lb) in a collection of coin sets that contains 9.00 lb of gold coins? (e) Determine the number of silver and gold atoms in the collection from part (d). (Assume that the coins are pure silver and pure gold.)

Answers to In-Chapter Materials

Answers to Practice Problems

5.1A (a) 4.40×10^{24} atoms, (b) 148 mol. **5.1B** (a) 6.32×10^{49} atoms, (b) 3.87×10^{-3} mol. **5.2A** (a) 3.066×10^{-1} mol, (b) 1.72×10^{-3} mol, (c) 2.98×10^{-1} mol. **5.2B** (a) 1.10×10^2 g, (b) 3.0×10^{-1} g, (c) 4.125×10^{-3} g. **5.3A** (a) 3.225×10^{23} atoms, (b) 5.374×10^{-10} g. **5.3B** (a) 8.116×10^2 g, (b) 1.439×10^{31} atoms. **5.4A** (a) 137.32 g/mol, (b) 98.09 g/mol, (c) 68.15 g/mol, (d) 146.07 g/mol. **5.4B** (a) 5, (b) 2, (c) 2, (d) 4. **5.5A** (a) 495 g, (b) 0.704 mol. **5.5B** 1.626 mol. **5.6A** (a) 6.68×10^{23} molecules, 1.34×10^{24} atoms; (b) 1.32×10^{-7} g. **5.6B** (a) 1.11×10^{25} N atoms, 2.79×10^{25} O atoms; (b) 33.4 g. **5.7A** 5.82×10^{23} Mg^{2+} ions, 1.16×10^{24} CN^- ions. **5.7B** 406.5 g. **5.8A** 57.13% C, 6.165% H, 9.52% N, 27.18% O. **5.8B** 70.95% C, 6.315% H, 3.401% F, 5.016% N, 14.32% O. **5.9A** C_2H_6O. **5.9B** $C_9H_8O_4$. **5.10A** C_2H_5O, $C_4H_{10}O_2$. **5.10B** C_3H_4, C_9H_{12}.

Answers to Checkpoints

5.1.1 a. **5.1.2** d. **5.1.3** b. **5.1.4** c. **5.2.1** e. **5.2.2** a. **5.2.3** b. **5.2.4** d. **5.3.1** b. **5.3.2** e. **5.3.3** d. **5.3.4** a. **5.4.1** a. **5.4.2** e. **5.5.1** e. **5.5.2** d. **5.5.3** d.

CHAPTER 6

Molecular Shape

6.1 **Drawing Simple Lewis Structures**
- Lewis Structures of Simple Molecules
- Lewis Structures of Molecules with a Central Atom
- Lewis Structures of Simple Polyatomic Ions

6.2 **Lewis Structures Continued**
- Lewis Structures with Less Obvious Skeletal Structures
- Lewis Structures with Multiple Bonds
- Exceptions to the Octet Rule

6.3 **Resonance Structures**

6.4 **Molecular Shape**
- Bond Angles

6.5 **Electronegativity and Polarity**
- Electronegativity
- Bond Polarity
- Molecular Polarity

6.6 **Intermolecular Forces**
- Dipole-Dipole Forces
- Hydrogen Bonding
- Dispersion Forces
- Intermolecular Forces in Review

The shapes of molecules, and the resulting attractive forces between them, give rise to some interesting observable phenomena, such as the beautiful and unique shapes of snowflakes.

Robin Treadwell/Science Source

In This Chapter, You Will Learn

How to draw Lewis structures for molecules and polyatomic ions; and how to predict molecular shape. Further, you will learn how to identify polar chemical bonds, how to determine whether a molecule is polar overall, and about the attractive forces that hold molecules and atoms together in a pure substance.

Things To Review Before You Begin

- How charged particles interact. [◀◀ Section 1.2]
- How to identify an atom's valence electrons. [◀◀ Section 2.5]
- Chapter 2 Key Skills [◀◀ pages 68–69]

We have seen how molecules can be represented with chemical formulas, with structural formulas, and with molecular models. In this chapter, we explain how to represent molecules in a way that can help us understand and predict some of their important properties, including their shapes and how they interact with one another.

6.1 Drawing Simple Lewis Structures

In Chapter 3, we saw how we could represent simple molecules such as Cl_2, H_2, and HCl by combining the Lewis symbols of their constituent atoms.

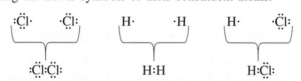

We also saw how the dots between atoms, which represent electrons shared by the two atoms, constitute a covalent bond and can also be drawn as dashes.

$$:\ddot{C}l—\ddot{C}l: \quad H—H \quad H—\ddot{C}l:$$

Each of these is an example of a Lewis dot structure, or more commonly, a *Lewis structure*—a representation we can draw of a chemical species that contains one or more covalent bonds. In this section, we present the steps involved in drawing Lewis structures for relatively simple chemical species, such as molecules and polyatomic ions.

Lewis Structures of Simple Molecules

We learned in Chapter 3 that the atoms of most main group elements achieve stability by becoming isoelectronic [◀◀ Section 2.7] with noble gas atoms. This tendency gives rise to something called the *octet rule,* which says that atoms will lose, gain, or share electrons so that each atom is surrounded by eight electrons. (A universal exception to this is hydrogen, which can accommodate only *two* electrons because of its extraordinarily small size.) Consider for a moment the simple Lewis structures shown previously. In the first, two chlorine atoms, each with seven valence electrons, have moved close enough together to share a pair of electrons. The total number of valence electrons, 14, is not sufficient for each Cl atom to possess the necessary eight electrons to obey the octet rule. However, when atoms form a covalent bond by *sharing* electrons, each atom counts the *shared* electrons as its own—which allows each atom to obey the octet rule. Although we have begun to show shared electron pairs as *dashes,* which is appropriate for the purpose of drawing Lewis structures, the way the octet rule applies to both atoms in the Cl_2 molecule may be easier to visualize if we temporarily revert to representing each electron as a dot:

$$:\overset{..}{\underset{..}{Cl}}\overset{..}{\underset{..}{Cl}}:$$

The left Cl atom counts all the   The right Cl atom counts all the
dots in the blue circle as its own.   dots in the pink circle as its own.

Thus, each Cl atom in the Cl_2 molecule is surrounded by eight electrons and has a complete octet. The steps involved in drawing Lewis structures for simple molecules are as follows:

1. Using the molecular formula, write the symbol for each of the atoms the way you expect them to be arranged in the molecule. This is referred to as drawing a **skeletal structure.** Then, using dashes, draw bonds between the atoms in the skeletal structure.
2. Count the total number of valence electrons possessed by all of the atoms in the molecule—remembering that the number of valence electrons possessed by any main-group atom is equal to the last digit in its group number. (Group 1: one valence electron, Group 2, two valence electrons, Group 13: three valence electrons, etc.)
3. For each bond (dash) in the structure drawn in step 1, subtract two electrons from the total number of valence electrons determined in step 2. The result is the number of electrons that must be distributed to complete the Lewis structure.
4. Distribute the number of electrons determined in step 3 as pairs of dots to complete the octet of each atom in the structure. (Remember that hydrogen gets only *two* electrons, not eight.)

Each of these steps is clarified in the examples that follow. Let's first look in detail at the implementation of these steps using the familiar examples of Cl_2 and HCl.

Step		Cl_2	HCl	
1	Arrange atoms in a skeletal structure, include dashes for bonds.	Cl—Cl	H—Cl	In each of these cases, there is only one way to arrange the atoms: side by side.
2	Sum the valence electrons for all atoms.	$7 + 7 = 14$	$1 + 7 = 8$	Each Cl, in Group 17, has seven valence electrons; each H, in Group 1, has one valence electron.
3	For each bond in the structure from step 1, subtract two electrons.	$14 - 2 = 12$	$8 - 2 = 6$	In each of these examples, there is just one bond in the skeletal structure, so we subtract 2 from the sum of valence electrons.
4	Distribute the number of electrons determined in step 3 as dots to give each atom a complete octet—except H, which gets just two electrons.	$:\overset{..}{\underset{..}{Cl}}-\overset{..}{\underset{..}{Cl}}:$	$H-\overset{..}{\underset{..}{Cl}}:$	With all of the electrons distributed, each Cl atom has a complete octet, and the H atom has two electrons.

Lewis Structures of Molecules with a Central Atom

In the case of a diatomic molecule such as Cl_2 or HCl, the first step in drawing a Lewis structure, drawing the *skeletal* structure, is fairly simple. The only way to arrange *two* atoms is by placing them next to each other. There are many cases, though, where a little more thought has to go into determining a skeletal structure. We have encountered polyatomic molecules such as CCl_4 and H_2O, in which there might be more than one possible way to arrange the atoms. Let's examine each of the steps again using the slightly less simple examples of CCl_4 and H_2O. Each of these is a binary compound with one *unique* atom, and more than one atom of another element. In CCl_4, the unique atom is *carbon;* in H_2O, the unique atom is *oxygen.* In cases such as these, we place the unique atom in the center, making it the **central atom.** We then arrange the other atoms symmetrically around the central atom, and draw a bond between the central atom and each of the other atoms. The atoms that surround a central atom in this way are referred to as **terminal atoms.**

Student Note: There will be cases where the rules we have learned so far do not enable you to determine the correct central atom. In these cases, you will be given more information regarding the arrangement of atoms.

Step		CCl_4	H_2O
1	Arrange atoms in a skeletal structure, include dashes for bonds—placing the unique atom at the center and arranging the other atoms around it symmetrically.	Cl | Cl—C—Cl | Cl	H—O—H
2	Sum the valence electrons for all atoms.	4 for C (Group 4) $4 + 4(7) = 32$ 7 for each Cl (Group 17)	6 for O (Group 16) $6 + 2(1) = 8$ 1 for each H (Group 1)
3	For each bond in the structure from step 1, subtract two electrons.	$32 - 4(2) = 24$ 2 for each of four bonds	$8 - 2(2) = 4$ 2 for each of two bonds
4	Distribute the number of electrons determined in step 3 as dots to give each atom a complete octet—except H, which gets just two electrons.	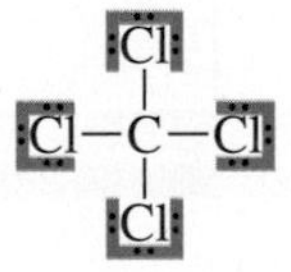 Cl | Cl—C—Cl | Cl	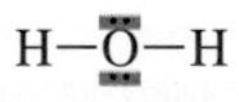 H—O—H

In each of the Lewis structures we have drawn so far, a dash represents a pair of *shared* electrons, also known as a **bond pair** or *bonding pair.* Each pair of dots represents a pair of electrons that is *not* shared—and is not involved in chemical bonding. An unshared pair of electrons is known as a **lone pair** or a *nonbonding pair.*

Lewis Structures of Simple Polyatomic Ions

We learned in Chapter 3 that the atoms in polyatomic ions are held together by covalent bonds. Therefore, as in the case of molecules, we can represent polyatomic ions with

Lewis structures. This requires two minor modifications of the steps we have learned so far for drawing Lewis structures:

- We must account for the polyatomic ion's charge in the count of valence electrons.
- We enclose the final Lewis structure in brackets and denote the ion's charge with a superscript.

To illustrate this, we will use two familiar examples of polyatomic ions: NH_4^+ and ClO^-.

Step		NH_4^+	ClO^-
1	Arrange atoms in a skeletal structure, include dashes for bonds. Where there is a unique atom, place it at the center and arrange the other atoms around it symmetrically.	$\begin{array}{c} H \\ \| \\ H-N-H \\ \| \\ H \end{array}$	$Cl-O$
2	Sum the valence electrons for all atoms, subtracting one electron for each positive charge and adding one electron for each negative charge.	5 for N (Group 5) Subtract 1 to account for +1 charge on ion. $5 + 4(1) - 1 = 8$ 1 for each H (Group 1)	7 for Cl (Group 17) Add 1 to account for −1 charge on ion. $7 + 6 + 1 = 14$ 6 for O (Group 16)
3	For each bond in the structure from step 1, subtract two electrons.	$8 - 4(2) = 0$ 2 for each of four bonds	$14 - 2 = 12$ 2 for just one bond
4	Distribute the number of electrons determined in step 3 as dots to give each atom a complete octet—except H, which gets just two electrons. Enclose structures of polyatomic ions in brackets with a superscript charge.	$\left[\begin{array}{c} H \\ \| \\ H-N-H \\ \| \\ H \end{array}\right]^+$	$[\ddot{\underset{..}{Cl}}-\ddot{\underset{..}{O}}]^-$

Figure 6.1 examines the Lewis structure of CCl_4 in some detail and calls attention to the important features.

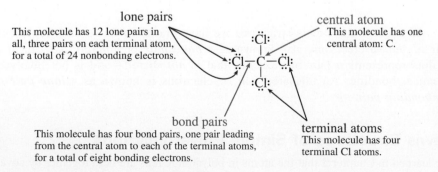

lone pairs
This molecule has 12 lone pairs in all, three pairs on each terminal atom, for a total of 24 nonbonding electrons.

central atom
This molecule has one central atom: C.

bond pairs
This molecule has four bond pairs, one pair leading from the central atom to each of the terminal atoms, for a total of eight bonding electrons.

terminal atoms
This molecule has four terminal Cl atoms.

Figure 6.1 Anatomy of a simple Lewis structure.

Although every Lewis structure contains at least two atoms and at least one bonding pair of electrons, not all of the Lewis structures we have seen so far contain all of the features illustrated in Figure 6.1. The Lewis structure for H_2, for example, contains no central or terminal atoms, and no lone pairs. That of HCl has lone pairs, but no central atom. Further, as we explain in the next section, some molecules have more than one central atom—and some bonds consist of more than just one shared pair of electrons. These things, along with some other more advanced aspects of Lewis structures, are covered later in the chapter; but for now it is important that you master the basic steps of drawing simple Lewis structures.

Sample Problem 6.1 lets you practice drawing Lewis structures of some simple molecules and polyatomic ions.

> **Student Note:** Often, the *first step* in solving a multistep chemistry problem is drawing a Lewis structure. If the Lewis structure you draw is not correct, the rest of the work that goes into solving such a problem—even if done correctly—may yield an incorrect answer. Therefore, it is *very* important that you learn to draw proper Lewis structures *now;* and the only way to do this is with *lots* of practice.

SAMPLE PROBLEM 6.1 Drawing Lewis Structures for Molecules and Polyatomic Ions

Draw the Lewis structure for each of the following substances: (a) $SeBr_2$, (b) H_3O^+, (c) PCl_3

Strategy Use the procedure described in steps 1–4 for drawing Lewis structures.

Setup & Solution

	(a)	(b)	(c)
Step 1: Determine the skeletal structure, with the unique atom at the center.	Br—Se—Br	$$\begin{array}{c} H \\ \mid \\ H-O-H \end{array}$$	$$\begin{array}{c} Cl \\ \mid \\ Cl-P-Cl \end{array}$$
Step 2: Sum the valence electrons.	$Se = 6$ $\underline{Br = 2(7) = 14}$ $20\ e^-$	$O = 6$ $H = 3(1) = 3$ $\underline{charge = -1}$ $8\ e^-$	$P = 5$ $\underline{Cl = 3(7) = 21}$ $26\ e^-$
Step 3: For each bond in the structure from step 1, subtract 2 electrons.	$20 - 2(2) = 16$	$8 - 3(2) = 2$	$26 - 3(2) = 20$
Step 4: Distribute the remaining electrons to give each atom a complete octet, except H.	$:\!\ddot{B}r\!-\!\ddot{S}e\!-\!\ddot{B}r\!:$	$\left[\begin{array}{c} H \\ \mid \\ H-\underset{\cdot\cdot}{O}-H \end{array} \right]^+$	$:\!\ddot{C}l\!: \\ \mid \\ :\!\ddot{C}l\!-\!P\!-\!\ddot{C}l\!:$

THINK ABOUT IT

Counting the total number of valence electrons should be relatively simple to do, but it is often done hastily and is therefore a potential source of error in this type of problem. Remember that the number of valence electrons for each element is equal to the last digit of the group number of that element.

Practice Problem **A**TTEMPT Draw the Lewis structure for each of the following substances:

(a) ClO_3^- (b) PCl_4^+ (c) $SiBr_4$

Practice Problem **B**UILD Draw the Lewis structure for each of the following substances:

(a) NF_4^+ (b) BrO_4^- (c) H_3S^+

Practice Problem **C**ONCEPTUALIZE Which of the elements listed *could* be represented by A in the Lewis structure shown? Which could be represented by X in the Lewis structure?

B, C, N, O, F, Al, Si, P, S, Cl, As, Se, Br, I, H

$$:\!\ddot{X}\!-\!\ddot{A}\!-\!\ddot{X}\!: \\ \mid \\ :\!\ddot{X}\!:$$

CHECKPOINT–SECTION 6.1   Drawing Simple Lewis Structures

6.1.1 Identify the correct Lewis structure for $AlCl_4^-$.

a)
$$\left[\begin{array}{c} Cl \\ | \\ Cl-Al-Cl \\ | \\ Cl \end{array}\right]^-$$

b)
$$\begin{array}{c} :\ddot{C}l: \\ | \\ :\ddot{C}l-Al-\ddot{C}l: \\ | \\ :\ddot{C}l: \end{array}$$

c)
$$\begin{array}{c} :\ddot{C}l: \\ | \\ :\ddot{C}l-Al-\ddot{C}l: \\ | \\ :\ddot{C}l: \end{array}$$

d)
$$\begin{array}{c} Cl \\ | \\ Cl-Al-Cl \\ | \\ Cl \end{array}$$

e)
$$\left[\begin{array}{c} :\ddot{C}l: \\ | \\ :\ddot{C}l-Al-\ddot{C}l: \\ | \\ :\ddot{C}l: \end{array}\right]^-$$

6.1.2 Identify the correct Lewis structure for ClO_2^-.

a) $\left[:\ddot{O}-Cl-\ddot{O}:\right]^-$

b) $:O-\ddot{C}l^-O:$

c) $\left[:\ddot{O}-\ddot{C}l-\ddot{O}:\right]^-$

d) $:\ddot{O}-\ddot{C}l^-\ddot{O}:$

e) $\left[:O-\ddot{C}l-O:\right]^-$

6.2 Lewis Structures Continued

Thus far, we have drawn Lewis structures for molecules that have, at most, one central atom; and in which each bond has consisted of a single pair of shared electrons. In this section, we describe how to draw Lewis structures for slightly more complex molecules and polyatomic ions.

Lewis Structures with Less Obvious Skeletal Structures

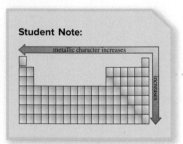

Student Note:

In some cases, it may not be immediately apparent how to arrange atoms in a skeletal structure for a molecule or polyatomic ion. For example, in the molecule CH_3Cl, there is not just *one* unique atom (one that appears only once in the formula) that the rules we have learned thus far tell us to place in the center of the structure—there are two: C and Cl. In such a case, we need to choose the central atom based on metallic character [◀◀ Section 2.6] and place the atom with the *greater* metallic character in the center of the skeletal structure. For CH_3Cl, because C has greater metallic character than Cl, this gives a skeletal structure of:

$$\begin{array}{c} H \\ | \\ H-C-Cl \\ | \\ H \end{array}$$

From this point, we apply the rules for drawing Lewis structures and get a final structure of:

$$\begin{array}{c} H \\ | \\ H-C-\ddot{C}l: \\ | \\ H \end{array}$$

Student Note: In this context, *central* atom refers to one that has bonds to more than one other atom—whereas a *terminal* atom has a bond to just *one* other atom.

There will also be cases where there is more than one "central" atom. An example is the molecule ethane, C_2H_6, where both C atoms are considered to be central atoms, and each of the H atoms is a terminal atom. Note that both C atoms have complete octets in the Lewis structure:

$$\begin{array}{c} H \quad H \\ | \quad | \\ H-C-C-H \\ | \quad | \\ H \quad H \end{array}$$

Lewis Structures with Multiple Bonds

Another common occurrence is running out of valence electrons before satisfying the octet rule for every atom in a molecule. Consider the O_2 molecule. If we follow steps 1 through 4 for drawing the Lewis structure of a molecule, we find the following:

1. Draw a skeletal structure.   O—O
2. Count valence electrons. $2(6) = 12$
3. Subtract 2 electrons per bond in the skeletal structure. $12 - 2 = 10$
4. Distribute remaining electrons.   :Ö—Ö:

Although we have distributed all of the available valence electrons, one of the oxygen atoms does not have a complete octet. In these cases, we must add a step to the process in which we move lone pairs on terminal atoms to positions in between atoms to create *multiple* bonds. A ***multiple bond*** is one in which more than one electron pair is shared between two atoms. Specifically, a ***double bond*** is one that consists of *two* shared pairs of electrons; and a ***triple bond*** is one that consists of *three* shared pairs of electrons. In this case, we take one of the lone pairs on the O atom that *does* have a complete octet, and turn it into a *shared* pair between the two O atoms:

5. If necessary, change lone pairs to bond pairs, creating multiple bonds as needed to complete octets.

:Ö—Ö: → :Ö=Ö: → (:Ö(:)Ö:)

In the resulting structure, :Ö=Ö:, there is a double bond between the O atoms, and each atom has a complete octet, which may be easier to see when shared electrons are represented with dots (right).

Sample Problem 6.2 lets you practice drawing Lewis structures that include multiple bonds.

> **Student Hot Spot**
>
> Student data indicate that you may struggle to draw the Lewis structure of a molecule with multiple bonds. Access the eBook to view additional Learning Resources on this topic.

SAMPLE PROBLEM 6.2 Drawing Lewis Structures with Double or Triple Bonds

Draw the Lewis structure for CO_2.

Strategy Use the procedure described in steps 1–5 for drawing Lewis structures.

Setup

Step 1: Place C, the unique element, in the center of the skeletal structure:

$$O-C-O$$

Step 2: The total number of valence electrons is $4 + 2(6) = 16$, four electrons from carbon and six from each oxygen atom.

Step 3: Subtract four electrons (two for each bond in the structure in step 1) from 16 and arrive at 12 remaining electrons.

Step 4: Distribute the 12 remaining electrons:

:Ö—C—Ö:

Step 5: Change lone pairs to bond pairs to complete octets.

Solution

:Ö=C=Ö:

THINK ABOUT IT

It is generally best to distribute the electrons in step 4 around the "outer" or *terminal* atoms first. After satisfying the octets of the terminal atoms, place any remaining electrons on the central atom.

Practice Problem **A**TTEMPT Draw the Lewis structure for HCN.

Practice Problem **B**UILD Draw the Lewis structure for NO^-.

Practice Problem **C**ONCEPTUALIZE Given the following Lewis structures, determine the charge on each:

(a) $[:N\equiv O:]^?$ (b) $[:Ö=N=Ö:]^?$

Exceptions to the Octet Rule

There are three common exceptions to the octet rule. The first involves molecules that have an odd number of electrons. NO, nitrogen monoxide (commonly known as *nitric oxide*), has a total of 11 valence electrons. There simply is no way to arrange this number of valence electrons in a way that gives a complete octet to both atoms. The best structure we can draw is one in which the O atom has a complete octet, and there are seven electrons around the N atom:

$$\cdot \ddot{N} = \ddot{O} :$$

The second exception involves molecules in which either beryllium (Be) or boron (B) is the central atom. Be and B both have greater metallic character than most of the elements that we encounter in molecular compounds. Neither needs to be surrounded by eight electrons; in fact, Be can have just *four* electrons around it, and B can have just *six*, as the structures of $BeCl_2$ and BF_3 illustrate:

The third exception involves molecules in which the central atom is an element from the third period or later. Because these atoms are larger than those in the second period, they can accommodate more than eight electrons around them—although they *can* and often *do* obey the octet rule. A central atom with more than eight electrons around it is said to have an **expanded octet.** Examples include PCl_5 and SF_6:

CHECKPOINT–SECTION 6.2 Lewis Structures Continued

6.2.1 Determine the Lewis structure of HSiN.

a) $H-\ddot{Si}=N:$

b) $H-Si\equiv N:$

c) $H-\ddot{Si}-\ddot{N}:$

d) $H-Si=\ddot{N}:$

e) $H-Si\equiv N$

6.2.2 Determine the Lewis structure of PO_3^-.

a) $:\ddot{O}-P=\ddot{O}:$ with $:\ddot{O}:$ below P

b) $\left[:\ddot{O}-P=\ddot{O}:\right]^-$ with $:\ddot{O}:$ below P

c) $\left[:\ddot{O}-\ddot{P}=\ddot{O}:\right]^-$ with $:\ddot{O}:$ below P

d) $\left[:\ddot{O}-\ddot{P}-\ddot{O}:\right]^-$ with $:\ddot{O}:$ below P

e) $\left[:\ddot{O}-P-\ddot{O}:\right]^-$ with $:\ddot{O}:$ below P

6.2.3 Determine the charge on the ion shown in the Lewis structure.

$$\left[:\ddot{O}-\ddot{A}s-\ddot{O}:\right]^?$$ with $:\ddot{O}:$ below As

a) 1– b) 2– c) 3– d) 1+ e) 2+

6.2.4 Determine the charge on the ion shown in the Lewis structure.

$$\left[:\ddot{O}-Br=\ddot{O}:\right]^?$$ with $:\ddot{O}:$ below Br

a) 1– b) 2– c) 3– d) 1+ e) 2+

Familiar Chemistry

Bleaching, Disinfecting, and Decontamination

Although it is not possible for us to draw completely satisfactory Lewis structures of molecules with odd numbers of electrons, many such substances exist. One such molecule of note is ClO_2, chlorine dioxide, which has a total of 19 valence electrons. The presence of an unpaired electron makes a molecule highly reactive—something that has enabled several industries to make use of ClO_2.

One industry that does this is the paper industry, which uses ClO_2 to bleach the wood pulp from which paper is made. The use of ClO_2 is part of an industry method known as elemental chlorine free (ECF) method, which produces fewer environmentally harmful by-products than the older method that involved the use of Cl_2 (elemental chlorine). Chlorination to disinfect municipal water supplies also uses ClO_2 rather than Cl_2 in many instances, for the purpose of minimizing by-products that are potentially harmful to human health.

Another common use of ClO_2 is decontamination—something for which it has been extraordinarily beneficial. In late September and October 2001, letters containing anthrax bacteria were mailed to several news media offices and to two U.S. senators. Of the 22 people who subsequently contracted anthrax, five died. Anthrax is a spore-forming bacterium (*Bacillus anthracis*) and is classified by the Centers for Disease Control and Prevention (CDC) as a potential bioterrorism agent. Spore-forming bacteria are notoriously difficult to kill, making the cleanup of the buildings contaminated by anthrax costly and time-consuming.

The American Media Inc. (AMI) building in Boca Raton, Florida, was not deemed safe to enter until July 2004, after it had been treated with

ClO_2, the only structural fumigant approved by the Environmental Protection Agency (EPA) for anthrax decontamination.

FBI/Getty Images

Joe Raedle/Getty Images

6.3 Resonance Structures

There are molecules and polyatomic ions for which more than one correct Lewis structure may be drawn. Let's consider the example of the ozone molecule, O_3. Following the stepwise procedure introduced in Section 6.1, we draw a skeletal structure:

$$O-O-O$$

We sum the valence electrons:

$$3(6) = 18$$

And subtract two electrons for each bond in the skeletal structure:

$$18 - 2(2) = 14$$

We then distribute the remaining electrons to satisfy the octet of each O atom, but we run out of electrons before completing all of the octets.

$$:\ddot{O}-\ddot{O}-\ddot{O}:$$

We can now satisfy the octets of all of the O atoms by moving one lone pair from a terminal atom to form a double bond between it and the central atom. However, there are two different ways for us to do this. One results in the double bond being on the right, and the other results in the double bond being on the left. Both of these structures are equally correct—and they differ only in placement of the valence electrons.

$$\ddot{\text{O}}\!-\!\ddot{\text{O}}\!=\!\ddot{\text{O}}\!: \qquad :\!\ddot{\text{O}}\!=\!\ddot{\text{O}}\!-\!\ddot{\text{O}}\!:$$

When two or more equally correct Lewis structures can be drawn for a molecule (or polyatomic ion), differing only in the placement of *electrons,* they are referred to as **resonance structures.**

It is important to realize that a sample of ozone gas does not consist of a mixture of two different molecules, some with double bonds on one side and some with double bonds on the other side. All of the ozone molecules in a sample have the *same* structure. Experimental evidence indicates that the two oxygen-oxygen bonds in each O_3 molecule are actually identical. The reason we must draw more than one structure to represent ozone has to do with the limitations of Lewis structures to represent some of the properties of real molecules. Despite their limitations, Lewis structures are an important and powerful tool in the interpretation of chemical formulas and in the explanation and prediction of many aspects of molecules and polyatomic ions.

The carbonate ion provides another example of a species for which resonance structures can be drawn, with a double bond that can occupy any one of three different positions:

$$\left[\begin{array}{c}:\ddot{\text{O}}:\\ \,|\\ :\text{O}\!=\!\text{C}\!-\!\ddot{\text{O}}:\end{array}\right]^{2-} \longleftrightarrow \left[\begin{array}{c}\cdot\ddot{\text{O}}\\ \,||\\ :\ddot{\text{O}}\!-\!\text{C}\!-\!\ddot{\text{O}}:\end{array}\right]^{2-} \longleftrightarrow \left[\begin{array}{c}:\ddot{\text{O}}:\\ \,|\\ :\ddot{\text{O}}\!-\!\text{C}\!=\!\text{O}:\end{array}\right]^{2-}$$

Again, experimental evidence indicates that all three of the carbon-oxygen bonds in CO_3^{2-} are identical. The actual structure of the carbonate ion is not perfectly represented by any one of the resonance structures, but by a combination of all three. When you draw Lewis structures with double bonds that can occupy more than one possible position in the structure—interchangeable by relocation of *electrons only,* you should recognize that two or more resonance structures are possible.

Sample Problem 6.3 gives you some practice drawing resonance structures when they are appropriate.

Student Note: A double-headed arrow is used to indicate that two or more Lewis structures are resonance structures, differing only in positions of electrons.

Student Note: Sometimes it is possible to draw multiple structures for a given chemical formula by arranging the *atoms* differently. Structures that differ in the positions of *atoms* are *not* resonance structures—they are *isomers.* Resonance structures must differ in the positions of electrons only.

SAMPLE PROBLEM (**6.3**) **Drawing Resonance Structures**

High oil and gasoline prices have renewed interest in alternative methods of producing energy, including the "clean" burning of coal. Part of what makes "dirty" coal *dirty* is its high sulfur content. Burning dirty coal produces sulfur dioxide (SO_2), among other pollutants. Sulfur dioxide is oxidized in the atmosphere to form sulfur trioxide (SO_3), which subsequently combines with water to produce sulfuric acid—a major component of acid rain. Draw all possible resonance structures of sulfur trioxide.

Strategy Draw two or more Lewis structures for SO_3 in which the atoms are arranged the same way but the electrons are arranged differently.

Setup Following the steps for drawing Lewis structures, we determine that a correct Lewis structure for SO_3 contains two sulfur-oxygen single bonds and one sulfur-oxygen double bond:

$$\begin{array}{c}:\ddot{\text{O}}:\\ \,|\\ :\text{O}\!=\!\text{S}\!-\!\ddot{\text{O}}:\end{array}$$

But the double bond can be put in any one of three positions in the molecule.

Solution

$$\begin{array}{c}:\ddot{\text{O}}:\\ \,|\\ :\text{O}\!=\!\text{S}\!-\!\ddot{\text{O}}:\end{array} \longleftrightarrow \begin{array}{c}\cdot\ddot{\text{O}}\\ \,||\\ :\ddot{\text{O}}\!-\!\text{S}\!-\!\ddot{\text{O}}:\end{array} \longleftrightarrow \begin{array}{c}:\ddot{\text{O}}:\\ \,|\\ :\ddot{\text{O}}\!-\!\text{S}\!=\!\text{O}:\end{array}$$

THINK ABOUT IT

Always make sure that resonance structures differ only in the positions of the electrons, not in the positions of the atoms.

Practice Problem (A)TTEMPT Draw all possible resonance structures for the nitrate ion (NO_3^-).

Practice Problem (B)UILD Draw two resonance structures for the acetate ion, $C_2H_3O_2^-$. The carbons are both central (CH_3—CO_2).

Practice Problem (C)ONCEPTUALIZE The Lewis structure of a molecule consisting of the hypothetical elements A, B, and C is shown here. Of the four other structures, identify any that is *not* a resonance structure of the original and explain why it is not a resonance structure.

$$:\ddot{B}:$$
$$|$$
$$:\ddot{C}—A{=}\ddot{C}:$$

$$:\ddot{A}:$$
$$|$$
$$:C{=}B—\ddot{C}:$$
$$\text{(i)}$$

$$\overset{\cdot\cdot}{B}\overset{\cdot}{}$$
$$\|$$
$$:\ddot{C}—A—\ddot{C}:$$
$$\text{(ii)}$$

$$:\ddot{B}:$$
$$|$$
$$:\ddot{C}—A—\ddot{C}:$$
$$\text{(iii)}$$

$$:\ddot{B}:$$
$$|$$
$$:\ddot{C}—A—\ddot{C}:$$
$$\text{(iv)}$$

CHECKPOINT–SECTION 6.3   Resonance Structures

6.3.1 Indicate which of the following are resonance structures of BrO_3^+.

a) $\left[:\ddot{O}—Br{=}\ddot{O}:\right]^+$
   $\qquad |$
   $\qquad :\ddot{O}:$

b) $\left[:\ddot{O}—\ddot{Br}{=}O:\right]^+$
   $\qquad |$
   $\qquad :\ddot{O}:$

c) $\left[:\ddot{O}—Br{=}\ddot{O}:\right]^+$
   $\qquad |$
   $\qquad :\ddot{O}:$

d) $\left[:\ddot{Br}—O{=}\ddot{O}:\right]^+$
   $\qquad |$
   $\qquad :\ddot{O}:$

e) $\left[:O{=}Br—\ddot{O}:\right]^+$
   $\qquad |$
   $\qquad :\ddot{O}:$

6.3.2 How many resonance structures can be drawn for the nitrite ion (NO_2^-)? (N and O must obey the octet rule.)

a) 1

b) 2

c) 3

d) 4

e) 5

6.4 Molecular Shape

Many of the chemical and biochemical processes that we experience depend on the three-dimensional shapes of the molecules and/or ions involved. Our sense of smell is one example; the effectiveness of a particular drug is another. Although the actual shapes must ultimately be determined experimentally, we can do a reasonably good job of predicting them using Lewis structures and our knowledge of how negative charges, in this case groups of electrons, repel one another. In this section, we examine how starting with a properly drawn Lewis structure can enable us to determine the shape of a molecule or polyatomic ion. The shapes we encounter most often are those illustrated in Figure 6.2.

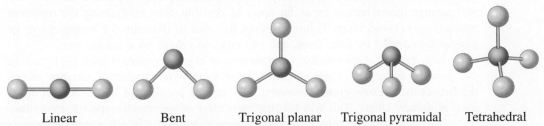

Linear   Bent   Trigonal planar   Trigonal pyramidal   Tetrahedral

Figure 6.2 Shapes of molecules and polyatomic ions with a central atom and two or more terminal atoms: linear, bent, trigonal planar, trigonal pyramidal, and tetrahedral.

TABLE 6.1	Partial Structures of Molecules/Polyatomic Ions Surrounded by 4, 3, or 2 Electron Groups			
Number of Electron Groups Around Central Atom		**Examples**		
4	 $-\overset{\mid}{\underset{\mid}{A}}-$	CH_4	NH_4^+	CH_2Cl_2
4	 $-\overset{\cdot\cdot}{\underset{\mid}{A}}-$	NH_3	H_3O^+	SO_3^{2-}
4	 $-\overset{\cdot\cdot}{\underset{\cdot\cdot}{A}}-$	H_2O	$HOCl$	ClO_2^-
3	 $-\overset{\mid}{A}=$	CO_3^{2-}	SO_3	H_2CO
3	 $-\overset{\cdot\cdot}{A}=$	HNO	NO_2^-	SO_2
3	 $-\overset{\mid}{\underset{\mid}{A}}-$	BCl_3	BF_3	BI_3
2	 $=A=$	CO_2	NO_2^+	CS_2
2	 $-A\equiv$	HCN	$NCCl$	$C_2H_2{}^*$
2	 $-A-$	$BeCl_2$	BeF_2	

*The C_2H_2 molecule has more than one central atom.

Student Note: The ability to visualize three-dimensional structures from two-dimensional drawings is an extremely important skill. It is highly recommended that you either purchase or borrow a model kit and build these molecules as you work through this section.

Recall that the electrons in the valence shell are the ones involved in the formation of covalent bonds [◄◄ Section 3.4]. The basis of the **valence-shell electron-pair repulsion (VSEPR) model** is that groups of electrons in the valence shell of a central atom *repel* one another. For the purpose of turning our two-dimensional Lewis structures into three-dimensional models, we will define an **electron group** on a central atom as either a *lone pair* or a *bond;* and the bond can be a *single* bond, a *double* bond, or a *triple* bond. The partial structures shown in Table 6.1 illustrate how electron groups on a central atom (represented in the partial structures as A) are counted.

The VSEPR model predicts that because electron groups repel one another, they will arrange themselves to be as far apart as possible, thus minimizing the repulsive interactions between them. Figure 6.3 uses balloons to illustrate the arrangements or *geometries* adopted by four, three, and two electron groups on a central atom.

How can just *three* possible arrangements of electron groups (Figure 6.3) result in *five* possible molecular shapes (Figure 6.2)? Although the number of electron groups determines the **electron-group geometry,** it is only the positions of the *atoms* that determine **molecular shape.** This will become more clear as we examine specific examples.

Let's consider the three different circumstances under which a central atom is surrounded by four electron groups. In the simplest case, all four electron groups are single bonds. The molecule methane, CH_4, is an example of this. The central atom is surrounded by four electron groups, and therefore has a tetrahedral electron-group

Thinking Outside the Box

Flavor, Molecular Shape, and Bond-Line Structures

Why is it so important to understand the type of bonding present within a molecule and, in turn, the molecule's shape? Not only does molecular shape help us to predict properties such as polarity and strength of intermolecular forces, but it also gives us important clues to other properties such as flavor, scent—even pharmaceutical efficacy.

Eugenol

Iconotec/Glow Images

You may have noticed that the structures of eugenol and isoeugenol shown here do not look quite like the Lewis structures you are learning to draw. These drawings are called *bond-line* structures and are a chemist's short-hand notation for larger, typically *organic* molecules. In bond-line drawings of molecular structures, when a line *ends* without another atom being shown, or a line *intersects* another line, it indicates the presence of a *carbon* atom along with whatever number of hydrogen atoms is necessary to give that carbon atom a total of four bonds. The complete

As an example, the distinctive compounds responsible for the characteristic flavors and aromas of two common spices, cloves and nutmeg, have the same chemical formula: $C_{10}H_{12}O_2$. The only difference between *eugenol* and *isoeugenol* is the placement of one double bond, which changes the flavor of the molecule from that of cloves to that of nutmeg. Flavor receptors on the tongue can sense this seemingly insignificant difference between these two molecules.

Isoeugenol

Iconotec/Glow Images

structure of isoeugenol is shown in the following illustration, in the more familiar form of a complete Lewis structure.

---

geometry. Further, each of the four electron groups is a *bond* to a terminal *atom;* therefore its molecular *shape* is also tetrahedral.

H–C–H (with H above and below)

Electron-group geometry: tetrahedral
-no lone pairs on the central atom-
Molecular shape: tetrahedral

> **Student Note:** Whenever there are no lone pairs on the central atom, the molecular shape is the same as the electron-group geometry.

The next case in which the central atom is surrounded by four electron groups is illustrated by the molecule ammonia, NH_3, where three of the electron groups are single bonds and one is a lone pair. Because there are four electron groups on the central atom, the electron-group geometry is tetrahedral. However, in this case, because only three of the electron groups are bonds leading to terminal *atoms,* the shape of the *molecule* is trigonal pyramidal.

H–N̈–H (with H below)

Electron-group geometry: tetrahedral
-one lone pair on the central atom-
Molecular shape: trigonal pyramidal

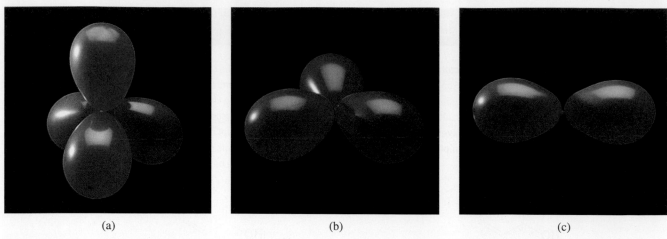

Figure 6.3 The geometries that result when (a) four, (b) three, and (c) two balloons are connected at their stems.

(all) Stephen Frisch/McGraw Hill

The last example of a central atom surrounded by four electron groups is illustrated by the water molecule, H_2O, where two of the electron groups are single bonds and two are lone pairs. Again, because there are *four* electron groups on the central atom, the electron-group geometry is tetrahedral. But because two of the electron groups are lone pairs and only two are bonds to terminal *atoms,* the molecular shape is *bent.*

$$H—\ddot{O}—H$$

Electron-group geometry: tetrahedral
-two lone pairs on the central atom-
Molecular shape: bent

So you can see how one electron-group geometry, tetrahedral, can result in *three* different molecular shapes: *tetrahedral, trigonal pyramidal,* and *bent*—depending on how many of the electron groups on the central atom are bonds to terminal atoms, and how many are lone pairs.

The other molecular shapes arise in the same way, being based on electron-group geometries and dependent on the number of lone pairs on the central atom. There are three ways that a central atom can have three electron groups around it: two single bonds and one double bond; one single bond, one double bond, and one lone pair; or three single bonds (an exception to the octet rule exhibited by boron). The two molecular shapes that are derived from the trigonal-planar electron-group geometry are trigonal planar, and bent—although *bent* molecules with trigonal-planar electron-group geometry, as we explain shortly, are slightly different from bent molecules with tetrahedral electron-group geometry (see Table 6.2). Examples that illustrate trigonal-planar electron-group geometry and the two molecular shapes that result are sulfur trioxide (SO_3), ozone (O_3), and boron trichloride (BCl_3):

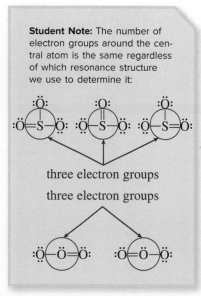

Student Note: The number of electron groups around the central atom is the same regardless of which resonance structure we use to determine it:

three electron groups

three electron groups

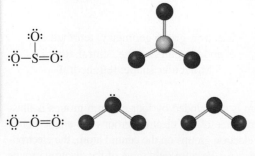

$$:\ddot{O}:$$
$$\ddot{O}—S=\ddot{O}:$$

Electron-group geometry: trigonal planar
-no lone pairs on the central atom-
Molecular shape: trigonal planar

$$\ddot{O}—\ddot{O}=\ddot{O}:$$

Electron-group geometry: trigonal planar
-one lone pair on the central atom-
Molecular shape: bent

$$:\ddot{Cl}:$$
$$:\ddot{Cl}—B—\ddot{Cl}:$$

Electron-group geometry: trigonal planar
-no lone pairs on the central atom-
Molecular shape: trigonal planar

TABLE 6.2	Electron-Group Geometries and Molecular Shapes with Bond Angles			
Number of Electron Groups	**Electron-Group Geometry**	**Number of Lone Pairs on the Central Atom**	**Molecular Shape and Bond Angle**	**Example**
4	tetrahedral	0	tetrahedral 109.5°	CH_4
4	tetrahedral	1	trigonal pyramidal ~109.5°	NH_3
4	tetrahedral	2	bent ~109.5°	H_2O
3	trigonal planar	0	trigonal planar 120°	SO_3
3	trigonal planar	1	bent ~120°	SO_2
2	linear	0	linear 180°	CO_2

*The ~ symbol is read as "approximately."

In the final case, when a central atom is surrounded by just two electron groups (and two terminal atoms), both the electron-group geometry and the molecular shape are *linear*. There are three ways that this can arise: both groups can be double bonds; one group can be a single bond and one can be a triple bond; or both can be single bonds (an exception to the octet rule exhibited by beryllium). These three possibilities are illustrated by carbon dioxide (CO_2), hydrogen cyanide (HCN), and beryllium difluoride, also commonly called *beryllium fluoride* (BeF_2):

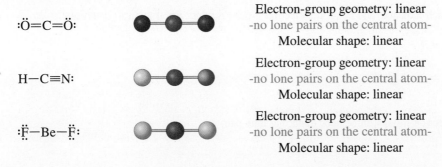

:Ö=C=Ö: Electron-group geometry: linear
-no lone pairs on the central atom-
Molecular shape: linear

H—C≡N: Electron-group geometry: linear
-no lone pairs on the central atom-
Molecular shape: linear

:F̈—Be—F̈:   Electron-group geometry: linear
-no lone pairs on the central atom-
Molecular shape: linear

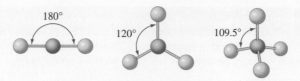

Figure 6.4 The three electron-group geometries associated with two, three, and four electron groups around a central atom—and the ideal angle between the electron groups in each case.

Bond Angles

The arrangements of electron groups around a central atom were illustrated in Figure 6.3 with balloons. Figure 6.4 shows these same arrangements with molecular models that make it easier to see the angles between electron groups in each case.

For molecules in which all of the electron groups on the central atom are identical, and lead to identical terminal atoms, the angles shown in Figure 6.4 are the angles between the bonds. For example, in the CH_4 molecule, where all four electron groups on the central atom are single bonds, and each leads to a hydrogen atom, the angles between any two of the C—H bonds is 109.5°.

In cases where the electron groups around the central atom are not all identical, as in the case of SO_2, the bond angles may differ slightly from those shown in Figure 6.4. Table 6.2 shows each of the three electron-group geometries we have discussed, along with the molecular shapes associated with them and the bond angles they contain.

With practice, determining the shape of a molecule or polyatomic ion is a relatively easy process. However, it is very important to follow the stepwise procedure. A review of the necessary steps follows:

1. Draw a correct Lewis structure. (See the steps for drawing Lewis structures in Section 6.1.)
2. Count the electron groups on the central atom.
3. Apply VSEPR theory to determine electron-group geometry.
4. Consider just the positions of *atoms* to determine molecular shape.

Sample Problem 6.4 lets you practice determining the shapes of molecules and polyatomic ions—and the bond angles they contain.

SAMPLE PROBLEM   **6.4**   **Determining Shapes and Bond Angles for Molecules and Polyatomic Ions**

Determine the electron-group geometry, molecular shape, and bond angles for (a) NO_3^- and (b) NCl_3.

Strategy Draw the Lewis structure for each molecule and count the electron groups on the central atom to determine electron-group geometry, molecular shape, and bond angles using Table 6.1.

Setup (a) $\left[\begin{array}{c} :\ddot{O}-N=\ddot{O}: \\ | \\ :\ddot{O}: \end{array}\right]^-$ The nitrogen has three electron groups, three separate bonds.

(b) $:\ddot{C}l-\underset{\underset{:\ddot{C}l:}{|}}{\ddot{N}}-\ddot{C}l:$ The nitrogen has four electron groups, three bonds and one lone pair.

Solution (a) The electron-group geometry for three electron groups is trigonal planar. The molecular shape is also trigonal planar because there is an atom attached to each electron group. The bond angles are all 120°.

(b) The electron-group geometry for four electron groups is tetrahedral. The molecular shape is also trigonal pyramidal because there is an atom attached to only three of the four electron groups. The bond angles are all about 109.5°.

THINK ABOUT IT

Remember that double and triple bonds only count as one electron group because they are attached to the same two atoms and must stay together.

Practice Problem **(A)TTEMPT** Determine the electron-group geometry, molecular shape, and bond angles for SO_3.

Practice Problem **(B)UILD** Consider the two hypothetical chemical species that appear to differ only in overall charge: AO_2 and AO_2^{2-}, where the central atom A is a member of Group 16. Determine the shape and the approximate O—A—O bond angle for each species. If the bond angles are different, explain why.

Practice Problem **(C)ONCEPTUALIZE** What *positive* charge on the species in Practice Problem B (AO_2^{2+}) would result in the shape being linear and the O—A—O bond angle being 180°?

Thinking Outside the Box

Molecular Shapes Resulting from Expanded Octets

One of the exceptions to the octet rule is that central atoms from the third period or below on the periodic table can have more than eight electrons around them—a so-called *expanded octet*.

This produces molecular shapes that we have not encountered before. Table 6.3 shows the arrangements of electron groups around central atoms with expanded octets, and the molecular shapes that result.

TABLE 6.3	Electron-Group Geometries and Molecular Shapes of Molecules with Expanded Octets					
Total Number of Electron Groups	**Type of Molecule**	**Electron-Group Geometry**	**Number of Lone Pairs**	**Placement of Lone Pairs**	**Molecular Shape**	**Example**
5	AB_5	Trigonal bipyramidal	0		Trigonal bipyramidal	PCl_5
5	AB_4	Trigonal bipyramidal	1		Seesaw-shaped	SF_4
5	AB_3	Trigonal bipyramidal	2		T-shaped	ClF_3

(Continued on next page)

TABLE 6.3   (Continued)

Total Number of Electron Groups	Type of Molecule	Electron-Group Geometry	Number of Lone Pairs	Placement of Lone Pairs	Molecular Shape	Example
5	AB_2	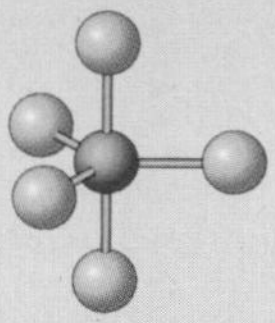 Trigonal bipyramidal	3	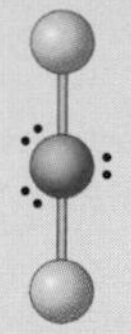	 Linear	IF_2^-
6	AB_6	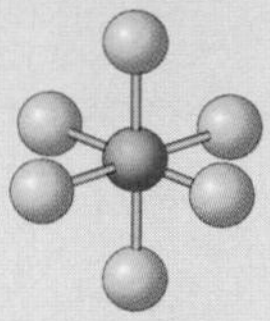 Octahedral	0		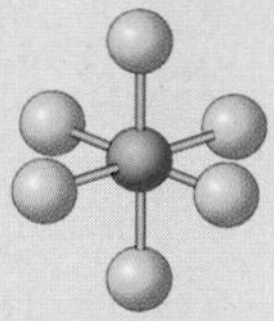 Octahedral	SF_6
6	AB_5	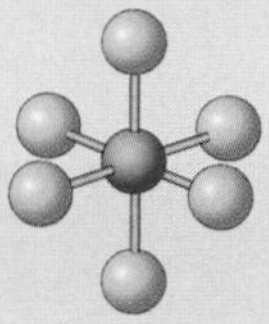 Octahedral	1	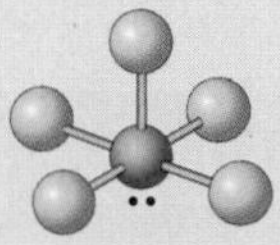	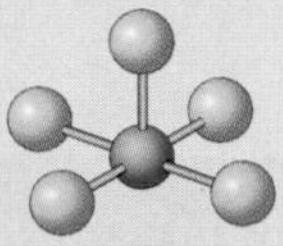 Square pyramidal	BrF_5
6	AB_4	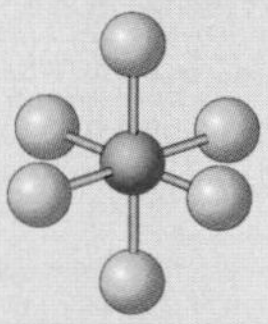 Octahedral	2	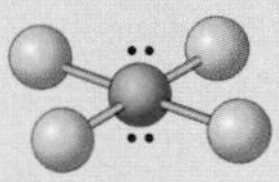	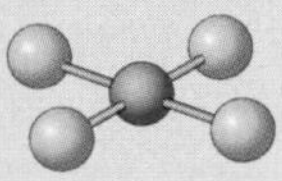 Square planar	XeF_4

CHECKPOINT–SECTION 6.4   Molecular Shape

6.4.1   What are the electron-group geometry and molecular geometry of CO_3^{2-}?

a) tetrahedral, trigonal planar

b) tetrahedral, trigonal pyramidal

c) trigonal pyramidal, trigonal pyramidal

d) trigonal planar, trigonal planar

e) tetrahedral, tetrahedral

6.4.2   What are the electron-group geometry and molecular geometry of ClO_3^-?

a) tetrahedral, trigonal planar

b) tetrahedral, trigonal pyramidal

c) trigonal pyramidal, trigonal pyramidal

d) trigonal planar, trigonal planar

e) tetrahedral, tetrahedral

6.4.3   What is the approximate value of the bond angle indicated?

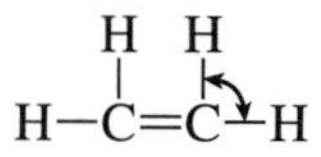

$$H-C=C-H$$

a) 45°

b) 90°

c) 109.5°

d) 120°

e) 180°

6.4.4 What is the approximate value of the bond angle indicated?

$$H-\overset{\displaystyle H}{\underset{\displaystyle H}{C}}-\ddot{O}-H$$

a) 45°   d) 120°

b) 90°   e) 180°

c) 109.5°

6.4.5 Which of the following combinations could the central atom in this molecule have? Select all that apply.

a) four electron groups total, two lone pairs

b) four electron groups total, one lone pair

c) four electron groups total, no lone pairs

d) three electron groups total, no lone pairs

e) three electron groups total, one lone pair

6.5   Electronegativity and Polarity

So far, we have talked about chemical bonds in terms of *ionic* bonding [◀◀ Section 3.2] and *covalent* bonding [◀◀ Section 3.4]. We have seen, for instance, when the elements sodium and chlorine are combined, each sodium atom will lose its only valence electron to become isoelectronic with the noble gas neon (Ne); and each chlorine atom will gain an electron to become isoelectronic with the noble gas argon (Ar). The resulting sodium cations (Na^+) and chloride anions (Cl^-) form an ionic solid in which the particles (in this case, ions) are held together by electrostatic attraction.

$$Na\cdot \quad \cdot\ddot{\underset{..}{Cl}}: \qquad Na \rightleftharpoons \leftarrow \left[:\ddot{\underset{..}{Cl}}:\right]^- \qquad Na^+ \left[:\ddot{\underset{..}{Cl}}:\right]^-$$

And we have seen that when chlorine atoms are not in the presence of a ready source of electrons (such as sodium atoms) they themselves will form diatomic *molecules,* connected by covalent bonds that each consists of a pair of shared electrons. By sharing a pair of electrons, each Cl atom in a Cl_2 molecule is also *effectively* isoelectronic with Ar:

$$:\ddot{\underset{..}{Cl}}\cdot \rightarrow \quad \leftarrow \cdot\ddot{\underset{..}{Cl}}: \qquad \textcircled{:Cl:}\textcircled{:Cl:}$$

In fact, these two scenarios represent extreme cases—opposite ends of a *spectrum* along which all chemical bonds lie. In the first scenario, an electron is transferred outright; and in the second, a pair of electrons is shared equally by two identical atoms. As it turns out, most chemical bonds fall somewhere between these two extremes.

Electronegativity

Let's consider the HCl molecule, which was represented as a Lewis structure along with those of Cl_2 and H_2 in Section 6.1. Unlike the shared electron pairs that constitute the covalent bonds in Cl_2 and H_2, the electron pair shared by H and Cl in the HCl molecule is *not* shared equally. The reason for this is that different elements have different *electronegativities.* **Electronegativity** is the ability of an atom to draw the electrons it shares with another atom toward *itself.* An element with higher electronegativity is said to be *more electronegative,* and an element with lower electronegativity is said to be *less electronegative.* In main group elements, electronegativity decreases from top to bottom and from right to left. Figure 6.5 shows the commonly used electronegativity values of the elements, and the periodic trends.

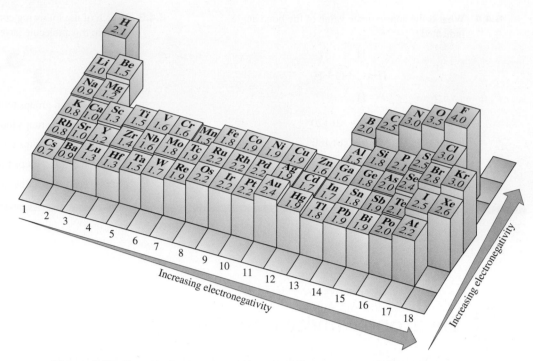

Figure 6.5 Electronegativity values for elements through period 6. These values were developed by Linus Pauling (Section 6.6), who assigned an arbitrary value of 4.0 to fluorine (the most electronegative element) and determined the other values relative to fluorine.

Sample Problem 6.5 lets you practice using the periodic trends to compare electronegativity values of various main group elements.

SAMPLE PROBLEM   6.5   Comparing Electronegativity Values Using the Periodic Table

For each pair of elements, indicate which one you would expect to be the *most* electronegative: (a) Rb or Li, (b) Ca or Br.

Strategy Consider the periodic trend where electronegativity increases, from left to right, within a period (row) and increases bottom to top within a group (column).

Setup (a) Li is in the same group as Rb and lies above it on the periodic table.

(b) Ca and Br are in the same period, with Br farther to the right.

Solution (a) Li has a higher electronegativity than Rb.

(b) Br has a higher electronegativity than Ca.

THINK ABOUT IT

The closer an element is to the top right of the periodic table, excluding the noble gases He and Ne (Figure 6.5), the higher its electronegativity.

Practice Problem **A**TTEMPT For each pair of elements, indicate which one you would expect to be the more electronegative: (a) Mg or Sr, (b) Rb or Te.

Practice Problem **B**UILD Without consulting Figure 6.5, place the following elements in order of decreasing electronegativity: (a) Li, C, and K; (b) Ba, Ca, and As.

Practice Problem **C**ONCEPTUALIZE Is it possible to rank the elements N, S, and Br in order of increasing electronegativity using only the periodic trend? If not, explain why.

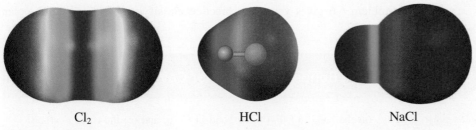

Cl₂   HCl   NaCl

Figure 6.6 Electron density maps show the distribution of negative charge in a covalent species (Cl_2), a polar covalent species (HCl), and an ionic species (NaCl). The highest density of negative charge in each model is the red region.

Bond Polarity

In the HCl example, chlorine is the more electronegative element, with a value of 3.0; and hydrogen is less electronegative, with a value of 2.1. Because of this difference in electronegativity, the electron pair between the two atoms is more tightly held by the Cl atom than by the H atom. Another way to say this is that the shared electron pair lies *closer* to the Cl atom—or that, on average, it spends more *time* near the Cl atom. However we choose to phrase it, in this example, the Cl atom has greater power to attract electrons to itself than the H atom. The difference is not so great that the Cl atom actually *removes* the H atom's electron—that would constitute the formation of an *ionic* bond, which the bond in HCl is *not*. But neither is it a purely *covalent* bond, because it consists of an unequally shared pair of electrons. A bond such as this one is known as a ***polar covalent bond.***

The mapping of electron density enables us to visualize the distribution of electrons in the bonds in NaCl, Cl_2, and HCl. Figure 6.6 shows what are known as *electrostatic potential models* of these three species. These models show regions where electrons spend a lot of time in red, and regions where electrons spend very little time in blue. (Regions where electrons spend a moderate amount of time appear in green.)

Figure 6.7 illustrates the spectrum of chemical bond types, ranging from ionic, to polar covalent, to purely covalent—which occurs only when the electronegativity values of the two bonded atoms are *equal*.

There really is no sharp distinction between ionic and very polar, or between somewhat polar, slightly polar, and nonpolar. What we typically do is classify bonds as ionic, polar, or nonpolar based on the difference between electronegativity values of the atoms involved. The following guidelines are used to classify the nature of bonds:

- A bond is considered ionic if the difference between electronegativity values of the two atoms is equal to or greater than 2.0.
- A bond is considered polar if the difference between electronegativity values of the two atoms is at least 0.5, but less than 2.0.
- A bond is considered nonpolar if the difference between electronegativity values of the two atoms is less than 0.5.

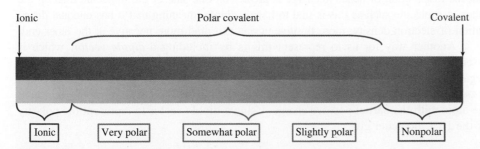

Figure 6.7 Spectrum of chemical bonding from ionic to covalent.

Sample Problem 6.6 gives you some practice classifying bonds as ionic, polar covalent, or covalent.

SAMPLE PROBLEM **6.6** Classifying Bonds as Ionic, Polar Covalent, or Covalent

Classify the following bonds as nonpolar, polar, or ionic: (a) the bond in ClF, (b) the bond in CsBr, and (c) the carbon-carbon double bond in C_2H_4.

Strategy Using the information in Figure 6.5, determine which bonds have identical, similar, and widely different electronegativities.

Strategy Electronegativity values from Figure 6.5 are Cl (3.0), F (4.0), Cs (0.7), Br (2.8), C (2.5).

Solution (a) The difference between the electronegativities of F and Cl is $4.0 - 3.0 = 1.0$, making the bond in ClF polar.

(b) In CsBr, the difference is $2.8 - 0.7 = 2.1$, making the bond ionic.

(c) In C_2H_4, the two atoms are identical. (Not only are they the same element, but each C atom is bonded to two H atoms.) The carbon-carbon double bond in C_2H_4 is nonpolar.

THINK ABOUT IT

By convention, the difference in electronegativity is always calculated by subtracting the smaller number from the larger one, so the result is always positive.

Practice Problem **A**TTEMPT Classify the following bonds as nonpolar, polar, or ionic: (a) the bonds in H_2S, (b) the H—O bonds in H_2O_2, and (c) the O—O bond in H_2O_2.

Practice Problem **B**UILD In order of increasing polarity, list the bonds between carbon and each of the nonmetal Group 17 elements.

Practice Problem **C**ONCEPTUALIZE Electrostatic potential maps are shown for HF and LiH. Determine which diagram is which. (The H atom is shown on the left in both.)

When we first learned to represent ionic compounds, we did so by placing the Lewis dot symbols for the individual ions next to each other—and these representations showed the charges on the ions explicitly. We know that molecules are electrically neutral, and that the individual atoms within a molecule do not have discrete charges, as was the case with the ions in ionic compounds. However, in a molecule made up of atoms with significantly different electronegativities, the atoms do bear *partial* charges because of the unequal sharing of the electron pairs in the covalent bonds. To denote this in a molecular formula, we use the lowercase Greek letter delta (δ) along with a positive or negative sign. The HCl molecule, for example, can be represented as

$$\delta+ \text{ H—Cl } \delta-$$

where the δ+ and δ− symbols indicate a *partial positive* and *partial negative* charge, respectively. A diatomic molecule in which the *bond* is polar is a *polar* molecule, meaning that the distribution of negative charge is greater on one side of the molecule than on the other. A molecule such as this is said to have a **dipole**, meaning that it has unequal distribution of electron density: more positive at one end, and more negative at the other end.

Another way for us to represent this is by including a **dipole vector**, which is essentially an arrow drawn parallel to the bond, and directed at the atom with the partial negative charge (the more electronegative of the two atoms). We also draw a crosshatch at the end near the *less* electronegative atom, making what looks like a plus sign near the atom with the partial positive charge. Using a dipole vector to represent the polarity of the HCl molecule gives:

$$\overset{\longrightarrow}{\text{H—Cl}}$$

As we explain in the next section, this uneven distribution of charge has an important impact on how molecules interact with one another.

Molecular Polarity

Although it is always true that diatomic molecules containing polar bonds are polar, there are many cases where *polyatomic* molecules contain polar bonds but are not themselves polar molecules. To see how this is so, we must consider both the polarity of a molecule's bonds and the molecule's *shape*. Consider the example of carbon dioxide, CO_2. Carbon and oxygen have different electronegativity values (see Figure 6.5) that differ by 1.0—making the carbon-oxygen bonds *polar*. We can represent the molecule with dipole vectors as:

$$\overset{\longleftarrow \;+\; \longrightarrow}{O=C=O}$$

However, when we determined the shape of the CO_2 molecule, we found it to be linear—with an angle between the two CO bonds of 180°. With two equally polar bonds pointing in opposite directions, the dipoles cancel one another, and the molecule overall is actually *nonpolar*. We can apply essentially the same analysis of whether or not the polarity of a molecule's bonds cancel one another with all of the molecular shapes that we have learned. Figure 6.8 shows how this type of analysis applies to each of the other molecular shapes we have encountered. Note that in each of the models shown in Figure 6.8, the bonds between the central and terminal atoms are identical.

It is important to recognize that when determining the polarity of a molecule, we must consider not just the *distribution* of bonds, but whether or not they are the same type of bond. For example, the molecule CH_4 is very similar to the molecule CH_3Cl. However, one is *nonpolar* and one is *polar*. In CH_4, there is a tetrahedral arrangement of identical C—H bonds, resulting in the bond dipoles all cancelling one another. In CH_3Cl, although there is the same tetrahedral arrangement of bonds, the bond dipoles are not all equal. There is a slightly greater difference in electronegativity between C and Cl (2.5 and 3.0, respectively) than there is between C and H (2.5 and 2.1, respectively). Moreover, the dipole vector we draw for the C—Cl bond in CH_3Cl

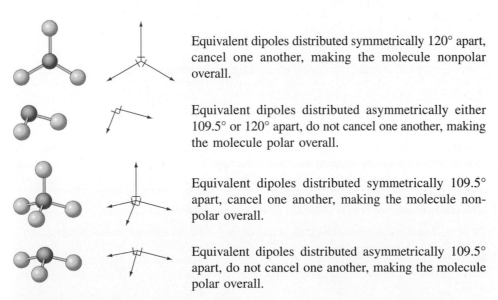

Equivalent dipoles distributed symmetrically 120° apart, cancel one another, making the molecule nonpolar overall.

Equivalent dipoles distributed asymmetrically either 109.5° or 120° apart, do not cancel one another, making the molecule polar overall.

Equivalent dipoles distributed symmetrically 109.5° apart, cancel one another, making the molecule nonpolar overall.

Equivalent dipoles distributed asymmetrically 109.5° apart, do not cancel one another, making the molecule polar overall.

Figure 6.8 Four of the molecular shapes we learned to draw in Section 6.4. When the bonds in molecules of these shapes are identical and distributed symmetrically, the molecules are nonpolar—even though the bonds they contain are polar. The two molecular shapes in this figure that are polar are those in which there are lone pairs on the central atom: bent and trigonal pyramidal. Note that this does not mean that every molecule in which there are lone pairs on the central atom is necessarily polar.

will not point in the same direction as the one we draw for the corresponding C—H bond in CH₄.

dipole vectors
sum to zero

dipole vectors do
not sum to zero

Student Note: Two criteria must be met in order for a molecule to be polar.
• The molecule must contain polar bonds.
• The polar bonds must be arranged asymmetrically—such that they do not cancel one another's contributions to the overall polarity.

Thus, despite the similarity in the appearance of their structures, CH₄ is nonpolar and CH₃Cl is polar.

Sample Problem 6.7 gives you some practice determining whether or not a molecule is polar.

SAMPLE PROBLEM 6.7 Assessing Molecular Polarity

Determine whether or not the following molecules are polar: (a) BCl₃, (b) AsCl₃.

Strategy Remember that you must start by drawing the correct Lewis structure in order to arrive at the correct answer. Then use the VSEPR model to determine its molecular geometry. Once the shape is known, determine if the dipoles cancel one another. IF they do, it is a nonpolar molecule.

Setup (a) The Lewis structure of BCl₃ is (remember B is an exception to the octet rule):

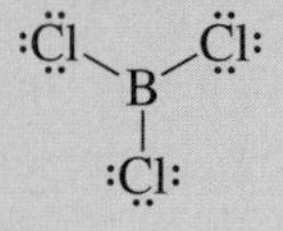

With three identical electron groups on the central atom, this molecule is trigonal planar and has 120 angles between the atoms. This means that all three chlorine atoms are "pulling" on the electrons with the same but opposite forces, which cancel them out.

(b) The Lewis structure of AsCl₃ is:

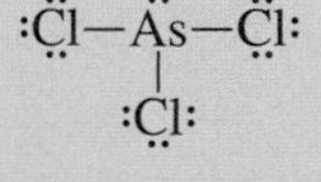

With four electron groups on the central atom, this molecule is trigonal pyramidal. All three chlorine atoms are "pulling" the shared electrons toward them, but are not opposing one another. The dipoles do not cancel one another.

Solution (a) BCl₃ is nonpolar. (b) AsCl₃ is polar.

THINK ABOUT IT

Without a properly drawn Lewis structure, it is unlikely that you will arrive at a correct answer in these types of problems. Never try to take a shortcut on problems such as these by skipping the Lewis structure!

Practice Problem **A**TTEMPT Determine whether or not the following molecules are polar: (a) BI₃, (b) SF₂.

Practice Problem **B**UILD Which of the following molecules could be polar? (Assume that all terminal atoms are identical.)

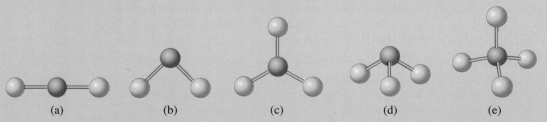

(a)   (b)   (c)   (d)   (e)

Practice Problem **C**ONCEPTUALIZE Explain why CO₂ is nonpolar but SO₂ is polar, despite the similarities in their formulas.

Thinking Outside the Box

How Bond Dipoles Sum to Determine Molecular Polarity

In AB_x molecules where $x \geq 3$, it may be less obvious whether the individual bond dipoles cancel one another. Consider the molecule BF_3, for example, which has a trigonal planar geometry:

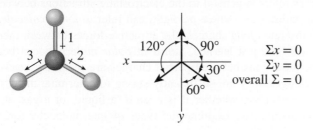

We can simplify the math in this analysis by assigning the vectors representing the three identical B—F bonds an arbitrary magnitude of 1.00. The x, y coordinates for the end of arrow 1 are (0, 1.00). Determining the coordinates for the ends of arrows 2 and 3 requires the use of trigonometric functions. You may have learned the mnemonic SOH CAH TOA, where the letters stand for

Sin = Opposite over Hypotenuse
Cos = Adjacent over Hypotenuse
Tan = Opposite over Adjacent

The x coordinate for the end of arrow 2 corresponds to the length of the line *opposite* the 60° angle. The hypotenuse of the triangle has a length of 1.00 (the arbitrarily assigned value). Therefore, using SOH,

$$\sin 60° = 0.866 = \frac{\text{opposite}}{\text{hypotenuse}} = \frac{\text{opposite}}{1}$$

so the x coordinate for the end of arrow 2 is 0.866.

The magnitude of the y coordinate corresponds to the length of the line *adjacent* to the 60° angle. Using TOA,

$$\tan 60° = 1.73 = \frac{\text{opposite}}{\text{adjacent}} = \frac{0.866}{\text{adjacent}}$$

$$\text{adjacent} = \frac{0.866}{1.73} = 0.500$$

so the y coordinate for the end of arrow 2 is −0.500. (The trigonometric formula gives us the length of the side. We know from the diagram that the sign of this y component is negative.)

Arrow 3 is similar to arrow 2. Its x component is equal in magnitude but opposite in sign, and its y component is the same magnitude and

sign as that for arrow 2. Therefore, the x and y coordinates for all three vectors are

	x	y
Arrow 1	0	1
Arrow 2	0.866	−0.500
Arrow 3	−0.866	−0.500
Sum =	0	0

Because the individual bond dipoles (represented here as the vectors) sum to zero, the molecule is nonpolar overall. Although it is somewhat more complicated, a similar analysis can be done to show that all x, y, and z coordinates sum to zero when there are four identical polar bonds arranged in a tetrahedron about a central atom. In fact, any time there are identical bonds symmetrically distributed around a central atom, with no lone pairs on the central atom, the molecule will be nonpolar overall, even if the bonds themselves are polar. In cases where the bonds are distributed symmetrically around the central atom, the nature of the atoms surrounding the central atom determines whether the molecule is polar overall. For example, CCl_4 and $CHCl_3$ have the same molecular geometry (tetrahedral), but CCl_4 is nonpolar because the bond dipoles cancel one another. In $CHCl_3$, however, the bonds are not all identical, and therefore the bond dipoles do not sum to zero. The $CHCl_3$ molecule is polar.

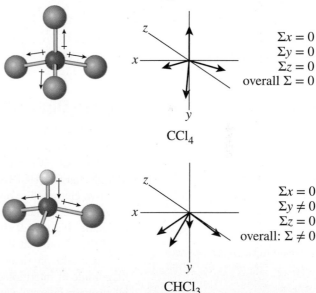

CHECKPOINT–SECTION 6.5 Electronegativity and Polarity

6.5.1 Identify the polar molecules in the following group: HBr, CH_4, CS_2.

a) HBr only

b) HBr and CS_2

c) HBr, CH_4, and CS_2

d) CH_4 and CS_2

e) CH_4 only

6.5.2 Identify the polar molecules in the following group: SO_2, NH_3, SeF_2.

a) SO_2, NH_3, and SeF_2

b) SO_2 only

c) SeF_2 only

d) SO_2 and SeF_2

e) SO_2 and NH_3

6.6 Intermolecular Forces

An important consequence of molecular polarity is the existence of attractive forces between molecules. We have seen how the oppositely charged ions in an ionic compound are held together by electrostatic attraction [◄◄ Section 3.2]. The ion-ion attractions that hold such substances together are one example of "intermolecular" forces. The term *intermolecular forces* refers in general to the electrostatic attractions between the particles that make up any substance—where *particles* can refer to *ions, molecules,* or *atoms.* Because ions have discrete (*full*) charges, the attractive forces between them are relatively strong. But as we have just learned, atoms in *polar molecules* can bear *partial* charges. Although the attractions between partial charges are relatively weaker than those between discrete charges, they are sufficiently strong to determine many of the properties of a substance—including whether it is a solid, a liquid, or a gas, at a particular temperature. In this section, we examine the types of intermolecular forces that exist between molecules and atoms.

As the graphic here illustrates, the atoms in a molecule are held together by covalent bonds. The molecules in a liquid or solid *substance* are held together by relatively weaker, *intermolecular* forces. Although both covalent bonds and intermolecular forces are the result of electrostatic attraction, because intermolecular forces are *weaker* forces, they are "broken" much more easily than covalent bonds. This is why we can warm a solid sample of a molecular substance and *melt* it, continue heating and *vaporize* it—without changing the identity of the substance. The same covalent bonds hold the atoms of the molecules together regardless of the physical state of the substance.

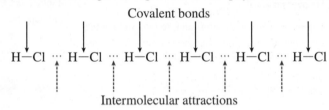

Dipole-Dipole Forces

Dipole-dipole forces are attractive forces that act between polar molecules. Recall that a diatomic molecule containing elements of significantly different electronegativity values, such as HCl, has an unequal distribution of electron density and therefore has partial charges. HCl has a partial positive charge ($\delta+$) on the H atom and a partial negative charge ($\delta-$) on the Cl atom. In a sample of HCl molecules, the partial positive charge on one molecule is attracted to the partial negative charge on a neighboring molecule. Figure 6.9 shows the orientations of polar molecules in a liquid (a) and in a solid (b). (Note that the arrangement of molecules is more orderly in the solid.)

Certain physical properties such as melting point and boiling point reflect the magnitude of intermolecular forces. A substance in which the particles are held together by stronger intermolecular forces will require more energy to separate the particles and

Student Note: We go into much greater detail about the melting and boiling processes in Chapter 7.

(a) liquid   (b) solid

Figure 6.9 Arrangement of polar diatomic molecules, such as HCl, (a) in a liquid and (b) in a solid. (Although HCl is a gas at room temperature, it is a liquid at temperatures below −85°C and a solid at temperatures below −114°C.)

will therefore melt at a higher temperature and boil at a higher temperature than a substance with weaker intermolecular forces. Consider the following comparison between a nonpolar molecule and a polar molecule of similar molar mass:

Student Note: The importance of comparing substances of similar molar mass will become clear later in this section.

Compound	Formula	Molar Mass (g/mol)	Boiling Point (°C)
propane	$CH_3CH_2CH_3$	44.09	−42
acetonitrile	CH_3CN	41.05	82

Because of the partial positive and negative charges on atoms in the acetonitrile molecules, there are relatively strong electrostatic attractions that hold the molecules together, making acetonitrile a liquid at room temperature. Propane, being nonpolar, has weaker intermolecular forces and is a gas at room temperature and ordinary pressures.

Sample Problem 6.8 lets you practice identifying molecules that exhibit dipole-dipole forces.

Student Note: As we will see shortly, there *are* attractive forces between nonpolar molecules such as propane. It is because of these intermolecular forces that we are able to *liquefy* propane under certain conditions.

SAMPLE PROBLEM 6.8   Identifying Dipole-Dipole Forces in Molecules

Determine whether or not these molecules exhibit dipole-dipole forces: (a) Br_2, (b) $SeCl_2$, (c) BF_3.

Strategy You will need to determine whether each molecule is polar or not by first drawing the proper Lewis structure for each. Any polar molecule exhibits dipole-dipole forces.

Setup Determine the Lewis structure and polarity of each molecule.

(a) :B̈r—B̈r: This molecule contains no dipole since both atoms are identical. It is therefore nonpolar.

(b) :C̈l—S̈e—C̈l: This molecule is bent where the dipole for each bond does not cancel the other, making this a polar molecule.

(c) :F̈⟍ ⟋F̈: This molecule is symmetrical, and the dipoles for each bond cancel one another out, making it nonpolar.
  ⎮ B
  :F̈:

Solution (a) no dipole-dipole forces present (b) dipole-dipole forces present (c) no dipole-dipole forces present

THINK ABOUT IT

Make sure that you understand how critical it is to draw a proper Lewis structure as the first step to answering these types of questions. Without the Lewis structure and knowledge of molecular shape, which you can determine only by doing a VSEPR analysis of the Lewis structure, you will not be able to answer these correctly.

Practice Problem **A**TTEMPT Determine if each of these molecules exhibits dipole-dipole forces: (a) HBr, (b) N_2, (c) NF_3.

Practice Problem **B**UILD Determine if each of these molecules exhibits dipole-dipole forces: (a) CH_3Cl, (b) OCS, (c) CS_2.

Practice Problem **C**ONCEPTUALIZE How could you make the following nonpolar compounds exhibit dipole-dipole forces by changing the identity of *one* atom in each molecule? (a) BCl_3, (b) CO_2

Hydrogen Bonding

Hydrogen bonding is a special type of dipole-dipole force that is exhibited only by substances that contain specific bonds N—H, O—H, or F—H. For example, hydrogen bonding is exhibited by HF, which is shown in Figure 6.10. Because the F atom is very small and has a very high electronegativity value (see Figure 6.5), it draws the shared electron pair in the F—H bond toward itself very effectively, resulting in relatively *large* partial charges— positive ($\delta+$) on the H atom and negative ($\delta-$) on the F atom. The large partial positive charge on H is powerfully attracted to the large partial negative charge on the F atom of a neighboring HF molecule. The result is an especially strong dipole-dipole attraction.

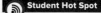 **Student Hot Spot**

Student data indicate you may struggle to identify hydrogen bonding in a substance. Access the eBook to view additional Learning Resources on this topic.

Figure 6.10 Hydrogen bonding between HF molecules. The electrostatic potential maps show the locations of the partial positive and partial negative charges and how they are drawn together.

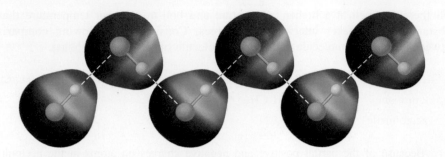

Sample Problem 6.9 gives you some practice identifying molecules that exhibit hydrogen bonding.

SAMPLE PROBLEM 6.9   Identifying Hydrogen Bonding in Molecules

Which of the following molecules exhibit hydrogen bonding?

(a) CH_2F_2 (b) NH_3 (c) H_2Se

Strategy You will need to determine whether each molecule is polar or not by first drawing the proper Lewis structure for each. Polar molecules that contain F—H, O—H, or N—H bonds exhibit hydrogen bonding.

Setup Draw the Lewis structure for each molecule and determine if it is polar or not, and whether it contains a bond from F, O, or N directly to a hydrogen atom.

(a)

$$\ddot{F}:$$
$$H-C-H$$
$$:\ddot{F}:$$

This molecule is polar, but it does not contain F—H, O—H, or N—H bonds.

(b) $H-\ddot{N}-H$ This molecule is polar and does contain at least one N—H bond.
         |
         H

(c) $H-\ddot{S}e-H$ This molecule is polar but does not contain F—H, O—H, or N—H bonds.

Solution (a) no hydrogen bonding (b) hydrogen bonding (c) no hydrogen bonding because it does not contain F, O, or N atoms.

THINK ABOUT IT

Note that molecules containing H and F, O, or N, but do not have the hydrogen bonded directly to the F, O, or N atom, do NOT exhibit hydrogen bonding. CH_2F_2 is a good example of this.

Practice Problem ATTEMPT Which of the following molecules exhibit hydrogen bonding?

(a) NCl_3 (b) PH_2F (c) CH_3OH

Practice Problem BUILD One of these molecules exhibits hydrogen bonding and the other does not, even though they have the same chemical formula of C_2H_6O. Which one exhibits hydrogen bonding? Explain why the other does not.

(a)
$$\begin{array}{ccc} H & & H \\ | & & | \\ H-C-\ddot{O}-C-H \\ | & & | \\ H & & H \end{array}$$

(b)
$$\begin{array}{ccc} H & H \\ | & | \\ H-C-C-\ddot{O}-H \\ | & | \\ H & H \end{array}$$

Practice Problem CONCEPTUALIZE The molecule HCl actually does exhibit some degree of hydrogen bonding, but it is insignificant compared to that exhibited by HF. Explain why this is.

Dispersion Forces

Nonpolar gases, such as N_2 and O_2, can be liquefied at certain combinations of temperature and pressure. This indicates that there must be some kinds of intermolecular forces strong enough to attract the molecules to one another under the right conditions.

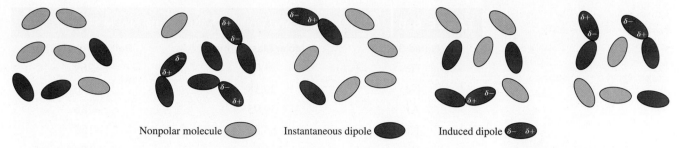

Nonpolar molecule ⬭   Instantaneous dipole ⬬   Induced dipole ⬬ δ− δ+

Figure 6.11 Instantaneous dipoles in ordinarily nonpolar molecules can induce temporary dipoles in neighboring molecules, causing the molecules to be attracted to one another. (This type of interaction is responsible for our ability to condense nonpolar gases.)

These intermolecular forces are electrostatic in nature (as are all intermolecular forces), but they differ from other intermolecular forces because they arise from the movement of electrons in nonpolar molecules.

On average, the distribution of electron density in a nonpolar molecule is uniform and symmetrical, which is what makes the molecule nonpolar. However, because electrons in a molecule have some freedom to move about, at any given point in time, the molecule may have a nonuniform distribution of electron density, giving it a fleeting, temporary dipole—called an ***instantaneous dipole.*** An instantaneous dipole in one molecule can cause a temporary ***induced dipole*** in a nearby molecule. For example, the temporary partial negative charge on a molecule repels the electrons in a molecule next to it. This repulsion causes a temporary dipole in the second molecule and so on, leaving a collection of ordinarily nonpolar molecules with partial positive and negative charges and electrostatic attractions between them. Figure 6.11 illustrates how instantaneous and induced dipoles in nonpolar molecules result in intermolecular attractive forces known as ***dispersion forces.*** Dispersion forces, although generally considered to be the weakest of the intermolecular forces, can be strong enough to allow us to liquefy such nonpolar substances as N_2 and O_2—and even to allow many nonpolar substances to be liquids or solids at room temperature.

The magnitude of dispersion forces depends on how easily the electrons in the molecule can move around. In small molecules, such as F_2, the electrons are relatively close to the atomic nuclei and cannot move about very freely; thus, the electron distribution in F_2 tends to remain fairly uniform, and F_2 does not easily achieve an instantaneous dipole. In a larger molecule, such as Cl_2, the electrons are somewhat farther away from the nuclei, are less tightly held in position, and therefore move about more freely. The electron density in Cl_2 is more easily *polarized,* meaning that the Cl_2 molecule achieves an instantaneous dipole more easily than does the smaller F_2 molecule. This results in *more* instantaneous dipoles, more *induced* dipoles, and ultimately *stronger* dispersion forces between Cl_2 molecules than between F_2 molecules. Br_2 and I_2, the next two halogens, are larger still and exhibit even greater tendencies to form instantaneous dipoles, induced dipoles, and stronger dispersion forces. Table 6.4 lists the halogens, their molar masses, and their boiling points—as a comparison of the magnitudes of their dispersion forces. All molecules exhibit dispersion forces, whether or not they are polar. In general, the greater the molar mass, the greater the strength of dispersion forces between the molecules.

TABLE 6.4	Molar Masses and Boiling Points of the Halogens	
Molecule	**Molar Mass (g/mol)**	**Boiling Point (°C)**
F_2	38.0	−188
Cl_2	70.9	−34
Br_2	159.8	59
I_2	253.8	184

As we mentioned early in this section, the term inter*molecular* forces can refer to attractive forces between ions, molecules, or *atoms.* The noble gases, which exist as individual atoms, also exhibit dispersion forces—and serve as another illustration of how the magnitude of dispersion forces increases with increasing molar mass. Table 6.5 lists the noble gases along with their molar masses and boiling points.

TABLE 6.5	Molar Masses and Boiling Points of the Noble Gases	
Noble Gas	**Molar Mass (g/mol)**	**Boiling Point (°C)**
He	4.003	−269
Ne	20.18	−246
Ar	39.95	−186
Kr	83.80	−153
Xe	131.3	−108
Rn	222	−62

Sample Problem 6.10 lets you practice comparing the magnitudes of dispersion forces in different substances.

SAMPLE PROBLEM 6.10 Comparing the Magnitudes of Dispersion Forces

In each pair of nonpolar molecules, choose the substance with the stronger dispersion forces:

(a) N_2 and Br_2 (b) CO_2 and I_2 (c) SiF_4 and $SiBr_4$

Strategy All of these are nonpolar molecules. (If it is not immediately obvious that they are nonpolar, draw the Lewis structures.) Simply determine the molar masses for comparison, where the higher the molar mass, the stronger the dispersion forces.

Setup (a) The molar masses for N_2 and Br_2 are 28.02 and 159.80 g/mol, respectively.

(b) The molar masses for CO_2 and I_2 are 44.01 and 253.81 g/mol, respectively.

(c) The molar masses for SiF_4 and $SiBr_4$ are 104.08 and 343.65 g/mol, respectively.

Solution (a) Br_2 has the larger molar mass and therefore the stronger dispersion forces.

(b) I_2 has the larger molar mass and therefore the stronger dispersion forces.

(c) $SiBr_4$ has the larger molar mass and therefore the stronger dispersion forces.

THINK ABOUT IT

Remember that we can simply compare molar masses here because we are comparing all nonpolar substances or atoms. If any of these molecules were polar, we would have to consider the strength of the dipole-dipole forces as well.

Practice Problem ATTEMPT In each pair, choose the substance with the stronger dispersion forces:

(a) Ar and Cl_2 (b) CCl_4 and O_2 (c) $SiCl_4$ and Kr

Practice Problem BUILD Which noble gas would you expect to have dispersion forces similar in strength to F_2? Why?

Practice Problem CONCEPTUALIZE The blue ovals represent nonpolar diatomic molecules. The black dots represent the bonding electrons in each molecule. Which of the following sequences best illustrates how dispersion forces cause intermolecular attractions between nonpolar molecules?

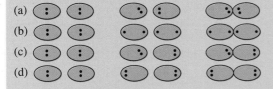

Having established that dispersion forces exist between all molecules, and indeed all *atoms,* and that substances with higher molar masses exhibit larger attractive dispersion forces—as evidenced by their higher boiling points—let's revisit the important concept of hydrogen bonding. Figure 6.12 shows the boiling points of the binary hydrogen compounds of Groups 14 through 17. Within the series of hydrogen compounds

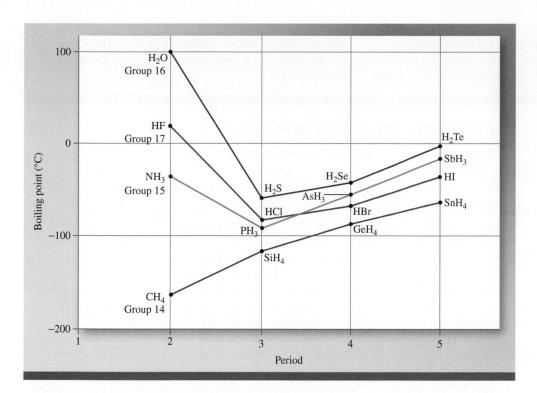

Figure 6.12 Boiling points of the hydrogen compounds of elements from Groups 14 through 17. Although normally the boiling point increases with increasing mass within a group, the lightest compound has the highest boiling point in Groups 15, 16, and 17. This departure from the observed trend is due to hydrogen bonding.

of Group 14, the boiling point increases with increasing molar mass. For Groups 15 through 17, the same trend is observed for all but the smallest member of each series, which we might expect to have the lowest boiling point in the series. Because of hydrogen bonding, however, the *smallest* binary hydrogen compounds of Groups 15, 16, and 17 have the *highest* boiling points. Note that the smallest member of the Group 14 series does not exhibit hydrogen bonding and therefore does not depart from the trend. Only molecules containing F—H, O—H, or N—H bonds exhibit hydrogen bonding.

Profiles in Science

Linus Pauling

The electronegativity values shown in Figure 6.5 were developed by Linus Pauling (1901–1994), American chemist, author, and social activist. Much of Pauling's work involved elucidation of the nature of chemical bonds; it was he who first recognized that ionic and covalent bonding were simply extremes on a continuum along which all chemical bonds lie. He is considered one of the founders of the fields of quantum chemistry and molecular biology. One of his most influential contributions to molecular biology was the determination that sickle-cell disease, a debilitating and often fatal condition prevalent in populations originating in tropical and subtropical regions, is the result of an inherited *molecular* abnormality.

In his later years, Pauling became a vociferous advocate of the use of large doses of vitamin C to ward off and/or cure maladies ranging from the common cold to cancer. Many of his findings and strongly held views in this vein were and *are* controversial—and unfortunately have served to taint his legacy. Regardless, Linus Pauling was truly a remarkable person—generally considered to be one of the most influential scientists of the twentieth century. The scope of his contributions to science is vast, and he is the only person ever to have won two *unshared* Nobel Prizes: the 1954 Nobel Prize in Chemistry for his research into the nature of the chemical bond, and the 1962 Nobel Peace Prize for his campaign against

the continued testing of nuclear weapons.

In 2008, the United States Postal Service issued a set of stamps honoring American scientists including theoretical physicist John Bardeen, biochemist Gerty Cori, astronomer Edwin Hubble, and Linus Pauling. The graphics on Pauling's stamp depict normal and deformed red blood cells, symbolizing his discovery of the molecular nature of sickle-cell disease.

Interestingly, Pauling was a high school dropout. When, at age 15, he had taken every science course offered by Washington High School in Portland, Oregon, he opted to leave without his diploma rather

Linus Pauling

than spend another year there to complete a two-course American history sequence, which he had petitioned unsuccessfully to take concurrently during his last semester. He was admitted to Oregon State University, then known as Oregon Agricultural College, without

having satisfied the requirements for high school graduation. In 1962, his high school (by then named Washington-Monroe High School) awarded Pauling his diploma—after he had received the first of his two Nobel Prizes.

Intermolecular Forces in Review

In general, the strength of intermolecular attractive forces in molecular substances is considered to be:

$$\text{dispersion forces} < \text{dipole-dipole forces} < \text{hydrogen bonding}$$

However, this generalization applies strictly only to substances of comparable molar mass and can be misleading because of the enormous range of molar masses in known compounds. For example, although molecular bromine (M = 159.8 g/mol) is nonpolar, exhibiting only dispersion forces, it is a *liquid* at room temperature. Molecular iodine (M = 253.8 g/mol) is nonpolar, also exhibiting only dispersion forces, and yet is a *solid* at room temperature. Thus, although dispersion forces are typically described as the *weakest* of the intermolecular forces, in very large molecules they are sufficiently strong to hold molecules together in a liquid or even a solid at room temperature.

An elegant example of the effect of molar mass and its effect on the magnitude of dispersion forces is the straight-chain alkanes. These are nonpolar compounds that consist of nothing but carbon atoms and hydrogen atoms, connected by only single bonds. The simplest of the alkanes is methane, CH_4, which we have seen in the context of this chapter. The next in the series of alkanes is ethane, C_2H_6, the next is propane, C_3H_8, and the next is butane, C_4H_{10}.

| Methane (gas) | Ethane (gas) | Propane (gas) | Butane (gas) |
| 16.04 g/mol | 30.07 g/mol | 44.09 g/mol | 58.12 g/mol |

Student Note: Both propane and butane are routinely liquefied under pressure for use as fuels in furnaces, stoves, and lighters.

All of these compounds are nonpolar, exhibiting only dispersion forces, and are gases at room temperature and ordinary pressures. As the number of carbons in this series of compounds grows, and the molar mass increases, the strength of the dispersion forces between molecules increases. The next compound in the series, pentane, is a chain with *five* carbon atoms. Pentane, C_5H_{12}, is a *liquid* at room temperature. The molar mass of pentane is sufficiently large, and the resulting dispersion forces strong enough, to hold pentane molecules together in the liquid state.

Pentane (liquid)
72.15 g/mol

Analogous molecules with carbon-chain lengths of six carbons through 17 carbons are all, likewise, liquids at room temperature—with the boiling point of each compound increasing along with molar mass. When the length of the carbon chain grows to 18, in the compound known as *octadecane,* the molar mass is sufficiently large, and the resulting dispersion forces sufficiently strong, that the compound is a *solid* at room temperature.

Octadecane, $C_{18}H_{38}$ (solid)
254.5 g/mol

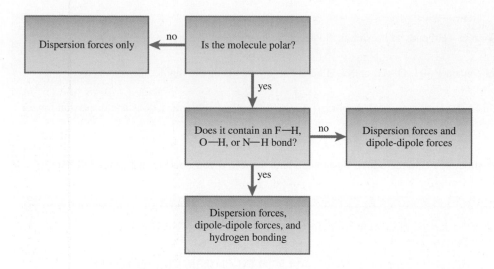

Figure 6.13 Flowchart for determining the type or types of intermolecular forces exhibited by a compound.

Straight-chain hydrocarbons with *more* than 18 carbons are all, likewise, solids—with melting points that increase with increasing molar mass.

You now have the skills necessary to predict the type or types of intermolecular forces a compound will exhibit, based on its chemical formula—and to compare the magnitudes of intermolecular forces in different compounds. It is important to remember, though, that this is not something you can do simply by *looking* at chemical formulas. You must use the following set of steps every time to determine whether or not a molecule is polar:

1. Use the formula to draw a Lewis structure.
2. Apply the VSEPR model to determine the electron-group geometry.
3. Use the electron-group geometry and the resulting positions of atoms to determine molecular shape.
4. Consider the arrangement and types of *bond* dipoles to determine whether or not the molecule is polar overall.

Once you have completed this process, you are ready to make predictions about intermolecular forces, and to compare their relative strengths in various compounds. Figure 6.13 summarizes the steps used to determine what type or types of intermolecular forces a compound exhibits.

Sample Problem 6.11 lets you practice determining what kinds of forces exist between particles in liquids.

Student Note: The takeaway message from this is that, although dispersion forces are generally listed as the weakest of the intermolecular forces, they can be very strong in very large molecules. Further, dispersion forces are exhibited by *all* molecules, not just nonpolar ones.

SAMPLE PROBLEM (6.11) Identifying Intermolecular Forces in Liquids

What kind(s) of intermolecular forces exist in (a) $CCl_4(l)$, (b) $CH_3COOH(l)$, (c) $CH_3COCH_3(l)$, and (d) $H_2S(l)$?

Strategy Draw Lewis dot structures and apply VSEPR theory [◄◄ Section 6.4] to determine whether each molecule is polar or nonpolar. Nonpolar molecules exhibit dispersion forces only. Polar molecules exhibit both dipole-dipole interactions and dispersion forces. Polar molecules with F—H, O—H, or N—H bonds exhibit dipole-dipole interactions (including hydrogen bonding) and dispersion forces.

Setup The Lewis dot structures for molecules (a) to (d) are:

(a)   (b)   (c)   (d)

Solution (a) CCl_4 is nonpolar, so the only intermolecular forces are dispersion forces.

(Continued on next page)

(b) CH_3COOH is polar and contains an O—H bond, so it exhibits dipole-dipole interactions (including hydrogen bonding) and dispersion forces.

(c) CH_3COCH_3 is polar but does not contain F—H, O—H, or N—H bonds, so it exhibits dipole-dipole interactions and dispersion forces.

(d) H_2S is polar but does not contain F—H, O—H, or N—H bonds, so it exhibits dipole-dipole interactions and dispersion forces.

THINK ABOUT IT

Being able to draw correct Lewis structures is, once again, vitally important. Review, if you need to, the procedure for drawing them [◄◄ Section 6.1].

Practice Problem Ⓐ**TTEMPT**   What kind(s) of intermolecular forces exist in (a) $CH_3CH_2CH_2CH_2CH_3(l)$, (b) $CH_3CH_2OH(l)$, (c) $H_2CO(l)$, and (d) $O_2(l)$?

Practice Problem Ⓑ**UILD**   What kind(s) of intermolecular forces exist in (a) $CH_2Cl_2(l)$, (b) $CH_3CH_2CH_2OH(l)$, (c) $H_2O_2(l)$, and (d) $N_2(l)$?

Practice Problem Ⓒ**ONCEPTUALIZE**   Samples of CH_3OH and CH_3SH are both at room temperature. One is a gas and the other a liquid under these conditions. Determine which is the gas and explain why.

CHECKPOINT–SECTION 6.6   Intermolecular Forces

6.6.1   What kind(s) of intermolecular forces exist between benzene molecules (C_6H_6)? (Select all that apply.)

　　a) Dispersion forces

　　b) Dipole-dipole interactions

　　c) Hydrogen bonding

　　d) Ion-dipole interactions

　　e) Ionic bonding

6.6.2   Which of the following exhibits significant hydrogen bonding? (Select all that apply.)

a) HBr	d) H_2O_2
b) H_2CF_2	e) CH_3CN
c) H_2	

6.6.3   Rank the set of molecules in order of *increasing* intermolecular forces.

$$Br_2 \qquad CF_4 \qquad NF_3$$

a) $NF_3 < Br_2 < CF_4$	d) $Br_2 < NF_3 < CF_4$
b) $Br_2 < CF_4 < NF_3$	e) $CF_4 < Br_2 < NF_3$
c) $NF_3 < CF_4 < Br_2$	

6.6.4   Rank the set of substances in order of *decreasing* intermolecular forces.

$$Xe \qquad CCl_3OH \qquad PCl_3$$

a) $PCl_3 > CCl_3OH > Xe$	d) $Xe > CCl_3OH > PCl_3$
b) $Xe > PCl_3 > CCl_3OH$	e) $PCl_3 > Xe > CCl_3OH$
c) $CCl_3OH > PCl_3 > Xe$	

Chapter Summary

Section 6.1

- *Lewis dot structures* or *Lewis structures* are drawn to represent species that contain covalent bonds; i.e., molecules and polyatomic ions.

- In most Lewis structures, atoms (with the exception of hydrogen) obey the *octet rule,* which says they must be surrounded by eight electrons. Hydrogen needs only two electrons.

- Lewis structures are drawn by counting valence electrons on all atoms, adding or subtracting electrons to adjust for charge

on a polyatomic ion, arranging atoms in a *skeletal structure,* and distributing the available valence electrons to complete the octets of all atoms in the structure.

- Many molecules for which we will draw Lewis structures have a *central atom* and two or more *terminal atoms.*

- In a Lewis structure, we generally use a dash to represent a *bond pair* or *bonding pair* of electrons, which is shared by two atoms. An unshared pair of electrons is represented by dots and is known as a *lone pair* or a *nonbonding pair.*

Section 6.2

- In Lewis structures of molecules for which there is more than one possible central atom, we generally choose the more metallic atom as the central atom for the skeletal structure. Some molecules have more than one central atom.

- If in the distribution of valence electrons in a Lewis structure, we run out of electrons before completing the octets of all atoms, we can move lone pairs between atoms to form *multiple bonds* (*double bonds* or *triple bonds*), thereby completing the octets of every atom.

- Exceptions to the octet rule include species with Be or B as the central atom, which can have just *four* and *six* electrons around them, respectively; any species with an odd number of electrons; and any species in which the central atom is from the third period or below on the periodic table. Atoms from the third period or below can have an *expanded octet,* with as many as 12 electrons around them.

Section 6.3

- Two or more equally correct Lewis structures that can be drawn for a single species, but that differ only in their placement of electrons, are called *resonance structures.*

- Resonance structures typically are necessary for molecules or polyatomic ions with double bonds that can be placed in two or more positions in the structure.

Section 6.4

- Molecular shape is important to chemical and biochemical processes—including our sense of smell and drug functionality.

- We can determine molecular shape by drawing a Lewis structure and applying the *valence-shell electron-pair repulsion (VSEPR) model.* This requires us to count the number of *electron groups* (bonds or lone pairs) on the central atom and determine the *electron-group geometry* based on the number of electron groups.

- Four electron groups on a central atom adopt a tetrahedral arrangement, three electron groups adopt a trigonal planar arrangement, and two electron groups adopt a linear arrangement.

- Electron-group geometry is determined by the arrangement of all electron groups on the central atom. *Molecular shape* is determined by the resulting arrangement of *atoms*.

Section 6.5

- Ionic bonding and covalent bonding are actually the extremes on a continuum of chemical bonding. Most bonds are neither completely ionic nor purely covalent.

- *Electronegativity* is the ability of an atom to attract electrons toward itself in a covalent bond. A covalent bond joining two atoms with different electronegativity values is called a *polar covalent bond.*

- Atoms with very large differences in electronegativity tend to form ionic bonds rather than covalent bonds; atoms with very small differences in electronegativity tend to form covalent bonds; and atoms with intermediate differences in electronegativity tend to form polar covalent bonds.

- A molecule that is polar is said to have a *dipole,* which can be represented by a *dipole vector.* A molecule can contain polar bonds and still be nonpolar overall if its dipole vectors are symmetrically distributed and of equal magnitude.

Section 6.6

- *Intermolecular forces* are attractive forces resulting from electrostatic attraction between molecules. The magnitude of intermolecular forces in a substance determines whether it is a solid, a liquid, or a gas at a given temperature.

- *Dipole-dipole forces* are intermolecular attractions between polar molecules.

- *Hydrogen bonding* is a particularly strong type of dipole-dipole force that is exhibited only by molecules in which hydrogen is bonded to F, O, or N.

- *Dispersion forces* are the only attractive forces between non-polar molecules and atoms—although all molecules, polar and nonpolar, exhibit dispersion forces.

- Dispersion forces are the result of random movement of electrons within molecules and atoms that result in a temporary *instantaneous dipole* in an ordinarily nonpolar species. An instantaneous dipole in one molecule or atom causes a temporary *induced dipole* in a neighboring molecule or atom, resulting in electrostatic attraction between the temporary dipoles.

- The magnitude of dispersion forces increases with increasing molar mass because the electrons in larger species have more freedom to move about, allowing larger molecules and atoms to become polarized more easily.

Key Terms

Bond pair 203

Central atom 203

Dipole 222

Dipole-dipole forces 226

Dipole vector 222

Dispersion forces 229

Double bond 207

Electron group 212

Electronegativity 219

Electron-group geometry 212

Expanded octet 208

Hydrogen bonding 227

Induced dipole 229

Instantaneous dipole 229

Intermolecular forces 226

Lewis structure 201

Lone pair 203

Molecular shape 212

Multiple bond 207

Octet rule 201

Polar covalent bond 221

Resonance structures 210

Skeletal structure 202

Terminal atom 203

Triple bond 207

Valence-shell electron-pair repulsion (VSEPR) model 212

KEY SKILLS   **Molecular Shape and Polarity**

Molecular polarity is tremendously important in determining the physical and chemical properties of a substance. Indeed, molecular polarity is one of the most important consequences of molecular shape. To determine the shape of a molecule, we use a stepwise procedure:

1. Draw a correct Lewis structure [◄◄ Sections 6.1 and 6.2].
2. Count electron groups on the central atom. Remember that an electron group can be a lone pair or a bond, and that a *bond* may be a *single* bond, a *double* bond, or a *triple* bond.
3. Apply the VSEPR model [◄◄ Section 6.4] to determine electron-group geometry.
4. Consider the positions of the atoms to determine the molecular shape, which may or may not be the same as the electron-group geometry.

Consider the examples of SO_2, C_2H_2, and CH_2Cl_2. We determine the molecular shape of each as follows:

Draw the Lewis structure	:Ö̈–S̈=Ö:	H–C≡C–H	Cl | H–C–Cl | H
Count the electron groups on the central atom(s)	3 electron groups: • 1 double bond • 1 single bond • 1 lone pair	2 electron groups on each central atom: • 1 single bond • 1 triple bond	4 electron groups: • 4 single bonds
Apply VSEPR to determine electron-group geometry	3 electron groups arrange themselves in a trigonal plane.	2 electron groups arrange themselves linearly.	4 electron groups arrange themselves in a tetrahedron.
	O⤴S̈⤵O	H–C≡C–H	Cl | C H⟋ ⟍Cl | H
Consider positions of atoms to determine molecular shape.	With 1 lone pair on the central atom, the molecular shape is bent.	With no lone pairs on the central atom, the molecular shape is linear.	With no lone pairs on the central atom, the molecular shape is tetrahedral.

Having determined molecular shape, we determine overall molecular polarity of each molecule by examining the individual bond dipoles and their arrangement:

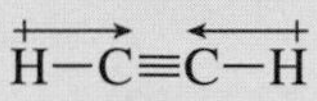

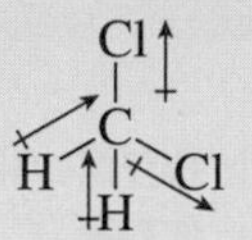

| Determine whether or not the individual bonds are polar. | S and O have electronegativity values of 2.5 and 3.5, respectively. Therefore, the bonds are polar. | C and H have electronegativity values of 2.5 and 2.1, respectively. Therefore, the bonds are considered nonpolar. | The C—H bonds are nonpolar. C and Cl have electronegativity values of 2.5 and 3.0, respectively. Therefore, the C—Cl bonds are polar. |

Only in C_2H_2 do the dipole-moment vectors cancel each other. C_2H_2 is nonpolar, SO_2 and CH_2Cl_2 are polar.

Even with polar bonds, a molecule may be nonpolar if it consists of equivalent bonds that are distributed symmetrically. Molecules with equivalent bonds that are *not* distributed symmetrically—or with bonds that are not *equivalent,* even if they *are* distributed symmetrically—are generally polar.

Key Skills Problems

6.1
Determine the molecular shape of selenium dibromide.
a) linear
b) bent
c) trigonal planar
d) trigonal pyramidal
e) tetrahedral

6.2
Determine the molecular shape of phosphorus triiodide.
a) linear
b) bent
c) trigonal planar
d) trigonal pyramidal
e) tetrahedral

6.3
Which of the following species is polar?
a) OBr_2
b) $GeCl_4$
c) $AlCl_3$
d) BH_3
e) BeF_2

6.4
Which of the following species is nonpolar?
a) NCl_3
b) $SeCl_2$
c) SO_2
d) CF_4
e) $AsBr_3$

Questions and Problems

SECTION 6.1: DRAWING SIMPLE LEWIS STRUCTURES

6.1 Why are valence electrons the only ones counted when predicting the bonding in a molecule?

6.2 List the steps necessary to draw an accurate Lewis structure for a molecule.

6.3 How does drawing the Lewis structure for a molecule differ from drawing the Lewis structure for a polyatomic ion?

6.4 How many *total* valence electrons are present in each of the following molecules?
(a) NCl_3 (c) $SiBr_4$
(b) OCl_2 (d) PF_3

6.5 How many *total* valence electrons are present in each of the following molecules?
(a) CF_4 (c) NI_3
(b) SCl_2 (d) $SeBr_2$

6.6 Draw the Lewis structure for each of the following molecules:
(a) Cl_2 (c) $SiCl_4$
(b) SF_2 (d) I_2

6.7 Draw the Lewis structure for each of the following molecules:
(a) OCl_2 (c) CI_4
(b) NI_3 (d) Br_2

6.8 Draw the Lewis structure for each of the following polyatomic ions:
(a) ClO_3^- (c) SiF_3^-
(b) $AsCl_4^+$ (d) BrS_2^-

6.9 Draw the Lewis structure for each of the following polyatomic ions:
(a) PF_4^+ (c) NO_2^-
(b) CO_3^{2-} (d) CCl_3^-

SECTION 6.2: LEWIS STRUCTURES CONTINUED

6.10 What are double and triple bonds? How are they different from a single bond?

6.11 Draw the Lewis structure for each of the following compounds:
(a) CS   (c) PCl_3
(b) CS_2 (d) HCN (C is central)

6.12 Draw the Lewis structure for each of the following compounds:
(a) H_2CO (C is central)   (c) NBr_3
(b) O_2 (d) OF_2

6.13 Draw the Lewis structure for each of the following compounds:
(a) N_2O (N is central)   (c) SO_2
(b) CCl_2F_2 (d) N_2

6.14 Draw the Lewis structure for each of the following compounds. The carbon atoms are central in each one.
(a) C_2H_6 (c) C_2H_2
(b) C_2H_4 (d) C_2Cl_2

6.15 Draw the Lewis structure for each of the following polyatomic ions:
(a) NO_2^- (c) SiO_3^{2-}
(b) SO_3^{2-} (d) SO_4^{2-}

6.16 Draw the Lewis structure for each of the following polyatomic ions:
(a) NO_3^- (c) AsO_3^-
(b) PO_3^{3-} (d) OH^-

6.17 Identify and correct any errors in the following Lewis structures:

6.18 Identify and correct any errors in the following Lewis structures:

6.19 Identify the seven elements that exist as diatomic molecules and use theory to explain why they exist in this form.

6.20 Using Lewis theory, determine which molecule in each pair is more likely to exist:
(a) CH_4/CH_2 (c) SiF_3/SiF_4
(b) NH_4/NH_3 (d) SCl_2/SCl_3

SECTION 6.3: RESONANCE STRUCTURES

6.21 What is a resonance structure? When is it necessary to draw resonance structures?

6.22 Draw the Lewis structure for each of the following ions. Include resonance structures when necessary.
(a) NO_2^- (c) ClO_4^-
(b) SO_3^{2-} (d) CO_3^{2-}

6.23 Draw the Lewis structure for each of the following ions. Include resonance structures when necessary.
(a) ClO_2^- (b) SiO_3^{2-} (c) NO_3^- (d) SO_4^{2-}

6.24 Draw all of the possible resonance structures for the following substances:
(a) SO_2 (c) N_2O (N is central)
(b) OCN^- (d) SiO_2^{2-}

6.25 Draw all of the possible resonance structures for the following substances:
(a) PO_2^+ (c) SO_3
(b) AsO_3^- (d) O_3

SECTION 6.4: MOLECULAR SHAPE

6.26 What is the difference between electron-group geometry and molecular shape? Explain why they sometimes are the same and sometimes are different.

6.27 Fill in the missing information in the table.

Generic Lewis Structure	Number of Electron Groups on Central Atom	Electron-Group Geometry	Molecular Shape	Bond Angles
(a) X—A—X				
(b) X—Ä—X				
(c) X—Ä—X				

6.28 Fill in the missing information in the table.

Generic Lewis Structure	Number of Electron Groups on Central Atom	Electron-Group Geometry	Molecular Shape	Bond Angles
(a) X—A—X X				
(b) X—Ä—X X				
(c) X X—A—X X				

6.29 Using VSEPR theory, determine the electron-group geometry and molecular shape of the following substances:
(a) BF_3 (c) CF_4
(b) NF_3 (d) CF_2H_2

6.30 Using VSEPR theory, determine the electron-group geometry and molecular shape of the following substances:
(a) $SeBr_2$ (c) SO_2
(b) CO_2 (d) $BeCl_2$

6.31 Using VSEPR theory, determine the electron-group geometry and molecular shape of the following substances:
(a) BF_4^- (c) $SiBr_4$
(b) SeO_2 (d) NO_3^-

6.32 Using VSEPR theory, determine the electron-group geometry and molecular shape of the following substances:
(a) NO_2^+ (c) BCl_2H
(b) PCl_3 (d) ClO_3^-

6.33 Using VSEPR theory, determine the electron-group geometry and molecular shape of the following substances:
(a) OF_2 (c) CF_3Cl
(b) H_2CO (C is central)   (d) PCl_2F

6.34 Using VSEPR theory, determine the electron-group geometry and molecular shape of the following substances. *Hint:* The carbon atoms are central, and you must consider each one separately.
(a) C_2H_6 (c) C_2H_2
(b) C_2H_4 (d) C_2Cl_2

6.35 Predict the bond angles for each molecule in Problem 29.

6.36 Predict the bond angles for each molecule in Problem 30.

6.37 Predict the bond angles for each substance in Problem 31.

6.38 Predict the bond angles for each substance in Problem 32.

6.39 Predict the bond angles for each substance in Problem 33.

6.40 Label the bond angles on each Lewis structure you drew in Problem 34.

SECTION 6.5: ELECTRONEGATIVITY AND POLARITY

6.41 Define the term *electronegativity* and describe the trends in electronegativity on the periodic table.

6.42 What makes a bond polar?

6.43 What makes a molecule polar?

6.44 Describe the importance of molecular shape in the determination of a molecule's polarity.

6.45 Can a molecule contain polar bonds, but be nonpolar? Explain.

6.46 Which of the following bonds are polar?
(a) C—F   (c) O—O
(b) P—F   (d) Si—O

6.47 Which of the following bonds are polar?
(a) As—Cl   (c) N—I
(b) F—F   (d) C—Cl

6.48 Label each of the polar bonds in Problem 46 with a dipole vector.

6.49 Label each of the polar bonds in Problem 47 with a dipole vector.

6.50 Which of the following molecules are polar?
(a) CCl_4 (c) BI_3
(b) PF_3 (d) OCl_2

6.51 Which of the following molecules are polar?
(a) $AsCl_3$ (b) NI_3 (c) CBr_4 (d) BCl_3

6.52   Which of the following molecules are polar?
   (a) CO   (c) SO_2
   (b) CO_2 (d) $BeCl_2$

6.53   Which of the following molecules are polar?
   (a) CH_2Cl_2 (c) $BHCl_2$
   (b) BeIF   (d) SCl_2

6.54   Elements X and A have electronegativity values that differ by 1.0. Determine which of the following molecules is polar:

   (a)   X   (c) X—Ä—X
         |
      X—A—X   (d) X—Ä—X
         |
         X

   (b) X—A—X
         |
         X

SECTION 6.6: INTERMOLECULAR FORCES

6.55   List the different types of intermolecular forces. What determines the type of intermolecular forces present in a substance?

6.56   List the steps necessary to determine what types of intermolecular forces are present in a substance, given its chemical formula.

6.57   Which of the following substances exhibit ONLY dispersion forces?
   (a) F_2 (c) Kr
   (b) SF_2 (d) SiS_2

6.58   Which of the following substances exhibit ONLY dispersion forces?
   (a) Xe   (c) OCl_2
   (b) O_2 (d) BCl_3

6.59   Which of the following substances exhibit dipole-dipole forces?
   (a) NF_3 (c) CH_2F_2
   (b) CF_4 (d) BF_3

6.60   Which of the following substances exhibit dipole-dipole forces?
   (a) $SiCl_4$ (c) $BeCl_2$
   (b) OF_2 (d) PCl_3

6.61   Which of the following substances exhibit hydrogen bonding?
   (a) CH_3CH_2OH (c) $N(CH_3)_2H$
   (b) CH_3F (d) NCl_3

6.62   Which of the following substances exhibit hydrogen bonding?
   (a) $HO—CH_2CH_2—OH$ (c) CH_3NH_2
   (b) $N(CH_3)_2F$ (d) SH_2

6.63   Identify the strongest type of intermolecular force present in each of the following substances:
   (a) CS_2 (c) CCl_2F_2
   (b) Kr   (d) O_2

6.64   Identify the strongest type of intermolecular force present in each of the following substances:
   (a) NH_3 (c) SO_3
   (b) PH_3 (d) BF_3

6.65   Place each set of molecules in order of increasing strength of intermolecular forces:
   (a) PCl_3, NCl_3, SO_2
   (b) CH_4, C_3H_8, C_2H_6
   (c) Xe, Cl_2, CO_2

6.66   Place each set of substances in order of increasing strength of intermolecular forces:
   (a) CH_3OH, C_4H_{10}, $NHCl_2$
   (b) CF_2H_2, CF_2Cl_2, CF_2I_2
   (c) NCl_3, CO_2, Kr

6.67   Place each set of molecules in order of decreasing strength of intermolecular forces:
   (a) BCl_3, OF_2, H_2O
   (b) NH_3, SeO_2, CO_2
   (c) BeF_2, PF_3, SF_2

6.68   Place each set of molecules in order of decreasing strength of intermolecular forces:
   (a) $AsCl_3$, CS_2, Kr
   (b) $BeCl_2$, BCl_3, CCl_4
   (c) HF, CF_4, CF_2H_2

CUMULATIVE PROBLEMS

6.69   Write the Lewis symbol for each ion present in MgO. How does this differ from the process for drawing the Lewis structure for the molecular compound CO?

6.70   Explain how molecules with very similar formulas, such as CO_2 and SeO_2, can exhibit different shapes, polarities, and intermolecular forces.

6.71   Explain how a substance can contain hydrogen atoms but not exhibit hydrogen bonding. Give examples.

Answers to In-Chapter Materials

Answers to Practice Problems

6.1A (a) $\left[:\overset{..}{O}-\overset{..}{\underset{\underset{:O:}{|}}{Cl}}-\overset{..}{O}: \right]^{-}$ (b) $\left[:\overset{:\overset{..}{Cl}:}{\underset{:\overset{..}{Cl}:}{\overset{|}{\underset{|}{Cl}-P-\overset{..}{Cl}:}}} \right]^{+}$

(c) $:\overset{:\overset{..}{Br}:}{\underset{:\overset{..}{Br}:}{\overset{|}{\underset{|}{Br}-Si-\overset{..}{Br}:}}}$ **6.1B** (a) $\left[:\overset{:\overset{..}{F}:}{\underset{:\overset{..}{F}:}{\overset{|}{\underset{|}{F}-N-\overset{..}{F}:}}} \right]^{+}$ (b) $\left[:\overset{..}{O}-Br-\overset{..}{O}: \atop \underset{:O:}{\overset{..}{O}:} \right]^{-}$

(c) $\left[\overset{H-\overset{..}{S}-H}{\underset{H}{\overset{|}{|}}} \right]^{+}$ **6.2A** H—C≡N: **6.2B** $\left[:N=\overset{..}{O}: \right]^{-}$

6.3A $\left[\overset{..}{O}=N-\overset{..}{O}: \atop \underset{:O:}{} \right]^{-} \longleftrightarrow \left[:\overset{..}{O}-N=\overset{..}{O} \atop \underset{:O:}{} \right]^{-} \longleftrightarrow \left[:\overset{..}{O}-N-\overset{..}{O}: \atop \underset{.\overset{..}{O}.}{\overset{||}{}} \right]^{-}$

6.3B $\left[\overset{H\ \ \overset{..}{O}}{\underset{H}{\overset{|\ \ ||}{H-C-C-\overset{..}{O}:}}} \right]^{-} \longleftrightarrow \left[\overset{H\ \ :\overset{..}{O}:}{\underset{H}{\overset{|\ \ |}{H-C-C=\overset{..}{O}:}}} \right]^{-}$

6.4A Trigonal planar, trigonal planar, 120°. **6.4B** (a) Both species are bent. The O—A—O bond angle in AO_2 is approximately 120°, whereas that in AO_2^{2-} is approximately 109.5°. The bond angles are different because there are different numbers of lone pairs on the central atom. **6.5A** (a) Mg, (b) Te. **6.5B** (a) C > Li > K, (b) As > Ca > Ba. **6.6A** (a) nonpolar, (b) polar, (c) nonpolar. **6.6B** C—I < C—Br < C—Cl < C—F. **6.7A** (a) nonpolar, (b) polar. **6.7B** b, d. **6.8A** (a) yes, (b) no, (c) yes. **6.8B** (a) yes, (b) yes, (c) no. **6.9A** c. **6.9B** (a) has no F—H, O—H, or N—H bonds. (b) exhibits H-bonding. **6.10A** (a) Cl_2, (b) CCl_4, (c) $SiCl_4$. **6.10B** Ar, similar molar mass. **6.11A** (a) dispersion; (b) dispersion, dipole-dipole, hydrogen bonding; (c) dispersion, dipole-dipole; (d) dispersion. **6.11B** (a) dispersion, dipole-dipole; (b) dispersion, dipole-dipole, hydrogen bonding; (c) dispersion, dipole-dipole, hydrogen bonding; (d) dispersion.

Answers to Checkpoints

6.1.1 e. **6.1.2** c. **6.2.1** b. **6.2.2** b. **6.2.3** c. **6.2.4** d. **6.3.1** a, e. **6.3.2** b. **6.4.1** d. **6.4.2** b. **6.4.3** d. **6.4.4** c. **6.4.5** a, e. **6.5.1** a. **6.5.2** a. **6.6.1** a. **6.6.2** d. **6.6.3** e. **6.6.4** c.

CHAPTER 7

Solids, Liquids, and Phase Changes

7.1 **General Properties of the Condensed Phases**

7.2 **Types of Solids**
- Ionic Solids
- Molecular Solids
- Atomic Solids
- Network Solids

7.3 **Physical Properties of Solids**
- Vapor Pressure
- Melting Point

7.4 **Physical Properties of Liquids**
- Viscosity
- Surface Tension
- Vapor Pressure
- Boiling Point

7.5 **Energy and Physical Changes**
- Temperature Changes
- Solid-Liquid Phase Changes: Melting and Freezing
- Liquid-Gas Phase Changes: Vaporization and Condensation
- Solid-Gas Phase Changes: Sublimation

Under specific atmospheric conditions, snow can fall through a layer of air warm enough to melt it, and then remain liquid until it contacts a surface cold enough to refreeze it. Ice storms can be dramatic reminders of some of water's extraordinary properties.

Tony Campbell/Shutterstock

In This Chapter, You Will Learn

About the attractive forces that hold atoms, molecules, or ions together in liquids and solids; and how the magnitudes of those attractive forces affect the observable physical properties of substances in the condensed phases.

Things To Review Before You Begin

- Molecular shape [◄◄ Section 6.4]
- Electronegativity and Polarity [◄◄ Section 6.5]
- Chapter 6 Key Skills [◄◄ pages 236–237]

We have learned that all substances, ionic, molecular, and atomic, can exhibit intermolecular forces. We have also seen that in many cases, the intermolecular forces are strong enough to hold the particles of a substance together—even at room temperature, resulting in many substances being solids or liquids (collectively known as the *condensed phases*). Moreover, we know that intermolecular forces are what enable us to liquefy gases under the right conditions. In this chapter, we examine the role that intermolecular forces play in determining the properties of solids and liquids, and the role that energy plays in the transitions between the solid, liquid, and gaseous states of pure substances.

7.1 General Properties of the Condensed Phases

Recall from the discussion of states of matter in Chapter 3 that the particles in a solid are close together and are typically held rigidly in position relative to one another, giving the solid a *fixed shape*. In addition, most solids are not compressible because their particles are already packed as tightly together as possible. This results in solids also having a *fixed volume*.

The particles that make up a liquid are also very close to one another but are not so tightly held that they cannot move past one another. This is what enables liquids to *flow*. Thus, a liquid also has a fixed *volume,* but *not* a fixed shape. For most substances, the solid state has a higher density than the liquid state, and as we will see in the next section, the particles in a solid may be very highly ordered (crystalline) or less ordered (amorphous).

In contrast to the condensed phases, gases consist of particles that are separated by large distances. Gases are easily compressed, and a sample of gas will assume both the shape and the *volume* of its container. For any substance, the gaseous state has (by far) the lowest density of the three states. Figure 7.1 shows the solid, liquid, and gaseous states of a typical substance.

Student Note: A notable exception to this is ice (solid *water*), which we examine in some detail later in this chapter.

Animation
States of Matter

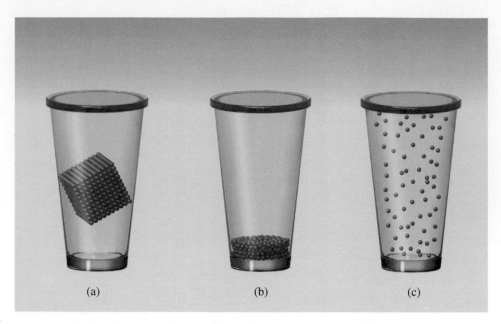

Figure 7.1 A typical substance in (a) solid, (b) liquid, and (c) gaseous states.

7.2 Types of Solids

We begin by distinguishing two different types of solids: amorphous and crystalline. An **amorphous solid** is one in which the constituent molecules are not arranged in a highly ordered way. Amorphous solids typically form when a liquid cools and solidifies too quickly to allow the molecules to move into the positions they would otherwise occupy, given sufficient time. An example of this phenomenon is the manufacture of glass—a very familiar amorphous solid. Quartz, the most common mineral in Earth's continental crust, is *crystalline* silicon dioxide, SiO_2. **Glass,** a term that usually refers to an optically transparent solid substance, is essentially the *amorphous* form of SiO_2. Figure 7.2 shows macroscopic and molecular-level pictures of quartz and glass.

There are many different types of glass, most of which contain additives used to impart specific properties such as durability and color. Glass is generally made by melting sand, a common form of quartz, and then shaping and allowing the molten mixture to cool. Table 7.1 shows the composition of three common types of glass.

Solids in which the ions, molecules, or atoms they contain are highly ordered are **crystalline solids,** and their properties depend on the nature of the particles they contain. What follows are descriptions of the four types of crystalline solids.

Ionic Solids

The strongest "intermolecular" force we have encountered is *ionic bonding* [◄◄ Section 3.2]. Most ionic substances are solids at room temperature because of the strength of the electrostatic forces that hold them together—typically in a highly ordered three-dimensional array of alternating cations and anions. Recall that the smallest unit of an ionic compound is referred to as a *formula unit* [◄◄ Section 5.2] and that ionic substances also exhibit dispersion forces.

Molecular Solids

The molecules in *all* molecular solids are held together by *dispersion* forces. Those in which the molecules are polar are *also* held together by *dipole-dipole* forces (including *hydrogen bonding* in substances that contain F—H, O—H, or N—H bonds). An example of a molecular substance that is a solid at room temperature is the element iodine (I_2). Iodine is nonpolar, so the only forces holding the molecules together are dispersion

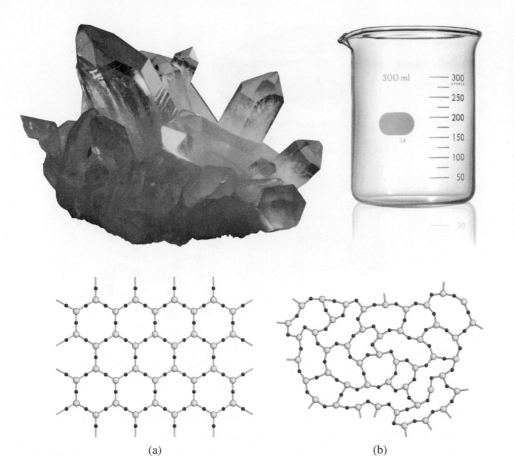

(a)   (b)

Figure 7.2 (a) Quartz, crystalline SiO_2, along with a two-dimensional representation of its structure; and (b) glass, amorphous SiO_2, along with a two-dimensional representation of its structure. Both solids actually exist as three-dimensional arrays, but are represented here in two dimensions to simplify the illustration of how they differ in degree of long-range order.

(a) Siede Preis/Photodisc/Getty Images; (b) Studiocasper/iStockphoto/Getty Images

TABLE 7.1	Composition and Properties of Three Types of Glass	
Pure quartz glass	100% SiO_2	Low thermal expansion, transparent to a wide range of wavelengths. Used in optical research.
Pyrex glass	60%–80% SiO_2, 10%–25% B_2O_3, some Al_2O_3	Low thermal expansion; transparent to visible and infrared, but not to ultraviolet light. Used in cookware and laboratory glassware.
Soda-lime glass	75% SiO_2, 15% Na_2O, 10% CaO	Easily attacked by chemicals and sensitive to thermal shocks. Transmits visible light but absorbs ultraviolet light. Used in windows and bottles.

forces. The dispersion forces in I_2 are sufficiently strong to make it a solid at room temperature because the molecules are large ($\mathcal{M} = 253.8$ g/mol). In general, the molecules that make up a molecular crystalline solid are packed together as closely as their size and shape will allow. An exception to this is perhaps the most familiar of all molecular solids: ice.

We have seen that water has an anomalously high boiling point for so small a molecule—because of the hydrogen bonding that it exhibits [◀◀ Section 6.6]. The hydrogen bonding in water is also what causes it to crystallize, when sufficiently cooled, in a structure with the molecules in a distinct hexagonal arrangement as shown in Figure 7.3. This hexagonal arrangement prevents the molecules in ice from being as close together as they would otherwise be. In fact, the water molecules in ice are farther apart than the molecules in *liquid* water, making the solid form of water less dense than the liquid form—a *very* unusual phenomenon.

Figure 7.3 The three-dimensional structure of ice. The covalent bonds that hold the atoms together in each water molecule are shown as short, solid lines; and the hydrogen bonds are shown as longer, dashed lines. The empty space in the hexagonal structure accounts for the low density of ice relative to liquid water.

Brian Hagiwara/Brand X Pictures/PunchStock

Student Note: Only one metal, mercury, is a liquid at room temperature—the reasons for which are beyond the scope of this book.

The forces that hold molecular solids together are generally much weaker than those that hold ionic solids together. In fact, most of the substances we encountered while learning to draw Lewis structures and determine molecular polarity (e.g., CCl_4, SO_2, CO, C_2H_6, C_2H_4) exist as solids only at temperatures far colder than room temperature.

Atomic Solids

Crystalline solids in which the smallest particles are *atoms* come in two varieties: *metallic* and *nonmetallic*. A metallic crystalline solid consists of a highly ordered three-dimensional array of metal atoms. What holds a solid metal together is something called *metallic bonding,* which we have not encountered before. In metallic bonding, each metal atom basically shares its valence electrons with all of the other atoms in the solid. The result is an array of neatly arranged, positively charged metal ions in a negatively charged "sea" of valence electrons. In fact, this description of metallic bonding is known as the "electron sea" model—and it is the free movement of the valence electrons in a metal that allow it to conduct electricity. Figure 7.4 illustrates the electron sea model of *metallic bonding.*

The strength of bonding in metallic solids is highly variable. Nearly all metals are solids at room temperature but have melting points ranging from just above room temperature for rubidium (Rb), cesium (Cs), and gallium (Ga) to thousands of degrees Celsius for tungsten (W), rhenium (Re), and osmium (Os). (We explain more about melting points and the factors that influence them in Section 7.3.)

Although gallium is a solid at room temperature, it will melt if you hold it in your hand [Figure 7.5(a)]. It has famously been used by scientists in a prank whereby a spoon made of gallium metal is used by an unsuspecting guest to stir a cup of tea. Hot tea melts the spoon, causing the bottom part of it to disappear [Figure 7.5(b)].

The second type of atomic solid, the nonmetallic, is less common because only the noble gases exist as isolated atoms under ordinary conditions. Moreover, because the only types of intermolecular forces they exhibit are relatively weak *dispersion* forces, the noble gases can exist as crystalline solids only at extraordinarily low temperatures.

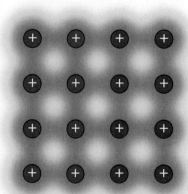

Figure 7.4 A cross section of a metallic crystalline solid. Each circled positive charge represents a metal cation, without its valence electrons. The grey area surrounding the positive metal ions represents the mobile "sea" of electrons that holds the metallic solid together.

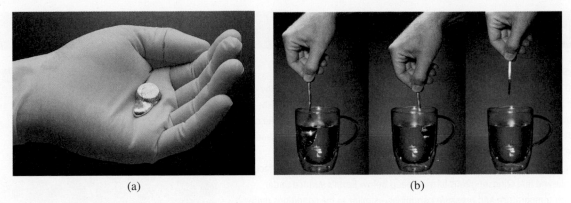

Figure 7.5 Gallium (a) melting just above room temperature, and (b) seeming to disappear when immersed in hot liquid.

(a) Stephen Frisch/McGraw Hill; (b) David A. Tietz/McGraw Hill

Network Solids

A *network solid* (or *covalent solid*) is one in which the atoms are connected by co-valent bonds, which are even stronger than ionic bonds. Familiar examples of network solids include diamond and graphite, both of which consist of carbon atoms. In dia-mond, each carbon atom is bonded covalently to four other carbon atoms in a tetra-hedral arrangement. In graphite, each carbon atom is bonded covalently to *three* other carbon atoms in a trigonal planar arrangement. The network of carbon atoms in graph-ite produces extensive sheets of carbon that are held to one another by relatively weak dispersion forces. This allows the sheets to slide past each other, giving graphite a slippery feel. It is used in many industrial applications as a lubricant and is the "lead" in pencils. Quartz (SiO_2), which we discussed earlier, is also a network solid, in which each silicon atom is bonded covalently to four oxygen atoms in a tetrahedral arrange-ment, and each oxygen atom is bonded to two silicon atoms. Figure 7.6 shows diamond, graphite, and quartz along with atomic-level illustrations of their structures.

Student Note: A single sheet of graphite is known as *graphene*.

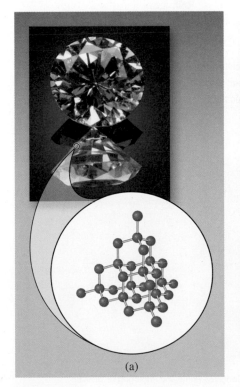
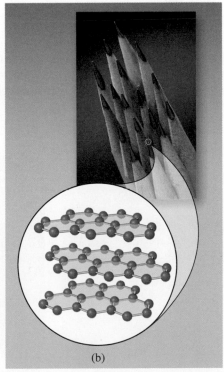
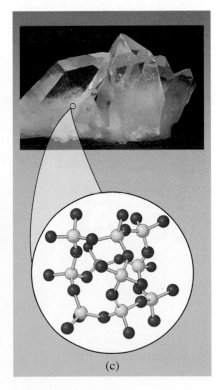

(a) (b) (c)

Figure 7.6 Structures of the network solids (a) diamond, (b) graphite, and (c) quartz (SiO_2).

(a) Charles D. Winters/Timeframe Photography/McGraw Hill; (b) Charles D. Winters/Timeframe Photography/McGraw Hill; (c) Geofile/Doug Sherman

Thinking Outside the Box

A Network Solid as Hard as Diamond

In August 2013, the International Mineralogical Association officially approved the name *Qingsongite* for a naturally occurring boron mineral discovered in 2009 in the Tibetan mountains of China. For over 50 years, the very same substance, a network solid called *cubic boron nitride* (shown here), has been manufactured as an inexpensive, sometimes superior alternative to industrial diamond abrasives.

   Although the other known boron-containing minerals are found at the surface of Earth, it is believed that Qingsongite is formed deep below Earth's surface under extremely high temperature and pressure conditions—similar to the conditions under which diamond forms. Further, the mineral is isoelectronic [◄◄ Section 2.7] with diamond and has the same tetrahedral structure, with alternating B and N atoms in place of C atoms throughout the network.

Courtesy of Worldwide Superabrasives, LLC

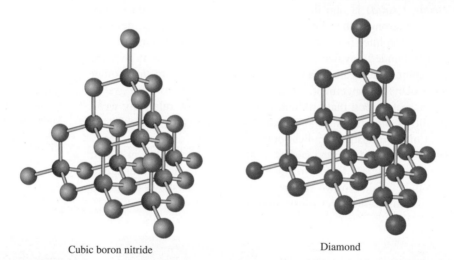

Cubic boron nitride   Diamond

   A more common natural form of boron is the mineral *borax*, also known as *sodium borate*, which has a somewhat complex chemical formula that includes boron, sodium, oxygen, and hydrogen. Borax is a relatively soft, water-soluble, *ionic* solid that has been used for a variety of industrial and household applications since the late nineteenth century when it was hauled from the borax deposits in Death Valley to the nearest rail line in California by a 20-mule team.

David A. Tietz/McGraw Hill

Harry Taylor/Dorling Kindersley/Getty Images

Library of Congress Prints and Photographs Division
[LC-USZ62-20305]

TABLE 7.2	Types of Crystalline Solids and the Intermolecular Forces They Exhibit	
Type of Crystalline Solids	**Intermolecular Forces**
Ionic	Dispersion forces and ionic bonding (electrostatic attractions)
Molecular	
nonpolar molecules	Dispersion forces
polar molecules	Dispersion forces and dipole-dipole forces
polar molecules with F—H, O—H, or N—H bonds	Dispersion forces, dipole-dipole forces, and hydrogen bonding
Atomic	
metallic	Dispersion forces and metallic bonding
nonmetallic	Dispersion forces
Network	Dispersion forces and covalent bonding

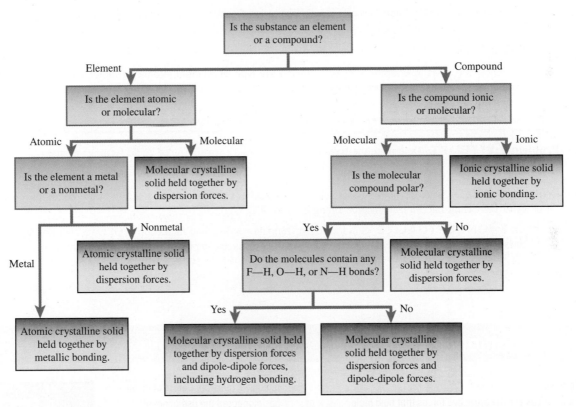

Figure 7.7 Flowchart showing how to determine what kind of crystalline solid a substance will form under appropriate conditions.

Table 7.2 summarizes the types of crystalline solids and the relative strengths of their intermolecular forces, and Figure 7.7 summarizes the steps in determining what kind of crystalline solid a substance can form under appropriate conditions. Note that either an element or a molecular compound may be a network solid, but this is not something that you can determine simply by looking at the chemical formula.

Sample Problem 7.1 gives you some practice determining what kind of crystalline solid a substance will form and determining the types of intermolecular forces that hold the particles in solids together.

SAMPLE PROBLEM 7.1 Using Chemical Formulas to Determine Crystalline Solid Type

Determine the type of crystalline solid each substance would form under the right conditions, and list the intermolecular forces that would hold the solid together.

(a) Fe   (b) $CaBr_2$ (c) I_2 (d) SCl_2

Strategy Use the flowchart in Figure 7.7 to determine the type of crystalline solid each substance could form, and use the information in Table 7.2 to determine what intermolecular forces are present.

Setup (a) Fe is an element, it is atomic, and it is a metal.

(b) $CaBr_2$ consists of more than one type of atom and is, therefore, a compound. Because it contains a metal (Ca) and a nonmetal (Br), it is an ionic compound.

(c) I_2 is an element that exists as molecules.

(d) SCl_2 is a compound that consists of two different types of nonmetal atoms. Therefore, it is a molecular compound. Further, it is a polar molecule. (Remember that to determine this, you must draw and analyze its Lewis structure.)

Solution (a) Type of crystalline solid: metallic. Intermolecular forces: dispersion forces, metallic bonding.

(b) Type of crystalline solid: ionic. Intermolecular forces: dispersion forces, ionic bonding.

(c) Type of crystalline solid: molecular. Intermolecular forces: dispersion forces.

(d) Type of crystalline solid: molecular. Intermolecular forces: dispersion forces and dipole-dipole forces.

THINK ABOUT IT

You may want to review the rules in Chapter 3 for determining the type of element, atomic or molecular; or the type of compound, molecular or ionic. Also recall that the types of intermolecular forces exhibited by molecular compounds depend on their polarity and their ability to exhibit hydrogen bonding.

Practice Problem Ⓐ**TTEMPT**   Determine the type of crystalline solid each substance could form, given the right conditions.

(a) Br_2 (b) NH_3 (c) Kr

Practice Problem Ⓑ**UILD**   Each of the following substances exists in the solid phase, under the right conditions. Write the substances in order of increasing intermolecular-force strength: Ar, $CaCl_2$, CO_2.

Practice Problem Ⓒ**ONCEPTUALIZE**   Solid nitrogen (N_2) and solid carbon monoxide (CO) have intermolecular forces of very similar magnitude. Predict which one has the stronger intermolecular forces, explain why, and explain why they are so similar.

Profiles in Science

Carol V. Robinson

Intermolecular forces are the attractive forces that hold molecules, ions, or atoms together in a substance. Intramolecular forces are attractive forces between different *parts* of a large molecule. Carol V. Robinson (born 1956) is a British chemist who has done ground-breaking research in the field of protein folding. Proteins are long molecules made up of hundreds—sometimes thousands—of amino acids. There are 20 different amino acids involved in protein formation, each with a different "side chain." The side chains vary in polarity, and the attractive forces between side chains at different locations along the molecule (intramolecular forces) cause the protein to fold into a specific three-dimensional structure. The three-dimensional structure of the molecule is vital for its biological function. Carol Robinson has pioneered the use of novel analytical approaches to the study of chemical biology. Her research has earned her numerous awards, including the prestigious Davy Medal, the Aston Medal, and the FEBS/EMBO Women in Science Award. Further, she was the first female professor of chemistry at both the University of Cambridge and the University of Oxford.

Carol V. Robinson
Courtesy of Carol Robinson

7.2.1 Determine the type of crystalline solid that $SiCl_4$ forms.

a) Metallic   d) Ionic

b) Nonmetallic   e) Network

c) Molecular

7.2.2 Determine the type of crystalline solid that NH_4Cl forms.

a) Metallic   d) Ionic

b) Nonmetallic   e) Network

c) Molecular

7.2.3 Place the following in order of *decreasing* intermolecular-force strength.

$$F_2 \qquad CH_2Cl_2 \qquad CF_4$$

a) $CH_2Cl_2 > F_2 > CF_4$ d) $F_2 > CH_2Cl_2 > CF_4$

b) $CF_4 > CH_2Cl_2 > F_2$ e) $F_2 > CF_4 > CH_2Cl_2$

c) $CH_2Cl_2 > CF_4 > F_2$

7.2.4 Place the following in order of *increasing* intermolecular-force strength.

$$BF_3 \qquad Ar \qquad OCl_2$$

a) $OCl_2 < Ar < BF_3$ d) $Ar < OCl_2 < BF_3$

b) $OCl_2 < BF_3 < Ar$ e) $Ar < BF_3 < OCl_2$

c) $BF_3 < Ar < OCl_2$

7.3 Physical Properties of Solids

Two of the physical properties that distinguish different solids are *vapor pressure* and *melting point*. In this section, we examine how intermolecular forces influence these two quantitative properties.

Vapor Pressure

Qualitatively, the term ***vapor pressure*** refers to the tendency of molecules of a substance to move from a condensed phase to the gas phase. The weaker the intermolecular forces holding the particles of a solid together, the greater the tendency for those particles to leave the condensed phase (solid or liquid) and become part of the gas phase. Because intermolecular forces in solids tend to be stronger than those in liquids, solids typically have *very* low vapor pressures. In the case of ionic solids, there is essentially *no* tendency for the solid particles to become gaseous. However, there are some familiar examples of solids with high vapor pressures pictured in Figure 7.8.

Ingram Publishing

(a)   (b)   (c)

Figure 7.8 Solids with relatively high vapor pressures. (a) Naphthalene, $C_{10}H_8$; (b) iodine, I_2; and (c) carbon dioxide, CO_2.

Solid naphthalene [Figure 7.8(a)] has a familiar, acrid, tarlike smell; and it is precisely because of its high vapor pressure and the toxicity of its vapor that it has been used as a pesticide. (Although naphthalene "moth balls" are still available, they have been replaced in many of their traditional uses by other substances that are less toxic to humans and pets.) Solid iodine [Figure 7.8(b)], with its high vapor pressure, readily becomes a visible purple gas. Solid carbon dioxide, commonly referred to as dry ice [Figure 7.8(c)], has an *extremely* high vapor pressure. Its vapor pressure is so high, in fact, that if sealed in a container such as the plastic water bottle shown in the figure, it will quickly cause the container to explode. Carbon dioxide is used in fire extinguishers, which are specially designed to safely contain carbon dioxide's extraordinarily high vapor pressure. Inside a CO_2 fire extinguisher, the vapor pressure of carbon dioxide is more than 50 times greater than the ordinary air pressure to which we are accustomed.

Sample Problem 7.2 lets you practice comparing the vapor pressures of solids.

> **Student Note:** Under such extraordinary pressure, CO_2 actually exists as a *liquid*—which is something that does not happen at ordinary pressure, regardless of temperature.

SAMPLE PROBLEM (7.2)   Comparing Vapor Pressures of Different Solids

At −125°C, both CHF_3 and CS_2 are solids. Determine which compound has the lower vapor pressure at that temperature.

Strategy Vapor pressure is lower when the intermolecular forces holding molecules together are stronger. Determine the type and relative magnitude of intermolecular forces exhibited by the two compounds. Because the compounds both consist entirely of nonmetals, we know that they are both molecular. The next step will be to determine whether or not the molecules are polar. To do this, you may wish to review the material in Section 6.5.

Setup Start by drawing the Lewis structures of both compounds. Following the steps introduced in Section 6.1, we get the following structures:

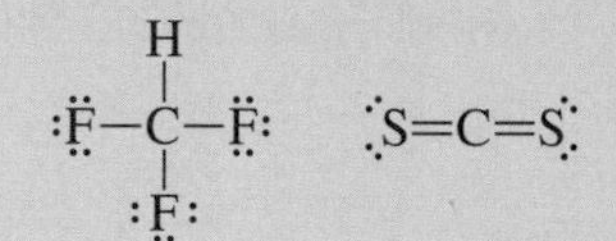

H
|
:F̈—C—F̈:   :S̈=C=S̈:
:F̈:

> **Student Note:** C and S have the same electronegativity value—see Figure 6.5.

Because of its linear symmetry and its nonpolar bonds, CS_2 is nonpolar. CHF_3, on the other hand, contains three polar C—F bonds and one nonpolar C—H bond. Using VSEPR, we see that with four electron groups around the central C atom, CHF_3 has a tetrahedral electron-group geometry. Further, because none of the electron groups around the central atom is a lone pair, the molecular shape is also tetrahedral. The polar C—F bonds are arranged such that their dipole moments do not cancel one another, making CHF_3 polar overall.

Solution These two compounds have similar molar masses (76.15 g/mol and 70.02 g/mol, respectively), so we would expect them to exhibit dispersion forces of similar magnitude. Only CHF_3 is polar and can also exhibit dipole-dipole forces. Therefore, we would expect it to exhibit *stronger* intermolecular forces—and to have the *lower* vapor pressure.

THINK ABOUT IT

In this case, we were comparing two molecular compounds with similar molar masses and very different polarities. In other cases, you may have to compare molecular compounds with similar *polarities* and very *different* molar masses—in which case the magnitude of dispersion forces would be the determining factor.

Practice Problem ATTEMPT At −195°C, both CF_4 and CH_2Cl_2 are solids. Determine which compound has the lower vapor pressure at this temperature.

Practice Problem BUILD Place the following solid substances in order of increasing vapor pressure: SiF_4, K_2SO_4, NCl_3.

Practice Problem CONCEPTUALIZE Solid CO_2 has a much higher vapor pressure than the other two substances shown in Figure 7.8. Explain why.

Melting Point

Although we know that the particles that make up a solid are in fixed positions, they do vibrate in position and therefore possess some energy. When we warm a solid, we are adding energy to the particles, thereby increasing their energy. When the energy of the particles exceeds the forces that hold the particles together in the solid state, the individual

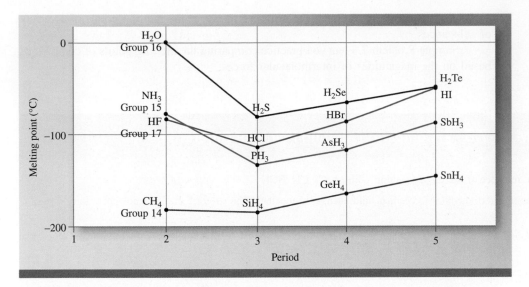

Figure 7.9 Melting points of the hydrogen compounds of elements from Groups 14 through 17. As with boiling point, melting point generally increases with increasing mass within a group—as illustrated by the hydrogen compounds of Periods 3, 4, and 5. The magnitude of intermolecular forces is stronger in compounds that exhibit hydrogen bonding (those containing F—H, O—H, or N—H bonds), making melting point higher than would be predicted by the trend.

particles can break free of the intermolecular forces that hold them in place, and the solid melts—becoming a liquid in which the particles are still in contact but can flow (move past one another). The temperature at which a solid melts is called the **melting point.**

Solid substances with stronger intermolecular forces melt at higher temperatures than those with weaker intermolecular forces. Figure 7.9 shows the melting points of the binary hydrogen compounds of the elements of Groups 14 through 17 (the same compounds used to illustrate the effects of molar mass and hydrogen bonding on *boiling* point in Figure 6.12. The trends in melting point are nearly identical to the trends in boiling point because both properties depend on the magnitude of intermolecular forces exhibited by a substance.

In general, attractive forces are greater between larger molecules than between smaller molecules; greater between polar molecules than between nonpolar molecules (of roughly equal size); and greater between molecules that can form hydrogen bonds than between molecules of similar size that cannot form hydrogen bonds. And because ion-ion interactions are typically much stronger than other types of intermolecular forces, the melting points of ionic solids tend to be very high. Table 7.3 shows how

TABLE 7.3	Intermolecular Forces and Melting Points		
Substance	**$\mathcal{M}$ (g/mol)**	**Types of IM Forces**	**Melting Point (°C)**
He	4.003	Dispersion	−272.2
H_2O	18.02	Dispersion, dipole-dipole, hydrogen bonding	0
HF	20.01	Dispersion, dipole-dipole, hydrogen bonding	−83.6
LiF	25.94	Dispersion, ionic bonding	845
NaF	41.99	Dispersion, ionic bonding	993
HCl	36.46	Dispersion, dipole-dipole	−114.2
F_2	38.00	Dispersion	−219.6
Ar	39.95	Dispersion	−189.3
NaCl	58.44	Ionic bonding	801
KI	166.0	Ionic bonding	680
$C_6H_{12}O_6$ (fructose)	180.2	Dispersion, dipole-dipole, hydrogen bonding	103
$C_{12}H_{22}O_{11}$ (sucrose)	342.3	Dispersion, dipole-dipole, hydrogen bonding	186

molar mass and types of intermolecular forces influence the melting points of a variety of substances.

Sample Problem 7.3 lets you practice comparing the melting points of compounds based on the magnitudes of intermolecular forces.

SAMPLE PROBLEM 7.3 Comparing Melting Points of Different Solids

Place the following substances in order of increasing melting point: $LiCl$, Cl_2, NCl_3

Strategy Determine the types of intermolecular forces that hold each solid together. The stronger the intermolecular forces, the higher the melting point.

Setup LiCl is an ionic compound and is held together by ionic bonds—these are generally much stronger than any other type of intermolecular force.

Cl_2 is a nonpolar substance whose solid form is held together by dispersion forces.

NCl_3 is a polar molecule whose solid form is held together by both dispersion and dipole-dipole forces.

Solution In order of increasing melting point (from lowest to highest): $Cl_2 < NCl_3 < LiCl$

THINK ABOUT IT

Remember that the attractions *between* molecules are broken during the process of melting, not the bonds between the atoms *within* molecules.

Practice Problem **A**TTEMPT Place the following compounds in order of increasing melting point: C_2H_6, C_3H_8, C_4H_{10}.

Practice Problem **B**UILD Place the following compounds in order of *decreasing* melting point: PBr_3, BCl_3, PCl_3.

Practice Problem **C**ONCEPTUALIZE Undecanoic acid and octadecane both have melting points near 30°C. Explain how they can have similar melting points.

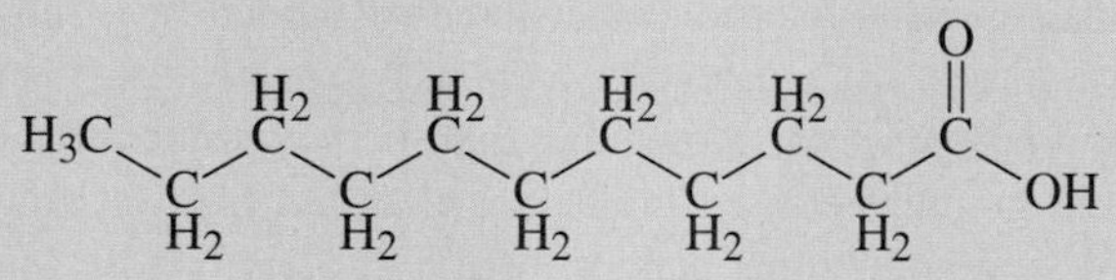

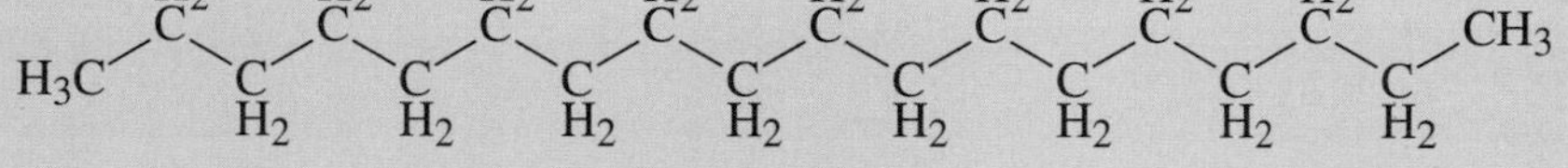

CHECKPOINT–SECTION 7.3 Physical Properties of Solids

7.3.1 At a temperature sufficiently low for all of the following substances to be solids, arrange the substances in order of increasing vapor pressure.

HCN   H₂O   O₂

a) $H_2O < HCN < O_2$

b) $O_2 < HCN < H_2O$

c) $HCN < H_2O < O_2$

d) $HCN < O_2 < H_2O$

e) $O_2 < H_2O < HCN$

7.3.2 At a temperature sufficiently low for all of the following substances to be solids, arrange the substances in order of decreasing vapor pressure.

CH₄   Br₂   Xe

a) $Br_2 > Xe > CH_4$

b) $CH_4 > Br_2 > Xe$

c) $Br_2 > CH_4 > Xe$

d) $CH_4 > Xe > Br_2$

e) $Xe > Br_2 > CH_4$

7.3.3 Place the following compounds in order of increasing melting point.

$$MgS \quad CS_2 \quad BeF_2$$

a) $CS_2 < BeF_2 < MgS$

b) $CS_2 < MgS < BeF_2$

c) $BeF_2 < CS_2 < MgS$

d) $BeF_2 < MgS < CS_2$

e) $MgS < CS_2 < BeF_2$

7.3.4 Place the following substances in order of decreasing melting point.

$$Al \quad SiCl_4 \quad BCl_3$$

a) $BCl_3 > SiCl_4 > Al$

b) $SiCl_4 > Al > BCl_3$

c) $Al > BCl_3 > SiCl_4$

d) $Al > SiCl_4 > BCl_3$

e) $BCl_3 > Al > SiCl_4$

7.4 Physical Properties of Liquids

As with solids, the magnitudes of intermolecular forces in liquids are responsible for some of their important physical properties, including *viscosity, surface tension, vapor pressure,* and *boiling point.*

Viscosity

The *viscosity* of a liquid is a measure of the liquid's resistance to flow. Consider two liquids: water and honey. If you had identical containers of both liquids, and tried to empty them by pouring out the contents, the container holding the water would be empty far sooner than the one holding the honey because honey has a greater resistance to flow than does water. Honey has a *higher viscosity* than water. A better example that compares two pure substances (honey is a mixture of water and a variety of dissolved substances) is the viscosity difference between ethylene glycol, $C_2H_6O_2$, a component of antifreeze, and glycerol, $C_3H_8O_3$, a common additive in the food, pharmaceutical, and personal-products industries. Figure 7.10 illustrates the difference in viscosity of these two liquids.

Surface Tension

A molecule within a liquid is pulled in all directions by the intermolecular forces between it and the other molecules that surround it. Because there is pull in *every* direction, there is no "net" pull in any *one* direction. Unlike the forces on molecules in the *interior* of a liquid, the forces on a molecule at the *surface* of a liquid do not all cancel one another. Although a molecule at the surface of the liquid is pulled down and to the sides by neighboring molecules, there is no *upward* pull. This causes a molecule at the surface of a liquid to experience a net pull downward or *inward* (into the bulk of the liquid) known as **surface tension.** This is what causes water to form beads (spherical droplets)

Figure 7.10 Ethylene glycol (left) and glycerol (right) being poured from graduated cylinders into beakers. Glycerol has a higher viscosity than ethylene glycol.

©Richard Megna/Fundamental Photographs, NYC

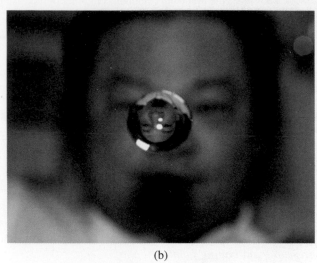

(a)   (b)

Figure 7.11 (a) Intermolecular forces acting on a molecule in the surface layer of a liquid and in the interior region of the liquid. (b) Water forming a sphere in zero gravity as a result of surface tension. (b) NASA

on a just-washed car—and what causes water to form spheres in the absence of gravity, as demonstrated on the International Space Station. Figure 7.11 shows the molecular-level illustration of the forces resulting in surface tension and a photograph from the International Space Station experiments with water in zero gravity.

Familiar Chemistry

Surface Tension and the Shape of Water Drops

You have no doubt seen droplets of water sitting on a surface, similar to the photo shown here. The droplets are never perfect spheres, although what you've just learned about surface tension might seem to suggest that they should be. The flattening of these approximately spherical shapes is due in large part to gravity. In space, when there is no gravity, it is possible to make large "drops" of water that take on nearly perfect spherical shapes. Note that the formation of spherical drops is the result of relatively strong intermolecular attractions between water molecules, and the liquid's resulting tendency to assume a shape that minimizes its surface area.

Water droplets on flower petals.
Holly Hildreth/McGraw Hill

A large "drop" of water in zero gravity.
NASA

It may not be obvious to you that a sphere has the smallest surface/volume ratio of all of the possible shapes. To demonstrate this, two sample calculations of surface area are shown comparing a rectangular shape to a sphere of the same volume, 10 cm³. Note how much smaller the surface area of the sphere is compared to that of the rectangular shape.

Rectangular shape: **Sphere:**

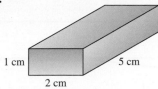

1 cm   5 cm
2 cm

$r = 1.34$ cm

volume $= l \times w \times h = 5$ cm $\times 2$ cm $\times 1$ cm $= 10.0$ cm³

volume $= \dfrac{4\pi r^3}{3} = \dfrac{4\pi}{3}(1.34$ cm$)^3 = 10.0$ cm³

surface area $= 2$ cm² $+ 2$ cm² $+ 5$ cm² $+ 5$ cm² $+ 10$ cm² $+ 10$ cm² $= 34$ cm²

surface area $= 4\pi r^2 = 4\pi(1.34$ cm$)^2 = 22.6$ cm²

Vapor Pressure

Vapor pressure is another property of liquids that depends on the magnitude of inter-molecular forces. Because the attractive forces between the particles that make up a liquid are generally much weaker than those between the particles in solids, liquids typically have higher vapor pressures than solids. Substances that have high vapor pressures at room temperature are said to be *volatile*. Those with very low vapor pressures are called *nonvolatile*.

Molecules in a liquid are in constant motion and therefore possess *kinetic energy,* the energy associated with motion. The molecules in a liquid do not all move at the same speed, and therefore do not all have the *same* kinetic energy. Figure 7.12 illustrates the distribution of kinetic energies of molecules in a liquid at a particular temperature. If a molecule at the surface of a liquid possesses enough kinetic energy to break free from the liquid and become part of the vapor phase, indicated by the colored portion of the curve in Figure 7.12, it can *evaporate* or *vaporize*. (Vaporization is one of the phase changes we discuss in detail in Section 7.5.) If we place a liquid in a closed container, molecules in the liquid phase with enough kinetic energy will escape the liquid surface and become part of the vapor phase. Once part of the vapor phase, a molecule will move about in the space above the liquid, and it may return to the liquid phase by colliding with the liquid surface. The process of returning to the liquid phase is called *condensation*—another of the phase changes we discuss in Section 7.5.

Figure 7.13 illustrates a simplified version of how the vapor pressure of a volatile liquid is established in a closed container. Within a closed container, evaporation of a given liquid at a given temperature occurs at a constant rate. Early in the process, the rate of condensation is very low because very few of the liquid molecules have entered the vapor phase. Over time, though, as evaporation continues and there are more molecules in the vapor phase, the rate of condensation increases. Eventually, the rate of condensation increases to the point where it is equal to the rate of evaporation. From that point on, although both processes (evaporation and condensation) continue to occur, because they are occurring at the *same rate,* the number of molecules in the vapor phase remains steady. This is a state known as a dynamic equilibrium, where a forward process (in this case, evaporation) and a reverse process (in this case, condensation) continue to occur at equal rates. The number of molecules in the vapor phase when this state has been achieved is a measure of the liquid's vapor pressure.

As temperature increases, the speed of molecules in a liquid increases, increasing the average kinetic energy of the molecules. Figure 7.14 shows the same liquid

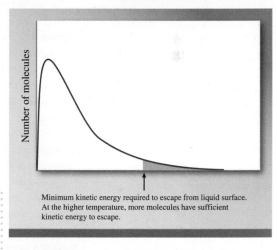

Number of molecules

Minimum kinetic energy required to escape from liquid surface. At the higher temperature, more molecules have sufficient kinetic energy to escape.

Figure 7.12 Distribution of kinetic energies in a sample of liquid at a specific temperature. Only molecules with the minimum required kinetic energy can evaporate.

Student Note: The terms *evaporation* and *vaporization* mean essentially the same thing, moving from the liquid phase to the gas phase; but the two terms are used in slightly different ways. Evaporation usually refers to a substance gradually becoming a vapor without being heated; vaporization generally refers to a substance becoming a vapor as a result of its being heated.

Student Note: We present more about the state of dynamic equilibrium in Chapter 13.

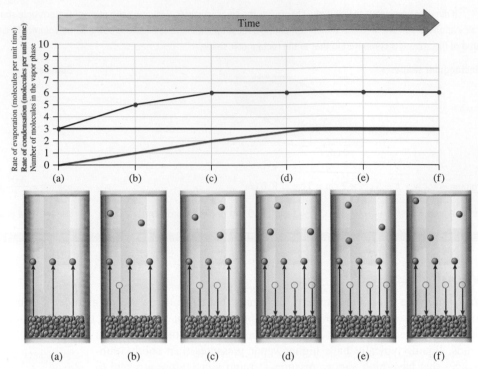

Figure 7.13 Simplified illustration of the establishment of vapor pressure over a volatile liquid in a closed container. (a) Initially, there are no liquid molecules in the vapor phase, so no condensation can occur. Three of the liquid molecules evaporate, making a total of three molecules in the vapor phase. (b) With three molecules in the vapor phase, one molecule condenses. Because the rate of evaporation is constant, three more molecules evaporate, leaving a total of five molecules in the vapor phase. (c) With five molecules in the vapor phase, two condense. Three more evaporate, leaving a total of six molecules in the vapor phase. (d) With six molecules in the vapor phase, three condense. Three more evaporate. (e) Again, with six molecules in the vapor phase, three more condense. Three more evaporate. (f) Again, three molecules evaporate, three condense, and there are six molecules in the vapor phase. Both processes continue at the same rate, and the number of molecules in the vapor phase has stopped changing. A dynamic equilibrium has been established whereby both processes (evaporation and condensation) continue to occur at a rate of three molecules per unit time, leaving a constant number of molecules (six) in the vapor phase. This represents the vapor pressure (sometimes called the *equilibrium* vapor pressure) of this liquid at this temperature.

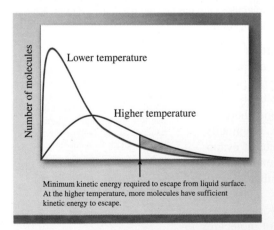

Figure 7.14 Distribution of kinetic energies in a sample of liquid at a higher temperature than that in Figure 7.12. At higher temperature, more molecules have the minimum required kinetic energy to evaporate.

as shown in Figure 7.12 at a higher temperature. Note that at the higher temperature, more molecules in the liquid have the amount of kinetic energy required to evaporate. This causes the vapor pressure of the liquid to increase, as it results in there being more molecules in the vapor phase when the state of dynamic equilibrium is achieved. Figure 7.15 illustrates the influence of temperature on vapor pressure at the molecular level.

Boiling Point

We have seen that the vapor pressure of a liquid increases with increasing temperature. Figure 7.16 shows how the vapor pressures of three different liquids, including water, change with temperature. When the vapor pressure of a liquid is equal to the surrounding atmospheric pressure, the liquid *boils*. When a liquid boils, molecules can enter the vapor phase throughout the liquid sample—not just from the surface as with ordinary evaporation. The **boiling point** of a substance is defined as the temperature at which its vapor pressure is equal to atmospheric pressure. As with other properties of liquids, boiling point depends on the magnitudes of intermolecular forces.

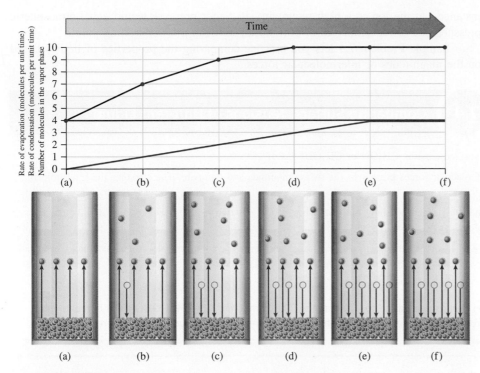

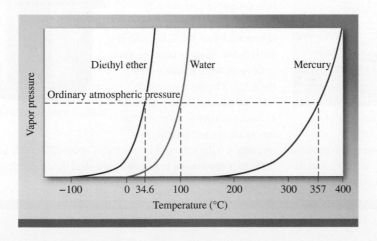

Figure 7.15 Establishment of vapor pressure over a volatile liquid in a closed container at a higher temperature than shown in Figure 7.13. (a) Initially, there are no liquid molecules in the vapor phase, so no condensation can occur. At this temperature, four of the liquid molecules evaporate, making a total of four molecules in the vapor phase. (b) With four molecules in the vapor phase, one molecule condenses. Because the rate of evaporation is constant, four more molecules evaporate, leaving a total of seven molecules in the vapor phase. (c) With seven molecules in the vapor phase, two condense. Four more evaporate, leaving a total of nine molecules in the vapor phase. (d) With nine molecules in the vapor phase, three condense. Four more evaporate, leaving ten in the vapor phase. (e) With ten molecules in the vapor phase, four condense, four more evaporate, and there are ten molecules in the vapor phase. (f) Again, four molecules evaporate, four condense, and there are ten molecules in the vapor phase. When dynamic equilibrium has been established, there are more molecules in the vapor phase at the higher temperature than there were at the lower temperature shown in Figure 7.13. As temperature increases, the vapor pressure of a liquid increases.

Recall from Chapter 6 [◀◀ Figure 6.12] that the hydrogen compounds of Groups 15, 16, and 17 have anomalously high boiling points because they exhibit hydrogen bonding.

As shown in Figure 7.16, the temperature at which water's vapor pressure is equal to ordinary atmospheric pressure (at sea level) is 100°C. If we were to heat water in an environment where the surrounding air pressure were significantly lower, for

Student Hot Spot

Student data indicate that you may struggle to understand the meaning of boiling point. Access the eBook to view additional Learning Resources on this topic.

Figure 7.16 Increase in vapor pressure with increasing temperature for three liquids.

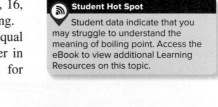

instance—at the top of a high mountain—its vapor pressure would equal atmospheric pressure at a lower temperature. This would result in a *lower* boiling point.

Sample Problem 7.4 lets you practice comparing the properties of liquids based on the magnitudes of intermolecular forces.

SAMPLE PROBLEM (**7.4**)   Comparing Surface Tensions of Different Liquids

Select the compound with the higher surface tension, CH_3OH or CH_2OHCH_2OH.

Strategy Identify the strongest intermolecular forces present in the two compounds and compare. The stronger the intermolecular forces, the higher the surface tension.

Setup Both of these compounds contain hydrogen bonding, but one of them contains two —OH groups, which strengthens the intermolecular attractions between molecules.

Solution The compound CH_2OHCH_2OH has stronger intermolecular forces and therefore a higher surface tension than CH_3OH.

THINK ABOUT IT

This trend continues as more —OH groups are added to a substance. The compound $CH_2OHCHOHCH_2OH$ contains more —OH groups and therefore has stronger intermolecular forces than either of the two compounds shown here.

Practice Problem **A**TTEMPT   Select the compound with the higher vapor pressure at room temperature, (i) CH_3OH or (ii) CH_2OHCH_2OH.

Practice Problem **B**UILD   Select the compound with the higher viscosity, (i) $CH_3CH_2CH_2CH_2CH_2CH_2CH_3$ or (ii) $CCl_3CCl_2CCl_2CCl_2CCl_2CCl_2CCl_3$.

Practice Problem **C**ONCEPTUALIZE   Suppose you have two compounds, CH_3X and CH_3Y. Select an atom or group of atoms for X and Y from among those shown here, such that the statement "CH_3X has a higher boiling point than CH_3Y" is true.

—H   —OH   —Cl   —$CH_2CH_2CH_2CH_2CH_2CH_2CH_2CH_3$

Familiar Chemistry

High Altitude and High-Pressure Cooking

You may have noticed that some packaged mixes and recipes for cakes and other baked goods have special *high altitude* directions on their labels. You have probably also seen or at least heard about *pressure cookers*, which shorten the time required to cook food. These are both related to the vapor pressure of water, and the dependence of water's boiling point on the ambient pressure.

At high altitude, the ambient air pressure is lower than at sea level. This results in the vapor pressure of water being equal to the ambient pressure at a *lower* temperature—lowering its boiling point to below the normal value of 100°C. Because the cooking process depends on the boiling point of water, this requires bakers at high altitudes to make special adjustments to their recipes so that they turn out correctly. An extreme example of this phenomenon, although no baking goes on

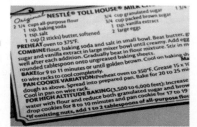

David A. Tietz/McGraw Hill

there, is at the top of Mount Everest—at 8848 meters, the highest elevation on Earth. The pressure at this altitude is only about one-third that at sea level, and water boils at just over 70°C.

Pressure cookers, on the other hand, are designed to contain the steam generated by boiling water in such a way as to increase the pressure inside the pot to higher than the ambient pressure. This results in water boiling at a higher temperature than 100°C, because in order to boil, its vapor pressure has to be equal to the pressure inside the pot. At a higher temperature, food cooks faster. Modern pressure cookers can achieve pressures that are double the ambient pressure,

David A. Tietz/McGraw Hill

resulting in water boiling at temperatures above 120°C. This enables cooks to prepare meals in as little as one-third the time ordinarily required.

CHECKPOINT–SECTION 7.4 Physical Properties of Liquids

7.4.1 Select the answer where the molecules are correctly arranged in order of increasing strength of intermolecular forces.

a) $BF_3 < BClF_2 < BH_3$

b) $BH_3 < BF_3 < BClF_2$

c) $BClF_2 < BH_3 < BF_3$

d) $BH_3 < BClF_2 < BF_3$

e) $BClF_2 < BF_3 < BH_3$

7.4.2 Select the answer that correctly shows the substances in order of decreasing boiling point.

a) $CaF_2 > SeF_2 > BeF_2$

b) $SeF_2 > BeF_2 > CaF_2$

c) $BeF_2 > CaF_2 > SeF_2$

d) $CaF_2 > BeF_2 > SeF_2$

e) $BeF_2 > SeF_2 > CaF_2$

7.4.3 Select the answer that correctly shows the substances in order of decreasing surface tension.

a) $PBr_3 > PCl_3 > BBr_3$ d) $PBr_3 > BBr_3 > PCl_3$

b) $PCl_3 > PBr_3 > BBr_3$ e) $BBr_3 > PBr_3 > PCl_3$

c) $BBr_3 > PCl_3 > PBr_3$

7.4.4 The contents of one of the bubbles in boiling water is represented here. Identify the molecules shown in the circle.

Stocksearch/Alamy Stock Photo

a) Air   d) H_2O

b) O_2 e) A mixture of H_2 and O_2

c) H_2

7.5 Energy and Physical Changes

In this section, we examine the energy involved in temperature and phase changes of pure substances. To begin, we must consider the nature of **energy**, the ability to produce heat or to do work, as it applies to macroscopic processes.

The SI unit of energy is the **joule (J)**, although another commonly used unit of energy is the **calorie (cal)**. Equation 7.1 expresses the relationship between these two units:

$$1 \text{ cal} \equiv 4.184 \text{ J} \qquad \textbf{Equation 7.1}$$

Also commonly used are the kilojoule (kJ), which is equal to 1000 joules, and the kilocalorie (Cal), which is equal to 1000 calories. [Note that the only difference between the abbreviations for calories and kilocalories is the *capitalization* in the case of kilocalories. The "calories" listed on food labels are actually kilocalories (Cal).]

Temperature Changes

At some point, you have probably put a pot of water on a stove and waited for the water to boil. You know now that the water boils when its *vapor pressure,* which increases as temperature increases, is equal to the ambient air pressure. But have you ever accidentally put an *empty* pan on a hot stove and noticed how much faster the empty pan gets hot than a pan with water in it? There are two reasons for this. The first is that the mass of water we typically bring to a boil to cook something is much greater than the mass of the pan in which we heat it. The second is that different substances require the input of different amounts of energy to increase their temperatures.

TABLE 7.4	Specific Heat Capacities of Several Substances
Substance	**Specific Heat (J/g · °C)**
Aluminum	0.900
Gold	0.129
Graphite	0.720
Copper	0.385
Iron	0.444
Mercury	0.139
Water (liquid)	4.184
Ethanol (C_2H_5OH)	2.46

Let's use aluminum (a common metal in cookware) and water to illustrate this. If we wanted to heat 2.000 kg of water from room temperature (25.0°C) to the boiling point (100.0°C), we would have to add more than 600 kJ of energy. To change the temperature of an equal mass of aluminum by the same amount would require significantly less energy. The reason for this is that different substances have different *specific heat capacities*.

Specific heat capacity or simply **specific heat** is the amount of energy required to increase the temperature of one gram of a substance by one degree Celsius (or by one kelvin). Water has an especially *high* specific heat, which makes it useful as a coolant—and is the reason it takes a sometimes frustratingly long time to boil. Table 7.4 gives the specific heat capacities of several familiar substances.

We express the amount of energy required for a process with the symbol q. The amount of energy required to change the temperature of a substance is calculated using Equation 7.2:

Equation 7.2 $$q = \text{specific heat capacity} \times m \times \Delta T$$

where m is the mass of the substance (in grams) and ΔT is the change in temperature (expressed either in degrees Celsius or in kelvins, which are numerically the same). ΔT is calculated as final temperature minus initial temperature.

Using Equation 7.2 and specific heat values from Table 7.4, we can calculate the amounts of energy (q) required to increase the temperature of 2.000 kg (2000 g) of water and of 2.000 kg of aluminum from room temperature (25.0°C) to 100.0°C ($\Delta T = 100.0°C - 25.0°C = 75.0°C$):

$$\text{Water} \qquad q = \frac{4.184 \text{ J}}{\text{g} \cdot °\text{C}} \times 2000 \text{ g} \times 75.0°\text{C} = 6.28 \times 10^2 \text{ kJ}$$

$$\text{Aluminum} \quad q = \frac{0.900 \text{ J}}{\text{g} \cdot °\text{C}} \times 2000 \text{ g} \times 75.0°\text{C} = 1.35 \times 10^2 \text{ kJ}$$

Note that it takes more than *four times* as much energy to increase the temperature of the water as it does to increase the temperature of the aluminum.

Sample Problem 7.5 shows how to use specific heat capacities to calculate a temperature change.

SAMPLE PROBLEM (**7.5**) Determining the Amount of Energy Required for a Specified Temperature Increase

Calculate the amount of energy (in kJ) required to increase the temperature of 255 g of water from 25.2°C to 90.5°C.

Strategy Use Equation 7.2 (q = specific heat capacity × m × ΔT) to calculate q.

Setup m = 255 g, s = 4.184 J/g · °C, and ΔT = 90.5°C − 25.2°C = 65.3°C.

Solution

$$q = \frac{4.184 \text{ J}}{\text{g} \cdot °\cancel{C}} \times 255 \text{ } \cancel{g} \times 65.3°\cancel{C} = 6.97 \times 10^4 \text{ J} \quad \text{or} \quad 69.7 \text{ kJ}$$

THINK ABOUT IT

Look carefully at the cancellation of units and make sure that the number of kilojoules is smaller than the number of joules. It is a common error to multiply by 1000 instead of dividing in conversions of this kind.

Practice Problem **A**TTEMPT Calculate the amount of energy (in kJ) required to increase the temperature of 1.01 kg water from 1.05°C to 35.81°C.

Practice Problem **B**UILD What will be the final temperature of a 514-g sample of water, initially at 10.0°C, after 90.8 kJ have been added to it?

Practice Problem **C**ONCEPTUALIZE Shown here are two samples of the same substance. When equal amounts of heat are added to both samples, the temperature of the sample on the left increases by 15.3°C. Determine the increase in temperature of the sample on the right.

 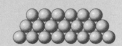

Solid-Liquid Phase Changes: Melting and Freezing

When an ice cube floats in a glass of water, the glass contains two phases of the same substance: ice (solid) and water (liquid). Although the substance *water* is the same in both cases, as we have seen, the different phases have very different properties. The transition from solid to liquid is generally referred to as *melting,* but is also known as *fusion.* The reverse process, the transition from liquid to solid, is known as *freezing.*

The process of fusion (melting) occurs when the molecules in a substance have enough energy to break free of the intermolecular forces that hold them rigidly in place—allowing them to move past one another.

Solid   Add energy   Solid and liquid phases   Add more energy   Liquid
in equilibrium

The energy required to melt a solid is called the *heat of fusion.* For the purpose of comparison, we generally report heats of fusion as *molar* quantities, thus the *molar heat of fusion (ΔH_{fus})* is the amount of energy (expressed in kJ/mol) required to melt one mole of a substance. Table 7.5 shows the molar heats of fusion of several substances. Note that the table includes melting points for all of the substances. The phase change from solid to liquid (or the reverse, from liquid to solid) for any substance always occurs at the melting point of the substance. The temperature of ice, for example, cannot be increased above 0°C. The temperature of the water can begin to increase only after all of the ice has melted. Phase changes always occur at *constant temperature.*

Changing a substance from the solid phase to the liquid phase is an **endothermic process,** meaning that to make it happen, we must add heat. This should make sense intuitively as you have undoubtedly witnessed this phenomenon in many circumstances.

TABLE 7.5	Molar Heats of Fusion	
Substance	**Melting Point (°C)**	**ΔH_{fus} (kJ/mol)**
Argon (Ar)	−190	1.3
Benzene (C_6H_6)	5.5	10.9
Ethanol (C_2H_5OH)	−117.3	7.61
Diethyl ether ($C_2H_5OC_2H_5$)	−116.2	6.90
Mercury (Hg)	−39	23.4
Methane (CH_4)	−183	0.84
Water (H_2O)	0	6.01

You know that an ice cube will melt faster in a hot pan on the stove than it would in a cold pan on the countertop. As we saw in Figure 7.5, gallium metal melts when we add heat to it—either by simply holding it, or by dipping it into hot liquid. Melting a solid is an *endothermic* process. What may be less intuitively obvious is that the reverse process, *freezing,* is an **exothermic process**—meaning that it gives *off* heat as it occurs. Perhaps you have had the experience of filling ice trays with water and placing them in your freezer to make ice cubes. If you take ice cream out of the freezer shortly thereafter, you will find that it has softened considerably. What's happened? The process of water freezing to become ice is *exothermic;* and the heat given off by the exothermic process softens the ice cream.

It is possible for us to calculate the amount of energy required to melt a substance at its melting point using Equation 7.3:

Equation 7.3	$q = \Delta H_{fus} \times$ moles of substance

Note that if we are given the *mass* of a substance, as is common, we will need to convert it to moles before using Equation 7.3 [◄◄ Section 5.2].

For example, we can calculate the amount of energy required to melt a 1.00-kg (1.00×10^3 g) block of ice at 0.00°C as follows:

$$q = \frac{6.01 \text{ kJ}}{1 \text{ mol } H_2O} \times 1.00 \times 10^3 \text{ g } H_2O \times \frac{1 \text{ mol } H_2O}{18.02 \text{ g } H_2O} = 334 \text{ kJ}$$

Conversely, if a 1.00-kg mass of water were to freeze at 0.00°C, the process would give off an equal amount of energy.

Liquid-Gas Phase Changes: Vaporization and Condensation

A tea kettle filled with boiling water contains the liquid and vapor phases of water in equilibrium. *Vaporization,* the transition from liquid to gas or vapor phase, occurs when the molecules in a liquid have enough energy to break free of the intermolecular forces that hold them together—allowing them to escape to the vapor phase and expand to fill their container. The reverse process, **condensation,** is the transition from the vapor phase to the liquid phase. Vaporization is endothermic; condensation is exothermic.

Vaporization

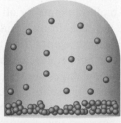

Liquid   Add energy   Liquid and vapor phases   Add more energy   Vapor
in equilibrium

Condensation

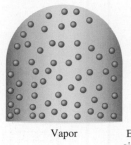 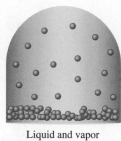 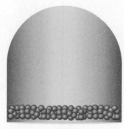

| Vapor | Energy given off | Liquid and vapor phases in equilibrium | More energy given off | Liquid |

The energy required to vaporize a mole of liquid is called the ***molar heat of vaporization (ΔH_{vap})***, which is also expressed in kJ/mol. Table 7.6 gives the molar heats of vaporization and boiling points of several substances.

Just as we were able to calculate the amount of energy required to melt a substance at its melting point using Equation 7.3, we can calculate the amount of energy required to vaporize a substance at its boiling point using Equation 7.4:

$$q = \Delta H_{vap} \times \text{moles of substance} \qquad \textbf{Equation 7.4}$$

Again, we need the amount of substance in *moles* and must convert to moles if we are given mass. The amount of energy required to vaporize 1.00 kg (1.00×10^3 g) of water at 100.0°C is calculated as:

$$q = \frac{40.79 \text{ kJ}}{1 \text{ mol } H_2O} \times 1.00 \times 10^3 \text{ g } H_2O \times \frac{1 \text{ mol } H_2O}{18.02 \text{ g } H_2O} = 2.26 \times 10^3 \text{ kJ}$$

And again, conversely, if 1.00 kg of water vapor were to *condense* at 100.0°C, the process would give off that same amount of energy.

Solid-Gas Phase Changes: Sublimation

Some substances, such as CO_2, do not have a liquid phase at ordinary pressures. Instead, they undergo **sublimation,** the transition directly from the solid phase to the vapor phase. The reverse process, also often called *sublimation* but commonly referred to as **deposition,** is the transition from the vapor phase directly to the solid phase. Sublimation is endothermic, deposition is exothermic. Figure 7.17 summarizes the six different phase changes for a hypothetical substance.

TABLE 7.6	Molar Heats of Vaporization	
Substance	**Boiling Point (°C)**	**ΔH_{vap} (kJ/mol)**
Argon (Ar)	−186	6.3
Benzene (C_6H_6)	80.1	31.0
Ethanol (C_2H_5OH)	78.3	39.3
Diethyl ether ($C_2H_5OC_2H_5$)	34.6	26.0
Mercury (Hg)	357	59.0
Methane (CH_4)	−164	9.2
Water (H_2O)	100	40.79

Figure 7.17 The six possible phase changes, melting (fusion), vaporization, sublimation, deposition, condensation, and freezing.

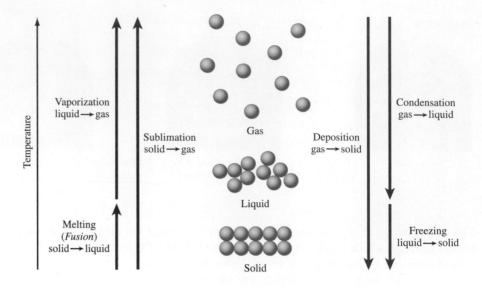

Sample Problem 7.6 lets you practice determining the amount of energy required for, or given off by, specific phase changes.

SAMPLE PROBLEM (**7.6**)   Calculating the Energy Given Off or Required for a Process

---

For each process, determine whether energy is given off or required—and the quantity of energy involved: (a) 1.0 mol of solid benzene is melted at 5.5°C; (b) 73.2 g of liquid methane is frozen at −183°C; (c) 1.32 mol of liquid ethanol is vaporized at 78.3°C.

Strategy Phase changes occur at constant temperature. The amount of energy given off or required depends on the amount of substance undergoing the phase change—and on the corresponding ΔH value. Energy is given off by gas-to-liquid, liquid-to-solid, and gas-to-solid phase changes. Energy is required for solid-to-liquid, liquid-to-gas, and solid-to-gas phase changes.

Setup (a) Melting or fusion (solid-to-liquid) *requires* energy to break apart the intermolecular attractions in the solid. The molar heat of fusion, ΔH_{fus}, of benzene at 5.5°C is 10.9 kJ/mol (see Table 7.5).

(b) Freezing (liquid-to-solid), the reverse of melting, *gives off* energy as intermolecular attractions are established in the solid phase. The molar heat of fusion, ΔH_{fus}, of methane at −183°C is 0.84 kJ/mol (see Table 7.5).

(c) Vaporization (liquid-to-gas) *requires* energy to break apart the intermolecular attractions in the liquid phase. The molar heat of vaporization, ΔH_{vap}, of ethanol at 78.3°C is 39.3 kJ/mol (see Table 7.6).

Solution (a) 1.0 mol benzene $\times \dfrac{10.9 \text{ kJ}}{\text{mol benzene}} = 10.9$ kJ (required)

(b) 73.2 g methane $\times \dfrac{1 \text{ mol methane}}{16.04 \text{ g methane}} \times \dfrac{0.84 \text{ kJ}}{\text{mol methane}} = 3.8$ kJ (required)

(c) 1.32 mol ethanol $\times \dfrac{39.3 \text{ kJ}}{\text{mol ethanol}} = 51.9$ kJ (given off)

THINK ABOUT IT

Molar heats of fusion and vaporization apply only at the temperatures at which phase changes occur. For example, because the melting point of benzene is 5.5°C, we can only calculate the amount of energy required to melt a given amount of benzene at *that* temperature.

Practice Problem ⒶTTEMPT For each process, determine whether energy is required or given off, and the quantity of energy involved: (a) 29.6 g of solid water melts at 0°C; (b) 2.44 mol of ethanol vapor condenses at 78.3°C; (c) 14.8 moles of argon melts at −190°C.

Practice Problem ⒷUILD Determine the amount of energy required to convert 45.5 g of liquid water at 25.0°C to water vapor at 100.0°C.

Practice Problem ⒸONCEPTUALIZE Which of the following best represents the process of sublimation?

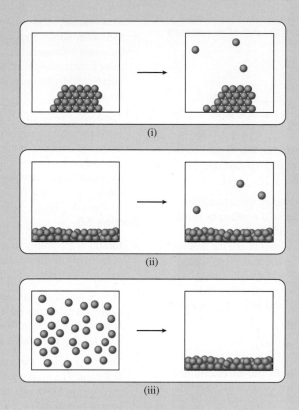

(i)

(ii)

(iii)

CHECKPOINT—SECTION 7.5 Energy and Physical Changes

7.5.1 Determine the amount of energy required to increase the temperature of 85.5 g Hg from room temperature (25.0°C) to 42.5°C.

a) 6.26×10^3 J

b) 505 J

c) 208 J

d) 297 J

e) 1.37×10^2 J

7.5.2 What mass of ice at 0°C could be melted by the heat given off when 100.0 g ethanol vapor condenses at 78.3 °C?

a) 2.17 g

b) 256 g

c) 85.3 g

d) 14.2 g

e) 0.601 g

Chapter Summary

Section 7.1

- Substances with sufficiently strong intermolecular forces to hold the particles together are solids or liquids. Solids and liquids are known collectively as the **condensed phases.**

- A solid has a fixed shape and a fixed volume. A liquid has a fixed volume but assumes the shape of the part of its container that it occupies.

Section 7.2

- Solids may be amorphous (lacking in long-range order) or crystalline (highly ordered). A familiar example of an **amorphous solid** is **glass.**

- **Crystalline solids** may be ionic, molecular, atomic, or network (covalent), depending on the particles they contain and the forces that hold the particles together.

- Ionic solids consist of ions held together by ionic bonding. Molecular solids consist of molecules held together by some combination of dispersion forces, dipole-dipole forces, and hydrogen bonding—depending on whether or not the molecules are polar and/or able to form hydrogen bonds. Atomic solids consist of atoms held together by dispersion forces (in the case of nonmetallic substances) or metallic bonding. A **network solid** or **covalent solid** consists of atoms connected by covalent bonds.

Section 7.3

- Two of the important properties of solids that are determined by the magnitude of intermolecular forces are vapor pressure and melting point.

- **Vapor pressure** is a measure of how readily the particles of a solid enter the vapor phase.

- **Melting point** is the temperature at which the particles that make up a solid have enough energy to overcome the intermolecular forces holding them together, and the solid becomes a liquid.

Section 7.4

- Important properties of liquids that are determined by the magnitude of intermolecular forces include viscosity, surface tension, vapor pressure, and boiling point.

- **Viscosity** is a measure of a liquid's resistance to flow. The greater the magnitude of intermolecular forces, the greater a liquid's viscosity.

- **Surface tension** is the result of the molecules at the surface of a liquid experiencing a net pull inward, toward the interior. Surface tension is what causes water to "bead."

- **Vapor pressure** of a liquid is a measure of its tendency to become a gas. Liquids with high vapor pressures are referred to as **volatile;** those with very low vapor pressures are referred to as **nonvolatile.** Vapor pressure increases as temperature increases because at higher temperatures, molecules have more kinetic energy and more of them can overcome the intermolecular forces holding them together and escape to the vapor phase.

- The **boiling point** of a liquid is the temperature at which its vapor pressure is equal to atmospheric pressure. At that point, the molecules have enough kinetic energy to overcome the intermolecular forces holding them together, and to escape to the vapor phase—not just from the surface, but throughout the liquid.

Section 7.5

- **Energy** is the ability to produce heat or to do work. The SI unit of energy is the **joule (J),** but another commonly used unit of energy is the **calorie (cal).**

- The amount of energy necessary to increase the temperature of a substance depends on the **specific heat capacity** or **specific heat** of the substance. Water has an especially high specific heat, making it useful as a coolant.

- The phase change from solid to liquid is known as **melting** or **fusion.** The reverse of melting is **freezing.** The amount of energy required to melt a mole of a substance is the **molar heat of fusion (ΔH_{fus}),** which is generally expressed in kJ/mol.

- Melting is an **endothermic process,** meaning that energy must be supplied to make it happen. Freezing is an **exothermic process,** meaning that energy is given off when the process occurs. Melting and freezing of a substance occur at the melting point of the substance.

- The phase change from liquid to gas is known as **vaporization.** The reverse of vaporization is **condensation.** The amount of energy required to vaporize a mole of a substance is the **molar heat of vaporization (ΔH_{vap}),** which is also generally expressed in kJ/mol. Vaporization and condensation of a substance occur at the boiling point of the substance.

- The phase change from solid to gas is known as **sublimation.** The reverse of sublimation is commonly referred to as **deposition.** Sublimation is endothermic; deposition is exothermic.

Key Terms

Amorphous solid 244	Endothermic process 263	Melting point 253	Specific heat capacity 262
Boiling point 258	Energy 261	Metallic bonding 246	Sublimation 265
Calorie (cal) 261	Exothermic process 264	Molar heat of fusion (ΔH_{fus}) 263	Surface tension 255
Condensation 264	Freezing 263	Molar heat of vaporization (ΔH_{vap}) 265	Vapor pressure 251
Condensed phases 243	Fusion 263		Vaporization 264
Covalent solid 247	Glass 244	Network solid 247	Viscosity 255
Crystalline solid 244	Joule (J) 261	Nonvolatile 257	Volatile 257
Deposition 265	Melting 263	Specific heat 262	

Key Equations

7.1	$1 \text{ cal} \equiv 4.184 \text{ J}$	One calorie (cal) is defined as 4.184 joules (J). This is the amount of energy required to raise the temperature of one gram of water by one degree Celsius.
7.2	$q = \text{specific heat capacity} \times m \times \Delta T$	The amount of energy (in J) required to increase the temperature of a known mass of a substance is calculated as the product of the specific heat of the substance, its mass in grams, and the temperature change.
7.3	$q = \Delta H_{fus} \times \text{moles of substance}$	The amount of energy (in kJ) required to melt a known mass of a substance is calculated as the product of the molar heat of fusion (ΔH_{fus}) and the number of moles of the substance. (Usually, we must calculate the number of moles by dividing the mass in grams by the molar mass of the substance.)
7.4	$q = \Delta H_{vap} \times \text{moles of substance}$	The amount of energy (in kJ) required to vaporize a known mass of a substance is calculated as the product of the molar heat of vaporization (ΔH_{fus}) and the number of moles of the substance. (Again, we typically must calculate the number of moles by dividing the mass in grams by the molar mass of the substance.)

Intermolecular Forces

The intermolecular forces discussed in Chapter 7 are those between particles (atoms, molecules, or ions) in a pure substance. However, our ability to predict how easily a substance can be dissolved in a particular solvent relies on our understanding of the forces between the particles of two *different* substances. The axiom "like dissolves like" refers to *polar* (or ionic) substances being more soluble in *polar* solvents, and *nonpolar* substances being more soluble in *nonpolar* solvents. To assess the solubility of a substance, we must identify it as ionic, polar, or nonpolar. The following flowchart illustrates this identification process and the conclusions we can draw about solubility.

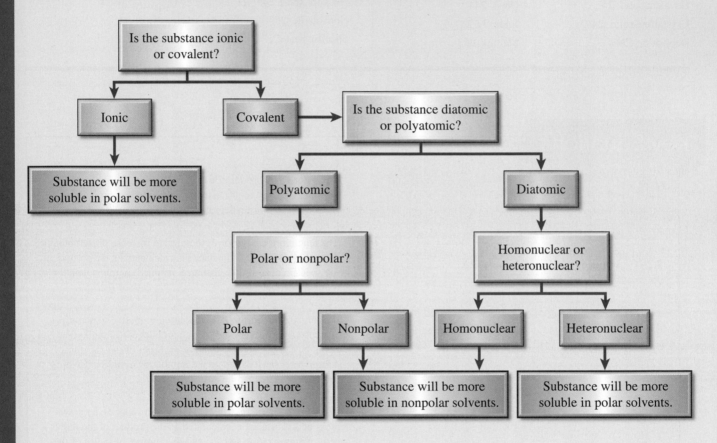

To determine whether or not a substance is ionic, we examine its chemical formula. If the formula contains either a metal cation or the ammonium ion (NH_4^+), it is ionic. There are some ionic substances that are considered insoluble in water. However, even these substances are more soluble in *water* than they are in *non*polar solvents.

If a substance is not ionic, it is covalent. Covalent substances may be polar or nonpolar, depending on the electronegativities of the atoms involved—and on molecular geometry [◄◄ Chapter 6 Key Skills]. Polar species will be more soluble in polar solvents, such as water; and nonpolar species will be more soluble in nonpolar solvents, such as benzene.

In addition to deciding whether a substance will be more soluble in polar or nonpolar solvents, it is possible to assess *relative* solubilities of two different solutes in the same solvent. For example, we may be asked to compare the solubilities (in water and in benzene) of two different molecules. If one of the molecules is polar and the other is nonpolar, we should expect the *polar* molecule to be more soluble in *water*—and the *nonpolar* one to be more soluble in *benzene*.

Further, some nonpolar substances do dissolve in water because of dispersion forces, which are stronger in larger molecules. Therefore, we should expect the larger of two nonpolar molecules to be more soluble in water.

Key Skills Problems

7.1
Which of the following would you expect to be more soluble in water than in benzene? (Select all that apply.)
a) CH_3OH
b) CCl_4
c)

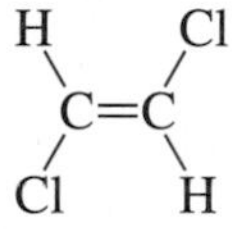

d)

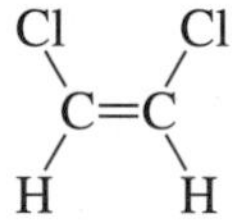

e) KI

7.2
Which of the following would you expect to be more soluble in benzene than in water? (Select all that apply.)
a) Br_2
b) KBr
c) NH_3
d)

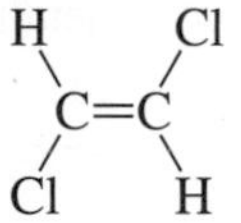

e)

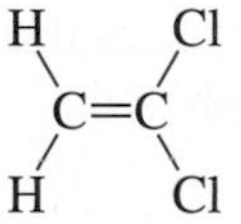

7.3
Arrange the following substances in order of *decreasing* solubility in water: Kr, O_2, N_2.
a) $Kr \approx O_2 > N_2$
b) $Kr > O_2 \approx N_2$
c) $Kr \approx N_2 > O_2$
d) $Kr > N_2 > O_2$
e) $Kr \approx N_2 \approx O_2$

7.4
Arrange the following substances in order of *increasing* solubility in water: C_2H_5OH, CO_2, N_2O.
a) $C_2H_5OH < CO_2 < N_2O$
b) $CO_2 < N_2O < C_2H_5OH$
c) $N_2O < C_2H_5OH < CO_2$
d) $CO_2 \approx N_2O < C_2H_5OH$
e) $CO_2 < C_2H_5OH < N_2O$

Questions and Problems

SECTION 7.1: GENERAL PROPERTIES OF THE CONDENSED PHASES

7.1   List the states of matter. Which of these states is/are compressible?

7.2   Which states of matter take on the shape of the container in which they are stored?

7.3   Classify each of the following samples as a solid, liquid, or gas.

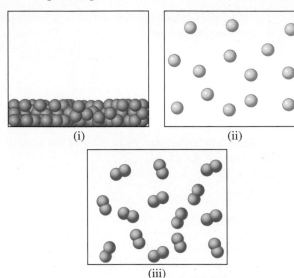

(i)   (ii)

(iii)

7.4   Classify each of the following samples as a solid, liquid, or gas.

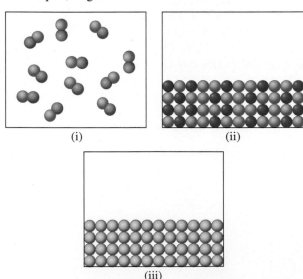

(i)   (ii)

(iii)

7.5   The diagram represents a gaseous molecular compound. Using the same style, sketch a picture of a gaseous atomic element, and a gaseous molecular element.

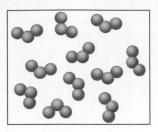

7.6   Using the same style as that shown in Problem 7.5, sketch a picture of a liquid atomic element, a liquid molecular element, and a liquid molecular compound.

7.7   Using the same style as that shown in Problem 7.5, sketch a picture of a solid atomic element, a solid molecular element, and a solid molecular compound.

SECTION 7.2: TYPES OF SOLIDS

7.8   Identify the type of solid each of the following substances can form. (Ionic, molecular, or atomic. For an atomic solid, indicate whether it is metallic or nonmetallic.)
(a) Al   (c) Xe
(b) O_2 (d) K_2O

7.9   Identify the type of solid each of the following substances can form. (Ionic, molecular, or atomic. For an atomic solid, indicate whether it is metallic or nonmetallic.)
(a) PCl_3 (c) N_2O
(b) N_2 (d) Cu

7.10   Identify the type of solid each of the following substances can form. (Ionic, molecular, or atomic. For an atomic solid, indicate whether it is metallic or nonmetallic.)
(a) NH_3 (c) Kr
(b) TiO_2 (d) C_2H_6

7.11   Identify the type(s) of intermolecular forces that hold the solid particles together in each of the substances in Problem 7.8.

7.12 Identify the type(s) of intermolecular forces that hold the solid particles together in each of the substances in Problem 7.9.

7.13 Identify the type(s) of intermolecular forces that hold the solid particles together in each of the substances in Problem 7.10.

7.14 Describe the three types of solids, ionic, molecular, and atomic, and explain how they are different.

7.15 Ionic solids have very high melting points compared to molecular solids. Explain why this is so, and give an example of each type of solid.

7.16 The Thinking Outside the Box feature in Section 7.2 describes boron nitride as a compound that mimics diamond. Given this, predict another form that boron nitride might take, and describe the structure and properties of this form.

SECTION 7.3: PHYSICAL PROPERTIES OF SOLIDS

7.17 Explain what the term *vapor pressure* means.

7.18 Using the style shown in Problem 7.5, sketch two diagrams of an ionic compound, such as NaCl, one showing the solid phase, and another showing how the compound would look when it is melted.

7.19 Using the style shown in Problem 7.5, sketch two diagrams of a molecular compound, such as CO, one showing the solid phase, and another showing how the compound would look when it is melted.

7.20 How is intermolecular-force strength related to vapor pressure?

7.21 Arrange each set of substances in order of increasing vapor pressure at the temperature indicated.
(a) CH_4, NH_3, HCl ($-200°C$, all three are solid)
(b) Ne, F_2, Kr ($-255°C$, all three are solid)
(c) $AsCl_3$, NCl_3, BCl_3 ($-120°C$, all three are solid)

7.22 Arrange each set of substances in order of decreasing vapor pressure at the temperature indicated.
(a) $BeCl_2$, $BeBr_2$, CBr_4 ($75°C$, all three are solid)
(b) SiF_4, CH_2Cl_2, Xe ($-125°C$, all three are solid)
(c) CF_4, $SiCl_4$, $NaCl$ ($-190°C$, all three are solid)

7.23 In each set, select the substance that would have the higher vapor pressure at the temperature indicated and explain why.
(a) BBr_3, $BeBr_2$ ($-50°C$, both are solid)
(b) BCl_3, BF_3 ($-130°C$, both are solid)
(c) HBr, Cl_2 ($-110°C$, both are solid)

7.24 In each set, select the substance that would have the higher vapor pressure at a given temperature and explain why. (These images are bond-line drawings, where each line indicates a bond or pair of electrons, and each "corner" where they meet represents a carbon atom with as many hydrogen atoms as needed to fill its octet.)

Cortisone

(a) Testosterone

(b)

SECTION 7.4: PHYSICAL PROPERTIES OF LIQUIDS

7.25 Explain the meaning of the term *viscosity*.

7.26 How are the intermolecular forces in a substance related to the viscosity of the substance?

7.27 How are the intermolecular forces in a substance related to the surface tension of the substance?

7.28 How are the vapor pressure and the boiling point of a substance related?

7.29 How are the intermolecular forces in a substance related to the vapor pressure of the substance?

7.30 How are the intermolecular forces in a substance related to the boiling point of the substance?

7.31 What is the effect of strong intermolecular forces in a liquid substance on each of the following properties of the liquid?
(a) Vapor pressure
(b) Boiling point
(c) Viscosity
(d) Surface tension

7.32 Does the vapor pressure of a liquid substance change as the liquid is heated? Explain.

7.33 Are bonds broken when a molecular substance boils? Explain.

7.34 Select the substance in each pair that you would expect to have the lower vapor pressure (at a temperature at which both are liquid) and explain why.
(a) CH_3OH, CH_3SH
(b) C_3H_8, H_2O
(c) CH_3OH, CH_4

7.35 Select the substance in each pair that you would expect to have the higher vapor pressure at a given temperature and explain why. Note that the structures shown do not convey information about bond angles, so you must take this into consideration.
(a) acetone (nail polish remover),

$$H_3C-\overset{\overset{\displaystyle O}{\|}}{C}-CH_3$$

isopropanol (rubbing alcohol)

$$H_3C-\underset{\underset{\displaystyle H}{|}}{\overset{\overset{\displaystyle OH}{|}}{C}}-CH_3$$

(b) $SeCl_2$, $BeCl_2$
(c) HF, HCl ($-50°C$, both liquid)

7.36 Arrange the following substances in order of decreasing boiling point.
(a) $CH_3CH_2CH_2CH_3$, $H_2NCH_2CH_2NH_2$, $CH_3CH_2CH_2NH_2$
(b) ICl, Br_2, N_2
(c) LiCl, CO_2, CS_2

7.37 Arrange the following substances in order of increasing boiling point.
(a) CH_3Cl, CH_3OH, CH_3F
(b) CaO, Xe, O_2
(c) Ar, CF_2H_2, $(CH_3)_2NH$

7.38 Select the substance in each pair that you would expect to have the lower boiling point and explain why.

(a)

$$H_3C-\overset{\overset{\displaystyle O}{\|}}{C}-OH$$

(CH_3COOH)
(i)

or

$$H-\underset{\underset{\displaystyle H}{|}}{\overset{\overset{\displaystyle H}{|}}{C}}-\underset{\underset{\displaystyle H}{|}}{\overset{\overset{\displaystyle H}{|}}{C}}-Cl$$

(C_2H_5Cl)
(ii)

(b)

$$H-\underset{\underset{\displaystyle H}{|}}{\overset{\overset{\displaystyle H}{|}}{C}}-\underset{\underset{\displaystyle H}{|}}{\overset{\overset{\displaystyle H}{|}}{C}}-OH$$

(C_2H_5OH)
(i)

or

$$H-\underset{\underset{\displaystyle H}{|}}{\overset{\overset{\displaystyle H}{|}}{C}}-O-\underset{\underset{\displaystyle H}{|}}{\overset{\overset{\displaystyle H}{|}}{C}}-H$$

(CH_3OCH_3)
(ii)

(c)

$$H-\underset{\underset{\displaystyle H}{|}}{\overset{\overset{\displaystyle H}{|}}{C}}-\underset{\underset{\displaystyle H}{|}}{\overset{\overset{\displaystyle H}{|}}{C}}-H$$

(C_2H_6)
(i)

or

$$H-\underset{\underset{\displaystyle H}{|}}{\overset{\overset{\displaystyle H}{|}}{C}}-\underset{\underset{\displaystyle H}{|}}{\overset{\overset{\displaystyle H}{|}}{C}}-\underset{\underset{\displaystyle H}{|}}{\overset{\overset{\displaystyle H}{|}}{C}}-\underset{\underset{\displaystyle H}{|}}{\overset{\overset{\displaystyle H}{|}}{C}}-H$$

(C_4H_{10})
(ii)

7.39 Select the substance in each pair that you would expect to have the higher boiling point and explain why. (These images are bond-line drawings, where each line indicates a bond or pair of electrons, and each "corner" where they meet represents a carbon atom with as many hydrogen atoms as needed to fill its octet.)

(a)

(i)   (ii)

(b) $H-\underset{\underset{\displaystyle H}{|}}{\overset{\overset{\displaystyle H}{|}}{C}}-\underset{\underset{\displaystyle H}{|}}{\overset{\overset{\displaystyle H}{|}}{C}}-H$ $H-\underset{\underset{\displaystyle H}{|}}{\overset{\overset{\displaystyle H}{|}}{C}}-\underset{\underset{\displaystyle H}{|}}{\overset{\overset{\displaystyle H}{|}}{C}}-OH$

(i)   (ii)

(c)

(i)   (ii)

SECTION 7.5: ENERGY AND PHYSICAL CHANGES

7.40 Explain what the term *specific heat* means. What are the units of specific heat?

7.41 Determine the amount of heat energy required to increase the temperature of 20.0 g of water from 15.5°C to 89.4°C.

7.42 Determine the amount of heat energy required to increase the temperature of 20.0 g of copper from 15.5°C to 89.4°C.

7.43 Determine the amount of heat energy required to increase the temperature of 55.0 g of each substance listed by 25.0°C. (See Table 7.4.)
(a) Aluminum   (c) Copper
(b) Ethanol

7.44 Determine the amount of heat energy given off when the temperature of a 23.9-g sample of each of the following substances drops by 15.8°C.
(a) Gold   (c) Water
(b) Iron

7.45 Determine the final temperature of a 49.8-g block of copper, initially at room temperature (25.0°C), after the addition of 228 J.

7.46 Determine the final temperature of a 296-g block of gold, initially at 39.8°C, after the addition of 1.05 kJ.

7.47 The temperature of a 5.75-g piece of an unknown metal increases from 20.6°C to 58.0°C with the addition of 65.1 J. Determine the specific heat capacity of the unknown metal.

7.48 The temperature of a 324-g piece of an unknown metal increases from 18.4°C to 48.5°C with the addition of 1.27 kJ. Determine the specific heat capacity of the unknown metal.

7.49 Indicate whether each of the following processes is exothermic or endothermic.
(a) Melting   (c) Vaporization
(b) Deposition

7.50 Indicate whether each of the following processes is exothermic or endothermic.
(a) Sublimation   (c) Condensation
(b) Freezing

7.51 What are the units of *molar heat of fusion* and *molar heat of vaporization?* At what temperature(s) is it appropriate to use these quantities to calculate the energy involved in a phase change?

7.52 Write the chemical equation that represents the melting of a substance. How is this equation related to the one representing the *freezing* of the same substance?

7.53 Draw diagrams to illustrate the two processes described in Problem 7.52.

7.54 Write the chemical equation that represents the vaporization of a substance. How is this equation related to the one representing the *condensation* of the same substance?

7.55 Draw diagrams to illustrate the two processes described in Problem 7.54.

7.56 Write the chemical equation that represents the sublimation of a substance. How is this equation related to the one representing the *deposition* of the same substance?

7.57 Draw diagrams to illustrate the two processes described in Problem 7.56.

7.58 Are covalent bonds broken when a solid molecular substance goes through a phase change? Explain.

7.59 Determine the amount of energy required to melt 55.8 g of ice at 0.0°C.

7.60 Determine the amount of energy required to vaporize 287 g of ethanol at 78.3°C.

7.61 Determine the amount of energy given off when 75.8 g of liquid diethyl ether freezes at −116.2°C.

7.62 Determine the amount of energy given off when 341 g of gaseous benzene condenses at 80.1°C.

7.63 Determine the amount of energy required to vaporize a 35.9-g sample of liquid water at 100.0°C.

7.64 Determine the amount of energy given off when a 98.2-g sample of gaseous ethanol condenses at 78.3°C.

Answers to In-Chapter Materials

Answers to Practice Problems

7.1A (a) molecular, (b) molecular, (c) atomic. **7.1B** Ar < CO_2 < $CaCl_2$. **7.2A** CH_2Cl_2. **7.2B** K_2SO_4 < NCl_3 < SiF_4. **7.3A** C_2H_6 < C_3H_8 < C_4H_{10}. **7.3B** PBr_3 > PCl_3 > BCl_3. **7.4A** i. **7.4B** ii. **7.5A** 1.5×10^2 kJ. **7.5B** 52.2°C. **7.6A** (a) 9.87 kJ required, (b) 95.9 kJ given off, (c) 19 kJ required. **7.6B** 117 kJ.

Answers to Checkpoints

7.2.1 c. **7.2.2** d. **7.2.3** c. **7.2.4** e. **7.3.1** a. **7.3.2** d. **7.3.3** a. **7.3.4** d. **7.4.1** b. **7.4.2** d. **7.4.3** a. **7.4.4** d. **7.5.1** c. **7.5.2** b.

CHAPTER 8

Gases

The ascent of spectacular hot air balloons is the result of hot air and cold air having different densities.

Eric Delmar/Vetta/Getty Images

8.1 Properties of Gases
- Gaseous Substances
- Kinetic Molecular Theory of Gases

8.2 Pressure
- Definition and Units of Pressure
- Measurement of Pressure

8.3 The Gas Equations
- The Ideal Gas Equation
- The Combined Gas Equation
- The Molar Mass Gas Equation

8.4 The Gas Laws
- Boyle's Law: The Pressure-Volume Relationship
- Charles's Law: The Temperature-Volume Relationship
- Avogadro's Law: The Moles-Volume Relationship

8.5 Gas Mixtures
- Dalton's Law of Partial Pressures
- Mole Fractions

In This Chapter, You Will Learn

About the properties of gases, and how our understanding of molecular properties helps us explain and predict the behavior of gaseous samples.

Things To Review Before You Begin

- Dimensional analysis [◄◄ Section 4.4]
- Chapter 4 Key Skills [◄◄ pages 160–161]

In Chapter 7, we examined the condensed phases and considered the role that intermolecular forces play in determining the physical properties of solids and liquids. In this chapter, we consider the physical properties of gases, substances in which the intermolecular forces are so weak that they are essentially negligible. We also discuss the laws that help us predict the behavior of gases and how to use some of the important equations that are derived from those laws. Gaseous substances may consist of atoms or molecules, but we refer to them collectively throughout this chapter as *molecules*.

8.1 Properties of Gases

Recall from Section 3.1 that, unlike the condensed phases, gases are substances whose molecules are not in contact with one another, but are separated by very large distances (Figure 8.1). Most substances that are solids or liquids can exist as gases under the

(a)   (b)   (c)

Figure 8.1 (a) Solid, (b) liquid, and (c) gaseous states of a substance.

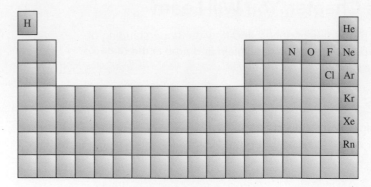

Figure 8.2 Elements that exist as gases at room temperature.

right conditions—generally meaning at higher temperatures. Water, for instance, vaporizes when heated; and water vapor is a gas. The molecular view of a gas shown in Figure 8.1(c) actually exaggerates the size of the gas molecules relative to the distances between them. In reality, the distances between gas molecules are so large as to make the size of individual molecules essentially negligible. To picture how far apart the molecules in a sample of water vapor really are, consider this: if we were to vaporize a teaspoon (5 mL) of water at the boiling point of water (100°C), the resulting water vapor would occupy approximately 8.5 L—greater than the volume of a 2-gallon gasoline can. The vapor occupies a volume 1700 *times* that occupied by the liquid.

Gaseous Substances

Relatively few elements exist as gases at room temperature. Those that do are hydrogen (H_2), nitrogen (N_2), oxygen (O_2), fluorine (F_2), chlorine (Cl_2), and the noble gases (Group 18). Of these, the noble gases exist as isolated atoms, whereas the others exist as diatomic molecules. Figure 8.2 shows where the gaseous elements appear in the periodic table.

Many molecular *compounds,* most often those with low molar masses, exist as gases at room temperature. Table 8.1 lists several gaseous compounds, some of which may be familiar to you.

Gases differ from the condensed phases (solids and liquids) in the following important ways:

1. *A sample of gas assumes both the shape and the volume of its container.* Like a liquid, a gas consists of molecules that do not have fixed positions in the sample. As a result, both liquids and gases are able to flow. (Gases and liquids are sometimes referred to collectively as *fluids.*) Although a sample of liquid will assume the shape of the part of its container that it occupies, a sample of gas will expand to fill the entire volume of its container.
2. *Gases are compressible.* Unlike a solid or a liquid, a gas consists of molecules with relatively large distances between them; that is, the distance between any two molecules in a gas is much larger than the size of the molecules themselves.

TABLE 8.1	Molecular Compounds That Are Gases at Room Temperature	
Molecular Formula	**Compound Name**
HCl	Hydrogen chloride
NH_3	Ammonia
CO_2	Carbon dioxide
N_2O	Dinitrogen monoxide or nitrous oxide
CH_4	Methane
HCN	Hydrogen cyanide

Because gas molecules are far apart, it is possible to move them closer together by confining them to a smaller volume.

3. *The densities of gases are much smaller than those of liquids and solids; and the density of a gaseous substance is highly variable depending on temperature and pressure.* The densities of gases are typically expressed in g/L, whereas those of liquids and solids are typically expressed in g/mL or the equivalent: g/cm^3.

4. *Gases form homogeneous mixtures with one another in any proportions.* Some liquids (e.g., oil and water) do not mix with one another. Gases, on the other hand, because their molecules are so far apart, do not interact with one another to any significant degree. This allows molecules of different gases to mix uniformly.

> **Student Note:** 1 mL = 1 cm^3
> [◀◀ Figure 4.7]

Each of these four characteristics is the result of the properties of gases at the molecular level.

Kinetic Molecular Theory of Gases

The *kinetic molecular theory* is a model [◀◀ Section 1.1] that explains how the nature of gases results in their physical properties. The kinetic molecular theory can be expressed by statement of the following four assumptions:

1. A gas is composed of molecules that are separated by relatively large distances. The volume occupied by individual molecules is negligible. Gases are compressible because molecules in the gas phase are separated by large distances and can be moved closer together by decreasing the volume occupied by a sample of gas.

In a container with a movable piston, a gas sample can be compressed into a smaller volume.

2. Gas molecules are constantly in random motion, moving in straight paths, colliding with the walls of their container and with one another, without any energy being lost in the collisions. (The collisions of gas molecules with the walls of their container is what constitutes *pressure.*)

Note that because the distances molecules travel between collisions is reduced when the sample is compressed, there are more frequent collisions, which constitutes higher gas pressure.

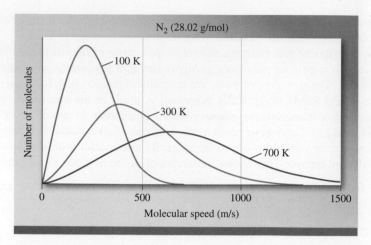

Figure 8.3 Distribution of speeds for molecules in a sample of nitrogen gas at three different temperatures. At higher temperatures, more molecules are moving faster—increasing the average kinetic energy of molecules in the sample.

3. Gas molecules do not exhibit intermolecular forces, either attractive or repulsive.
4. The average kinetic energy of gas molecules in a sample is proportional to the absolute temperature of the sample. (Recall that the *absolute* temperature is the temperature expressed in *kelvins*.) Figure 8.3 illustrates how the average speeds of molecules in the gas phase depend on temperature. (The greater a molecule's speed, the greater its kinetic energy.) Note that this is the same principle we have encountered in the context of liquids. [◀◀ Figure 7.14]

8.2 Pressure

According to kinetic molecular theory, gas molecules in a sample are in constant motion, colliding with one another and with the walls of their container. The collisions of gas molecules with the walls of a container constitute the pressure exerted by a sample of gas. For example, the air in the tires of your car exerts pressure by virtue of the molecules in the air colliding with the inside walls of the tires. In fact, gases exert pressure on everything they touch. Thus, while you may add enough air to increase the pressure inside a tire to 32 pounds per square inch (psi), per the manufacturer's recommendation, there is also a pressure of approximately 14.7 psi, called *atmospheric pressure,* acting on the *outside* of the tire—and on everything else, including your body. (The reason you don't feel the pressure of the atmosphere is that it is balanced by an equal pressure exerted from within your body.)

Atmospheric pressure, the pressure exerted by Earth's atmosphere, can be demonstrated using the empty metal container shown in Figure 8.4(a). Because the container is open to the atmosphere, atmospheric pressure acts on both the internal and external walls of the container. When we attach a vacuum pump to the opening of the container and draw air out of it, however, we reduce the pressure inside the container. When the pressure against the interior walls is reduced, atmospheric pressure crushes the container [Figure 8.4(b)].

Definition and Units of Pressure

Although you are probably at least casually familiar with the concept of pressure—you may have encountered the term *psi* before; and you may have heard "inches of mercury" used in a weather forecast on television—we need a specific definition of pressure to facilitate our discussion of the behavior of gases. When gas molecules collide

(a)   (b)

Charles D. Winters/Timeframe Photography/McGraw Hill   Charles D. Winters/Timeframe Photography/McGraw Hill

Figure 8.4 (a) An empty metal can. (b) When the air is removed by a vacuum pump, atmospheric pressure crushes the can.

with a surface, they exert a force on the surface. *Pressure* is defined as *force* exerted per unit *area:*

$$\text{pressure} = \frac{\text{force}}{\text{area}}$$

The SI unit of force is the ***newton (N),*** which in terms of SI base units is

$$1\text{ N} = 1\frac{1\text{ kg} \cdot \text{m}}{\text{s}^2}$$

The SI unit of pressure is the ***pascal (Pa),*** defined as 1 newton per square meter.

$$1\text{ Pa} = \frac{1\text{ N}}{\text{m}^2}$$

Although the pascal is the SI unit of pressure, there are other units of pressure that are still commonly used. Table 8.2 gives standard atmospheric pressure in several of the most commonly used units. Which of these units of pressure you encounter most often will depend on your specific field of study. Use of atmospheres (atm) is common in chemistry—although bar is also being used with increasing frequency. Millimeters of mercury (mmHg) is common in medicine and meteorology. We use atm predominantly in this text, but it is important that you be comfortable using and converting between the various pressure units in Table 8.2.

> **Student Hot Spot**
>
> Student data indicate that you may struggle to understand where gas pressure originates. Access the eBook to view additional Learning Resources on this topic.

TABLE 8.2	Standard Atmospheric Pressure Expressed in Various Units
Unit	**Typical Air Pressure Value at Sea Level**
pounds per square inch (psi)	14.7 psi
pascals (Pa)	101,325 Pa*
kilopascals (kPa)	101.325 kPa
atmospheres (atm)	1 atm
inches of mercury (in Hg)	29.92 in Hg
millimeters of mercury (mmHg)	760 mmHg
torr (equal to mmHg)	760 torr
bar (equal to 1×10^5 Pa)	1.01325 bar

*The pascal unit refers to a very small amount of pressure, which is why it is not commonly used—the necessary numerical values are unwieldy.

Sample Problem 8.1 gives you some practice converting pressures from one unit to another.

SAMPLE PROBLEM (8.1)   Converting Between Different Units of Pressure

Perform the following pressure conversions: (a) 695 mmHg to units of atm, (b) 3.45 atm to units of psi, (c) 2.87 bar to units of torr.

Strategy Use Table 8.2 to locate the conversion factors needed for each conversion. Remember that the values in the second column are all equal.

Setup

(a) 1 atm = 760 mmHg, so we can write two conversion factors: $\dfrac{1\ \text{atm}}{760\ \text{mmHg}}$ and $\dfrac{760\ \text{mmHg}}{1\ \text{atm}}$.

We select the conversion factor that will facilitate proper cancellation of units from mmHg to atm, $\dfrac{1\ \text{atm}}{760\ \text{mmHg}}$.

(b) 14.7 psi = 1 atm gives the conversion factors: $\dfrac{14.7\ \text{psi}}{1\ \text{atm}}$ and $\dfrac{1\ \text{atm}}{14.7\ \text{psi}}$. Again, we select the one that will cancel units correctly (converting from atm to psi), $\dfrac{14.7\ \text{psi}}{1\ \text{atm}}$.

(c) 1 atm = 1.01325 bar gives the conversion factors: $\dfrac{1\ \text{atm}}{1.01325\ \text{bar}}$ and $\dfrac{1.01325\ \text{bar}}{1\ \text{atm}}$. The conversion factor we need in this case, to convert from bar to atm, is $\dfrac{1\ \text{atm}}{1.01325\ \text{bar}}$.

Solution

(a) $695\ \cancel{\text{mmHg}} \times \dfrac{1\ \text{atm}}{760\ \cancel{\text{mmHg}}} = 0.914\ \text{atm}$ (b) $3.45\ \cancel{\text{atm}} \times \dfrac{14.7\ \text{psi}}{1\ \cancel{\text{atm}}} = 50.7\ \text{psi}$ (c) $2.87\ \cancel{\text{bar}} \times \dfrac{1\ \text{atm}}{1.01325\ \cancel{\text{bar}}} = 2.83\ \text{atm}$

THINK ABOUT IT

Pay careful attention to the cancellation of units. It is also important to consider if the size of each answer is reasonable. In part (a), the initial value of 695 mmHg is lower than atmospheric pressure, which results in the final answer being slightly less than 1 atm. In part (b), the answer is much larger than the initial value because there are many psi in each atmosphere. In part (c), the answer is about the same size as the initial value because atmospheres and bars are about the same size.

Practice Problem Ⓐ**TTEMPT** Perform the following pressure conversions: (a) 39.4 psi to units of atm, (b) 1.75 atm to units of torr, (c) 651 kPa to units of mmHg.

Practice Problem Ⓑ**UILD** Express each of the units found in Table 8.2 in terms of pascals. For example, determine how many pascals are equivalent to 1 psi.

Practice Problem Ⓒ**ONCEPTUALIZE** Rank the following gas samples from lowest to highest pressure.

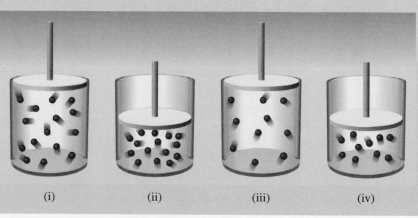

(i)   (ii)   (iii)   (iv)

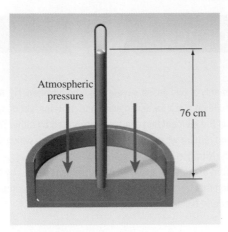

Figure 8.5 Barometer. A long glass tube, closed at one end, is filled with mercury and carefully inverted into a container of mercury. Some of the mercury flows out of the tube into the container. What remains in the tube is the column of mercury that is supported by atmospheric pressure.

Measurement of Pressure

One of the ways that pressure can be measured is by the use of a ***barometer.*** A barometer consists of a long glass tube, closed at one end and filled with mercury. The tube is carefully inverted in a container of mercury so that no air can enter the tube. When the tube is inverted, and the open end is submerged in the container of mercury, some of the mercury in the tube will flow out into the container—creating an empty space in the top of the tube (Figure 8.5). The mercury that remains in the tube is held there by atmospheric pressure pushing down on the surface of the mercury in the container. In other words, the pressure exerted by the column of mercury is equal to the pressure exerted by the atmosphere. Standard atmospheric pressure (1 atm) was originally defined as the pressure that would support a column of mercury exactly 76 cm high at 0°C—at sea level. Note that the mercury column is 760 mm high, and that the mmHg unit is also called the ***torr*** (see Table 8.2) after the Italian scientist Evangelista Torricelli, who invented the barometer.

A ***manometer*** is a device used to measure pressures other than atmospheric pressure. The principle of operation of a manometer is similar to that of a barometer. There are two types of manometers, both of which are shown in Figure 8.6. The closed-tube manometer [Figure 8.6(a)] is normally used to measure pressures that are lower than atmospheric pressure, whereas the open-tube manometer [Figure 8.6(b)] is used to measure pressures either equal to or greater than atmospheric pressure.

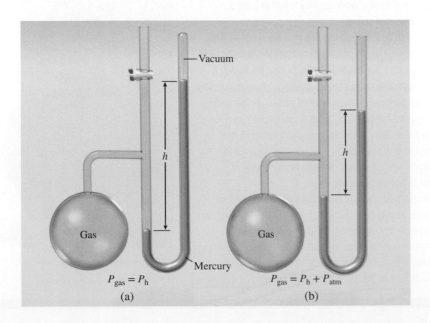

Figure 8.6 (a) Closed-tube manometer. (b) Open-tube manometer.

Profiles in Science

Fritz Haber

Our atmosphere consists primarily of two gases: oxygen, O_2, and nitrogen, N_2. Air is nearly 80 percent N_2. Nitrogen is a necessary building block for amino acids and proteins; yet despite its atmospheric abundance, N_2 is not a form of the element that can be incorporated by either plants or animals. Most of the plants we cultivate for food require what is known as *fixed nitrogen*, which refers to *compounds* that *can* be incorporated by plants. Nitrogen is fixed in nature by lightning strikes, and by certain bacteria in the soil. Historically, the amount of food that could be produced was limited by the amount of fixed nitrogen available—a condition that became critical late in the nineteenth century, when the world's human population first exceeded 1 billion.

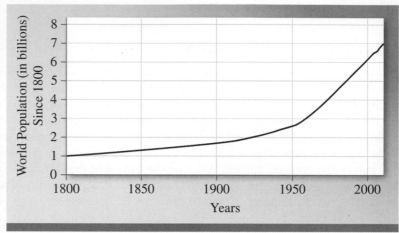

Fritz Jakob Haber
ullstein bild Dtl./Getty Images

Fritz Haber, working with Carl Bosch (both German chemists), developed a method for converting atmospheric nitrogen (N_2) into a form of fixed nitrogen (ammonia, NH_3), from which fertilizers are made. Without the process developed to fix nitrogen artificially, even using all of our arable land, it would be possible to produce enough food for only about half of Earth's current population—which was estimated to have reached 7 billion in October of 2011. Put another way, if all of the factories that produce NH_3 using the Haber-Bosch process were to stop working, nearly half of Earth's population would starve.

Despite having developed a process credited with literally saving the lives of *billions* of people, Haber's work with gases also had a sinister side. During World War I, he worked tirelessly to develop gases as chemical warfare agents for Germany—and personally oversaw the testing and implementation of the substances he produced for that purpose. Other aspects of his work involved the development of explosives, which were also used in Germany's war effort and resulted in the deaths of millions in both world wars.

Haber's first wife, Clara Immerwahr, was the first woman to earn a PhD in chemistry at the University of Breslau (now known as the University of Wroclaw).

Haber was awarded the 1918 Nobel Prize in Chemistry for his work on the nitrogen-fixation process—an award that was controversial because of his work with chemical warfare.

Clara Immerwahr Haber
H.S. Photos/Alamy Stock Photo

CHECKPOINT–SECTION 8.2 Pressure

8.2.1 Express the pressure 21.1 psi in units of atmospheres.

a) 310 atm

b) 0.697 atm

c) 2.37 atm

d) 1.44 atm

e) 0.00323 atm

8.2.2 Express the pressure 8.75 atm in units of mmHg.

a) 1.50×10^{-4} mmHg

b) 4.22×10^{-3} mmHg

c) 6.65×10^{3} mmHg

d) 86.9 mmHg

e) 1.15×10^{-2} mmHg

8.2.3 Express a pressure of 1.15 atm in units of bar.

a) 1.17 bar

b) 1.13 bar

c) 0.881 bar

d) 874 bar

e) 1.51×10^{-3} bar

8.2.4 Which of the following is true? (Select all that apply.)

a) 0.80 atm = 0.80 torr

b) 4180 mmHg = 5.573×10^{5} Pa

c) 433 torr = 433 mmHg

d) 2.300 atm = 1748 torr

e) 5.5 atm = 1.0×10^{5} Pa

8.3 The Gas Equations

Most of the calculations involved in understanding, describing, and predicting the behavior of gases are done using just a few equations. In this section, we examine these important equations and learn how to apply them to solve problems involving gases.

The Ideal Gas Equation

The *ideal gas equation* (Equation 8.1) relates several properties of a gaseous sample:

$$PV = nRT \qquad \text{Equation 8.1}$$

where P is pressure (usually in atm), V is volume (in L), n is the number of moles of the gaseous substance, T is the absolute temperature (in K), and R is a constant known as the *ideal gas constant*. When pressure is given in atm, the ideal gas constant, R, is expressed as 0.0821 L · atm/K · mol.

Algebraic manipulation of the ideal gas equation enables us to solve for any variable, as long as the others are known. For example, if we know the pressure, the number of moles, and the absolute temperature of a gas sample, we can calculate the volume it will occupy. We do so by dividing both sides of Equation 8.1 by P:

$$\frac{\cancel{P}V}{\cancel{P}} = \frac{nRT}{P} \quad \text{to give} \quad V = \frac{nRT}{P}$$

We can, in fact, solve Equation 8.1 in the same way for any of the variables:

$$P = \frac{nRT}{V} \quad n = \frac{PV}{RT} \quad T = \frac{PV}{nR}$$

Thus, if we know the values of any *three* of the four variables (P, V, n, and T), we can calculate the value of the one we do not know.

Sample Problems 8.2, 8.3, 8.4, and 8.5 let you practice solving problems using the ideal gas equation.

Student Note: Remember that R is a *constant*, not a variable.

SAMPLE PROBLEM (8.2)   Using the Ideal Gas Equation to Calculate Volume

Calculate the volume of a mole of ideal gas at room temperature (25°C) and 1.00 atm.

Strategy Convert the temperature in °C to temperature in kelvins, and use the ideal gas equation to solve for the unknown volume.

Setup The data given are $n = 1.00$ mol, $T = 298$ K, and $P = 1.00$ atm. Because the pressure is expressed in atmospheres, we use $R = 0.0821$ L · atm/K · mol to solve for volume in liters.

Solution

$$V = \frac{(1 \text{ mol})\left(0.0821 \frac{\text{L} \cdot \text{atm}}{\text{K} \cdot \text{mol}}\right)(298 \text{ K})}{1 \text{ atm}} = 24.5 \text{ L}$$

> **Student Note:** It is a very common mistake to fail to convert to absolute temperature when solving a gas problem. Most often, temperatures are given in degrees Celsius. The ideal gas equation only works when the temperature used is in kelvins. Remember: K = °C + 273.

THINK ABOUT IT

With the pressure held constant, we should expect the volume to increase with increased temperature. Room temperature is higher than the standard temperature for gases (0°C), so the molar volume at room temperature (25°C) should be higher than the molar volume at 0°C—and it is.

Practice Problem **A**TTEMPT   What is the volume of 5.12 mol of an ideal gas at 32°C and 1.00 atm?

Practice Problem **B**UILD   At what temperature (in °C) would 1 mole of ideal gas occupy 50.0 L ($P = 1.00$ atm)?

Practice Problem **C**ONCEPTUALIZE   The diagram on the left represents a sample of gas in a container with a movable piston. Which of the other diagrams [(i)–(iv)] best represents the sample (a) after the absolute temperature has been doubled; (b) after the volume has been decreased by half; and (c) after the external pressure has been doubled? (In each case, assume that the only variable that has changed is the one specified.)

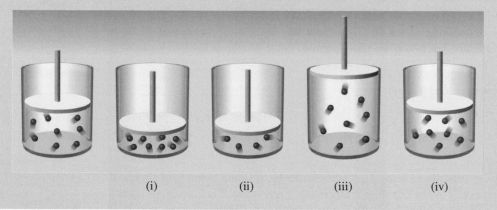

(i)   (ii)   (iii)   (iv)

SAMPLE PROBLEM (8.3)   Using the Ideal Gas Equation to Calculate Pressure

Calculate the pressure of 1.44 mol of an ideal gas in a 5.00-L container at 36°C.

Strategy Rearrange the ideal gas equation (Equation 8.1) to isolate pressure, P. Convert the temperature into kelvins, $36 + 273 = 309$ K.

Setup

$$P = \frac{nRT}{V}$$

Solution

$$P = \frac{nRT}{V} = \frac{1.44 \text{ mol} \times 0.0821 \frac{\text{L} \cdot \text{atm}}{\text{K} \cdot \text{mol}} \times 309 \text{ K}}{5.00 \text{ L}} = 7.31 \text{ atm}$$

THINK ABOUT IT

This pressure seems reasonably high when comparing to the volume (22.4 L) occupied by 1.00 mol of an ideal gas at STP. The volume in this problem is less than ¼ of the standard volume and contains a higher number of moles at a higher temperature, all leading to a pressure much higher than the standard 1 atm.

Practice Problem ATTEMPT Calculate the pressure of 3.15 mol of an ideal gas in a 4.33-L container at 55°C.

Practice Problem BUILD Calculate the pressure of 24.5 grams of He in a 15.5-L container at 25°C.

Practice Problem CONCEPTUALIZE Each of the diagrams represents a sample of gas at room temperature. Which of the samples has the highest pressure? Which has the lowest pressure?

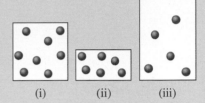

(i)　　　(ii)　　　(iii)

SAMPLE PROBLEM (8.4) Using the Ideal Gas Equation to Calculate Moles

Determine the number of moles of ideal gas that occupy 12.3 L at 298 K and 1.80 atm.

Strategy Rearrange the ideal gas equation (Eq 8.1) to isolate moles, n.

Setup

$$n = \frac{PV}{RT}$$

Solution

$$n = \frac{PV}{RT} = \frac{1.80 \ \cancel{atm} \times 12.3 \ \cancel{L}}{0.0821 \ \frac{\cancel{L} \cdot \cancel{atm}}{K \cdot mol} \times 298 \ \cancel{K}} = 0.905 \ mol$$

THINK ABOUT IT

This answer seems reasonable when comparing to the 22.4 L volume that an ideal gas occupies at STP. The volume in this problem is close to half the standard volume, but nearly double the standard pressure.

Practice Problem ATTEMPT Determine the number of moles of ideal gas that occupy 39.2 L at STP.

Practice Problem BUILD Determine the mass of Ne that occupies 21.8 L at 25°C and 0.885 atm.

Practice Problem CONCEPTUALIZE The diagram on the left represents a sample of gas in a sealed container. Which of the other diagrams [(i)–(iv)] could represent the sample after both the temperature and the volume have changed?

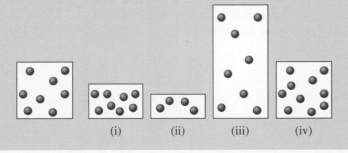

(i)　　　(ii)　　　(iii)　　　(iv)

SAMPLE PROBLEM **8.5** Using the Ideal Gas Equation to Calculate Absolute Temperature

At what temperature (in K) would 2.71 mol of an ideal gas occupy 30.0 L at standard pressure?

Strategy Rearrange the ideal gas equation (Equation 8.1) to isolate temperature, T. Standard pressure is 1.00 atm.

Setup

$$T = \frac{PV}{nR}$$

Solution

$$T = \frac{PV}{nR} = \frac{1.00 \text{ atm} \times 30.0 \text{ L}}{2.71 \text{ mol} \times 0.0821 \frac{\text{L} \cdot \text{atm}}{\text{K} \cdot \text{mol}}} = 135 \text{ K}$$

THINK ABOUT IT

Carefully cancelling units will stop you from making the most common errors in these types of problems.

Practice Problem **A** **TTEMPT** At what temperature (in K) would 0.181 mol of ideal gas occupy 15.0 L at a pressure of 1.60 atm?

Practice Problem **B** **UILD** At what temperature (in °C) would 1.05 grams of H_2 occupy 10.0 L at a pressure of 0.306 atm?

Practice Problem **C** **ONCEPTUALIZE** Is it possible for a sample of gas to undergo a change in pressure and volume without the temperature changing? Explain.

The *standard state* of a gas is defined as a temperature of 0°C and a pressure of 1 atm. These conditions are commonly referred to as **STP,** for standard temperature and pressure. Often, gas problems will refer to STP, so it is important that you understand what this term means. Using Equation 8.1, we can calculate the volume occupied by 1 mol of an ideal gas at STP:

Rearranging Equation 8.1 to solve for V, we get

$$V = \frac{nRT}{P}$$

And plugging in values for n, R, T, and P, we determine that the volume occupied by 1 mol of gas at STP is

$$V = \frac{(1 \text{ mol}) \left(\frac{0.0821 \text{ L} \cdot \text{atm}}{\text{K} \cdot \text{mol}} \right) (273 \text{ K})}{1 \text{ atm}} = 22.4 \text{ L}$$

That is, a mole of gas at STP occupies 22.4 L—regardless of the identity of the gas. Note that the number of significant figures in the result is not limited by either the 1 mol term or the 1 atm term. 1 mol is a specified quantity; our purpose is to calculate the volume of exactly one mole. Likewise, 1 atm is part of a definition—the P in STP is *defined* as 1 atm.

Thinking Outside the Box

Pressure Exerted by a Column of Fluid

The pressure exerted by a column of fluid, such as that in a barometer (Figure 8.5), can be calculated using the following equation:

$$P = hdg$$

where h is the height of the column in meters, d is the density of the fluid in kg/m^3, and g is the gravitational constant equal to 9.80665 m/s^2. (Using this equation yields pressure in units of Pascals.) This equation explains why barometers historically have been constructed using *mercury*. The height of a column of fluid supported by a given pressure is inversely proportional to the density of the fluid. (At a given pressure, as the density of the fluid goes down, h must go up—and vice versa.) Mercury's high density made it possible to construct barometers and manometers of manageable size. Traditional barometers, with mercury column heights in the neighborhood of 76 cm, may seem large and unwieldy. But consider this: Because the density of mercury is more than 13 times that of water, the same pressure measured using a barometer constructed with water would have to be more than 13 times as high—over 10 meters high!

Notice that the equation does not contain a term for the cross section of the column. The pressure exerted depends only on the *height* of the column, *not* on how big around it is. Thus, the three columns of fluid represented here all exert the same pressure (as long as they contain the same fluid).

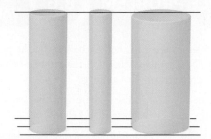

The Combined Gas Equation

When the quantity of a gas is fixed, meaning that n does not change, we can use a modified form of the ideal gas equation to determine changes in the remaining variables, P, T, and V. If we know that the number of moles is constant, we can rearrange the ideal gas equation as:

$$nR = \frac{PV}{T}$$

Moreover, because the product of n and R is constant (because both n and R are constant), we know that this equality must hold regardless of any changes in P, V, and T. So consider this sample of gas under one set of conditions, where $T = T_1$, $P = P_1$, and $V = V_1$:

$$nR = \frac{P_1V_1}{T_1}$$

And then consider the same sample of gas under a different set of conditions where $T = T_2$, $P = P_2$, and $V = V_2$:

$$nR = \frac{P_2V_2}{T_2}$$

Because the value of nR is constant, these two expressions are equal to each other, and we can also write

$$\frac{P_1V_1}{T_1} = \frac{P_2V_2}{T_2} \qquad \textbf{Equation 8.2}$$

which is known as the **combined gas equation.** The combined gas equation contains *three* variables and is useful when changes in *two* of the variables are known—and we wish to calculate the change in the *third* variable.

Consider the following example: We have a sample of gas that occupies 1.00 L at 1.00 atm and 25°C. If the pressure stays the same but the temperature increases to 50°C, what volume will the sample occupy? As before, we can rearrange Equation 8.2 algebraically to solve for any of the variables. So if we know P_1 (1.00 atm), V_1 (1.00 L), and T_1 (298 K); and we know P_2 (1.00 atm) and T_2 (323 K), we can solve for the new volume by multiplying both sides of Equation 8.2 by T_2 and dividing both sides by P_2:

$$\frac{T_2}{P_2} \times \frac{P_1V_1}{T_1} = \frac{P_2V_2}{T_2} \times \frac{T_2}{P_2} \quad \text{to give} \quad \frac{T_2P_1V_1}{P_2T_1} = V_2$$

Plugging in the values we know for T_2, P_1, V_1, P_2, and T_1 gives:

$$\frac{(323\ \cancel{K})(1.00\ \cancel{atm})(1.00\ L)}{(1.00\ \cancel{atm})(298\ \cancel{K})} = V_2 = 1.08\ L$$

Like the ideal gas equation, the combined gas equation can be solved for any of the variables it contains with simple algebraic manipulations.

Sample Problem 8.6 shows how the combined gas equation can be used to solve problems involving fixed amounts of gas.

SAMPLE PROBLEM (**8.6**)   Using the Combined Gas Equation to Calculate a New Volume as Temperature and/or Pressure Change

If a child releases a 6.25-L helium balloon in the parking lot of an amusement park where the temperature is 28°C and the air pressure is 757.2 mmHg, what will the volume of the balloon be when it has risen to an altitude where the temperature is −34°C and the air pressure is 366.4 mmHg?

Strategy In this case, because there is a fixed amount of gas, we use Equation 8.2. The only value we don't know is V_2. Temperatures must be expressed in kelvins. We can use any units of pressure, as long as we are consistent.

Setup $T_1 = 301$ K, $T_2 = 239$ K. Solving Equation 8.2 for V_2 gives

$$V_2 = \frac{P_1 T_2 V_1}{P_2 T_1}$$

Solution

$$V_2 = \frac{(757.2\ \cancel{mmHg})(239\ \cancel{K})(6.25\ L)}{(366.4\ \cancel{mmHg})(301\ \cancel{K})} = 10.3\ L$$

THINK ABOUT IT

Note that the solution is essentially multiplying the original volume by the ratio of P_1 to P_2 and by the ratio of T_2 to T_1. The effect of decreasing external pressure is to increase the balloon volume. The effect of decreasing temperature is to decrease the volume. In this case, the effect of decreasing external pressure predominates, and the balloon volume increases significantly.

Practice Problem **A**TTEMPT What would be the volume of the balloon in Sample Problem 8.6 if, instead of being released to rise in the atmosphere, it were submerged in a swimming pool to a depth where the pressure is 922.3 mmHg and the temperature is 26°C?

Practice Problem **B**UILD What would the temperature of the water in Practice Problem A have to be for the balloon to have the same volume when it is submerged as it had originally, prior to being released in Sample Problem 8.6?

Practice Problem **C**ONCEPTUALIZE Which of the following diagrams could represent a gas sample in a balloon before and after an increase in temperature and an increase in external pressure?

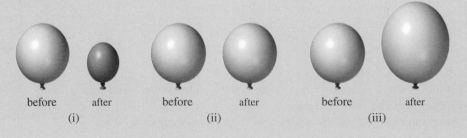

before   after   before   after   before   after
    (i)   (ii)   (iii)

The Molar Mass Gas Equation

Another minor manipulation of the ideal gas equation enables us to determine the density (d) of a gas if we know its molar mass. Starting with the ideal gas equation:

$$PV = nRT$$

we divide both sides by V and by RT:

$$\frac{P\cancel{V}}{\cancel{V}RT} = \frac{nRT}{V\cancel{RT}} \quad \text{to get} \quad \frac{P}{RT} = \frac{n}{V}$$

On the right side of the resulting equation, we have n (mol) divided by V (L). Recall that multiplying moles of a substance by its molar mass gives the mass of substance in grams [◄◄ Section 5.2]. Therefore, if we multiply both sides of the equation by molar mass, we get

$$\mathcal{M} \times \frac{P}{RT} = \frac{n}{V} \times \mathcal{M}$$

or

$$d(\text{in g/L}) = \frac{P\mathcal{M}}{RT}$$

Further, we can rearrange this equation to solve for molar mass—which we can determine if we know the density of a gas:

$$\mathcal{M} = \frac{dRT}{P} \qquad \textbf{Equation 8.3}$$

Equation 8.3 is used often to determine the molar mass of a gas whose density has been measured experimentally. Figure 8.7 illustrates the experimental method by which molar mass can be determined using Equation 8.3.

Sample Problem 8.7 shows how Equation 8.3 is used to determine the density of a gas from experimental data and molar mass.

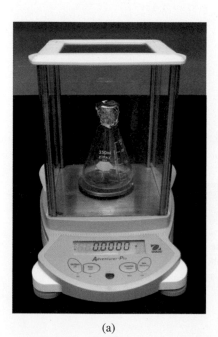

(a)

(b)

(c)

Figure 8.7 To determine the molar mass of a liquid, (a) we weigh an empty flask fitted with a foil cap with a tiny hole for air and vapor to escape. (b) We then put a small volume of the liquid into the flask and immerse it in boiling water long enough for all of the liquid to be vaporized. The vapor displaces the air that originally filled the flask. (c) After the liquid has all vaporized, we remove the flask from the boiling water, run cold water on the outside to cool it, carefully dry the outside, and reweigh it. The difference between the initial and final masses is the mass of the vapor. Knowing the temperature of the boiling water, the pressure in the room, and the volume of the flask, we can use Equation 8.3 to calculate the molar mass of the liquid.

SAMPLE PROBLEM 8.7   Calculating Gas Density from Molar Mass

Carbon dioxide is effective in fire extinguishers partly because its density is greater than that of air, so CO_2 can smother the flames by depriving them of oxygen. (Air has a density of approximately 1.2 g/L at room temperature and 1 atm.) Calculate the density of CO_2 at room temperature (25°C) and 1.0 atm.

Strategy Use Equation 8.3 to solve for density. Because the pressure is expressed in atm, we should use $R = 0.0821$ L · atm/K · mol. Remember to express temperature in kelvins.

Setup The molar mass of CO_2 is 44.01 g/mol.

Solution

$$d = \frac{P\mathcal{M}}{RT} = \frac{(1 \text{ atm})\left(44.01\frac{\text{g}}{\text{mol}}\right)}{\left(0.0821\frac{\text{L} \cdot \text{atm}}{\text{K} \cdot \text{mol}}\right)(298 \text{ K})} = 1.8 \text{ g/L}$$

THINK ABOUT IT

The calculated density of CO_2 is greater than that of air under the same conditions (as expected). Although it may seem tedious, it is a good idea to write units for each and every entry in a problem such as this. Unit cancellation is very useful for detecting errors in your reasoning or your solution setup.

Practice Problem **A**TTEMPT Calculate the density of air at 0°C and 1 atm. (Assume that air is 80 percent N_2 and 20 percent O_2.)

Practice Problem **B**UILD What pressure would be required for helium at 25°C to have the same density as carbon dioxide at 25°C and 1 atm?

Practice Problem **C**ONCEPTUALIZE Two samples of gas are shown at the same temperature and pressure. Which sample has the greater density? Which exerts the greater pressure?

     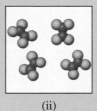

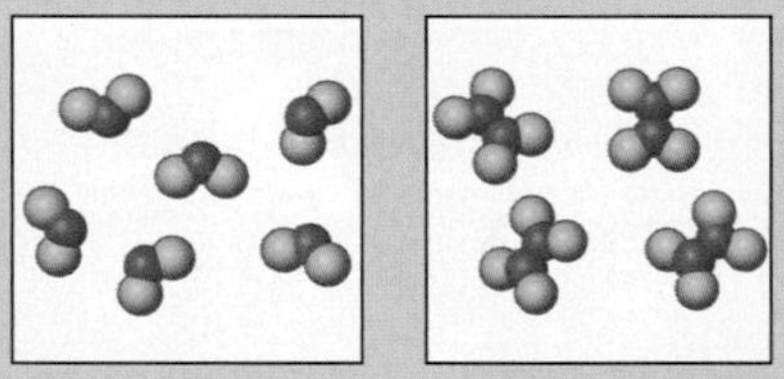

(i)   (ii)

CHECKPOINT–SECTION 8.3   The Gas Equations

8.3.1 Calculate the volume occupied by 8.75 mol of an ideal gas at STP.

a) 196 L

b) 268 L

c) 0.718 L

d) 18.0 L

e) 2.56 L

8.3.2 Calculate the pressure exerted by 10.2 mol of an ideal gas in a 7.5-L vessel at 150°C.

a) 17 atm

b) 31 atm

c) 0.72 atm

d) 1.3 atm

e) 47 atm

8.3.3 Determine the density of a gas with $\mathcal{M} = 146.07$ g/mol at 1.00 atm and 100.0°C.

a) 6.85×10^{-3} g/L

b) 4.77 g/L

c) 146 g/L

d) 30.6 g/L

e) 17.8 g/L

8.3.4 Determine the molar mass of a gas with $d = 1.963$ g/L at 1.00 atm and 100.0°C.

a) 0.0166 g/mol

b) 60.1 g/mol

c) 16.1 g/mol

d) 6.09×10^3 g/mol

e) 1.63×10^3 g/mol

The Gas Laws

The gas equations that we encountered in Section 8.3 are all derived from specific gas *laws* [◄◄ Section 1.1], which were developed by scientists in the seventeenth, eighteenth, and early nineteenth centuries. In this section, we describe these laws and how they led to the development of the gas equations.

Boyle's Law: The Pressure-Volume Relationship

Imagine that you have a plastic syringe filled with air. If you hold your finger tightly against the tip of the syringe and push the plunger with your other hand, decreasing the volume of the air, you will increase the pressure in the syringe. During the seventeenth century, Robert Boyle (British chemist, 1627–1691) conducted systematic studies of the relationship between gas volume and pressure using a simple apparatus like the one shown in Figure 8.8. The J-shaped tube contains a sample of gas confined by a column of mercury. The apparatus functions as an open-end manometer. When the mercury levels on both sides are equal [Figure 8.8(a)], the pressure of the confined gas is equal to atmospheric pressure. When more mercury is added through the open end, the pressure of the confined gas is increased by an amount proportional to the height of the added mercury—and the volume of the gas decreases. If, for example, as shown in Figure 8.8(b), we *double* the pressure on the confined gas by adding

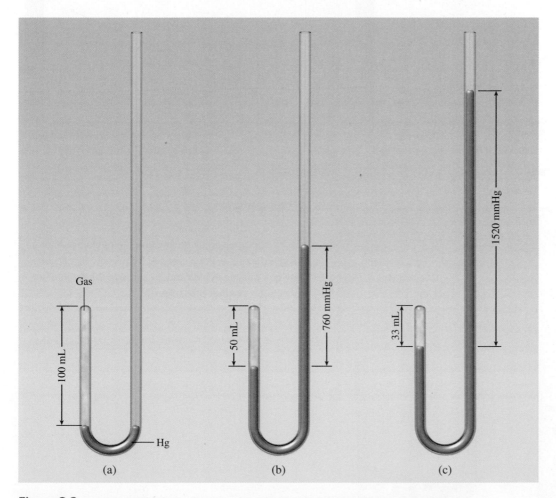

Figure 8.8 Demonstration of Boyle's law. The volume of a sample of gas is inversely proportional to its pressure. (a) $P = 760$ mmHg, $V = 100$ mL. (b) $P = 1520$ mmHg, $V = 50$ mL. (c) $P = 2280$ mmHg, $V = 33$ mL. Note that the total pressure exerted on the gas is the sum of atmospheric pressure (760 mmHg) and the difference in mercury heights.

TABLE 8.3	Typical Data from Experiments with the Apparatus of Figure 8.8								
P (mmHg)	760	855	950	1045	1140	1235	1330	1425	1520	2280
V (mL)	100	89	78	72	66	59	55	54	50	33
	Figure 8.8(a)								Figure 8.8(b)	Figure 8.8(c)

enough mercury to make the difference in mercury levels on the left and right 760 mm (the height of a mercury column that exerts a pressure equal to 1 atm), the volume of the gas is reduced by *half*. If we triple the original pressure on the confined gas by adding more mercury, the volume of the gas is reduced to one-third of its original volume [Figure 8.8(c)].

Table 8.3 gives a set of data typical of Boyle's experiments. Figure 8.9 shows some of the volume data plotted (a) as a function of pressure and (b) as a function of the inverse of pressure, respectively. These data illustrate **Boyle's law,** which states that the pressure of a fixed amount of gas at a constant temperature is inversely proportional

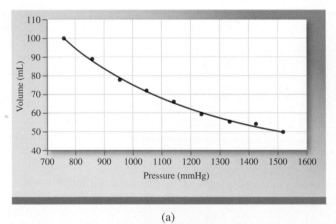

(a)

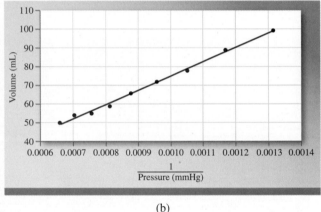

(b)

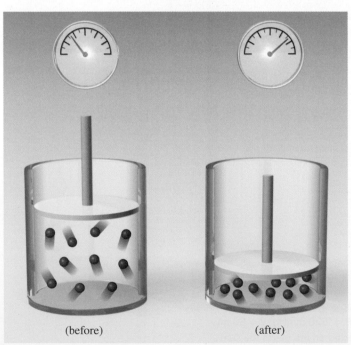

(before)   (after)

(c)

Figure 8.9 Plots of volume (a) as a function of pressure and (b) as a function of 1/pressure. (c) Gases can be compressed by decreasing their volume. After a volume decrease, the increased frequency of collisions between gas molecules and the walls of their container constitutes a higher pressure.

to the volume of the gas. This inverse relationship between pressure and volume can be expressed mathematically as

$$V \propto \frac{1}{P}$$

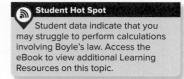

Student Hot Spot

Student data indicate that you may struggle to perform calculations involving Boyle's law. Access the eBook to view additional Learning Resources on this topic.

where the symbol $\propto$ means "is proportional to." Another way to express this relationship is that the *product* of V and P ($V \times P$) is constant at constant temperature.

Sample Problem 8.8 illustrates the use of Boyle's law.

SAMPLE PROBLEM **8.8** Using Boyle's Law to Calculate Volume as Pressure Changes at Constant Temperature

If a skin diver takes a breath at the surface, filling his lungs with 5.82 L of air, what volume will the air in his lungs occupy when he dives to a depth where the pressure is 1.92 atm? (Assume constant temperature and that the pressure at the surface is exactly 1 atm.)

Strategy Use Equation 8.2 to solve for V_2.

Setup $P_1 = 1.00$ atm, $V_1 = 5.82$ L, and $P_2 = 1.92$ atm.

Solution

$$V_2 = \frac{P_1 \times V_1}{P_2} = \frac{1.00 \text{ atm} \times 5.82 \text{ L}}{1.92 \text{ atm}} = 3.03 \text{ L}$$

THINK ABOUT IT

At higher pressure, the volume should be smaller. Therefore, the answer makes sense.

Practice Problem **A**TTEMPT Calculate the volume of a sample of gas at 5.75 atm if it occupies 5.14 L at 2.49 atm. (Assume constant temperature.)

Practice Problem **B**UILD At what pressure would a sample of gas occupy 7.86 L if it occupies 3.44 L at 4.11 atm? (Assume constant temperature.)

Practice Problem **C**ONCEPTUALIZE Which of the following diagrams could represent a gas sample in a balloon at constant temperature before and after an increase in external pressure?

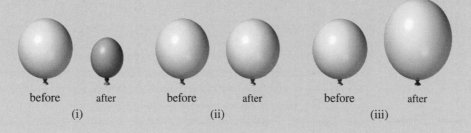

before	after	before	after	before	after
(i)		(ii)		(iii)	

Charles's Law: The Temperature-Volume Relationship

If you took a helium-filled Mylar balloon outdoors on a cold day, the balloon would shrink somewhat when it came into contact with the cold air. This would occur because the volume of a sample of gas depends on the temperature. A more dramatic illustration is shown in Figure 8.10, where liquid nitrogen is being poured over an air-filled balloon. The large drop in temperature of the air in the balloon (the boiling liquid nitrogen has a temperature of $-196°C$) results in a significant decrease in its volume, causing the balloon to shrink. Note that the pressure inside the balloon is roughly equal to the external pressure.

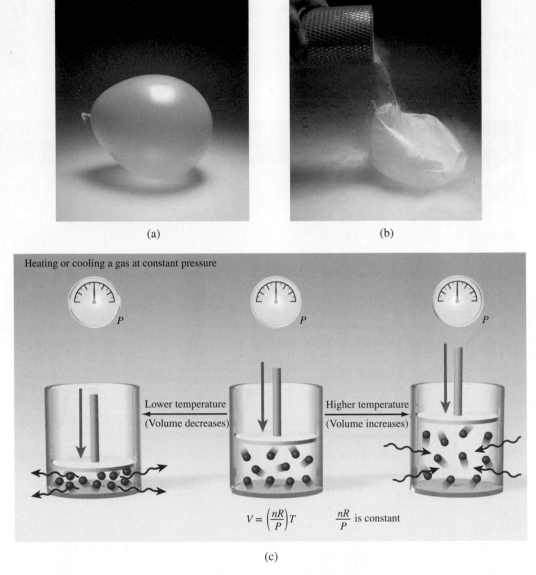

(a)

(b)

(c)

Figure 8.10 (a) Air-filled balloon. (b) Lowering the temperature with liquid nitrogen causes a dramatic volume decrease. The pressure inside the balloon, which is roughly equal to the external pressure, remains constant in this process. (c) Molecular-level illustration of Charles's law.

(both) Charles D. Winters/Timeframe Photography/McGraw Hill

During the eighteenth century, French scientists Jacques Charles (1746–1823) and Joseph Gay-Lussac (1778–1850) studied the relationship between temperature and volume of gas samples at constant pressure. Their studies showed that at constant pressure, the volume of a gas sample increases when heated and decreases when cooled. Figure 8.11(a) shows a plot of data typical of Charles's and Gay-Lussac's experiments. Note that at constant pressure, the volume of a gas is directly proportional to its absolute temperature. This relationship is known as ***Charles's law.*** These experiments were carried out at several different pressures [Figure 8.11(b)], with each yielding a different straight line. Interestingly, when all of the lines are extrapolated (continued beyond the data points that define the lines), they converge at the *x*-axis at −273°C. The temperature at which the temperature-volume lines meet is significant: it is what we call ***absolute zero;*** that is, zero on the Kelvin scale.

Student Note: The implication is that a gas sample occupies *zero* volume at −273°C. This is not actually observed—gases condense to form liquids typically at temperatures well above that point.

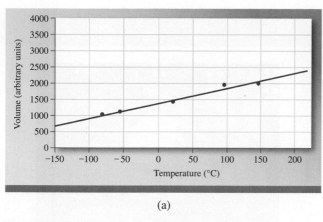
(a)

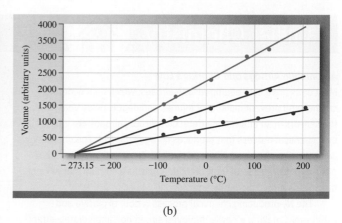
(b)

Figure 8.11 (a) Plot of the volume of a sample of gas as a function of temperature. (b) Plot of the volume of a sample of gas as a function of temperature at three different pressures. Recall that the Celsius and Kelvin scales are actually offset by 273.15. [◄◄ Section 4.1]

The relationship between volume and temperature can be expressed as

$$V \propto T$$

Another way to express this relationship is that the *quotient* of V and T (V/T) is constant at constant pressure.

Sample Problem 8.9 shows how to use Charles's law.

SAMPLE PROBLEM   8.9   Using Charles's Law to Calculate a New Volume as Temperature Changes at Constant Pressure

A sample of argon gas that originally occupied 14.6 L at 25°C was heated to 50°C at constant pressure. What is its new volume?

Strategy Use Equation 8.3 to solve for V_2. Remember that temperatures must be expressed in kelvins.

Setup $T_1 = 298$ K, $V_1 = 14.6$ L, and $T_2 = 323$ K.

Solution

$$V_2 = \frac{V_1 \times T_2}{T_1} = \frac{14.6 \text{ L} \times 323 \text{ K}}{298 \text{ K}} = 15.8 \text{ L}$$

THINK ABOUT IT

When temperature increases at constant pressure, the volume of a gas sample increases.

Practice Problem ⒶTTEMPT A sample of gas originally occupies 29.1 L at 0°C. What is its new volume when it is heated to 15°C? (Assume constant pressure.)

Practice Problem ⒷUILD At what temperature (in °C) will a sample of gas occupy 82.3 L if it occupies 50.0 L at 75°C? (Assume constant pressure.)

Practice Problem ⒸONCEPTUALIZE The first diagram shows a sample of gas at 50°C is contained in a cylinder with a movable piston. Which of the other diagrams [(i)–(iv)] best represents the system when the temperature of the sample has been increased to 100°C?

(i)

(ii)

(iii)

(iv)

Familiar Chemistry

Automobile Airbags and Charles's Law

Most of us drive cars equipped with airbags, which deploy during crashes and prevent millions of injuries and deaths every year. When the sensors in an automobile detect a collision, airbags are inflated by a chemical process that produces nitrogen gas (N_2). The photo shows an explosively deployed driver-side airbag. The inflation, which takes place over the course of 0.06 second, produces a cushion that prevents the driver's body from hitting the steering wheel.

The N_2-producing chemical process is both explosive and exothermic, making a newly deployed airbag hot. Immediately after deployment, the airbag begins to shrink, making it possible for the driver or passenger to move freely and escape the vehicle. One cause of the shrinkage is gas escaping from the bag through holes or vents that are part of its design. The other cause is the rapid drop in temperature that begins immediately following the inflation.

fStop Images - Caspar Benson/Brand X Pictures/Getty Images

As a gas cools at constant pressure, its volume decreases. The volume decrease resulting from cooling at constant pressure is an illustration of Charles's law.

Avogadro's Law: The Moles-Volume Relationship

Early in the nineteenth century, Amedeo Avogadro (Italian scientist, 1776–1856) proposed that equal volumes of different gases (at the same temperature and pressure) contain the same numbers of molecules. This hypothesis gave rise to *Avogadro's law,* which states that, at constant temperature and pressure, the volume of a sample of gas is directly proportional to the number of moles in the sample:

$$V \propto n$$

Figure 8.12 Avogadro's law. The volume of a gas at constant temperature and pressure is proportional to the number of moles.

Another way to express this relationship is that the quotient of V and n (V/n) is constant at constant temperature and pressure. Figure 8.12 illustrates Avogadro's law at the molecular level.

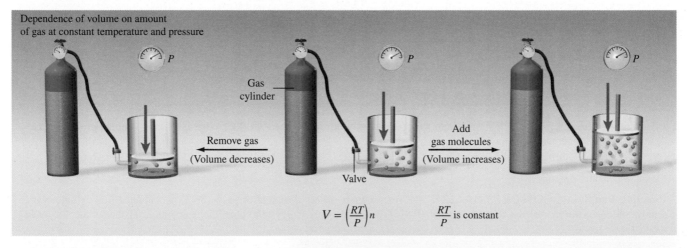

Dependence of volume on amount of gas at constant temperature and pressure

Gas cylinder

Remove gas
(Volume decreases)

Add gas molecules
(Volume increases)

Valve

$$V = \left(\frac{RT}{P}\right)n \qquad \frac{RT}{P} \text{ is constant}$$

Profiles in Science

Amanda Theodosia Jones

Amanda Theodosia Jones (1835–1914) was a writer and inventor in the late nineteenth and early twentieth centuries. She was also an active supporter of women's rights and women's suffrage. One of her patented inventions was an apparatus used for "vacuum canning" of food—sometimes known as *the Jones method*. Although her invention may look complicated, it constitutes a fairly simple use of the gas laws. Food to be preserved was placed in canning jars with rubber gaskets and sealed loosely with glass tops. The chamber holding the jars was then heated using steam—increasing the temperature of the jars and their contents. The temperature increase did two things: it killed bacteria in the food, and it caused an increase in pressure inside the jars, causing nearly all the air in the jars to escape through the loose seal. After a prescribed period of sustained high temperature, the jars were allowed to cool. The cooling caused a decrease in volume and pressure of what little air remained in the jars, and caused the condensation of water vapor, further decreasing pressure inside the jars. The decreased pressure drew the glass tops downward, compressing the rubber gaskets and forming a vacuum seal. Jones's vacuum method of canning was a marked improvement over existing food-preservation methods in the nineteenth century.

Amanda Theodosia Jones
Steve Beach Miller

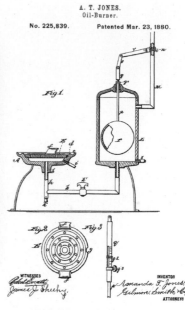

United States Patent and Trademark Office,
https://www.uspto.gov/

Sample Problem 8.10 shows how to use Avogadro's law.

SAMPLE PROBLEM 8.10 Using Avogadro's Law to Calculate the Volume of a Gas Sample

A 1.50-mol sample of an ideal gas occupies 10.0 L under certain temperature and pressure conditions. Determine the volume that 3.00 mol of an ideal gas would occupy under these same conditions.

Strategy Avogadro's law tells us that if the temperature and pressure are constant, the volume of an ideal gas is proportional to the moles of gas present. This means that the ratio of V/n is constant if T and P are constant.

Setup Avogadro's law tells us the volume is proportional to the number of moles of gas, at constant temperature and pressure. We can set up a proportionality to express this relationship: $\dfrac{V_1}{n_1} = \dfrac{V_2}{n_2}$

Rearrange this relationship to isolate V_2. $V_2 = \dfrac{V_1 n_2}{n_1}$

Solution

$$V_2 = \frac{10.0 \text{ L} \times 3.00 \text{ mol}}{1.50 \text{ mol}} = 20.0 \text{ L}$$

THINK ABOUT IT

This answer makes sense as the volume of gas doubled when the moles of gas were doubled under the same temperature and pressure conditions.

Practice Problem (A)TTEMPT A 4.89-mol sample of Ar occupies 57.5 L under certain temperature and pressure conditions. Determine the volume that 1.43 mol of Ar would occupy under the same conditions.

Practice Problem (B)UILD A 2.33-mol sample of CO occupies 14.5 L at a certain temperature and pressure. What is the new volume of the CO gas sample if 3.67 mol of CO is added? Assume constant T and P.

Practice Problem (C)ONCEPTUALIZE Would the answer to Practice Problem B be different if the identity of the gas were changed from CO to CO_2? Why or why not?

To summarize the mathematical expressions of the gas laws:

Boyle's law: $V \propto \frac{1}{P}$, which can also be written as $V = a \times \frac{1}{P}$, where a is a constant.

Charles's law: $V \propto T$, which can also be written as $V = b \times T$, where b is a constant.

Avogadro's law: $V \propto n$, which can also be written as $V = c \times n$, where c is a constant.

We can combine these three equations, each of which describes the relationship between volume and another variable, to get

$$V = a \times \frac{1}{P} \times b \times T \times c \times n$$

Combining the constants to give a single constant, and calling the new constant R,

$$V = R \times \frac{nT}{P}$$

which we see, when we multiply both sides by P, is the same as Equation 8.1:

$$P \times V = R \times \frac{nT}{\cancel{P}} \times \cancel{P} \quad \text{or} \quad PV = nRT$$

CHECKPOINT–SECTION 8.4   The Gas Laws

8.4.1 Given $P_1 = 1.50$ atm, $V_1 = 37.3$ mL, and $P_2 = 1.18$ atm, calculate V_2. Assume that n and T are constant.

a) 0.0211 mL

b) 0.0341 mL

c) 29.3 mL

d) 12.7 mL

e) 47.4 mL

8.4.2 Given $T_1 = 21.5°C$, $V_1 = 50.0$ mL, and $T_2 = 316°C$, calculate V_2. Assume that n and P are constant.

a) 100 mL

b) 73.5 mL

c) 25.0 mL

d) 3.40 mL

e) 26.5 mL

8.4.3 At what temperature will a gas sample occupy 100.0 L if it originally occupies 76.1 L at 89.5°C? Assume constant P.

a) 276°C

b) 118°C

c) 203°C

d) 68.1°C

e) 99.6°C

8.4.4 Place the following samples of gas in order of increasing volume if they are compared at the same temperature and pressure.

    I. 10.0 g He   II. 10.0 g Ne   III. 10.0 g CO_2

a) I = II = III

b) III < II < I

c) I < II < III

d) III < I < II

e) II < I < III

8.4.5 Which diagram could represent the result of increasing the temperature and decreasing the external pressure on a fixed amount of gas in a balloon?

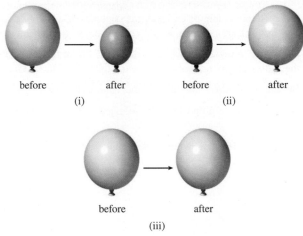

   before   after   before   after

       (i)   (ii)

      before   after

        (iii)

a) i only

b) ii only

c) i and ii

d) i, ii, and iii

e) iii only

8.4.6 Which diagram in question 8.4.5 could represent the result of decreasing both the temperature and the external pressure?

a) i only

b) ii only

c) i and ii

d) i, ii, and iii

e) iii only

8.5 Gas Mixtures

So far our discussion of the physical properties of gases has focused on the behavior of *pure* gaseous substances, even though the gas laws were all developed based on observations of samples of air, which is a *mixture* of gases. In this section, we consider gas mixtures and their physical behavior.

Dalton's Law of Partial Pressures

When two or more gaseous substances are placed in a container, each gas behaves as though it occupies the container alone. For example, if we place 1.00 mol of N_2 gas in a 5.00-L container at 0°C, it exerts a pressure that we can calculate using the ideal gas equation:

$$P = \frac{(1.00 \text{ mol})(0.0821 \text{ L} \cdot \text{atm/K} \cdot \text{mol})(273 \text{ K})}{5.00 \text{ L}} = 4.48 \text{ atm}$$

If we then add a mole of another gas, such as O_2, the pressure exerted by N_2 does not change. It remains at 4.48 atm. The O_2 exerts its own pressure, also 4.48 atm. Neither gas is affected by the presence of the other. In a mixture of gases, the pressure exerted by each gas is known as the **partial pressure (P_i)** of the gas, where the subscript identifies each individual component of the gas mixture.

$$P_{N_2} = \frac{(1.00 \text{ mol})(0.0821 \text{ L} \cdot \text{atm/K} \cdot \text{mol})(273 \text{ K})}{5.00 \text{ L}} = 4.48 \text{ atm}$$

$$P_{O_2} = \frac{(1.00 \text{ mol})(0.0821 \text{ L} \cdot \text{atm/K} \cdot \text{mol})(273 \text{ K})}{5.00 \text{ L}} = 4.48 \text{ atm}$$

Dalton's law of partial pressures states that the total pressure exerted by a gas mixture is the sum of the partial pressures exerted by each component of the mixture (Figure 8.13). In the case of a mixture consisting only of N_2 and O_2, the total pressure is the sum of the partial pressures of those two gases. Thus, the total pressure exerted by a mixture of 1.00 mol N_2 and 1.00 mol O_2 in a 5.00-L container at 0°C is

$$P_{\text{total}} = P_{N_2} + P_{O_2} = 4.48 \text{ atm} + 4.48 \text{ atm} = 8.96 \text{ atm}$$

Sample Problem 8.11 illustrates the use of Dalton's law of partial pressures.

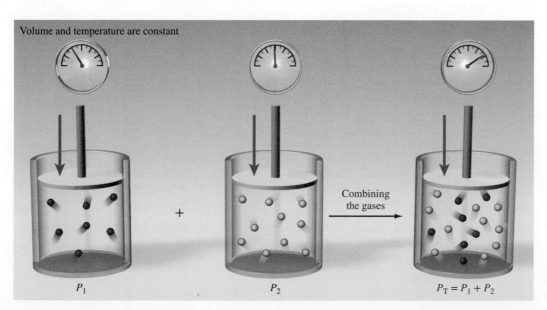

Volume and temperature are constant

P_1 + P_2 Combining the gases → $P_T = P_1 + P_2$

Figure 8.13 Dalton's law of partial pressures. Each component of a gas mixture exerts a pressure independent of the other components. The total pressure is the sum of the individual components' partial pressures.

SAMPLE PROBLEM (**8.11**)   Using Dalton's Law of Partial Pressures to Determine Total Pressure in a Gas Mixture

A 1.00-L vessel contains 0.215 mol of N_2 gas and 0.0118 mol of H_2 gas at 25°C. Determine the partial pressure of each component and the total pressure in the vessel.

Strategy Use the ideal gas equation to find the partial pressure of each component of the mixture, and sum the two partial pressures to find the total pressure.

Setup $T = 298$ K

Solution mol of H_2 = total moles − mol N_2 = 6.029 − 6.022 = 0.007 mol

$$P_{N_2} = \frac{(0.215 \text{ mol})\left(0.0821 \frac{L \cdot atm}{K \cdot mol}\right)(298 \text{ K})}{1.00 \text{ L}} = 5.26 \text{ atm}$$

$$P_{H_2} = \frac{(0.0118 \text{ mol})\left(0.0821 \frac{L \cdot atm}{K \cdot mol}\right)(298 \text{ K})}{1.00 \text{ L}} = 0.289 \text{ atm}$$

$$P_{total} = P_{N_2} + P_{H_2} = 5.26 \text{ atm} + 0.289 \text{ atm} = 5.55 \text{ atm}$$

THINK ABOUT IT

The total pressure in the vessel can also be determined by summing the number of moles of mixture components (0.215 + 0.0118 = 0.227 mol) and solving the ideal gas equation for P_{total}:

$$P_{total} = \frac{(0.227 \text{ mol})\left(0.0821 \frac{L \cdot atm}{K \cdot mol}\right)(298 \text{ K})}{1.00 \text{ L}} = 5.55 \text{ atm}$$

Practice Problem **A**TTEMPT Determine the partial pressures and the total pressure in a 2.50-L vessel containing the following mixture of gases at 16°C: 0.0194 mol He, 0.0411 mol H_2, and 0.169 mol Ne.

Practice Problem **B**UILD Determine the number of moles of each gas present in a mixture of CH_4 and C_2H_6 in a 2.00-L vessel at 25°C and 1.50 atm, given that the partial pressure of CH_4 is 0.39 atm.

Practice Problem **C**ONCEPTUALIZE The diagram represents a mixture of three different gases. The partial pressure of the gas represented by red spheres is 1.25 atm. Determine the partial pressures of the other gases and determine the total pressure.

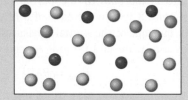

Mole Fractions

The relative amounts of the components of a gas mixture can be specified using mole fractions. The *mole fraction* (χ_i) of a component of a mixture is the number of moles of that component divided by the total number of moles (of *all* components) in the mixture:

Equation 8.4 $$\chi_i = \frac{n_i}{n_{total}}$$

There are three important things to remember about mole fractions:

1. The mole fraction of a mixture component is always less than 1.
2. The sum of mole fractions for all components of a mixture is always 1.
3. Mole fractions are dimensionless quantities—they have no units.

In addition, because of the proportional relationship between n and P at a specified temperature and volume, we can determine mole fraction of a component by dividing its partial pressure by the total pressure:

Equation 8.5 $$\chi_i = \frac{P_i}{P_{total}}$$

Like the other equations we have encountered in this chapter, Equations 8.4 and 8.5 can be manipulated to solve for any of the variables they contain. For example, if we know a component's mole fraction and the total pressure of the mixture, we can rearrange Equation 8.5 and solve for the component's partial pressure.

Sample Problem 8.12 lets you practice calculations involving mole fractions, partial pressures, and total pressures in gas mixtures.

Student Hot Spot

Student data indicate that you may struggle to perform calculations involving mole fractions and partial pressures. Access the eBook to view additional Learning Resources on this topic.

SAMPLE PROBLEM 8.12 Calculating Mole Fraction of a Gas-Mixture Component

In 1999, the FDA approved the use of nitric oxide (NO) to treat and prevent lung disease, which occurs commonly in premature infants. The nitric oxide used in this therapy is supplied to hospitals in the form of an N_2/NO mixture. Calculate the mole fraction of NO in a 10.00-L gas cylinder at room temperature (25°C) that contains 6.022 mol N_2 and in which the total pressure is 14.75 atm.

Strategy Use the ideal gas equation to calculate the total number of moles in the cylinder. Subtract moles of N_2 from the total to determine moles of NO. Divide moles NO by total moles to get mole fraction (Equation 8.5).

Setup The temperature is 298 K.

Solution Rearranging the ideal gas equation, $PV = nRT$, to solve for total moles gives:

$$n_{total} = \frac{P_{total} \cdot V}{RT} = \frac{14.75 \text{ atm} \cdot 10.00 \text{ L}}{\left(0.0821 \frac{L \cdot atm}{K \cdot mol}\right) \cdot 298 \text{ K}} = 6.029 \text{ mol}$$

mol NO = total moles − mol N_2 = 6.029 − 6.022 = 0.007 mol NO

$$\chi_{NO} = \frac{n_{NO}}{n_{total}} = \frac{0.007 \text{ mol NO}}{6.029 \text{ mol}} = 0.001$$

THINK ABOUT IT

To check your work, determine χ_{N_2} by subtracting χ_{NO} from 1. Using each mole fraction and the total pressure, calculate the partial pressure of each component using Equation 8.4 and verify that they sum to the total pressure.

Practice Problem A TTEMPT Determine the mole fractions and partial pressures of CO_2, CH_4, and He in a sample of gas that contains 0.250 mol of CO_2, 1.29 mol of CH_4, and 3.51 mol of He, and in which the total pressure is 5.78 atm.

Practice Problem B UILD Determine the partial pressure and number of moles of each gas in a 15.75-L vessel at 30°C containing a mixture of xenon and neon gases only. The total pressure in the vessel is 6.50 atm, and the mole fraction of xenon is 0.761.

Practice Problem C ONCEPTUALIZE A mixture of gases can be represented with red, yellow, and green spheres. The diagram shows such a mixture, but the green spheres are missing. Determine the number of green spheres missing, the mole fraction of yellow, and the mole fraction of green, given that the mole fraction of red is 0.28.

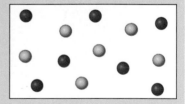

One place where Dalton's law of partial pressures is used is in the measurement of the volume of gas produced by a chemical process. This experimental technique is illustrated in Figure 8.14. As gas is produced in the test tube, it flows through the tubing and bubbles out under an inverted graduated cylinder that has been filled with water. The volume of gas collected is equal to the volume of water displaced. However, because the measured volume contains both the gas produced by the chemical process *and* water vapor, the pressure exerted inside the graduated cylinder is the *sum* of the two partial pressures:

$$P_{total} = P_{gas} + P_{water}$$

By subtracting the partial pressure of water from the total pressure, which is equal to atmospheric pressure, we can determine the partial pressure of the collected gas—and

Familiar Chemistry

Natural Gas

Many people in the United States live in homes where natural gas is used to heat the house, heat the water, and cook. Natural gas is predominantly methane, CH_4, which we first encountered in Section 6.4. Depending on the source of the natural gas, there are varying small amounts of other gases, including larger alkanes such as ethane, C_2H_6, and propane, C_3H_8. If you have lived in a home with natural gas, you are probably familiar with the telltale sign of a gas leak—a very unpleasant smell reminiscent of rotten eggs. But this odor is not that of methane. None of the hydrocarbons in natural gas has a detectable smell. In fact, the odor is the result of the addition of an "oderant," the main ingredient of which is a *mercaptan,* a sulfur-containing hydrocarbon. The smell of a mercaptan is extremely potent. It takes the addition of only a *tiny* amount for it to be detectable. Typically, natural gas contains less than one part per million of the smelly molecule—although it can actually be detected by humans at *far* smaller concentrations.

Jupiterimages/Photos.com/Getty Images

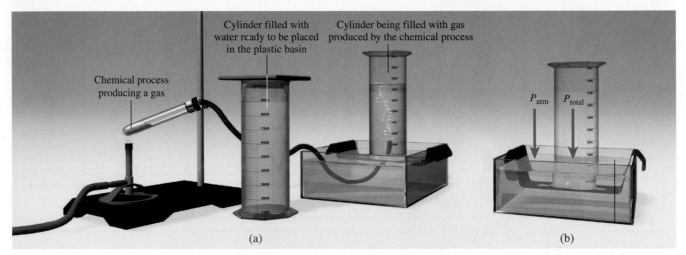

Figure 8.14 (a) Apparatus for measuring the amount of gas produced by a chemical process. (b) When the water levels inside and outside the collection vessel are the same, the pressure inside the cylinder is equal to atmospheric pressure.

thereby determine how many moles of gas were actually collected. We get the partial pressure of water, which depends on temperature, from a table of values. Table 8.4 lists the partial pressure of water at different temperatures.

Sample Problem 8.13 shows how to use Dalton's law of partial pressure to determine the amount of gas produced by a chemical process and collected over water.

TABLE 8.4	Vapor Pressure of Water (P_{H_2O}) as a Function of Temperature				
T (°C)	P (torr)	T (°C)	P (torr)	T (°C)	P (torr)
0	4.6	35	42.2	70	233.7
5	6.5	40	55.3	75	289.1
10	9.2	45	71.9	80	355.1
15	12.8	50	92.5	85	433.6
20	17.5	55	118.0	90	525.8
25	23.8	60	149.4	95	633.9
30	31.8	65	187.5	100	760.0

SAMPLE PROBLEM (8.13) Calculating the Mass of a Gaseous Product in a Chemical Reaction

When calcium metal is placed in water, hydrogen gas is produced. Determine the mass of H_2 produced at 25°C and 0.967 atm when 525 mL of the gas is collected over water as shown in Figure 8.14.

Strategy Use Dalton's law of partial pressures to determine the partial pressure of H_2, use the ideal gas equation to determine moles of H_2, and then use the molar mass of H_2 to convert to mass. (Pay careful attention to units. Atmospheric pressure is given in atmospheres, whereas the vapor pressure of water is tabulated in torr.)

Setup $V = 0.525$ L and $T = 298$ K. The partial pressure of water at 25°C is 23.8 torr (Table 8.4). $23.8 \text{ torr} \times \dfrac{1 \text{ atm}}{760 \text{ torr}} = 0.0313$ atm. The molar mass of H_2 is 2.016 g/mol.

Solution

$$P_{H_2} = P_{total} - P_{H_2O} = 0.967 \text{ atm} - 0.0313 \text{ atm} = 0.936 \text{ atm}$$

$$\text{mol } H_2 = \frac{(0.936 \text{ atm})(0.525 \text{ L})}{\left(0.0821 \dfrac{L \cdot atm}{K \cdot mol}\right)(298 \text{ K})} = 2.01 \times 10^{-2} \text{ mol}$$

$$\text{mass of } H_2 = (2.008 \times 10^{-2} \text{ mol})(2.016 \text{ g/mol}) = 0.0405 \text{ g } H_2$$

THINK ABOUT IT

Check unit cancellation carefully, and remember that the densities of gases are relatively low. The mass of approximately half a liter of hydrogen at or near room temperature and 1 atm should be a very small number.

Practice Problem **A**TTEMPT Calculate the mass of O_2 produced by the decomposition of $KClO_3$ when 821 mL of gas is collected over water at 30°C and 1.015 atm.

Practice Problem **B**UILD Determine the volume of gas collected over water when 0.501 g O_2 is produced by the decomposition of $KClO_3$ at 35°C and 1.08 atm.

Practice Problem **C**ONCEPTUALIZE The diagram on the top represents the result of an experiment in which the oxygen gas produced by a chemical reaction is collected over water at typical room temperature. Which of the diagrams [(i)–(iv)] best represents the result of the same experiment on a day when the temperature in the laboratory is significantly warmer?

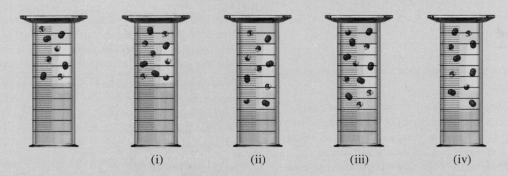

(i)   (ii)   (iii)   (iv)

CHECKPOINT–SECTION 8.5 Gas Mixtures

8.5.1 What is the partial pressure of He in a 5.00-L vessel at 25°C that contains 0.0410 mol of He, 0.121 mol of Ne, and 0.0922 mol of Ar?

a) 1.24 atm   d) 2.87 atm

b) 0.248 atm   e) 0.201 atm

c) 0.117 atm

8.5.2 What is the mole fraction of CO_2 in a mixture of 0.756 mol of N_2, 0.189 mol of O_2, and 0.0132 mol of CO_2?

a) 0.789   d) 1.003

b) 0.0138   e) 0.798

c) 0.0140

(Continued on next page)

8.5.3 What is the partial pressure of oxygen in a gas mixture that contains 4.10 mol of oxygen, 2.38 mol of nitrogen, and 0.917 mol of carbon dioxide and that has a total pressure of 2.89 atm?

a) 1.60 atm

d) 0.705 atm

b) 3.59 atm

e) 0.624 atm

c) 0.391 atm

8.5.4 In the diagram, each color represents a different gas molecule. Calculate the mole fraction of each gas.

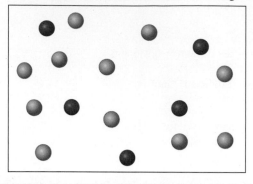

a) $\chi_{red} = 0.5$, $\chi_{blue} = 0.4$, $\chi_{green} = 0.7$

b) $\chi_{red} = 0.05$, $\chi_{blue} = 0.07$, $\chi_{green} = 0.07$

c) $\chi_{red} = 0.3125$, $\chi_{blue} = 0.25$, $\chi_{green} = 0.4375$

d) $\chi_{red} = 0.4167$, $\chi_{blue} = 0.333$, $\chi_{green} = 0.5833$

e) $\chi_{red} = 0.333$, $\chi_{blue} = 0.333$, $\chi_{green} = 0.333$

8.5.5 Calculate the partial pressure of each gas in the diagram in question 8.5.4 if the total pressure is 8.21 atm.

a) $P_{red} = 0.5$ atm, $P_{blue} = 0.25$ atm, $P_{green} = 4.5$ atm

b) $P_{red} = 2.57$ atm, $P_{blue} = 2.05$ atm, $P_{green} = 3.59$ atm

c) $P_{red} = 3.13$ atm, $P_{blue} = 2.50$ atm, $P_{green} = 2.58$ atm

d) $P_{red} = 2.74$ atm, $P_{blue} = 2.74$ atm, $P_{green} = 2.74$ atm

e) $P_{red} = 3.125$ atm, $P_{blue} = 2.500$ atm, $P_{green} = 4.375$ atm

Chapter Summary

Section 8.1

- Substances with intermolecular forces too weak to hold the molecules together are gases.

- Gases assume both the shape and volume of their containers, are compressible, have low densities, and form homogeneous mixtures with other gases.

- The **kinetic molecular theory** relates the molecular nature of gases to their macroscopic behaviors. Its four essential assumptions are that gas molecules are separated by relatively large distances; are in constant, random motion and collide with one another and the walls of their container; exhibit no intermolecular forces; and have average kinetic energy proportional to absolute temperature.

Section 8.2

- **Pressure** is defined as force exerted per unit area. Gas pressure is the result of collisions of gas molecules. The SI unit of force is the **newton (N),** and the SI unit of pressure is the **pascal (Pa).** 1 Pa = 1 N/m².

- Other commonly used units of pressure include the atmosphere (atm), millimeters of mercury (mmHg), pounds per square inch (psi), and bar.

- Pressure can be measured using a **barometer** or a **manometer.** The pressure exerted by a column of liquid depends on the height of the liquid, its density, and the gravitational constant.

Section 8.3

- The ideal gas equation relates the pressure, volume, quantity (number of moles), and absolute temperature of a sample of gas using the **ideal gas constant (R).** We can solve for any of the four variables (P, V, n, or T) if we know the other three.

- The **combined gas equation** relates the pressure, volume, and temperature of a fixed quantity of gas (n is constant) under two different sets of conditions. Like the ideal gas equation, the combined gas equation can be solved for any of the variables it contains.

- The ideal gas equation can be manipulated to solve for the density of a gas, if its molar mass is known; or to solve for the molar mass of a gas, if its density is known.

Section 8.4

- The gas equations are derived from the gas laws, which were determined empirically during the seventeenth, eighteenth, and early nineteenth centuries.

- **Boyle's law** says that the volume of a sample of gas at constant temperature is inversely proportional to pressure: $V \propto 1/P$.

- **Charles's law** says that the volume of a sample of gas at constant pressure is directly proportional to temperature: $V \propto T$.

- **Avogadro's law** says that the volume of a sample of gas at constant temperature and pressure is directly proportional to the number of moles: $V \propto n$.

Section 8.5

- **Dalton's law of partial pressures** says that gases in a mixture behave independently of one another. Each component of a gaseous mixture exerts the same pressure that it would if the other components of the mixture were not present. The pressure exerted by a gaseous component of a mixture is called the **partial pressure (P_i)**.

- The relative amounts of components in a gas mixture can be expressed in terms of mole fraction. The **mole fraction (χ_i)** of

a gas in a mixture is the number of moles of that gas divided by the total number of moles of gas in the mixture: $\chi_i = n_i/n_{total}$. Mole fraction can also be calculated using partial pressures: $\chi_i = P_i/P_{total}$.

- When gas is collected over water using an inverted graduated cylinder, the partial pressure of water must be subtracted from the total pressure to determine the pressure of the collected gas.

Key Terms

Absolute zero 296	Combined gas equation 289	Kinetic molecular theory 279	Pascal (Pa) 281
Avogadro's law 298	Dalton's law of partial	Manometer 283	Pressure 281
Barometer 283	pressures 301	Mole fraction (χ_i) 302	STP 288
Boyle's law 294	Ideal gas constant 285	Newton (N) 281	Torr 283
Charles's law 296	Ideal gas equation 285	Partial pressure (P_i) 301	

Key Equations

8.1	$PV = nRT$	The ideal gas equation. Four variables of a gas (pressure, volume, number of moles, and absolute temperature) are related by the ideal gas constant (R). We can solve for any of the variables as long as the others are known. Generally, P is expressed in atm, and V is expressed in L. T is always expressed in K, and R is 0.0821 L · atm/K · mol.
8.2	$\dfrac{P_1V_1}{T_1} = \dfrac{P_2V_2}{T_2}$	The combined gas equation. Three variables (pressure, volume, and absolute temperature) of a fixed quantity of gas (n is constant) are related under two different sets of conditions. We can calculate the change in any of the variables as long as we know the changes in the other variables. Any units of pressure and volume can be used—as long as they are consistent; but temperature *must* be expressed as K.
8.3	$\mathcal{M} = \dfrac{dRT}{P}$	The ideal gas equation can be solved for molar mass. We can determine the molar mass of a gas from its experimentally determined density. Density is expressed as g/L; R is 0.0821 L · atm/K · mol; pressure is expressed as atm; and temperature, as always, must be expressed as K.
8.4	$\chi_i = \dfrac{n_i}{n_{total}}$	The relative amounts of the components of a gaseous mixture can be expressed using mole fractions. The mole fraction of an individual component, (χ_i), is the number of moles of the component, n_i, divided by the total number of moles in the mixture n_{total}. Mole fractions are dimensionless quantities, meaning that they have no units.
8.5	$\chi_i = \dfrac{P_i}{P_{total}}$	The mole fraction of an individual component in a gaseous mixture can also be calculated using partial pressures. The mole fraction of an individual component, (χ_i), is the partial pressure of the component, P_i, divided by the total pressure of the mixture P_{total}.

Mole Fractions

Most of the gases that we encounter are mixtures of two or more different gases. The concentrations of gases in a mixture are typically expressed using mole fractions, which are calculated using Equation 8.4:

$$\chi_i = \frac{n_i}{n_{\text{total}}}$$

Depending on the information given in a problem, calculating mole fractions may require you to determine molar masses and carry out mass-to-mole conversions [◄◄ Section 5.2].

For example, consider a mixture that consists of known masses of three different gases: 5.50 g He, 7.75 g N_2O, and 10.00 g SF_6. Molar masses of the components are

He: $4.003 = \boxed{\dfrac{4.003\ \text{g}}{\text{mol}}}$ N_2O: $2(14.01) + (16.00) = \boxed{\dfrac{44.02\ \text{g}}{\text{mol}}}$ SF_6: $32.07 + 6(19.00) = \boxed{\dfrac{146.1\ \text{g}}{\text{mol}}}$

We convert each of the masses given in the problem to moles by dividing each by the corresponding molar mass:

$\dfrac{5.50\ \text{g He}}{4.003\ \text{g/mol}} = 1.374\ \text{mol He}$ $\dfrac{7.75\ \text{g } N_2O}{44.02\ \text{g/mol}} = 0.1761\ \text{mol } N_2O$ $\dfrac{10.00\ \text{g } SF_6}{146.1\ \text{g/mol}} = 0.06846\ \text{mol } SF_6$

We then determine the total number of moles in the mixture:

$1.374\ \text{mol He} + 0.1761\ \text{mol } N_2O + 0.06846\ \text{mol } SF_6 = 1.619\ \text{mol}$

We divide the number of moles of each component by the total number of moles to get each component's mole fraction.

$\chi_{He} = \dfrac{1.374\ \text{mol He}}{1.619\ \text{mol}} = 0.849$ $\chi_{N_2O} = \dfrac{0.1761\ \text{mol } N_2O}{1.619\ \text{mol}} = 0.109$ $\chi_{SF_6} = \dfrac{0.06846\ \text{mol } SF_6}{1.619\ \text{mol}} = 0.0423$

The resulting mole fractions have no units; and for any mixture, the sum of mole fractions of all components is 1. Rounding error may result in the overall sum of mole fractions not being exactly 1. In this case, to the appropriate number of significant figures [◄◄ Section 4.3], the sum is 1.00. (Note that we kept an extra digit throughout the calculations.)

Because at a given temperature, pressure is proportional to the number of moles, mole fractions can also be calculated using the partial pressures of the gaseous components using Equation 8.5:

$$\chi_i = \frac{P_i}{P_{\text{total}}}$$

Although we have learned to determine mole fractions in gaseous mixtures, they can also be determined for mixtures containing liquids and solids. When liquids are involved, it is typically necessary to convert from a given volume (using the liquid's density) to mass, and then to moles (using molar mass).

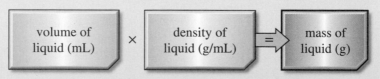

volume of liquid (mL) × density of liquid (g/mL) = mass of liquid (g)

Consider the following example: 5.75 g of sugar (sucrose, $C_{12}H_{22}O_{11}$) is dissolved in 100.0 mL of water at 25°C. We first determine the molar masses of sucrose and water.

$$H_2O: 2(1.008) + 16.00 = \boxed{\dfrac{18.02\ g}{mol}} \qquad C_{12}H_{22}O_{11}: 12(12.01) + 11(16.00) = \boxed{\dfrac{342.3\ g}{mol}}$$

Then we use the density of water to convert the volume given to a mass. The density of water at 25°C is 0.9970 g/mL.

$$100.0\ mL\ H_2O \quad \times \quad \dfrac{0.9970\ g}{mL} \quad = \quad 99.70\ g\ H_2O$$

We convert the masses of both solution components to moles:

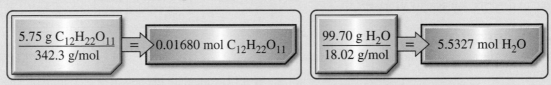

$$\dfrac{5.75\ g\ C_{12}H_{22}O_{11}}{342.3\ g/mol} = 0.01680\ mol\ C_{12}H_{22}O_{11} \qquad \dfrac{99.70\ g\ H_2O}{18.02\ g/mol} = 5.5327\ mol\ H_2O$$

We then sum the number of moles and divide moles of each component by the total.

$$0.01680\ mol\ C_{12}H_{22}O_{11} + 5.5327\ mol\ H_2O = 5.5495\ mol$$

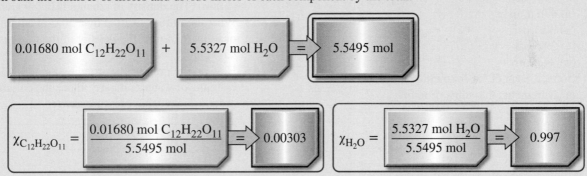

$$\chi_{C_{12}H_{22}O_{11}} = \dfrac{0.01680\ mol\ C_{12}H_{22}O_{11}}{5.5495\ mol} = 0.00303 \qquad \chi_{H_2O} = \dfrac{5.5327\ mol\ H_2O}{5.5495\ mol} = 0.997$$

To the appropriate number of significant figures, the mole fractions sum to 1.

Key Skills Problems

8.1
Determine the mole fraction of helium in a gaseous mixture consisting of 0.524 g He, 0.275 g Ar, and 2.05 g CH$_4$.
a) 0.0069
b) 0.0259
c) 0.481
d) 0.493
e) 0.131

8.2
Determine the mole fraction of argon in a gaseous mixture in which the partial pressures of H$_2$, N$_2$, and Ar are 0.01887 atm, 0.3105 atm, and 1.027 atm, respectively.
a) 0.01391
b) 0.2289
c) 0.7572
d) 0.01887
e) 1.027

8.3
Determine the mole fraction of *water* in a solution consisting of 5.00 g glucose (C$_6$H$_{12}$O$_6$) and 250.0 g water.
a) 0.00200
b) 0.998
c) 0.0278
d) 1.00
e) 0.907

8.4
Determine the mole fraction of ethanol in a solution containing 15.50 mL ethanol (C$_2$H$_5$OH) and 110.0 mL water. (The density of ethanol is 0.789 g/mL; the density of water is 0.997 g/mL.)
a) 0.0436
b) 6.08
c) 0.265
d) 0.958
e) 0.0418

Questions and Problems

SECTION 8.1: PROPERTIES OF GASES

8.1   Compare the physical properties of a gas with those of a substance in a condensed phase.

8.2   List the elements that exist as gases at room temperature.

8.3   Explain each of the four assumptions of the kinetic molecular theory.

8.4   What happens to a gas if you heat it?

8.5   Why is the density of a gas so much lower than that of a solid or liquid? What units are typically used to describe the density of gases?

SECTION 8.2: PRESSURE

8.6   Define pressure, and give the most common units of pressure.

8.7   Convert each of the following pressure measurements into units of atmospheres.
(a) 475 mmHg
(b) 32.1 psi
(c) 3.85 bar
(d) 744 torr

8.8   Convert each of the following pressure measurements into units of mmHg.
(a) 9.01 atm
(b) 10.8 psi
(c) 635 torr
(d) 1.71 kPa

8.9   Convert each of the following pressure measurements into units of torr.
(a) 13.6 psi   (c) 2.08 bar
(b) 5.5×10^{-3} kPa   (d) 11.7 atm

8.10   The air pressure in Death Valley, CA, the lowest point in North America, can be as high as 33.5 inches of mercury. Convert this pressure to each of the following units:
(a) mmHg
(b) atm
(c) Pa
(d) psi
(e) torr
(f) bar

8.11   The air pressure at the top of Denali, the highest mountain peak in North America, typically is near 572 mmHg. Convert this pressure to the following units:
(a) inches Hg
(b) atm
(c) Pa
(d) psi
(e) torr
(f) bar

8.12   Perform the necessary unit conversions and complete the following table.

psi	Pa	kPa	atm	mmHg	in Hg	torr	bar
28.6							
			1.22				
							2.94

8.13   Perform the necessary unit conversions and complete the following table.

psi	Pa	kPa	atm	mmHg	in Hg	torr	bar
	245,289						
				895			
						544	

8.14   Calculate the height of a column of ethanol that would be supported by atmospheric pressure (1 atm). The density of ethanol is 0.789 g/cm³.

8.15   What pressure is exerted by a 50.0-m column of water? Assume the density of the water is 1.00 g/cm³.

8.16   How do the pressures exerted by the following columns of water compare? Place the columns in order of increasing pressure exerted.

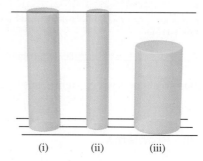

(i)   (ii)   (iii)

SECTION 8.3: THE GAS EQUATIONS

8.17   Write the ideal gas equation.

8.18   What units of temperature must be used in the ideal gas equation? Explain.

8.19   Rearrange the ideal gas equation algebraically to solve for volume.

8.20   Rearrange the ideal gas equation algebraically to solve for pressure.

8.21   Rearrange the ideal gas equation algebraically to solve for temperature.

8.22   Rearrange the ideal gas equation algebraically to solve for moles.

8.23   Does the ideal gas equation depend on the identity of the gas? Explain your answer.

8.24   Explain why a hot air balloon floats.

8.25   Determine the volume (in L) of a 0.235-mol ideal gas sample at 1.10 atm and 25°C.

8.26 Determine the volume (in L) of a 1.35-mol ideal gas sample at 1.59 atm and 125°C.

8.27 Determine the volume (in L) of each of the following gas samples at 811 mmHg and 373 K.
(a) 1.33 mol He
(b) 6.88 g He
(c) 35.0 g O_2

8.28 Determine the volume (in L) of each of the following gas samples at 639 torr and 38°C.
(a) 65.5 g CH_4
(b) 81.2 g CO_2
(c) 29.6 g OF_2

8.29 Determine the volume (in L) of 1.54 mol of an ideal gas under each set of conditions.
(a) 695 torr and 85°C
(b) 735 mmHg and 35°C
(c) 23.9 psi and 48°C

8.30 Determine the volume (in L) of each of the following gas samples at STP.
(a) 30.7 g Ar
(b) 12.4 g CO
(c) 8.56 mol Cl_2
(d) 4.13×10^{23} molecules SO_2

8.31 Determine the pressure (in atm) of each of the following gas samples in a 12.3-L container at 35°C.
(a) 1.22 mol SO_2
(b) 2.44 mol Kr
(c) 3.66 mol N_2

8.32 Determine the pressure (in atm) of each of the following gas samples in a 22.4-L container at 373 K.
(a) 1.55 g H_2
(b) 2.88 g He
(c) 3.27 mol OCl_2
(d) 5.83×10^{23} molecules O_2

8.33 Determine the pressure (in atm) that a 23.7-g sample of NH_3 would exert in the following containers at the temperatures given.
(a) 13.2-L container at 45°C
(b) 1244-mL container at 24°C
(c) 2455-cm^3 container at 37°C

8.34 Determine the pressure (in atm) that a 10.0-mol sample of Ar would exert in each of the following containers at standard temperature.
(a) 10.0-L container
(b) 20.0-L container
(c) 30.0-L container

8.35 Determine the temperature (in K) of each gas sample under the given conditions.
(a) 4.39 mol Ar in a 10.5-L container at 1.76 atm
(b) 2.66 mol C_2H_6 in a 73.4-L container at 1.93 atm
(c) 12.7 g O_2 in a 33.7-L container at 1.57 atm

8.36 Determine the temperature (in °C) of each gas sample under the given conditions.
(a) 1.22 mol He in a 27.3-L container at 1.05 atm
(b) 1.94 mol CH_4 in a 13.6-L container at 1.43 atm
(c) 2.48 mol CO_2 in a 12.9-L container at 1.66 atm

8.37 Determine the temperature (in °C) of a 1.32-mol gas sample in a 22.0-L container at the following pressures.
(a) 1.22 atm   (c) 4.88 atm
(b) 2.44 atm

8.38 Determine the temperature (in K) of each gas sample given that it is placed in a 15.5-L container at standard pressure.
(a) 0.735 mol Cl_2 (c) 23.8 g He
(b) 23.8 g SF_6

8.39 Determine the number of moles of gas present in each of the following samples under the specified conditions.
(a) 2.33 atm at 355 K in a 22.4-L container
(b) 1.12 atm at 298 K in a 9.77-L container
(c) 1.84 atm at 469 K in a 17.6-L container

8.40 Determine the number of moles of gas present in each of the following samples under the specified conditions.
(a) 407 torr at 15°C in a 5.00-L container
(b) 42.0 psi at 285 K in a 15.0-L container
(c) 2.52 bar at 225°C in a 10.0-L container

8.41 Determine the number of moles of gas present in a 44.0-L container under the specified conditions.
(a) 175 kPa at 27.9°C
(b) 853 mmHg at 40.1°C
(c) 37.2 psi at 134°C

8.42 Determine the number of moles of gas present at STP in containers with the following volumes.
(a) 789 mL   (c) 27.8 L
(b) 2.78 L

8.43 Refer to the diagram that shows heavy gas molecules (larger) and light gas molecules (smaller) in identical containers at the same temperature. Identify the gas sample(s) that has/have the

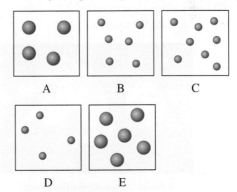

A   B   C

D   E

(a) highest pressure
(b) lowest pressure

8.44 A 2.11-g sample of gas exerts a pressure of 1.08 atm in a 225-mL container at 298 K. Determine the molar mass of the gas.

8.45 A 0.943-g sample of gas exerts a pressure of 1.13 atm in a 495-mL container at 45°C. Determine the molar mass of the gas.

8.46 Determine the molar mass of a gas with a density of 1.905 g/L at 80.0°C and 1.00 atm.

8.47 Determine the molar mass of a gas with a density of 6.52 g/L at STP.

8.48 Determine the density of a gas with a molar mass of 146.07 g/mol at 1.00 atm and 100.0°C.

8.49 Determine the density of a gas with a molar mass of 70.1 g/mol at 1.33 atm and 175.0°C.

8.50 A company claims to have developed a new synthetic "alcohol" for nonalcoholic beverages. The chemical formula is proprietary. You analyze a sample of the new product by placing a small volume of it in a round-bottomed flask with a volume of 511.1 mL and a mass of 131.918 g when empty. You submerge the flask in a water bath at 100.0°C and allow the volatile liquid to vaporize. You then cap the flask and remove it from the water bath. You reweigh it and determine the mass of the vapor in the flask to be 0.768 g. What is the molar mass of the volatile liquid? (Assume the pressure in the laboratory is 1.00 atm.) Comment on your result.

8.51 A sample of the volatile liquid propyl acetate ($C_5H_{10}O_2$) is analyzed using the procedure and equipment described in Problem 8.50. What will the mass of the 511.1-mL flask be after evaporation of the propyl acetate?

8.52 Wish lanterns have become a popular addition to wedding celebrations. The rice paper lantern has a small candle attached in the opening, and when it is lit, it warms the air inside the lantern, causing it to rise. Calculate the difference in the density of the gas contained inside the lantern before the candle is lit ($T = 298$ K, $P = 1.00$ atm, $V = 35.0$ L) and after the candle is lit ($T = 348$ K, $P = 1.00$ atm, $V = 35.0$ L). The average molar mass of dry air is 28.97 g/mol. Remember that the pressure inside and outside the lantern remains constant because it is not a sealed container.

Guillermo Vazquez/EyeEm/Getty Images

SECTION 8.4: THE GAS LAWS

8.53 Which of the following samples of gas will exert the highest pressure in a 10.0-L container at 298 K?
(a) 20.0 g H_2
(b) 20.0 g He
(c) 20.0 g Ne
(d) 20.0 g CH_4

8.54 Write the equation for Boyle's law and explain what it describes. What conditions are required for Boyle's law to apply?

8.55 A sample of gas occupies 2.94 L at a pressure of 794 mmHg. Determine the new pressure of the sample when the volume expands to 3.88 L at constant temperature.

8.56 A sample of gas occupies 6.57 L at a pressure of 1.75 atm. Determine the new pressure of the sample when the volume contracts to 1.63 L at constant temperature.

8.57 Write the equation for Charles's law and explain what it describes. What conditions are necessary for Charles's law to apply?

8.58 A latex balloon has a volume of 3.00 L at 298 K. To what volume will the balloon expand if it gets stuck over a heating vent that causes its temperature to increase to 308 K?

8.59 A balloon has a volume of 3.00 L at 298 K. To what volume will the balloon shrink if its temperature drops to 273 K?

8.60 Write the equation for Avogadro's law and explain what it describes. What conditions are necessary for Avogadro's law to apply?

8.61 Use equations to show how the combined gas law can be used to solve Boyle's law problems. State any assumptions that must be made.

8.62 Use equations to show how the combined gas law can be used to solve Charles's law problems. State any assumptions that must be made.

8.63 Why is it a bad idea for scuba divers to hold their breath as they rapidly ascend to the water's surface?

8.64 Newer cars have tire pressure measuring system (TPMS) indicators that sometimes light up on the first cold day of the season. Often, the light will go off after a bit of driving. The TPMS indicator light in a specific car model comes on when the pressure in a tire falls below 28.8 psi. At what temperature will the TPMS indicator light come on if the pressure in the tires is 30.0 psi at 273 K? (Assume the volume of the tire stays constant.)

David A. Tietz/McGraw Hill

SECTION 8.5: GAS MIXTURES

8.65 Explain what is meant by the term *partial pressure*. When is partial pressure important?

8.66 The box on the left represents a container in which the total pressure is 1.00 atm. Fill in the boxes labeled He and Ar to represent each component of the gas mixture. What pressure does the helium exert? What pressure does the argon exert?

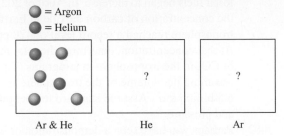

● = Argon
● = Helium

Ar & He   He   Ar

8.67 The box on the left represents a container in which the total pressure is 2.00 atm. Fill in the boxes labeled H_2 and O_2 to represent each component of the gas mixture. What pressure does the hydrogen exert? What pressure does the oxygen exert?

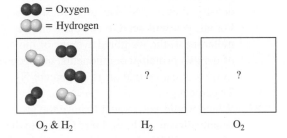

●● = Oxygen
○○ = Hydrogen

O_2 & H_2 H_2 O_2

8.68 The boxes on the right represent containers of H_2 and O_2 in which the pressures are 2.00 atm and 0.50 atm, respectively. Draw the O_2 and H_2 molecules in the empty box on the left to represent a mixture of these two gas samples. What is the total pressure of the gas mixture?

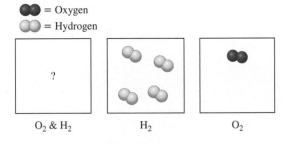

●● = Oxygen
○○ = Hydrogen

O_2 & H_2 H_2 O_2

8.69 The box on the right represents a sample of F_2 at 0.50 atm. The middle box represents a sample of neon at the same temperature—in a container of the same volume. Draw the molecules in the empty box to represent a mixture of these two gas samples. What is the pressure of the neon? What is the total pressure of the gas mixture?

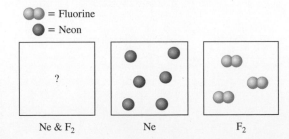

○○ = Fluorine
● = Neon

Ne & F_2 Ne F_2

8.70 Determine the mole fraction of each gas in the mixture shown in Problem 8.66.

8.71 Determine the mole fraction of each gas in the mixture shown in Problem 8.67.

8.72 Determine the mole fraction of each gas in the mixture drawn in Problem 8.68.

8.73 Determine the mole fraction of each gas in the mixture drawn in Problem 8.69.

8.74 Consider the three containers shown, all of which have the same volume and are at the same temperature. (a) Which container has the smallest mole fraction of gas A (red)? (b) Which container has the highest partial pressure of gas B (green)? (c) Which container has the highest total pressure?

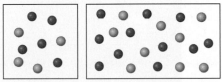

(i)   (ii)   (iii)

8.75 The volume of the box on the right is twice that of the box on the left. The boxes contain helium atoms (red) and hydrogen molecules (green) at the same temperature. (a) Which box has a higher total pressure? (b) Which box has a higher partial pressure of helium?

8.76 A gas mixture is composed of 2.55 mol of SCl_2 and 3.89 mol of NO_2 in a 15.2-L container at 398 K.
(a) Determine the partial pressure of SCl_2 in the mixture.
(b) Determine the partial pressure of NO_2 in the mixture.
(c) Determine the total pressure of the mixture.
(d) Determine the mole fraction of SCl_2 and the mole fraction of NO_2.

8.77 A gas mixture is composed of 24.8 g of Ar and 24.8 g of N_2 in a 31.3-L container at 298 K.
(a) Determine the partial pressure of argon in the mixture.
(b) Determine the partial pressure of nitrogen in the mixture.
(c) Determine the total pressure of the mixture.
(d) Determine the mole fraction of Ar and the mole fraction of N_2.

8.78 Determine the partial pressure of Cl_2 in a gas mixture containing Cl_2 and F_2. The gas mixture is in an 8.55-L container at STP. The mole fraction (χ) of F_2 is 0.22.

8.79 Determine the partial pressure of NCl_3 in a gas mixture containing NCl_3 and Cl_2. The gas mixture is in a 25.0-L container at 25°C and 1.55 atm. The mole fraction (χ) of NCl_3 is 0.69.

8.80   A popular demonstration is performed in which a small amount of liquid water is placed in the otherwise empty soda can as shown in the top photo. The water is then heated until steam is observed escaping the can. The can is then placed, opening down, into a beaker of cold water where it implodes. Explain why there must be water in the can for this demonstration to work.

David A. Tietz/McGraw Hill

8.81   With the advent of the Industrial Revolution, during the first half of the nineteenth century, carbon dioxide emissions resulting from the burning of fossil fuels began to increase. In April of 2021, the concentration of carbon dioxide in Earth's troposphere reached a record level of 420 ppm. At that concentration, determine the total mass of CO_2 in the troposphere in teragrams (Tg), assuming the volume of the troposphere is 6.5 billion km^3. Assume standard temperature and pressure.

8.82   Perhaps you have seen a demonstration of how inhaling helium makes a person's voice squeaky and high-pitched. This is because the density of He gas is significantly lower than that of air. Sulfur hexafluoride is a gas with a density significantly *higher* than that of air. Inhaling SF_6 gas will cause your voice to deepen. Calculate the density of each gas—He, air, and SF_6—at STP. For air, you will need to calculate an *effective* molar mass, the weighted average of molar masses of the two principal components, nitrogen and oxygen. Assume that air is 79 percent N_2 and 21 percent O_2.

8.83   Which would you expect to have the greater density, dry air or humid air? Calculate the density of each at STP, assuming that dry air is 79 percent nitrogen and 21 percent oxygen, whereas humid air is 69 percent nitrogen, 11 percent oxygen, and 20 percent water vapor. You will have to calculate an *effective* molar mass, the weighted average of molar masses, for each gas mixture.

Answers to In-Chapter Materials

Answers to Practice Problems

8.1A (a) 2.68 atm, (b) 1.33×10^3 torr, (c) 4.88×10^3 mmHg.
8.1B 1 psi = 6.89×10^3 Pa, 1 kPa = 1000 Pa, 1 atm = 101,325 Pa,
1 in Hg = 3.387×10^3 Pa, 1 mmHg = 133 Pa, 1 torr = 133 Pa,
1 bar = 1×10^5 Pa. **8.2A** 128 L. **8.2B** 336°C. **8.3A** 19.6 atm.
8.3B 9.66 atm. **8.4A** 1.75 mol. **8.4B** 15.9 g. **8.5A** 1.62×10^3 K.
8.5B −201°C. **8.6A** 5.10 L. **8.6B** 93.6°C. **8.7A** 1.286 g/L.
8.7B 11 atm. **8.8A** 2.23 L. **8.8B** 1.80 L. **8.9A** 30.7 L. **8.9B** 300°C.
8.10A 16.8 L. **8.10B** 37.3 L. **8.11A** P_{He} = 0.184 atm,
P_{H_2} = 0.390 atm, P_{Ne} = 1.60 atm, P_{total} = 2.18 atm.

8.11B 0.032 mol CH_4, 0.0907 mol C_2H_6. **8.12A** χ_{CO_2} = 0.0495,
P_{CO_2} = 0.286 atm; χ_{CH_4} = 0.255, P_{CH_4} = 1.48 atm; χ_{He} = 0.695,
P_{He} = 4.02 atm. **8.12B** P_{Xe} = 4.95 atm, n_{Xe} = 3.13 mol;
P_{Ne} = 1.55 atm, n_{Ne} = 0.984 mol. **8.13A** 1.03 g O_2. **8.13B** 386 mL.

Answers to Checkpoints

8.2.1 d. **8.2.2** c. **8.2.3** a. **8.2.4** b, c, d. **8.3.1** a. **8.3.2** e. **8.3.3** b.
8.3.4 b. **8.4.1** e. **8.4.2** a. **8.4.3** c. **8.4.4** b. **8.4.5** b. **8.4.6** d. **8.5.1** e.
8.5.2 b. **8.5.3** a. **8.5.4** c. **8.5.5** b.

CHAPTER 9

Physical Properties of Solutions

Fruit-flavored drinks such as Kool-Aid® are aqueous solutions. Concentrated drops for preparing these colorful drinks have grown very popular in recent years. The difference between the commercially available drops and the prepared beverages is simply concentration.

Brian Rayburn/McGraw Hill

9.1   General Properties of Solutions

9.2   Aqueous Solubility

9.3   Solution Concentration
- Percent by Mass
- Molarity
- Molality
- Comparison of Concentration Units

9.4   Solution Composition

9.5   Solution Preparation
- Preparation of a Solution from a Solid
- Preparation of a More Dilute Solution from a Concentrated Solution

9.6   Colligative Properties
- Freezing-Point Depression
- Boiling-Point Elevation
- Osmotic Pressure

Most of our planet is water, but most water is not pure. Instead, it has other substances dissolved in it. Seawater, for example, contains dissolved salts, including sodium chloride—as well as other dissolved substances. The water we drink, although sometimes labeled "pure," also contains small amounts of dissolved substances. Water that contains dissolved substances is an example of a homogeneous mixture [◀◀ Section 3.1]. In this chapter, we explore the properties of homogeneous mixtures, which are also known as *solutions*.

9.1 General Properties of Solutions

We learned about homogeneous mixtures in Section 3.1, and that they consist of two or more substances—each of which retains its individual chemical identity. The term *solution* is simply another word for *homogeneous mixture,* but we now introduce specific names for the mixture components. *Solvent* refers to the component present in greatest amount. *Solute* refers to a substance *dissolved* in the solvent. Every solution consists of *one* solvent, and one or *more* solutes. Using salt water to illustrate: water is the *solvent* and salt (NaCl) is the *solute*. Solutions such as this, in which *water* is the solvent, are called *aqueous* solutions. Figure 9.1 includes several familiar aqueous solutions that you may have encountered.

Water is a very common solvent, so common that it is sometimes called the *universal* solvent, but there are many others. Some of the other solvents are liquids; such as alcohol in a tincture, or oil in an essential-oil preparation. Some are solids; such as silver in sterling silver, or copper in brass. And some are gases; such as nitrogen in air, or helium in Heliox®—compressed gas for scuba diving. So, although most familiar examples of solutions are liquids, the term solution can refer to any homogeneous

Student Note: Metal solutions are known as *alloys*.

Figure 9.1 Familiar products that are aqueous solutions.
Brian Rayburn/McGraw Hill

Familiar Chemistry

Honey – A Supersaturated Solution

You may have enjoyed honey on a slice of toast or in a cup of hot tea. If you have a pantry, you probably have a jar of honey. You may also have seen a jar of honey begin to "crystallize" if it has been in your pantry for a long period of time. What was once a clear, golden liquid turns into a grainy, unsightly mixture that can be difficult to dispense. Although it may be inconvenient, crystallization does not mean that the honey has spoiled. Restoring the honey's texture and appearance simply requires some knowledge of the properties of solutions.

Bees produce honey as a sort of emergency nourishment source—feeding on it when nectar and pollen are scarce. During the seasons when plants are in bloom, worker bees tirelessly collect nectar from blossoms and process it to produce an aqueous solution of several different sugars. The bees then fan the sweet solution with their wings to evaporate most of the water, rendering the final product less than 20 percent water by mass. As the percentage of water decreases, the solution becomes

supersaturated in sugars, making it viscous and naturally resistant to bacterial spoilage.

Bee hives are typically hubs of activity and are very warm. The elevated temperature in a hive helps to stabilize the supersaturated sugar solution. When we remove the honey to a cooler location, such as a pantry, the solution may begin to crystallize. Because the aqueous solubility of sugars increases with increasing temperature, we can redissolve the crystallized sugars simply by warming the honey.

James Ross/Getty Images

mixture, and a solution may consist of nearly any combination of phases. Table 9.1 lists and gives examples of the various combinations of phases that constitute solutions.

Despite the many possible combinations that can constitute solutions, *aqueous* solutions are especially important—in part because of their importance in biological systems. For this reason, most of the rest of this chapter focuses on *aqueous* solutions.

In most cases, there is a limit to how much of a substance we can dissolve in a given amount of water. For example, if we were to add 75 grams of sodium acetate ($NaC_2H_3O_2$) to a beaker containing 100 mL water at room temperature (25°C) and stir, all of the sodium acetate would dissolve. (The resulting solution would resemble water.) However, if we were to add 150 grams of sodium acetate to 100 mL water at 25°C, no matter how long we stirred, it would not all dissolve. In fact, only 121 g of sodium acetate will dissolve in this volume of water at room temperature. The rest will remain undissolved at the bottom of the beaker. The first solution, in which *all* of the solute dissolves, is known as an ***unsaturated*** solution. An unsaturated solution is one in which more solute *could* be dissolved. The second, in which some of the solute remains *undissolved,* is known as a ***saturated*** solution. Further, the amount of solute dissolved in a saturated solution is called the ***solubility.*** Thus, the solubility of sodium acetate in water at 25°C is 121 g/100 mL. Note that the *solute,* the *solvent,* and the *temperature* are all specified. The solubility of sodium acetate in a different solvent, or at a different temperature, would be a different number of grams.

In certain circumstances, it is possible to prepare a solution that contains *more* dissolved solute than a saturated solution. Such solutions are known as ***supersaturated.*** Supersaturated solutions can be tricky to prepare, and they are typically unstable. They easily revert to saturated solutions, with the excess solute coming out of solution. Consider

Student Note: The process of coming out of solution is referred to as *precipitation.* [▶▶ Section 10.4]

TABLE 9.1	Examples of Solutions				
Phase Combination			**Example**	**Solute(s)**	**Solvent**
Gas	dissolved in	Gas	Air	O_2, CO_2 (and others)	N_2
Gas	dissolved in	Liquid	Water prepared to fill a fish bowl	O_2 (and others)	H_2O
Liquid	dissolved in	Liquid	3% Hydrogen peroxide	H_2O_2	H_2O
Solid	dissolved in	Liquid	Simple syrup (used to sweeten drinks)	sugar	H_2O
Liquid	dissolved in	Solid	Dental amalgam (used for fillings)	Hg	Ag
Solid	dissolved in	Solid	Bronze	Sn	Cu

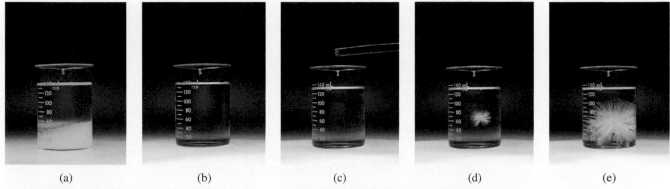

(a) (b) (c) (d) (e)

Figure 9.2 (a) A saturated solution of sodium acetate. (b) Upon heating, all of the solid dissolves. Subsequent, careful cooling yields a supersaturated solution. (c) The addition of a tiny crystal of sodium acetate initiates the precipitation of excess dissolved solute. (d) and (e) Precipitation of solute continues until the remaining solution is saturated.

(all) Charles D. Winters/McGraw Hill

Familiar Chemistry

Instant Hot Packs

One type of instant hot pack works using a supersaturated solution of sodium acetate ($NaC_2H_3O_2$), the solid shown in Figure 9.2. The packs are prepared by heating a solution of sodium acetate to dissolve a large amount of the solid. The hot sodium acetate solution is then sealed in plastic packs, each with a metal "clicker" disk, and the plastic packs are cooled to room temperature. A pack is activated when the clicker disk is squeezed, and crystallization of sodium acetate is initiated. As the excess sodium acetate crystallizes out of solution, the heat used to dissolve it is released and the pack gets warm.

This type of hot pack can be regenerated and reused. Regeneration is done by heating the packs, typically by immersing them in boiling water, to redissolve the sodium acetate. When the solid has all redissolved, the packs can be cooled to room temperature and are ready to be used again.

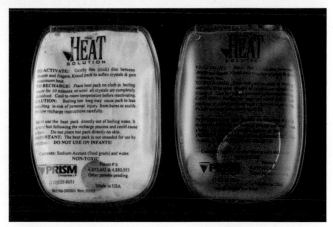

David A. Tietz/McGraw Hill

the saturated solution of sodium acetate, in which 121 g $NaC_2H_3O_2$ is dissolved, and the remainder (29 g) remains undissolved. We can make the excess solid dissolve if we warm the saturated solution to 80°C. (At 80°C, the solubility of sodium acetate is 150 g/100 mL.) Having done this, it is then possible to cool the solution to room temperature with all of the sodium acetate still dissolved. The result, a solution at room temperature consisting of 150 g $NaC_2H_3O_2$ dissolved in 100 mL water, is *supersaturated*. Figure 9.2 illustrates the preparation of a supersaturated solution, and how the extra solute can be made to come back out of solution.

9.2 Aqueous Solubility

We learned about the various types of intermolecular forces in the context of pure substances in Chapter 7. However, intermolecular forces are also important in solutions. When a substance dissolves in water, its particles become dispersed throughout the solution, with each solute particle surrounded by water molecules. The substances that are most soluble in water are those with intermolecular forces similar in magnitude to those of water; that is, *polar* substances—especially those capable of forming *hydrogen bonds*. Ionic substances, with some exceptions, also tend to be soluble in water. There is a saying

TABLE 9.2	Water-Soluble Ionic Compounds	
Water-Soluble Compounds	**Insoluble Exceptions**	
Compounds containing an alkali metal cation (Li^+, Na^+, K^+, Rb^+, Cs^+) or the ammonium ion (NH_4^+)		
Compounds containing the nitrate ion (NO_3^-), acetate ion ($C_2H_3O_2^-$), or chlorate ion (ClO_3^-)		
Compounds containing the chloride ion (Cl^-), bromide ion (Br^-), or iodide ion (I^-)	Compounds containing Ag^+, Hg_2^{2+}, or Pb^{2+}	
Compounds containing the sulfate ion (SO_4^{2-})	Compounds containing Ag^+, Hg_2^{2+}, Pb^{2+}, Ca^{2+}, Sr^{2+}, or Ba^{2+}	

TABLE 9.3	Water-Insoluble Ionic Compounds	
Water-Insoluble Compounds	**Soluble Exceptions**	
Compounds containing the carbonate ion (CO_3^{2-}), phosphate ion (PO_4^{3-}), chromate ion (CrO_4^{2-}), or sulfide ion (S^{2-})	Compounds containing Li^+, Na^+, K^+, Rb^+, Cs^+, or NH_4^+	
Compounds containing the hydroxide ion (OH^-)	Compounds containing Li^+, Na^+, K^+, Rb^+, Cs^+, or Ba^{2+}, or NH_4^+	

Narcisopa/Shutterstock

Student Hot Spot

Student data indicate that you may struggle to identify soluble ionic compounds. Access the eBook to view additional Learning Resources on this topic.

regarding solubility: "Like dissolves like." What this means is that a solute is typically soluble in a solvent when the two exhibit the same types of intermolecular forces. Thus, *polar* (and ionic) substances tend to dissolve best in *polar* solvents, such as water; and *nonpolar* substances tend to dissolve best in *nonpolar* solvents. This explains why oil (which consists of nonpolar carbon-hydrogen chains) and water do not mix, but rather settle into separate layers.

Although many ionic substances are soluble in water, there are some important exceptions. Tables 9.2 and 9.3 give guidelines for determining whether or not an ionic substance is soluble in water.

Sample Problem 9.1 lets you practice naming ionic compounds and determining whether or not they are water soluble.

SAMPLE PROBLEM 9.1

Name each ionic compound and determine whether or not it is soluble in water.

(a) Ag_2SO_4 (b) $SrCO_3$ (c) KNO_3 (d) Li_2S (e) $PbBr_2$

Strategy We first name the compound using the procedure summarized in Figure 3.8 (page 90). Then we locate one of the ions in the compound in Table 9.2 or Table 9.3 to determine if it is generally found in water-*soluble* compounds, or generally found in water-*insoluble* compounds. Then we check to see if the *other* ion constitutes an exception to water-solubility or to water-insolubility.

Setup

(a) Ag_2SO_4 consists of silver ions, Ag^+, and sulfate ions, SO_4^{2-}. The sulfate ion appears in Table 9.2 because most ionic compounds containing it are water soluble.

(b) $SrCO_3$ consists of strontium ions, Sr^{2-}, and carbonate ions, CO_3^{2-}. The carbonate ion appears in Table 9.3 because most ionic compounds containing it are water insoluble.

(c) KNO_3 consists of potassium ions, K^+, and nitrate ions, NO_3^-. The nitrate ion appears in Table 9.2 because all ionic compounds that contain it are water soluble.

(d) Li_2S consists of lithium ions, Li^+, and sulfide ions, S^{2-} ions. The sulfide ion appears in Table 9.3 because most ionic compounds containing it are water insoluble.

(e) $PbBr_2$ consists of lead(II) ions, Pb^{2+}, and bromide ions, Br^-. The bromide ion appears in Table 9.2 because most ionic compounds containing it are water soluble.

Solution (a) Ag_2SO_4 is silver sulfate. Although most compounds containing the sulfate ion are soluble, Table 9.2 lists the silver ion as an insoluble exception. Ag_2SO_4 is *not* soluble in water.

(b) $SrCO_3$ is strontium carbonate. Most compounds containing the carbonate ion are insoluble, and the strontium ion is not listed as an exception in Table 9.3. $SrCO_3$ is *not* soluble in water.

(c) KNO_3 is potassium nitrate. All ionic compounds containing the nitrate ion are soluble—there are no exceptions. KNO_3 is soluble in water.

(d) Li_2S is lithium sulfide. Although most compounds containing the sulfide ion are insoluble, Table 9.3 lists the lithium ion as a soluble exception. Li_2S is soluble in water.

(e) $PbBr_2$ is lead(II) bromide. Although most compounds containing the bromide ion are soluble, Table 9.2 lists the lead(II) ion as an insoluble exception. $PbBr_2$ is *not* soluble.

THINK ABOUT IT

In part (c), note that both ions in the compound, K^+ and NO_3^-, are listed as soluble with no insoluble exceptions. Any ionic compound that contains either of these ions is water soluble. Also, in part (e), remember that because lead can form ions of different charges, we must specify the charge in the compound's name using a Roman numeral in parentheses. We know that the charge on lead in $PbBr_2$ is $+2$ because the charge on the bromide ion is always -1, and there are two bromide ions for every lead ion.

Practice Problem **TTEMPT** Name and determine the aqueous solubility for each compound.

(a) K_2CO_3 (b) $Ba(NO_3)_2$ (c) $(NH_4)_2CO_3$ (d) $BaSO_4$ (e) Hg_2I_2

Practice Problem **B**UILD Name and determine the aqueous solubility for each compound.

(a) $CaCrO_4$ (b) Na_3PO_4 (c) $Ba(C_2H_3O_2)_2$ (d) KOH (e) $Sr(CrO_4)_2$

Practice Problem **C**ONCEPTUALIZE Determine which, if any, of the images shown here could be used to represent the following ionic compounds: iron(III) chloride, barium nitrate, ammonium bromide, lithium carbonate.

(i)   (ii)   (iii)

Profiles in Science

Alice Ball

Alice Ball (1892–1916) earned bachelor's degrees in chemistry and pharmacy at the University of Washington. After graduating, she was offered a number of scholarships for post-graduate study. She elected to attend the College of Hawaii (now the University of Hawaii) and was the first African American and the first woman to complete a master's degree in chemistry at that institution. Further, at the age of 23, she became the first African American research chemist and instructor at the College of Hawaii. Ball's research included developing a revolutionary treatment for leprosy, a debilitating and highly stigmatizing disease. Her work involved use of the oil of the chaulmoogra tree, which had actually been used topically as a leprosy treatment for hundreds of years, albeit with very limited success. The oil is viscous and sticky, making it difficult to apply to the skin. And the extremely acrid taste of the oil made oral administration impossible. Ball devised a way to isolate and chemically alter the therapeutic components of the oil, making them sufficiently water soluble to be administered by injection. Unfortunately, Ball died very young, and never got to see how successful the therapy she developed would ultimately be. Apallingly, after her death, a colleague stole her research and published it under his own name—giving her no credit—and naming the method after himself. In 1922, another colleague attempted to correct this injustice by publishing a paper in which he referred to the treatment as the "Ball method." Since then, several other scientists and historians have studied Ball's research, and have cultivated the proper recognition of her extraordinary achievements.

Alice Ball

CHECKPOINT–SECTION 9.2   Aqueous Solubility

9.2.1 Determine the aqueous solubility of $AgNO_3$, $CaCO_3$, and $FePO_3$.

a) soluble, insoluble, soluble

b) insoluble, soluble, insoluble

c) soluble, insoluble, insoluble

d) All three compounds are soluble.

e) All three compounds are insoluble.

9.2.2 Determine the correct name and aqueous solubility of Hg_2SO_4.

a) mercury(II) sulfate, soluble

b) mercury(II) sulfide, insoluble

c) mercury(I) sulfide, insoluble

d) mercury(I) sulfate, insoluble

e) mercury(I) sulfate, soluble

9.2.3 Determine the correct name and aqueous solubility of $CaBr_2$.

a) calcium bromide, insoluble

b) calcium bromide, soluble

c) cadmium bromide, insoluble

d) cadmium bromide, soluble

e) None of these is correct.

9.2.4 Which of the following is the correct name and aqueous solubility of AgS_2?

a) silver disulfate, soluble

b) silver sulfate, insoluble

c) silver sulfate, soluble

d) silver sulfide, soluble

e) silver sulfide, insoluble

9.2.5 Which of the following is the correct name and aqueous solubility of $Mg(OH)_2$?

a) magnesium hydroxide, insoluble

b) manganese hydroxide, soluble

c) magnesium hydroxide, soluble

d) manganese hydroxide, insoluble

e) None of these is correct.

9.2.6 Which of the following is the correct name and aqueous solubility of $(NH_4)_2CrO_4$?

a) diammonia chromate, insoluble

b) ammonium chromate, insoluble

c) diammonia chromate, soluble

d) ammonium chromate, soluble

e) None of these is correct.

9.3 Solution Concentration

In general, a solution containing a *small* amount of solute is said to be **dilute;** and a solution containing a *large* amount of solute is said to be **concentrated.** The terms dilute and concentrated are simply relative terms, though, and do not indicate the actual amount of solute contained in a solution. Chemists can use several different methods for quantitatively describing the amount of solute a solution contains. Which method they choose depends on the particular circumstance and how the information is to be used. In this section, we discuss three distinct ways to specify a solution's concentration: *percent by mass, molarity,* and *molality.*

Percent by Mass

Percent by mass, also known as *percent by weight,* is the ratio of the mass of a *solute* to the mass of the *solution,* multiplied by 100 percent:

Equation 9.1 $$\text{percent by mass} = \frac{\text{mass of solute}}{\text{mass of solution (solute + solvent)}} \times 100\%$$

Because the units of mass cancel on the top and bottom of the fraction, any units of mass can be used—provided they are used consistently.

Sample Problems 9.2 and 9.3 show how to calculate the percent by mass concentration of an aqueous solution, and how to calculate mass amounts of solutes in solutions of known compositions.

SAMPLE PROBLEM 9.2 Calculating Mass Percent Composition of an Aqueous Solution

Determine the percent by mass concentration of each of the following aqueous solutions: (a) 29.5 g $MgCl_2$ dissolved in 100.0 g of water, (b) 3.72 g CO_2 dissolved in 500.00 g of water, (c) 278 mg LiF dissolved in 475 mg of water.

Strategy Using Equation 9.1, identify the quantities needed to determine the percent by mass.

Setup (a) mass solute = 29.5 g $MgCl_2$, mass of solution = 29.5 g $MgCl_2$ + 100.0 g H_2O = 129.5 g

(b) mass solute = 3.72 g CO_2, mass of solution = 3.72 g CO_2 + 500.00 g H_2O = 503.72 g

(c) mass solute = 278 mg LiF, mass of solution = 278 mg LiF + 475 mg H_2O = 753 mg

Solution (a) percent by mass = $\dfrac{\text{mass of solute}}{\text{mass of solution (solute + solvent)}} \times 100\% = \dfrac{29.5\ g}{129.5\ g} \times 100\% = 22.8\%\ MgCl_2$

(b) percent by mass = $\dfrac{\text{mass of solute}}{\text{mass of solution (solute + solvent)}} \times 100\% = \dfrac{3.72\ g}{503.72\ g} \times 100\% = 0.739\%\ CO_2$

(c) percent by mass = $\dfrac{\text{mass of solute}}{\text{mass of solution (solute + solvent)}} \times 100\% = \dfrac{278\ mg}{753\ mg} \times 100\% = 36.9\%\ LiF$

THINK ABOUT IT

Note that any units of mass can be used, provided that they are consistent within the mass percent calculation you are performing.

Practice Problem ATTEMPT Determine the percent by mass concentration of each of the following solutions: (a) 143 mg $C_6H_{12}O_6$ dissolved in 700.0 mg of water, (b) 54.8 g CCl_4 dissolved in 500.0 g of hexane, (c) 169 g $BaCl_2$ dissolved in 975 g of water.

Practice Problem BUILD Determine the percent by mass concentration of each of the following solutions: (a) 762 mg LiF dissolved in 1.00 kg water, (b) 0.00331 kg K_2CO_3 dissolved in 1975 g water, (c) 59.8 mg $Sr(ClO_3)_2$ dissolved in 192 g water.

Practice Problem CONCEPTUALIZE The diagram represents two aqueous solutions. Each sphere in the container on the left represents a dissolved molecule. How many spheres would have to be in the container on the right for it to represent a solution with the same percent by mass concentration as the solution on the left?

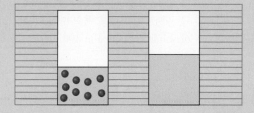

Thinking Outside the Box

Trace Concentrations

Often we hear about very low or *trace* levels of contaminants in the context of environmental concerns. The Federal Drug Administration (FDA), for example, has established the maximum allowable level of mercury in fish intended for human consumption as 1 part per million (ppm); and the Environmental Protection Agency (EPA) has established the maximum allowable level of arsenic in drinking water as 10 parts per billion (ppb). The determination of ppm and ppb concentrations is analogous to that of

(Continued on next page)

percent composition. Consider this: percent *means* parts per *hundred*. We determine percent by dividing the mass of a mixture component by the total mass of the mixture, and multiplying by 100:

$$\frac{\text{mass of component}}{\text{total mass of mixture}} \times 100 = \textit{parts per hundred or percent}$$

If we want the concentration in parts per million or parts per billion, we simply multiply by 1 million, or 1 billion, respectively:

$$\frac{\text{mass of component}}{\text{total mass of mixture}} \times 1,000,000 = \textit{parts per million or ppm}$$

$$\frac{\text{mass of component}}{\text{total mass of mixture}} \times 1,000,000,000 = \textit{parts per billion or ppb}$$

Javier Larrea/Pixtal/age fotostock

Jacques Cornell/McGraw Hill

SAMPLE PROBLEM 9.3 Calculating Amounts of Solutes in Solutions of Known Composition

Determine the mass of solute present in each of the following aqueous solutions: (a) 250.0 g of solution that is 11.8% $NaNO_3$ by mass, (b) 150.0 mg of solution that is 5.44% CH_3OH by mass, (c) 375 g of solution that is 1.89% Na_2S by mass.

Strategy Percentages should be written as the mass of solute contained in 100 grams (or the mass units used in the problem) of solution.

Setup (a) $\dfrac{11.8 \text{ g } NaNO_3}{100.0 \text{ g solution}}$ (b) $\dfrac{5.44 \text{ mg } CH_3OH}{100.0 \text{ mg solution}}$ (c) $\dfrac{1.89 \text{ g } Na_2S}{100.0 \text{ g solution}}$

Solution (a) $250.0 \text{ g solution} \times \dfrac{11.8 \text{ g } NaNO_3}{100.0 \text{ g solution}} = 29.5 \text{ g } NaNO_3$

(b) $150.0 \text{ mg solution} \times \dfrac{5.44 \text{ mg } CH_3OH}{100.0 \text{ mg solution}} = 8.16 \text{ mg } CH_3OH$

(c) $375 \text{ g solution} \times \dfrac{1.89 \text{ g } Na_2S}{100.0 \text{ g solution}} = 7.09 \text{ g } Na_2S$

THINK ABOUT IT

Writing the percent by mass values as fractions with 100 in the denominator makes it much easier to use them as conversion factors. This allows for the proper cancellation of units.

Practice Problem **A**TTEMPT Determine the mass of solute present in each of the following aqueous solutions: (a) 750.0 g of solution that is 3.91% $Cu(NO_3)_2$ by mass, (b) 275 g of solution that is 25.6% NH_4CN by mass, (c) 75 g of solution that is 0.0668% NaCl by mass.

Practice Problem **B**UILD Determine the mass of each solution that would contain 15.0 g of solute: (a) 25.1% KCl, (b) 9.77% $Mg(C_2H_3O_2)_2$, (c) 2.11% LiCN.

Practice Problem **C**ONCEPTUALIZE The diagram represents two aqueous solutions with the *same* percent-by-mass concentration. One of the solutions contains methanol (CH_3OH) and the other contains propanol ($CH_3CH_2CH_2OH$). Determine in which solution the solute is methanol and in which it is propanol.

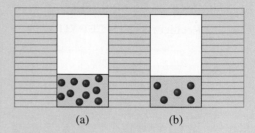

(a)   (b)

Molarity

Molarity (M), also called *molar concentration,* is determined by dividing the number of *moles* of solute by the *volume* of the solution (in L):

$$\text{molarity} = \frac{\text{moles solute}}{\text{liters solution}}$$ **Equation 9.2**

Molarity is more convenient than percent by mass when laboratory quantities are measured using *volume,* rather than mass. However, calculating molarity typically requires the conversion of a *mass* of solute to *moles* of solute [◄◄ Section 5.2].

Sample Problem 9.4 shows how to use Equation 9.2 to determine the molarity of a solution.

Student Hot Spot

Student data indicate that you may struggle to perform calculations involving molarity. Access the eBook to view additional Learning Resources on this topic.

SAMPLE PROBLEM ⬤ 9.4   Calculating Molarity of a Solution

Determine the molarity of the following glucose ($C_6H_{12}O_6$) solutions: (a) 0.223 mol glucose in 1.50 L of solution, (b) 50.0 g glucose in 2.00 L of solution, (c) 136 g glucose in 750.0 mL of solution.

Strategy Using Equation 9.2, identify all of the required values.

Setup (a) moles solute = 0.223 mol glucose, volume of solution = 1.50 L

(b) moles solute = convert 50.0 g glucose to moles using molar mass $\left(50.0 \text{ g glucose} \times \dfrac{1 \text{ mol glucose}}{180.16 \text{ g glucose}} = 0.278 \text{ mol glucose}\right)$, volume of solution = 2.00 L

(c) moles solute = convert 136 g glucose to moles using molar mass $\left(136 \text{ g glucose} \times \dfrac{1 \text{ mol glucose}}{180.16 \text{ g glucose}} = 0.755 \text{ mol glucose}\right)$, volume of solution = 0.750 L

Solution (a) molarity $= \dfrac{0.223 \text{ mol glucose}}{1.50 \text{ L solution}} = 0.149 \ M$ glucose

(b) molarity $= \dfrac{0.278 \text{ mol glucose}}{2.00 \text{ L solution}} = 0.139 \ M$ glucose

(c) molarity $= \dfrac{0.755 \text{ mol glucose}}{0.750 \text{ L solution}} = 1.01 \ M$ glucose

---

THINK ABOUT IT

Make sure to pay careful attention to the values given in these types of problems. Many students will fall into the habit of always using the molar mass here and you can see that part (a) did not require it.

(Continued on next page)

Practice Problem **A**TTEMPT Determine the molarity of the following HF solutions: (a) 0.118 mol HF in 1.50 L solution, (b) 2.99 g HF in 1.25 L solution, (c) 14.2 g HF in 844 mL solution.

Practice Problem **B**UILD Determine the molarity of each solution in Practice Problem A, replacing the HF solute with ethanol (C_2H_5OH).

Practice Problem **C**ONCEPTUALIZE Which of the following solutions has the same molarity as the one on the left?

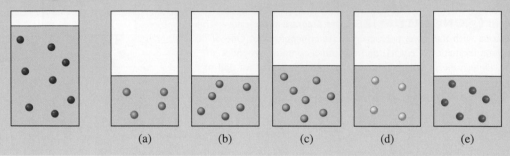

(a)   (b)   (c)   (d)   (e)

Students sometimes have difficulty seeing how units cancel in these equations. It may help to write M as mol/L until you become completely comfortable with these calculations.

Sample Problem 9.5 lets you practice converting from moles of solute to molarity of solution.

SAMPLE PROBLEM **9.5**   Converting Between Moles Solute and Molarity of Solution

Determine the volume (in L) of each solution that contains 0.313 mole of solute: (a) 0.0448 M Na_3PO_4, (b) 0.105 M $LiClO_4$, (c) 0.268 M $Fe(NO_3)_3$.

Strategy Use Equation 9.2 to create a conversion factor from molarity.

Setup (a) $\dfrac{\text{1 L solution}}{0.0448 \text{ mol } Na_3PO_4}$ (b) $\dfrac{\text{1 L solution}}{0.105 \text{ mol } LiClO_4}$ (c) $\dfrac{\text{1 L solution}}{0.268 \text{ mol } Fe(NO_3)_3}$

Solution (a) $0.313 \text{ mol } \cancel{Na_3PO_4} \times \dfrac{\text{1 L solution}}{0.0448 \cancel{\text{ mol } Na_3PO_4}} = 6.99$ L Na_3PO_4 solution

(b) $0.313 \text{ mol } \cancel{LiClO_4} \times \dfrac{\text{1 L solution}}{0.105 \cancel{\text{ mol } LiClO_4}} = 2.98$ L $LiClO_4$ solution

(c) $0.313 \text{ mol } \cancel{Fe(NO_3)_3} \times \dfrac{\text{1 L solution}}{0.268 \cancel{\text{ mol } Fe(NO_3)_3}} = 1.17$ L $Fe(NO_3)_3$ solution

---

THINK ABOUT IT

Remember that conversion factors can be "flipped" (written with the numerator and denominator switched). We use whichever form is needed for proper cancellation of units.

Practice Problem **A**TTEMPT Determine the volume (in mL) of each solution that contains 0.0570 mole of solute: (a) 0.199 M glucose, (b) 0.211 M NaCl, (c) 0.322 M MgF_2.

Practice Problem **B**UILD Determine the mass of solute present in each of the following solutions: (a) 1.25 L of 0.229 M $(NH_4)_2S$, (b) 25.0 mL of 2.63 M HBr, (c) 50.0 mL of 0.119 M NaCl.

Practice Problem **C**ONCEPTUALIZE The diagrams represent solutions of two different concentrations. What volume of solution 2 contains the same amount of solute as 5.00 mL of solution 1? What volume of solution 1 contains the same amount of solute as 30.0 mL of solution 2?

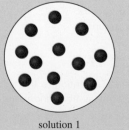

solution 1   solution 2

Molality

Molality (m), also called *molal concentration,* is determined by dividing the number of moles of solute by the *mass* of the *solvent* (in kg):

$$\text{molality} = \frac{\text{moles solute}}{\text{kg solvent}} \qquad \textbf{Equation 9.3}$$

As with molarity, calculation of molality generally requires the conversion of a mass of solute to moles of solute—using molar mass.

Sample Problem 9.6 shows how to use Equation 9.3 to determine the molality of a solution.

SAMPLE PROBLEM **9.6** **Calculating Molality of a Solution**

Determine the molality of each of the following solutions: (a) 0.253 mol sucrose dissolved in 1.75 kg water, (b) 0.172 mol CH_3OH in 195 g water, (c) 12.1 g CH_3OH in 275 g water.

Strategy Using Equation 9.3, determine the quantities required from the given values.

Setup (a) moles solute = 0.253 mol sucrose, kg solvent = 1.75 kg

(b) moles solute = 0.172 mol CH_3OH, kg solvent = 0.195 kg

(c) convert mass of CH_3OH into moles solute using molar mass $(12.1 \text{ g } CH_3OH \times \dfrac{1 \text{ mol } CH_3OH}{32.04 \text{ g } CH_3OH} = 0.378 \text{ mol } CH_3OH)$,
kg solvent = 0.275 kg water

Solution (a) $\dfrac{0.253 \text{ mol sucrose}}{1.75 \text{ kg } H_2O} = 0.145 \, m$ sucrose

(b) $\dfrac{0.172 \text{ mol } CH_3OH}{0.195 \text{ kg } H_2O} = 0.882 \, m \, CH_3OH$

(c) $\dfrac{0.378 \text{ mol } CH_3OH}{0.275 \text{ kg } H_2O} = 1.37 \, m \, CH_3OH$

THINK ABOUT IT

It is tempting to add the mass of the solute and solvent together here, but think carefully about what concentration units you are using and how they are defined. Remember, molality is defined as moles solute per kg of solvent—*not* per kg of *solution*.

Practice Problem Ⓐ**TTEMPT** Determine the molality of each of the following solutions: (a) 1.17 mol KF dissolved in 1.99 kg water, (b) 0.0787 mol CO_2 dissolved in 2755 g water, (c) 59.6 g CO_2 dissolved in 4999 g water.

Practice Problem Ⓑ**UILD** Determine the molality of each of the following solutions: (a) 2.42 mol KF dissolved in water to make 1.99 kg of solution, (b) 0.787 mol CH_3OH dissolved in enough water to make 2755 g of solution, (c) 596 g CH_3OH dissolved in enough water to make 4999 g solution.

Practice Problem Ⓒ**ONCEPTUALIZE** How much do the values of molality and percent by mass differ? Calculate the percent by mass for each solution in Practice Problem B to confirm your answer.

Comparison of Concentration Units

Often it is necessary to convert the concentration of a solution from one unit to another. For example, the same solution may be used for different experiments that require different concentration units for calculations. Suppose we want to express the concentration of a 0.396 m aqueous glucose $(C_6H_{12}O_6)$ solution (at 25°C) in molarity. We know there is 0.396 mole of glucose in 1000 g of the solvent. We need to determine the *volume* of this solution to calculate molarity. To determine volume, we must first calculate its mass:

$$0.396 \; \cancel{\text{mol } C_6H_{12}O_6} \times \frac{180.2 \text{ g}}{1 \; \cancel{\text{mol } C_6H_{12}O_6}} = 71.4 \text{ g } C_6H_{12}O_6$$

$$71.4 \text{ g } C_6H_{12}O_6 + 1000 \text{ g } H_2O = 1071 \text{ g solution}$$

Once we have determined the mass of the solution, we use the *density* of the solution, which is typically determined experimentally, to determine its volume. The density of a 0.396 *m* glucose solution is 1.16 g/mL at 25°C. Therefore, its volume is

$$\text{volume} = \frac{\text{mass}}{\text{density}}$$

$$= \frac{1071 \text{ g} \times 1 \; \cancel{\text{mL}}}{1.16 \text{ g}} \times \frac{1 \text{ L}}{1000 \; \cancel{\text{mL}}}$$

$$= 0.923 \text{ L}$$

Having determined the volume of the solution, the molarity is given by

$$\text{molarity} = \frac{\text{moles of solute}}{\text{liters of solution}}$$

$$= \frac{0.396 \text{ mol}}{0.923 \text{ L}}$$

$$= 0.429 \text{ mol/L} = 0.429 \; M$$

Sample Problem 9.7 shows how to convert from one unit of concentration to another.

SAMPLE PROBLEM (9.7)　Conversion Between Different Concentration Units

"Rubbing alcohol" is a mixture of isopropyl alcohol (C_3H_7OH) and water that is 70 percent isopropyl alcohol by mass (density = 0.79 g/mL at 20°C). Express the concentration of rubbing alcohol in (a) molarity and (b) molality.

Strategy (a) Use density to determine the total mass of a liter of solution, and use percent by mass to determine the mass of isopropyl alcohol in a liter of solution. Convert the mass of isopropyl alcohol to moles, and divide moles by liters of solution to get molarity. We can choose to start with any volume in a problem like this. Choosing 1 L simplifies the math.

(b) Subtract the mass of C_3H_7OH from the mass of solution to get the mass of water. Divide moles of C_3H_7OH by the mass of water (in kg) to get molality.

Setup The mass of a liter of rubbing alcohol is 790 g, and the molar mass of isopropyl alcohol is 60.09 g/mol.

Solution

(a) $$\frac{790 \text{ g } \cancel{\text{solution}}}{\text{L solution}} \times \frac{70 \text{ g } C_3H_7OH}{100 \text{ g } \cancel{\text{solution}}} = \frac{553 \text{ g } C_3H_7OH}{\text{L solution}}$$

$$\frac{553 \text{ g } \cancel{C_3H_7OH}}{\text{L solution}} \times \frac{1 \text{ mol}}{60.09 \text{ g } \cancel{C_3H_7OH}} = \frac{9.20 \text{ mol } C_3H_7OH}{\text{L solution}} = 9.2 \; M$$

(b) 790 g solution − 553 g C_3H_7OH = 237 g water = 0.237 kg water

$$\frac{9.20 \text{ mol } C_3H_7OH}{0.237 \text{ kg water}} = 39 \; m$$

Rubbing alcohol is 9.2 *M* and 39 *m* in isopropyl alcohol.

THINK ABOUT IT

Note the large difference between molarity and molality in this case. Molarity and molality are the same (or similar) only for very dilute aqueous solutions.

Practice Problem **A**TTEMPT   An aqueous solution that is 16 percent sulfuric acid (H_2SO_4) by mass has a density of 1.109 g/mL at 25°C. Determine (a) the molarity and (b) the molality of the solution at 25°C.

Practice Problem **B**UILD   Determine the percent sulfuric acid by mass of a 1.49-m aqueous solution of H_2SO_4.

Practice Problem **C**ONCEPTUALIZE   The diagrams represent solutions of a solid substance that is soluble in both water (density 1 g/cm^3) and chloroform (density 1.5 g/cm^3). For which of these solutions will the numerical value of molarity be closest to that of the molality? For which will the values of molarity and molality be most different?

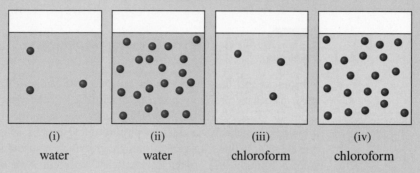

(i)	(ii)	(iii)	(iv)
Solvent: water	water	chloroform	chloroform

CHECKPOINT–SECTION 9.3   Solution Concentration

9.3.1 Determine the percent by mass percent concentration of a solution formed by dissolving 10.9 g $AgClO_4$ in 375 g water.

a) 2.91%   d) 1.78%

b) 3.44%   e) 2.82%

c) 5.85%

9.3.2 Calculate the mass of solute present in 95.0 g of a 0.0847% HCl solution.

a) 8.05 g   d) 0.0805 g

b) 1.12 g   e) 1.24 g

c) 0.892 g

9.3.3 Determine the molarity of a KOH solution prepared by dissolving 2.09 moles of KOH in enough water to make 4.00 L of solution.

a) 8.36 M d) 9.31×10^{-3} M

b) 0.523 M e) 0.209 M

c) 1.91 M

9.3.4 Calculate the molar concentration of a solution prepared by dissolving 58.5 g NaOH in enough water to yield 1.25 L of solution.

a) 1.46 M d) 1.17 M

b) 46.8 M e) 0.855 M

c) 2.14×10^{-2} M

9.3.5 The diagram represents three aqueous solutions formed from the same solute. Place the solutions in order of increasing molarity.

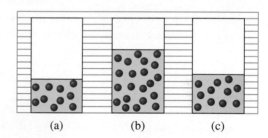

(a)	(b)	(c)

a) $a < c < b$ d) $c < a < b$

b) $b < c < a$ e) $c < b < a$

c) $b < a < c$

9.3.6 Determine the molality of a solution prepared by dissolving 0.441 mole of $Cu(NO_3)_2$ in 275 g water.

a) 1.60 m d) 4.32 m

b) 0.624 m e) 2.45 m

c) 0.231 m

9.3.7 Determine the molality of a solution prepared by dissolving 6.44 g of naphthalene ($C_{10}H_8$) in 80.1 g benzene.

a) 1.13 m d) 0.627 m

b) 80.4 m e) 11.7 m

c) 0.804 m

(Continued on next page)

9.3.8 At 20.0°C, a 0.258-*m* aqueous solution of glucose ($C_6H_{12}O_6$) has a density of 1.0173 g/mL. Calculate the molarity of this solution.

a) 0.258 *M*   d) 0.251 *M*

b) 0.300 *M*   e) 0.448 *M*

c) 0.456 *M*

9.3.9 At 25.0°C, an aqueous solution that is 25.0 percent H_2SO_4 by mass has a density of 1.178 g/mL. Calculate the molarity and the molality of this solution.

a) 3.00 *M* and 3.40 *m*   d) 3.00 *M* and 2.98 *m*

b) 3.40 *M* and 3.40 *m*   e) 3.44 *M* and 3.14 *m*

c) 3.00 *M* and 3.00 *m*

9.4 Solution Composition

Substances that dissolve in water fall into two categories: *electrolytes* and *nonelectrolytes*. You have probably heard of electrolytes in the context of sports drinks such as Gatorade. Electrolytes in body fluids are necessary for the transmission of electrical impulses, which are critical to physiological processes such as nerve impulses and muscle contractions. In general, an **electrolyte** is a substance that dissolves in water to yield a solution that conducts electricity. By contrast, a **nonelectrolyte** is a substance that dissolves in water to yield a solution that does *not* conduct electricity. Familiar examples of each are salt (NaCl, electrolyte) and sugar ($C_{12}H_{22}O_{11}$, nonelectrolyte).

The critical difference between an aqueous solution of salt and one of sugar is that the salt solution contains *ions;* the sugar solution does not. When salt (NaCl) dissolves in water, it undergoes **dissociation,** meaning that the ions it contains (Na^+ and Cl^-) separate from one another. When sugar dissolves, the $C_{12}H_{22}O_{11}$ molecules remain intact. The ions in the sodium chloride solution are what enable a salt solution to conduct electricity. Soluble ionic compounds such as NaCl are strong electrolytes, meaning that they undergo *complete* dissociation when they dissolve and exist in solution *entirely* as individual ions. Figure 9.3 shows the process of NaCl dissolving in water.

We can distinguish electrolytes and nonelectrolytes experimentally using an apparatus like the one pictured in Figure 9.4. A lightbulb is connected to a battery using a circuit that includes the contents of the beaker. For the bulb to light, electric current

Student Hot Spot

Student data indicate that you may struggle to understand what occurs when an ionic compound dissolves in water. Access the eBook to view additional Learning Resources on this topic.

Student Note: Electrolytes can be divided into two types, *strong* and *weak*. Strong electrolytes dissociate completely and exist entirely as ions when dissolved, whereas weak electrolytes exist in solution only partly as ions. In this chapter, we discuss only *strong* electrolytes.

Figure 9.3 Sodium chloride dissolving in water. Water molecules surround each Cl^- ion with their partial positive charges (on H atoms) oriented toward the negatively charged anion; and they surround each Na^+ ion with their partial negative charges (on O atoms) oriented toward the positively charged cation.

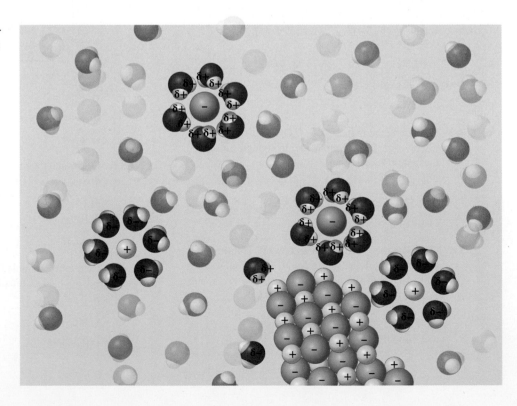

must flow through the beaker's contents. In order for electric current to flow, the beaker's contents must include *ions*. Pure water is a very poor conductor of electricity because H_2O consists of molecules, not ions. Therefore, water is a *nonelectrolyte,* and when the beaker contains pure water [Figure 9.4(a)], the bulb does not light. If we add some salt (NaCl) to the water, because NaCl is a strong electrolyte and dissociates into ions in solution, the bulb begins to glow [Figure 9.4(b)]. We can represent the dissociation of NaCl by writing:

$$NaCl(s) \rightarrow Na^+(aq) + Cl^-(aq)$$

where NaCl(s) represents the solid salt, and $Na^+(aq)$ and $Cl^-(aq)$ represent the aqueous sodium and chloride ions.

If we add sugar to the water instead of salt [Figure 9.4(c)], the bulb does not light. Sugar (sucrose, $C_{12}H_{22}O_{11}$) is a *nonelectrolyte*. Although it *dissolves,* it does *not* dissociate. Instead, sugar simply dissolves to give aqueous sugar molecules:

$$C_{12}H_{22}O_{11}(s) \rightarrow C_{12}H_{22}O_{11}(aq)$$

Because a solution of sugar contains no ions, it does not conduct electricity.

> **Student Note:** This is an example of a *chemical equation*—a sort of chemical shorthand used to represent physical and chemical processes. We examine chemical equations in some detail in Chapter 10.

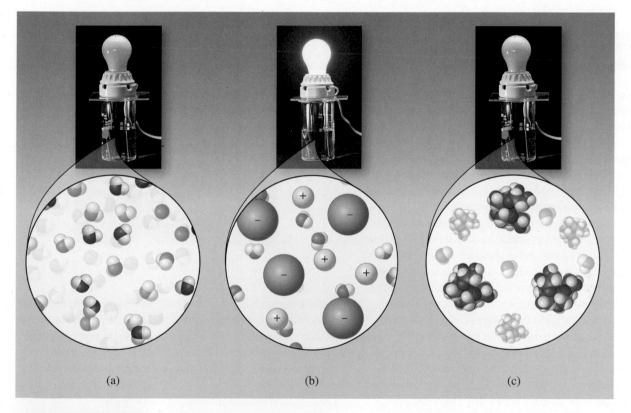

(a)   (b)   (c)

Figure 9.4 An apparatus for distinguishing electrolytes from nonelectrolytes. (a) Pure water consists of water molecules and does not conduct electricity; therefore, the lightbulb does not light. (b) When sodium chloride is dissolved in the water, it dissociates to give sodium ions and chloride ions. (Water molecules surrounding each ion have been omitted for clarity.) Because this solution contains ions, it conducts electricity and the bulb lights up. (c) When sugar is dissolved in the water instead, it does not dissociate. This solution does not contain ions, so it does not conduct electricity—the bulb does not light.

(light bulb photos) Stephen Frisch/McGraw Hill;
(salt and sugar) David A. Tietz/McGraw Hill

Figure 9.5 Solutions of
(a) barium nitrate [Ba(NO₃)₂],
(b) aluminum chloride (AlCl₃), and
(c) sodium phosphate (Na₃PO₄).

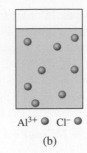

Ba^{2+} ○ NO_3^- ● Al^{3+} ● Cl^- ● Na^+ ● PO_4^{3-} ○
(a)   (b)   (c)

Student Hot Spot

Student data indicate that you may struggle to understand how an ionic compound dissociates in solution. Access the eBook to view additional Learning Resources on this topic.

When some ionic compounds dissolve, they dissociate into *more* than two ions. Sodium sulfate (Na₂SO₄), for example, consists of twice as many sodium ions as sulfate ions. When sodium sulfate dissociates in solution, the resulting concentration of sodium ion is twice that of sulfate ion. Figure 9.5 compares solutions of several soluble ionic compounds at the atomic level. It is important that you recognize what ions (and how many of each) an ionic compound contains. Now is a good time to review the common polyatomic ions [◄◄ Table 3.6].

Profiles in Science

Robert Cade, M.D.

In 1965, University of Florida (UF) assistant coach Dwayne Douglas was concerned about the health of Gators football players. He noted that during practices and games in hot weather the players (1) lost a great deal of weight, (2) seldom needed to urinate, and (3) had limited stamina, especially during the second half of a practice or game. He consulted Dr. Robert Cade, researcher and kidney-disease specialist at UF's medical college, who embarked on a project to identify the cause of the athletes' lack of endurance. It was found that after a period of intense activity accompanied by profuse sweating, the players had low blood sugar, low blood volume, and an imbalance of electrolytes—all of which contributed to heat exhaustion. Cade and his research fellows theorized that the depletion of sugar, water, and electrolytes might be remedied by having the athletes drink a solution containing just the right amounts of each. Using this theory, they developed a beverage containing water, sugar, and sodium and potassium salts similar to those present in sweat. By all accounts, the beverage tasted so bad that no one would drink it. Mary Cade, Robert Cade's wife, suggested adding lemon juice to make the concoction more palatable—and the drink that would become Gatorade was born. In their 1966 season the Gators

David A. Tietz/McGraw Hill

earned a reputation as the "second-half" team, often coming from behind in the third or fourth quarter. Gators coach Ray Graves attributed his team's newfound late-in-the-game strength to the newly developed sideline beverage that replenished blood sugar, blood volume, and electrolyte balance. Sports drinks are now a multibillion-dollar industry, and there are several popular brands, although Gatorade still maintains a large share of the market.

Sample Problem 9.8 shows how to determine the concentrations of individual ions in a solution of a strong electrolyte.

SAMPLE PROBLEM **9.8** **Calculating Concentrations of Individual Ions in Electrolyte Solutions**

Determine the concentration of chloride ions (in *M*) in each of the following solutions: (a) 0.150 *M* LiCl, (b) 0.150 *M* MgCl₂, (c) 0.150 *M* AlCl₃.

Strategy Write each molarity as *x* moles solute/1 L solution so units can be canceled. Determine the mole ratio of chloride ions to compound for each substance.

Setup (a) $\dfrac{0.150 \text{ mol LiCl}}{1 \text{ L solution}}$ and $\dfrac{1 \text{ mol Cl}^-}{1 \text{ mol LiCl}}$

(b) $\dfrac{0.150 \text{ mol MgCl}_2}{1 \text{ L solution}}$ and $\dfrac{2 \text{ mol Cl}^-}{1 \text{ mol MgCl}_2}$

(c) $\dfrac{0.150 \text{ mol AlCl}_3}{1 \text{ L solution}}$ and $\dfrac{3 \text{ mol Cl}^-}{1 \text{ mol AlCl}_3}$

Solution (a) $\dfrac{0.150 \text{ mol LiCl}}{1 \text{ L solution}} \times \dfrac{1 \text{ mol Cl}^-}{1 \text{ mol LiCl}} = \dfrac{0.150 \text{ mol Cl}^-}{1 \text{ L solution}}$ or $0.150 \ M \ \text{Cl}^-$

(b) $\dfrac{0.150 \text{ mol MgCl}_2}{1 \text{ L solution}} \times \dfrac{2 \text{ mol Cl}^-}{1 \text{ mol MgCl}_2} = \dfrac{0.300 \text{ mol Cl}^-}{1 \text{ L solution}}$ or $0.300 \ M \ \text{Cl}^-$

(c) $\dfrac{0.150 \text{ mol AlCl}_3}{1 \text{ L solution}} \times \dfrac{3 \text{ mol Cl}^-}{1 \text{ mol AlCl}_3} = \dfrac{0.450 \text{ mol Cl}^-}{1 \text{ L solution}}$ or $0.450 \ M \ \text{Cl}^-$

THINK ABOUT IT

Notice how all three solutions have the same molarity of each compound, but differ in the concentration of chloride ions. This depends on the chemical formula.

Practice Problem Ⓐ **TTEMPT** Determine the concentration of nitrate ions (in M) in each of the following solutions: (a) $0.200 \ M \ \text{NaNO}_3$, (b) $0.200 \ M \ \text{Mg(NO}_3)_2$, (c) $0.200 \ M \ \text{Al(NO}_3)_3$.

Practice Problem Ⓑ **UILD** If a solution is $0.333 \ M$ in NH_4^+, what is the concentration of $(\text{NH}_4)_3\text{PO}_4$ in the same solution?

Practice Problem Ⓒ **ONCEPTUALIZE** Determine which diagram, if any, could represent an aqueous solution of each of the following compounds: LiCl, CuSO_4, K_2SO_4, H_2CO_3, $\text{Al}_2(\text{SO}_4)_3$, AlCl_3, Na_3PO_4. (Red and blue spheres represent different chemical species.)

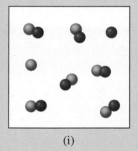
(i)

(ii)

(iii)

(iv)

CHECKPOINT–SECTION 9.4 Solution Composition

9.4.1 A solution that is $0.18 \ M$ in Na_2CO_3 is also _____. (Choose all that apply.)

a) $0.18 \ M$ in CO_3^{2-} b) $0.18 \ M$ in Na^+ c) $0.09 \ M$ in Na^+ d) $0.09 \ M$ in CO_3^{2-} e) $0.36 \ M$ in Na^+

9.4.2 Which diagram best represents an aqueous solution of sodium sulfate? $\text{Na}^+ = \bullet$ $\text{SO}_4^{2-} = \circ$

(a)

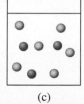
(b)

(c)

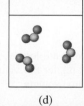
(d)

(e)

9.5 Solution Preparation

There are two methods by which aqueous solutions are commonly prepared. One involves dissolving a solid in water to prepare a solution of specific concentration, and the other involves diluting (adding water to) a previously prepared *concentrated* solution to prepare a more dilute solution.

Preparation of a Solution from a Solid

The procedure for preparing a solution of known molar concentration from a solid is as follows:

1. The solute is weighed accurately and transferred to a volumetric flask of the desired volume.
2. Water is added to the flask, which is then swirled to dissolve the solute.
3. Additional water is added to the flask to bring the volume up to the mark on the flask's neck.
4. The flask is capped and inverted to ensure thorough mixing and uniform composition throughout the solution.

This process is illustrated in Figure 9.6. Knowing the volume of the solution in the flask and the amount of solid substance dissolved, we can determine the molarity of the solution using Equation 9.2. (Note that we will first need to convert the measured *mass* of solute to *moles* using the molar mass.)

Sample Problem 9.9 shows how to determine the molar concentration of a solution prepared by dissolving a solid in water.

SAMPLE PROBLEM   9.9   Calculating Molarity of an Aqueous Solution Prepared by Dissolving a Solid in Water

For an aqueous solution of glucose ($C_6H_{12}O_6$), determine (a) the molarity of 2.00 L of a solution that contains 50.0 g of glucose, (b) the volume of this solution that would contain 0.250 mole of glucose, and (c) the number of moles of glucose in 0.500 L of this solution.

Strategy Convert the mass of glucose given to moles, and use the equations for interconversions of *M*, liters, and moles to calculate the answers.

Setup The molar mass of glucose is 180.2 g.

$$\text{moles of glucose} = \frac{50.0 \text{ g} \times 1 \text{ mol}}{180.2 \text{ g}} = 0.277 \text{ mol}$$

Solution (a) molarity $= \dfrac{0.277 \text{ mol } C_6H_{12}O_6}{2.00 \text{ L solution}} = 0.139 \text{ mol/L}$

A common way to state the concentration of this solution is to say, "This solution is 0.139 *M* in glucose."

(b) volume $= \dfrac{0.250 \text{ mol } C_6H_{12}O_6 \times 1 \text{ L}}{0.139 \text{ mol}} = 1.80 \text{ L}$

(c) moles of $C_6H_{12}O_6$ in 0.500 L $= 0.500 \text{ L} \times \dfrac{0.139 \text{ mol}}{\text{L}} = 0.0695 \text{ mol}$

THINK ABOUT IT

Check to see that the magnitudes of your answers are logical. For example, the mass given in the problem corresponds to 0.277 mol of solute. If you are asked, as in part (b), for the volume that contains a number of moles smaller than 0.277, make sure your answer is smaller than the original volume.

Practice Problem **A**TTEMPT For an aqueous solution of sucrose ($C_{12}H_{22}O_{11}$), determine (a) the molarity of 5.00 L of a solution that contains 235 g of sucrose, (b) the volume of this solution that would contain 1.26 mol of sucrose, and (c) the number of moles of sucrose in 1.89 L of this solution.

Practice Problem **B**UILD For an aqueous solution of sodium chloride (NaCl), determine (a) the molarity of 3.75 L of a solution that contains 155 g of sodium chloride, (b) the volume of this solution that would contain 4.58 mol of sodium chloride, and (c) the number of moles of sodium chloride in 22.75 L of this solution.

Practice Problem **C**ONCEPTUALIZE What would be the effect on the final concentration of a prepared solution if you were to first fill the volumetric flask to the mark and then add and dissolve the solid?

Preparation of a More Dilute Solution from a Concentrated Solution

The other common procedure for preparing solutions of the desired concentrations is through the use of concentrated "stock" solutions that are typically kept in laboratory stockrooms. *Dilution* is the process of preparing a less concentrated solution from a more concentrated one. This also requires the use of a volumetric flask and is done using the following steps:

1. A precise volume of concentrated stock solution is measured using a volumetric pipet and is transferred to a volumetric flask.
2. Water is added to the volumetric flask to bring the volume up to the mark on the flask's neck.
3. The flask is capped and inverted to ensure thorough mixing and uniform composition throughout the solution.

In carrying out a dilution, it is important to remember that adding more water to a given amount of concentrated stock solution changes (decreases) the *concentration* of the solution—but does *not* change the number of *moles* of solute present in the solution. We can calculate the number of moles in a given volume of a solution of known concentration by rearranging Equation 9.2 to solve for moles solute:

$$\text{moles solute} = \text{molarity} \times \text{liters solution} \quad \text{or} \quad \text{moles solute} = M \cdot L$$

And because we know that moles of solute does not change with dilution, we can write

$$M_c \times L_c = M_d \times L_d \qquad \textbf{Equation 9.4}$$

where the subscripts c and d refer to *concentrated* (before dilution) and *dilute* (after dilution). Equation 9.4 enables us to determine how much concentrated stock solution (L_c) is needed to prepare a known volume of solution (L_d) with a specific concentration (M_d):

$$L_c = \frac{M_d \times L_d}{M_c}$$

Conversely, we might know the volume of concentrated stock solution to be diluted, know its concentration, and know the final volume—and we can calculate the concentration of the resulting diluted solution:

$$L_d = \frac{M_c \times L_c}{M_d}$$

Further, because laboratory volume measurements are more often done using mL than L, Equation 9.4 and any rearrangements can also be written with the volumes in mL:

$$M_c \times mL_c = M_d \times mL_d$$

$$mL_c = \frac{M_d \times mL_d}{M_c} \qquad\qquad mL_d = \frac{M_c \times mL_c}{M_d}$$

Student Note: When we write Equation 9.4 using mL instead of L, the product on each side of the equation is millimoles (mmol) rather than moles.

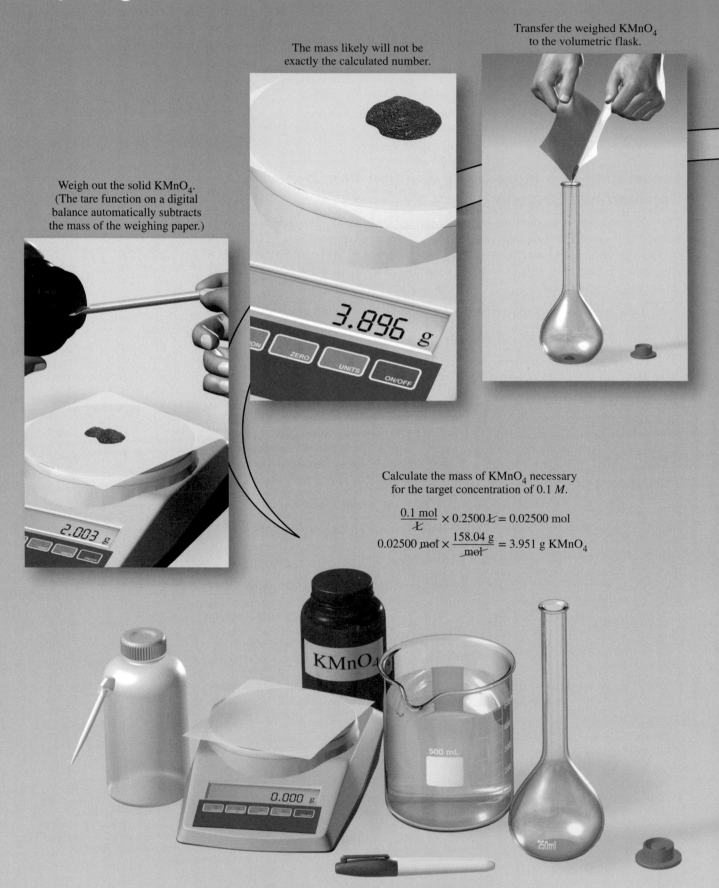

Figure 9.6
Preparing a Solution from a Solid

Weigh out the solid $KMnO_4$. (The tare function on a digital balance automatically subtracts the mass of the weighing paper.)

The mass likely will not be exactly the calculated number.

Transfer the weighed $KMnO_4$ to the volumetric flask.

Calculate the mass of $KMnO_4$ necessary for the target concentration of 0.1 M.

$$\frac{0.1\ \text{mol}}{\cancel{L}} \times 0.2500\ \cancel{L} = 0.02500\ \text{mol}$$

$$0.02500\ \cancel{\text{mol}} \times \frac{158.04\ \text{g}}{\cancel{\text{mol}}} = 3.951\ \text{g}\ KMnO_4$$

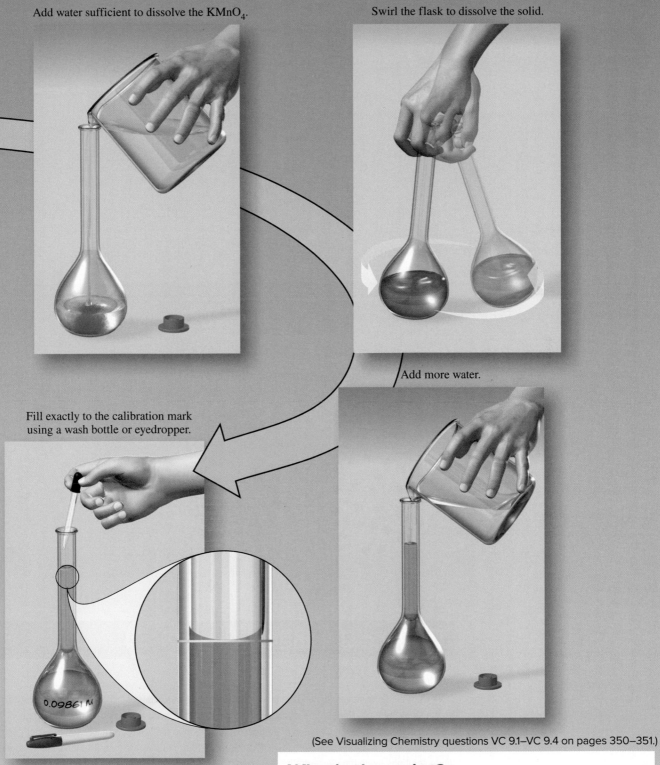

Add water sufficient to dissolve the KMnO₄.

Swirl the flask to dissolve the solid.

Add more water.

Fill exactly to the calibration mark using a wash bottle or eyedropper.

0.09861 M

After capping and inverting the flask to ensure complete mixing, we calculate the actual concentration of the prepared solution.

$$3.896 \text{ g KMnO}_4 \times \frac{1 \text{ mol}}{158.04 \text{ g}} = 0.024652 \text{ mol}$$

$$\frac{0.024652 \text{ mol}}{0.2500 \text{ L}} = 0.09861 \text{ M}$$

(See Visualizing Chemistry questions VC 9.1–VC 9.4 on pages 350–351.)

What's the point?

The goal is to prepare a solution of precisely known concentration, with that concentration being very close to the target concentration of 0.1 M. Note that because 0.1 is a *specified* number, it does not limit the number of significant figures in our calculations.

337

Sample Problem 9.10 lets you practice using Equation 9.4 to determine the amount of concentrated stock solution needed to prepare a solution by the dilution method.

SAMPLE PROBLEM **9.10** Calculating the Amount of Concentrated Stock Solution Necessary to Prepare a Dilute Solution with a Specific Concentration

What volume of 12.0 M HCl, a common laboratory stock solution, must be used to prepare 250.0 mL of 0.125 M HCl?

Strategy Use Equation 9.4 to determine the volume of 12.0 M HCl required for the dilution.

Setup M_c = 12.0 M, M_d = 0.125 M, mL_d = 250.0 mL.

Solution 12.0 M × mL_c = 0.125 M × 250.0 mL

$$mL_c = \frac{0.125\ M \times 250.0\ mL}{12.0\ M} = 2.60\ mL$$

THINK ABOUT IT

Plug the answer into Equation 9.4, and make sure that the product of concentration and volume is the same on both sides of the equation.

Practice Problem **A**TTEMPT What volume of 6.0 M H_2SO_4 is needed to prepare 500.0 mL of a solution that is 0.25 M in H_2SO_4?

Practice Problem **B**UILD What volume of 0.20 M H_2SO_4 can be prepared by diluting 127 mL of 6.0 M H_2SO_4?

Practice Problem **C**ONCEPTUALIZE The diagrams represent a concentrated stock solution (left) and a dilute solution (right) that can be prepared by dilution of the stock solution. How many milliliters of the concentrated stock solution are needed to prepare solutions of the same concentration as the dilute solution of each of the following final volumes? (a) 50.0 mL, (b) 100.0 mL, (c) 250.0 mL

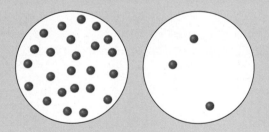

Thinking Outside the Box

Serial Dilution

A series of dilutions may be used in the laboratory to prepare a number of increasingly dilute solutions from a stock solution. The method involves preparing a solution, either from a solid or from a concentrated stock solution, and diluting a portion of the prepared solution to make a *more* dilute solution. For example, we could use the 0.400 M $KMnO_4$ solution described previously to prepare a series of five increasingly dilute solutions—with the concentration decreasing by a factor of 10 at each stage. Using a volumetric pipette, we withdraw 10.00 mL of the 0.400-M solution and deliver it into a 100.00-mL volumetric flask as shown in the figure. We then dilute to the volumetric mark and cap and invert the flask to ensure complete mixing. The concentration of the newly prepared solution is determined using Equation 9.4, where M_c is 0.400 M, and mL_c and mL_d are 10.00 mL and 100.00 mL, respectively.

$$0.400\ M \times 10.00\ mL = M_d \times 100.00\ mL$$
$$M_d = 0.0400\ M\ \text{or}\ 4.00 \times 10^{-2}\ M$$

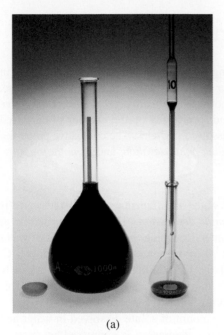

(a)

Charles D. Winters/Timeframe Photography/
McGraw Hill

(b)

Charles D. Winters/Timeframe Photography/
McGraw Hill

Serial dilution. (a) A solution of precisely known concentration is prepared in a volumetric flask. A precise volume of the solution is transferred to a second volumetric flask and subsequently diluted. (b) A precise volume of the second solution is transferred to a third volumetric flask and diluted. The process is repeated several times, each time producing a more dilute solution. In this example, the concentration is reduced by a factor of 10 at each stage.

CHECKPOINT—SECTION 9.5   Solution Preparation

9.5.1 What mass of $Mg(NO_3)_2$ must be used to prepare 250.0 mL of a 0.228 M $Mg(NO_3)_2$ solution?

a) 163 g

b) 13.5 g

c) 2.60 g

d) 28.1 g

e) 8.45 g

9.5.2 What mass of glucose ($C_6H_{12}O_6$) in grams must be used in order to prepare 500 mL of a solution that is 2.50 M in glucose?

a) 225 g

b) 125 g

c) 200 g

d) 1.25 g

e) 625 g

9.5.3 What volume in milliliters of a 1.20 M HCl solution must be diluted in order to prepare 1.00 L of 0.0150 M HCl?

a) 15.0 mL

b) 12.5 mL

c) 12.0 mL

d) 85.0 mL

e) 115 mL

9.5.4 Which best represents the before-and-after molecular-level view of the dilution of a concentrated stock solution?

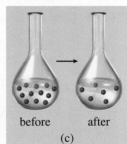

before   after

(a)

before   after

(c)

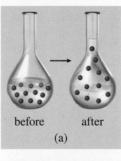

before   after

(b)

before   after

(d)

9.6 Colligative Properties

You may be familiar with the practice of salting roads to prevent their becoming icy in winter—or that of adding antifreeze to the water in a car's radiator for the same purpose. These practices take advantage of what are known as *colligative properties*. **Colligative properties** are properties of solutions that depend specifically on the *number* of dissolved particles in a solution—but not on the *type* of dissolved particles. The colligative properties that we discuss are *freezing-point depression, boiling-point elevation,* and *osmotic pressure.*

Freezing-Point Depression

When a substance such as salt is dissolved in water, the freezing point of the water is lowered—a phenomenon known as **freezing-point depression.** This is the principle at work when salt is spread on roads during the winter. Lowering the freezing point of water results in roads being wet rather than icy—even when the temperature is at or below 0°C, the normal freezing point of pure water. Similarly, the addition of antifreeze (typically ethylene glycol) to the water in an automobile's radiator lowers the freezing point. This allows the coolant to remain liquid at temperatures below the freezing point of water.

The quantitative relationship between the amount of solute dissolved in water and the amount by which the freezing point of water is depressed is expressed as:

Equation 9.5 $\qquad\qquad \Delta T_f = K_f m$

Student Note: Remember that *molality* is moles of solute divided by kilograms of solvent:

$$M = \frac{\text{mol solute}}{\text{kg solvent}}$$

where ΔT_f is the number of degrees (Celsius) by which the freezing point is lowered; K_f is the **freezing-point-depression constant** of water, which is equal to 1.86°C/*m*; and *m* is the concentration of the solution, expressed as molality. Note that Equation 9.5 does not contain any term specific to the solute. It is the *molal concentration*—not the *identity* of the solute—that determines by how much freezing point is depressed.

Sample Problem 9.11 shows how to use Equation 9.5 to determine the freezing point of an aqueous solution.

SAMPLE PROBLEM 9.11 Calculating Freezing Point of an Aqueous Solution

Ethylene glycol [$CH_2(OH)CH_2(OH)$] is a common automobile antifreeze. It is water soluble and fairly nonvolatile (B.P. 197°C). Calculate the freezing point of a solution containing 11.04 moles of ethylene glycol in 2075 g of water.

Strategy Using Equation 9.5, the molality of the solution and the K_f value for water, 1.86°C/*m*, must be used.

Setup $\dfrac{11.04 \text{ mol ethylene glycol}}{2.075 \text{ kg water}} = 5.321 \ m$ ethylene glycol

Solution $\Delta T_f = K_f m = \dfrac{1.86°C}{m} \times 5.321 \ m = 9.90°C$

The freezing point of water is 0°C, so this *change* in freezing point must be subtracted from 0°C.

$$T_f = 0°C - 9.90°C = -9.90°C$$

THINK ABOUT IT

Remember that the result of Equation 9.5 (ΔT_f), which is always a positive number, must be *subtracted* from the normal freezing point of water (0°C) to determine the new freezing point. The freezing point of an aqueous solution is always a *negative* number.

Practice Problem ATTEMPT Determine the freezing point of an aqueous solution formed by dissolving 4.33 moles of ethylene glycol in 638 g of water.

Practice Problem BUILD Determine the freezing point of an aqueous solution formed by dissolving 47.3 g ethylene glycol in 168 g water.

Practice Problem ⓒONCEPTUALIZE The diagrams represent four different aqueous solutions of the same solute. Which of the solutions has the lowest freezing point?

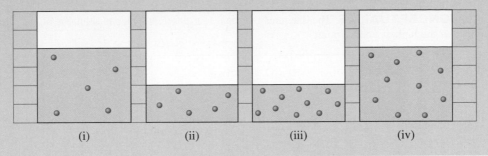

(i) (ii) (iii) (iv)

Boiling-Point Elevation

The presence of a solute also affects boiling point. When a nonvolatile substance [◄◄ Section 7.4] is dissolved in water, the resulting solution boils at a *higher* temperature than pure water, which boils at 100°C. This phenomenon is known as *boiling-point elevation.* In fact, the purpose of using ethylene glycol in car radiators is twofold: it both lowers the freezing point and raises the boiling point—thereby increasing the range of temperatures over which the coolant remains *liquid* and, therefore, *effective.*

The quantitative relationship between solution concentration and boiling-point elevation is:

$$\Delta T_b = K_b m \qquad \text{Equation 9.6}$$

where ΔT_b is the number of degrees (Celsius) by which the boiling point is raised; K_b is the *boiling-point-elevation constant* of water, which is equal to 0.512°C/*m*; and *m* is the concentration of the solution, expressed as molality. Note that as with freezing-point depression, boiling-point elevation depends on the concentration of the solute, not on its identity.

Sample Problem 9.12 shows how to use Equation 9.6 to determine the boiling point of an aqueous solution.

> **Student Note:** Remember that *molality* is moles of solute divided by kilograms of solvent:
>
> $$m = \frac{\text{mol solute}}{\text{kg solvent}}$$

SAMPLE PROBLEM (**9.12**) Calculating the Boiling Point of an Aqueous Solution

Glycerol [$CH_2(OH)CH(OH)CH_2(OH)$] is a common ingredient in pharmaceutical and personal-care products. Calculate the boiling point of a solution containing 7.75 moles of glycerol in 1895 g of water.

Strategy Using Equation 9.6, the molality of the solution and the K_b value for water, 0.52°C/*m*, must be used.

Setup

$$\frac{7.75 \text{ mol glycerol}}{1.895 \text{ kg water}} = 4.0897 \ m \text{ glycerol}$$

Solution

$$\Delta T_b = K_b m = \frac{0.512°C}{\cancel{m}} \times 4.0897 \ \cancel{m} = 2.09°C$$

$$T_b = 100.00°C + 2.09°C = 102.09°C$$

THINK ABOUT IT

The result of Equation 9.6 must be *added* to the normal boiling point of water to get the solution's boiling point. Aqueous solutions of nonvolatile solutes boil at temperatures *greater* than 100°C.

(Continued on next page)

Practice Problem ATTEMPT Determine the boiling point of a solution containing 3.65 moles of glycerol in 575 g water.

Practice Problem BUILD Determine the boiling point of a solution containing 3165 g of glycerol in 2.50 kg of water.

Practice Problem CONCEPTUALIZE The diagrams represent four different aqueous solutions of the same solute. Which of the solutions has the highest boiling point?

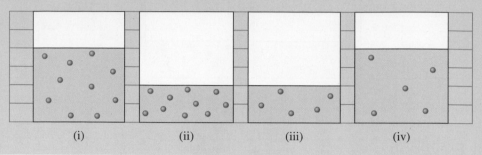

(i)   (ii)   (iii)   (iv)

Familiar Chemistry

Ice Melters

We learned in Section 9.4 that strong electrolytes, such as sodium chloride, dissociate and exist entirely as ions in solution. When a mole of sodium chloride dissolves in a kilogram of water, it results in a solution with *twice* as many dissolved solute particles as a comparable aqueous solution containing a mole of *sucrose*. Because colligative properties depend solely on the concentration of solute particles, the sodium chloride solution will exhibit *twice* the freezing-point depression and *twice* the boiling-point elevation of the sucrose solution. Further, there are strong electrolytes that dissociate in solution to give *more* than two ions. Calcium chloride ($CaCl_2$), for

example, dissociates to give *three* ions in solution: one Ca^{2+} ion and two Cl^- ions.

$$CaCl_2(s) \longrightarrow Ca^{2+}(aq) + 2Cl^-(aq)$$

Therefore, on a mole-for-mole basis, calcium chloride results in a greater freezing-point depression and a greater boiling-point elevation than sodium chloride. There are several soluble ionic compounds used to melt or prevent ice—including calcium chloride, magnesium chloride, and sodium acetate.

David A. Tietz/McGraw Hill

DigitalVues/Alamy Stock Photo

Osmotic Pressure

If an aqueous solution and pure water are separated by a **semipermeable membrane,** one through which water molecules can pass but solute particles cannot, water molecules will move through the membrane *from* the pure water side *to* the solution side. This process is called **osmosis.** Figure 9.7 illustrates this process. It is possible to prevent the flow of water through the semipermeable membrane by the application of pressure to the solvent side of the apparatus shown in Figure 9.7. The pressure necessary to stop the flow of water across the membrane is known as the *osmotic pressure* of the solution. The greater a solution's concentration, the higher its osmotic pressure.

Osmosis and osmotic pressure are very important in biological systems. Human blood, for example, consists in part of red blood cells (*erythrocytes*) suspended in plasma, an aqueous solution containing a variety of solutes, including various proteins. Each red blood cell is encased by a protective semipermeable membrane. The concentration of dissolved substances must be the same inside the membrane (in the cells) and outside the membrane (in the plasma) so that there is no net passage of water either into the cells or out of them. Figure 9.8 shows what happens to red blood cells when they are placed in a solution with a different concentration of dissolved substances than they themselves contain. The osmotic pressure of human plasma must be maintained within a very narrow range to prevent damage to red blood cells. It is for this reason that most fluids that are given intravenously (injected directly into the bloodstream) must be carefully prepared to have the same concentration of dissolved substances as the plasma itself.

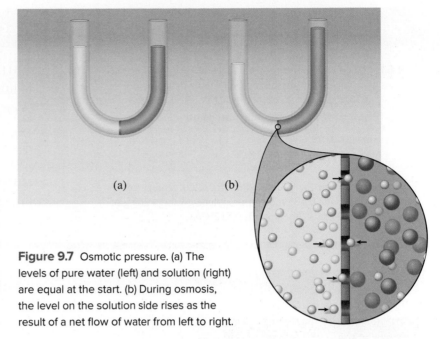

(a)   (b)

Figure 9.7 Osmotic pressure. (a) The levels of pure water (left) and solution (right) are equal at the start. (b) During osmosis, the level on the solution side rises as the result of a net flow of water from left to right.

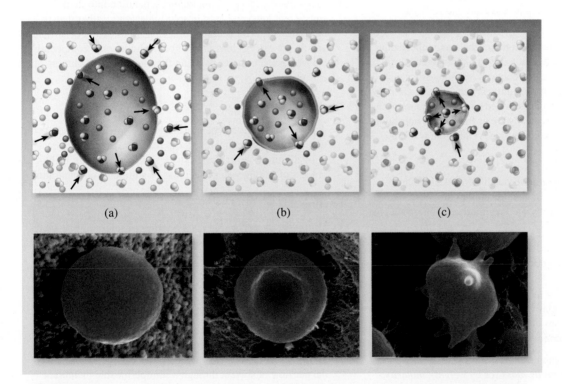

(a)   (b)   (c)

Figure 9.8 A red blood cell in a solution with a concentration lower than that inside the cell. Water flows through the membrane into the cell, causing the cell to swell and potentially burst. (b) A red blood cell in a solution with a concentration equal to that inside the cell. There is no net flow of water into or out of the cell. (c) A red blood cell in a solution with a concentration higher than that inside the cell. Water flows through the membrane out of the cell, causing the cell to shrink and potentially collapse.

(all) David M. Phillips/Science Source

CHECKPOINT–SECTION 9.6   Colligative Properties

9.6.1 Calculate the freezing point of a sucrose solution that is prepared by dissolving 0.918 mole of sucrose in 635 g of water. The K_f for water is 1.86°C/m.

a) 1.71°C

b) −1.71°C

c) −3.22°C

d) 2.69°C

e) −2.69°C

9.6.2 Determine the boiling point of a solution prepared by dissolving 0.589 mole of propylene glycol in 269 g of water. The K_b for water is 0.512°C/m.

a) 111.0°C

b) 104.0°C

c) 101.1°C

d) 96.0°C

e) 98.9°C

9.6.3 Determine the freezing point of a solution prepared by dissolving 678 g of glucose (M = 180.16 g/mol) in 2.0 kg of water. The K_f for water is 1.86°C/m.

a) 3.5°C

b) −3.5°C

c) 7.0°C

d) −7.0°C

e) −6.3°C

Chapter Summary

Section 9.1

- A *solution* is a homogeneous mixture consisting of a *solvent* and one or more *solutes.*

- Solutions in which the solvent is water are referred to as *aqueous* solutions.

- A *saturated* solution is one that contains the maximum amount of solute that can be dissolved at a particular temperature. An *unsaturated* solution is one that contains *less* solute than that amount; and a *supersaturated* solution is one that contains *more* solute than that amount. Supersaturated solutions require special conditions to prepare and typically are unstable. The amount of solute dissolved in a saturated solution is called the *solubility.*

Section 9.2

- Solubility depends on the intermolecular forces exhibited by the solvent and the solute. In general, solutes are most soluble in solvents with similar types and magnitudes of intermolecular forces: (Like dissolves like).

- Ionic compounds tend to be soluble in water, which is a polar solvent, but there are exceptions.

Section 9.3

- Solutions containing a relatively small amount of solute are known as *dilute.* Solutions containing a relatively large amount of solute are known as *concentrated.*

- Solution concentration can be expressed in several different ways, including as *percent by mass* (or by *weight*), *molarity,* and *molality.*

Section 9.4

- Solutes in aqueous solutions fall into two categories: electrolytes and nonelectrolytes. An *electrolyte* is a substance that dissolves in water to give a solution that conducts electricity. A *nonelectrolyte* is one that dissolves in water to give a solution that does not conduct electricity.

- Electrolytes undergo *dissociation* (separation into their constituent ions) and exist in solution as *ions,* rather than as molecules or formula units. Nonelectrolytes do not dissociate and do not exist as ions in solution. Soluble ionic compounds are *strong* electrolytes, meaning that they exist in solution *entirely* as ions.

Section 9.5

- Aqueous solutions can be prepared by dissolving a solid in water and adding enough water to reach the desired volume, or by diluting a concentrated stock solution. *Dilution* is the process of adding water to decrease a solution's concentration.

Section 9.6

- *Colligative properties* are the properties of solutions that depend solely on the number of dissolved particles in solution. They include *freezing-point depression, boiling-point elevation,* and *osmotic pressure.*

- The magnitude of freezing-point depression and boiling-point elevation of aqueous solutions depends on the molal concentration of solute.

- A *semipermeable membrane* is one through which water molecules can pass, but solute particles cannot. *Osmosis* is the net flow of solvent through a semipermeable membrane separating an aqueous solution from pure water. Solvent flows *from* the pure-water side *to* the solution side. *Osmotic pressure* is the pressure that must be applied to the solution side in order to prevent the flow of water through the semipermeable membrane.

Key Terms

Aqueous 317

Boiling-point elevation 341

Boiling-point-elevation
constant 341

Colligative properties 340

Concentrated 322

Dilute 322

Dilution 335

Dissociation 330

Electrolyte 330

Freezing-point depression 340

Freezing-point-depression
constant 340

Molal concentration 327

Molality (*m*) 327

Molar concentration 325

Molarity (*M*) 325

Nonelectrolyte 330

Osmosis 342

Osmotic pressure 343

Percent by mass 322

Percent by weight 322

Saturated 318

Semipermeable membrane 342

Solubility 318

Solute 317

Solution 317

Solvent 317

Supersaturated 318

Unsaturated 318

Key Equations

9.1	$\text{percent by mass} = \dfrac{\text{mass of solute}}{\text{mass of solution (solute + solvent)}} \times 100\%$	The concentration of a solution can be expressed as percent by mass, which is determined by dividing the solute mass by total solution mass and multiplying by 100. Any mass units can be used as long as the use is consistent.
9.2	$\text{molarity} = \dfrac{\text{moles solute}}{\text{liters solution}}$	Solution concentration can also be expressed as *molarity* or *molar concentration*. Molarity, calculated as moles of solute divided by liters of solution, is more convenient for laboratory work where solution quantities are most often measured as volumes, rather than as masses.
9.3	$\text{molality} = \dfrac{\text{moles solute}}{\text{kg solvent}}$	Solution concentration can also be expressed as *molality* or *molal concentration*. Molality, calculated as moles of solute divided by kilograms of solvent, is necessary for calculations of freezing-point depression and boiling-point elevation.
9.4	$M_c \times L_c = M_d \times L_d$	The concentration of a solution before and after dilution is calculated by multiplying concentration by volume for both the concentrated solution and the dilute solution. Dilution of a concentrated stock solution does not change the number of moles of solute.
9.5	$\Delta T_f = K_f m$	The number of degrees by which the freezing point of water is depressed by the presence of a solute is calculated as the product of the freezing-point-depression constant, K_f, for water ($1.86°C/m$) and the molal concentration of the dissolved substance. The calculated value of ΔT_f must be *subtracted* from the normal freezing point of water.
9.6	$\Delta T_b = K_b m$	The number of degrees by which the boiling point of water is elevated by the presence of a solute is calculated as the product of the boiling-point-elevation constant, K_b, for water ($0.512°C/m$) and the molal concentration of the dissolved substance. The calculated value of ΔT_b must be *added* to the normal boiling point of water.

The unit of concentration used most often throughout the rest of this book is *molarity*. It is important that you be able to calculate and express the molar concentration of an aqueous solution; and it is equally important that you be able to determine how many moles of solute a given amount of solution contains—given its molarity.

When the concentration is given as a molarity, remember that the necessary unit cancellation will be more obvious if you write the concentration as moles per liter, rather than simply writing *M*. For example:

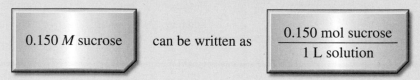

The concentration 0.150 *M* has *three* significant figures. Note that when we write the concentration out as moles per liter, it still has three significant figures. The 1 L in the denominator is an exact number.

If we were asked to determine how many moles of sucrose there are in 2.80 L of this solution, we would multiply the molar concentration by the volume:

$$\frac{0.150 \text{ mol sucrose}}{1 \text{ L solution}} \times 2.80 \text{ L solution} = 0.42 \text{ mol sucrose}$$

In the laboratory, volume measurements are more commonly given in mL than in L. This requires us to convert the volume in mL to L before multiplying by molar concentration. Consider another example using the same sucrose solution: How many moles of sucrose are in 152 mL of 0.150 *M* sucrose? We begin by converting the volume 152 mL to L:

$$152 \text{ mL solution} \times \frac{1 \text{ L}}{1000 \text{ mL}} = 0.152 \text{ L solution}$$

We then proceed as before, by multiplying the molar concentration by the volume in liters.

$$\frac{0.150 \text{ mol sucrose}}{1 \text{ L solution}} \times 0.152 \text{ L solution} = 0.0228 \text{ mol sucrose}$$

Remember that we can also solve this type of unit-conversion problem by combining all of the steps on a single line:

We can also use these mole-volume-concentration relationships to determine what volume of solution contains a specified amount of solute. For example: What volume of this sucrose solution contains 0.765 mol sucrose? In this case, we divide the number of moles by the molar concentration, which is the same as multiplying by the reciprocal of molar concentration:

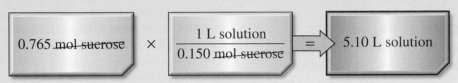

$$0.765 \text{ mol sucrose} \times \frac{1 \text{ L solution}}{0.150 \text{ mol sucrose}} = 5.10 \text{ L solution}$$

And if we needed the answer in milliliters, we would perform one more step (multiplication by 1000 mL/1 L) to convert to mL.

Key Skills Problems

9.1
How many moles of NaCl are there in 125 mL of a 0.0180-M solution of NaCl?
a) 2.25 mol
b) 6.94 mol
c) 0.00225 mol
d) 0.00694 mol
e) 0.0180 mol

9.2
How many moles of glucose are there in 500.0 mL of a 0.112-M solution of glucose?
a) 0.0560 mol
b) 56.0 mol
c) 4460 mol
d) 0.000224 mol
e) 2.24 mol

9.3
What volume of a 0.250-M solution of glucose contains 0.0379 mol glucose?
a) 6.60 mL
b) 660 mL
c) 0.152 mL
d) 9.48 mL
e) 152 mL

9.4
How many grams of NaCl are there in 37.5 mL of a 0.0180-M solution of NaCl? (*Hint:* You will need the molar mass of NaCl.)
a) 67.5 g
b) 58.4 g
c) 0.0394 g
d) 39.4 g
e) 675 g

Questions and Problems

SECTION 9.1: GENERAL PROPERTIES OF SOLUTIONS

9.1 Define the term *solution*. Are all solutions liquid? Explain.

9.2 Explain what the terms *solvent* and *solute* mean.

9.3 Which of the following are solutions?
(a) Gravel
(b) Stainless steel (a metal alloy of iron, chromium, and carbon)
(c) Hot tea
(d) Tap water

9.4 Which of the following are solutions?
(a) Fresh-squeezed orange juice
(b) Black coffee
(c) Maple syrup
(d) Nail polish remover

9.5 Identify the solute and solvent in each of the following solutions:
(a) Everclear (a 190-proof liquor, 95% ethanol by 5% water by volume)
(b) Rose gold (25% copper and 75% gold alloy)
(c) Rubbing alcohol (70% isopropanol and 30% water by volume)

9.6 Identify the solute and solvent in each of the following solutions:
(a) Bronze (12% tin and 88% copper)
(b) Saline solution (NaCl in water)
(c) Heliox (compressed scuba-diving gas, 21% oxygen and 79% helium)

9.7 Identify each of the images as an unsaturated solution, saturated solution, or supersaturated solution.

   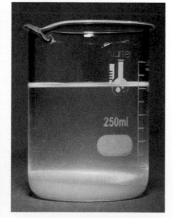

David A. Tietz/McGraw Hill   David A. Tietz/McGraw Hill

SECTION 9.2: AQUEOUS SOLUBILITY

9.8 Explain the meaning of the term *water soluble*.

9.9 If a substance such as NaCl is *soluble* in water, does that mean there is no limit to the amount of NaCl that can be dissolved in a particular quantity of water? Explain.

9.10 What does the phrase "Like dissolves like" mean? Give an example.

9.11 Substances that exhibit similar intermolecular forces are generally soluble in one another. In which of the following solvents would $CH_3CH_2CH_2CH_2CH_3$ most likely be soluble?
(a) H_2O (b) NH_3 (c) C_6H_{14}

9.12 Which of the following pairs of substances would you expect to be soluble in one another? Explain.
(a) $I_2(s)$ and $CCl_4(l)$
(b) $CH_3OH(l)$ and $H_2O(l)$
(c) $LiNO_3$ and C_6H_{14}

9.13 Indicate whether you expect each of the following substances to be soluble in water.
(a) NaCl   (b) KNO_3 (c) NH_4CN (d) AgCl

9.14 Indicate whether you expect each of the following substances to be soluble in water.
(a) PbI_2 (b) $Fe(OH)_3$ (c) $CaCO_3$ (d) $KClO_4$

9.15 Indicate whether you expect each of the following substances to be soluble in water.
(a) $AlPO_4$ (c) KBr
(b) MgS   (d) $AgC_2H_3O_2$

9.16 Indicate whether you expect each of the following substances to be soluble in water.
(a) $BaSO_4$ (b) Li_2CO_3 (c) KOH   (d) $Ni(NO_3)_2$

9.17 Which of the following anions, when combined with Na^+, form a water-soluble compound?
(a) $C_2H_3O_2^-$ (b) SO_4^{2-} (c) CO_3^{2-} (d) OH^-

9.18 Which of the following anions, when combined with Ca^{2+}, form a water-soluble compound?
(a) PO_4^{3-} (b) Br^- (c) NO_3^- (d) S^{2-}

9.19 Which of the following cations, when combined with OH^-, form a water-insoluble compound? In each case where you expect a water-insoluble compound to form, write the chemical formula for the compound.
(a) NH_4^+ (b) Ni^{2+} (c) Mg^{2+} (d) Li^+

9.20 Which of the following cations, when combined with PO_4^{3-}, form a water-insoluble compound? In each case where you expect a water-insoluble compound to form, write the chemical formula for the compound.
(a) Na^+ (b) Cu^{2+} (c) Zn^{2+} (d) NH_4^+

SECTION 9.3: SOLUTION CONCENTRATION

9.21 Define the term *mass percent* in the context of solution concentrations.

9.22 Describe how *molarity* is different from *molality*.

9.23 Calculate the mass percent composition of each of the following solutions:
(a) 34.2 g KF in 275 g H_2O
(b) 10.0 g CH_3CH_2OH in 75.0 g H_2O
(c) 12.8 g LiOH in 100.0 g H_2O

9.24   Calculate the mass percent of each of the following solutions:
(a) 1.5 mol $Mg(NO_3)_2$ in 200.0 g water
(b) 1.5 mol CH_3CH_2OH in 200.0 g water
(c) 1.5 mol I_2 in 200.0 g hexane (C_6H_{14})

9.25   Calculate the mass percent of each of the following solutions. (The density of water at room temperature is 1.0 g/mL.)
(a) 2.1 mol NaF in 1.0 L H_2O
(b) 2.1 mol K_3PO_4 in 1.0 L H_2O
(c) 2.1 mol CH_3OH in 1.0 L H_2O

9.26   Determine the mass of solute dissolved in 250.0 grams of each of the following aqueous solutions:
(a) 2.95% NaCl
(b) 21.7% $C_{12}H_{22}O_{11}$ (sucrose)
(c) 60.0% glycerol

9.27   Determine the mass of solute dissolved in 375.0 grams of each of the following aqueous solutions:
(a) 13.6% ethylene glycol
(b) 9.65% sugar
(c) 4.55% sodium bicarbonate ($NaHCO_3$)

9.28   Determine the mass of each aqueous solution that contains 25.0 grams of solute.
(a) 3.70% $NiCl_2$ (c) 12.8% K_2CO_3
(b) 4.75% $AgNO_3$

9.29   Determine the mass of each aqueous solution that contains 93.7 grams of solute.
(a) 7.44% HCl   (c) 8.33% H_2O_2
(b) 0.351% $Fe(NO_3)_3$

9.30   Determine what mass of each of the following aqueous solutions contains 2.50 g of F^-.
(a) 1.54% NaF   (b) 16.8% SrF_2 (c) 9.88% AlF_3

9.31   Determine what mass of each of the following aqueous solutions contains 10.0 g of sodium.
(a) 23.1% $NaC_2H_3O_2$ (c) 13.7% Na_2SO_4
(b) 18.3% Na_3PO_4

9.32   Determine the molality of each of the following solutions:
(a) 1.55 g LiF in 150.0 g water
(b) 45.5 g sucrose ($C_{12}H_{22}O_{11}$) in 500.0 g water
(c) 31.9 g CH_3OH in 285 g water

9.33   Determine the molality of each of the following solutions:
(a) 10.0 g $C_3H_8O_2$ (propylene glycol) in 100.0 g of water
(b) 24.3 g glucose ($C_6H_{12}O_6$) in 250.0 g of water
(c) 24.3 g $Sr_3(PO_4)_2$ in 250.0 g of water

9.34   Determine the molality of each of the following solutions:
(a) 0.444 mol C_2H_5OH in exactly 1500 g water
(b) 0.189 mol $(NH_4)_2CO_3$ in exactly 200 g water
(c) 0.189 mol $Ba(ClO_4)_2$ in exactly 200 g water

9.35   Determine the molality of each of the following solutions:
(a) 0.359 mol CH_3OH in 355 g ethanol
(b) 0.293 mol $C_6H_{12}O_6$ in 479 g water
(c) 0.293 mol NaCl in 479 g water

9.36   Determine the molality of each of the following solutions. Recall that water has a density of 1.0 g/mL at room temperature.
(a) 1.83 mol solute in 250.0 mL water
(b) 0.448 mol solute in 475 mL water
(c) 2.39 mol solute in 125 mL water

9.37   Determine the molality of each of the following solutions. (Assume the density of water is 1.0 g/mL.)
(a) 3.48 mol glucose in 750.0 mL water
(b) 1.75 mol glycerol in 675 mL water
(c) 4.76 mol solute in 325 mL water

9.38   Determine the molarity of each solution formed by dissolving the given solute in enough water to make 500.0 mL of solution.
(a) 2.33 mol calcium chloride
(b) 1.55 mol zinc carbonate
(c) 0.434 mol potassium nitrate

9.39   Determine the molarity of each solution formed by dissolving the indicated amount of solute in enough water to make 500.0 mL of solution.
(a) 0.119 mol sucrose
(b) 0.497 mol ethanol
(c) 0.296 mol glycerol

9.40   Determine the molarity of each solution formed by dissolving the indicated amount of solute in enough water to make 250.0 mL of solution.
(a) 21.3 g strontium chlorate
(b) 45.9 g barium phosphate
(c) 121 g ethanol (C_2H_5OH)

9.41   Determine the molarity of each solution formed by dissolving the indicated amount of solute in enough water to make 100.0 mL of solution.
(a) 2.79 g methanol (CH_3OH)
(b) 24.2 g iron(III) chloride
(c) 345 g $Sr(BrO_3)_2$

9.42   Determine the volume of each solution that contains 0.342 mol $ZnBr_2$.
(a) 1.43 M $ZnBr_2$ (c) 3.29 M $ZnBr_2$
(b) 0.221 M $ZnBr_2$

9.43   Determine the volume of each solution that contains 1.55 mol I_2.
(a) 0.113 M I_2 (b) 0.0273 M I_2 (c) 0.0786 M I_2

9.44   Determine the volume of each solution that contains 1.89 g HCl.
(a) 1.55 M HCl   (b) 2.11 M HCl   (c) 0.219 M HCl

9.45   Determine the volume of each solution that contains 25.0 g K_3PO_4.
(a) 0.0759 M K_3PO_4 (c) 0.664 M K_3PO_4
(b) 0.118 M K_3PO_4

9.46   Determine the mass of solute dissolved in 250.0 mL of each solution.
(a) 1.33 M AlF_3 (c) 0.766 M CH_3CH_2OH
(b) 0.441 M $LiNO_3$

9.47   Determine the mass of solute dissolved in 750.0 mL of each solution.
(a) 2.45 M $C_6H_{12}O_6$ (c) 0.00211 M CO_2
(b) 1.05 M $Ca(C_2H_3O_2)_2$

9.48   Determine the mass of water in a 24-oz jar of honey that is 18.2% water by mass. (Assume the jar contains exactly 1000 grams of honey.)

9.49   Outline the steps for converting from molarity to molality.

9.50   Outline the steps for converting from molality to percent by mass.

9.51   A sulfuric acid (H_2SO_4) solution is 18.4 M with a density of 1.84 g/mL.
(a) Determine the molality of the solution.
(b) Determine the percent by mass of H_2SO_4 in the solution.

9.52   A phosphoric acid (H_3PO_4) solution is 14.8 M with a density of 1.71 g/mL.
(a) Determine the molality of the solution.
(b) Determine the percent by mass of H_3PO_4 in the solution.

9.53   A solution of nitric acid (HNO_3) is 36.5 m with a density of 1.41 g/mL.
(a) Determine the molarity of the solution.
(b) Determine the percent by mass of HNO_3 in the solution.

9.54   A solution of formic acid (HCO_2H) is 539 m with a density of 1.13 g/mL.
(a) Determine the molarity of the solution.
(b) Determine the percent by mass of HCO_2H in the solution.

9.55   The FDA allows no more than 1 ppm mercury in fish meant for human consumption. Determine the percent by mass that is represented by 1 ppm.

9.56   The Environmental Protection Agency has placed a limit of 15 ppb on lead in drinking water. What is this concentration in percent by mass?

9.57   The Environmental Protection Agency has placed a limit of 4 ppm on fluoride in drinking water. (Municipal water supplies generally fluoridate the water at levels of 1 ppm to be safe.) What are each of these concentrations in percent by mass?

9.58   The Environmental Protection Agency has placed a limit of 10 ppb on arsenic in drinking water. What is this concentration in percent by mass?

SECTION 9.4: SOLUTION COMPOSITION

9.59   How are electrolytes and nonelectrolytes different from one another?

9.60   Indicate whether each substance is an electrolyte or a nonelectrolyte.
(a) methanol (CH_3OH)   (c) $LiNO_3$
(b) ethylene glycol ($C_2H_6O_2$)   (d) NH_4Cl

9.61   Identify each substance as an electrolyte or a nonelectrolyte.
(a) $Ca(ClO_4)_2$ (c) SrF_2
(b) Na_2SO_4 (d) glycerol ($C_3H_8O_3$)

9.62   Do a 0.25-m solution of KCl and a 0.25-m solution of $CaCl_2$ have the same concentration of chloride ions? Explain.

9.63   Which of the following sketches correctly represent the formation of a solution from the given solute? For any that do not, describe why they do not.

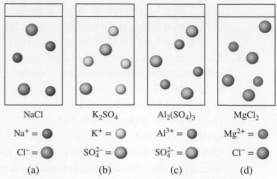

NaCl	K_2SO_4	$Al_2(SO_4)_3$	$MgCl_2$
$Na^+ = $	$K^+ = $	$Al^{3+} = $	$Mg^{2+} = $
$Cl^- = $	$SO_4^{2-} = $	$SO_4^{2-} = $	$Cl^- = $
(a)	(b)	(c)	(d)

9.64   Using the style shown in Problem 9.63, draw molecular-art diagrams to represent LiCl solid dissolving in water. (Represent each formula unit of LiCl with spheres of two different colors, and use two diagrams—one representing before dissolution and one representing after.)

9.65   Using the style shown in Problem 9.63, draw molecular-art diagrams to represent $Al_2(SO_4)_3$ solid dissolving in water. (Represent each formula unit of $Al_2(SO_4)_3$ with spheres of two different colors, and use two diagrams—one representing before dissolution and one representing after.)

9.66   Draw a diagram to represent the process of sugar dissolving in water (represent each individual sugar molecule with a single sphere).

9.67   Determine the molar concentration of ammonium ions in each of the following solutions:
(a) 0.10 M NH_4Cl (c) 0.10 M $(NH_4)_3PO_4$
(b) 0.10 M $(NH_4)_2S$

9.68   Determine the molar concentration of magnesium ions in each of the following solutions:
(a) 0.25 M $Mg(NO_3)_2$ (c) 0.25 M $Mg_3(PO_4)_2$
(b) 0.25 M $MgSO_4$

9.69   Determine the molar concentration of magnesium ions in a solution formed by mixing 100.0 mL of 0.100 M $MgCl_2$ solution with 100.0 mL of 0.100 M $Mg_3(PO_4)_2$ solution.

9.70   Determine the molar concentration of chloride ions in a solution formed by mixing 250.0 mL of 0.155 M $CaCl_2$ with 150.0 mL of 0.240 M $AlCl_3$.

SECTION 9.5: SOLUTION PREPARATION

Visualizing Chemistry
Figure 9.6

VC 9.1   Which of the following would result in the actual concentration of the prepared solution being higher than the final, calculated value?
a) Loss of some of the solid during transfer to the volumetric flask.
b) Neglecting to add the last bit of water with the wash bottle to fill to the volumetric mark.
c) Neglecting to tare the balance with the weigh paper on the pan.

VC 9.2   Why can't we prepare the solution by first filling the volumetric flask to the mark and then adding the solid?
a) The solid would not all dissolve.
b) The solid would not all fit into the flask.
c) The final volume would not be correct.

VC 9.3   What causes the concentration of the prepared solution not to be exactly 0.1 M?
a) Rounding error in the calculations.
b) The volume of the flask is not exactly 250 mL.
c) The amount of solid weighed out is not exactly the calculated mass.

VC 9.4   The volumetric flask used to prepare a solution from a solid is shown before and after the last of the water has been added.

(i)   (ii)

Which of the following statements is true?
a) The concentration of solute is greater in (i) than in (ii).
b) The concentration of solute is smaller in (i) than in (ii).
c) The concentration of solute in (i) is equal to the concentration of solute in (ii).

9.71   Explain what is meant by the terms *concentrated* and *dilute*.

9.72   When a concentrated solution is diluted, its volume increases. What remains constant?

9.73   Describe in detail how you would prepare 25.0 mL of 0.233 M aqueous NaCl starting with solid NaCl.

9.74   Describe in detail how you would prepare 100.0 mL of 0.119 M aqueous $Al(NO_3)_3$ starting with solid $Al(NO_3)_3$.

9.75   Calculate the mass of $RbNO_3$ necessary to prepare 500.0 mL of 0.355 M $RbNO_3$.

9.76   Calculate the mass of Na_2SO_3 necessary to prepare 750.0 mL of 0.355 M Na_2SO_3.

9.77   Describe in detail how you would prepare 25.0 mL of 0.133 M aqueous HNO_3 starting with an aqueous stock solution that is 1.00 M in HNO_3.

9.78   The diagram on the left represents a concentrated stock solution of a strong electrolyte. Which of the solutions represented on the right could be prepared by diluting a sample of the stock solution? Select all that apply.

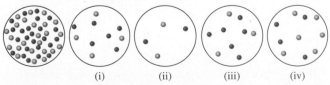

(i)   (ii)   (iii)   (iv)

9.79   What volume of 15.8 M HNO_3 stock solution must be used to prepare 100.0 mL of 0.255 M HNO_3?

9.80   What volume of 17.4 M acetic acid stock solution must be used to prepare 250.0 mL of 0.750 M $HC_2H_3O_2$?

9.81   Starting with a 2.00-M stock solution of HCl, four standard solutions (a–d) are prepared by sequentially diluting 10.00 mL of each solution to 250.0 mL. Determine (a) the concentrations of all four standard solutions and (b) the number of moles of HCl in each solution.

9.82   Starting with a 16.0-M stock solution of HNO_3, four standard solutions (a–d) are prepared by sequentially diluting 25.0 mL of each solution to 250.0 mL. Determine (a) the concentrations of all four standard solutions and (b) the number of moles of HNO_3 in each solution.

SECTION 9.6: COLLIGATIVE PROPERTIES

9.83   Explain the term *colligative property* and list the colligative properties covered in this chapter. On what do colligative properties depend?

9.84   Is it possible for an aqueous solution to have a boiling point below 100.0°C under standard conditions? Explain.

9.85   Is it possible for an aqueous solution to have a freezing point above 0.0°C under standard conditions? Explain.

9.86   Explain why two 0.10-m aqueous solutions, one of NaCl and the other of glucose ($C_6H_{12}O_6$), do not boil at the same temperature.

9.87   Explain why two 0.23-m aqueous solutions, one of $CaCl_2$ and the other of NaCl, do not freeze at the same temperature.

9.88   Arrange the following aqueous solutions in order of decreasing freezing point: 2.3 m glucose ($C_6H_{12}O_6$), 2.3 m KI, 2.3 m K_3PO_4.

9.89   Arrange the following aqueous solutions in order of increasing boiling point: 3.9 m $SrCl_2$, 3.9 m $MgSO_4$, 3.9 m ethanol (CH_3OH).

9.90   Determine the freezing point of an aqueous solution that is 0.24 m in sucrose.

9.91   Determine the freezing point of an aqueous solution that is 2.95 m in isopropanol.

9.92   Determine the freezing point of a solution composed of 415 g sucrose ($C_{12}H_{22}O_{11}$) dissolved in 800.0 g water.

9.93   Determine the freezing point of a solution composed of 355 g glycerol ($C_3H_8O_3$) dissolved in 789 g water.

9.94   Determine the boiling point of an aqueous solution that is 0.95 m in ethanol.

9.95   Determine the boiling point of an aqueous solution that is 6.99 m in glycerol.

9.96   Determine the boiling point of a solution prepared by dissolving 119 g ethanol (CH_3CH_2OH) in 425 g water.

9.97   Determine the boiling point of a solution prepared by dissolving 745 g isopropanol (C_3H_8O) in 1955 g water.

9.98 The first diagram represents an aqueous solution. Which of the other pictures represents a solution that has the same osmotic pressure as the first?

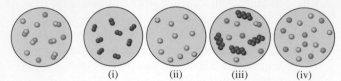

    (i)   (ii)   (iii)   (iv)

9.99 The first diagram represents one aqueous solution separated from another by a semipermeable membrane. Which of the other diagrams could represent the same system after the passage of some time?

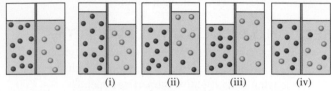

    (i)   (ii)   (iii)   (iv)

9.100 Osmotic pressure (π) can be calculated using the equation $\pi = MRT$, where M represents molar concentration, R is the gas constant, and T represents the temperature in kelvins. Determine the osmotic pressure of a solution at 25°C that is 3.07 M in glycerol.

9.101 Osmotic pressure (π) can be calculated using the equation $\pi = MRT$, where M represents molar concentration, R is the gas constant, and T represents the temperature in kelvins. Determine the osmotic pressure of a solution at 25°C that is 1.50 M in ethanol and 2.00 M in sucrose.

Cumulative

9.102 The hydrogen peroxide (H_2O_2) that can be purchased at a drug store typically is 3.0% hydrogen peroxide by mass, with the remainder of the solution being water. What mass (in g) of hydrogen peroxide is present in a 32-ounce container of such a solution? (Assume the density of the solution to be the same as that of pure water: 1.00 g/mL.)

9.103 A prescription treatment for sensitive teeth is a 0.63% aqueous solution of tin(II) fluoride by mass. [Tin(II) fluoride is also commonly called *stannous* fluoride.] What mass (in g) of tin(II) fluoride is present in the 283.5-g bottle shown in the photo? How many moles of fluoride ion (F^-) does the bottle contain?

9.104 Determine the molarity of fluoride ions in the fluoride treatment described in Problem 9.103. (Assume the density of the solution to be the same as that of pure water: 1.00 g/mL.)

9.105 The *carat* scale is used to describe the amount of gold in the metal alloys used to make jewelry. Pure gold is referred to as 24-carat. The carat scale is similar to a mass-percent scale, except that instead of specifying how many parts per *hundred*, the carat scale specifies how many parts per 24. It can be calculated as

David A. Tietz/McGraw Hill

$$\text{carat weight} = \frac{\text{mass of gold}}{\text{total mass of metal alloy}} \times 24$$

(Note the similarity of this equation to that used to calculate percent by mass.)

(a) White gold is an alloy of gold and one or more white metals, such as platinum, palladium, silver, or nickel. If a ring is described as 14-carat white gold, what percent by mass of gold is the ring?

(b) Rose gold is 75 percent gold, 22.25 percent copper, and 2.75 percent silver by mass. What is this gold on the carat scale?

(c) Standard 18-carat yellow gold is an alloy of gold blended with equal masses of silver and copper. Determine the percent by mass of silver in 18-carat yellow gold.

Answers to In-Chapter Materials

Answers to Practice Problems

9.1A (a) potassium carbonate, soluble. (b) barium nitrate, soluble. (c) ammonium carbonate, soluble. (d) barium sulfate, insoluble. (e) mercury(I) iodide, insoluble. **9.1B** (a) calcium chromate, insoluble. (b) sodium phosphate, soluble. (c) barium acetate, soluble. (d) potassium hydroxide, soluble. (e) strontium chromate, insoluble. **9.2A** (a) 17.0% $C_6H_{12}O_6$, (b) 9.88% CCl_4, (c) 14.8% $BaCl_2$. **9.2B** (a) 0.0761% LiF, (b) 0.167% K_2CO_3, (c) 0.0311% $Sr(ClO_3)_2$. **9.3A** (a) 29.3 g $Cu(NO_3)_2$, (b) 70.4 g NH_4CN, (c) 0.050 g NaCl. **9.3B** (a) 59.8 g, (b) 154 g, (c) 711 g. **9.4A** (a) 0.0787 M, (b) 0.120 M, (c) 0.841 M. **9.4B** (a) 0.0787 M, (b) 0.0519 M, (c) 0.365 M. **9.5A** (a) 286 mL, (b) 270 mL, (c) 177 mL.

9.5B (a) 19.5 g, (b) 5.32 g, (c) 0.348 g. **9.6A** (a) 0.588 m, (b) 0.0286 m, (c) 0.271 m. **9.6B** (a) 1.31 m, (b) 0.288 m, (c) 4.22 m. **9.7A** (a) 1.8 M, (b) 1.9 m. **9.7B** 12.8%. **9.8A** (a) 0.200 M, (b) 0.400 M, (c) 0.600 M. **9.8B** 0.111 M. **9.9A** (a) 0.137 M, (b) 9.18 L, (c) 0.260 mol. **9.9B** (a) 0.707 M, (b) 6.48 L, (c) 16.1 mol. **9.10A** 21 mL. **9.10B** 3.8 L. **9.11A** $-12.6°C$. **9.11B** $-8.44°C$. **9.12A** 103.25°C. **9.12B** 107.04°C.

Answers to Checkpoints

9.2.1 c. **9.2.2** d. **9.2.3** b. **9.2.4** e. **9.2.5** a. **9.2.6** d. **9.3.1** e. **9.3.2** d. **9.3.3** b. **9.3.4** d. **9.3.5** d. **9.3.6** a. **9.3.7** d. **9.3.8** d. **9.3.9** a. **9.4.1** a, e. **9.4.2** e. **9.5.1** e. **9.5.2** a. **9.5.3** b. **9.5.4** a. **9.6.1** e. **9.6.2** c. **9.6.3** b.

CHAPTER 10

Chemical Reactions and Chemical Equations

10.1 Recognizing Chemical Reactions

10.2 Representing Chemical Reactions with Chemical Equations
- Metals
- Nonmetals
- Noble Gases
- Metalloids

10.3 Balancing Chemical Equations

10.4 Types of Chemical Reactions
- Precipitation Reactions
- Acid-Base Reactions
- Oxidation-Reduction Reactions

10.5 Chemical Reactions and Energy

10.6 Chemical Reactions in Review

One of the most easily recognized chemical reactions, fire, has been part of the human experience for over a million years.

Lindsay Upson/Image Source/Getty Images

In the early chapters of this book, we learned about the structure of atoms and how the number and arrangement of subatomic particles give rise to the properties of atoms. The properties of atoms determine how they interact with one another to become the matter that we encounter every day. In this chapter, we investigate the chemical changes that matter can undergo, and how we represent these changes, known as *chemical reactions,* with *chemical equations.*

10.1 Recognizing Chemical Reactions

How do we know when a chemical reaction is occurring—or has occurred? Let's consider the two examples used to illustrate chemical changes in Chapter 1: the rusting of iron, and the rising of cakes or cookies during baking. When an iron object, such as the horseshoe shown in Figure 10.1(a) rusts, both its color and its surface texture change dramatically. A new horseshoe is grey and smooth to the touch; and has a relatively shiny, hard surface. A rusted horseshoe appears brown and its surface is rough and flaky. Unlike the iron of the new shoe, we could easily scrape off some of the surface of the rusted shoe. The matter that makes up the surface of the horseshoe has undergone a chemical change. Its identity has changed from *iron* to *rust.* The change is due to a **chemical reaction** involving iron, oxygen, and water. Figure 10.1(b) shows cake batter before and after baking. Baking soda in the batter causes a chemical reaction that produces numerous tiny gas bubbles—causing the cake to rise. These examples illustrate some of the ways that we can recognize a chemical reaction. When the properties of matter change, including changes in color and texture; or when a gas is produced when different types of matter are combined, we can tell that a chemical reaction has occurred. Other evidence of chemical reactions includes the production of heat and light, such as a flame. The burning of wood, for example, is a chemical reaction.

In Chapters 1 and 3, we encountered the first two hypotheses of Dalton's atomic theory. They were:

1. Matter is composed of tiny, indivisible particles called atoms; all atoms of a given element are identical; and atoms of one element are different from atoms of any other element. [◄◄ Section 1.2]
2. Compounds are made up of atoms of more than one element; and in any given compound, the same types of atoms are always present in the same relative numbers. [◄◄ Section 3.4]

We now add Dalton's third hypothesis:

3. Chemical reactions cause the *rearrangement* of atoms, but do not cause either the creation or the destruction of atoms.

Student Note: It is not always obvious to us that a chemical reaction has occurred. The combination of certain aqueous acid and base solutions is an example of a chemical reaction that is not accompanied by any apparent changes in properties, and we must use techniques other than simple observation to determine that a reaction has taken place.

Rewind: We have since learned that the atoms of a given element are not all identical. Atoms whose nuclei contain the same number of protons, but different numbers of neutrons, are different isotopes of the same element.

Figure 10.1 (a) Rusting of iron and (b) rising of baked goods caused by baking soda are familiar examples of chemical reactions.

(a, left) Charles Mann/E+/Getty Images;
(a, right) Davies and Starr/Getty Images;
(b, both) David A. Tietz/McGraw Hill

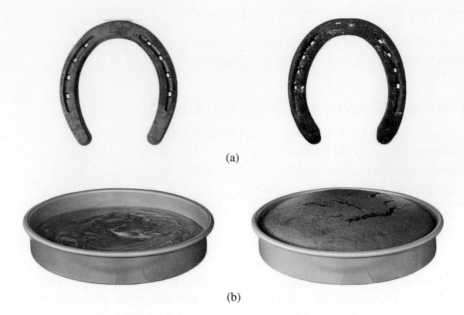

(a)

(b)

Put another way: The same atoms are present *after* a chemical reaction that were present *before* the reaction. Further, because atoms have specific masses, the mass of all of the atoms involved in a chemical reaction is the same before and after the reaction. This is a statement of the **law of conservation of mass.**

Sample Problem 10.1 lets you practice identifying evidence of a chemical reaction.

SAMPLE PROBLEM **10.1** **Identifying Chemical Reactions**

The diagram in (a) shows a compound made up of atoms of two elements (represented by the green and red spheres) in the liquid state. Which of the diagrams in (b) to (d) represent a physical change, and which diagrams represent a chemical reaction?

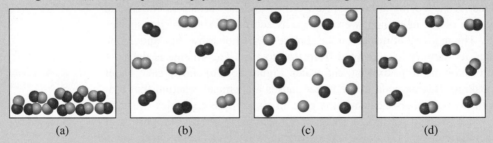

(a)   (b)   (c)   (d)

Strategy We review the discussion of physical change and chemical reactions. A physical change does not change the *identity* of a substance, whereas a chemical reaction *does* change the identity of a substance.

Setup The diagram in (a) shows a substance that consists of molecules of a compound, each of which contains two different atoms, represented by green and red spheres. Diagram (b) contains the same numbers of red and green spheres, but they are not arranged the same way as in diagram (a). In (b), each molecule is made up of two identical atoms. These are molecules of *elements,* rather than molecules of a compound. Diagram (c) also contains the same numbers of red and green spheres as diagram (a). In (c), however, all the atoms are shown as isolated spheres. These are atoms of elements, rather than molecules of a compound. In diagram (d), the spheres are arranged in molecules, each containing one red and one green sphere. Although the molecules are farther apart in diagram (d), they are the same molecules as shown in diagram (a).

Solution Diagrams (b) and (c) represent chemical reactions. Diagram (d) represents a physical change.

THINK ABOUT IT

A chemical reaction changes the *identity* of matter. A physical change does not.

Practice Problem **A**TTEMPT   Which of the following represents a chemical reaction? (a) evaporation of water; (b) combination of hydrogen and oxygen gas to produce water; (c) dissolution of sugar in water; (d) separation of sodium chloride (table salt) into its constituent elements, sodium and chlorine; (e) combustion of sugar to produce carbon dioxide and water.

Practice Problem **B**UILD   The first diagram shows a system prior to a process taking place. Which of the other diagrams (i to iv) could represent the system following a chemical reaction?

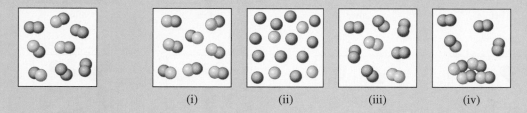

                (i)   (ii)   (iii)   (iv)

Practice Problem **C**ONCEPTUALIZE   The diagram on the left represents the result of a process. Which of the diagrams (i to iii) could represent the starting material if the process were a chemical reaction?

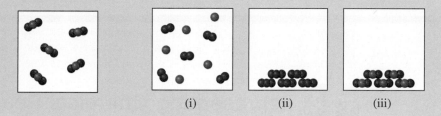

                (i)   (ii)   (iii)

CHECKPOINT–SECTION 10.1   Recognizing Chemical Reactions

10.1.1   Which of the following (a–f) represent chemical reactions?

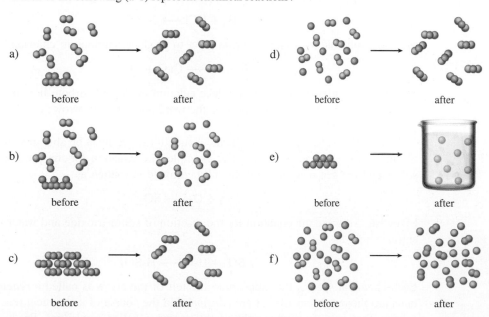

(Continued on next page)

10.1.2 Which of the following sets of before and after diagrams represent chemical reactions? Each sphere color represents a different element. (Select all that apply.)

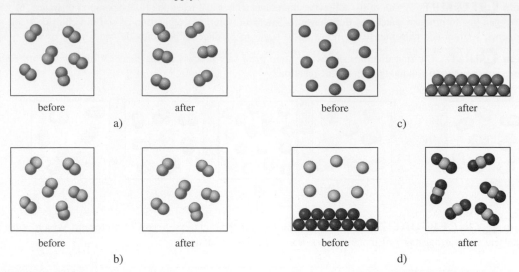

before	after
a)	

before	after
c)	

before	after
b)	

before	after
d)	

10.2 Representing Chemical Reactions with Chemical Equations

A ***chemical equation*** uses chemical symbols to denote what occurs in a chemical reaction. We have seen how chemists represent elements and compounds using chemical symbols. Now we consider how chemists represent chemical reactions using chemical equations.

A chemical equation represents a *chemical statement*. When you encounter a chemical equation, you may find it useful to read it as though it were a sentence. Read

$$NH_3 + HCl \longrightarrow NH_4Cl$$

as "Ammonia and hydrogen chloride react to produce ammonium chloride." Read

$$CaCO_3 \longrightarrow CaO + CO_2$$

as "Calcium carbonate reacts to produce calcium oxide and carbon dioxide." Thus, the plus signs can be interpreted simply as the word *and,* and the arrows can be interpreted as the phrase "react(s) to produce."

In addition to interpreting chemical equations, you must be able to write chemical equations to represent reactions. For example, the equation for the process by which sulfur and oxygen react to produce sulfur dioxide is written as

$$S + O_2 \longrightarrow SO_2$$

Likewise, we write the equation for the reaction of sulfur trioxide and water to produce sulfuric acid as

$$SO_3 + H_2O \longrightarrow H_2SO_4$$

Each chemical species that appears to the left of the arrow is called a ***reactant***. Reactants are those substances that are *consumed* in the course of a chemical reaction. Each species that appears to the right of the arrow is called a ***product***. Products are the substances that *form* during the course of a chemical reaction.

Chemists usually indicate the physical states of reactants and products with italicized letters in parentheses following each species in the equation. Gases, liquids, and

Student Note: Some problems in later chapters can be solved only if the states of reactants and products are specified. It is a good idea to get in the habit now of including the physical states of reactants and products in the chemical equations that you write.

solids are labeled with (g), (l), and (s), respectively. Chemical species that are dissolved in water are said to be *aqueous* and are labeled (aq). The equation examples given previously can be written as follows:

$$NH_3(g) + HCl(g) \longrightarrow NH_4Cl(s) \qquad S(s) + O_2(g) \longrightarrow SO_2(g)$$

$$CaCO_3(s) \longrightarrow CaO(s) + CO_2(g) \qquad SO_3(g) + H_2O(l) \longrightarrow H_2SO_4(l)$$

At this point, it is useful to discuss how substances are represented in chemical equations. Compounds, whether they are ionic or molecular, are represented with their chemical formulas. However, because *free* elements (those that are *uncombined*) exist in a variety of forms, and because some elements can exist in multiple forms, there are two different ways to represent them.

Metals

Because metals usually do not exist in discrete molecular units but rather in complex, three-dimensional networks of atoms, we always use their empirical formulas in chemical equations. The empirical formulas are the same as the symbols that represent the elements. For example, the empirical formula for iron is Fe, the same as the symbol for the element.

Nonmetals

There is no single rule regarding the representation of nonmetals in chemical equations. Carbon, for example, exists in several different forms. Regardless of the form, we use its empirical formula C to represent elemental carbon in chemical equations. Often the symbol C will be followed by the specific form in parentheses. Thus, we represent two of carbon's forms as C(graphite) and C(diamond).

For nonmetals that exist as polyatomic molecules, we generally use the molecular formula in equations: H_2, N_2, O_2, F_2, Cl_2, Br_2, I_2, and P_4, for example. In the case of sulfur, however, we usually use the empirical formula S rather than the molecular formula S_8—although the molecular formula S_8 is equally correct and is used sometimes.

Noble Gases

All the noble gases exist as isolated atoms, so we use their symbols: He, Ne, Ar, Kr, Xe, and Rn.

Metalloids

The metalloids, like the metals, generally have complex three-dimensional networks, so we also represent them with their empirical formulas—that is, their symbols: B, Si, Ge, and so on.

Chemical equations are also used to represent *physical* processes. Sucrose $(C_{12}H_{22}O_{11})$ dissolving in water, for example, is a physical process [◄ Section 3.1] that can be represented with the following chemical equation:

$$C_{12}H_{22}O_{11}(s) \xrightarrow{H_2O} C_{12}H_{22}O_{11}(aq)$$

The H_2O over the arrow in the equation denotes the process of dissolving a substance in water. Although formulas or symbols are sometimes omitted for simplicity, they can be written over an arrow in a chemical equation to indicate the conditions under which the reaction takes place. For example, in the chemical equation

$$2KClO_3(s) \xrightarrow{\Delta} 2KCl(s) + 3O_2(g)$$

the symbol Δ indicates that the addition of heat energy is necessary to make $KClO_3$ react to form KCl and O_2.

Student Note: Students sometimes fear that they will be asked to determine the products of an unfamiliar chemical reaction—and will be unable to do so. In this chapter and in Chapter 11, we explain how to recognize *patterns* in chemical reactions, and how to deduce the products of several different types of reactions.

Student Note: Recall that for ionic compounds, the chemical formula is usually the *empirical* formula [◄◄ Section 3.4].

10.3 Balancing Chemical Equations

Based on what you have learned so far, the chemical equation for the explosive reaction of hydrogen gas with oxygen gas to form liquid water would be

$$H_2(g) \quad + \quad O_2(g) \quad \longrightarrow \quad H_2O(l)$$

This equation as it is written, however, violates the *law of conservation of mass* because four atoms (two H and two O) react to produce only three atoms (two H and one O). The law of conservation of mass is another way of stating Dalton's third hypothesis, which is that atoms can be neither created nor destroyed.

The equation must be *balanced* so that the same number of each kind of atom appears on both sides of the reaction arrow. Balancing is achieved by writing appropriate ***stoichiometric coefficients*** (often referred to simply as ***coefficients***) to the left of the chemical formulas. In this case, we write a coefficient of 2 to the left of both the $H_2(g)$ and the $H_2O(l)$:

$$2H_2(g) \quad + \quad O_2(g) \quad \longrightarrow \quad 2H_2O(l)$$

There are now four H atoms and two O atoms on each side of the arrow. When balancing a chemical equation, we can change only the coefficients that precede the chemical formulas, *not* the subscripts within the chemical formulas. Changing the subscripts would change the formulas for the species involved in the reaction. For example, changing the product from H_2O to H_2O_2 would result in equal numbers of each kind of atom on both sides of the equation, but the equation we set out to balance represented the combination of hydrogen gas and oxygen gas to form water (H_2O), not hydrogen peroxide (H_2O_2). Additionally, we cannot add reactants or products to the chemical equation for the purpose of balancing it. To do so would result in an equation that represents the wrong reaction. The chemical equation must be made quantitatively correct without changing its qualitative chemical statement.

Balancing a chemical equation requires something of a trial-and-error approach. You may find that you change the coefficient for a particular reactant or product, only to have to change it again later in the process. In general, it will facilitate the balancing process if you do the following:

1. Change the coefficients of compounds (e.g., CO_2) before changing the coefficients of elements (e.g., O_2).
2. Treat polyatomic ions that appear on both sides of the equation (e.g., CO_3^{2-}) as units, rather than counting their constituent atoms individually.
3. Count atoms and/or polyatomic ions carefully, and track their numbers each time you change a coefficient.

Student Note: *Combustion* refers to burning in the presence of oxygen. Combustion of a hydrocarbon such as butane produces carbon dioxide and water.

To balance the chemical equation for the combustion of butane, we first take an inventory of the numbers of each type of atom on each side of the arrow.

$$C_4H_{10}(g) + O_2(g) \longrightarrow CO_2(g) + H_2O(g)$$

$$4-C-1$$

$$10-H-2$$

$$2-O-3$$

Initially, there are 4 C atoms on the left and 1 on the right; 10 H atoms on the left and 2 on the right; and 2 O atoms on the left with 3 on the right. As a first step, we will place a coefficient of 4 in front of $CO_2(g)$ on the product side.

$$C_4H_{10}(g) + O_2(g) \longrightarrow 4CO_2(g) + H_2O(g)$$

$$4-C-4$$
$$10-H-2$$
$$2-O-9$$

This changes the tally of atoms as shown. Thus, the equation is balanced for carbon, but not for hydrogen or oxygen. Next, we place a coefficient of 5 in front of $H_2O(g)$ on the product side and tally the atoms on both sides again.

$$C_4H_{10}(g) + O_2(g) \longrightarrow 4CO_2(g) + 5H_2O(g)$$

$$4-C-4$$
$$10-H-10$$
$$2-O-13$$

Now the equation is balanced for carbon and hydrogen. Only oxygen remains to be balanced. There are 13 O atoms on the product side of the equation (8 in CO_2 molecules and another 5 in H_2O molecules), so we need 13 O atoms on the reactant side. Because each oxygen molecule contains 2 O atoms, we will have to place a coefficient of $\frac{13}{2}$ in front of $O_2(g)$:

$$C_4H_{10}(g) + \tfrac{13}{2}O_2(g) \longrightarrow 4CO_2(g) + 5H_2O(g)$$

$$4-C-4$$
$$10-H-10$$
$$13-O-13$$

With equal numbers of each kind of atom on both sides of the equation, this equation is now balanced. For now, however, you should practice balancing equations with the smallest possible *whole* number coefficients. Multiplying each coefficient by 2 gives all whole numbers and a final balanced equation:

$$2C_4H_{10}(g) + 13O_2(g) \longrightarrow 8CO_2(g) + 10H_2O(g)$$

$$8-C-8$$
$$20-H-20$$
$$26-O-26$$

Student Note: A balanced equation is, in a sense, a mathematical equality. We can multiply or divide through by any number, and the equality will still be valid.

Sample Problems 10.2 and 10.3 let you practice writing and balancing chemical equations.

SAMPLE PROBLEM (**10.2**) Writing and Balancing Chemical Equations

Write and balance the chemical equation for the aqueous reaction of barium hydroxide and perchloric acid to produce aqueous barium perchlorate and water.

Determine the formulas and physical states of all reactants and products, and use them to write a chemical equation that makes the correct chemical statement. Finally, adjust coefficients in the resulting chemical equation to ensure that there are identical numbers of each type of atom on both sides of the reaction arrow.

The reactants are $Ba(OH)_2$ and $HClO_4$, and the products are $Ba(ClO_4)_2$ and H_2O. Because the reaction is aqueous, all species except H_2O will be labeled (*aq*) in the equation. Being a liquid, H_2O will be labeled (*l*).

Student Note: You may wish to review how to deduce a compound's formula from its name [◄◄ Sections 3.3 and 3.5].

Solution The chemical statement "barium hydroxide and perchloric acid react to produce barium perchlorate and water" can be represented with the following unbalanced equation:

$$Ba(OH)_2(aq) + HClO_4(aq) \longrightarrow Ba(ClO_4)_2(aq) + H_2O(l)$$

(Continued on next page)

Perchlorate ions (ClO_4^-) appear on both sides of the equation, so count them as units, rather than counting the individual atoms they contain. Thus, the tally of atoms and polyatomic ions is

$$Ba(OH)_2(aq) + HClO_4(aq) \longrightarrow Ba(ClO_4)_2(aq) + H_2O(l)$$

$$1-Ba-1$$

$$2-O-1 \text{ (not including O atoms in } ClO_4^- \text{ ions)}$$

$$3-H-2$$

$$1-ClO_4^--2$$

The barium atoms are already balanced, and placing a coefficient of 2 in front of $HClO_4(aq)$ balances the number of perchlorate ions.

$$Ba(OH)_2(aq) + 2HClO_4(aq) \longrightarrow Ba(ClO_4)_2(aq) + H_2O(l)$$

$$1-Ba-1$$

$$2-O-1 \text{ (not including O atoms in } ClO_4^- \text{ ions)}$$

$$4-H-2$$

$$2-ClO_4^--2$$

Placing a coefficient of 2 in front of $H_2O(l)$ balances both the O and H atoms, giving us the final balanced equation:

$$Ba(OH)_2(aq) + 2HClO_4(aq) \longrightarrow Ba(ClO_4)_2(aq) + 2H_2O(l)$$

$$1-Ba-1$$

$$2-O-2 \text{ (not including O atoms in } ClO_4^- \text{ ions)}$$

$$4-H-4$$

$$2-ClO_4^--2$$

THINK ABOUT IT

Check to be sure the equation is balanced by counting all the atoms individually.

$$1-Ba-1$$
$$10-O-10$$
$$4-H-4$$
$$2-Cl-2$$

Practice Problem **A**TTEMPT   Write and balance the chemical equation that represents the combustion of propane (i.e., the reaction of propane gas, C_3H_8, with oxygen gas to produce carbon dioxide gas and water vapor).

Practice Problem **B**UILD   Write and balance the chemical equation that represents the aqueous reaction of sulfuric acid with sodium hydroxide to form water and sodium sulfate.

Practice Problem **C**ONCEPTUALIZE   Write a balanced equation for the combustion reaction shown here.

SAMPLE PROBLEM (10.3) Writing a Balanced Chemical Equation for a Combustion Reaction

Butyric acid (also known as butanoic acid, $C_4H_8O_2$) is one of many compounds found in milk fat. First isolated from rancid butter in 1869, butyric acid has received a great deal of attention in recent years as a potential anticancer agent. Write a balanced equation for the metabolism of butyric acid. Assume that the overall processes of metabolism and combustion are the same (i.e., reaction with oxygen to produce carbon dioxide and water).

Strategy Begin by writing an unbalanced equation to represent the combination of reactants and formation of products as stated in the problem, and then balance the equation.

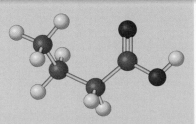

Butyric acid

Setup Metabolism in this context refers to the combination of aqueous $C_4H_8O_2$ with O_2 gas to produce CO_2 gas and H_2O vapor. (Remember that the word "vapor" is used to describe the gaseous state of a substance that ordinarily exists in a condensed state. Because water is ordinarily a liquid, we refer to its gaseous state as water *vapor.*)

Solution

$$C_4H_8O_2(aq) + O_2(g) \longrightarrow CO_2(g) + H_2O(g)$$

Balance the number of C atoms by changing the coefficient for CO_2 from 1 to 4.

$$C_4H_8O_2(aq) + O_2(g) \longrightarrow 4CO_2(g) + H_2O(g)$$

Balance the number of H atoms by changing the coefficient for H_2O from 1 to 4.

$$C_4H_8O_2(aq) + O_2(g) \longrightarrow 4CO_2(g) + 4H_2O(g)$$

Finally, balance the number of O atoms by changing the coefficient for O_2 from 1 to 5.

$$C_4H_8O_2(aq) + 5O_2(g) \longrightarrow 4CO_2(g) + 4H_2O(g)$$

THINK ABOUT IT

Count the number of each type of atom on each side of the reaction arrow to verify that the equation is properly balanced. There are 4 C, 8 H, and 12 O in the reactants and in the products, so the equation is balanced.

Practice Problem ATTEMPT Another compound found in milk fat that appears to have anticancer and anti-obesity properties is conjugated linoleic acid (CLA; $C_{18}H_{32}O_2$). Assuming again that the only products are CO_2 gas and H_2O vapor, write a balanced equation for the metabolism of CLA.

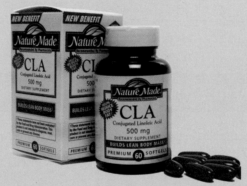

John Flournoy/McGraw Hill

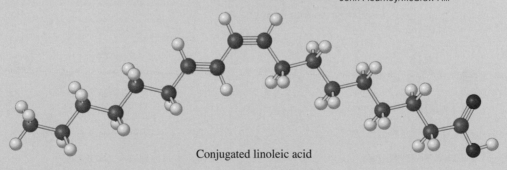

Conjugated linoleic acid

Practice Problem BUILD Write a balanced equation for the combination of ammonia gas with solid copper(II) oxide to form copper metal, nitrogen gas, and liquid water.

Practice Problem CONCEPTUALIZE The compound shown on the left reacts with nitrogen dioxide to form the compound shown on the right and iodine. Write a balanced equation for the reaction.

Methyl iodide

Nitromethane

Familiar Chemistry

The Stoichiometry of Metabolism

The carbohydrates and fats we eat are broken down into small molecules in the digestive system. Carbohydrates are broken down into simple sugars such as glucose ($C_6H_{12}O_6$), and fats are broken down into fatty acids and glycerol ($C_3H_8O_3$). The small molecules produced in the digestion process are subsequently consumed by a series of complex biochemical reactions. Although the metabolism of simple sugars and fatty acids involves relatively complex processes, the results are essentially the same as that of combustion—that is, simple sugars and fatty acids react with oxygen to produce carbon dioxide, water, and energy. The balanced chemical equation for the metabolism of glucose is

Glycerol

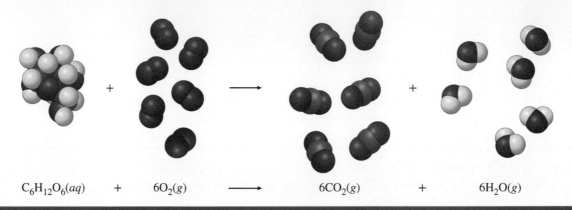

$$C_6H_{12}O_6(aq) \quad + \quad 6O_2(g) \quad \longrightarrow \quad 6CO_2(g) \quad + \quad 6H_2O(g)$$

CHECKPOINT–SECTION 10.3   Balancing Chemical Equations

10.3.1 What are the stoichiometric coefficients in the following equation when it is balanced?

$$CH_4(g) + H_2O(g) \longrightarrow H_2(g) + CO_2(g)$$

    a) 1, 2, 2, 2   b) 2, 1, 1, 2   c) 1, 2, 2, 1   d) 2, 2, 2, 1   e) 1, 2, 4, 1

10.3.2 Which chemical equation represents the reaction shown?

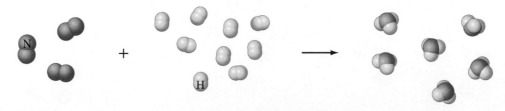

    a) $6N(g) + 18H(g) \longrightarrow 6NH_3(g)$ d) $N_2(g) + 3H_2(g) \longrightarrow 2NH_3(g)$

    b) $3N_2(g) + 3H_2(g) \longrightarrow 2NH_3(g)$ e) $3N_2(g) + 6H_2(g) \longrightarrow 6NH_3(g)$

    c) $2N_2(g) + 3H_2(g) \longrightarrow 2NH_3(g)$

10.3.3 Which is the correctly balanced form of the given equation?

    a) $S(s) + O_3(g) \longrightarrow SO_2(g)$ d) $3S(s) + 2O_3(g) \longrightarrow SO_2(g)$

    b) $3S(s) + 6O_3(g) \longrightarrow 3SO_2(g)$ e) $3S(s) + 2O_3(g) \longrightarrow 3SO_2(g)$

    c) $3S(s) + O_3(g) \longrightarrow 3SO_2(g)$

10.3.4 Carbon monoxide reacts with oxygen to produce carbon dioxide according to the following balanced equation:

$$2CO(g) + O_2(g) \longrightarrow 2CO_2(g)$$

A reaction vessel containing the reactants is pictured.

Which of the following represents the contents of the reaction vessel when the reaction is complete?

a)   b)   c)   d)

10.4 Types of Chemical Reactions

In this section, we look at several types of reactions that occur in aqueous solution, and we see how balanced chemical equations are used to represent them.

Precipitation Reactions

One of the more obvious indicators of a chemical reaction is the formation of a solid when two solutions are combined. When an aqueous solution of lead(II) nitrate [Pb(NO$_3$)$_2$] is added to an aqueous solution of sodium iodide (NaI), a yellow insoluble solid—lead(II) iodide (PbI$_2$)—forms. Sodium nitrate (NaNO$_3$), the other reaction product, remains in solution. Figure 10.2 shows this reaction in progress. An insoluble ionic

Figure 10.2 A colorless aqueous solution of NaI is added to a colorless aqueous solution of Pb(NO$_3$)$_2$. A yellow precipitate, PbI$_2$, forms. Na$^+$ and NO$_3^-$ ions remain in solution.

Charles D. Winters/Timeframe Photography/McGraw Hill

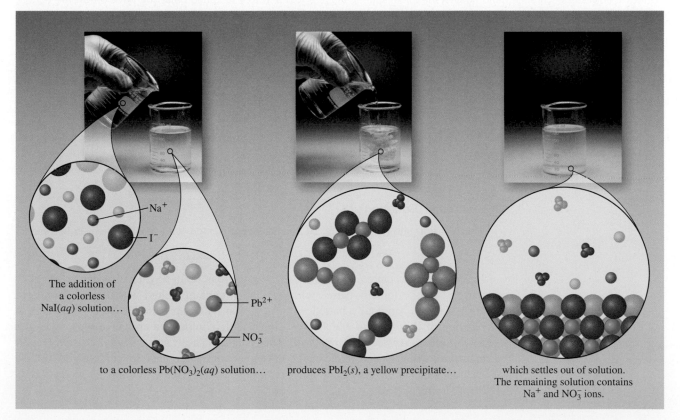

The addition of a colorless NaI(aq) solution...

to a colorless Pb(NO$_3$)$_2$(aq) solution...

Na$^+$

I$^-$

Pb^{2+}

NO$_3^-$

produces PbI$_2$(s), a yellow precipitate...

which settles out of solution. The remaining solution contains Na$^+$ and NO$_3^-$ ions.

solid product that separates from a solution is called a ***precipitate,*** and a chemical reaction in which a precipitate forms is called a ***precipitation reaction.*** Precipitation reactions usually involve ionic compounds, but a precipitate does not form every time two solutions of electrolytes are combined. Instead, whether or not a precipitate forms when two solutions are mixed depends on the solubility of the products. Because being able to predict when a precipitate will form requires you to use the solubility guidelines from Chapter 9, we review them here.

Solubility Guidelines Revisited

Figure 10.3 shows the solubility guidelines for ionic compounds that typically are soluble; and Figure 10.4 shows the solubility guidelines for ionic compounds that typically are not soluble.

Sample Problem 10.4 lets you practice determining whether or not an ionic compound is water soluble.

SAMPLE PROBLEM Determining Aqueous Solubility/Insolubility of Ionic Compounds

Classify each of the following compounds as soluble or insoluble in water: (a) $AgNO_3$, (b) $CaSO_4$, (c) K_2CO_3.

Strategy Use the guidelines in Figures 10.3 and 10.4 to determine whether or not each compound is expected to be water soluble.

Setup (a) $AgNO_3$ contains the nitrate ion (NO_3^-). According to Figure 10.3, *all* compounds containing the nitrate ion are soluble. (b) $CaSO_4$ contains the sulfate ion (SO_4^{2-}). According to Figure 10.3, compounds containing the sulfate ion are soluble unless the cation is Ag^+, Hg_2^{2+}, Pb^{2+}, Ca^{2+}, Sr^{2+}, or Ba^{2+}. Thus, the Ca^{2+} ion is one of the insoluble exceptions. (c) K_2CO_3 contains an alkali metal cation (K^+) for which, according to Figure 10.3, there are no insoluble exceptions. Alternatively, Figure 10.4 shows that most compounds containing the carbonate ion (CO_3^{2-}) are insoluble—but compounds containing a Group 1 cation such as K^+ are soluble exceptions.

Solution (a) soluble, (b) insoluble, (c) soluble

THINK ABOUT IT

Check the ions in each compound against the information in Figures 10.3 and 10.4 to confirm that you have drawn the right conclusions.

Practice Problem **A**TTEMPT   Classify each of the following compounds as soluble or insoluble in water: (a) $PbCl_2$, (b) $(NH_4)_3PO_4$, (c) $Fe(OH)_3$.

Practice Problem **B**UILD   Classify each of the following compounds as soluble or insoluble in water: (a) $MgBr_2$, (b) $Ca_3(PO_4)_2$, (c) $KClO_3$.

Practice Problem **C**ONCEPTUALIZE   Using Figures 10.3 and 10.4, identify a compound that will cause precipitation of two different insoluble ionic compounds when an aqueous solution of it is added to an aqueous solution of iron(III) sulfate.

Molecular Equations

The reaction shown in Figure 10.2 can be represented with the chemical equation

$$Pb(NO_3)_2(aq) + 2NaI(aq) \longrightarrow 2NaNO_3(aq) + PbI_2(s)$$

Based on this chemical equation, the metal cations seem to exchange anions. That is, the Pb^{2+} ion, originally paired with NO_3^- ions, ends up paired with I^- ions; similarly, each Na^+ ion, originally paired with an I^- ion, ends up paired with an NO_3^- ion. A reaction of this type, in which two soluble ionic compounds exchange ions, is known as a ***double-displacement reaction.*** This equation, as written, is called a ***molecular equation,*** which is a chemical equation written with all compounds represented by their chemical formulas, making it look as though they exist in solution as molecules or formula units.

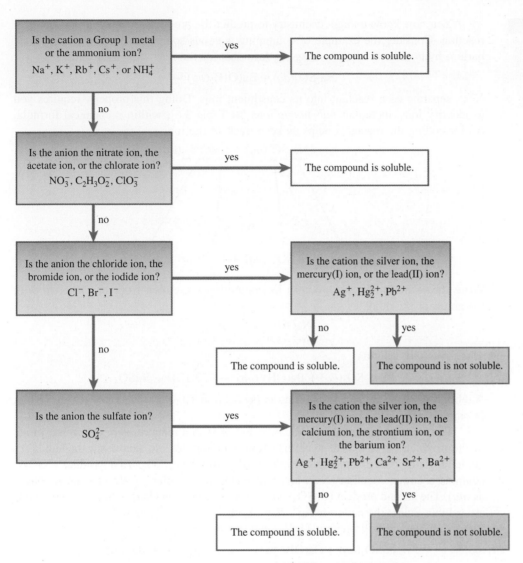

Figure 10.3 Flowchart for typically *soluble* compounds.

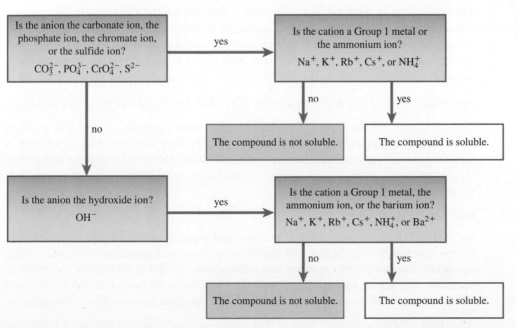

Figure 10.4 Flowchart for typically *insoluble* compounds.

You now know enough chemistry to predict the products of this type of chemical reaction. Consider the example of combining aqueous solutions of sodium sulfate and barium hydroxide. Start by writing the formulas for the reactants [◄◄ Section 3.3].

$$Na_2SO_4(aq) + Ba(OH)_2(aq) \longrightarrow$$

Next, separate each reactant into its constituent ions. Doing this properly requires you to identify ions, including *polyatomic* ions [◄◄ Table 3.6], within a chemical formula. (Color coding the reactants helps us keep track of the ions.)

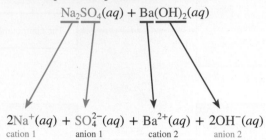

$$\underset{\text{cation 1}}{2Na^+(aq)} + \underset{\text{anion 1}}{SO_4^{2-}(aq)} + \underset{\text{cation 2}}{Ba^{2+}(aq)} + \underset{\text{anion 2}}{2OH^-(aq)}$$

Write the formulas for the products by combining the cation from one reactant with the anion of the other.

$$\underset{\text{cation 1 \quad anion 2}}{NaOH} + \underset{\text{cation 2 \quad anion 1}}{BaSO_4}$$

This gives the equation:

$$Na_2SO_4(aq) + Ba(OH)_2(aq) \longrightarrow 2NaOH + BaSO_4$$

Although we have balanced the equation [◄◄ Section 10.3], we have not yet put phases in parentheses for the products.

The final step in predicting the outcome of such a reaction is to determine which of the products, if any, will precipitate from solution. We do this using the solubility guidelines for ionic compounds (Figures 10.3 and 10.4). The first product (NaOH) contains a Group 1 cation (Na^+) and will therefore be soluble. We indicate its phase as (aq). The second product ($BaSO_4$) contains the sulfate ion (SO_4^{2-}). Sulfate compounds are soluble unless the cation is Ag^+, (Hg_2^{2+}), Pb^{2+}, Ca^{2+}, Sr^{2+}, or Ba^{2+}. $BaSO_4$ is therefore insoluble and will precipitate. We indicate its phase as (s):

$$Na_2SO_4(aq) + Ba(OH)_2(aq) \longrightarrow 2NaOH(aq) + BaSO_4(s)$$

Complete Ionic Equations

Although molecular equations are useful, especially from the standpoint of knowing which solutions to combine in the laboratory, they are not the most realistic representation of what's actually happening in solution. Because soluble ionic compounds are *strong electrolytes* [◄◄ Section 9.4], they exist in solution as hydrated *ions,* rather than as formula units. Thus, it would be more realistic to represent the aqueous species in the reaction of $Na_2SO_4(aq)$ with $Ba(OH)_2(aq)$ as separate, aqueous ions:

$$2Na^+(aq) + SO_4^{2-}(aq) + Ba^{2+}(aq) + 2OH^-(aq) \longrightarrow 2Na^+(aq) + 2OH^-(aq) + BaSO_4(s)$$

This version of the equation is called a **complete ionic equation,** a chemical equation in which any compound that exists completely or predominantly as ions in solution is represented as those ions. Species that are insoluble or that exist in solution completely or predominantly as molecules are represented with their chemical formulas, as they were in the molecular equation.

Net Ionic Equations

$Na^+(aq)$ and $OH^-(aq)$ both appear as reactants and products in the complete ionic equation for the reaction of $Na_2SO_4(aq)$ with $Ba(OH)_2(aq)$. Ions that appear on both sides of the equation arrow are called **spectator ions** because they do not participate in the reaction. Spectator ions cancel one another, just as identical terms on both sides of an algebraic equation cancel one another, so we need not show spectator ions in chemical equations.

$$2\cancel{Na^+(aq)} + SO_4^{2-}(aq) + Ba^{2+}(aq) + 2\cancel{OH^-(aq)} \longrightarrow 2\cancel{Na^+(aq)} + 2\cancel{OH^-(aq)} + BaSO_4(s)$$

Eliminating the spectator ions yields the following equation:

$$Ba^{2+}(aq) + SO_4^{2-}(aq) \longrightarrow BaSO_4(s)$$

> **Student Note:** Although the reactants may be written in either order in the net ionic equation, it is common for the cation to be shown first and the anion second.

This version of the equation is called a ***net ionic equation,*** which is a chemical equation that includes only the species that are actually involved in the reaction. The net ionic equation, in effect, tells us what actually happens when we combine solutions of sodium sulfate and barium hydroxide.

The steps necessary to determine the molecular, complete ionic, and net ionic equations for a precipitation reaction are as follows:

1. Write and balance the molecular equation, predicting the products by assuming that the cations trade anions.
2. Write the complete ionic equation by separating strong electrolytes into their constituent ions.
3. Write the net ionic equation by identifying and canceling spectator ions on both sides of the equation.

If both products of a reaction are strong electrolytes, all the ions in solution are spectator ions. In this case, there is no net ionic equation and no reaction takes place.

Sample Problem 10.5 illustrates the stepwise determination of molecular, complete ionic, and net ionic equations.

SAMPLE PROBLEM 10.5   Writing Molecular, Complete Ionic, and Net Ionic Equations

Write the molecular, complete ionic, and net ionic equations for the reaction that occurs when aqueous solutions of lead(II) acetate [$Pb(C_2H_3O_2)_2$] and calcium chloride ($CaCl_2$) are combined.

Strategy Predict the products by exchanging ions and balance the equation. Determine which product will precipitate based on the solubility guidelines in Figures 10.3 and 10.4. Rewrite the equation showing strong electrolytes as ions. Identify and cancel spectator ions.

Setup The products of the reaction are $PbCl_2$ and $Ca(C_2H_3O_2)_2$. $PbCl_2$ is insoluble because Pb^{2+} is one of the insoluble exceptions for chlorides, which are generally soluble. $Ca(C_2H_3O_2)_2$ is soluble because all acetates are soluble.

Solution Molecular equation:

$$Pb(C_2H_3O_2)_2(aq) + CaCl_2(aq) \longrightarrow PbCl_2(s) + Ca(C_2H_3O_2)_2(aq)$$

Complete ionic equation:

$$Pb^{2+}(aq) + 2C_2H_3O_2^-(aq) + Ca^{2+}(aq) + 2Cl^-(aq) \longrightarrow PbCl_2(s) + Ca^{2+}(aq) + 2C_2H_3O_2^-(aq)$$

Net ionic equation:

$$Pb^{2+}(aq) + 2\cancel{C_2H_3O_2^-}(aq) + \cancel{Ca^{2+}}(aq) + 2Cl^-(aq) \longrightarrow PbCl_2(s) + \cancel{Ca^{2+}}(aq) + 2\cancel{C_2H_3O_2^-}(aq)$$

Canceling spectator ions in the complete ionic equation gives

$$Pb^{2+}(aq) + 2Cl^-(aq) \longrightarrow PbCl_2(s)$$

THINK ABOUT IT

Remember that the charges on ions in a compound must sum to zero. Make sure that you have written correct formulas for the products and that each of the equations you have written is balanced. If you find that you are having trouble balancing an equation, check to make sure you have correct formulas for the products.

Practice Problem A **TTEMPT** Write the molecular, complete ionic, and net ionic equations for the combination of $Sr(NO_3)_2(aq)$ and $Li_2SO_4(aq)$.

Practice Problem B **UILD** Write the molecular, complete ionic, and net ionic equations for the combination of $KNO_3(aq)$ and $BaCl_2(aq)$.

(Continued on next page)

Practice Problem CONCEPTUALIZE Which diagram best represents the result when equal volumes of equal-concentration aqueous solutions of barium nitrate and potassium phosphate are combined?

$PO_4^{3-} = $ ○ $Ba^{2+} = $ ● $NO_3^- = $ ● $K^+ = $ ●

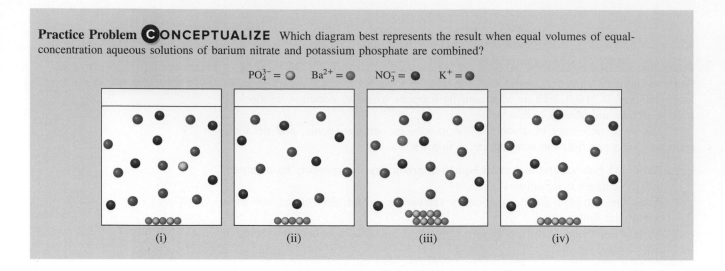

(i)   (ii)   (iii)   (iv)

Acid-Base Reactions

We learned in Chapter 3 that an acid is a substance that produces hydrogen ions, H^+, when dissolved in water. A ***base*** is a substance that produces *hydroxide* ions, OH^-, when dissolved in water. We frequently encounter acids and bases in everyday life (Figure 10.5). Ascorbic acid, for instance, is also known as vitamin C; acetic acid is the component responsible for the sour taste and characteristic smell of vinegar.

Figure 10.5 Some common acids and bases. From left to right: Sodium hydroxide (NaOH), ascorbic acid ($H_2C_6H_6O_6$), hydrochloric acid (HCl), acetic acid ($HC_2H_3O_2$), and ammonia (NH_3). Although it is not an ionic compound, HCl is a strong electrolyte and exists in solution entirely as H^+ and Cl^- ions.

David A. Tietz/Editorial Image, LLC

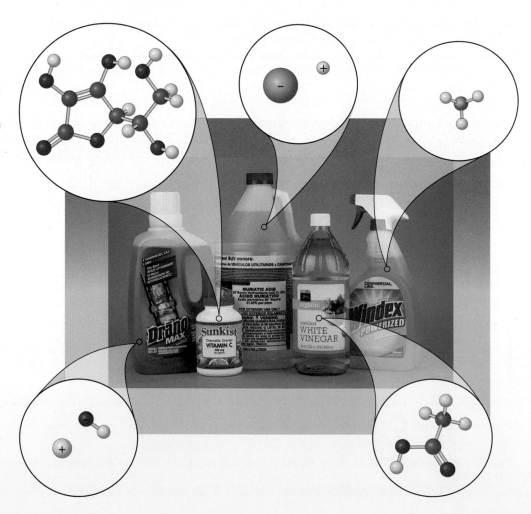

Hydrochloric acid is the acid in muriatic acid, commonly used to clean swimming pools; and is also the principal ingredient in gastric juice (stomach acid). Ammonia, found in many cleaning products, and sodium hydroxide, found in drain cleaner, are common bases.

A *neutralization reaction* is a reaction between an acid and a base. In general, an aqueous acid-base reaction produces water and a *salt,* which is an ionic compound made up of the cation from a base and the anion from an acid. The substance we know as table salt, NaCl, is a familiar example. It is a product of the following acid-base reaction:

$$HCl(aq) + NaOH(aq) \longrightarrow H_2O(l) + NaCl(aq)$$

However, because the acid, base, and salt are all strong electrolytes, they exist entirely as ions in solution. The complete ionic equation is

$$H^+(aq) + Cl^-(aq) + Na^+(aq) + OH^-(aq) \longrightarrow H_2O(l) + Na^+(aq) + Cl^-(aq)$$

The net ionic equation is

$$H^+(aq) + OH^-(aq) \longrightarrow H_2O(l)$$

Both Na^+ and Cl^- are spectator ions. If we were to carry out the preceding reaction using stoichiometric amounts [◄◄ Section 10.3] of HCl and NaOH, the result would be neutral salt water with no leftover acid or base.

The following are also examples of acid-base neutralization reactions, represented by molecular equations:

$$HNO_3(aq) + KOH(aq) \longrightarrow H_2O(l) + KNO_3(aq)$$

$$H_2SO_4(aq) + 2NaOH(aq) \longrightarrow 2H_2O(l) + Na_2SO_4(aq)$$

$$2HC_2H_3O_2(aq) + Ba(OH)_2(aq) \longrightarrow 2H_2O(l) + Ba(C_2H_3O_2)_2(aq)$$

Like a precipitation reaction, acid-base neutralization is a double-displacement reaction where two aqueous compounds exchange ions.

Familiar Chemistry

Oxygen Generators

Although most people have heard the routine safety instructions given by flight attendants before every flight, relatively few of us have actually had the yellow oxygen masks drop down because of a sudden change in cabin air pressure. When the masks do drop down, they deliver supplemental oxygen generated by a chemical reaction. A common ingredient in aircraft oxygen generators is sodium chlorate, $NaClO_3$. Solid $NaClO_3$ reacts to produce sodium chloride and oxygen gas. The balanced equation for the reaction is:

$$2NaClO_3(s) \longrightarrow 2NaCl(s) + 3O_2(g)$$

The $NaClO_3$ is stored in canisters approximately the size of a soda can. The oxygen-producing reaction begins when passengers pull the masks toward themselves. This action breaks the canister's release pins and causes detonation of a small blasting cap that initiates the reaction. Because the reaction is exothermic, the canisters get hot when activated.

Although the canisters do not ordinarily pose a safety risk, in one tragic case a mislabeled box of canisters was being transported in the cargo hold of a commercial aircraft. The canisters caught fire and set fire to the rest of the cargo hold's contents. The fire burned extraordinarily fast and hot because of the oxygen produced by the canisters, and the plane crashed into the Everglades, killing all 110 people on board.

withGod/Shutterstock

Peter A. Harris/AP Images

Sample Problem 10.6 involves an acid-base neutralization reaction.

SAMPLE PROBLEM (**10.6**)   Writing Molecular, Complete Ionic, and Net Ionic Equations for an Acid-Base Reaction

Milk of magnesia, an over-the-counter laxative, is a mixture of magnesium hydroxide [$Mg(OH)_2$] and water. Because $Mg(OH)_2$ is insoluble in water (see Figure 10.4), milk of magnesia is a suspension rather than a solution. The undissolved solid is responsible for the milky appearance of the product. When acid such as HCl is added to milk of magnesia, the suspended $Mg(OH)_2$ dissolves, and the result is a clear, colorless solution. Write and balance the molecular equation, and then give the complete ionic and net ionic equations for this reaction.

> **Student Note:** Most suspended solids will settle to the bottom of the bottle, making it necessary to "shake well before using." Shaking redistributes the solid throughout the liquid.

(a) Milk of magnesia

(b) Addition of HCl

(c) Resulting clear solution

(all) Charles D. Winters/Timeframe Photography/McGraw Hill

Strategy Determine the products of the reaction; then write and balance the equation. Remember that one of the reactants, $Mg(OH)_2$, is a solid. Identify any strong electrolytes and rewrite the equation showing strong electrolytes as ions. Identify and cancel the spectator ions.

Setup Because this is an acid-base neutralization reaction, one of the products is water. The other product is a salt comprising the cation from the base, Mg^{2+}, and the anion from the acid, Cl^-. For the formula to be neutral, these ions combine in a 1:2 ratio, giving $MgCl_2$ as the formula of the salt.

Solution

$$Mg(OH)_2(s) + 2HCl(aq) \longrightarrow 2H_2O(l) + MgCl(aq)$$

Of the species in the molecular equation, only HCl and $MgCl_2$ are strong electrolytes. Therefore, the complete ionic equation is

$$Mg(OH)_2(s) + 2H^+(aq) + 2Cl^-(aq) \longrightarrow 2H_2O(l) + Mg^{2+}(aq) + 2Cl^-(aq)$$

Cl^- is the only spectator ion. The net ionic equation is

$$Mg(OH)_2(s) + 2H^+(aq) \longrightarrow 2H_2O(l) + Mg^{2+}(aq)$$

THINK ABOUT IT

Make sure your equation is balanced and that you only show strong electrolytes as ions. $Mg(OH)_2$ is *not* shown as aqueous ions because it is insoluble.

Practice Problem (**A**)**TTEMPT**   Write and balance the molecular equation and then give the complete ionic and net ionic equations for the neutralization reaction between $Ba(OH)_2(aq)$ and $HI(aq)$.

Practice Problem (**B**)**UILD**   Write and balance the molecular equation and then give the complete ionic and net ionic equations for the neutralization reaction between $LiOH(aq)$ and $H_2SO_4(aq)$.

Practice Problem **CONCEPTUALIZE** Which diagram best represents the ions remaining in solution after stoichiometric amounts of aqueous barium hydroxide and hydrobromic acid are combined?

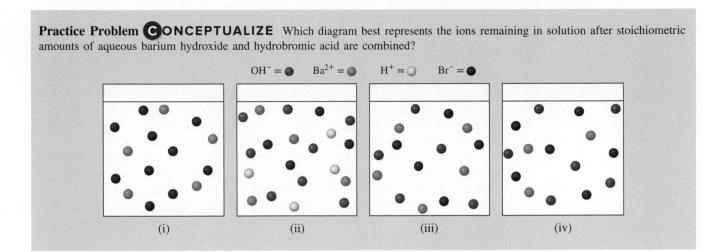

OH⁻ = ●   Ba²⁺ = ●   H⁺ = ○   Br⁻ = ●

(i)   (ii)   (iii)   (iv)

Oxidation-Reduction Reactions

When a piece of zinc metal is immersed in an aqueous solution of copper(II) sulfate, as shown in Figure 10.6, the zinc appears to darken as it becomes coated with copper metal. We can see the copper metal that has deposited on the zinc. What we cannot see in the figure is that some of the zinc metal has been transformed to zinc *ions* (Zn^{2+}) and has become part of the solution. The overall equation for this process is:

$$Zn(s) + CuSO_4(aq) \longrightarrow ZnSO_4(aq) + Cu(s)$$

or, written as a net ionic equation (sulfate is a spectator ion):

$$Zn(s) + Cu^{2+}(aq) \longrightarrow Zn^{2+}(aq) + Cu(s)$$

Figure 10.6 Oxidation of zinc in a solution of copper(II) sulfate.

(both) Charles D. Winters/Timeframe Photography/McGraw Hill

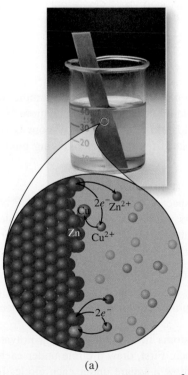

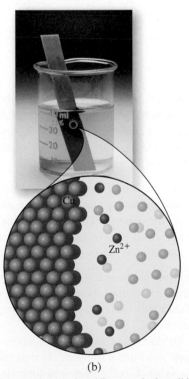

(a)

Zinc atoms enter the solution as zinc ions (Zn^{2+}). Copper ions are reduced to copper atoms on the surface of the metal.

(b)

Cu atoms have replaced Zn atoms in the solid, and Zn^{2+} ions have replaced Cu^{2+} ions in solution.

This is an example of an ***oxidation-reduction reaction,*** commonly called a ***redox reaction.*** In a redox reaction, one reactant *loses* electrons and the other reactant *gains* electrons. Losing electrons is known as ***oxidation,*** whereas gaining electrons is known as ***reduction.*** In the reaction of $Zn(s)$ and $Cu^{2+}(aq)$, even though the equation doesn't *show* the electrons, each Zn atom loses two electrons to become a Zn^{2+} ion (zinc is oxidized); and each Cu^{2+} ion *gains* two electrons to become a Cu atom (copper is reduced). Redox reactions happen because different elements have different natural tendencies to lose and gain electrons. In this case, copper has a greater tendency to gain electrons than zinc. Put another way: Zinc has a greater tendency to *lose* electrons than copper.

Because the reactants have such different tendencies to lose and gain electrons, the reaction of $Zn(s)$ and $Cu^{2+}(aq)$ involves an actual *transfer* of electrons from one species to the other. But there are many redox reactions in which the chemical species have more similar tendencies to lose or gain electrons. In these cases, the reactants typically combine to form molecular compounds in which the electrons are *shared.* Consider the reaction of the gaseous elements hydrogen (H_2) and fluorine (F_2):

$$H_2(g) + F_2(g) \longrightarrow 2HF(g)$$

There is no actual *transfer* of electrons from one reactant to the other, but this is a redox reaction in which H_2 is oxidized and F_2 is reduced. Although the electrons involved in a redox reaction are not shown in the chemical equation, *oxidation numbers* provide a way for us to keep track of them.

Oxidation Numbers

We can think of the ***oxidation number*** or ***oxidation state*** of an atom as the charge the atom *would* have if electrons were actually transferred from one reactant to the other. The oxidation number of an atom is essentially its contribution to the overall charge of the species. For example, we can rewrite the equations for our two redox examples as follows:

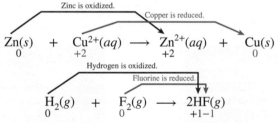

The numbers below each element are the oxidation numbers. In the first example, where electrons actually are transferred from one reactant to the other, it may be more obvious that the process is a redox reaction, and the oxidation number of each atom is simply equal to its charge. In the second example, we need to apply the following guidelines for determining oxidation numbers:

1. The oxidation number of any element, in its elemental form, is zero.
2. The oxidation numbers in any chemical species must sum to the overall charge on the species. That is, oxidation numbers must sum to zero for any molecule, and must sum to the charge on any polyatomic ion. The oxidation number of a monatomic ion is equal to the charge on the ion.

In addition to these two rules, it is necessary to know the oxidation numbers of the "reliable" elements—those that always, or nearly always, have the same oxidation number. Table 10.1 lists elements whose oxidation numbers are reliable, in order of decreasing reliability.

To determine oxidation numbers of the atoms in a compound or a polyatomic ion, you must use a stepwise, systematic approach. First, draw a circle under each element's symbol in the chemical formula. Then draw a square under each circle. In the circle, write the oxidation number of the element; in the square, write the total contribution of that element to the overall charge of the species. Start with the oxidation numbers you know, using the information in Table 10.1, and use the ones you know

TABLE 10.1	Elements with Reliable Oxidation Numbers in Compounds or Polyatomic Ions	
Element	**Oxidation Number**	**Exceptions**
Fluorine	-1	
Group 1 or 2 metal	$+1$ or $+2$, respectively	
Hydrogen	$+1$	Any combination with a Group 1 or 2 metal to form a metal hydride. Examples: LiH and CaH_2—the oxidation number of H is -1 in both examples.
Oxygen	-2	Any combination with something higher on the list that necessitates its having a different oxidation number (see rule 2 for assigning oxidation numbers). Examples: H_2O_2 and KO_2—the oxidation number of O for H_2O_2 is -1 and for KO_2 is $-\frac{1}{2}$.
Group 17 (other than fluorine)	-1	Any combination with something higher on the list that necessitates its having a different oxidation number (see rule 2 for assigning oxidation numbers). Examples: ClF, BrO_4^-, and IO_3^-—the oxidation numbers of Cl, Br, and I are $+1$, $+7$, and $+5$, respectively. Remember that these exceptions do not apply to fluorine, which *always* has an oxidation state of -1 when it is part of a compound.

to figure out the ones you don't know. We use the compound potassium permanganate ($KMnO_4$) to illustrate this process:

$$KMnO_4$$

Oxidation number
Total contribution to charge

Fill in the oxidation number first for the element that appears highest on the list in Table 10.1. Potassium (K) is a Group 1 metal. In its compounds, it always has the oxidation number $+1$. We write $+1$ in the circle beneath the K. Because there is only one K atom in this formula, the total contribution to charge is also $+1$, so we also write $+1$ in the square beneath the K.

$$KMnO_4$$

Oxidation number
Total contribution to charge

Next on the list is oxygen (O). In compounds, O usually has the oxidation number -2, so we assign it -2. Because there are four O atoms in the formula, the total contribution to charge by O atoms is $4(-2) = -8$.

$$KMnO_4$$

Oxidation number
Total contribution to charge

The numbers in the squares, all the contributions to overall charge, must sum to zero. This requires putting $+7$ in the box beneath the Mn atom. Because there is just one Mn atom in this formula, the contribution to charge is the same as the oxidation number. Thus, $(+1) + (+7) + (-8) = 0$.

$$KMnO_4$$

Oxidation number
Total contribution to charge

Sample Problem 10.7 lets you determine oxidation numbers in three more compounds and a polyatomic ion.

SAMPLE PROBLEM **10.7** Determining Oxidation Numbers in Compounds and Polyatomic Ions

Determine the oxidation number of each atom in the following compounds and ion: (a) SO_2, (b) NaH, (c) CO_3^{2-}, (d) N_2O_5.

Strategy For each compound, assign an oxidation number first to the element that appears higher in Table 10.1. Then use rule 2 to determine the oxidation number of the other element.

Setup (a) O appears in Table 10.1 but S does not, so we assign oxidation number −2 to O. Because there are two O atoms in the molecule, the total contribution to charge by O is 2(−2) = −4. The lone S atom must therefore contribute +4 to the overall charge. (b) Both Na and H appear in Table 10.1, but Na appears higher in the table, so we assign the oxidation number +1 to Na. This means that H must contribute −1 to the overall charge. (H^- is the hydride ion.) (c) We assign the oxidation number −2 to O. Because there are three O atoms in the carbonate ion, the total contribution to charge by O is −6. To have the contributions to charge sum to the charge on the ion (−2), the C atom must contribute +4. (d) We assign the oxidation number −2 to O. Because there are five O atoms in the N_2O_5 molecule, the total contribution to charge by O is −10. To have the contributions to charge sum to zero, the contribution by N must be +10, and because there are two N atoms, each one must contribute +5. Therefore, the oxidation number of N is +5.

Solution (a) In SO_2, the oxidation numbers of S and O are +4 and −2, respectively.

(b) In NaH, the oxidation numbers of Na and H are +1 and −1, respectively.

(c) In CO_3^{2-}, the oxidation numbers of C and O are +4 and −2, respectively.

(d) In N_2O_5, the oxidation numbers of N and O are +5 and −2, respectively.

THINK ABOUT IT

Use the circle and square system to verify that the oxidation numbers you have assigned do indeed sum to the overall charge on each species.

Practice Problem **A**TTEMPT Assign oxidation numbers to each atom in the following compounds: H_2O_2, MnO_2, H_2SO_4.

Practice Problem **B**UILD Assign oxidation numbers to each atom in the following polyatomic ions: O_2^{2-}, ClO^-, ClO_3^-.

Practice Problem **C**ONCEPTUALIZE Write the balanced equation for the reaction represented by the models and determine oxidation states for each element before and after the reaction.

Types of Redox Reactions

Redox reactions come in several forms. In what follows, we describe four types of reactions that involve the oxidation of one reactant, and the reduction of the other: *single displacement, combination, decomposition,* and *combustion.*

Our first redox example:

$$Zn(s) + CuSO_4(aq) \longrightarrow ZnSO_4(aq) + Cu(s)$$

is an example of a ***single-displacement reaction,*** in which one element replaces another in a compound. In this case, the element zinc replaces copper in *copper(II) sulfate* to produce *zinc sulfate*. Other examples of single-displacement reactions include:

$$2Ag(s) + Pt(NO_3)_2(aq) \longrightarrow 2AgNO_3(aq) + Pt(s)$$

and

$$2Al(s) + 3NiCl_2(aq) \longrightarrow 2AlCl_3(aq) + 3Ni(s)$$

A ***combination reaction*** is the combination of reactants to form a single product. Examples include the combination of the elements nitrogen and hydrogen to form ammonia:

$$N_2(g) \quad + \quad 3H_2(g) \quad \longrightarrow \quad 2NH_3(g)$$

the combination of sodium and chlorine to form sodium chloride:

$$2Na \quad + \quad Cl_2 \quad \longrightarrow \quad 2NaCl$$

and the reaction of hydrogen and oxygen to form water:

$$2H_2 \quad + \quad O_2 \quad \longrightarrow \quad 2H_2O$$

A ***decomposition reaction*** is essentially the opposite of a combination: One reactant produces two or more products. Examples include the decomposition of sodium hydride into its constituent elements:

$$2NaH(s) \quad \longrightarrow \quad 2Na(s) \quad + \quad H_2(g)$$

the decomposition of potassium chlorate to give potassium chloride and oxygen:

$$2KClO_3(s) \quad \longrightarrow \quad 2KCl(s) \quad + \quad 3O_2(g)$$

and the decomposition of hydrogen peroxide to give water and oxygen:

$$2H_2O_2(aq) \quad \longrightarrow \quad 2H_2O(l) \quad + \quad O_2(g)$$

Profiles in Science

Antoine Lavoisier

Antoine Lavoisier (1743–1794) often is referred to as the *father of modern chemistry*. His work is credited with changing chemistry from a qualitative science to a quantitative one. He did numerous experiments, many involving combustion, in which he carefully measured the masses of substances both before and after the reaction—and produced a large body of work in support of the law of conservation of mass. (In Lavoisier's native France, the law of conservation of mass is called Lavoisier's law.)

Prior to Lavoisier's contributions, combustion had been something of a mystery. The theory used at the time to explain combustion was known as "phlogiston theory." Phlogiston was believed to be a substance contained by combustible substances. As substances burned, they were thought to give off the phlogiston they contained, which caused them to lose mass. However, although many substances lost mass when they burned, others actually *gained* mass—a reality that Lavoisier demonstrated with a large number of meticulously performed experiments—and one that could not be explained using phlogiston theory. Lavoisier is responsible for our current understanding of combustion as the reaction of a substance with oxygen.

Lavoisier is also responsible for proving beyond any doubt that water is a compound, and not an element; and that sulfur is an element and not a compound—as had been believed by scientists for millennia. He recognized and named the elements oxygen and hydrogen, constructed the first comprehensive list of elements known at the time, and predicted the existence of an element that had not yet been discovered (silicon). He further proposed a system of chemical nomenclature that helped to organize and simplify the voluminous and rapidly growing body of knowledge in chemistry, and he helped establish the metric system.

Antoine Lavoisier
Fine Art Images/Heritage Images/Getty Images

Tragically, Lavoisier was executed in the Reign of Terror during the French Revolution. He and many other scientists and intellectuals were seen by violent political factions as "enemies of the revolution," and were guillotined. A year and a half after his execution, the French government exonerated Lavoisier, saying that he had been wrongly convicted.

Finally, a **combustion reaction** is one in which a substance burns in the presence of oxygen. The type of combustion reaction we will encounter most often is the burning of carbon-containing substances in oxygen. These combustion reactions produce carbon dioxide and water vapor. Examples include the burning of natural gas (primarily methane) in gas stoves:

$$CH_4(g) + 2O_2(g) \longrightarrow CO_2(g) + 2H_2O(g)$$

$$\boxed{-4}\boxed{+1} \quad \boxed{0} \quad \boxed{+4}\boxed{-2} \quad \boxed{+1}\boxed{-2}$$
$$\boxed{-4}\boxed{+4} \quad \boxed{0} \quad \boxed{+4}\boxed{+4} \quad \boxed{+2}\boxed{-2}$$

and the burning of coal, which is done to generate much of the electricity used in the United States:

$$C(s) + O_2(g) \longrightarrow CO_2(g)$$

$$\boxed{0} \quad \boxed{0} \quad \boxed{+4}\boxed{-2}$$
$$\boxed{0} \quad \boxed{0} \quad \boxed{+4}\boxed{+4}$$

and the burning of acetylene in torches:

$$2C_2H_2(g) + 5O_2(g) \longrightarrow 4CO_2(g) + 2H_2O(g)$$

$$\boxed{+1}\boxed{+1} \quad \boxed{0} \quad \boxed{+4}\boxed{-2} \quad \boxed{+1}\boxed{-2}$$
$$\boxed{-2}\boxed{+2} \quad \boxed{0} \quad \boxed{+4}\boxed{-4} \quad \boxed{+2}\boxed{-2}$$

Sample Problem 10.8 lets you practice identifying types of redox reactions.

SAMPLE PROBLEM 10.8 Identifying Various Types of Redox Reactions

Determine whether each of the following equations represents a combination reaction, a decomposition reaction, or a combustion reaction:

(a) $H_2(g) + Br_2(g) \longrightarrow 2HBr(g)$

(b) $2HCO_2H(l) + O_2(g) \longrightarrow 2CO_2(g) + 2H_2O(g)$

(c) $2KClO_3(s) \longrightarrow 2KCl(s) + 3O_2(g)$

(d) $2Ag(s) + Cu(NO_3)_2(aq) \longrightarrow Cu(s) + 2AgNO_3(aq)$

(e) $2C_2H_6(g) + 7O_2(g) \longrightarrow 4CO_2(g) + 6H_2O(g)$

(f) $CaCO_3(s) \longrightarrow CaO(s) + CO_2(g)$

(g) $Mg(s) + ZnCl_2(aq) \longrightarrow MgCl_2(aq) + Zn(s)$

(h) $P_4(s) + 6Cl_2(g) \longrightarrow 4PCl_3(l)$

Strategy Look at the reactants and products in each balanced equation to see if two or more reactants combine into one product (a combination reaction), if one reactant splits into two or more products (a decomposition reaction), if one metal replaces another in an ionic compound (a single-displacement reaction), or if the main products formed are carbon dioxide gas and water (a combustion reaction).

Setup The equations in parts (a) and (h) depict two reactants and one product. The equations in parts (b) and (e) represent a combination with O_2 of a compound containing C and H to produce CO_2 and H_2O. The equations in parts (c) and (f) represent two products being formed from a single reactant. The equations in parts (d) and (g) depict a metal replacing another metal within an ionic compound.

Solution The equations in parts (a) and (h) represent combination reactions; those in parts (b) and (e) represent combustion reactions; those in parts (c) and (f) represent decomposition reactions; and those in parts (d) and (g) represent single-displacement reactions.

THINK ABOUT IT

Make sure that a reaction identified as a single displacement has one metal replacing another in an ionic compound, a reaction identified as a combination has only one product, a reaction identified as a decomposition has only one reactant, and a reaction identified as a combustion produces only CO_2 and H_2O.

Practice Problem Ⓐ**TTEMPT** Identify each of the following as a combination, decomposition, or combustion reaction:

(a) $C_2H_4O_2(l) + 2O_2(g) \longrightarrow 2CO_2(g) + 2H_2O(g)$

(b) $2Na(s) + Cl_2(g) \longrightarrow 2NaCl(s)$

(c) $2NaH(s) \longrightarrow 2Na(s) + H_2(g)$

(d) $2NO(g) + O_2(g) \longrightarrow 2NO_2(g)$

(e) $(CH_3)_2O(g) + O_2(g) \longrightarrow CO_2(g) + H_2O(g)$

(f) $2NaCl(l) \longrightarrow 2Na(l) + Cl_2(g)$

Practice Problem Ⓑ**UILD** Using the chemical species A_2, X, and AX, write a balanced equation for a combination reaction.

Practice Problem Ⓒ**ONCEPTUALIZE** Each of the diagrams represents a reaction mixture before and after a chemical reaction. Identify each of the reactions shown as combination, decomposition, combustion, or single displacement.

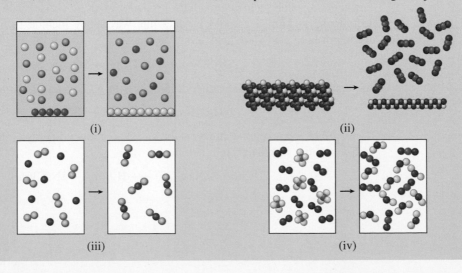

(i)   (ii)

(iii)   (iv)

Thinking Outside the Box

Dental Pain and Redox

Anyone who accidentally bites on a piece of aluminum foil in such a way that the foil touches an old-fashioned dental filling will experience a momentary sharp pain. The pain is actually the result of electrical stimulation of the nerve of a tooth caused by a current flowing between the aluminum foil and the metal in the filling. Historically, the material most commonly used to fill cavities is known as *dental amalgam*. (An *amalgam* is a substance made by combining mercury with one or more other metals.) Dental amalgam consists of liquid mercury mixed in roughly equal parts with an alloy powder containing silver, tin, copper, and sometimes smaller amounts of other metals such as zinc.

What happens is this: Aluminum has a greater tendency to lose electrons than any of the metals in the amalgam. Therefore, when it comes into contact with the amalgam filling, it loses electrons to the other metals. The flow of electrons stimulates the nerve of the tooth, causing a *very* unpleasant sensation.

The pain caused by having aluminum foil contact an amalgam filling is the result of a redox reaction.

Gas-Producing Reactions

Some of the reaction types we have encountered in this section can form gaseous products, which bubble out of solution as they form. An acid-base example of a gas-producing reaction is the combination of hydrochloric acid and sodium carbonate:

$$2HCl(aq) + Na_2CO_3(aq) \longrightarrow 2NaCl(aq) + H_2O(l) + CO_2(g)$$

> **Student Note:** Note that none of the oxidation numbers changes for any of the elements in this reaction.

This is a double-displacement reaction that may be easier to visualize as a two-step process. In the first step, the two reactants exchange ions, with the H^+ ions from the acid combining with the carbonate ion to form carbonic acid:

$$2HCl(aq) + Na_2CO_3(aq) \longrightarrow 2NaCl(aq) + H_2CO_3(aq)$$

Aqueous carbonic acid is unstable and decomposes to form water and carbon dioxide:

$$H_2CO_3(aq) + 2NaCl(aq) \longrightarrow H_2O(l) + CO_2(g) + 2NaCl(aq)$$

Note that the aqueous sodium chloride consists of spectator ions and can be left out of the last equation:

$$H_2CO_3(aq) \longrightarrow H_2O(l) + CO_2(g)$$

So the overall equation for the reaction of hydrochloric acid and sodium carbonate can be written as it first appeared:

$$2HCl(aq) + Na_2CO_3(aq) \longrightarrow 2NaCl(aq) + H_2O(l) + CO_2(g)$$

There are several such *intermediate* products of double-displacement reactions that decompose to form a gaseous product this way. They include sulfurous acid, which decomposes to give water and sulfur dioxide:

$$H_2SO_3(aq) \longrightarrow H_2O(l) + SO_2(g)$$

and ammonium hydroxide, which decomposes to give water and ammonia:

$$NH_4OH(aq) \longrightarrow H_2O(l) + NH_3(g)$$

A gas-producing reaction that forms the gaseous product directly is the reaction of acid with a sulfide, such as sodium sulfide:

$$2HCl(aq) + Na_2S(aq) \longrightarrow H_2S(aq) + 2NaCl(aq)$$

Sample Problem 10.9 gives you some practice identifying the products of gas-producing reactions.

SAMPLE PROBLEM (10.9) Identifying the Products of a Gas-Producing Reaction

Identify the products of the following reaction and write a balanced equation:

$$NaHSO_3(s) + HCl(aq) \longrightarrow ?$$

Strategy Approach this reaction just as you would a precipitation or acid-base reaction, meaning you should take note of the ions present in solution at the moment that the two reactants are mixed together and then predict the two new products that *could* form. If you see one of the three intermediate compounds (shown above) as a product, you should break it down into the gas and water that it forms.

Setup The ions present are $Na^+(aq)$, $HSO_3^-(aq)$, $H^+(aq)$, and $Cl^-(aq)$. The two possible products are NaCl and H_2SO_3 ($H^+ + HSO_3^-$). Recognize that H_2SO_3 breaks down right away into $H_2O(l)$ and $SO_2(g)$.

Solution Your process should look something like this:

$$NaHSO_3(s) + HCl(aq) \longrightarrow NaCl(aq) + H_2SO_3(aq) \longrightarrow NaCl(aq) + H_2O(l) + SO_2(g)$$

The final answer and overall equation is:

$$NaHSO_3(s) + HCl(aq) \longrightarrow NaCl(aq) + H_2O(l) + SO_2(g)$$

THINK ABOUT IT

As a beginner, it is easiest to predict the intermediate product, recognize it, and then rewrite the equation as shown here. When you get better at recognizing these gas-producing reactions, you will be able to do it in one step.

Practice Problem ATTEMPT Identify the products of the following reaction and write a balanced equation:

$$CaCO_3(s) + HNO_3(aq) \longrightarrow ?$$

Practice Problem BUILD Identify the products of the following reaction and write the net ionic equation:

$$NH_4Cl(s) + LiOH(aq) \longrightarrow ?$$

Practice Problem CONCEPTUALIZE Give two compounds (other than those given in the sample problem) that produce $H_2O(l)$ and $SO_2(g)$ when their aqueous solutions are mixed.

CHECKPOINT–SECTION 10.4 Types of Chemical Reactions

10.4.1 Determine the oxidation number of sulfur in each of the following species: H_2S, HSO_3^-, SCl_2, and S_8.

a) $+2, +6, -2, +2, +\frac{1}{4}$ b) $-2, +3, +2, 0$ c) $-2, +5, +2, +2, -\frac{1}{4}$ d) $-1, +4, +2, 0$ e) $-2, +4, +2, 0$

10.4.2 What species gets oxidized in the following equation?

$$Mg(s) + 2HCl(aq) \longrightarrow MgCl_2(aq) + H_2(g)$$

a) $Mg(s)$ b) $H^+(aq)$ c) $Cl^-(aq)$ d) $Mg^{2+}(aq)$ e) $H_2(g)$

10.4.3 Which of the following equations represent redox reactions? (Choose all that apply.)

a) $2Mg(s) + O_2(g) \longrightarrow 2MgO(s)$ d) $2NaN_3(s) \longrightarrow 2Na(s) + 3N_2(g)$

b) $Cu(s) + PtCl_2(aq) \longrightarrow CuCl_2(aq) + Pt(s)$ e) $CaCO_3(s) \longrightarrow CaO(s) + CO_2(g)$

c) $NH_4Cl(aq) + AgNO_3(aq) \longrightarrow NH_4NO_3(aq) + AgCl(s)$

(Continued on next page)

Refer to the following diagrams to answer questions 10.4.4–10.4.6.

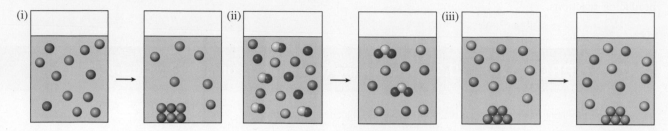

10.4.4 Which of the diagrams depict acid-base neutralization reactions?

a) i

b) ii

c) iii

d) i and ii

e) ii and iii

10.4.5 Which of the diagrams depict precipitation reactions?

a) i

b) ii

c) iii

d) ii and iii

e) i and iii

10.4.6 Which of the diagrams depict redox reactions?

a) i

b) ii

c) iii

d) i and iii

e) i, ii, and iii

10.5 Chemical Reactions and Energy

Like physical processes, chemical processes can be either *endothermic* or *exothermic* [◄◄ Section 7.5]. When we write a chemical equation to represent an endo- or exothermic reaction, we can include heat as part of the equation. For an endothermic process, one to which heat must be *added,* we can show heat as a *reactant.* An example of an endothermic reaction is the decomposition of solid calcium carbonate to give solid calcium oxide and carbon dioxide gas. We can write the equation as:

$$heat + CaCO_3(s) \longrightarrow CaO(s) + CO_2(g)$$

An example of a highly *exothermic* reaction is combustion, which is precisely why we use combustion to generate the energy we need to function as a society. For example, we can write the equation for combustion of natural gas as:

$$CH_4(g) + 2O_2(g) \longrightarrow CO_2(g) + 2H_2O(g) + heat$$

In the case of an exothermic process, we show heat as a *product.* We explain more about the energy changes that accompany chemical reactions in Chapter 11.

10.6 Chemical Reactions in Review

In order for a chemical reaction to occur, there must be a ***driving force*** that makes it happen. The driving force can be the formation of a precipitate, the formation of water, the transfer of electrons, or the formation of a gas. We have seen examples of all of these in the preceding sections. Figure 10.7 shows how to classify the reaction types that we have learned. Figures 10.8 and 10.9 show how to further refine the classification of acid-base and redox reactions.

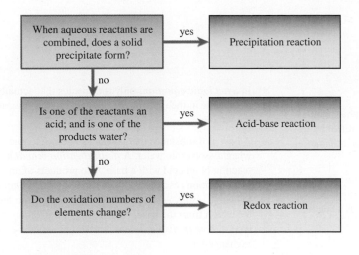

Figure 10.7 Flowchart for classifying reactions as precipitation, acid-base, or redox. Both precipitation and acid-base reactions are double-displacement reactions.

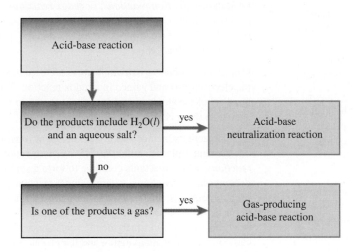

Figure 10.8 Flowchart for classifying acid-base reactions as neutralization or gas producing.

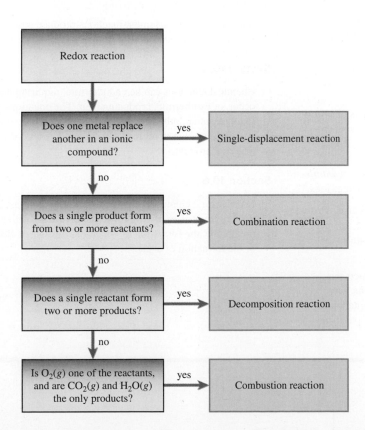

Figure 10.9 Flowchart for classifying redox reactions as single displacement, combination, decomposition, combustion, or gas producing.

Chapter Summary

Section 10.1

- *Chemical reactions* are processes that result in a change in the identity of matter. They can sometimes be identified by a change in the appearance of matter, the presence of a flame, or the production of a gas.

- The third hypothesis of Dalton's atomic theory states: *Chemical reactions cause the rearrangement of atoms, but do not cause either the creation or the destruction of atoms.* This is a statement of the *law of conservation of mass.*

Section 10.2

- A *chemical equation* is used to represent a chemical reaction with chemical symbols. A *reactant* is a species present before the reaction occurs, and a *product* is a species present after the reaction occurs. Reactants are written on the left, followed by an arrow pointing to the right; products are written on the right in a chemical equation.

- The species involved in a chemical reaction should be labeled with parenthetical italicized letters to indicate their phase: (*s*) for solid, (*l*) for liquid, (*g*) for gas, and (*aq*) for aqueous.

Section 10.3

- Chemical equations must be balanced so that the same number of each type of atom is represented on both sides of the equation. The balancing of chemical equations is done by changing the *stoichiometric coefficients* or *coefficients*—the numbers that appear to the left of each chemical species. When a coefficient is 1, it is not shown.

- To simplify the balancing of chemical equations, the coefficients of compounds should be changed before those of elements.

Section 10.4

- A *precipitation reaction* occurs when two solutions of strong electrolytes are combined, resulting in the formation of an insoluble salt. The insoluble salt that results is called the *precipitate,* and it typically settles to the bottom of the reaction vessel. A precipitation reaction is one type of *double-displacement reaction,* in which the reactants exchange ions.

- The equation for a double-displacement reaction can be written in three different forms: molecular, complete ionic, and net ionic. In a *molecular equation,* reactants and products are represented using their chemical formulas.

- In a *complete ionic equation,* reactants and products that are strong electrolytes are shown as individual ions. Ions that appear on both sides of the complete ionic equation are *spectator ions,* meaning that they do not actually participate in the reaction.

- In a *net ionic equation,* only the species that actually participate in the reaction are shown; i.e., the spectator ions are removed.

- A *base* is a substance that produces hydroxide ions, OH^-, when dissolved in water. A *neutralization reaction* is a reaction of an acid with a base. The products of an acid-base neutralization reaction are *water* and an ionic compound known as a *salt.* A salt consists of the anion from the acid and the cation from the base. Acid-base neutralization is another type of *double-displacement reaction,* where the reactants exchange ions.

- *Oxidation-reduction reactions,* or *redox reactions,* are those in which electrons are lost by one reactant and gained by the other. The loss of electrons is *oxidation,* and the gain of electrons is *reduction.*

- *Oxidation numbers* or *oxidation states* are used to keep track of the electrons lost and gained in a redox reaction. The oxidation numbers of the atoms in any species must sum to the overall charge on the species.

- Redox reactions include *single-displacement reactions,* where one element replaces another in a compound; *combination reactions,* where reactants combine to form a single product; *decomposition reactions,* where a single reactant forms two or more products; and *combustion reactions,* where a reactant burns in the presence of oxygen. The most common combustion reactions are those in which the reactants are oxygen and a carbon-containing compound—and the products are carbon dioxide and water.

- Some double-displacement reactions produce unstable intermediate products that decompose to produce a gas.

Section 10.5

- Chemical reactions can be endothermic, requiring heat to occur, or exothermic, producing heat. The equations written to represent endothermic reactions can show heat as a reactant; and those written to represent exothermic reactions can show heat as a product.

Section 10.6

- In order for a chemical reaction to occur, there must be a *driving force.* The driving force can be the formation of a precipitate, the formation of water, the transfer of electrons, or the formation of a gas.

Key Terms

Base 370

Chemical equation 358

Chemical reaction 355

Coefficients 360

Combination reaction 377

Combustion reaction 378

Complete ionic equation 368

Decomposition reaction 377

Double-displacement
reaction 366

Driving force 382

Law of conservation
of mass 356

Molecular equation 366

Net ionic equation 369

Neutralization reaction 371

Oxidation 374

Oxidation number 374

Oxidation-reduction
reaction 374

Oxidation state 374

Precipitate 366

Precipitation reaction 366

Product 358

Reactant 358

Redox reaction 374

Reduction 374

Salt 371

Single-displacement
reaction 377

Spectator ion 368

Stoichiometric coefficients 360

Net Ionic Equations

Molecular equations can be useful for stoichiometric calculations (Chapter 11), but molecular equations are not always the best way to represent the species that are actually in solution.

Net ionic equations are preferable in many instances because they indicate more succinctly the species in solution and the actual chemical process that a chemical equation represents. Writing net ionic equations is an important part of solving a variety of problems, including those involving precipitation reactions, redox reactions, and acid-base neutralization reactions. To write net ionic equations, you must draw on several of the skills that you have learned.

- Recognition of the common polyatomic ions [◄◄ Section 3.6]
- Balancing chemical equations and labeling species with (s), (l), (g), or (aq) [◄◄ Section 10.2]
- Identification of strong electrolytes and nonelectrolytes [◄◄ Section 9.4]

Writing a net ionic equation begins with writing and balancing the molecular equation. For example, consider the precipitation reaction that occurs when aqueous solutions of lead(II) nitrate and sodium iodide are combined.

$$\text{Pb(NO}_3)_2(aq) + \text{NaI}(aq) \longrightarrow$$

Exchanging the ions of the two aqueous reactants gives us the formulas of the products. The phases of the products are determined by considering the solubility guidelines [◄◄ Figures 10.3 and 10.4].

$$\text{Pb(NO}_3)_2(aq) + \text{NaI}(aq) \longrightarrow \text{PbI}_2(s) + \text{NaNO}_3(aq)$$

We balance the equation and separate the soluble strong electrolytes to get the ionic equation.

$$\text{Pb(NO}_3)_2(aq) + 2\,\text{NaI}(aq) \longrightarrow \text{PbI}_2(s) + 2\,\text{NaNO}_3(aq)$$

$$\text{Pb}^{2+}(aq) + 2\,\text{NO}_3^-(aq) + 2\,\text{Na}^+(aq) + 2\,\text{I}^-(aq) \longrightarrow \text{PbI}_2(s) + 2\,\text{Na}^+(aq) + 2\,\text{NO}_3^-(aq)$$

We then identify the spectator ions, those that are identical on both sides of the equation, and eliminate them.

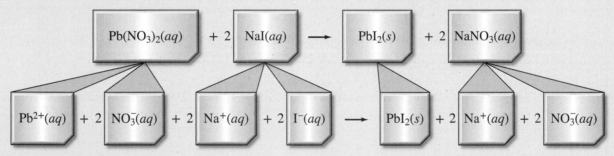

$$\text{Pb}^{2+}(aq) + 2\,\text{I}^-(aq) \longrightarrow \text{PbI}_2(s)$$

Consider now the reaction that occurs when aqueous solutions of hydrochloric acid and potassium hydroxide are combined.

$$\text{HCl}(aq) + \text{KOH}(aq) \longrightarrow$$

Again, exchanging the ions of the two aqueous reactants gives us the formulas of the products.

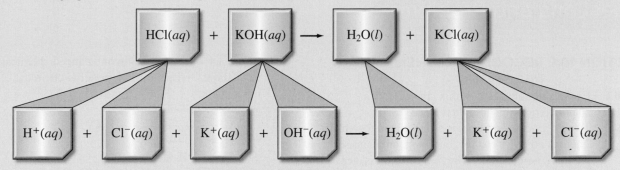

In this case, the aqueous product is a strong electrolyte. The other product, H_2O, is a nonelectrolyte.

We identify the spectator ions and eliminate them.

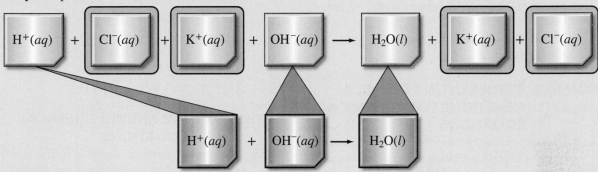

What remains is the net ionic equation.

You must be able to identify each species in solution as a strong electrolyte or a nonelectrolyte so that you know which should be separated into ions and which should be left as molecular or formula units.

Key Skills Problems

10.1
What is the balanced net ionic equation for the precipitation of $FeSO_4(s)$ when aqueous solutions of K_2SO_4 and $FeCl_2$ are combined?
a) $2K^+(aq) + SO_4^{2-}(aq) + Fe^{2+}(aq) + 2Cl^-(aq) \longrightarrow$
$$FeSO_4(s) + 2K^+(aq) + 2Cl^-(aq)$$
b) $Fe^{2+}(aq) + SO_4^{2-}(aq) \longrightarrow FeSO_4(s)$
c) $K_2SO_4(aq) + FeCl_2(aq) \longrightarrow FeSO_4(s) + 2KCl(aq)$
d) $Fe^{2+}(aq) + 2SO_4^{2-}(aq) \longrightarrow FeSO_4(s)$
e) $2K^+(aq) + SO_4^{2-}(aq) + Fe^{2+}(aq) + 2Cl^-(aq) \longrightarrow FeSO_4(s)$

10.2
Consider the following net ionic equation: $Cd^{2+}(aq) + 2OH^-(aq) \longrightarrow Cd(OH)_2(s)$. If the spectator ions in the ionic equation are $NO_3^-(aq) + K^+(aq)$, what is the molecular equation for this reaction?
a) $CdNO_3(aq) + KOH(aq) \longrightarrow Cd(OH)_2(s) + KNO_3(aq)$
b) $Cd^{2+}(aq) + NO_3^-(aq) + 2k^+(aq) + OH^-(aq) \longrightarrow$
$$Cd(OH)_2(s) + 2K^+(aq) + NO_3^-(aq)$$
c) $Cd(NO_3)_2(aq) + 2KOH(aq) \longrightarrow Cd(OH)_2(s) + 2KNO_3(aq)$
d) $Cd(OH)_2(s) + 2KNO_3(aq) \longrightarrow Cd(NO_3)_2(aq) + 2KOH(aq)$
e) $Cd^{2+}(aq) + NO_3^-(aq) + K^+(aq) + OH^-(aq) \longrightarrow$
$$Cd(OH)_2(s) + K^+(aq) + NO_3^-(aq)$$

10.3
The net ionic equation for the neutralization of hydroiodic acid (HI) with lithium hydroxide [$LiOH(aq)$] is
a) $H^+(aq) + OH^-(aq) \longrightarrow H_2O(l)$
b) $H^+(aq) + I^-(aq) \longrightarrow HI(aq)$
c) $HI(aq) + OH^-(aq) \longrightarrow H_2O(l) + I^-(aq)$
d) $HI(aq) + OH^-(aq) \longrightarrow H_2O(l) + HI(aq)$
e) $H^+(aq) + I^-(aq) + OH^-(aq) \longrightarrow H_2O(l) + I^-(aq)$

10.4
When steel wool [$Fe(s)$] is placed in a solution of $CuSO_4(aq)$, the steel becomes coated with copper metal and the characteristic blue color of the solution fades. What is the net ionic equation for this reaction?
a) $Fe(s) + CuSO_4(aq) \longrightarrow FeSO_4(aq) + Cu(s)$
b) $Fe^{2+}(aq) + Cu(s) \longrightarrow Fe(s) + Cu^{2+}(aq)$
c) $FeSO_4(aq) + Cu(s) \longrightarrow Fe(s) + CuSO_4(aq)$
d) $Fe(s) + Cu^{2+}(aq) \longrightarrow Fe^{2+}(aq) + Cu(s)$
e) $Fe(s) + Cu(aq) \longrightarrow Fe(aq) + Cu(s)$

Questions and Problems

SECTION 10.1: RECOGNIZING CHEMICAL REACTIONS

10.1 List observable changes that can indicate a chemical reaction has taken place.

10.2 Two clear, colorless solutions are mixed together. There is no color change, bubbling, or change in temperature observed. The new solution is clear and colorless. Has a chemical reaction occurred? Explain your reasoning.

10.3 The Statue of Liberty is made of cast iron, but has a substantial coating of copper. Based on the appearance of the statue, has a chemical reaction taken place on the surface of the statue? Explain.

SECTION 10.2: REPRESENTING CHEMICAL REACTIONS WITH CHEMICAL EQUATIONS

10.4 Use the formation of water from hydrogen and oxygen to explain the following terms: *chemical reaction, reactant,* and *product.*

10.5 What is the difference between a chemical reaction and a chemical equation?

10.6 Write the symbols used to represent gas, liquid, solid, and the aqueous phase in chemical equations.

10.7 Write an unbalanced equation to represent each of the following reactions:
  (a) Nitrogen and oxygen react to form nitrogen dioxide.
  (b) Dinitrogen pentoxide reacts to form dinitrogen tetroxide and oxygen.
  (c) Ozone reacts to form oxygen.
  (d) Chlorine and sodium iodide react to form iodine and sodium chloride.
  (e) Magnesium and oxygen react to form magnesium oxide.

10.8 Write an unbalanced equation to represent each of the following reactions:
  (a) Potassium hydroxide and phosphoric acid react to form potassium phosphate and water.
  (b) Zinc and silver chloride react to form zinc chloride and silver.
  (c) Sodium hydrogen carbonate reacts to form sodium carbonate, water, and carbon dioxide.
  (d) Ammonium nitrite reacts to form nitrogen and water.
  (e) Carbon dioxide and potassium hydroxide react to form potassium carbonate and water.

10.9 For each of the following unbalanced chemical equations, write the corresponding chemical statement.
  (a) $S_8 + O_2 \longrightarrow SO_2$
  (b) $CH_4 + O_2 \longrightarrow CO_2 + H_2O$
  (c) $N_2 + H_2 \longrightarrow NH_3$
  (d) $P_4O_{10} + H_2O \longrightarrow H_3PO_4$
  (e) $S + HNO_3 \longrightarrow H_2SO_4 + NO_2 + H_2O$

10.10 For each of the following unbalanced chemical equations, write the corresponding chemical statement:
  (a) $K + H_2O \longrightarrow KOH + H_2$
  (b) $Ba(OH)_2 + HCl \longrightarrow BaCl_2 + H_2O$
  (c) $Cu + HNO_3 \longrightarrow Cu(NO_3)_2 + NO_2 + H_2O$
  (d) $Al + H_2SO_4 \longrightarrow Al_2(SO_4)_3 + H_2$
  (e) $HI \longrightarrow H_2 + I_2$

SECTION 10.3: BALANCING CHEMICAL EQUATIONS

10.11 Why must a chemical equation be balanced? What law is obeyed by a balanced chemical equation?

10.12 Sketch the molecules involved in the reaction between hydrogen gas and oxygen gas to form water to show how the equation is balanced using coefficients.

10.13 Balance the chemical equations in Problem 10.7.

10.14 Balance the chemical equations in Problem 10.8.

10.15 Balance the chemical equations in Problem 10.9.

10.16 Balance the chemical equations in Problem 10.10.

10.17 Balance the following chemical equations:
  (a) $C + O_2 \longrightarrow CO$
  (b) $CO + O_2 \longrightarrow CO_2$
  (c) $H_2 + Br_2 \longrightarrow HBr$
  (d) $Ca + O_2 \longrightarrow CaO$
  (e) $O_2 + Cl_2 \longrightarrow OCl_2$

10.18 Balance the following chemical equations:
  (a) $H_2O_2 \longrightarrow H_2O + O_2$
  (b) $Zn + AgCl \longrightarrow ZnCl_2 + Ag$
  (c) $NaOH + H_2SO_4 \longrightarrow Na_2SO_4 + H_2O$
  (d) $Cl_2 + NaI \longrightarrow NaCl + I_2$
  (e) $KOH + H_3PO_4 \longrightarrow H_2O + K_3PO_4$

10.19 Balance the following chemical equations:
  (a) $CH_4 + Br_2 \longrightarrow CBr_4 + HBr$
  (b) $N_2H_4 + HNO_3 \longrightarrow N_2 + H_2O$
  (c) $KNO_3 \longrightarrow KNO_2 + O_2$
  (d) $NH_4NO_3 \longrightarrow N_2O + H_2O$
  (e) $NH_4NO_2 \longrightarrow N_2 + H_2O$

10.20 Balance the following chemical equations:
  (a) $NaHCO_3 \longrightarrow Na_2CO_3 + H_2O + CO_2$
  (b) $HCl + CaCO_3 \longrightarrow CaCl_2 + H_2O + CO_2$
  (c) $CO_2 + KOH \longrightarrow K_2CO_3 + H_2O$
  (d) $Be_2C + H_2O \longrightarrow Be(OH)_2 + CH_4$
  (e) $NH_3 + CuO \longrightarrow Cu + N_2 + H_2O$

10.21 Which of the following equations best represents the reaction shown in the diagram?
(a) $8A + 4B \longrightarrow C + D$
(b) $4A + 8B \longrightarrow 4C + 4D$
(c) $2A + B \longrightarrow C + D$
(d) $4A + 2B \longrightarrow 4C + 4D$
(e) $2A + 4B \longrightarrow C + D$

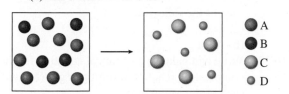

- A
- B
- C
- D

10.22 Which of the following equations best represents the reaction shown in the diagram?
(a) $A + B \longrightarrow C + D$
(b) $6A + 4B \longrightarrow C + D$
(c) $A + 2B \longrightarrow 2C + D$
(d) $3A + 2B \longrightarrow 2C + D$
(e) $3A + 2B \longrightarrow 4C + 2D$

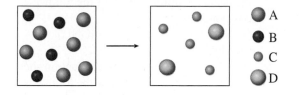

- A
- B
- C
- D

SECTION 10.4: TYPES OF CHEMICAL REACTIONS

10.23 What is the difference between a complete ionic equation and a molecular equation?

10.24 What is the advantage of writing a net ionic equation?

10.25 How can a net ionic equation be used to show that no reaction occurs when two solutions are mixed?

10.26 Classify the following reactions as precipitation, acid-base, or redox.
(a) $4NH_3(g) + 5O_2(g) \longrightarrow 4NO(g) + 6H_2O(g)$
(b) $H_2SO_4(aq) + 2LiOH(aq) \longrightarrow$
$\qquad Li_2SO_4(aq) + 2H_2O(l)$
(c) $2Na(s) + 2H_2O(l) \longrightarrow 2NaOH(aq) + H_2(g)$
(d) $CaCl_2(aq) + Li_2SO_4(aq) \longrightarrow$
$\qquad 2LiCl(aq) + CaSO_4(s)$

10.27 Classify the following reactions as precipitation, acid-base, or redox.
(a) $2Al(s) + 3CuO(s) \longrightarrow Al_2O_3(s) + 3Cu(s)$
(b) $3LiOH(aq) + Fe(NO_3)_3(aq) \longrightarrow$
$\qquad Fe(OH)_3(s) + 3LiNO_3(aq)$
(c) $2Li(s) + Br_2(l) \longrightarrow 2LiBr(s)$
(d) $2HF(aq) + Mg(OH)_2(aq) \longrightarrow$
$\qquad MgF_2(aq) + 2H_2O(l)$

10.28 Classify the following reactions as precipitation, acid-base, or redox.
(a) $C_2H_5OH(l) + 3O_2(g) \longrightarrow 2CO_2(g) + 3H_2O(g)$
(b) $HClO_4(aq) + LiOH(aq) \longrightarrow$
$\qquad H_2O(l) + LiClO_4(aq)$
(c) $3(NH_4)_2CO_3(aq) + 2Al(CN)_3(aq) \longrightarrow$
$\qquad Al_2(CO_3)_3(s) + 6NH_4CN(aq)$
(d) $3Mg(s) + Fe_2O_3(s) \longrightarrow 3MgO(s) + 2Fe(s)$

10.29 Classify the following reactions as precipitation, acid-base, or redox.
(a) $2Si_2S_4(s) \longrightarrow 4Si(s) + S_8(s)$
(b) $3BaCl_2(aq) + 2Li_3PO_4(aq) \longrightarrow$
$\qquad 6LiCl(aq) + Ba_3(PO_4)_2(s)$
(c) $Ni(ClO_3)_2(aq) + Zn(s) \longrightarrow$
$\qquad Zn(ClO_3)_2(aq) + Ni(s)$
(d) $HF(aq) + LiOH(aq) \longrightarrow H_2O(l) + LiF(aq)$

10.30 Classify the reactions in Problem 10.26 as single displacement, double displacement, combination, decomposition, or combustion.

10.31 Classify the reactions in Problem 10.27 as single displacement, double displacement, combination, decomposition, or combustion.

10.32 Classify the reactions in Problem 10.28 as single displacement, double displacement, combination, decomposition, or combustion.

10.33 Classify the reactions in Problem 10.29 as single displacement, double displacement, combination, decomposition, or combustion.

10.34 Sketch a beaker of aqueous $AgNO_3$ and a second beaker of aqueous LiCl. Make sure to show the substances as they exist in solution. What happens when you mix the two solutions together in one beaker? Sketch the result.

10.35 Sketch a beaker of aqueous HCl and a second beaker of aqueous $Ca(OH)_2$. Make sure to show the substances as they exist in solution. What happens when you mix the two solutions together in one beaker? Sketch the result.

10.36 For each of the combinations given, determine whether or not a reaction occurs when the two aqueous solutions are combined. For each reaction that does happen, determine the products and balance the equation.
(a) $NaCl(aq) + LiNO_3(aq) \longrightarrow ?$
(b) $Ca(ClO_4)_2(aq) + Na_2CO_3(aq) \longrightarrow ?$
(c) $AgC_2H_3O_2(aq) + KCl(aq) \longrightarrow ?$
(d) $(NH_4)_3PO_4(aq) + MgCl_2(aq) \longrightarrow ?$

10.37 For each of the combinations given, determine whether or not a reaction occurs when the two aqueous solutions are combined. For each reaction that does happen, determine the products and balance the equation.
(a) $Li_2S(aq) + Mg(C_2H_3O_2)_2(aq) \longrightarrow ?$
(b) $(NH_4)_2SO_4(aq) + BaBr_2(aq) \longrightarrow ?$
(c) $LiOH(aq) + NaCl(aq) \longrightarrow ?$
(d) $H_3PO_4(aq) + LiOH(aq) \longrightarrow ?$

10.38 For each of the combinations given, determine whether or not a reaction occurs when the two aqueous solutions are combined. For each reaction that does happen, determine the products and balance the equation.
(a) $HCl(aq) + NaOH(aq) \longrightarrow$?
(b) $Ca(OH)_2(aq) + HNO_3(aq) \longrightarrow$?
(c) $HC_2H_3O_2(aq) + LiOH(aq) \longrightarrow$?
(d) $Sr(OH)_2(aq) + H_3PO_4(aq) \longrightarrow$?

10.39 For each of the combinations given, determine whether or not a reaction occurs when the two aqueous solutions are combined. For each reaction that does happen, determine the products and balance the equation.
(a) $K_2CO_3(aq) + HNO_3(aq) \longrightarrow$?
(b) $NH_4Cl(aq) + LiOH(aq) \longrightarrow$?
(c) $Na_2SO_3(aq) + HBr(aq) \longrightarrow$?
(d) $NaNO_3(aq) + K_2CO_3(aq) \longrightarrow$?

10.40 Write the balanced molecular, complete ionic, and net ionic equations for each of the following:
(a) Aqueous copper(II) chloride reacts with solid zinc to form aqueous zinc chloride and solid copper.
(b) Aqueous nitric acid reacts with solid calcium carbonate to form gaseous carbon dioxide, liquid water, and aqueous calcium nitrate.
(c) Solid iron reacts with aqueous tin(II) chloride to form solid tin and aqueous iron(III) chloride.

10.41 Write the balanced molecular, complete ionic, and net ionic equations for each of the following:
(a) Aqueous silver nitrate reacts with solid lead to form aqueous lead(II) nitrate and solid silver.
(b) Solid calcium carbide (CaC_2) reacts with water to form gaseous acetylene (C_2H_2) and aqueous calcium hydroxide.
(c) Aqueous barium hydroxide reacts with sulfuric acid to form solid barium sulfate and water.

10.42 Write the complete ionic and net ionic equations for each equation from Problem 10.36.

10.43 Write the complete ionic and net ionic equations for each equation from Problem 10.37.

10.44 Write the complete ionic and net ionic equations for each equation from Problem 10.38.

10.45 Write the complete ionic and net ionic equations for each equation from Problem 10.39.

10.46 List a pair of aqueous solutions that would form $AlPO_4(s)$ when combined.

10.47 List a pair of aqueous solutions that would form $PbSO_4(s)$ when combined.

10.48 Identify a soluble ionic compound that will cause a precipitation reaction to occur when it is added to an aqueous solution of Li_2SO_4, but not when it is added to aqueous solutions of either KCl or NaCN.

10.49 Determine the oxidation number of the underlined atom in each of the following substances:
(a) $\underline{N}O_2$ (d) $\underline{C}_2H_6$ (g) $\underline{P}_2O_4$
(b) $\underline{C}H_4$ (e) $\underline{Si}Cl_4$ (h) $\underline{S}O_3$
(c) $\underline{C}O_2$ (f) $\underline{O}_2$ (i) $\underline{P}Cl_3$

10.50 Determine the oxidation number of the underlined atom in each of the following polyatomic ions:
(a) $\underline{N}O_3^-$ (d) $\underline{S}O_4^{2-}$ (g) $\underline{N}O_2^-$
(b) $\underline{Cl}O_4^-$ (e) $\underline{P}O_4^{3-}$ (h) $\underline{O}_2^{2-}$
(c) $\underline{Mn}O_4^-$ (f) $\underline{C}_2O_4^{2-}$ (i) $\underline{H}_3O^+$

10.51 Write and balance a chemical equation for the combination reaction of chlorine gas (Cl_2) with each of the following:
(a) $K(s)$ (c) $Sr(s)$
(b) $Al(s)$

10.52 Write and balance a chemical equation for the combination reaction of nitrogen gas (N_2) with each of the following:
(a) $Mg(s)$ (c) $Li(s)$
(b) $Rb(s)$

10.53 Sketch a molecular-art diagram for each balanced equation in Problem 10.51.

10.54 Sketch a molecular-art diagram for each balanced equation in Problem 10.52.

10.55 Write and balance a chemical equation showing the decomposition of each of the following compounds:
(a) $Na_2O(s)$ (c) $K_3N(s)$
(b) $MgS(s)$

10.56 Write and balance a chemical equation showing the decomposition of each of the following compounds:
(a) $KCl(s)$ (c) $Li_2S(s)$
(b) $SrO(s)$

10.57 Sketch a molecular-art diagram for each balanced equation in Problem 10.55.

10.58 Sketch a molecular-art diagram for each balanced equation in Problem 10.56.

10.59 Write and balance a chemical equation showing the single-displacement reaction that occurs when solid aluminum is combined with each of the following:
(a) $Zn(NO_3)_2(aq)$ (c) $Pb(ClO_4)_2(aq)$
(b) $FeCl_3(aq)$

10.60 Write and balance a chemical equation showing the single-displacement reaction that occurs when solid barium is combined with each of the following:
(a) $Ca(CN)_2(aq)$ (c) $Ni(C_2H_3O_2)_2(aq)$
(b) $AgClO_3(aq)$

10.61 Sketch a molecular-art diagram for each balanced equation in Problem 10.59.

10.62 Sketch a molecular-art diagram for each balanced equation in Problem 10.60.

10.63 Identify the substance being reduced in each of the following unbalanced reactions:
(a) $K(s) + H_2O(l) \longrightarrow KOH(aq) + H_2(g)$
(b) $Zn(s) + HCl(aq) \longrightarrow ZnCl_2(aq) + H_2(g)$
(c) $SO_2(g) + O_2(g) \longrightarrow SO_3(g)$

10.64 Identify the substance being oxidized in each of the following unbalanced reactions:
(a) $F_2(g) + H_2O(l) \longrightarrow HF(aq) + O_2(g)$
(b) $KNO_3(s) \longrightarrow KNO_2(s) + O_2(g)$
(c) $Mg(l) + TiCl_4(g) \longrightarrow MgCl_2(s) + Ti(s)$

SECTION 10.5: CHEMICAL REACTIONS AND ENERGY

10.65 Explain what is meant by the terms *exothermic* and *endothermic*.

10.66 Write a chemical equation to represent the melting of water. Is heat a reactant or a product in this process?

SECTION 10.6: CHEMICAL REACTIONS IN REVIEW

10.67 Identify the driving force for each of the reactions represented here.

(a) $2HgO(s) \longrightarrow 2Hg(l) + O_2(g)$

(b) $AgClO_4(aq) + LiCl(aq) \longrightarrow$
$$AgCl(s) + LiClO_4(aq)$$

(c) $K_2CO_3(aq) + 2HNO_3(aq) \longrightarrow$
$$H_2O(l) + CO_2(g) + 2KNO_3(aq)$$

(d) $C_6H_{12}O_6(s) + 6O_2(g) \longrightarrow 6CO_2(g) + 6H_2O(g)$

10.68 Identify the driving force for each of the reactions represented here.

(a) $2HBr(aq) + Ca(OH)_2(aq) \longrightarrow$
$$2H_2O(l) + CaBr_2(aq)$$

(b) $Cl_2(g) + 2KBr(s) \longrightarrow 2KCl(s) + Br_2(l)$

(c) $2Na_3PO_4(aq) + 3Mg(NO_3)_2(aq) \longrightarrow$
$$Mg_3(PO_4)_2(s) + 6NaNO_3(aq)$$

(d) $Mg(s) + Cl_2(g) \longrightarrow MgCl_2(s)$

10.69 Classify each of the reactions in Problem 10.67 in as many of the categories in Figures 10.7, 10.8, and 10.9 as possible.

10.70 Classify each of the reactions in Problem 10.68 in as many of the categories in Figures 10.7, 10.8, and 10.9 as possible.

Answers to In-Chapter Materials

Answers to Practice Problems

10.1A b, d, e. **10.1B** ii, iii.

10.2A $C_3H_8(g) + 5O_2(g) \longrightarrow 3CO_2(g) + 4H_2O(g)$.

10.2B $H_2SO_4(aq) + 2NaOH(aq) \longrightarrow 2H_2O(l) + Na_2SO_4(aq)$.

10.3A $C_{18}H_{32}O_2(l) + 25O_2(g) \longrightarrow 18CO_2(g) + 16H_2O(l)$.

10.3B $2NH_3(g) + 3CuO(s) \longrightarrow 3Cu(s) + N_2(g) + 3H_2O(l)$.

10.4A (a) insoluble, (b) soluble, (c) insoluble.

10.4B (a) soluble, (b) insoluble, (c) soluble.

10.5A $Sr(NO_3)_2(aq) + Li_2SO_4(aq) \longrightarrow SrSO_4(s) + 2LiNO_3(aq)$
$Sr^{2+}(aq) + 2NO_3^-(aq) + 2Li^+(aq) + SO_4^{2-}(aq) \longrightarrow$
$$SrSO_4(s) + 2Li^+(aq) + 2NO_3^-(aq)$$
$Sr^{2+}(aq) + SO_4^{2-}(aq) \longrightarrow SrSO_4(s)$.

10.5B $2KNO_3(aq) + BaCl_2(aq) \longrightarrow 2KCl(aq) + Ba(NO_3)_2(aq)$
$2K^+(aq) + 2NO_3^-(aq) + Ba^{2+}(aq) + 2Cl^-(aq) \longrightarrow$
$$2K^+(aq) + 2Cl^-(aq) + Ba^{2+}(aq) + 2NO_3^-(aq)$$
No net ionic equation, no reaction. All ions are spectator ions.

10.6A $Ba(OH)_2(aq) + 2HI(aq) \longrightarrow 2H_2O(l) + BaI_2(aq)$
$Ba^{2+}(aq) + 2OH^-(aq) + 2H^+(aq) + 2I^-(aq) \longrightarrow$
$$2H_2O(l) + Ba^{2+}(aq) + 2I^-(aq)$$
$OH^-(aq) + H^+(aq) \longrightarrow H_2O(l)$.

10.6B $2LiOH(aq) + H_2SO_4(aq) \longrightarrow 2H_2O(l) + Li_2SO_4(aq)$
$2Li^+(aq) + 2OH^-(aq) + 2H^+(aq) + SO_4^{2-}(aq) \longrightarrow$
$$2H_2O(l) + 2Li^+(aq) + SO_4^{2-}(aq)$$
$OH^-(aq) + H^+(aq) \longrightarrow H_2O(l)$.

10.7A H: +1, O: −1; Mn: +4, O: −2, H: +1, S: +6, O: −2.

10.7B O: −1; Cl: +1, O: −2; Cl: +5, O: −2.

10.8A (a) combustion, (b) combination, (c) decomposition, (d) combination, (e) combustion, (f) decomposition.

10.8B $A_2 + 2X \longrightarrow 2AX$. **10.9A** $CaCO_3(s) + 2HNO_3(aq) \longrightarrow$ $Ca(NO_3)_2(aq) + H_2O(l) + CO_2(g)$.

10.9B $NH_4Cl(s) + LiOH(aq) \longrightarrow LiCl(aq) + H_2O(l) + NH_3(g)$.

Answers to Checkpoints

10.1.1 a, d, f. **10.1.2** a, d. **10.3.1** e. **10.3.2** d. **10.3.3** e. **10.3.4** d. **10.4.1** e. **10.4.2** a. **10.4.3** a, b, d. **10.4.4** b. **10.4.5** a. **10.4.6** c.

CHAPTER 11

Using Balanced Chemical Equations

11.1 **Mole to Mole Conversions**

11.2 **Mass to Mass Conversions**

11.3 **Limitations on Reaction Yield**
- Limiting Reactant
- Percent Yield

11.4 **Aqueous Reactions**

11.5 **Gases in Chemical Reactions**
- Predicting the Volume of a Gaseous Product
- Calculating the Required Volume of a Gaseous Reactant

11.6 **Chemical Reactions and Heat**

Breathalyzer tests are sometimes used to gauge a driver's sobriety. The test relies on precise measurement of one of the products of an aqueous reaction to determine blood alcohol level.
Piotr290/iStock/Getty Images

In This Chapter, You Will Learn

How balanced chemical equations are used to solve quantitative problems.

Things To Review Before You Begin

- Molar mass [◄◄ Section 5.2]
- Ideal gas equation [◄◄ Section 8.3]
- Balancing chemical equations [◄◄ Section 10.3]
- Chapter 5 Key Skills [◄◄ pages 192–193]

In the last chapter, we learned how to represent chemical reactions with chemical equations; and how to balance chemical equations. Often we would like to predict how much of a particular product will form from a given amount of a reactant. Other times, we need to produce a specific quantity of a product, so we need to determine how much of the reactants are necessary. Balanced chemical equations can be powerful tools for this type of problem solving. In this chapter, we present several ways that balanced chemical equations are used to solve problems.

11.1 Mole to Mole Conversions

We first encountered the term *stoichiometric coefficient* in the context of balancing chemical equations. ***Stoichiometry*** refers to the numerical relationship between the amounts of reactants and products in a balanced equation. Consider the following example:

Based on the equation for the reaction of dinitrogen dioxide with oxygen to produce nitrogen dioxide,

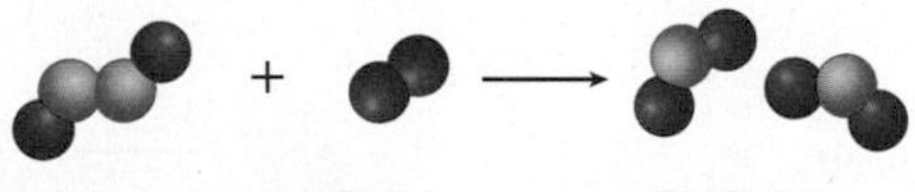

$$N_2O_2(g) + O_2(g) \longrightarrow 2NO_2(g)$$

1 mole of N_2O_2 combines with 1 mole of O_2 to produce 2 moles of NO_2. In stoichiometric calculations, we say that 1 mole of N_2O_2 is *equivalent* to 2 moles of NO_2, which can be represented as

$$1 \text{ mol } N_2O_2 \simeq 2 \text{ mol } NO_2$$

where the symbol $\simeq$ means "is stoichiometrically equivalent to" or simply "is equivalent to." The ratio of moles of N_2O_2 consumed to moles of NO_2 produced is 1:2. Whatever number of moles of N_2O_2 are consumed in the reaction, twice that number of moles of NO_2 will be produced. We can use this constant ratio as a conversion factor that can be written as

$$\frac{1 \text{ mol } N_2O_2}{2 \text{ mol } NO_2}$$

The ratio can also be written as the reciprocal,

$$\frac{2 \text{ mol } NO_2}{1 \text{ mol } N_2O_2}$$

These conversion factors enable us to determine how many moles of NO_2 will be produced upon reaction of a given amount of N_2O_2, or how much N_2O_2 is necessary to produce a specific amount of NO_2. Consider the complete reaction of 3.82 moles of N_2O_2 to form NO_2. To calculate the number of moles of NO_2 produced, we use the conversion

Student Note: When reactants are combined in exactly the mole ratio specified by the balanced chemical equation, they are said to be combined *in stoichiometric amounts.*

Figure 11.1 Starting with moles of reactant, use the stoichiometric conversion factor derived from the balanced chemical equation to determine moles of product. The example shows the result of starting with 3.80 moles of N_2O_2 in the reaction of N_2O_2 with O_2 to produce NO_2.

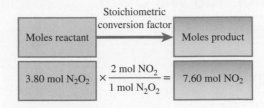

factor with moles of NO_2 in the numerator and moles of N_2O_2 in the denominator. Figure 11.1 illustrates the process of converting from moles reactant to moles product.

$$\text{moles } NO_2 \text{ produced} = 3.82 \text{ mol } N_2O_2 \times \frac{1 \text{ mol } NO_2}{1 \text{ mol } N_2O_2} = 3.82 \text{ mol } NO_2$$

Similarly, we can use other ratios represented in the balanced equation as conversion factors. For example, we have 1 mol $O_2 \simeq 2$ mol NO_2 and 1 mol $N_2O_2 \simeq 1$ mol O_2. The corresponding conversion factors allow us to calculate the amount of NO_2 produced upon reaction of a given amount of O_2, and the amount of one reactant necessary to react completely with a given amount of the other. Using the preceding example, we can determine the *stoichiometric amount* of O_2 (how many moles of O_2 are needed to react with 3.82 moles of N_2O_2).

$$\text{moles } O_2 \text{ needed} = 3.82 \text{ mol } N_2O_2 \times \frac{1 \text{ mol } O_2}{1 \text{ mol } N_2O_2} = 3.82 \text{ mol } O_2$$

Sample Problem 11.1 illustrates how to determine reactant and product amounts using a balanced chemical equation.

SAMPLE PROBLEM   **11.1**   Using a Balanced Equation to Convert from Moles Reactant to Moles Product

Urea [$(NH_2)_2CO$] is a by-product of protein metabolism. This waste product is formed in the liver and then filtered from the blood and excreted in the urine by the kidneys. Urea can be synthesized in the laboratory by the combination of ammonia and carbon dioxide according to the equation shown at the right.

(a) Calculate the amount of urea that will be produced by the complete reaction of 5.25 moles of ammonia. (b) Determine the stoichiometric amount of carbon dioxide required to react with 5.25 moles of ammonia.

Strategy Use the balanced chemical equation to determine the correct stoichiometric conversion factors, and then multiply by the number of moles of ammonia given.

$$2NH_3(g) + CO_2(g) \longrightarrow (NH_2)_2CO(aq) + H_2O(l)$$

Setup According to the balanced chemical equation, the conversion factor for ammonia and urea is either

$$\frac{2 \text{ mol } NH_3}{1 \text{ mol } (NH_2)_2CO} \quad \text{or} \quad \frac{1 \text{ mol } (NH_2)_2CO}{2 \text{ mol } NH_3}$$

To multiply by moles of NH_3 and have the units cancel properly, we use the conversion factor with moles of NH_3 in the denominator.

Similarly, the conversion factor for ammonia and carbon dioxide can be written as

$$\frac{2 \text{ mol } NH_3}{1 \text{ mol } CO_2} \quad \text{or} \quad \frac{1 \text{ mol } CO_2}{2 \text{ mol } NH_3}$$

Again, we select the conversion factor with ammonia in the denominator so that moles of NH_3 will cancel in the calculation.

Solution

(a) moles $(NH_2)_2CO$ produced $= 5.25 \text{ mol } NH_3 \times \dfrac{1 \text{ mol } (NH_2)_2CO}{2 \text{ mol } NH_3} = 2.63$ mol $(NH_2)_2CO$

(b) moles CO_2 required $= 5.25 \text{ mol } NH_3 \times \dfrac{1 \text{ mol } CO_2}{2 \text{ mol } NH_3} = 2.63$ mol CO_2

THINK ABOUT IT

As always, check to be sure that units cancel properly in the calculation. Also, the balanced equation indicates that there will be *fewer* moles of urea produced than ammonia consumed. Therefore, your calculated number of moles of urea (2.63) should be *smaller* than the number of moles given in the problem (5.25). Similarly, the stoichiometric coefficients in the balanced equation are the same for carbon dioxide and urea, so your answers to this problem should also be the same for both species.

Practice Problem Ⓐ**TTEMPT**   Nitrogen and hydrogen react to form ammonia according to the following balanced equation: $N_2(g) + 3H_2(g) \rightarrow 2NH_3(g)$. Calculate the number of moles of hydrogen required to react with 0.0880 mole of nitrogen, and the number of moles of ammonia that will form.

Practice Problem Ⓑ**UILD**   Tetraphosphorus decoxide (P_4O_{10}) reacts with water to produce phosphoric acid. Write and balance the equation for this reaction, and determine the number of moles of each reactant required to produce 5.80 moles of phosphoric acid.

Practice Problem Ⓒ**ONCEPTUALIZE**   The models depict the balanced equation for the reaction of nitric acid with tin metal to form metastannic acid (H_2SnO_3), water, and nitrogen dioxide. Determine how many moles of nitric acid must react to produce 8.75 mol H_2SnO_3.

CHECKPOINT–SECTION 11.1   Mole to Mole Conversions

11.1.1   How many moles of LiOH will be produced if 0.550 mol Li reacts according to the following equation?

$$2Li(s) + 2H_2O(l) \longrightarrow 2LiOH(aq) + H_2(g)$$

a) 0.550 mol   d) 2.20 mol

b) 1.10 mol   e) 2.00 mol

c) 0.275 mol

11.1.2   How many moles of sulfur form from the reaction of 4.00 moles of H_2S, according to the equation? (Don't forget to balance the equation first.)

$$H_2S(g) + SO_2(g) \longrightarrow S(s) + H_2O(l)$$

a) 1.00 mol S   d) 3.00 mol S

b) 4.00 mol S   e) 1.33 mol S

c) 6.00 mol S

11.1.3   How many moles of oxygen are required to react with 3.95 moles of C_6H_{14}, according to the following equation? (Don't forget to balance the equation first.)

$$C_6H_{14}(l) + O_2(g) \longrightarrow CO_2(g) + H_2O(g)$$

a) 19.0 mol O_2 c) 3.95 mol O_2 e) 26.8 mol O_2

b) 0.416 mol O_2 d) 37.5 mol O_2

11.1.4   Using the molecular art shown, write a balanced chemical equation. How many moles of oxygen form from the decomposition of 3.50 moles of H_2O_2?

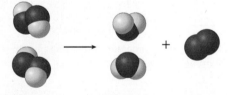

a) 7.00 mol O_2 c) 3.50 mol O_2

b) 1.00 mol O_2 d) 1.75 mol O_2 e) 2.00 mol O_2

11.2   Mass to Mass Conversions

Balanced chemical equations give us the relative amounts of reactants and products in terms of *moles*. However, because we measure reactants and products in the laboratory by weighing them, most often such calculations start with *mass* rather than the number of moles. Figure 11.2 illustrates the general process of converting from mass of reactant to mass of product.

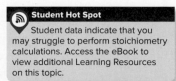

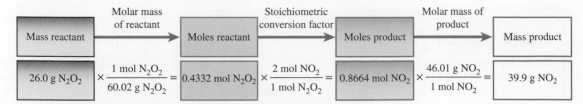

Figure 11.2 Starting with mass of reactant (usually given in grams), use the reactant's molar mass to convert to moles. Then use the stoichiometric conversion factor derived from the balanced chemical equation to determine moles of product; and finally use the product's molar mass to convert to mass of product. The example shows the result of starting with 26.0 grams of N_2O_2 in the reaction of N_2O_2 with O_2 to produce NO_2.

Sample Problem 11.2 illustrates how to determine amounts of reactants and products in terms of grams.

SAMPLE PROBLEM **11.2**   Using a Balanced Equation to Convert Between Reactant Mass and Product Mass

Nitrous oxide (N_2O), also known as "laughing gas," is commonly used as an anesthetic in dentistry. It is manufactured by heating ammonium nitrate. The balanced equation is

$$NH_4NO_3(s) \xrightarrow{\Delta} N_2O(g) + 2H_2O(g)$$

(a) Calculate the mass of ammonium nitrate that must be heated in order to produce 10.0 g of nitrous oxide. (b) Determine the corresponding mass of water produced in the reaction.

Strategy For part (a), use the molar mass of nitrous oxide to convert the given mass of nitrous oxide to moles, use the appropriate stoichiometric conversion factor to convert to moles of ammonium nitrate, and then use the molar mass of ammonium nitrate to convert to grams of ammonium nitrate. For part (b), use the molar mass of nitrous oxide to convert the given mass of nitrous oxide to moles, use the stoichiometric conversion factor to convert from moles of nitrous oxide to moles of water, and then use the molar mass of water to convert to grams of water.

Setup The molar masses are as follows: 80.05 g/mol for NH_4NO_3, 44.02 g/mol for N_2O, and 18.02 g/mol for H_2O. The conversion factors from nitrous oxide to ammonium nitrate and from nitrous oxide to water are, respectively:

$$\frac{1 \text{ mol } NH_4NO_3}{1 \text{ mol } N_2O} \quad \text{and} \quad \frac{2 \text{ mol } H_2O}{1 \text{ mol } N_2O}$$

Solution

(a) $10.0 \text{ g } N_2O \times \dfrac{1 \text{ mol } N_2O}{44.02 \text{ g } N_2O} = 0.227 \text{ mol } N_2O$

$0.227 \text{ mol } N_2O \times \dfrac{1 \text{ mol } NH_4NO_3}{1 \text{ mol } N_2O} = 0.227 \text{ mol } NH_4NO_3$

$0.227 \text{ mol } NH_4NO_3 \times \dfrac{80.05 \text{ g } NH_4NO_3}{1 \text{ mol } NH_4NO_3} = 18.2 \text{ g } NH_4NO_3$

Thus, 18.2 g of ammonium nitrate must be heated in order to produce 10.0 g of nitrous oxide.

(b) Starting with the number of moles of nitrous oxide determined in the first step of part (a),

$$0.227 \text{ mol } N_2O \times \frac{2 \text{ mol } H_2O}{1 \text{ mol } N_2O} = 0.454 \text{ mol } H_2O$$

$$0.454 \text{ mol } H_2O \times \frac{18.02 \text{ g } H_2O}{1 \text{ mol } H_2O} = 8.18 \text{ g } H_2O$$

Therefore, 8.18 g of water will also be produced in the reaction.

THINK ABOUT IT

Use the law of conservation of mass to check your answers. Make sure that the combined mass of both products is equal to the mass of reactant you determined in part (a). In this case (rounded to the appropriate number of significant figures), 10.0 g + 8.18 g = 18.2 g. Remember that small differences may arise as the result of rounding.

Practice Problem **A**TTEMPT   Calculate the mass of water produced by the metabolism of 56.8 g of glucose. (See the Familiar Chemistry Box in Section 10.3 for the necessary equation.)

Practice Problem **B**UILD   What mass of glucose must be metabolized in order to produce 175 g of water?

Practice Problem CONCEPTUALIZE The models here represent the reaction of nitrogen dioxide with water to form nitrogen monoxide and nitric acid. What mass of nitrogen dioxide must react for 100.0 g HNO_3 to be produced? (Don't forget to balance the equation.)

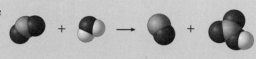

CHECKPOINT–SECTION 11.2   Mass to Mass Conversions

Use the following chemical equation for the first two questions.

$$3NO_2(g) + H_2O(l) \longrightarrow 2HNO_3(aq) + NO(g)$$

11.2.1 Determine the mass (in grams) of HNO_3 formed when 15.5 g NO_2 react with excess H_2O.

    a) 18.7 g   b) 10.3 g   c) 21.2 g   d) 11.3 g   e) 14.2 g

11.2.2 Determine the mass (in grams) of NO_2 that must react to form 15.5 g HNO_3.

    a) 24.6 g   b) 7.54 g   c) 17.0 g   d) 11.3 g   e) 23.3 g

11.2.3 Determine the mass (in grams) of O_2 necessary to react with 5.71 g Al according to the following equation:

$$4Al(s) + 3O_2(g) \longrightarrow 2Al_2O_3(s)$$

    a) 5.08 g   b) 9.03 g   c) 2.54 g   d) 4.28 g   e) 7.61 g

11.2.4 Using the molecular art shown, write a balanced chemical equation. Determine the mass of product (in grams) that forms if 20.0 grams of the blue molecules react with excess of the other reactant. The blue atoms represent nitrogen and the red atoms represent oxygen atoms.

    a) 131 g   b) 120 g   c) 32.8 g   d) 65.7 g   e) 16.4 g

Profiles in Science

Marie-Anne Paulze Lavoisier

Antoine Lavoisier is known as the *father of modern chemistry*. His numerous experiments, including meticulous measurements of the masses of combustion reactants and products, are well known. Less well known is his wife, Marie-Anne Paulze Lavoisier (1758–1836). Madame Lavoisier, herself a chemist, worked tirelessly alongside her husband designing and conducting experiments and making painstakingly exacting measurements. Further, unlike her husband, she was fluent in several languages. It is believed that without his wife's translations of scientific works to French, which made them accessible to him, Antoine Lavoisier's work could not have been as exhaustive as it was—and he would likely be less prominent in the history of chemistry.

In addition to being a chemist, Madame Lavoisier was a classically trained artist. This made it possible for her to illustrate experimental equipment and techniques, which was instrumental to communicating their scientific results to their contemporaries. Together, they produced a large body of work in support of the law of conservation of mass. (In France, this law is known as *Lavoisier's law*.)

Tragically, Antoine Lavoisier was executed in the Reign of Terror during the French Revolution. The government also seized all of the Lavoisiers' assets, notebooks, laboratory equipment, and property. Their property was eventually returned to her after the French government exonerated her husband, a year and a half after his execution. Before her own death, Madame Lavoisier managed to recover most of the confiscated notebooks and equipment, enabling her to publish a comprehensive memoir.

Marie-Anne Paulze Lavoisier
Fine Art/Alamy Stock Photo

11.3 Limitations on Reaction Yield

When a chemist carries out a reaction, the reactants usually are not present in stoichiometric amounts. Because the goal of a reaction is usually to produce the maximum quantity of a useful compound from the starting materials, an excess of one reactant is commonly supplied to ensure that the more expensive or more important reactant is converted completely to the desired product. Consequently, some of the reactant supplied in excess will be left over at the end of the reaction. Moreover, for a variety of reasons, many reactions simply do not yield as much product as the balanced equation predicts. In this section, we consider the limitations on reaction yield, and how we take them into account.

Limiting Reactant

Student Note: Limiting reactants and excess reactants are also referred to as *limiting reagents* and *excess reagents*.

The reactant used up first in a reaction is called the **limiting reactant,** because the amount of this reactant *limits* the amount of product that can form. When all the limiting reactant has been consumed, no more product can be formed. **Excess reactants** are those present in quantities *greater* than necessary to react with the quantity of the limiting reactant.

The concept of a limiting reactant applies to everyday tasks, too, such as making beef and mushroom kabobs. Suppose you want to make the maximum number of kabobs possible, each of which will consist of 1 skewer, 4 mushrooms, and 3 pieces of beef. If you have 4 skewers, 12 mushrooms, and 15 pieces of beef, how many kabobs can you make? The answer, as Figure 11.3 illustrates, is 3. After making 3 kabobs, your supply of mushrooms will be exhausted. Although you will still have 1 skewer and 6 pieces of beef remaining, you will not have the necessary ingredients to make any more kabobs according to the recipe. In this example, because the mushrooms will run out first, mushrooms are the limiting "reactant." The total amount of product, in this case the total number of complete kabobs, is limited by the amount of one ingredient.

In problems involving limiting reactants, the first step is to determine which is the limiting reactant. After the limiting reactant has been identified, the rest of the problem can be solved using the approach outlined in Section 11.1. Consider the formation of methanol (CH_3OH) from carbon monoxide and hydrogen:

$$CO(g) + 2H_2(g) \longrightarrow CH_3OH(l)$$

Suppose that initially we have 5 moles of CO and 8 moles of H_2, the ratio shown in Figure 11.4(a). We can use the stoichiometric conversion factors to determine how many moles of H_2 are necessary in order for all the CO to react. From the balanced equation, we have 1 mol CO $\coloneqq$ 2 mol H_2. Therefore, the amount of H_2 necessary to react with 5 mol CO is

$$\text{moles of } H_2 = 5 \; \cancel{\text{mol CO}} \times \frac{2 \text{ mol } H_2}{1 \; \cancel{\text{mol CO}}} = 10 \text{ mol } H_2$$

Because there are only 8 moles of H_2 available, there is insufficient H_2 to react with all the CO. Therefore, H_2 is the limiting reactant and CO is the excess reactant. H_2 will be used up first, and when it is gone, the formation of methanol will cease and there will be some CO left over, as shown in Figure 11.4(b). To determine how much CO will be left over when the reaction is complete, we must first calculate the amount of

4 skewers, 12 mushrooms, 15 pieces of beef
(a)

3 assembled kabobs, 1 skewer, and 6 pieces of beef
(b)

Figure 11.3 Part (a) 4 skewers, 12 mushrooms, 15 pieces of beef, (b) 3 assembled kabobs, 1 skewer, and 6 pieces of beef. The number of mushrooms limits the number of kabobs that can be assembled according to the recipe.

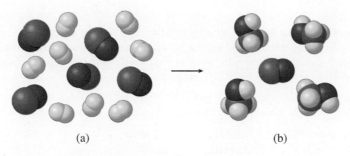

(a)   (b)

Figure 11.4 The reaction of
(a) CO and H_2 to form (b) CH_3OH.
Each molecule represents 1 mole
of substance. In this case, H_2 is
the limiting reactant and there is
1 mole of CO remaining when the
reaction is complete.

CO that will react with all 8 moles of H_2:

$$\text{moles of CO} = 8 \, \text{mol } H_2 \times \frac{1 \text{ mol CO}}{2 \text{ mol } H_2} = 4 \text{ mol CO}$$

Thus, there will be 4 moles of CO consumed and 1 mole (5 mol − 4 mol) left over.

 Another way to approach the determination of limiting reactant is to calculate the amount of product produced for *each* reactant amount given in the problem. In this case, if we use the amount of CO for one calculation, and the amount of H_2 for another calculation, we get two different product amounts:

$$5 \, \text{mol CO} \times \frac{1 \text{ mol } CH_3OH}{1 \text{ mol CO}} = 5 \text{ mol } CH_3OH$$

$$8 \, \text{mol } H_2 \times \frac{1 \text{ mol } CH_3OH}{2 \text{ mol } H_2} = 4 \text{ mol } CH_3OH$$

Student Note: We apparently
have enough CO to produce
5 moles of product; but we only
have enough H_2 to produce
4 moles of product. The amount
of H_2 limits the amount of
product formed.

The calculation that results in the *smaller* amount of product is the one we must use. In this case, the calculation beginning with the amount of H_2 given in the problem produces a smaller amount of CH_3OH than the one beginning with the amount of CO given in the problem. In other words, we have enough CO to produce 5 mol CH_3OH—provided that we have sufficient H_2 for the reaction—but only enough H_2 to produce 4 mol CH_3OH. This makes H_2 the limiting reactant. Figure 11.5 illustrates the steps for this type of calculation.

Student Hot Spot

Student data indicate that you
may struggle to perform limiting
reactant calculations. Access the
eBook to view additional Learning
Resources on this topic.

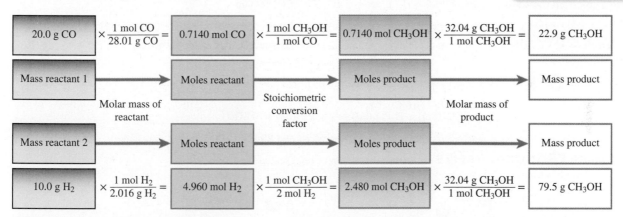

Figure 11.5 Determine the number of moles of each reactant and use each result to determine moles of product. The reactant that results in the smaller amount of product is the limiting reactant. The example shows the result of starting with 20.0 grams of CO and 10.0 grams of H_2 in the reaction of CO with H_2 to produce CH_3OH. Because the calculation starting with 20.0 g CO yields a smaller mass of CH_3OH, CO is the limiting reactant.

 Sample Problem 11.3 illustrates how to combine the concept of a limiting reactant with the conversion between mass and moles.

SAMPLE PROBLEM **11.3**   Determining Limiting Reactant, Mass of Product, and Mass of Excess Reactant Remaining After Reaction

Ammonia is produced by the reaction of nitrogen and hydrogen according to the equation shown. Determine the limiting reactant and calculate the mass of ammonia produced when 35.0 g of nitrogen react with 12.5 g of hydrogen. Which is the excess reactant and how much of it will be left over when the reaction is complete?

$$N_2(g) + 3H_2(g) \longrightarrow 2NH_3(g)$$

(Continued on next page)

Strategy Convert each of the reactant masses to moles. Use the balanced equation to write the necessary stoichiometric conversion factors and determine which reactant is limiting. Again, using the balanced equations, write the stoichiometric conversion factors to determine the number of moles of excess reactant remaining and the number of moles of NH_3 produced. Finally, use the appropriate molar masses to convert moles of excess reactant and moles of NH_3 to grams.

Setup The required molar masses are 28.02 g/mol for N_2, 2.02 g/mol for H_2, and 17.03 g/mol for NH_3. From the balanced equation we have 1 mol $N_2 \simeq$ 2 mol NH_3, 3 mol $H_2 \simeq$ 2 mol NH_3. The necessary stoichiometric conversion factors are therefore:

$$\frac{1 \text{ mol } N_2}{2 \text{ mol } NH_3} \qquad \frac{2 \text{ mol } NH_3}{1 \text{ mol } N_2} \qquad \frac{3 \text{ mol } H_2}{2 \text{ mol } NH_3} \qquad \frac{2 \text{ mol } NH_3}{3 \text{ mol } H_2}$$

Solution

$$35.0 \text{ g } N_2 \times \frac{1 \text{ mol } N_2}{28.02 \text{ g } N_2} = 1.249 \text{ mol } N_2$$

$$12.5 \text{ g } H_2 \times \frac{1 \text{ mol } H_2}{2.02 \text{ g } H_2} = 6.188 \text{ mol } H_2$$

To determine the limiting reactant, calculate the mass of ammonia that each starting amount of the reactants can produce.

$$1.249 \text{ mol } N_2 \times \frac{2 \text{ mol } NH_3}{1 \text{ mol } N_2} \times \frac{17.03 \text{ g } NH_3}{1 \text{ mol } NH_3} = 42.5 \text{ g } NH_3$$

$$6.188 \text{ mol } H_2 \times \frac{2 \text{ mol } NH_3}{3 \text{ mol } H_2} \times \frac{17.03 \text{ g } NH_3}{1 \text{ mol } NH_3} = 70.3 \text{ g } NH_3$$

The smallest amount of ammonia that is made is 42.5 grams, which is the amount that forms from nitrogen. This makes nitrogen the limiting reactant. This also means that hydrogen is the excess reactant.

To determine the mass of hydrogen left, determine how much hydrogen will react.

$$1.249 \text{ mol } N_2 \times \frac{3 \text{ mol } H_2}{1 \text{ mol } N_2} \times \frac{2.02 \text{ g } H_2}{1 \text{ mol } H_2} = 7.57 \text{ g } H_2 \text{ reacted}$$

Subtract this mass from the starting mass of 12.5 g H_2 to determine 4.9 g of H_2 remain.

Practice Problem Ⓐ**TTEMPT** Calculate the mass of ammonia produced when 35.0 g of nitrogen react with 2.5 g hydrogen. Which is the excess reactant and how much of it will be left over when the reaction is complete?

Practice Problem Ⓑ**UILD** Potassium hydroxide and phosphoric acid react to form potassium phosphate and water according to the equation:

$$3KOH(aq) + H_3PO_4(aq) \longrightarrow K_3PO_4(aq) + 3H_2O(l)$$

Determine the starting mass of each reactant if 55.7 g K_3PO_4 are produced and 89.8 g H_3PO_4 remain unreacted.

Practice Problem Ⓒ**ONCEPTUALIZE** A chemical reaction is represented by:

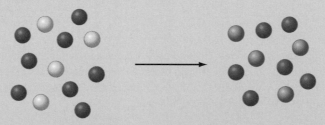

Use the diagram to write the corresponding balanced chemical equation and determine which of the images (i–iv) could represent the product of the combination shown here on the left:

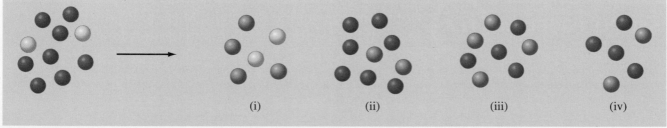

(i) (ii) (iii) (iv)

Percent Yield

When you use stoichiometry to calculate the amount of product formed in a reaction, you are calculating the *theoretical yield* of the reaction. The theoretical yield is the amount of product that forms when *all* the limiting reactant reacts to form the desired product. It is the *maximum* obtainable yield, predicted by the balanced equation. In practice, the *actual yield*—the amount of product actually obtained from a reaction—is almost always less than the theoretical yield. There are many reasons for the difference between the actual and theoretical yields. For instance, some of the reactants may not react to form the desired product. They may react to form different products, in something known as *side reactions,* or they may simply remain unreacted. In addition, it may be difficult to isolate and recover all the product at the end of the reaction. Chemists often determine the efficiency of a chemical reaction by calculating its *percent yield,* which tells *what percentage the actual yield is of the theoretical yield.* It is calculated as follows:

$$\% \text{ yield} = \frac{\text{actual yield}}{\text{theoretical yield}} \times 100\% \qquad \textbf{Equation 11.1}$$

Percent yields may range from a tiny fraction to 100 percent. (They cannot exceed 100 percent.) Chemists try to maximize percent yield in a variety of ways. Factors that can affect percent yield, including temperature and pressure, are discussed in Chapter 13.

Sample Problem 11.4 shows how to use a balanced equation and actual yield to determine percent yield of a reaction.

SAMPLE PROBLEM **11.4** Using a Balanced Equation and Actual Yield to Determine Percent Yield of Reaction

Phosphorus pentachloride (PCl_5) decomposes to form phosphorus trichloride (PCl_3) and chlorine (Cl_2) according to the equation:

$$PCl_5(g) \longrightarrow PCl_3(g) + Cl_2(g)$$

Determine the percent yield if 3.45 grams of PCl_5 decomposes to yield 2.10 grams of PCl_3.

Strategy Convert grams of PCl_5 to moles of PCl_5 and use the balanced equation to determine the number of moles of PCl_3 that we expect to be produced. Convert moles of PCl_3 to grams of PCl_3, and compare to the number of grams actually produced.

Setup The molar mass of PCl_5 is 208.2 g/mol and that of PCl_3 is 137.3 g/mol. The balanced equation indicates that 1 mol PCl_3 is produced for every 1 mol PCl_5 that decomposes. The necessary stoichiometric conversion factors are:

$$\frac{1 \text{ mol } PCl_3}{1 \text{ mol } PCl_5} \qquad \frac{1 \text{ mol } PCl_5}{1 \text{ mol } PCl_3}$$

Solution

$$3.45 \text{ g } PCl_5 \times \frac{1 \text{ mol } PCl_5}{208.2 \text{ g } PCl_5} = 0.01657 \text{ mol } PCl_5$$

$$0.01657 \text{ mol } PCl_5 \times \frac{1 \text{ mol } PCl_3}{1 \text{ mol } PCl_5} = 0.01657 \text{ mol } PCl_3$$

$$0.01657 \text{ mol } PCl_3 \times \frac{137.3 \text{ g } PCl_3}{1 \text{ mol } PCl_3} = 2.275 \text{ g } PCl_3$$

Thus, the theoretical yield is 2.275 grams of PCl_3. The percent yield is

$$\frac{2.10 \text{ g } PCl_3}{2.275 \text{ g } PCl_3} \times 100\% = 92.3\% \text{ yield}$$

THINK ABOUT IT

Percent yield can never exceed 100%. If you determine a percent yield in excess of 100%, check your work for errors.

(Continued on next page)

Practice Problem ATTEMPT Diethyl ether is produced from ethanol according to the following equation:

$$2CH_3CH_2OH(l) \longrightarrow CH_3CH_2OCH_2CH_3(l) + H_2O(l)$$

Calculate the percent yield if 68.6 g of ethanol react to produce 16.1 g of ether.

Practice Problem BUILD What mass of ether will be produced if 207 g of ethanol react with a 73.2 percent yield?

Practice Problem CONCEPTUALIZE Hydrogen peroxide (H_2O_2) decomposes to yield water and oxygen gas. The diagram represents this decomposition before and after the reaction. Determine the percent yield of oxygen in this reaction.

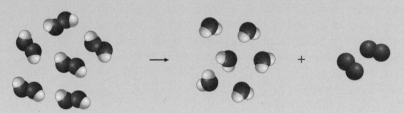

In reactions with more than one reactant, we must be sure to determine theoretical yield using the *limiting* reactant. Sample Problem 11.5 shows how to calculate the percent yield of a pharmaceutical manufacturing process in which there is a limiting reactant.

SAMPLE PROBLEM (**11.5**) **Determining Percent Yield of Reaction with a Limiting Reactant**

Aspirin, acetylsalicylic acid ($C_9H_8O_4$), is the most commonly used pain reliever in the world. It is produced by the reaction of salicylic acid ($C_7H_6O_3$) and acetic anhydride ($C_4H_6O_3$) according to the following equation:

$C_7H_6O_3$	+	$C_4H_6O_3$	$\longrightarrow$	$C_9H_8O_4$	+	$HC_2H_3O_2$
salicylic acid		acetic anhydride		acetylsalicylic acid		acetic acid

In a certain aspirin synthesis, 104.8 g of salicylic acid and 102.3 g of acetic anhydride are combined. Calculate the percent yield of the reaction if 105.6 g of aspirin are produced.

Strategy Convert reactant grams to moles for both reactants and determine which reactant is limiting. Use the moles of limiting reactant and the balanced equation to determine the number of moles of aspirin that can be produced. Convert this number of moles to grams for the theoretical yield, and compare it to the actual yield (given in the problem) to calculate the percent yield.

Setup The necessary molar masses are 138.12 g/mol for salicylic acid, 102.09 g/mol for acetic anhydride, and 180.15 g/mol for aspirin.

Solution

$$104.8 \text{ g } C_7H_6O_3 \times \frac{1 \text{ mol } C_7H_6O_3}{138.12 \text{ g } C_7H_6O_3} = 0.7588 \text{ mol } C_7H_6O_3$$

$$102.3 \text{ g } C_4H_6O_3 \times \frac{1 \text{ mol } C_4H_6O_3}{102.09 \text{ g } C_4H_6O_3} = 1.002 \text{ mol } C_4H_6O_3$$

The two reactants combine in a 1:1 mole ratio, so the reactant present in the smallest molar amount, salicylic acid ($C_7H_6O_3$), is the limiting reactant. Using moles of limiting reactant, according to the balanced equation, one mole of aspirin is produced for every mole of salicylic acid consumed.

$$1 \text{ mol salicylic acid } (C_7H_6O_3) \simeq 1 \text{ mol aspirin } (C_9H_8O_4)$$

Therefore, the theoretical yield of aspirin is 0.7588 mol. We convert this to grams using the molar mass of aspirin:

$$0.7588 \text{ mol } C_9H_8O_4 \times \frac{180.15 \text{ g } C_9H_8O_4}{1 \text{ mol } C_9H_8O_4} = 136.7 \text{ g } C_9H_8O_4$$

Thus, the theoretical yield is 136.7 g. If the actual yield is 105.6 g, the percent yield is

$$\% \text{ yield} = \frac{105.6 \text{ g}}{136.7 \text{ g}} \times 100\% = 77.25\% \text{ yield}$$

THINK ABOUT IT

In most instances, reactants are not combined in stoichiometric amounts. Be sure to convert to moles for each reactant and determine if there is a limiting reactant. If there is, you must use the limiting reactant amount to determine theoretical yield.

Practice Problem ATTEMPT Determine the percent yield of the reaction in Sample Problem 11.3 if the actual yield is 31.6 g NH_3.

Practice Problem BUILD What mass of NH_3 would be produced if the percent yield in Sample Problem 11.3 were 85.0%?

Practice Problem CONCEPTUALIZE The diagrams show a mixture of reactants (before) and the mixture of recovered products (after) for an experiment using the chemical reaction introduced in Practice Problem 11.3C. Identify the limiting reactant and determine the percent yield of carbon dioxide.

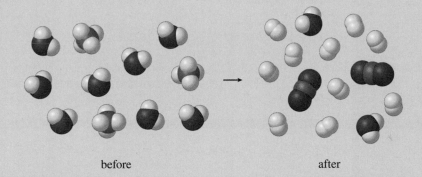

before   after

Thinking Outside the Box

Combustion Analysis

Often chemists need to determine the empirical formula of a compound they have isolated or synthesized. When a carbon-containing compound such as glucose is burned in a combustion analysis apparatus, carbon dioxide (CO_2) and water (H_2O) are produced.

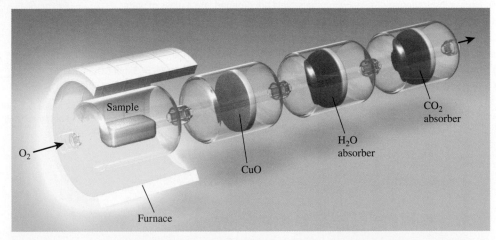

Schematic of a combustion analysis apparatus. CO_2 and H_2O produced in combustion are trapped and weighed. The amounts of these products are used to determine how much carbon and hydrogen the combusted sample contained. (CuO is used to ensure complete combustion of all carbon to CO_2.)

The combustion of glucose can be represented by the balanced chemical equation:

$$C_6H_{12}O_6(s) + 6O_2(g) \longrightarrow 6CO_2(g) + 6H_2O(g)$$

Because only oxygen gas is added to the reaction, the carbon and hydrogen present in the products must have come from the glucose. The oxygen in the products *may* have come from the glucose, but it may also have come from the added oxygen. Suppose that in one such experiment the combustion of 18.8 g of glucose produced 27.6 g of CO_2 and 11.3 g of H_2O. We can calculate the mass of carbon and hydrogen in the original 18.8-g sample of glucose as follows:

$$\text{mass of C} = 27.6 \ \cancel{g\ CO_2} \times \frac{1 \ \cancel{mol\ CO_2}}{44.01 \ \cancel{g\ CO_2}} \times \frac{1 \ \cancel{mol\ C}}{1 \ \cancel{mol\ CO_2}} \times \frac{12.01 \text{ g C}}{1 \ \cancel{mol\ C}} = 7.53 \text{ g C}$$

$$\text{mass of H} = 11.3 \ \cancel{g\ H_2O} \times \frac{1 \ \cancel{mol\ H_2O}}{18.02 \ \cancel{g\ H_2O}} \times \frac{2 \ \cancel{mol\ H}}{1 \ \cancel{mol\ H_2O}} \times \frac{1.008 \text{ g H}}{1 \ \cancel{mol\ H}} = 1.26 \text{ g H}$$

Thus, 18.8 g of glucose contains 7.53 g of carbon and 1.26 g of hydrogen. The remaining mass [18.8 g − (7.53 g + 1.26 g) = 10.0 g] is oxygen.

The number of moles of each element present in 18.8 g of glucose is

$$\text{moles of C} = 7.53 \ \cancel{g\ C} \times \frac{1 \text{ mol C}}{12.01 \ \cancel{g\ C}} = 0.627 \text{ mol C}$$

$$\text{moles of H} = 1.26 \ \cancel{g\ H} \times \frac{1 \text{ mol H}}{1.008 \ \cancel{g\ H}} = 1.25 \text{ mol H}$$

$$\text{moles of O} = 10.0 \ \cancel{g\ O} \times \frac{1 \text{ mol O}}{16.00 \ \cancel{g\ O}} = 0.626 \text{ mol O}$$

The empirical formula of glucose can therefore be written $C_{0.627}H_{1.25}O_{0.626}$. Because the numbers in an empirical formula must be integers, we divide each of the subscripts by the smallest subscript, 0.626 (0.627/0.626 ≈ 1, 1.25/0.626 ≈ 2, and 0.626/0.626 = 1), and obtain CH_2O for the empirical formula.

Familiar Chemistry

Alka-Seltzer

Alka-Seltzer tablets contain aspirin, sodium bicarbonate, and citric acid. When they come into contact with water, the sodium bicarbonate ($NaHCO_3$) and citric acid ($H_3C_6H_5O_7$) react to form carbon dioxide gas, among other products.

$$3NaHCO_3(aq) + H_3C_6H_5O_7(aq) \longrightarrow 3CO_2(g) + 3H_2O(l) + Na_3C_6H_5O_7(aq)$$

The formation of CO_2 causes the trademark fizzing when the tablets are dropped into a glass of water. An Alka-Seltzer tablet contains 1.700 g of sodium bicarbonate and 1.000 g of citric acid. We can determine which ingredient is the limiting reactant and what mass of CO_2 forms when a single tablet dissolves as follows: The required molar masses are 84.01 g/mol for $NaHCO_3$, 192.12 g/mol for $H_3C_6H_5O_7$, and 44.01 g/mol for CO_2. From the balanced equation we have 3 mol $NaHCO_3$ ≏ 1 mol $H_3C_6H_5O_7$, 3 mol $NaHCO_3$ ≏ 3 mol CO_2, and 1 mol $H_3C_6H_5O_7$ ≏ 3 mol CO_2. The necessary stoichiometric conversion factors are therefore:

Charles D. Winters/Timeframe Photography/McGraw Hill

$$\frac{3 \text{ mol } NaHCO_3}{1 \text{ mol } H_3C_6H_5O_7} \quad \frac{1 \text{ mol } H_3C_6H_5O_7}{3 \text{ mol } NaHCO_3} \quad \frac{3 \text{ mol } CO_2}{3 \text{ mol } NaHCO_3} \quad \frac{3 \text{ mol } CO_2}{1 \text{ mol } H_3C_6H_5O_7}$$

Solution

$$1.700 \ \cancel{g\ NaHCO_3} \times \frac{1 \text{ mol } NaHCO_3}{84.01 \ \cancel{g\ NaHCO_3}} = 0.02024 \text{ mol } NaHCO_3$$

$$1.000 \ \cancel{g\ H_3C_6H_5O_7} \times \frac{1 \text{ mol } H_3C_6H_5O_7}{192.12 \ \cancel{g\ H_3C_6H_5O_7}} = 0.005205 \text{ mol } H_3C_6H_5O_7$$

To determine which reactant is limiting, we calculate the amount of citric acid necessary to react completely with 0.02024 mol sodium bicarbonate.

$$0.02024 \ \cancel{mol\ NaHCO_3} \times \frac{1 \text{ mol } H_3C_6H_5O_7}{3 \ \cancel{mol\ NaHCO_3}} = 0.006745 \text{ mol } H_3C_6H_5O_7$$

The amount of $H_3C_6H_5O_7$ required to react with 0.02024 mol $NaHCO_3$ is more than a tablet contains. Therefore, citric acid is the limiting reactant and sodium bicarbonate is the excess reactant.

To determine the mass of CO_2 produced, we first calculate the number of moles of CO_2 produced from the number of moles of limiting reactant ($H_3C_6H_5O_7$) consumed:

$$0.005205 \text{ mol } H_3C_6H_5O_7 \times \frac{3 \text{ mol } CO_2}{1 \text{ mol } H_3C_6H_5O_7} = 0.01562 \text{ mol } CO_2$$

We convert this amount to grams as follows:

$$0.01562 \text{ mol } CO_2 \times \frac{44.01 \text{ g } CO_2}{1 \text{ mol } CO_2} = 0.6874 \text{ g } CO_2$$

CHECKPOINT–SECTION 11.3   Limitations on Reaction Yield

11.3.1 What mass of $CaSO_4$ is produced according to the given equation when 5.00 g of each reactant are combined?

$$CaF_2(s) + H_2SO_4(aq) \longrightarrow CaSO_4(s) + 2HF(g)$$

    a) 10.0 g   b) 11.6 g   c) 6.94 g   d) 8.72 g   e) 5.02 g

11.3.2 What is the percent yield for a process in which 10.4 g CH_3OH reacts and 10.1 g CO_2 forms according to the following equation?

$$2CH_3OH(g) + 3O_2(g) \longrightarrow 2CO_2(g) + 4H_2O(l)$$

    a) 97.1%   b) 70.7%   c) 52.1%   d) 103%   e) 37.9%

11.3.3 How many moles of NH_3 can be produced by the combination of 3.0 mol N_2 and 1.5 mol H_2?

    a) 2.0 mol   b) 1.5 mol   c) 0.50 mol   d) 6.0 mol   e) 1.0 mol

11.3.4 What mass of water can be produced by the reaction of 50.0 g CH_3OH with an excess of O_2; and what mass would be produced if the reaction yield were only 53.2 percent?

$$2CH_3OH(g) + 3O_2(g) \longrightarrow 2CO_2(g) + 4H_2O(l)$$

    a) 28.1 g, 14.9 g   b) 75.0 g, 39.9 g   c) 56.2 g, 29.9 g   d) 29.9 g, 15.0 g   e) 112 g, 59.8 g

11.3.5 Reactants A (red) and B (blue) combine to form a single product C (purple) according to the equation $2A + B \rightarrow C$. What is the limiting reactant in the reaction vessel shown?

    a) A

    b) B

    c) C

    d) None. Reactants are present in stoichiometric amounts.

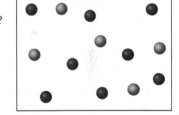

11.3.6 Which of the following represents the contents of the reaction vessel in Checkpoint 11.3.5 after the reaction is complete?

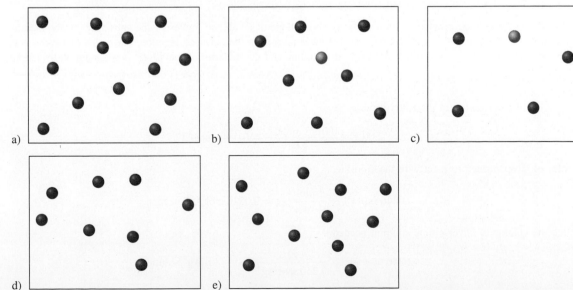

11.4 Aqueous Reactions

Rewind: Molarity is defined as moles solute per liter of solution. [◄◄ Section 9.3] To determine moles of an aqueous reactant, we multiply the volume in L by the molarity:

Volume (L) × M = moles reactant

When aqueous reactants are combined, we determine the number of moles of each reactant using volume and molar concentration (molarity, M [◄◄ Section 9.3]). We have encountered three types of aqueous reactions for which this method is used commonly: precipitation, acid-base, and redox. For each type, it is important to start with a balanced chemical equation, and to keep solution composition [◄◄ Section 9.4] in mind. Figure 11.6 illustrates the general steps for solving stoichiometric problems with aqueous reactions.

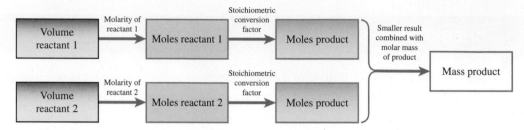

Figure 11.6 When reactant volumes and molar concentrations are given, use this information to determine moles of each reactant. Identify the limiting reactant, if any, and use it to determine the product amount.

Sample Problems 11.6 through 11.8 show how to solve problems involving aqueous reactions.

SAMPLE PROBLEM **11.6** **Determining the Volume of One Aqueous Reactant Necessary to React Completely with Another**

The amount of chloride ion in water can be determined by adding silver nitrate to precipitate the chloride as silver chloride:

$$Cl^-(aq) + Ag^+(aq) \longrightarrow AgCl(s)$$

Recall that silver chloride (AgCl) is insoluble in water [◄◄ Table 9.2] and will precipitate from solution as it is produced. (Nitrate, NO_3^-, is a spectator ion. [◄◄ Section 10.4])

(a) Determine how much of a 0.250-M solution of silver nitrate is necessary to precipitate all of the chloride from 2.00 L of water in which the chloride ion concentration is 0.0173 M. (b) Determine how much 0.250 M silver nitrate is necessary to precipitate all of the chloride from 1.50 L of solution that is 0.0435 M in barium chloride.

Strategy Whenever you are provided with a molarity and volume of a solution, moles of the substance can be determined. Using the moles of chloride ion in the solution and the mole ratio from the balanced reaction, the moles of silver nitrate required can be found. The final step is to use the moles of silver nitrate and the molarity of the silver nitrate solution to determine the volume needed.

Setup
(a) Find moles of chloride ion in the water sample:

$$2.00 \text{ L} \times \frac{0.0173 \text{ mol Cl}^-}{1 \text{ L}} = 0.0346 \text{ mol Cl}^-$$

Find moles of silver needed to react with the chloride:

$$0.0346 \text{ mol Cl}^- \times \frac{1 \text{ mol Ag}^+}{1 \text{ mol Cl}^-} = 0.0346 \text{ mol Ag}^+ \text{ needed}$$

Determine the volume of silver nitrate solution needed:

$$0.0346 \text{ mol Ag}^+ \times \frac{1 \text{ L}}{0.250 \text{ mol Ag}^+} = 0.138 \text{ L Ag}^+ \text{ solution}$$

(b) Find moles of chloride ion in the solution:

$$1.50 \text{ L } \cancel{\text{BaCl}_2} \times \frac{0.0435 \text{ mol } \cancel{\text{BaCl}_2}}{1 \cancel{L}} \times \frac{2 \text{ mol Cl}^-}{1 \text{ mol } \cancel{\text{BaCl}_2}} = 0.1305 \text{ mol Cl}^-$$

Find the moles of silver needed to react with the chloride:

$$0.1305 \cancel{\text{mol Cl}^-} \times \frac{1 \text{ mol Ag}^+}{1 \cancel{\text{mol Cl}^-}} = 0.1305 \text{ mol Ag}^+ \text{ needed}$$

Determine the volume of silver nitrate solution needed:

$$0.1350 \cancel{\text{mol Ag}^+} \times \frac{1 \text{ L}}{0.250 \cancel{\text{mol Ag}^+}} = 0.522 \text{ L Ag}^+ \text{ solution}$$

Solution If we connect all of these separate calculations, it is more efficient and will look like the following:

(a) $2.00 \cancel{L} \times \dfrac{0.0173 \cancel{\text{mol Cl}^-}}{1 \cancel{L}} \times \dfrac{1 \cancel{\text{mol Ag}^+}}{1 \cancel{\text{mol Cl}^-}} \times \dfrac{1 \text{ L}}{0.250 \cancel{\text{mol Ag}^+}} = 0.138 \text{ L Ag}^+ \text{ solution}$

(b) $1.50 \text{ L } \cancel{\text{BaCl}_2} \times \dfrac{0.0435 \cancel{\text{mol BaCl}_2}}{1 \cancel{L}} \times \dfrac{2 \cancel{\text{mol Cl}^-}}{1 \text{ mol } \cancel{\text{BaCl}_2}} \times \dfrac{1 \cancel{\text{mol Ag}^+}}{1 \cancel{\text{mol Cl}^-}} \times \dfrac{1 \text{ L}}{0.250 \cancel{\text{mol Ag}^+}} = 0.522 \text{ L Ag}^+ \text{ solution}$

THINK ABOUT IT

If you carefully label each value with specific units, you can avoid any errors in your calculations. Make sure that your units all cancel and you arrive at the units that correspond to what you are trying to find.

Practice Problem ⒶTTEMPT The amount of lead ion in water can be determined by adding sodium chloride to precipitate the lead as lead(II) chloride:

$$\text{Pb}^+(aq) + 2\text{Cl}^-(aq) \longrightarrow \text{PbCl}_2(s)$$

Recall that lead(II) chloride (PbCl_2) is insoluble in water [◄◄ Table 9.2] and will precipitate from solution as it is produced. (Sodium, Na^+, is a spectator ion. [◄◄ Section 10.4])

(a) Determine how much of a 0.119-M solution of sodium chloride is necessary to precipitate all of the lead from 1.00 L of water in which the lead(II) ion concentration is 0.00215 M. (b) Determine how much 0.119 M sodium chloride is necessary to precipitate all of the lead from 4.50 L of solution that is 0.00616 M in lead(II) nitrate [$\text{Pb(NO}_3)_2$].

Practice Problem ⒷUILD Determine the mass, in grams, of lead(II) chloride (PbCl_2) that will precipitate when 34.5 mL of 0.129 M $\text{Pb(NO}_3)_2$ reacts with 50.0 mL of 0.0533 M NaCl.

Practice Problem ⒸONCEPTUALIZE Which diagram best represents the solution (originally containing sodium chloride) from which the chloride has been removed by the addition of excess silver nitrate?

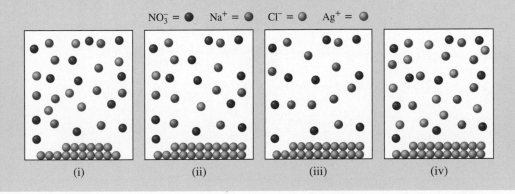

$\text{NO}_3^- = $ ⚫ $\text{Na}^+ = $ ⚫ $\text{Cl}^- = $ ⚫ $\text{Ag}^+ = $ ⚫

(i)   (ii)   (iii)   (iv)

SAMPLE PROBLEM 11.7 · Determining What Volume of Base Is Required to Neutralize a Known Quantity of Acid

How much 0.150-M NaOH is necessary to completely neutralize the acid in (a) 275 mL of a solution that is 0.0750 M in HCl; (b) 1.95 L of a solution that is 0.225 M in H_2SO_4; and (c) 50.0 mL of a solution that is 0.0583 M in H_3PO_4?

Strategy The moles of acid in each part can be found using the molarity and volume. The balanced chemical equations must be written for all three reactions. The mole ratio between the base and acid in the reactions is needed to determine the moles of NaOH needed and, finally, the volume of NaOH solution.

(a) $NaOH(aq) + HCl(aq) \longrightarrow NaCl(aq) + H_2O(l)$

(b) $2NaOH(aq) + H_2SO_4(aq) \longrightarrow Na_2SO_4(aq) + 2H_2O(l)$

(c) $3NaOH(aq) + H_3PO_4(aq) \longrightarrow Na_3PO_4(aq) + 3H_2O(l)$

Setup The moles of acid in each part can be found using the molarity and volume.

(a) $275 \text{ mL HCl solution} \times \dfrac{L}{1000 \text{ mL}} \times \dfrac{0.0750 \text{ mol HCl}}{L} = 0.02063 \text{ mol HCl}$

(b) $1.95 \text{ L } H_2SO_4 \text{ solution} \times \dfrac{0.225 \text{ mol } H_2SO_4}{L} = 0.4388 \text{ mol } H_2SO_4$

(c) $50.0 \text{ mL } H_3PO_4 \text{ solution} \times \dfrac{L}{1000 \text{ mL}} \times \dfrac{0.0583 \text{ mol } H_3PO_4}{L} = 0.002915 \text{ mol } H_3PO_4$

Determine the moles of NaOH needed to react with the moles of acid present, using the mole ratios from the balanced reactions.

(a) $0.02063 \text{ mol HCl} \times \dfrac{1 \text{ mol NaOH}}{1 \text{ mol HCl}} = 0.02063 \text{ mol NaOH}$

(b) $0.4388 \text{ mol } H_2SO_4 \times \dfrac{2 \text{ mol NaOH}}{1 \text{ mol } H_2SO_4} = 0.8776 \text{ mol NaOH}$

(c) $0.002915 \text{ mol } H_3PO_4 \times \dfrac{3 \text{ mol NaOH}}{1 \text{ mol } H_3PO_4} = 0.008745 \text{ mol NaOH}$

Determine the volume of NaOH solution needed:

(a) $0.02063 \text{ mol NaOH} \times \dfrac{1 \text{ L NaOH}}{0.150 \text{ mol NaOH}} = 0.138 \text{ L NaOH}$

(b) $0.8776 \text{ mol NaOH} \times \dfrac{1 \text{ L NaOH}}{0.150 \text{ mol NaOH}} = 5.85 \text{ L NaOH}$

(c) $0.008745 \text{ mol NaOH} \times \dfrac{1 \text{ L NaOH}}{0.150 \text{ mol NaOH}} = 0.0583 \text{ L NaOH}$

Solution If we string together all of the separate calculations we find:

(a) $275 \text{ mL HCl solution} \times \dfrac{L}{1000 \text{ mL}} \times \dfrac{0.0750 \text{ mol HCl}}{L} \times \dfrac{1 \text{ mol NaOH}}{1 \text{ mol HCl}} \times \dfrac{1 \text{ L NaOH}}{0.150 \text{ mol NaOH}} = 0.138 \text{ L NaOH}$

(b) $1.95 \text{ L } H_2SO_4 \text{ solution} \times \dfrac{0.225 \text{ mol } H_2SO_4}{L} \times \dfrac{2 \text{ mol NaOH}}{1 \text{ mol } H_2SO_4} \times \dfrac{1 \text{ L NaOH}}{0.150 \text{ mol NaOH}} = 5.85 \text{ L NaOH}$

(c) $50.0 \text{ mL } H_3PO_4 \text{ solution} \times \dfrac{L}{1000 \text{ mL}} \times \dfrac{0.0583 \text{ mol } H_3PO_4}{L} \times \dfrac{3 \text{ mol NaOH}}{1 \text{ mol } H_3PO_4} \times \dfrac{1 \text{ L NaOH}}{0.150 \text{ mol NaOH}} = 0.0583 \text{ L NaOH}$

Practice Problem Ⓐ**TTEMPT** How many milliliters of a 1.42 M H_2SO_4 solution are needed to neutralize 95.5 mL of a 0.336 M KOH solution?

Practice Problem Ⓑ**UILD** How many milliliters of a 0.211 M HCl solution are needed to neutralize 275 mL of a 0.0350 M $Ba(OH)_2$ solution?

Practice Problem **C**ONCEPTUALIZE Which diagram best represents the solution that would result from the neutralization of a solution of $Ba(OH)_2$ with a solution of HCl?

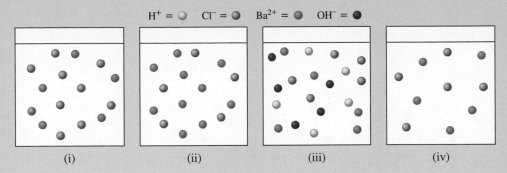

$H^+ = $ ○ $Cl^- = $ ● $Ba^{2+} = $ ● $OH^- = $ ●

      (i)   (ii)   (iii)   (iv)

SAMPLE PROBLEM (11.8) Determining the Amount of a Specific Solute in Solution by Reaction with Another Solution

The vitamin C (ascorbic acid, $C_6H_8O_6$) content of Gatorade and other sports beverages can be measured by reaction with aqueous triiodide (I_3^-). The reaction can be represented with the equation

$$I_3^-(aq) + C_6H_8O_6(aq) \longrightarrow 3I^-(aq) + C_6H_6O_6(aq) + 2H^+(aq)$$

Determine the mass of vitamin C (in mg) contained in a 350.0-mL bottle of Gatorade if a 25.0-mL sample requires 29.25 mL of 0.00125 M I_3^- solution to react with all of the vitamin C.

Strategy Use the volume and concentration of the triiodide solution to determine the number of moles of triiodide reacted; then use the balanced equation to determine the number of moles of vitamin C reacted. (In this case, the ratio of combination is 1:1.) Use this number of moles and the molar mass of vitamin C to determine the mass of vitamin C in the 25.0-mL sample; and then determine the mass of vitamin C in the total volume (350.0 mL).

Setup The molar mass of vitamin C is 176.1 g/mol. The volume of I_3^- solution in liters is 0.02925 L.

Solution

$$0.02925 \; \cancel{L} \times \frac{0.00125 \; \cancel{mol\,I_3^-}}{\cancel{L}} \times \frac{1 \; mol \; C_6H_8O_6}{1 \; \cancel{mol\,I_3^-}} = 3.656 \times 10^{-5} \; mol \; C_6H_8O_6$$

$$3.656 \times 10^{-5} \; \cancel{mol\,C_6H_8O_6} \times \frac{176.1 \; g \; C_6H_8O_6}{\cancel{mol\,C_6H_8O_6}} = 6.44 \times 10^{-3} \; g \; C_6H_8O_6$$

$$\frac{6.44 \times 10^{-3} \; \cancel{g\,C_6H_8O_6}}{25.0 \; \cancel{mL}} \times 350.0 \; \cancel{mL} \times \frac{1000 \; mg}{1 \; \cancel{g}} = 90.2 \; mg$$

THINK ABOUT IT

In this particular problem, the mole ratio between I_3^- and vitamin C is 1:1, but that is not always the case. Make sure to always take the mole ratio into account.

Practice Problem **A**TTEMPT Using the same reaction shown in the sample problem, determine the mass of vitamin C (in mg) contained in a 295-mL bottle of orange juice if a 50.0-mL sample of juice requires 41.6 mL of 0.00250 M I_3^- solution to react with all of the vitamin C.

Practice Problem **B**UILD The iron content of drinking water can be measured by reaction with potassium permanganate. This reaction is represented by the equation:

$$5Fe^{2+}(aq) + KMnO_4^-(aq) + 8H^+(aq) \longrightarrow 5Fe^{3+}(aq) + Mn^{2+}(aq) + 4H_2O(l)$$

Determine the concentration of iron in ppm (mg/L) of a sample of water if 25.00 mL of the water requires 21.30 mL of 2.175×10^{-5} M $KMnO_4$ to react with all of the iron in the water.

(Continued on next page)

Practice Problem Ⓒ**ONCEPTUALIZE** Because iodine itself is not very soluble in water, "iodine" solutions used in redox titrations generally contain the triiodide ion (I_3^-). Thus, the equation for the redox titration of vitamin C with iodine can be written as

$$C_6H_8O_6(aq) + I_3^-(aq) \longrightarrow C_6H_6O_6(aq) + 3I^-(aq) + 2H^+(aq)$$

Which diagram best represents the result when the $C_6H_8O_6$ (vitamin C) in a solution has all reacted with I_3^-? (Spectator ions are not shown.)

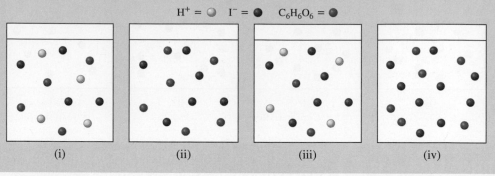

$$H^+ = \bigcirc \quad I^- = \bullet \quad C_6H_6O_6 = \bullet$$

(i)   (ii)   (iii)   (iv)

CHECKPOINT–SECTION 11.4   Aqueous Reactions

11.4.1 What mass of AgCl will be recovered if a solution containing 5.00 g of NaCl is treated with enough $AgNO_3$ to precipitate all the chloride ion?

    a) 12.3 g   d) 9.23 g

    b) 5.00 g   e) 10.0 g

    c) 3.03 g

11.4.2 A 10.0-g sample of an unknown ionic compound is dissolved, and the solution is treated with enough $AgNO_3$ to precipitate all the chloride ion. If 30.1 g of AgCl is recovered, which of the following compounds could be the unknown?

    a) NaCl   d) $MgCl_2$

    b) $NaNO_3$ e) KCl

    c) $BaCl_2$

11.4.3 Which of the following best represents the contents of a beaker in which equal volumes of 0.10 M $BaCl_2$ and 0.10 M $AgNO_3$ were combined?

$$Ag^+ = \bullet \quad NO_3^- = \bullet \quad Ba^{2+} = \bullet \quad Cl^- = \bullet$$

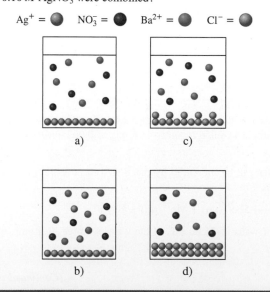

a)   c)

b)   d)

11.4.4 If 25.0 mL of an H_2SO_4 solution requires 39.9 mL of 0.228 M NaOH to neutralize, what is the concentration of the H_2SO_4 solution?

    a) 0.728 M d) 0.228 M

    b) 0.364 M e) 0.910 M

    c) 0.182 M

11.4.5 What volume of 0.144 M H_2SO_4 is required to neutralize 25.0 mL of 0.0415 M $Ba(OH)_2$?

    a) 7.20 mL   d) 50.0 mL

    b) 3.60 mL   e) 12.5 mL

    c) 14.4 mL

11.4.6 Which of the following best represents the contents of a beaker in which equal volumes of 0.10 M NaCl and 0.10 M $Pb(NO_3)_2$ were combined?

$$Na^+ = \bullet \quad NO_3^- = \bullet \quad Pb^{2+} = \bullet \quad Cl^- = \bullet$$

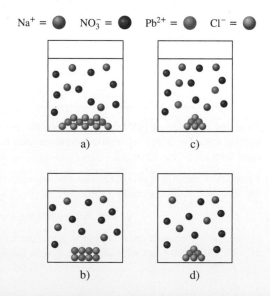

a)   c)

b)   d)

11.5 Gases in Chemical Reactions

Throughout this chapter, we have used balanced chemical equations to calculate amounts of reactants or products in chemical reactions. Because the coefficients in a balanced chemical equation correspond to *moles* of reactants and products, it has been necessary for us to determine *molar* amounts of reactants and/or products. Most often, we were given the masses of reactants and had to use molar masses for the necessary conversions to moles. In the case of aqueous reactions, we were given the volume and concentration of each reactant—again, requiring conversion to moles. In this section, we consider gaseous reactions, for which reactant and product amounts are generally given as volumes. This requires conversions between volume and moles and is done using the ideal gas equation [◄◄ Section 8.3].

> **Rewind:** Recall that the ideal gas equation is
>
> $$PV = nRT$$
>
> where P is pressure (usually in atm), V is volume (in L), n is the number of moles, R is the ideal gas constant (usually 0.0821 L·atm/K·mol), and T is absolute temperature (in kelvins).

Predicting the Volume of a Gaseous Product

Using a combination of stoichiometry and the ideal gas equation, we can calculate the volume of gas that we expect to be produced in a chemical reaction. We first use stoichiometry to determine the number of moles produced, and then apply the ideal gas equation to determine what volume will be occupied by that number of moles under the specified conditions.

Sample Problem 11.9 shows how to predict the volume of a gaseous product.

SAMPLE PROBLEM (11.9) Using a Balanced Chemical Equation to Determine Volume of a Gaseous Product

The airbags in cars are inflated when a collision triggers the explosive, highly exothermic decomposition of sodium azide (NaN_3):

$$2NaN_3(s) \longrightarrow 2Na(s) + 3N_2(g)$$

A typical driver-side airbag contains about 50 g of NaN_3. Determine the volume of N_2 gas that would be generated by the decomposition of 50.0 g of sodium azide at 85.0°C and 1.00 atm.

Strategy Convert the given mass of NaN_3 to moles, use the ratio of the coefficients from the balanced chemical equation to determine the corresponding number of moles of N_2 produced, and then use the ideal gas equation to determine the volume of that number of moles at the specified temperature and pressure.

Setup The molar mass of NaN_3 is 65.02 g/mol.

Solution

$$50.0 \text{ g NaN}_3 \times \frac{1 \text{ mol NaN}_3}{65.02 \text{ g NaN}_3} = 0.769 \text{ mol NaN}_3$$

$$0.769 \text{ mol NaN}_3 \times \left(\frac{3 \text{ mol N}_2}{2 \text{ mol NaN}_3}\right) = 1.15 \text{ mol N}_2$$

$$V_{N_2} = \frac{(1.15 \text{ mol N}_2)(0.0821 \text{ L} \cdot \text{atm/K} \cdot \text{mol})(358 \text{ K})}{1 \text{ atm}} = 33.8 \text{ L N}_2$$

THINK ABOUT IT

The calculated volume represents the space between the driver and the steering wheel and dashboard that must be filled by the airbag to prevent injury. Airbags also contain an oxidant that consumes the sodium metal produced in the reaction. Note that the final answer will be slightly different if we don't round any of the intermediate answers. If we retain all of the digits in the first and second calculations, the final answer will be 33.9 L N_2.

Practice Problem **TTEMPT** The chemical equation for the metabolic breakdown of glucose ($C_6H_{12}O_6$) is the same as that for the combustion of glucose [◄◄ Section 10.3—Familiar Chemistry box]:

$$C_6H_{12}O_6(aq) + 6O_2(g) \longrightarrow 6CO_2(g) + 6H_2O(l)$$

Calculate the volume of CO_2 produced at normal human body temperature (37°C) and 1.00 atm when 10.0 g of glucose is consumed in the reaction.

(Continued on next page)

Practice Problem **B**UILD   The passenger-side airbag in a typical car must fill a space approximately four times as large as the driver-side airbag to be effective. Calculate the mass of sodium azide required to fill a 125-L airbag at 85.0°C and 1.00 atm.

Practice Problem **C**ONCEPTUALIZE   The unbalanced decomposition reactions of two solid compounds are represented here. Both decompose to form the same two gaseous products, in different amounts. Which compound will produce the greater volume of products when equal numbers of moles decompose? Which compound will produce the greater volume of products when equal numbers of grams decompose?

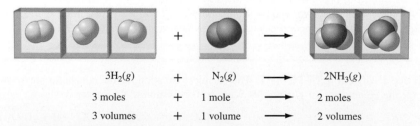

Calculating the Required Volume of a Gaseous Reactant

According to Avogadro's law [◄◄ Section 8.4], the volume of a gas at a given temperature and pressure is proportional to the number of moles. This means that balanced chemical equations not only specify the ratios in which *moles* of reactants combine, they also specify the ratios in which *volumes* of gaseous reactants combine—as illustrated in Figure 11.7. Therefore, if we know the volume of one reactant in a gaseous reaction, we can determine the required amount of another reactant (at the same temperature and pressure).

Let's consider the example of the reaction of carbon monoxide with oxygen to yield carbon dioxide:

$$2CO(g) + O_2(g) \longrightarrow 2CO_2(g)$$

The ratio of combination of CO and O_2 is 2:1, whether we are talking about moles or units of volume. Thus, if we want to determine the stoichiometric amount [◄◄ Section 11.1] of O_2 required to combine with a particular volume of CO, we simply use the conversion factor provided by the balanced equation, which can be expressed as any of the following:

$$\frac{1 \text{ mol } O_2}{2 \text{ mol CO}} \quad \text{or} \quad \frac{1 \text{ L } O_2}{2 \text{ L CO}} \quad \text{or} \quad \frac{1 \text{ mL } O_2}{2 \text{ mL CO}}$$

Let's say we want to determine what volume of O_2 is required to react completely with 65.8 mL of CO at STP. We could use the ideal gas equation to convert the volume of CO to moles, use the stoichiometric conversion factor to convert to moles O_2, and then use the ideal gas equation again to convert moles O_2 to volume. But this method involves several unnecessary steps. We get the same result simply by using the conversion factor expressed in milliliters:

$$65.8 \text{ mL CO} \times \frac{1 \text{ mL } O_2}{2 \text{ mL CO}} = 32.9 \text{ mL } O_2$$

3H₂(g)	+	N₂(g)	⟶	2NH₃(g)
3 moles	+	1 mole	⟶	2 moles
3 volumes	+	1 volume	⟶	2 volumes

Figure 11.7 Illustration of Avogadro's law. Because the volume of a gas is proportional to the number of moles, the coefficients in a balanced equation can be interpreted as volumes of gaseous reactants and products—typically expressed in liters (L) or milliliters (mL).

Recall, for example, the reaction of sodium and chlorine to produce sodium chloride [◄◄ Section 3.2]. Sodium metal is a solid, chlorine is a gas, and sodium chloride is a solid. This reaction has just one gaseous species, the chlorine gas reactant, and is represented by the following balanced equation:

$$2Na(s) + Cl_2(g) \longrightarrow 2NaCl(s)$$

Given moles (or more commonly the *mass*) of Na, and information regarding temperature and pressure, we can determine the volume of Cl_2 required to react completely:

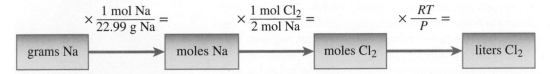

Sample Problem 11.10 shows how to use the ideal gas equation in a stoichiometric analysis.

SAMPLE PROBLEM (11.10) Using a Balanced Chemical Equation to Determine How Much Reactant Is Consumed in a Reaction

Sodium peroxide (Na_2O_2) is used to remove carbon dioxide from (and add oxygen to) the air supply in spacecrafts. It works by reacting with CO_2 in the air to produce sodium carbonate (Na_2CO_3) and O_2.

$$2Na_2O_2(s) + 2CO_2(g) \longrightarrow 2Na_2CO_3(s) + O_2(g)$$

What volume (in liters) of CO_2 (at STP) will react with 1.00 kg of Na_2O_2?

Strategy Convert the given mass of Na_2O_2 to moles, use the balanced equation to determine the stoichiometric amount of CO_2, and then use the ideal gas equation to convert moles of CO_2 to liters.

Setup The molar mass of Na_2O_2 is 77.98 g/mol (1.00 kg = 1.00×10^3 g).

Solution

$$1.00 \times 10^3 \text{ g } Na_2O_2 \times \frac{1 \text{ mol } Na_2O_2}{77.98 \text{ g } Na_2O_2} = 12.8 \text{ mol } Na_2O_2$$

$$12.8 \text{ mol } Na_2O_2 \times \frac{2 \text{ mol } CO_2}{2 \text{ mol } Na_2O_2} = 12.8 \text{ mol } CO_2$$

$$V_{CO_2} = \frac{(12.8 \text{ mol } CO_2)(0.0821 \text{ L} \cdot \text{atm/K} \cdot \text{mol})(273 \text{ K})}{1 \text{ atm}} = 287 \text{ L } CO_2$$

THINK ABOUT IT

The answer may seem like an enormous volume of CO_2. If you check the cancellation of units carefully in ideal gas equation problems, however, with practice you will develop a sense of whether such a calculated volume is reasonable.

Practice Problem ⒶTTEMPT What volume (in liters) of CO_2 can be consumed at STP by 525 g Na_2O_2?

Practice Problem ⒷUILD What mass (in grams) of Na_2O_2 is necessary to consume 1.00 L CO_2 at STP?

Practice Problem ⒸONCEPTUALIZE The decomposition reactions of two solid compounds are represented here.

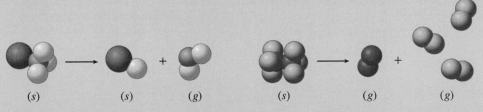

If an equal number of moles of each solid reactant were to decompose, how would the volume of products of the second decomposition compare to the volume of products of the first decomposition?

Profiles in Science

Joseph Louis Gay-Lussac

Joseph Gay-Lussac (1778–1850) made a number of important contributions to the fields of chemistry and physics. Some of his more enduring contributions are the experimental proof that water is composed of two parts hydrogen and one part oxygen; the discovery that the composition of air does not vary with altitude; the demonstration that the pressure of a gas is proportional to its absolute temperature; the naming of the element iodine; the discovery of the element boron (along with two of his contemporaries: Humphry Davy and Louis Jacques Thenard); and the introduction of copper(II) oxide to improve the quantitative accuracy of elemental analysis via combustion (see Thinking Outside the Box—Combustion Analysis in Section 11.3).

One of Gay-Lussac's more memorable series of experiments was done in a hydrogen balloon in which he ascended to an altitude of over 23,000 ft—an altitude record that went unchallenged for half a century. During that flight, he established the constant composition of air in Earth's atmosphere and that the intensity of Earth's magnetism is undiminished at high elevations.

Gay-Lussac and his wife (Geneviève-Marie-Joseph Rojot) had five children. Gay-Lussac's name is one of 72 names engraved on the Eiffel Tower in Paris. The list consists of historically important French scientists, engineers, and mathematicians—although it does not include any women. (Perhaps the most egregious omission is that of Sophie Germain, the French mathematician and physicist without whose pioneering work in elasticity theory the Eiffel Tower could not have been built.)

Joseph Louis Gay-Lussac
Library of Congress Prints & Photographs Division [LC-DIG-ppmsca-02237]

CHECKPOINT–SECTION 11.5   Gases in Chemical Reactions

11.5.1   Determine the volume of Cl_2 gas at STP that will react with 1.00 mole of Na solid to produce NaCl according to the equation,

$$2Na(s) + Cl_2(g) \longrightarrow 2NaCl(s)$$

a) 22.4 L

b) 44.8 L

c) 15.3 L

d) 11.2 L

e) 30.6 L

11.5.2   Determine the mass of NaN_3 required for an airbag to produce 100.0 L of N_2 gas at 85.0°C and 1.00 atm according to the equation,

$$2NaN_3(s) \longrightarrow 2Na(s) + 3N_2(g)$$

a) 177 g

b) 147 g

c) 221 g

d) 664 g

e) 442 g

11.5.3   Determine the mass (in grams) of NO needed to react with 13.9 L of O_2 at STP.

$$2NO(g) + O_2(g) \longrightarrow 2NO_2(g)$$

a) 18.6 g

b) 37.2 g

c) 57.1 g

d) 9.31 g

e) 28.5 g

Use the molecular art shown here to answer Questions 11.5.4 and 11.5.5.

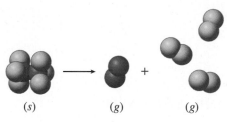

   (s)   (g)   (g)

11.5.4   Determine the volume of gas (in L) that would be formed from the decomposition of 4.33 moles of the reactant at 375°C and 1.50 atm.

a) 154 L

b) 307 L

c) 355 L

d) 614 L

e) 88.9 L

11.5.5   Determine the volume of gas (in L) that would be formed from the decomposition of 2.80 moles of the reactant at 263°C and 3.20 atm.

a) 154 L

b) 307 L

c) 355 L

d) 614 L

e) 88.9 L

11.6 Chemical Reactions and Heat

In addition to enabling us to determine amounts of products produced or reactants consumed, balanced chemical equations enable us to determine the amount of heat produced (or consumed) by a chemical reaction. We have seen that *heat* can be included in a chemical equation either as a reactant (for an endothermic reaction) or as a product (for an exothermic reaction) [◀◀ Section 10.5]. Recall the examples of endothermic and exothermic reactions that were given in Chapter 10:

endothermic: $heat + CaCO_3(s) \longrightarrow CaO(s) + CO_2(g)$

exothermic: $CH_4(g) + 2O_2(g) \longrightarrow CO_2(g) + H_2O(g) + heat$

The amount of heat required for the endothermic decomposition of one mole of $CaCO_3(s)$ is 177.8 kJ; and the amount of heat produced by the combustion of one mole of $CH_4(g)$ is 560.5 kJ. We can write each of these equations to include the **heat of reaction,** which specifies the exact amount of heat involved and indicates whether the heat is *consumed* or *produced*. For the first reaction, we can write:

$CaCO_3(s) \longrightarrow CaO(s) + CO_2(g)$ heat of reaction: +177.8 kJ

(The heat of reaction for an *endothermic* process is *positive*.) And for the second reaction, we can write:

$CH_4(g) + 2O_2(g) \longrightarrow CO_2(g) + H_2O(g)$ heat of reaction: −560.5 kJ

(The heat of reaction for an *exothermic* process is *negative*.)

Sample Problem 11.11 illustrates how heat of reaction is used to determine the amount of heat required for an endothermic reaction to produce a specific amount of product.

Student Note: The *heat of reaction* is also known as the *enthalpy* of reaction, and refers specifically to *the amount of energy consumed by a reaction carried out at constant pressure*—which is the case for most laboratory experiments. A negative sign indicates that energy is *produced*, rather than consumed.

SAMPLE PROBLEM 11.11 Using a Balanced Chemical Equation to Determine the Energy Required to Make a Particular Mass of Product

Given the chemical equation representing photosynthesis and its heat of reaction:

$6H_2O(l) + 6CO_2(g) \longrightarrow C_6H_{12}O_6(s) + 6O_2(g)$ heat of reaction = +2803 kJ

calculate the solar energy required to produce 75.0 g of $C_6H_{12}O_6$.

Strategy The equation shows that for every mole of $C_6H_{12}O_6$ produced, 2803 kJ is absorbed. We need to find out how much energy is absorbed for the production of 75.0 g of $C_6H_{12}O_6$. We must first find out how many moles there are in 75.0 g of $C_6H_{12}O_6$.

Setup The molar mass of $C_6H_{12}O_6$ is 180.2 g/mol, so 75.0 g of $C_6H_{12}O_6$ is

$$75.0 \text{ g } C_6H_{12}O_6 \times \frac{1 \text{ mol } C_6H_{12}O_6}{180.2 \text{ g } C_6H_{12}O_6} = 0.461 \text{ mol}$$

We will multiply the equation, including the heat of reaction, by 0.416, in order to write the equation in terms of the appropriate amount of $C_6H_{12}O_6$.

Solution

$$(0.416 \text{ mol})[6H_2O(l) + 6CO_2(g) \longrightarrow C_6H_{12}O_6(s) + 6O_2(g)]$$

and (0.416 mol)(heat of reaction) = (0.416 mol)(2803 kJ/mol) gives

$$2.50 H_2O(l) + 2.50 CO_2(g) \longrightarrow 0.416 C_6H_{12}O_6(s) + 2.50 O_2(g)   \Delta H = 1.17 \times 10^3 \text{ kJ}$$

Therefore, 1.17×10^3 kJ of energy in the form of sunlight is consumed in the production of 75.0 g of $C_6H_{12}O_6$.

(Continued on next page)

THINK ABOUT IT

The specified amount of $C_6H_{12}O_6$ is less than half a mole. Therefore, we should expect the heat consumed to be less than half that specified in the equation for the production of 1 mole of $C_6H_{12}O_6$. Consider, too, that the heat of reaction can be used in a stoichiometric conversion factor. In this example, one such conversion factor would be

$$\frac{2803 \text{ kJ consumed}}{1 \text{ mol } C_6H_{12}O_6 \text{ produced}} \qquad \text{which can also be written as} \qquad \frac{1 \text{ mol } C_6H_{12}O_6 \text{ produced}}{2803 \text{ kJ consumed}}$$

Practice Problem ATTEMPT Calculate the solar energy required to produce 5255 g of $C_6H_{12}O_6$.

Practice Problem BUILD Calculate the mass (in grams) of O_2 that is produced by photosynthesis when 2.490×10^4 kJ of solar energy is consumed.

Practice Problem CONCEPTUALIZE The diagrams represent systems before and after two related chemical reactions. The heat of reaction for the first is −1755.0 kJ. Determine the heat of reaction for the second.

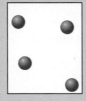      heat of reaction = −1755.0 kJ

before   after

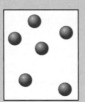      heat of reaction = ?

before   after

CHECKPOINT–SECTION 11.6 Chemical Reactions and Heat

11.6.1 Given the equation:

$$H_2(g) + Br_2(l) \longrightarrow 2HBr(g)$$

with heat of reaction = −72.4 kJ, calculate the amount of heat released when a kilogram of $Br_2(l)$ is consumed in this reaction.

a) 7.24×10^4 kJ   d) 227 kJ

b) 453 kJ   e) 724 kJ

c) 906 kJ

11.6.2 Given the equation:

$$2Cu_2O(s) \longrightarrow 4Cu(s) + O_2(g)$$

with heat of reaction = +333.8 kJ, calculate the mass of copper produced when 1.47×10^4 kJ is consumed in this reaction.

a) 11.2 kg   d) 334 kg

b) 176 kg   e) 782 kg

c) 44.0 kg

Chapter Summary

Section 11.1

- *Stoichiometry* refers to the numerical relationships between the amounts of reactants and products in a chemical reaction.

- The stoichiometric coefficients in a balanced equation are used to construct stoichiometric conversion factors. A balanced chemical equation can be used to determine how many moles of product can be produced, given molar amounts of reactants; or how many moles of reactants are necessary to produce a particular amount of product.

- The *stoichiometric amount* of a reactant is the precise amount necessary to react with the other reactant(s) in a chemical equation—according to the balanced chemical equation.

Section 11.2

- Using molar masses of reactants and products, we can use balanced chemical equations to determine masses of reactants corresponding to given masses of products, and vice versa.

Section 11.3

- The *limiting reactant* is the reactant that is consumed completely in a chemical reaction. An *excess reactant* is a reactant that is not consumed completely. The maximum amount of product that can form in a chemical reaction depends on the amount of limiting reactant.

- The *theoretical yield* of a reaction is the amount of product that will form if all of the limiting reactant is consumed by the desired reaction. The *actual yield* is the amount of product that actually forms in a reaction.

- *Percent yield* [(actual/theoretical) × 100%] is a measure of the efficiency of a chemical reaction.

Section 11.4

- When aqueous substances react with one another, we use their molar concentrations and volumes to relate reactant and product amounts.

Section 11.5

- For a reaction occurring at constant temperature and pressure, and involving only gases, the coefficients in the balanced chemical equation apply to units of volume, as well as to numbers of molecules or moles.

- A balanced chemical equation and the ideal gas equation can be used to determine volumes of gaseous reactants and/or products in a reaction.

Section 11.6

- *Heat of reaction* refers to the amount of heat produced (exothermic) or consumed (endothermic) by a chemical reaction. Heats of reaction are expressed in kJ. A negative heat of reaction indicates an exothermic reaction; a positive heat of reaction indicates an endothermic reaction.

- A balanced chemical equation and the corresponding heat of reaction can be used to determine how much heat will be produced or consumed by specified amounts of reactants or products.

Key Terms

Actual yield 401	Heat of reaction 415	Percent yield 401	Stoichiometry 393
Excess reactant 398	Limiting reactant 398	Stoichiometric amounts 394	Theoretical yield 401

Key Equation

11.1 $\% \text{ yield} = \dfrac{\text{actual yield}}{\text{theoretical yield}} \times 100\%$

The amount of product actually produced in a reaction will nearly always be less than that predicted by the balanced equation. We use the actual (measured) amount of product and the calculated amount of product to determine the percent yield of a reaction.

KEY SKILLS Limiting Reactant

The amount of product that can be produced in a chemical reaction typically is limited by the amount of *one* of the reactants—known as the *limiting* reactant. The practice of identifying the limiting reactant, calculating the maximum possible amount of product, and determining the percent yield and remaining amount of an excess reactant requires several skills:

- Balancing chemical equations [◄◄ Section 10.3]
- Determining molar mass [◄◄ Section 5.2]
- Converting between mass and moles [◄◄ Section 5.2]
- Using stoichiometric conversion factors [◄◄ Section 11.1]

Consider the following example. Hydrazine (N_2H_4) reacts with dinitrogen tetroxide (N_2O_4) to form nitrogen monoxide (NO) and water. Determine the mass of NO that can be produced when 10.45 g of N_2H_4 and 53.68 g of N_2O_4 are combined. The unbalanced equation is

$$N_2H_4 + N_2O_4 \longrightarrow NO + H_2O$$

We first balance the equation.

$$N_2H_4 + 2N_2O_4 \longrightarrow 6NO + 2H_2O$$

Next, we determine the necessary molar masses.

N_2H_4: $2(14.01) + 4(1.008) =$ $\dfrac{32.05\ g}{mol}$ N_2O_4: $2(14.01) + 4(16.00) =$ $\dfrac{92.02\ g}{mol}$ NO: $14.01 + 16.00 =$ $\dfrac{30.01\ g}{mol}$

We convert the reactant masses given in the problem to moles. Then we determine the mole amount of NO that could be produced from the mole amount of each reactant by multiplying each of the *reactant* mole amounts by the appropriate stoichiometric conversion factor, which we derive from the balanced equation. According to the balanced equation:

$$1\ mol\ N_2H_4 \simeq 6\ mol\ NO \quad and \quad 2\ mol\ N_2O_4 \simeq 6\ mol\ NO$$

The reactant that produces the smaller amount of product is the limiting reactant; in this case, N_2O_4.

We continue the problem using the mole amount of NO produced by reaction of the given amount of N_2O_4. To convert from moles to mass (grams), we multiply the number of moles NO by the molar mass of NO:

Thus, 52.52 g NO can be produced by the reaction. Note that we retained an extra significant figure until the end of the calculation.

To determine the mass of remaining excess reactant, we must first determine what amount was consumed in the reaction. To do this, we multiply the mole amount of limiting reactant (N_2O_4) by the appropriate stoichiometric conversion factor. According to the balanced equation:

$$1 \text{ mol } N_2H_4 \rightleftharpoons 2 \text{ mol } N_2O_4$$

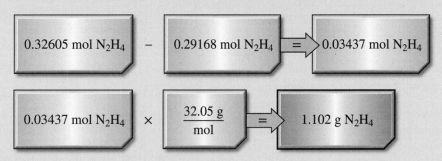

This is the amount of N_2H_4 consumed. The amount remaining is the difference between this and the original amount. We convert the remaining mole amount to grams using the molar mass of N_2H_4.

Thus, 1.102 g N_2H_4 remain when the reaction is complete.

We can check our work in a problem such as this by also calculating the mass of the other product, in this case water. The mass of all products plus the mass of any remaining reactant must equal the sum of starting reactant masses.

Key Skills Problems

11.1
Calculate the mass of water produced in the example above.
a) 21.02 g
b) 10.51 g
c) 11.61 g
d) 11.75 g
e) 5.400 g

Use the following information to answer questions 11.2, 11.3, and 11.4.

Calcium phosphide (Ca_3P_2) and water react to form calcium hydroxide and phosphine (PH_3). In a particular experiment, 225.0 g Ca_3P_2 and 125.0 g water are combined.

$$Ca_3P_2(s) + H_2O(l) \longrightarrow Ca(OH)_2(aq) + PH_3(g)$$

(Don't forget to balance the equation.)

11.2
How much PH_3 can be produced?
a) 350.0 g
b) 235.0 g
c) 78.59 g
d) 83.96 g
e) 41.98 g

11.3
How much $Ca(OH)_2$ can be produced?
a) 91.51 g
b) 274.5 g
c) 513.8 g
d) 85.63 g
e) 257.0 g

11.4
How much of the excess reactant remains when the reaction is complete?
a) 14.37 g
b) 235.0 g
c) 78.56 g
d) 83.96 g
e) 41.98 g

Questions and Problems

SECTION 11.1: MOLE TO MOLE CONVERSIONS

11.1 Why is it essential to use balanced equations when solving stoichiometric problems?

11.2 How would you explain to a friend that 10.0 g of N_2 does not react with 30.0 g of H_2 to form 40.0 g of NH_3?

$$N_2(g) + 3H_2(g) \longrightarrow 2NH_3(g)$$

11.3 Given the chemical equation shown, write down all possible conversion factors that show the relationships between the moles of reactants and products.

$$4NH_3(g) + 3O_2(g) \longrightarrow 2N_2(g) + 6H_2O(g)$$

(a) Which of these conversion factors would you use to determine the moles of nitrogen formed from 1 mole of ammonia (NH_3)?

(b) Which of these conversion factors would you use to determine the moles of oxygen needed to react with 1 mole of ammonia?

(c) Which of these conversion factors would you use to calculate the moles of water formed when 3 moles of nitrogen form?

11.4 If the image shows the magnesium solid prior to reaction, sketch the oxygen molecules that would be needed to exactly use up all of the magnesium according to the balanced chemical equation shown.

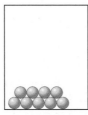

$$2Mg(s) + O_2(g) \longrightarrow 2MgO(s)$$

11.5 Consider the combustion of carbon monoxide (CO) in oxygen gas:

$$2CO(g) + O_2(g) \longrightarrow 2CO_2(g)$$

Starting with 3.60 moles of CO, calculate the number of moles of CO_2 produced if there is enough oxygen gas to react with all of the CO. How many moles of oxygen gas are required to react with the CO?

11.6 Silicon tetrachloride ($SiCl_4$) can be prepared by heating Si in chlorine gas:

$$Si(s) + 2Cl_2(g) \longrightarrow SiCl_4(l)$$

In one reaction, 0.507 mol $SiCl_4$ is produced. How many moles of molecular chlorine were used in the reaction? How many moles of Si?

11.7 Consider the following balanced chemical equation:

$$2N_2H_4(g) + N_2O_4(g) \longrightarrow 3N_2(g) + 4H_2O(g)$$

(a) How many moles of water form from 2.0 moles of N_2O_4?

(b) How many moles of water form from 2.0 moles of N_2H_4?

(c) How many moles of nitrogen form from 2.0 moles of N_2O_4?

(d) How many moles of nitrogen form from 2.0 moles of N_2H_4?

11.8 Consider the following balanced chemical equation:

$$2H_2S(g) + SO_2(g) \longrightarrow 3S(s) + 2H_2O(g)$$

(a) How many moles of H_2S must react to form 4.0 moles of S?

(b) How many moles of SO_2 must react to form 4.0 moles of S?

(c) How many moles of H_2S must react to form 4.0 moles of H_2O?

(d) How many moles of SO_2 must react to form 4.0 moles of H_2O?

11.9 Consider the following balanced chemical equation:

$$C_2H_5OH(l) + 3O_2(g) \longrightarrow 2CO_2(g) + 3H_2O(l)$$

(a) How many moles of CO_2 form from the reaction of 3.33 moles of C_2H_5OH?

(b) How many moles of H_2O form from the reaction of 3.33 moles of C_2H_5OH?

(c) How many moles of O_2 are required to form 8.65 moles of H_2O?

(d) How many moles of C_2H_5OH are required to react with 7.44 moles of O_2?

(e) How many moles of CO_2 form when 2.94 moles of H_2O form?

(f) How many molecules of CO_2 form when 2.94 moles of H_2O form?

11.10 Consider the following balanced chemical equation:

$$4Fe(s) + 3O_2(g) \longrightarrow 2Fe_2O_3(s)$$

(a) How many moles of iron are required to react with 24.0 moles of O_2?

(b) How many moles of oxygen are required to react with 24.0 moles of iron?

(c) How many moles of iron are required to produce 12.5 moles of Fe_2O_3?

(d) How many moles of oxygen are required to produce 12.5 moles of Fe_2O_3?

(e) How many molecules of oxygen are required to react with 9.55 moles of iron?

(f) How many formula units of iron(III) oxide are formed from 7.55 moles of iron?

SECTION 11.2: MASS TO MASS CONVERSIONS

11.11 How much oxygen (in grams) will form from the given quantities of N_2O_5?

$$2N_2O_5(s) \longrightarrow 4NO_2(g) + O_2(g)$$

(a) 23.3 mg
(b) 23.3 g
(c) 23.3 kg

11.12 How much methanol (CH_3OH, in grams) can be formed from the given quantities of hydrogen? Assume excess CO.

$$CO(g) + 2H_2(g) \longrightarrow CH_3OH(g)$$

(a) 65.4 mg
(b) 65.4 g
(c) 65.4 kg

11.13 What mass, in grams, of NO_2 is needed to form the given quantities of NO? Assume excess water.

$$3NO_2(g) + H_2O(l) \longrightarrow HNO_3(aq) + NO(g)$$

(a) 197.4 g
(b) 35.8 mg
(c) 144 ng

11.14 What mass, in grams, of C_2H_2 is needed to form the given quantities of CO_2? Assume excess oxygen.

$$2C_2H_2(g) + 5O_2(g) \longrightarrow 4CO_2(g) + 2H_2O(g)$$

(a) 569 mg
(b) 29.4 g
(c) 0.227 kg

11.15 In each of the following chemical equations, determine the moles and mass (in g) of the product formed when 1.50 moles of the substance highlighted in blue reacts. Assume an excess of the other reactant.
(a) $Ti(s) + 2Cl_2(g) \longrightarrow TiCl_4(s)$
(b) $2Mn(s) + 3O_2(g) \longrightarrow 2MnO_3(s)$
(c) $4Cr(s) + 3O_2(g) \longrightarrow 2Cr_2O_3(s)$

11.16 In each of the following chemical equations, determine the moles and mass (in g) of the reactant used when 3.50 moles of the substance highlighted in blue forms.
(a) $3N_2H_4(l) \longrightarrow 4NH_3(g) + N_2(g)$
(b) $4PCl_5(s) \longrightarrow P_4(s) + 10Cl_2(g)$
(c) $4PCl_3(g) \longrightarrow P_4(s) + 6Cl_2(g)$

11.17 Consider the following reaction:

$$B_2H_6(g) + 3O_2(l) \longrightarrow 2HBO_2(g) + 2H_2O(l)$$

(a) Determine the mass (in g) of HBO_2 formed from 40.0 g B_2H_6.
(b) Determine the mass (in g) of H_2O formed from 40.0 g B_2H_6.
(c) Determine the mass (in g) of oxygen needed to react with 40.0 g B_2H_6.

11.18 Consider the following reaction:

$$CS_2(g) + 3Cl_2(g) \longrightarrow CCl_4(g) + S_2Cl_2(g)$$

(a) Determine the mass (in kg) of CCl_4 formed from 39.5 g Cl_2.
(b) Determine the number of Cl_2 molecules required to form 45.9 grams of S_2Cl_2.
(c) Determine the mass (in grams) of CS_2 required to react with 3.45×10^{23} molecules of Cl_2.

11.19 Determine the mass of Al_4C_3, in grams, that is required to react with 98.9 grams of water.

$$Al_4C_3(s) + 12H_2O(l) \longrightarrow 4Al(OH)_3(s) + 3CH_4(g)$$

11.20 Determine the mass of carbon, in kg, that is required to react with 433 grams of Fe_2O_3.

$$Fe_2O_3(s) + 3C(s) \longrightarrow 2Fe(s) + 3CO(g)$$

SECTION 11.3: LIMITATIONS ON REACTION YIELD

11.21 The reaction to form ammonia from nitrogen and hydrogen is shown. Suppose you place 10.0 g of N_2 and 5.0 g of H_2 in a flask and start the reaction. Is it correct to say that H_2 is the limiting reactant because there is less of it? Prove your answer by calculating the limiting reactant.

$$N_2(g) + 3H_2(g) \longrightarrow 2NH_3(g)$$

11.22 Consider the following balanced chemical equation:

$$4KO_2(s) + 2H_2O(l) \longrightarrow 3O_2(g) + 4KOH(s)$$

(a) Determine the mass (in g) of KOH formed if 10.0 g of KO_2 reacts with 10.0 g of H_2O. Identify the limiting reactant.
(b) Determine the mass (in g) of KOH formed when 20.0 g of KO_2 reacts with 10.0 g of H_2O. Identify the limiting reactant.
(c) Determine the mass (in g) of O_2 formed when 25.0 g of KO_2 reacts with 5.00 g of H_2O. Identify the limiting reactant.

11.23 Consider the following balanced chemical equation:

$$2Cu_2O(s) + O_2(g) \longrightarrow 4CuO(s)$$

(a) Determine the mass (in g) of CuO formed from the reaction of 92.0 g Cu_2O and 12.0 g O_2. Identify the limiting reactant.
(b) Determine the mass (in g) of CuO formed from the reaction of 12.0 g Cu_2O and 92.0 g O_2. Identify the limiting reactant.
(c) Determine the mass (in g) of CuO formed from the reaction of 12.0 g Cu_2O and 12.0 g O_2. Identify the limiting reactant.

11.24 Consider the following balanced chemical equation:

$$CS_2(g) + 3O_2(g) \longrightarrow CO_2(g) + 2SO_2(g)$$

(a) Determine the mass (in g) of CO_2 formed if 8.75 g of CS_2 reacts with 8.75 g of O_2.
(b) Determine the mass (in g) of SO_2 formed if 8.75 g of CS_2 reacts with 9.22 g of O_2.
(c) Determine the mass (in g) of CO_2 formed if 4.89 g of CS_2 reacts with 21.1 g of O_2.

11.25 Consider the following balanced chemical equation:

$$2Al(s) + 3Cl_2(g) \longrightarrow 2AlCl_3(s)$$

(a) Determine the mass (in kg) of $AlCl_3$ formed when 375 g of aluminum reacts with 256 g of chlorine.
(b) Determine the mass (in g) of $AlCl_3$ formed when 39.9 g of aluminum reacts with 39.9 g of chlorine.
(c) Determine the moles of $AlCl_3$ formed when 47.6 g of aluminum reacts with 83.2 grams of chlorine.

11.26 Determine the theoretical yield of HF, in grams, when 49.8 g of NH_3 reacts with 49.8 g of F_2. What is the percent yield if 7.22 g of HF is actually formed during the reaction?

$$2NH_3(g) + 5F_2(g) \longrightarrow N_2F_4(g) + 6HF(g)$$

11.27 Determine the theoretical yield of HNO_3, in grams, when 122 g of H_2O_2 reacts with 74.6 g of N_2H_4. What is the percent yield if 59.5 g of HNO_3 is actually formed during the reaction?

$$7H_2O_2(l) + N_2H_4(g) \longrightarrow 2HNO_3(aq) + 8H_2O(l)$$

11.28 Determine the limiting reactant, the mass of water formed (in g), and the mass that remains of the excess reactant when 20.0 g of HCl reacts with 20.0 grams of O_2.

$$4HCl(g) + O_2(g) \longrightarrow 2Cl_2(g) + 2H_2O(g)$$

11.29 Determine the limiting reactant, the mass of $Ca(OH)_2$ formed (in g), and the mass that remains of the excess reactant when 24.7 g of Ca_3P_2 reacts with 28.5 grams of water.

$$Ca_3P_2(s) + 6H_2O(l) \longrightarrow 3Ca(OH)_2(aq) + 2PH_3(g)$$

11.30 Determine the limiting reactant and mass of oxygen formed, in grams, when 20.0 g KO_2 reacts with 20.0 g H_2O and 20.0 g CO_2.

$$4KO_2(s) + 2H_2O(l) + 4CO_2(g) \longrightarrow$$
$$4KHCO_3(s) + 3O_2(g)$$

SECTION 11.4: AQUEOUS REACTIONS

11.31 What mass of copper (in g) will form when 125.0 mL of 0.0455 M $CuSO_4$ reacts with excess aluminum?

$$2Al(s) + 3CuSO_4(aq) \longrightarrow 3Cu(s) + Al_2(SO_4)_3(aq)$$

11.32 Determine the mass, in grams, of precipitate that forms when 78.8 mL of 0.0223 M Na_2CO_3 reacts with excess calcium hydroxide.

$$Ca(OH)_2(aq) + Na_2CO_3(aq) \longrightarrow$$
$$2NaOH(aq) + CaCO_3(s)$$

11.33 Determine the mass, in kilograms, of precipitate that forms when 345 mL of 0.232 M KI reacts with 265 mL of 0.173 M $Pb(NO_3)_2$.

$$2KI(aq) + Pb(NO_3)_2(aq) \longrightarrow 2KNO_3(aq) + PbI_2(s)$$

11.34 Determine the mass, in grams, of precipitate that will form when 25.8 mL of 0.133 M $Ba(ClO_4)_2$ reacts with 45.4 mL of 0.0528 M K_2SO_4.

$$Ba(ClO_4)_2(aq) + K_2SO_4(aq) \longrightarrow$$
$$BaSO_4(s) + 2KClO_4(aq)$$

11.35 What volume, in mL, of 0.0551 M LiOH solution is required to produce 25.5 grams of water? Assume excess sulfuric acid.

$$H_2SO_4(aq) + 2LiOH(aq) \longrightarrow 2H_2O(l) + Li_2SO_4(aq)$$

11.36 What volume, in mL, of 0.109 M $NiCl_2$ solution is required to form 78.8 grams of precipitate? Assume excess Na_3PO_4.

$$3NiCl_2(aq) + 2Na_3PO_4(aq) \longrightarrow$$
$$Ni_3(PO_4)_2(s) + 6NaCl(aq)$$

11.37 What volume, in mL, of 0.236 M Li_2CO_3 is required to react with 68.7 mL of 0.308 M HNO_3?

$$Li_2CO_3(aq) + 2HNO_3(aq) \longrightarrow$$
$$H_2O(l) + LiNO_3(aq) + CO_2(g)$$

11.38 What mass of magnesium (in grams) is required to react with 29.5 mL of 0.118 M $AgNO_3$ solution?

$$Mg(s) + 2AgNO_3(aq) \longrightarrow Mg(NO_3)_2(aq) + 2Ag(s)$$

11.39 What mass of aluminum (in grams) is required to react with 0.228 L of 0.0811 M $AgNO_3$ solution?

$$3AgNO_3(aq) + Al(s) \longrightarrow 3Ag(s) + Al(NO_3)_3(aq)$$

11.40 What volume (in mL) of 0.844 M NaOH solution is required to react exactly with 48.5 mL of 1.39 M H_2SO_4 solution?

11.41 What volume (in mL) of 0.622 M KBr is required to react with 55.9 mL of 0.118 M $Pb(ClO_4)_2$?

SECTION 11.5: GASES IN CHEMICAL REACTIONS

11.42   What volume (in liters, at 355 K and 1.10 atm) of chlorine gas is required to react with 39.4 grams of P?

$$2P(s) + 5Cl_2(g) \longrightarrow 2PCl_5(s)$$

11.43   Determine the volume of chlorine gas (in liters, at 298 K and 1.50 atm) required to react with 13.7 L of O_2 gas at 425 K and 1.00 atm.

$$2Cl_2(g) + O_2(g) \longrightarrow 2OCl_2(g)$$

11.44   Determine the mass, in grams, of Li_3N that would produce 38.8 L N_2 gas at STP.

$$2Li_3N(s) \longrightarrow 6Li(s) + N_2(g)$$

11.45   Determine the volume (in L) of oxygen that forms from the decomposition of 98.5 g $KClO_3$ under each set of conditions given.

$$2KClO_3(s) \longrightarrow 2KCl(s) + 3O_2(g)$$

(a) 273 K and 1.00 atm
(b) 373 K and 1.00 atm
(c) 373 K and 4.00 atm
(d) 473 K and 2.00 atm

11.46   Determine the mass (in g) of HgO that must decompose to form 45.3 L of oxygen gas under each set of conditions given.

$$2HgO(s) \longrightarrow 2Hg(l) + O_2(g)$$

(a) STP
(b) 573 K and 1.00 atm
(c) 298 K and 1.50 atm
(d) 355 K and 2.50 atm

11.47   Determine the volume of oxygen gas (in mL) required to react with 195 g of aluminum under each set of conditions given.

$$4Al(s) + 3O_2(g) \longrightarrow 2Al_2O_3(s)$$

(a) STP
(b) 399 K and 3.00 atm
(c) 25°C and 855 mmHg
(d) 499°C and 1.68 atm

11.48   Determine the pressure of CO gas formed (in atm) when 76.6 g of C react under each set of conditions given.

$$SiO_2(s) + 3C(s) \longrightarrow SiC(s) + 2CO(g)$$

(a) 35°C in a 4.50-L container
(b) 555 K in a 5950-mL container
(c) 75°C in a 3444-mL container
(d) 298 K in a 7.39-L container

11.49   What volume of *gas,* in liters at 305 K and 1.10 atm, is formed from the reaction of 50.0 grams of carbon with excess SO_2?

$$5C(s) + 2SO_2(g) \longrightarrow CS_2(g) + 4CO(g)$$

11.50   The airbags in cars inflate through the explosive decomposition of a solid to produce a gas. The simplified version of the reaction that takes place in most airbags is shown.

$$2NaN_3(s) \longrightarrow 2Na(s) + 3N_2(g)$$

(a) If a typical airbag has a volume of 140.0 L when inflated, how much NaN_3 (in grams) is required to inflate such an airbag at STP?
(b) Would you want to rely on an airbag that contained only 100.0 g NaN_3? Explain.

11.51   Hydrogen can be prepared in the laboratory using the reaction shown. What volume, in mL, of hydrogen gas can be collected at 305 K and 740.0 mmHg when 185.0 g of magnesium reacts?

$$Mg(s) + 2HCl(aq) \longrightarrow H_2(g) + MgCl_2(aq)$$

SECTION 11.6: CHEMICAL REACTIONS AND HEAT

11.52   Define the term *heat of reaction.*

11.53   Write all of the heat to mole conversion factors from the balanced chemical equation.

$$C_3H_8(g) + 5O_2(g) \longrightarrow 3CO_2(g) + 4H_2O(g)$$
$$\text{heat of reaction} = -2044 \text{ kJ}$$

11.54   Consider the balanced chemical equation:

$$2Al(s) + Fe_2O_3(s) \longrightarrow 2Fe(s) + Al_2O_3(s)$$
$$\text{heat of reaction} = -851.5 \text{ kJ}$$

(a) How many kilojoules of heat are produced when 4.88 moles of Al react?
(b) How many kilojoules of heat are produced when 4.88 moles of Fe_2O_3 react?
(c) What mass of Fe_2O_3 must react to produce 387 kJ of heat?
(d) What mass of Fe is produced at the same time 387 kJ of heat is produced?

11.55   Consider the following balanced chemical equation:

$$4NO_2(g) + O_2(g) \longrightarrow 2N_2O_5(g)$$
$$\text{heat of reaction} = -110.2 \text{ kJ}$$

(a) How much heat is produced when 9.22 moles of N_2O_5 are produced?
(b) How much heat is produced when 39.5 grams of NO_2 react?
(c) How much heat is produced when 39.5 g of O_2 reacts?
(d) How much heat is produced when 39.5 g of NO_2 reacts with 39.5 g of O_2?

11.56   Consider the following balanced chemical equation:

$$4NH_3(g) + 5O_2(g) \longrightarrow 4NO(g) + 6H_2O(g)$$
$$\text{heat of reaction} = -906 \text{ kJ}$$

(a) How much heat is produced when 3.44 moles of oxygen react?

(b) How much heat is produced when 3.44 L of oxygen at STP reacts?

(c) How much heat is produced when 3.44 g of NO form?

(d) How much heat is produced when 3.44 g of water form?

11.57   Consider the following balanced chemical equation:

$$2SO_2(g) + O_2(g) \longrightarrow 2SO_3(g)$$
$$\text{heat of reaction} = -198 \text{ kJ}$$

(a) How much heat is produced when 75.6 moles of oxygen react?

(b) How much heat is produced when 75.6 L of SO_2 at STP reacts?

(c) How much heat is produced when 75.6 g of SO_3 form?

(d) How much heat is produced when 75.6 g of SO_2 react?

11.58   Consider the following balanced chemical equation:

$$H_2(g) + 2C(s) + N_2(g) \longrightarrow 2HCN(g)$$
$$\text{heat of reaction} = 270. \text{ kJ}$$

(a) How much heat is used when 20.0 g of carbon reacts?

(b) How much heat is used when 20.4 L of nitrogen reacts at STP?

(c) How much heat is used when 100.0 g of hydrogen reacts?

(d) How much heat is used when 63.3 g of HCN is formed?

11.59   Consider the following balanced chemical equation:

$$N_2(g) + O_2(g) \longrightarrow 2NO(g)$$
$$\text{heat of reaction} = 181 \text{ kJ}$$

(a) How much heat is used when 79.0 L of oxygen reacts at 455 K and 1.10 atm?

(b) How much heat is used when 226 g of NO forms?

(c) How much heat is used when 2.99×10^{23} nitrogen molecules react?

(d) How much heat is used when 25 kg of oxygen reacts?

CUMULATIVE PROBLEMS

11.60   Consider the reaction pictured, where each red sphere represents an oxygen atom and each blue sphere represents a nitrogen atom. Write the balanced equation and identify the limiting reactant.

11.61   Consider the reaction

$$N_2 + 3H_2 \longrightarrow 2NH_3$$

Assuming each model represents one mole of the substance, show the number of moles of the product and the excess reactant left after the complete reaction.

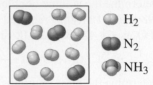

11.62   Consider the reaction of hydrogen gas with oxygen gas:

$$2H_2(g) + O_2(g) \longrightarrow 2H_2O(g)$$

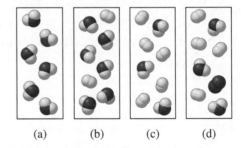

Assuming a complete reaction, which of the diagrams (a–d) shown here represents the amounts of reactants and products left after the reaction?

| (a) | (b) | (c) | (d) |

11.63   Baking soda (sodium bicarbonate) is used frequently to neutralize acid spills in the laboratory. What mass of sodium bicarbonate is needed to neutralize a 100.0-mL puddle of 0.120 M HNO_3? What volume of CO_2 gas is produced at STP during this neutralization?

11.64   A miner's lamp burns acetylene (C_2H_2) that is produced from water dripping onto solid CaC_2. The other product formed from the reaction is calcium hydroxide. What volume of water ($d = 1.00$ g/mL) is required to react with 12.5 g of calcium carbide (CaC_2) in the lamp?

An acetylene gas miner's lamp

Charles D. Winters/Science Source

11.65 Barbecue grills typically burn propane (C_3H_8) gas from a 20.0-lb tank. How many kilojoules of heat energy are released by the combustion of 20.0 lb of propane?

$$C_3H_8(g) + 5O_2(g) \longrightarrow 3CO_2(g) + 4H_2O(g)$$
$$\text{heat of reaction} = -2045 \text{ kJ}$$

11.66 Milk of magnesia is a suspension of magnesium hydroxide. It is sometimes taken to reduce heartburn caused by excess acid in the stomach. How many milliliters of milk of magnesia [1200 mg $Mg(OH)_2$ per 15 mL] are required to neutralize 50.0 mL of 0.113 M HCl?

11.67 What mass of CO, in grams, forms from 2.00 moles of carbon and 13.2 L of SO_2 at 325 K and 1.00 atm?

$$5C(s) + 2SO_2(g) \longrightarrow CS_2(g) + 4CO(g)$$

11.68 How many atoms of cadmium form from the reaction of 25.0 mL of 0.0277 M $Cd(NO_3)_2$ with 10.5 g of aluminum? The equation is not balanced.

$$Cd(NO_3)_2(aq) + Al(s) \longrightarrow Cd(s) + Al(NO_3)_3(aq)$$

11.69 How many molecules of Cl_2 are required to react with 74.9 g of S_8? The equation is not balanced.

$$S_8(l) + Cl_2(g) \longrightarrow S_2Cl_2(l)$$

11.70 What volume of chlorine is formed at 345 K and 1.22 atm from the reaction of 95.0 g of MnO_2 with 5.75×10^{23} molecules of HCl? The reaction is not balanced.

$$MnO_2(s) + HCl(g) \longrightarrow MnCl_2(s) + Cl_2(g) + H_2O(l)$$

11.71 Determine the concentration of lead ions that remain in solution after 50.0 mL of 0.0611 M $Pb(NO_3)_2$ is mixed with 50.0 mL of 0.0556 M KI.

11.72 How much heat (in kJ) is produced when 400 L of acetylene (C_2H_2) reacts at 425 K and 1.20 atm? The equation is not balanced.

$$C_2H_2(g) + O_2(g) \longrightarrow CO_2(g) + H_2O(g)$$
$$\text{heat of reaction} = -2600 \text{ kJ}$$

11.73 Determine the volume of 0.116 M Li_3PO_4 and 0.0871 M $Mg(C_2H_3O_2)_2$ solutions needed to form 135 g of precipitate when mixed.

11.74 How much heat is used when 7.99×10^{24} CO molecules react? Assume excess Fe_3O_4. The equation is not balanced.

$$Fe_3O_4(s) + CO(g) \longrightarrow FeO(s) + CO_2(g)$$
$$\text{heat of reaction} = 35.9 \text{ kJ}$$

11.75 What volume of CO_2 (in L at 298 K and 1.50 atm) is formed from 189 mL of 0.0667 M HNO_3 solution? Assume excess potassium carbonate. The equation is not balanced.

$$K_2CO_3(aq) + HNO_3(aq) \longrightarrow$$
$$H_2O(l) + KNO_3(aq) + CO_2(g)$$

Answers to In-Chapter Materials

Answers to Practice Problems

11.1A 0.264 mol H_2, 0.176 mol NH_3. **11.1B** $P_4O_{10} + 6H_2O \longrightarrow$ $4H_3PO_4$, 1.45 mol P_4O_{10}, 8.70 mol H_2O. **11.2A** 34.1 g. **11.2B** 292 g. **11.3A** 14.1 g NH_3, N_2, 23.4 g left over. **11.3B** 44.2 g KOH, 116 g H_3PO_4. **11.4A** 29.2% yield. **11.4B** 122 g. **11.5A** 74.4%. **11.5B** 36.1 g. **11.6A** (a) 36.1 mL, (b) 466 mL. **11.6B** 0.371 g. **11.7A** 11.3 mL. **11.7B** 91.2 mL. **11.8A** 0.108 g. **11.8B** 5.175 ppm. **11.9A** 8.48 L. **11.9B** 184 g. **11.10A** 151 L. **11.10B** 3.48 g. **11.11A** 8.174×10^4 kJ. **11.11B** 1706 g O_2.

Answers to Checkpoints

11.1.1 a. **11.1.2** c. **11.1.3** d. **11.1.4** d. **11.2.1** e. **11.2.2** c. **11.2.3** a. **11.2.4** d. **11.3.1** c. **11.3.2** b. **11.3.3** e. **11.3.4** c. **11.3.5** a. **11.3.6** c. **11.4.1** a. **11.4.2** d. **11.4.3** b. **11.4.4** c. **11.4.5** a. **11.4.6** c. **11.5.1** d. **11.5.2** b. **11.5.3** b. **11.5.4** d. **11.5.5** a. **11.6.1** b. **11.6.2** a.

CHAPTER 12

Acids and Bases

12.1 Properties of Acids and Bases

12.2 Definitions of Acids and Bases
- Arrhenius Acids and Bases
- Brønsted Acids and Bases
- Conjugate Acid-Base Pairs

12.3 Water as an Acid; Water as a Base

12.4 Strong Acids and Bases

12.5 pH and pOH Scales

12.6 Weak Acids and Bases

12.7 Acid-Base Titrations

12.8 Buffers

Acids are present in many familiar foods. Citrus fruits are sour because of the acids they contain. Bases in food are less common, but impart a bitter flavor when they are present.
Aflo Co., Ltd./Alamy Stock Photo

In This Chapter, You Will Learn

About some of the properties and reactions of acids and bases, the pH scale, which is used to quantify acidity, and the use of buffers to control pH.

Things To Review Before You Begin

- Balancing chemical equations [◄◄ Section 10.3]
- Acid-base reactions [◄◄ Section 10.4]
- Chapter 11 Key Skills [◄◄ pages 418–419]

Acids and bases are present in many of the substances we encounter every day. Examples include some of the key ingredients in vitamin C, vinegar, and gastric juice (acids); and in drain cleaner and window cleaner (bases). In this chapter, we examine the nature of acids and bases, and look more closely at the neutralization reactions that occur when they are combined. We also explain the pH scale, which is how acidity generally is expressed.

David A. Tietz/McGraw Hill

12.1 Properties of Acids and Bases

Acids are common ingredients in many foods and beverages, and they typically impart a sour taste to foods that contain them. If you have ever taken a chewable vitamin C tablet, you will recognize the sour taste of ascorbic acid, $C_6H_8O_6$.

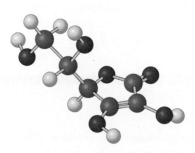

Acids are compounds that produce H^+ ions in aqueous solution [◄◄ Section 3.7]. For a compound to do this, it must have one or more hydrogen atoms that can separate from the molecules when dissolved in water. Such hydrogen atoms are called ***ionizable hydrogen atoms.*** Many acids have just one ionizable H atom per molecule but some, including ascorbic acid, have more than one. (Ascorbic acid has two ionizable H atoms, shown below in red.)

$$
\begin{array}{c}
\text{H—O \quad O—H} \\
\text{C=C \quad\quad H} \\
\text{O=C \quad\quad C—H \quad H \quad H} \\
\text{O \quad\quad C—C—OH} \\
\text{OH \quad H}
\end{array}
$$

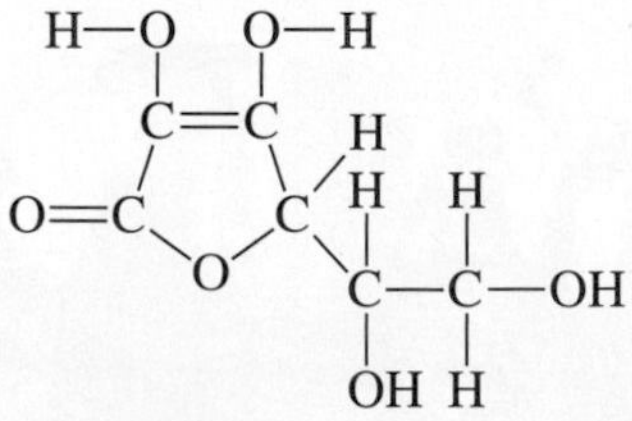

Natural sources of ascorbic acid include citrus fruits, such as oranges, lemons, and limes.

Student Note: When a hydrogen atom detaches from an acid molecule, it does not take its electron with it. Because a hydrogen atom consists only of a proton and an electron, *without* its electron a hydrogen atom is simply a *proton*. We represent protons in solution as H^+ ions; and we refer to H^+ ions that separate from a molecule as *protons*.

427

Profiles in Science

James Lind

James Lind (1716–1794) was a Scottish-born physician and surgeon. He served as ship's surgeon on several British Naval expeditions—most famously in 1747 on *HMS Salisbury* of the Channel Fleet. It was on HMS *Salisbury* that Lind conducted the world's first clinical trial of treatment for a disease, namely scurvy.

Scurvy, a disease now known to be the result of vitamin C deficiency, had been known since antiquity and was a particular problem for sailors on long voyages. The symptoms of scurvy include failure of connective tissues, chronic bleeding of the gums, loss of teeth, anemia, and debilitating pain and weakness. Many more British sailors are believed to have died from scurvy than from any other cause—including battles with enemy military forces. During the ill-fated and ultimately unsuccessful circumnavigation attempt by Admiral of the Fleet George Anson in the middle of the eighteenth century, 1400 of Anson's original crew of 1900 reportedly died of scurvy.

James Lind

H.S. Photos/Alamy Stock Photo

The susceptibility of sailors to scurvy resulted from their diets, which typically consisted of salted pork, dried peas, oatmeal, weevil-infested hard tack biscuits, and rum. (Interestingly, the issuance of a daily ration of rum to British sailors persisted well into the twentieth century, having been discontinued only in 1970.) None of the rations included any food sources of vitamin C, which is found primarily in fresh fruits and vegetables.

Lind designed and implemented an experiment using a dozen sailors on HMS *Salisbury,* all of whom were suffering from scurvy. He divided the men into six groups and gave each group a different supplement to the regular daily ration. The six different supplements he provided were cider, sulfuric acid, vinegar, seawater, barley water and spices, and citrus fruit (two oranges and one lemon). Only the last group, the sailors who received citrus, showed significant improvement after just one week. From this, Lind deduced that something in the citrus was in part responsible for curing scurvy—although the concept of a vitamin was unknown in Lind's time. Despite the success of his experiment, the practice of providing citrus to sailors in the British Navy did not become routine until early in the nineteenth century.

Bases, which are compounds that produce OH⁻ ions in aqueous solution [◀◀ Section 10.4], are less common in foods, partly because those that are edible have a *bitter* flavor. Many bases are not edible. In fact, an unpalatable bitter taste is thought by many to be nature's way of preventing us from consuming poisonous plants. Two bases found in most homes are ammonia (NH_3), in household cleaners; and sodium hydroxide (NaOH), in drain cleaner.

David A. Tietz/McGraw Hill

12.2 Definitions of Acids and Bases

Arrhenius Acids and Bases

The definitions of acid and base that we learned in Chapters 3 and 10:

> acid: substance that produces H^+ ions in aqueous solution

> base: substance that produces OH^- ions in aqueous solution

are the definitions of an ***Arrhenius acid*** and an ***Arrhenius base***—developed by Swedish chemist Svante Arrhenius in the nineteenth century. One compound that acts as an Arrhenius acid is hydrogen chloride, HCl, which is called *hydrochloric acid* when it is dissolved in water. HCl is the acid in gastric juice, making it the *acid* in *acid reflux*. The process by which HCl acts as an Arrhenius acid in water can be represented by the equation

$$HCl(aq) \xrightarrow{H_2O} H^+(aq) + Cl^-(aq)$$

Similarly, the process by which sodium hydroxide, NaOH, acts as an Arrhenius base in water can be represented by the equation

$$NaOH(s) \xrightarrow{H_2O} Na^+(aq) + OH^-(aq)$$

Although the Arrhenius definitions of acid and base are useful, they are not inclusive enough to identify compounds that act as acids or bases other than in aqueous solutions. For example, the reaction of ammonia gas and hydrogen chloride gas to produce ammonium chloride, which can be represented by the equation

$$NH_3(g) + HCl(g) \longrightarrow NH_4Cl(s)$$

is an acid-base reaction that is not described as such by the Arrhenius definitions.

Student Hot Spot

Student data indicate that you may struggle to understand what happens to an acid when it is ionized in water. Access the eBook to view additional Learning Resources on this topic.

Brønsted Acids and Bases

To encompass more acid-base reactions, we need a more general definition of acid-base behavior. This more general definition was developed by Danish chemist Johannes Brønsted, early in the twentieth century. Brønsted acids and bases are defined in terms of substances *donating* and *accepting* protons. (Remember that in this context, a *proton* is a hydrogen atom without its electron.) A ***Brønsted acid*** is a substance that donates a proton, and a ***Brønsted base*** is a substance that accepts a proton. If we look carefully at the reaction of ammonia gas and hydrogen chloride gas, we can see that HCl donates its only proton to the NH_3 molecule—making it into an ammonium ion (NH_4^+):

Student Note: Some texts refer to Brønsted acids and bases as Brønsted–Lowry acids and bases.

Brønsted base accepts a proton   Brønsted acid donates a proton

The resulting oppositely charged ions, NH_4^+ and Cl^-, are pulled together by electrostatic attraction [◄◄ Section 3.2] to form the ionic solid ammonium chloride, NH_4Cl. It is important to recognize that the Brønsted definition of acid-base behavior is more inclusive than the Arrhenius definition. Any reaction that was described as acid-base using the *Arrhenius* definition is *also* described as acid-base using the *Brønsted* definition. As an example, consider again the process by which HCl produces H^+ ions in aqueous solution. To illustrate this point, we will write the equation slightly differently, showing water as a reactant:

$$HCl(aq) + H_2O(l) \longrightarrow H_3O^+(aq) + Cl^-(aq)$$

Brønsted acid donates a proton   Brønsted base accepts a proton

TABLE 12.1	Conjugate Bases of Some Common Species		**TABLE 12.2**	Conjugate Acids of Some Common Species
Species	**Conjugate Base**		**Species**	**Conjugate Acid**
CH_3COOH	CH_3COO^-		NH_3	NH_4^+
H_2O	OH^-		H_2O	H_3O^+
HNO_2	NO_2^-		OH^-	H_2O
H_2SO_4	HSO_4^-		H_2NCONH_2 (urea)	$H_2NCONH_3^+$

Written this way, the products are $Cl^-(aq)$ and $H_3O^+(aq)$, rather than $Cl^-(aq)$ and $H^+(aq)$. Although H_3O^+ and H^+ may *look* different, they actually represent the same species. When ionizable hydrogen atoms separate from acid molecules as protons, they become attached to water molecules. The formula H_3O^+ simply represents the aqueous proton or *hydrogen ion* in a somewhat more realistic way—as attached to a water molecule. When we represent the aqueous proton with the formula H_3O^+, we call it the **hydronium ion.**

Conjugate Acid-Base Pairs

When a Brønsted acid donates a proton, what remains of the acid is known as a **conjugate base.** For example, in the ionization of HCl in water,

$$\underset{\text{acid}}{HCl(aq)} + H_2O(l) \longrightarrow H_3O^+(aq) + \underset{\text{conjugate base}}{Cl^-(aq)}$$

HCl donates a proton to water, producing the hydronium ion (H_3O^+) and the chloride ion (Cl^-), which is the conjugate base of HCl. The two species, HCl and Cl^-, are known as a *conjugate acid-base pair* or simply a **conjugate pair.** Table 12.1 lists the conjugate bases of several familiar species.

Conversely, when a Brønsted base *accepts* a proton, the newly formed *protonated* species is known as a **conjugate acid.** When ammonia (NH_3) ionizes in water,

Student Note: The double arrow in this chemical equation indicates that the reaction does not go to completion. We explain more about the use of this type of arrow in Section 12.6.

$$\underset{\text{base}}{NH_3(aq)} + H_2O(l) \rightleftharpoons \underset{\text{conjugate acid}}{NH_4^+(aq)} + OH^-(aq)$$

NH_3 accepts a proton from water to become the ammonium ion $(NH)_4^+$. The ammonium ion is the conjugate acid of ammonia. Table 12.2 lists the conjugate acids of several common species.

Any reaction that we describe using Brønsted acid-base theory involves an acid and a base. The acid donates the proton, and the base accepts it. Furthermore, the products of such a reaction are always a conjugate base and a conjugate acid. It is useful to identify and label each species in a Brønsted acid-base reaction. For the ionization of HCl in water, the species are labeled as follows:

$$\underset{\text{acid}}{HCl(aq)} + \underset{\text{base}}{H_2O(l)} \longrightarrow \underset{\substack{\text{conjugate} \\ \text{acid}}}{H_3O^+(aq)} + \underset{\substack{\text{conjugate} \\ \text{base}}}{Cl^-(aq)}$$

And for the ionization of NH_3 in water,

$$\underset{\text{base}}{NH_3(aq)} + \underset{\text{acid}}{H_2O(l)} \rightleftharpoons \underset{\substack{\text{conjugate} \\ \text{acid}}}{NH_4^+(aq)} + \underset{\substack{\text{conjugate} \\ \text{base}}}{OH^-(aq)}$$

Sample Problems 12.1 and 12.2 let you practice identifying conjugate pairs and the species in a Brønsted acid-base reaction.

SAMPLE PROBLEM **12.1** Determining Formulas for Conjugate Acids and Bases

What is (a) the conjugate base of HNO_3, (b) the conjugate acid of O^{2-}, (c) the conjugate base of HSO_4^-, and (d) the conjugate acid of HCO_3^-?

Strategy To find the conjugate base of a species, *remove* a proton from the formula. To find the conjugate acid of a species, *add* a proton to the formula.

Setup The word *proton*, in this context, refers to H^+. Thus, the formula and the charge will both be affected by the addition or removal of H^+.

Solution (a) NO_3^- (b) OH^- (c) SO_4^{2-} (d) H_2CO_3

> **THINK ABOUT IT**
>
> A species does not need to be what we think of as an acid for it to have a conjugate base. For example, we would not refer to the hydroxide ion (OH^-) as an acid—but it does have a conjugate base, the oxide ion (O^{2-}). Furthermore, a species that can either lose or gain a proton, such as HCO_3^-, has both a conjugate base (CO_3^{2-}) and a conjugate acid (H_2CO_3).

Practice Problem ATTEMPT What is (a) the conjugate acid of ClO_4^-, (b) the conjugate acid of S^{2-}, (c) the conjugate base of H_2S, and (d) the conjugate base of $H_2C_2O_4$?

Practice Problem BUILD HSO_3^- is the conjugate acid of what species? HSO_3^- is the conjugate base of what species?

Practice Problem CONCEPTUALIZE Which of the models could represent a species that has a conjugate base? Which could represent a species that is the conjugate base of another species?

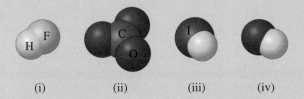

    (i)   (ii)   (iii)   (iv)

SAMPLE PROBLEM **12.2** Identifying Chemical Species as Acids, Bases, Conjugate Bases, and Conjugate Acids

Label each of the species in the following equations as an acid, base, conjugate base, or conjugate acid:

(a) $HF(aq) + NH_3(aq) \rightleftharpoons F^-(aq) + NH_4^+(aq)$

(b) $CH_3COO^-(aq) + H_2O(l) \rightleftharpoons CH_3COOH(aq) + OH^-(aq)$

Strategy In each equation, the reactant that loses a proton is the acid and the reactant that gains a proton is the base. Each product is the conjugate of one of the reactants. Two species that differ only by a proton constitute a conjugate pair.

Setup (a) HF loses a proton and becomes F^-; NH_3 gains a proton and becomes NH_4^+

(b) CH_3COO^- gains a proton to become CH_3COOH; H_2O loses a proton to become OH^-

Solution

(a) $HF(aq) + NH_3(aq) \rightleftharpoons F^-(aq) + NH_4^+(aq)$
    acid   base   conjugate   conjugate
                               base   acid

(b) $CH_3COO^-(aq) + H_2O(l) \rightleftharpoons CH_3COOH(aq) + OH^-(aq)$
     base   acid   conjugate acid   conjugate base

(Continued on next page)

THINK ABOUT IT

In a Brønsted acid-base reaction, there is always an acid and a base, and whether a substance behaves as an acid or a base depends on what it is combined with. Water, for example, behaves as a base when combined with HCl but behaves as an acid when combined with NH_3.

Practice Problem **A**TTEMPT   Identify and label the species in each reaction.

(a) $NH_4^+(aq) + H_2O(l) \rightleftharpoons NH_3(aq) + H_3O^+(aq)$

(b) $CN^-(aq) + H_2O(l) \rightleftharpoons HCN(aq) + OH^-(aq)$

Practice Problem **B**UILD   (a) Write an equation in which HSO_4^- reacts (with water) to form its conjugate base (b) Write an equation in which HSO_4^- reacts (with water) to form its conjugate acid.

Practice Problem **C**ONCEPTUALIZE   Write the formula and charge for each species in this reaction and identify each as an acid, a base, a conjugate acid, or a conjugate base.

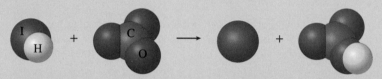

CHECKPOINT–SECTION 12.2   Definitions of Acids and Bases

12.2.1 Which of the following pairs of species are conjugate pairs? (Select all that apply.)

a) H_2S and S^{2-}

b) NH_2^- and NH_3

c) O_2 and H_2O_2

d) HBr and Br^-

e) HCl and OH^-

12.2.2 Which of the following species does *not* have a conjugate base? (Select all that apply.)

a) $HC_2O_4^-$

b) OH^-

c) O^{2-}

d) CO_3^{2-}

e) HClO

12.3 Water as an Acid; Water as a Base

Water is often referred to as the "universal solvent," because it is so common and so important to life on Earth. In addition, most of the acid-base chemistry that we describe takes place in aqueous solution. In this section, we take a closer look at water's ability to act as either a Brønsted acid (as in the ionization of NH_3) or a Brønsted base (as in the ionization of HCl). A species that can behave as either a Brønsted acid or a Brønsted base is called **amphoteric.**

Even in pure water, a very small percentage of the molecules act as Brønsted acids and bases, with a small number of water molecules donating protons to other water molecules. This process is known as the **autoionization of water,** and is responsible for there being a *very* small number of hydronium ions—and an equally small number of hydroxide ions in "pure" water.

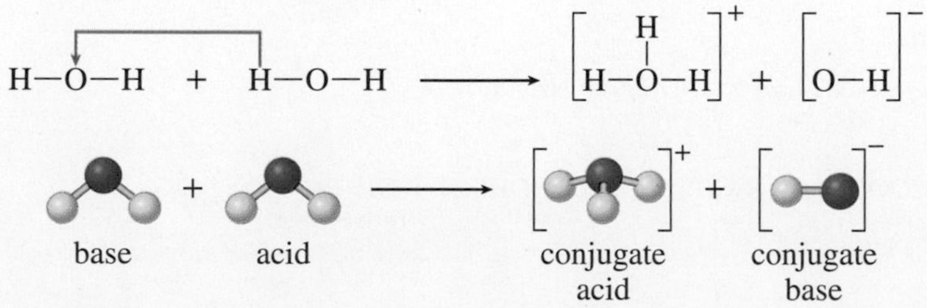

$$H-\overset{|}{\underset{|}{O}}-H \; + \; H-\overset{|}{\underset{|}{O}}-H \longrightarrow \left[H-\overset{\overset{\displaystyle H}{|}}{\underset{|}{O}}-H \right]^+ + \left[O-H \right]^-$$

base   acid   conjugate   conjugate
                                        acid   base

Note that for every H_2O molecule that acts as a Brønsted acid (*donates* a proton), there is another that acts as a Brønsted base (*accepts* a proton)—making the concentrations of hydronium ion and hydroxide ion equal to each other. In pure water at room temperature (25°C), the concentration of hydronium ion and the concentration of hydroxide ion are both 1.0×10^{-7} *M*—*extremely* small. All aqueous solutions contain some hydronium ion and some hydroxide ion. Solutions in which the hydronium ion concentration, $[H_3O^+]$, exceeds the hydroxide ion concentration, $[OH^-]$, are acidic. Those in which the hydroxide ion concentration exceeds the hydronium ion concentration are basic or *alkaline*. Those in which the concentrations of hydronium and hydroxide ions are equal are neutral.

Student Note: This is true only in *pure* water, where autoionization is the *only* source of H_3O^+ and OH^- ions.

- If $[H_3O^+] > [OH^-]$, the solution is acidic.
- If $[OH^-] > [H_3O^+]$, the solution is basic.
- If $[H_3O^+] = [OH^-]$, the solution is neutral.

Moreover, the molar concentrations of H_3O^+ and OH^- ions are related. In any aqueous solution at room temperature (25°C), the *product* of the two concentrations is constant—and is equal to 1.0×10^{-14}. Thus, if we know the concentration of *one* of the ions, we can *calculate* the concentration of the other.

$$\text{at } 25°C \quad [H_3O^+] \times [OH^-] = 1.0 \times 10^{-14} \qquad \textbf{Equation 12.1}$$

Sample Problem 12.3 shows how to use Equation 12.1 to determine hydronium or hydroxide concentration in an aqueous solution at 25°C.

SAMPLE PROBLEM **12.3** Calculating Hydroxide Concentration from Hydronium Concentration

The concentration of hydronium ions in stomach acid is 0.10 *M*. Calculate the concentration of hydroxide ions in stomach acid at 25°C.

Strategy Use the relationship between hydronium and hydroxide ion concentrations given in Equation 12.1 to determine $[OH^-]$ when $[H_3O^+] = 0.10$ *M*.

Setup $K_w = [H_3O^+][OH^-] = 1.0 \times 10^{-14}$ at 25°C. Rearranging Equation 12.1 to solve for $[OH^-]$,

$$[OH^-] = \frac{1.0 \times 10^{-14}}{[H_3O^+]}$$

Solution

$$[OH^-] = \frac{1.0 \times 10^{-14}}{0.10 \ M} = 1.0 \times 10^{-13} \ M$$

THINK ABOUT IT

This is a good time to review how to enter scientific notation correctly with your calculator [◄◄ Section 4.2].

Practice Problem Ⓐ**TTEMPT** The concentration of hydroxide ions in the antacid milk of magnesia is 5.0×10^{-4} *M*. Calculate the concentration of hydronium ions at 25°C.

Practice Problem Ⓑ**UILD** The product of hydronium and hydroxide ion concentrations at normal body temperature (37°C) is 2.8×10^{-14}. Calculate the concentration of hydroxide ions in stomach acid at body temperature. ($[H_3O^+] = 0.10$ *M*.)

(Continued on next page)

Practice Problem **C**ONCEPTUALIZE Shown here are two blank graphs. One represents hydronium ion concentration versus hydroxide ion concentration in an aqueous solution, and the other represents the product of hydroxide and hydronium concentrations versus total volume of the solution. Indicate which of the red lines (i–v) is appropriate for each graph.

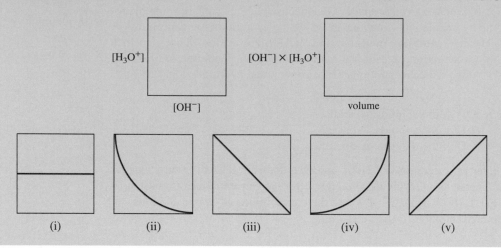

$[H_3O^+]$

$[OH^-]$

$[OH^-] \times [H_3O^+]$

volume

(i) (ii) (iii) (iv) (v)

CHECKPOINT–SECTION 12.3 Water as an Acid; Water as a Base

12.3.1 Calculate $[OH^-]$ in a solution in which $[H_3O^+] = 0.0012\ M$ at 25°C.

a) $1.2 \times 10^{-3}\ M$

d) $8.3 \times 10^{-12}\ M$

b) $8.3 \times 10^{-17}\ M$

e) $1.2 \times 10^{11}\ M$

c) $1.0 \times 10^{-14}\ M$

12.3.2 Calculate $[H_3O^+]$ in a solution in which $[OH^-] = 0.25\ M$ at 25°C.

a) $4.0 \times 10^{-14}\ M$

d) $1.0 \times 10^{-7}\ M$

b) $1.0 \times 10^{-14}\ M$

e) $4.0 \times 10^{-7}\ M$

c) $2.5 \times 10^{13}\ M$

12.3.3 Calculate $[OH^-]$ in a solution in which $[H_3O^+] = 3.4 \times 10^{-5}\ M$ at 25°C.

a) $2.9 \times 10^{-10}\ M$

d) $3.4 \times 10^{-19}\ M$

b) $3.4 \times 10^{9}\ M$

e) $1.0 \times 10^{-7}\ M$

c) $2.9 \times 10^{-9}\ M$

12.3.4 The hydronium ion and hydroxide ion concentrations are both $2.3 \times 10^{-7}\ M$ in pure water at 50°C. What is the value of the product of concentrations at this temperature?

a) 1.0×10^{-14}

d) 2.3×10^{-14}

b) 4.3×10^{-8}

e) 5.3×10^{-14}

c) 2.3×10^{7}

12.4 Strong Acids and Bases

A *strong acid* is one that ionizes completely in aqueous solution. HCl is a familiar example of a strong acid. Ionizing completely in aqueous solution means that when we dissolve HCl in water, all of the HCl molecules come apart—and the solution consists of aqueous hydronium ions and aqueous chloride ions. No HCl molecules remain intact in the solution. Thus, if we prepare a 1-*M* solution of HCl, the solution is actually 1 *M* in hydronium ion and 1 *M* in chloride ion—it contains essentially no HCl molecules, as depicted in Figure 12.1.

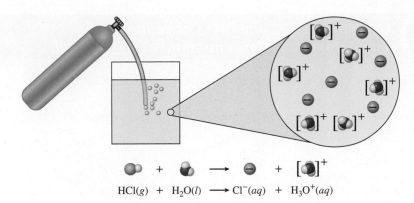

Figure 12.1 When HCl dissolves in water, all of the molecules ionize to produce aqueous hydronium ions and aqueous chloride ions. No HCl molecules remain in solution.

$$HCl(g) + H_2O(l) \longrightarrow Cl^-(aq) + H_3O^+(aq)$$

Among all of the many acids we encounter in everyday life, only a relatively small number are strong acids. The short list includes

- hydrochloric acid, HCl
- hydrobromic acid, HBr
- hydroiodic acid, HI
- nitric acid, HNO_3

- chloric acid, $HClO_3$
- perchloric acid, $HClO_4$
- sulfuric acid, H_2SO_4

Each of the strong acids (with the exception of H_2SO_4) is a ***monoprotic acid,*** meaning that it has just *one* ionizable H atom. Another way to say this is that it has just one *proton*. Sulfuric acid, H_2SO_4, is a ***diprotic acid,*** meaning that it has *two* protons. Although it has two protons, the vast majority of sulfuric acid molecules lose just *one* proton when dissolved in water. The chemical equations for the ionizations of strong acids are shown here.

Strong Acid	Ionization Reaction
Hydrochloric acid	$HCl(aq) + H_2O(l) \longrightarrow H_3O^+(aq) + Cl^-(aq)$
Hydrobromic acid	$HBr(aq) + H_2O(l) \longrightarrow H_3O^+(aq) + Br^-(aq)$
Hydroiodic acid	$HI(aq) + H_2O(l) \longrightarrow H_3O^+(aq) + I^-(aq)$
Nitric acid	$HNO_3(aq) + H_2O(l) \longrightarrow H_3O^+(aq) + NO_3^-(aq)$
Chloric acid	$HClO_3(aq) + H_2O(l) \longrightarrow H_3O^+(aq) + ClO_3^-(aq)$
Perchloric acid	$HClO_4(aq) + H_2O(l) \longrightarrow H_3O^+(aq) + ClO_4^-(aq)$
Sulfuric acid	$H_2SO_4(aq) + H_2O(l) \longrightarrow H_3O^+(aq) + HSO_4^-(aq)$

The list of ***strong bases*** is also fairly short. It consists of the hydroxides of alkali metals (Group 1) and the hydroxides of the heaviest alkaline earth metals (Group 2). The dissociation of a strong base is, for practical purposes, complete. Equations representing dissociations of the strong bases are as follows:

Group 1 hydroxides

$$LiOH(aq) \longrightarrow Li^+(aq) + OH^-(aq)$$
$$NaOH(aq) \longrightarrow Na^+(aq) + OH^-(aq)$$
$$KOH(aq) \longrightarrow K^+(aq) + OH^-(aq)$$
$$RbOH(aq) \longrightarrow Rb^+(aq) + OH^-(aq)$$
$$CsOH(aq) \longrightarrow Cs^+(aq) + OH^-(aq)$$

Group 2 hydroxides

$$Ca(OH)_2(aq) \longrightarrow Ca^{2+}(aq) + 2OH^-(aq)$$
$$Sr(OH)_2(aq) \longrightarrow Sr^{2+}(aq) + 2OH^-(aq)$$
$$Ba(OH)_2(aq) \longrightarrow Ba^{2+}(aq) + 2OH^-(aq)$$

Student Note: $Ca(OH)_2$ and $Sr(OH)_2$ are not very soluble, but what *does* dissolve dissociates completely [◄◄ Section 9.2, Table 9.3].

SAMPLE PROBLEM (12.4) Calculating H_3O^+ and OH^- Concentrations from Strong-Acid Concentration

Calculate the $[H_3O^+]$ and $[OH^-]$ of an aqueous solution that is (a) 0.0311 M HNO_3, (b) 4.51×10^{-5} M $HClO_4$, and (c) 8.74×10^{-6} M HI.

Strategy HNO_3, $HClO_4$, and HI are all strong acids, so the concentration of hydronium ions in each solution is the same as the stated concentration of the acid. Use Equation 12.1 to determine $[OH^-]$.

Setup

(a) $[H_3O^+] = 0.0311$ M

(b) $[H_3O^+] = 4.51 \times 10^{-5}$ M

(c) $[H_3O^+] = 8.74 \times 10^{-6}$ M

Solution

(a) $[OH^-] = \dfrac{1.0 \times 10^{-14}}{[H_3O^+]} = \dfrac{1.0 \times 10^{-14}}{0.0311\ M} = 3.2 \times 10^{-13}$ M

(b) $[OH^-] = \dfrac{1.0 \times 10^{-14}}{4.51 \times 10^{-5}\ M} = 2.2 \times 10^{-10}$ M

(c) $[OH^-] = \dfrac{1.0 \times 10^{-14}}{8.74 \times 10^{-6}\ M} = 1.1 \times 10^{-9}$ M

THINK ABOUT IT

You can check your answers by multiplying the $[OH^-]$ by the given $[H_3O^+]$. The product at 25°C should be 1.0×10^{-14}. You should also note that the higher the $[H_3O^+]$, the lower the $[OH^-]$ should be.

Practice Problem ATTEMPT Calculate the $[H_3O^+]$ and $[OH^-]$ of an aqueous solution that is (a) 0.118 M HNO_3, (b) 7.84×10^{-3} M $HClO_4$, and (c) 9.33×10^{-2} M HI.

Practice Problem BUILD Calculate the concentration of HBr required to give the following $[OH^-]$: (a) 1.2×10^{-8} M, (b) 3.75×10^{-9} M, and (c) 4.88×10^{-12} M.

Practice Problem CONCEPTUALIZE Which of the following diagrams could represent a diprotic acid where the first ionization is complete, but the second is not?

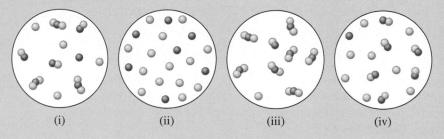

| (i) | (ii) | (iii) | (iv) |

SAMPLE PROBLEM (12.5) Calculating H_3O^+ and OH^- from Strong-Base Concentration

Calculate the $[OH^-]$ and $[H_3O^+]$ of an aqueous solution that is (a) 0.0311 M LiOH, (b) 4.15×10^{-5} M $Ca(OH)_2$, and (c) 8.74×10^{-6} M KOH.

Strategy LiOH, KOH, and $Ca(OH)_2$ are all strong bases, so the concentration of hydroxide ions in each solution can be easily determined from the stated concentration of the base. Note that in bases like $Ca(OH)_2$, two moles of OH^- form when one formula unit of $Ca(OH)_2$ dissolves in solution. Use Equation 12.1 to determine $[H_3O^+]$ from $[OH^-]$.

$$Ca(OH)_2(s) \longrightarrow Ca^{2+}(aq) + 2\ OH^-(aq)$$

Setup

(a) $[OH^-] = 0.0311\ M$

(b) $[OH^-] = 2 \times (4.15 \times 10^{-5}\ M) = 8.30 \times 10^{-5}\ M$

(c) $[OH^-] = 8.74 \times 10^{-6}\ M$

Solution

(a) $[H_3O^+] = \dfrac{1.0 \times 10^{-14}}{[OH^-]} = \dfrac{1.0 \times 10^{-14}}{0.0311\ M} = 3.2 \times 10^{-13}\ M$

(b) $[H_3O^+] = \dfrac{1.0 \times 10^{-14}}{8.30 \times 10^{-5}\ M} = 1.2 \times 10^{-10}\ M$

(c) $[H_3O^+] = \dfrac{1.0 \times 10^{-14}}{8.74 \times 10^{-6}\ M} = 1.1 \times 10^{-9}\ M$

THINK ABOUT IT

Again, you can check your answers by multiplying the $[H_3O^+]$ by the given $[OH^-]$. The product at 25°C should be 1.0×10^{-14}. And remember the reciprocal relationship between $[OH^-]$ and $[H_3O^+]$. As one goes up, the other must go down.

Practice Problem Ⓐ**TTEMPT** Calculate the $[OH^-]$ and $[H_3O^+]$ of an aqueous solution that is (a) 0.118 M NaOH, (b) $7.84 \times 10^{-3}\ M$ LiOH, and (c) $9.33 \times 10^{-2}\ M$ $Sr(OH)_2$.

Practice Problem Ⓑ**UILD** Calculate the concentration of NaOH required to give the following $[H_3O^+]$: (a) $1.21 \times 10^{-8}\ M$, (b) $3.75 \times 10^{-9}\ M$, and (c) $4.88 \times 10^{-12}\ M$.

Practice Problem Ⓒ**ONCEPTUALIZE** Calculate the concentration of $Ba(OH)_2$ required to give $[H_3O^+] = 9.6 \times 10^{-9}\ M$.

CHECKPOINT–SECTION 12.4   Strong Acids and Bases

12.4.1 Determine $[H_3O^+]$ and $[OH^-]$ in 0.015-M aqueous HCl at 25°C.

a) $[H_3O^+] = 0.015\ M$, $[OH^-] = 0.015\ M$

b) $[H_3O^+] = 0.015\ M$, $[OH^-] = 6.7 \times 10^{-13}\ M$

c) $[H_3O^+] = 6.7 \times 10^{-13}\ M$, $[OH^-] = 0.015\ M$

d) $[H_3O^+] = 0.030\ M$, $[OH^-] = 3.3 \times 10^{-13}\ M$

e) $[H_3O^+] = 3.3 \times 10^{-13}\ M$, $[OH^-] = 0.030\ M$

12.4.2 Determine $[OH^-]$ and $[H_3O^+]$ in 0.45-M aqueous $Ca(OH)_2$ at 25°C.

a) $[OH^-] = 0.45\ M$, $[H_3O^+] = 2.2 \times 10^{-14}\ M$

b) $[OH^-] = 0.90\ M$, $[H_3O^+] = 2.2 \times 10^{-14}\ M$

c) $[OH^-] = 0.90\ M$, $[H_3O^+] = 1.1 \times 10^{-14}\ M$

d) $[OH^-] = 1.1 \times 10^{-14}$, $[H_3O^+] = 1.1 \times 10^{-14}\ M$

e) $[OH^-] = 0.45\ M$, $[H_3O^+] = 0.90\ M$

12.5 pH and pOH Scales

The acidity of an aqueous solution depends on the concentration of hydronium ions, $[H_3O^+]$. This concentration can range over many orders of magnitude, which can make reporting the numbers cumbersome. To describe the acidity of a solution, rather than report the molar concentration of hydronium ions, we typically use the more convenient pH scale. The **pH** of a solution is defined as the negative base-10 logarithm of the hydronium ion concentration (in mol/L).

$$pH = -\log[H_3O^+] \quad \text{or} \quad pH = -\log[H^+] \quad \textbf{Equation 12.2}$$

Student Note: A word about significant figures: When we take the log of a number with two significant figures, we report the result to two places past the decimal point. Thus, pH 7.00 has two significant figures, not three.

Equation 12.2 converts numbers that can span an enormous range ($\sim 10^1$ to 10^{-14}) to numbers generally ranging from ~ 1 to 14. The pH of a solution is a dimensionless quantity, so the units of concentration must be removed from $[H_3O^+]$ before taking the logarithm. It is important that you be able to use the logarithm function on your calculator properly. Figure 12.2 shows how to calculate logarithms on several popular brands of calculator.

Because $[H_3O^+] = [OH^-] = 1.0 \times 10^{-7}$ *M* in pure water at 25°C, the pH of pure water at 25°C is

$$-\log{(1.0 \times 10^{-7})} = 7.00$$

The log base 10 function on the Sharp model EL-531X is the key labeled **log.** To calculate the pH of a solution with a hydronium ion concentration of 4.5×10^{-6} *M,* you must push the following sequence of keys:

Remember that to enter scientific notation, you must use the Exp function.

When you know the pH and wish to calculate the hydronium ion concentration ($[H_3O^+] = 10^{-pH}$), you must use the *antilog* (10^x) function. The antilog function on the Sharp model EL-531X is the second function of the **log** key. To calculate the $[H_3O^+]$ of a solution with pH 8.31, you must push the following sequence of keys:

The log base 10 function on the TI-30XIIS is the key labeled **LOG**. To calculate the pH of a solution with a hydronium ion concentration of 4.5×10^{-6} *M,* you must push the following sequence of keys:

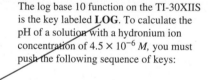

Remember that to enter scientific notation, you must use the EE function—which is the second function of the x^{-1} button. The calculator will automatically insert the left parenthesis—you only need to key in the right one.

When you know the pH and wish to calculate the hydronium ion concentration ($[H_3O^+] = 10^{-pH}$), you must use the *antilog* (10^x) function. The antilog function on the TI-30XIIS is the second function of the **LOG** key. To calculate the $[H_3O^+]$ of a solution with pH 8.31, you must push the following sequence of keys:

Figure 12.2 Visualizing Chemistry – Using Logarithmic Functions on Your Scientific Calculator

(all) David A. Tietz/McGraw Hill

The log base 10 function on the TI-30Xa is the key labeled **LOG.** To calculate the pH of a solution with a hydronium ion concentration of 4.5×10^{-6} *M,* you must push the following sequence of keys:

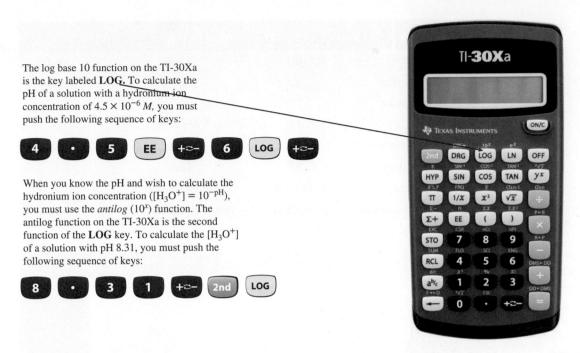

When you know the pH and wish to calculate the hydronium ion concentration ($[H_3O^+] = 10^{-pH}$), you must use the *antilog* (10^x) function. The antilog function on the TI-30Xa is the second function of the **LOG** key. To calculate the $[H_3O^+]$ of a solution with pH 8.31, you must push the following sequence of keys:

Figure 12.2, continued. Using Logarithmic Functions on Your Scientific Calculator

(all) David A. Tietz/McGraw Hill

Remember, too, that a solution in which $[H_3O^+] = [OH^-]$ is neutral. At 25°C, therefore, a neutral solution has pH 7.00. An acidic solution, one in which $[H_3O^+] > [OH^-]$, has pH < 7.00, whereas a basic solution, in which $[H_3O^+] < [OH^-]$, has pH > 7.00. Table 12.3 shows the calculation of pH for solutions ranging from 0.10 *M* to 1.0×10^{-14} *M.*

In the laboratory, pH is measured with a pH meter (Figure 12.3). Table 12.4 lists the pH values of a number of common fluids. Note that the pH of body fluids varies greatly, depending on the location and function of the fluid. The low pH (high acidity) of gastric juices is vital for digestion of food, whereas the higher pH of blood is required to facilitate the transport of oxygen.

TABLE 12.3	Benchmark pH Values for a Range of Hydronium Ion Concentrations at 25°C		
$[H_3O^+](M)$	$-\log[H_3O^+]$	pH	
0.10	$-\log (1.0 \times 10^{-1})$	1.00	
0.010	$-\log (1.0 \times 10^{-2})$	2.00	
1.0×10^{-3}	$-\log (1.0 \times 10^{-3})$	3.00	
1.0×10^{-4}	$-\log (1.0 \times 10^{-4})$	4.00	
1.0×10^{-5}	$-\log (1.0 \times 10^{-5})$	5.00	
1.0×10^{-6}	$-\log (1.0 \times 10^{-6})$	6.00	Acidic
1.0×10^{-7}	$-\log (1.0 \times 10^{-7})$	7.00	Neutral
1.0×10^{-8}	$-\log (1.0 \times 10^{-8})$	8.00	Basic
1.0×10^{-9}	$-\log (1.0 \times 10^{-9})$	9.00	
1.0×10^{-10}	$-\log (1.0 \times 10^{-10})$	10.00	
1.0×10^{-11}	$-\log (1.0 \times 10^{-11})$	11.00	
1.0×10^{-12}	$-\log (1.0 \times 10^{-12})$	12.00	
1.0×10^{-13}	$-\log (1.0 \times 10^{-13})$	13.00	
1.0×10^{-14}	$-\log (1.0 \times 10^{-14})$	14.00	

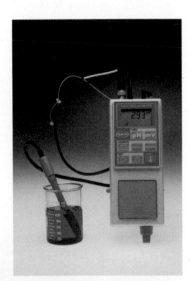

Figure 12.3 A pH meter is commonly used in the laboratory to determine the pH of a solution. Although many pH meters have a range of 1 to 14, pH values can actually be less than 1 and greater than 14.

Charles D. Winters/Timeframe Photography/McGraw Hill

TABLE 12.4	pH Values of Some Common Fluids		
Fluid	pH	Fluid	pH
Stomach acid	1.5	Saliva	6.4–6.9
Lemon juice	2.0	Milk	6.5
Vinegar	3.0	Pure water	7.0
Grapefruit juice	3.2	Blood	7.35–7.45
Orange juice	3.5	Tears	7.4
Urine	4.8–7.5	Milk of magnesia	10.6
Rainwater (in clean air)	5.5	Household ammonia	11.5

Figure 12.4 A change in one pH unit corresponds to a 10-fold change in hydronium ion concentration. A solution with pH equal to 1 has a hydronium ion concentration 1000 times that of a solution with pH equal to 4. Collections of blue dots to the right of the graph illustrate the relative amounts of hydronium ion in solutions at pH values 4, 3, 2, and 1.

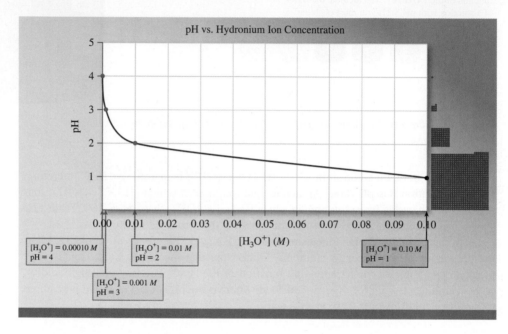

Because pH is a logarithmic scale, a change in one pH unit corresponds to a 10-fold change in hydronium ion concentration. Figure 12.4 illustrates this.

A measured pH can be used to determine experimentally the concentration of hydronium ion in solution. Solving Equation 12.2 for $[H_3O^+]$ gives

Student Note: 10^x is the inverse function of log. (It is usually the second function on the same key.) You must be comfortable performing these operations on your calculator.

Equation 12.3 $[H_3O^+] = 10^{-pH}$

Sample Problems 12.6 and 12.7 illustrate calculations involving pH.

SAMPLE PROBLEM (12.6) Calculating pH from H_3O^+ Concentration

Determine the pH of a solution at 25°C in which the hydronium ion concentration is (a) 3.5×10^{-4} M, (b) 1.7×10^{-7} M, and (c) 8.8×10^{-11} M.

Strategy Given $[H_3O^+]$, use Equation 12.2 to solve for pH.

Setup

(a) pH = $-\log (3.5 \times 10^{-4})$

(b) pH = $-\log (1.7 \times 10^{-7})$

(c) pH = $-\log (8.8 \times 10^{-11})$

Solution

(a) pH = 3.46

(b) pH = 6.77

(c) pH = 10.06

THINK ABOUT IT

When a hydronium ion concentration falls between two "benchmark" concentrations in Table 12.3, the pH falls between the two corresponding pH values. In part (c), for example, the hydronium ion concentration (8.8×10^{-11} M) is greater than 1.0×10^{-11} M but less than 1.0×10^{-10} M. Therefore, we expect the pH to be between 11.00 and 10.00.

$[H_3O^+](M)$	$-\log [H_3O^+]$	pH
1.0×10^{-10}	$-\log (1.0 \times 10^{-10})$	10.00
$8.8 \times 10^{-11*}$	$-\log (8.8 \times 10^{-11})$	10.06[†]
1.0×10^{-11}	$-\log (1.0 \times 10^{-11})$	11.00

*$[H_3O^+]$ between two benchmark values
[†]pH between two benchmark values

Recognizing the benchmark concentrations and corresponding pH values is a good way to determine whether or not your calculated result is reasonable.

Practice Problem **A**TTEMPT Determine the pH of a solution at 25°C in which the hydronium ion concentration is (a) 3.2×10^{-9} M, (b) 4.0×10^{-8} M, and (c) 5.6×10^{-2} M.

Practice Problem **B**UILD Determine the pH of a solution at 25°C in which the hydroxide ion concentration is (a) 8.3×10^{-8} M, (b) 3.3×10^{-4} M, and (c) 1.2×10^{-3} M.

Practice Problem **C**ONCEPTUALIZE Strong acid is added in 1-mL increments to a liter of water at 25°C. Which of the following graphs best approximates the result of plotting hydronium ion concentration as a function of mL acid added? Which graph best approximates the result of plotting pH as a function of mL acid added?

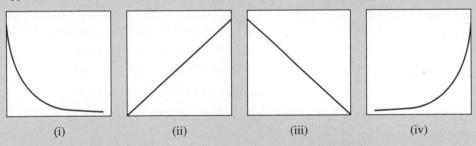

(i)   (ii)   (iii)   (iv)

SAMPLE PROBLEM 12.7 Calculating H_3O^+ Concentration from pH

Calculate the hydronium ion concentration in a solution at 25°C in which the pH is (a) 4.76, (b) 11.95, and (c) 8.01.

Strategy Given pH, use Equation 12.3 to calculate $[H_3O^+]$.

Setup

(a) $[H_3O^+] = 10^{-4.76}$

(b) $[H_3O^+] = 10^{-11.95}$

(c) $[H_3O^+] = 10^{-8.01}$

Solution

(a) $[H_3O^+] = 1.7 \times 10^{-5}$ M

(b) $[H_3O^+] = 1.1 \times 10^{-12}$ M

(c) $[H_3O^+] = 9.8 \times 10^{-9}$ M

(Continued on next page)

THINK ABOUT IT

If you use the calculated hydronium ion concentrations to recalculate pH, you will get numbers slightly different from those given in the problem. In part (a), for example, $-\log(1.7 \times 10^{-5}) = 4.77$. The small difference between this and 4.76 (the pH given in the problem) is due to a rounding error. Remember that a concentration derived from a pH with two digits to the right of the decimal point can have only two significant figures. Note also that the benchmarks can be used equally well in this circumstance. A pH between 4 and 5 corresponds to a hydronium ion concentration between 1×10^{-4} M and 1×10^{-5} M.

Practice Problem **A**TTEMPT Calculate the hydronium ion concentration in a solution at 25°C in which the pH is (a) 9.90, (b) 1.45, and (c) 7.01.

Practice Problem **B**UILD Calculate the hydroxide ion concentration in a solution at 25°C in which the pH is (a) 11.89, (b) 2.41, and (c) 7.13.

Practice Problem **C**ONCEPTUALIZE What is the value of the exponent in the hydronium ion concentration for solutions with pH values of 5.90, 10.11, and 1.25?

A *pOH* scale analogous to the pH scale can be defined using the negative base-10 logarithm of the *hydroxide* ion concentration of a solution, [OH⁻].

Equation 12.4
$$pOH = -\log[OH^-]$$

Rearranging Equation 12.4 to solve for hydroxide ion concentration gives

Equation 12.5
$$[OH^-] = 10^{-pOH}$$

Remember that the product of hydronium and hydroxide ions at 25°C is 1×10^{-14}:

$$[H_3O^+] \times [OH^-] = 1.0 \times 10^{-14}$$

Taking the negative logarithm of both sides, we obtain

$$-\log([H_3O^+][OH^-]) = -\log(1.0 \times 10^{-14})$$
$$-(\log[H_3O^+] + \log[OH^-]) = 14.00$$
$$-\log[H_3O^+] - \log[OH^-] = 14.00$$
$$(-\log[H_3O^+]) + (-\log[OH^-]) = 14.00$$

And from the definitions of pH and pOH, we see that at 25°C

Equation 12.6
$$pH + pOH = 14.00$$

Equation 12.6 provides another way to express the relationship between the hydronium ion concentration and the hydroxide ion concentration. On the pOH scale, 7.00 is neutral, numbers greater than 7.00 indicate that a solution is acidic, and numbers less than 7.00 indicate that a solution is basic. Table 12.5 lists pOH values for a range of hydroxide ion concentrations at 25°C.

TABLE 12.5	Benchmark pOH Values for a Range of Hydroxide Ion Concentrations at 25°C	
[OH⁻](M)	**pOH**	
0.10	1.00	↑
1.0×10^{-3}	3.00	
1.0×10^{-5}	5.00	Basic
1.0×10^{-7}	7.00	Neutral
1.0×10^{-9}	9.00	Acidic
1.0×10^{-11}	11.00	
1.0×10^{-13}	13.00	↓

Sample Problems 12.8 and 12.9 illustrate calculations involving pOH.

SAMPLE PROBLEM (12.8) Calculating pOH from OH⁻ Concentration

Determine the pOH of a solution at 25°C in which the hydroxide ion concentration is (a) 3.7×10^{-5} M, (b) 4.1×10^{-7} M, and (c) 8.3×10^{-2} M.

Strategy Given [OH⁻], use Equation 12.4 to calculate pOH.

Setup

(a) $pOH = -\log (3.7 \times 10^{-5})$

(b) $pOH = -\log (4.1 \times 10^{-7})$

(c) $pOH = -\log (8.3 \times 10^{-2})$

Solution

(a) $pOH = 4.43$

(b) $pOH = 6.39$

(c) $pOH = 1.08$

THINK ABOUT IT

Remember that the pOH scale is, in essence, the *reverse* of the pH scale. On the pOH scale, numbers below 7 indicate a basic solution, whereas numbers above 7 indicate an acidic solution. The pOH benchmarks (abbreviated in Table 12.5) work the same way the pH benchmarks do. In part (a), for example, a hydroxide ion concentration between 1×10^{-4} M and 1×10^{-5} M corresponds to a pOH between 4 and 5:

[OH⁻](M)	pOH
1.0×10^{-4}	4.00
$3.7 \times 10^{-5*}$	4.43†
1.0×10^{-5}	5.00

*[OH⁻] between two benchmark values

†pOH between two benchmark values

Practice Problem ATTEMPT Determine the pOH of a solution at 25°C in which the hydroxide ion concentration is (a) 5.7×10^{-12} M, (b) 7.3×10^{-3} M, and (c) 8.5×10^{-6} M.

Practice Problem BUILD Determine the pH of a solution at 25°C in which the hydroxide ion concentration is (a) 2.8×10^{-8} M, (b) 9.9×10^{-9} M, and (c) 1.0×10^{-11} M.

Practice Problem CONCEPTUALIZE Without doing any calculations, determine between which two whole numbers the pOH will be for solutions with OH⁻ concentrations of 4.71×10^{-5} M, 2.9×10^{-12} M, and 7.15×10^{-3} M.

SAMPLE PROBLEM (12.9) Calculating OH⁻ Concentration from pOH

Calculate the hydroxide ion concentration in a solution at 25°C in which the pOH is (a) 4.91, (b) 9.03, and (c) 10.55.

Strategy Given pOH, use Equation 12.5 to calculate [OH⁻].

Setup

(a) $[OH^-] = 10^{-4.91}$

(b) $[OH^-] = 10^{-9.03}$

(c) $[OH^-] = 10^{-10.55}$

(Continued on next page)

Solution

(a) $[\text{OH}^-] = 1.2 \times 10^{-5}\ M$

(b) $[\text{OH}^-] = 9.3 \times 10^{-10}\ M$

(c) $[\text{OH}^-] = 2.8 \times 10^{-11}\ M$

THINK ABOUT IT

Use the benchmark pOH values to determine whether these solutions are reasonable. In part (a), for example, the pOH between 4 and 5 corresponds to $[\text{OH}^-]$ between $1 \times 10^{-4}\ M$ and $1 \times 10^{-5}\ M$.

Practice Problem **TTEMPT** Calculate the hydroxide ion concentration in a solution at 25°C in which the pOH is (a) 13.02, (b) 5.14, and (c) 6.98.

Practice Problem **B**UILD Calculate the hydronium ion concentration in a solution at 25°C in which the pOH is (a) 2.74, (b) 10.31, and (c) 12.40.

Practice Problem **C**ONCEPTUALIZE What is the value of the exponent in the hydronium ion concentration for solutions with pOH values of 2.90, 8.75, and 11.86?

Familiar Chemistry

Commonly Encountered Acids and Bases

The list of familiar acids is pretty long. Citrus fruits, for example, contain a variety of acids, including ascorbic acid, which is vitamin C. Malic acid is found in stone fruits and some berries. Tartaric acid is found in grapes and supplemental tartaric acid is an important component of winemaking. The use of citric acid, found naturally in citrus fruits and produced commercially by the fermentation of molasses, is critical to safe home canning. Carbonated beverages contain carbonic acid—a result of the carbonation process. Phosphoric acid is added to give these beverages a tangy flavor—and as a preservative. It acts to inhibit mold and bacterial growth. If you have ever maintained a swimming pool, you are familiar with muriatic acid—the pool industry name for hydrochloric acid. And if you have ever smelled vinegar, you have encountered acetic acid. It is not quite so easy to identify a long list of familiar bases. Ammonia is common in window cleaners. Sodium hydroxide is used in drain cleaner. Beyond that, it may be difficult to point to very many familiar bases. But there is one base that most of us have in our pantries: sodium bicarbonate. Sodium bicarbonate, $NaHCO_3$, is commonly known as baking soda. It is also a common ingredient in over-the-counter antacids, including Alka Seltzer. And it is famously used in a popular science demonstration: the vinegar–baking soda volcano. In each of these applications, sodium bicarbonate reacts with an acid to produce water, carbon dioxide, and a salt—the identity of which depends on the acid.

In baking, sodium bicarbonate reacts with an acidic ingredient in the recipe. Common acidic ingredients in baked goods include cream of tartar (a derivative of tartaric acid), lemon juice, and buttermilk. In the case of cream of tartar (potassium bitartrate), the reaction can be represented as

$$NaHCO_3(s) + KC_4H_5O_6(s) \longrightarrow H_2O(l) + CO_2(g) + NaKC_4H_5O_6(s)$$

Because both reactants are solids, they must be combined in an aqueous solution, a batter, or dough, in order to react. The purpose of using this reaction in baked goods is the gaseous product. The production of many tiny CO_2 bubbles is what causes baked goods to "rise."

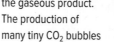

A Science fair volcano

Jamie Grill/Getty Images

The sodium bicarbonate in an antacid reacts with excess acid in gastric juice, which contains hydrochloric acid.

$$NaHCO_3(s) + HCl(aq) \longrightarrow H_2O(l) + CO_2(g) + NaCl(aq)$$

The production of carbon dioxide in this reaction can induce burping, which also aids in relieving an upset stomach.

In a vinegar–baking soda volcano, the reaction responsible for the "eruption" is that of sodium bicarbonate and acetic acid.

$$NaHCO_3(s) + CH_3COOH(aq) \longrightarrow H_2O(l) + CO_2(g) + NaCH_3COO(aq)$$

Generally, liquid detergent is added to make the escape of carbon dioxide gas more dramatic. The viscous, foamy mixture of products runs down the outside of the volcano gradually, prolonging the visual effect.

Thinking Outside the Box

Lake Natron

Natron is a naturally occurring mixture of salts consisting primarily of sodium carbonate, sodium bicarbonate, sodium chloride, and sodium sulfate. It is found in high concentrations in Lake Natron in Tanzania, where it causes the pH of the lake to be quite high (between 9 and 10.5). When animals die in the water, they do not decompose in the way we would expect because most microbes involved in decomposition do not thrive in this harshly alkaline environment.

Paul & Paveena McKenzie/Oxford Scientific/Getty Images

Why does the presence of carbonate, bicarbonate, and sulfate ions raise the pH of the lake to an extreme level? Remember that these ions are the conjugate bases of weak acids. Each of them reacts with water, pulling away a proton and leaving behind a hydroxide ion. This is what makes the water basic.

$$CO_3^{2-}(aq) + H_2O(l) \rightleftharpoons HCO_3^-(aq) + OH^-(aq)$$

$$HCO_3^-(aq) + H_2O(l) \rightleftharpoons H_2CO_3(aq) + OH^-(aq)$$

$$SO_4^{2-}(aq) + H_2O(l) \rightleftharpoons HSO_4^-(aq) + OH^-(aq)$$

Further, the high concentration of salts in the water causes significant dessication (drying) via osmosis. Water from once-living cells is drawn out into the more highly concentrated salt solution of the lake water, making the cells less susceptible to decomposition. (This is essentially the mechanism by which dried fruit, pickles, and cured meats are preserved.) The result is preservation of the bodies of dead animals and birds, similar to the preservation of human bodies in the ancient Egyptian mummification process.

CHECKPOINT–SECTION 12.5 pH and pOH Scales

12.5.1 Determine the pH of a solution at 25°C in which $[H_3O^+] = 6.35 \times 10^{-8} M$.

a) 7.651

b) 6.803

c) 7.197

d) 6.350

e) 8.000

12.5.2 Determine $[H_3O^+]$ in a solution at 25°C if pH = 5.75.

a) $1.8 \times 10^{-6} M$

b) $5.6 \times 10^{-9} M$

c) $5.8 \times 10^{-6} M$

d) $2.4 \times 10^{-9} M$

e) $1.0 \times 10^{-6} M$

12.5.3 Determine the pOH of a solution at 25°C in which $[OH^-] = 4.65 \times 10^{-3} M$.

a) 1.668

b) 3.677

c) 0.322

d) 4.647

e) 2.333

12.5.4 Determine $[OH^-]$ in a solution at 25°C if pH = 10.50.

a) $3.2 \times 10^{-11} M$

b) $3.2 \times 10^{-4} M$

c) $1.1 \times 10^{-2} M$

d) $7.1 \times 10^{-8} M$

e) $8.5 \times 10^{-7} M$

12.6 Weak Acids and Bases

In contrast to the seven strong acids listed in Section 12.4, most of the acids we encounter commonly are *weak* acids. A **weak acid** is one that ionizes only *partially* in water, leaving most of the acid molecules intact. We write the equation for the ionization of a weak acid using a double arrow as shown in Figure 12.5.

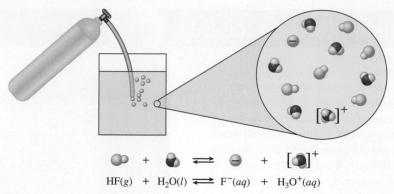

Figure 12.5 Unlike a strong acid, which ionizes completely, a weak acid ionizes only partially. Most weak acid molecules remain intact in solution.

Animation
Acid Ionization

Student Note: The carboxyl group is an example of a functional group—a specific arrangement of atoms that determines many of a molecule's chemical properties. We explain more about functional groups in Chapter 14.

Student Note: The structure of benzoic acid includes a circle within a hexagon. This is a type of shorthand used to represent a common group of atoms—and the bonds between them. The phenyl group consists of six carbon atoms and five hydrogen atoms. When we draw it showing every atom, we see that it has two resonance structures [◄◄ Section 6.3], both of which can be represented with the same shorthand.

The double arrow indicates that although the forward reaction (HF donating a proton to H_2O) occurs, the *reverse* reaction (H_3O^+ giving the proton *back* to F^-) also occurs—leaving most of the HF molecules intact, with only a very small percentage of them ionized in solution. Because only a small percentage of weak acid molecules are ionized in solution, the hydronium ion concentration is lower than it would be in a solution of strong acid of the same concentration. Table 12.6 lists several common weak acids along with their formulas and structures.

Until now, we have seen the chemical formulas of acids written with their ionizable hydrogen atoms first: HCl, HNO_3, HF, and so on. However, you will notice that for three of the five acids in Table 12.6 the formulas are not written this way. Instead, they are written with the same group of atoms at the *end* of the formula, COOH. This group of atoms is known as a carboxyl group, and an acid whose formula includes this group is a ***carboxylic acid.*** You should be able to recognize a carboxylic acid written either way: with its ionizable hydrogen atom(s) first, or with its carboxyl group(s) kept together at the end of the formula. Examples include:

formic acid   benzoic acid   acetic acid
HCOOH or $HCHO_2$ C_6H_5COOH or $HC_6H_5CO_2$ CH_3COOH or $HC_2H_3O_2$

(Ionizable hydrogen atoms are shown in pink.)

TABLE 12.6	Formulas and Structures of Some Weak Acids

Name of Acid	Formula	Structure
Hydrofluoric acid	HF	H—F
Nitrous acid	HNO_2	O=N—O—H
Formic acid	HCOOH	$\underset{H-C-O-H}{\overset{O}{\overset{\parallel}{}}}$
Benzoic acid	C_6H_5COOH	
Acetic acid	CH_3COOH	

Note: The ionizable hydrogen for each acid is shown in pink.

Profiles in Science

St. Elmo Brady

St. Elmo Brady (1916–1966) was the first African American to earn a Ph.D. in chemistry in the United States. He attended Fisk University in Nashville, Tennessee, and then did his graduate work at the University of Illinois Urbana-Champaign. There he did exhaustive studies on the impact that various substituents have on the strength of carboxylic acids. In this context, a "substituent" is an atom or *group* of atoms that replaces one or more of the nonionizable hydrogens in the structure of a carboxylic acid. For example, acetic acid, CH_3COOH, can have one or more of its nonionizable H atoms replaced by chlorine. Replacement of one H by Cl results in *chloroacetic* acid, which is a stronger acid than the original acetic acid. Replacement of two and three H atoms results in dichloroacetic acid and trichloroacetic acid, respectively. With each additional Cl substituent, the acidic strength increases.

St. Elmo Brady
Image courtesy of the University of Illinois Archives

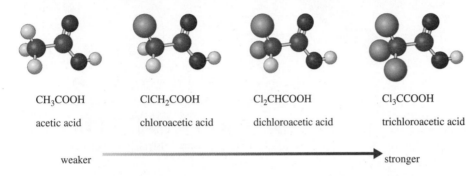

| CH_3COOH | $ClCH_2COOH$ | $Cl_2CHCOOH$ | Cl_3CCOOH |
| acetic acid | chloroacetic acid | dichloroacetic acid | trichloroacetic acid |

weaker ⟶ stronger

Brady went on to teach at the Tuskegee Normal and Industrial Institute, now Tuskegee University. He later became chair of the chemistry department at Howard University in Washington, DC. Howard University, founded in 1867, is the oldest HBCU (historically Black colleges and universities) in the United States. Later, Brady would return to lead the chemistry department at Fisk University, his undergraduate alma mater. There he designed the nation's first HBCU graduate program in chemistry. Eventually, he would go on to develop such programs at several other HBCUs. In 2019, Brady was honored posthumously with a National Historic Chemical Landmark by the American Chemical Society.

Sample Problem 12.10 shows how to estimate pH values when weak acids are involved.

SAMPLE PROBLEM (12.10) Estimating pH Value of a Weak-Acid Solution

Determine the pH value that a 0.10-*M* solution of $HC_2H_3O_2$ must be above.

Strategy Acetic acid, $HC_2H_3O_2$, is a weak acid and does not ionize completely. This means that while the concentration of the acid is 0.10 *M*, the concentration of H_3O^+ will be much smaller than 0.10 *M*. The lowest pH possible would occur if the acid were strong, so we can determine the pH of a 0.10-*M* solution of a strong acid and know that the pH of the acetic acid must be higher.

Setup Knowing that $[H_3O^+] < 0.10$ *M*, use Equation 12.2 to calculate pH.

Solution pH = −log(0.10) = 1.00 if this were a strong acid, so the pH of the weak acid must be above 1.00.

THINK ABOUT IT

Note that although the pH of the acetic acid is higher than it would be if it were a strong acid, the pH is still lower than that of pure water.

(Continued on next page)

Practice Problem **A** **TTEMPT**   Determine the pH value that a 0.078-*M* solution of HClO must be above.

Practice Problem **B** **UILD**   Place the following three solutions in order of increasing pH: 0.050 *M* HNO₂, 0.13 *M* HNO₂, and 0.093 *M* HNO₂.

Practice Problem **C** **ONCEPTUALIZE**   Which of the diagrams shown here could represent a solution of HF?

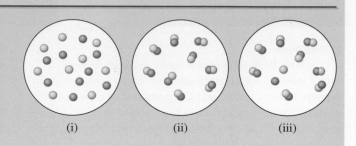

(i)   (ii)   (iii)

Figure 12.6 Weak bases, such as ammonia, ionize only partially in water. Most of the weak base molecules remain intact in solution.

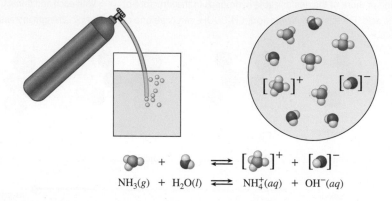

$$NH_3(g) + H_2O(l) \rightleftharpoons NH_4^+(aq) + OH^-(aq)$$

Like weak acids, ***weak bases*** are those that do not ionize completely in water. Weak bases, such as ammonia (NH_3), increase the concentration of hydroxide in water by accepting a proton from water, as shown in Figure 12.6.

Again, the double arrow indicates that the ionization is not complete, and most of the weak base molecules, in this case ammonia molecules, are intact in solution. Only a very small percentage of them ionizes to produce hydroxide ions in solution. Thus, a solution of weak base will have a lower pH than an equal concentration solution of a strong base.

Table 12.7 lists several common bases, along with their formulas and structures.

Note that the structures of the weak bases in Table 12.7 are all somewhat similar to that of ammonia—a nitrogen atom surrounded by three bonds.

Student Note: Strong bases are ionic compounds in which the anion is hydroxide, OH^-. They dissociate in water to produce aqueous hydroxide ions.

Animation
Base Ionization

TABLE 12.7	Formulas and Structures of Some Weak Bases	
Name of Base	**Formula**	**Structure**
Ethylamine	$C_2H_5NH_2$	$CH_3-CH_2-\ddot{N}-H$ with H below
Methylamine	CH_3NH_2	$CH_3-\ddot{N}-H$ with H below
Ammonia	NH_3	$H-\ddot{N}-H$ with H below
Pyridine	C_5H_5N	⬡N:
Aniline	$C_6H_5NH_2$	⬡$-\ddot{N}-H$ with H below
Dimethyl amine	C_2H_7N	$H-C-N-C-H$ with H atoms

Sample Problem 12.11 shows how to compare pH values for strong and weak bases.

SAMPLE PROBLEM 12.11 Estimating pH of a Weak-Base Solution

Determine the pH value that a 0.089-M solution of NH_3 must be below.

Strategy Ammonia, NH_3, is a weak base and does not ionize completely. This means that while the concentration of the base is 0.089 M, the concentration of OH^- will be much smaller than 0.089 M. The highest pH possible would be if the base were strong, so we can determine the pH of a 0.089-M solution of a strong base and know that the pH of the ammonia must be lower.

Setup Knowing that $[OH^-] < 0.089$ M, use Equation 12.1 to determine the concentration of hydronium ion and Equation 12.2 to calculate pH.

Solution $[H_3O^+][OH^-] = 1.0 \times 10^{-14}$

$$[H_3O^+] = \frac{1.0 \times 10^{-14}}{0.089\ M} = 1.12 \times 10^{-13}\ M$$

$pH = -\log(1.12 \times 10^{-13}) = 12.95$ if this were a strong base, so the pH of the weak base must be below 12.95.

THINK ABOUT IT

Note that although the pH of the ammonia is lower than it would be if it were a strong base, the pH is still higher than that of pure water.

Practice Problem ATTEMPT Determine the pH value that a 0.148-M solution of CH_3NH_2 must be below.

Practice Problem BUILD Place the following three solutions in order of increasing pH: 0.150 M $C_6H_5NH_2$, 0.0664 M $C_6H_5NH_2$, and 0.393 M $C_6H_5NH_2$.

Practice Problem CONCEPTUALIZE Place the weak bases shown in the diagram in order of increasing pOH.

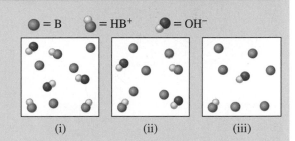

$\bullet$ = B $\bullet$ = HB^+ $\bullet$ = OH^-

(i)   (ii)   (iii)

CHECKPOINT–SECTION 12.6 Weak Acids and Bases

12.6.1 Place the solutions in order of decreasing pH: 0.067 M HCl, 0.067 M CH_3NH_2, 0.067 M HF.

a) HCl > HF > CH_3NH_2

b) HCl > CH_3NH_2 > HF

c) CH_3NH_2 > HCl > HF

d) HF > CH_3NH_2 > HCl

e) CH_3NH_2 > HF > HCl

12.6.2 Place the solutions in order of increasing pH: 0.155 M NaOH, 0.155 M $Sr(OH)_2$, and 0.155 M NH_3.

a) NH_3 < NaOH < $Sr(OH)_2$

b) NaOH < $Sr(OH)_2$ < NH_3

c) $Sr(OH)_2$ < NH_3 < NaOH

d) NaOH < NH_3 < $Sr(OH)_2$

e) NH_3 < NaOH = $Sr(OH)_2$

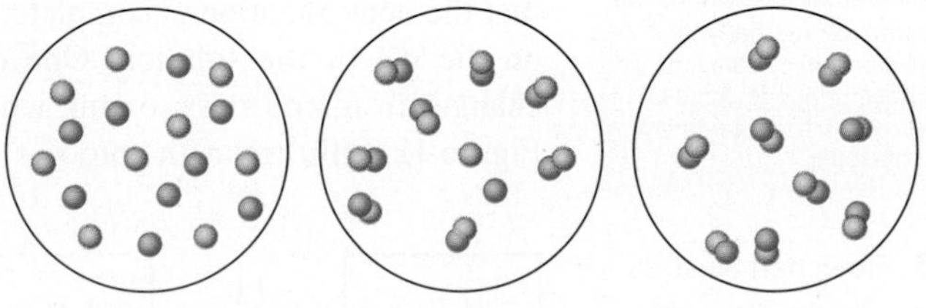

(i)   (ii)   (iii)

12.6.3 Identify the strong acid(s) in the diagram.

a) i   b) ii   c) iii   d) ii and iii   e) i, ii, and iii

12.6.4 Identify the weak acid(s) in the diagram.

a) i   b) ii   c) iii   d) ii and iii   e) i, ii, and iii

Figure 12.7 A base of known concentration is added from a buret to a flask containing acid of unknown concentration. When all of the acid has been neutralized, we determine its original concentration from the volume of base added.

(all) David A. Tietz/McGraw Hill

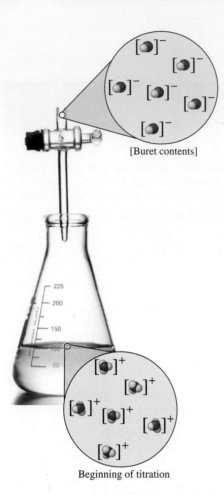

[Buret contents]

Beginning of titration

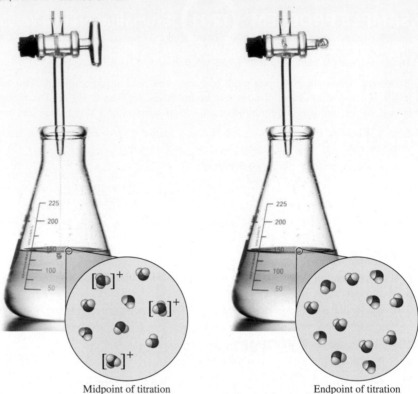

Midpoint of titration

Endpoint of titration

Animation
Acid-Base Titration

12.7 Acid-Base Titrations

We learned in Section 11.4 to relate the volumes and concentrations of solutions to the molar amounts of reactants they contain. Acid-base neutralization (combination of hydronium and hydroxide ions to form water) is used routinely in a type of chemical analysis known as *titration*. In a typical titration, a solution of base with a precisely known concentration is gradually added to a solution of acid with an unknown concentration. When we know the precise volume and concentration of base necessary to neutralize a sample of acid, we can calculate the amount of acid neutralized—and knowing *its* volume, we can calculate its original concentration. A typical titration apparatus is shown in Figure 12.7.

The point in a titration at which all of the acid has been neutralized is called the *equivalence point.* Most acid-base titrations require the use of an *indicator* to show that the neutralization is complete. An acid-base indicator is a dye whose color depends on the pH of the solution. One commonly used indicator is phenolphthalein, which changes from colorless to pink when the pH of a solution changes from acidic to basic. Figure 12.8 illustrates the process of determining acid concentration from titration data.

Figure 12.8 Flowchart illustrating the process of determining acid concentration from titration data.

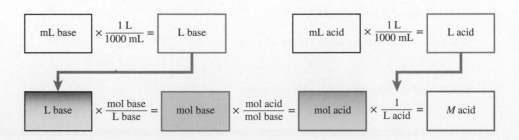

$$\boxed{\text{mL base}} \times \frac{1\ \text{L}}{1000\ \text{mL}} = \boxed{\text{L base}}$$

$$\boxed{\text{mL acid}} \times \frac{1\ \text{L}}{1000\ \text{mL}} = \boxed{\text{L acid}}$$

$$\boxed{\text{L base}} \times \frac{\text{mol base}}{\text{L base}} = \boxed{\text{mol base}} \times \frac{\text{mol acid}}{\text{mol base}} = \boxed{\text{mol acid}} \times \frac{1}{\text{L acid}} = \boxed{M\ \text{acid}}$$

Sample Problem 12.12 shows how acid-base titration is used to determine the concentration of an acid.

SAMPLE PROBLEM **12.12**   Using Titration Data to Calculate Acid Concentration

(a) If 25.0 mL of an HCl solution requires 46.3 mL of a 0.203 *M* NaOH solution to neutralize, what is the concentration of the HCl solution? (b) If 25.0 mL of an H_2SO_4 solution requires 46.3 mL of a 0.203 *M* NaOH solution to neutralize, what is the concentration of the H_2SO_4 solution?

Strategy First, write and balance the chemical equation that corresponds to each neutralization reaction:

$$NaOH(aq) + HCl(aq) \longrightarrow H_2O(l) + NaCl(aq)$$

The base and acid combine in a 1:1 ratio: NaOH ≃ HCl. Use the molarity and the volume of NaOH to determine the number of moles of NaOH consumed. Use the moles of NaOH to determine the moles of HCl neutralized. Using the moles of HCl neutralized and the volume given in the problem, determine the original concentration of HCl.

$$2NaOH(aq) + H_2SO_4(aq) \longrightarrow 2H_2O(l) + Na_2SO_4(aq)$$

H_2SO_4 is a diprotic acid. The base and diprotic acid combine in a 2:1 ratio: 2NaOH ≃ H_2SO_4. Determine moles of NaOH consumed. [The molarity and volume of base in part (b) are the same as those in part (a).] Use the moles of NaOH to determine the moles of H_2SO_4 neutralized. Using the moles of H_2SO_4 neutralized and the volume given in the problem, determine the original concentration of H_2SO_4.

Setup

(a) From the balanced equation, we derive the stoichiometric conversion factor: $\dfrac{1 \text{ mol HCl}}{1 \text{ mol NaOH}}$

(b) For the neutralization of H_2SO_4, the stoichiometric conversion factor is: $\dfrac{1 \text{ mol } H_2SO_4}{2 \text{ mol NaOH}}$

Solution

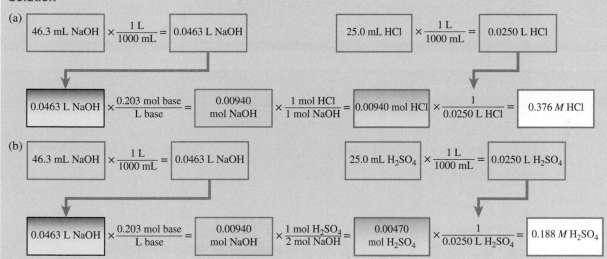

THINK ABOUT IT

Remember to use a balanced chemical equation to derive stoichiometric conversion factors. Many acid-base neutralizations involve acids and bases that combine in a 1:1 ratio, but this is not always the case.

Practice Problem Ⓐ**TTEMPT** How many milliliters of a 1.42 *M* HClO solution are needed to neutralize 95.5 mL of a 0.336 *M* KOH solution?

Practice Problem Ⓑ**UILD** How many milliliters of a 0.211 *M* HCl solution are needed to neutralize 275 mL of a 0.0350 *M* $Ba(OH)_2$ solution?

Practice Problem Ⓒ**ONCEPTUALIZE** Which diagram best represents the ions in solution at the equivalence point in the titration of $Ba(OH)_2$ with HCl? (*Hint:* Start by writing a balanced chemical equation.)

(Continued on next page)

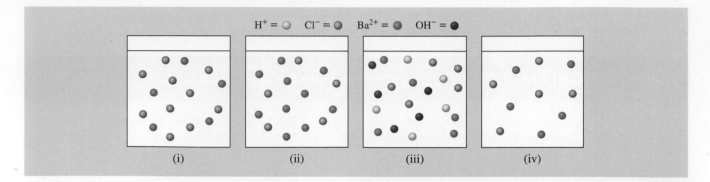

Thinking Outside the Box

Using Millimoles to Simplify Titration Calculations

It is possible to simplify the calculations in acid-base titration problems by using a slightly less familiar unit, the *millimole* (mmol).

We saw in Chapter 9 that molarity (*M*) is defined as $\dfrac{\text{mol solute}}{\text{L solution}}$. But it can also be expressed in the smaller units $\dfrac{\text{mmol solute}}{\text{mL solution}}$. The only

difference is that each of the new units is 1000 times smaller than the original. This eliminates the need to convert from mL, which is generally how volumes are given, to L for both acid and base—thus reducing the number of steps in the calculation. Consider the following solutions to parts (a) and (b) of Sample Problem 12.12.

$$\boxed{46.3 \text{ mL NaOH}} \times \frac{0.203 \text{ mmol base}}{\text{mL base}} = \boxed{\begin{array}{c}9.40\\ \text{mmol NaOH}\end{array}} \times \frac{1 \text{ mmol HCl}}{1 \text{ mmol NaOH}} = \boxed{9.40 \text{ mmol HCl}} \times \frac{1}{25.0 \text{ mL HCl}} = \boxed{0.376 \; M \text{ HCl}}$$

$$\boxed{46.3 \text{ mL NaOH}} \times \frac{0.203 \text{ mmol base}}{\text{mL base}} = \boxed{\begin{array}{c}9.40\\ \text{mmol NaOH}\end{array}} \times \frac{1 \text{ mmol H}_2\text{SO}_4}{2 \text{ mmol NaOH}} = \boxed{\begin{array}{c}4.70\\ \text{mmol H}_2\text{SO}_4\end{array}} \times \frac{1}{25.0 \text{ mL H}_2\text{SO}_4} = \boxed{0.188 \; M \text{ H}_2\text{SO}_4}$$

CHECKPOINT–SECTION 12.7   Acid-Base Titrations

12.7.1 If 15.0 mL of an HNO_3 solution requires 22.6 mL of 0.117 *M* LiOH to neutralize, what is the concentration of the HNO_3 solution?

   a) 0.264 *M*   d) 0.176 *M*

   b) 0.0777 *M*   e) 0.568 *M*

   c) 0.117 *M*

12.7.2 If 37.5 mL of an $HC_2H_3O_2$ solution requires 48.6 mL of 0.0578 *M* KOH to neutralize, what is the concentration of the $HC_2H_3O_2$ solution?

   a) 0.0749 *M*   d) 0.0356 *M*

   b) 0.0446 *M*   e) 0.134 *M*

   c) 0.00281 *M*

12.7.3 If 25.0 mL of an H_2SO_4 solution requires 39.9 mL of 0.228 *M* NaOH to neutralize, what is the concentration of the H_2SO_4 solution?

   a) 0.728 *M*   d) 0.228 *M*

   b) 0.364 *M*   e) 0.910 *M*

   c) 0.182 *M*

12.7.4 What volume of 0.144 *M* H_2SO_4 is required to neutralize 25.0 mL of 0.0415 *M* $Ba(OH)_2$?

   a) 7.20 mL   d) 50.0 mL

   b) 3.60 mL   e) 12.5 mL

   c) 14.4 mL

12.8   Buffers

It takes very little strong acid or strong base to change the pH of pure water. For example, consider the beaker of water and the bottle of 1.0 *M* HCl shown in Figure 12.9. A single drop of the acid in the bottle will reduce the pH in the beaker from 7.0 to 3.3. A ***buffer*** is

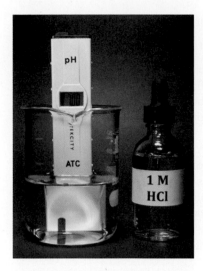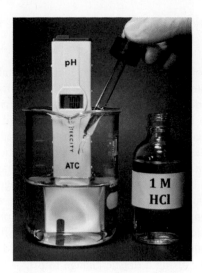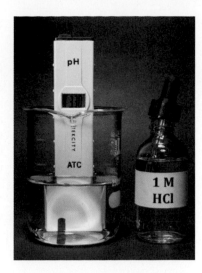

Figure 12.9 Adding a single drop of strong acid to pure water causes a large change in pH—from neutral (7.00) to strongly acidic (3.30).
(all) David A. Tietz/McGraw Hill

a solution that prevents drastic pH change upon the addition of acids or bases. The components of a buffer are a weak acid and its conjugate base. For example, we can prepare a buffer by combining 100 mL of 0.10 M acetic acid with 0.010 mol (0.82 g) sodium acetate. If we then add a single drop of 1.0 M HCl to this solution, the pH will drop by so little that our pH meter reading will not change! Even if we were to add *five* drops of the 1.0 M HCl (Figure 12.10), the pH would only drop from an initial value of 4.74 to 4.72.

Many important processes, including most biological processes, can occur only within a very narrow pH range. Human blood plasma, for example, must have a pH between 7.35 and 7.45. Plasma pH outside that range can result in severe lethargy, seizures, and death. The pH of blood plasma is maintained by a buffer system similar to that just described—although the weak acid principally involved is carbonic acid (H_2CO_3) rather than acetic acid.

A buffer works by containing one species that neutralizes strong acid and another species that neutralizes strong base. Consider the acetic acid–sodium acetate buffer described at the beginning of this section. The beaker in which we combine 100 mL of 0.10 M acetic acid and 0.010 mol sodium acetate contains 0.010 mol each of the weak acid and its conjugate base. When strong acid is added to the buffer, it reacts with the conjugate base:

$$H^+(aq) + C_2H_3O_2^-(aq) \longrightarrow HC_2H_3O_2(aq)$$

> **Student Note:** Acetate ion is the conjugate base of acetic acid. Sodium is a spectator ion—not an active ingredient of the buffer.

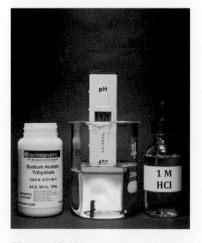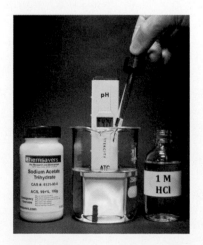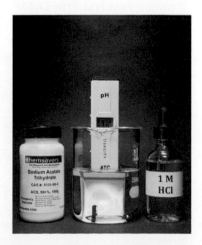

Figure 12.10 Adding a drop of strong acid to a buffered solution causes a much smaller pH change, in this case from 4.74 to 4.72.
(all) David A. Tietz/McGraw Hill

Animation
Buffers

Similarly, when strong base is added to the buffer, it reacts with the weak acid:

$$OH^-(aq) + HC_2H_3O_2(aq) \longrightarrow H_2O(l) + C_2H_3O_2^-(aq)$$

For a solution to be a buffer, it must contain comparable amounts of weak acid and conjugate base. If we were to add more than 0.010 mol of strong acid to the acetic acid–sodium acetate buffer, we would override the buffer's capacity to prevent a drastic drop in pH because we would have used up all of the conjugate base (neutralizing the added acid). Likewise, if we were to add more than 0.010 mol of strong base to the buffer, we would use up all of the acetic acid. A solution can act as a buffer only if it contains significant amounts of both members of the conjugate acid-base pair.

A buffer can prevent drastic pH change only as long as the necessary member of the conjugate pair remains. When all of the weak acid has been consumed, the solution is no longer buffered against pH change from the addition of base—and vice versa.

Sample Problems 12.13 and 12.14 show how to identify the components of a buffer, and how to determine how much acid or base can be added to a particular buffer without causing a drastic change in pH.

SAMPLE PROBLEM (**12.13**) Identifying Components Suitable for Preparing Buffers

Indicate which of the following pairs of substances can be used to prepare a buffer:

(a) HCl and NaCl, (b) NaF and KF, (c) HCN and NaCN.

Strategy A buffer needs to contain comparable amounts of a weak acid and conjugate base.

Setup Identify an acid present and determine its conjugate base.

(a) HCl is a strong acid, and its conjugate base is Cl^-.

(b) There is no acid present; both substances contain F^-, the conjugate base of HF.

(c) HCN is a weak acid, and its conjugate base is CN^-.

Solution (a) Since HCl is a strong acid, this cannot be a buffer system, even though the conjugate base is present.

(b) Since there is no acid present, only a conjugate base, this does not constitute a buffer system.

(c) Since a weak acid and its conjugate base are both present here, this combination of substances would form a buffer.

> **THINK ABOUT IT**
>
> Remember that the sodium and potassium ions in these substances are spectator ions and are not part of the buffer system.

Practice Problem ATTEMPT Indicate which of the following pairs of substances can be used to prepare a buffer:

(a) HF and HCN, (b) HNO_2 and $NaNO_2$, (c) HBr and KBr.

Practice Problem BUILD Indicate which of the following pairs of substances can be used to prepare a buffer. For any pair that cannot be used to prepare a buffer, explain why.

(a) $NaNO_2$ and $LiNO_2$, (b) HCOOH (formic acid) and KCOOH, (c) HF and KF.

Practice Problem CONCEPTUALIZE Which diagram best represents what happens in solution when a strong acid is added to a buffer solution?

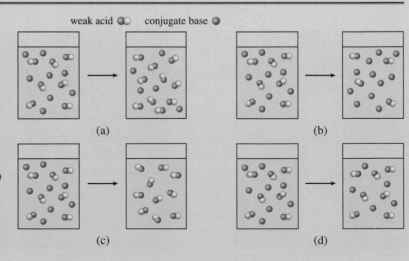

weak acid ⬤⬤   conjugate base ⬤

(a)   (b)

(c)   (d)

SAMPLE PROBLEM 12.14 Determining Buffer Capacity

How much strong acid can be added to a buffer prepared by combining 125 mL each of 0.10 M acetic acid and 0.10 M potassium acetate without causing a drastic pH change?

Strategy Determine the number of moles of conjugate base present. The amount of strong acid that can be absorbed by the buffer is limited by the number of moles of conjugate base present.

Setup

$$0.125 \, \cancel{L} \times \frac{0.10 \text{ mol } C_2H_3O_2^-}{\cancel{L}} = 0.0125 \text{ mol } C_2H_3O_2^-$$

Solution The largest amount of strong acid that could be absorbed by this buffer is 0.0125 moles, as limited by the number of moles of acetate ion (conjugate base) present.

THINK ABOUT IT

Strong acid will react with the base present in the buffer, which is the acetate ion in this case.

Practice Problem **A**TTEMPT How much strong acid can be added to a buffer prepared by combining 75.0 mL each of 0.12 M acetic acid and 0.15 M potassium acetate without causing a drastic pH change?

Practice Problem **B**UILD How much strong base can be added to the buffer in Practice Problem A without causing a drastic pH change?

Practice Problem **C**ONCEPTUALIZE Which diagram represents the buffer to which the largest amount of strong acid can be added without causing a drastic pH change? Which represents the buffer to which the largest amount of strong base can be added without causing a drastic pH change?

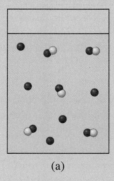
(a)

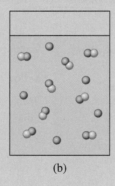
(b)

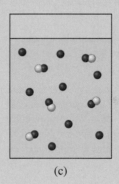
(c)

CHECKPOINT–SECTION 12.8 Buffers

12.8.1 Which of the following combinations can be used to prepare a buffer?

a) HCl/NaCl

b) HF/KF

c) NH_3/NH_4Cl

d) HNO_3/HNO_2

e) $NaNO_3/HNO_3$

12.8.2 How many moles of strong acid can be added to a buffer containing 0.223 mole of HCOOH and 0.184 mole of HCOONa without causing a drastic change in pH?

a) 0.0920 mole

b) 0.184 mole

c) 0.112 mole

d) 0.223 mole

e) 0.446 mole

Chapter Summary

Section 12.1

- *Acids* are compounds that produce H^+ ions in aqueous solutions. Acids are common in foods and beverages and typically have a sour taste.

- For a compound to produce H^+ ions, it must have at least one *ionizable hydrogen atom* or *proton*. An ionizable hydrogen atom is one that separates from the molecule in solution to become an aqueous proton.

- Bases are compounds that produce OH^- ions in aqueous solutions. Bases are less common in foods and beverages in part because they taste bitter, but they are common in household cleaning products.

Section 12.2

- Acids and bases can be defined in more than one way. A compound that produces H^+ ions in aqueous solution is an *Arrhenius acid.* A compound that produces OH^- ions in aqueous solution is an *Arrhenius base.* The Arrhenius definitions of acids and bases are limited to the behavior of compounds dissolved in water.

- The Brønsted definitions of acids and bases are broader and include nonaqueous substances. A *Brønsted acid* is a substance that donates a proton; a *Brønsted base* is a substance that accepts a proton.

- Aqueous protons $[H^+(aq)]$ are actually attached to water molecules and are therefore better represented by the formula $H_3O^+(aq)$, which is referred to as the *hydronium ion.*

- When a Brønsted acid donates a proton, what remains of the original acid is the acid's *conjugate base.* When a Brønsted base accepts a proton, the protonated species is the base's *conjugate acid.* Together, an acid and its conjugate base—or a base and its conjugate acid—constitute a *conjugate pair.*

Section 12.3

- Depending on the substance with which it is combined, water can act either as a Brønsted acid or as a Brønsted base. A substance, such as water, that can either donate or accept a proton is known as *amphoteric.*

- Water undergoes *autoionization,* resulting in very small concentrations of both hydronium and hydroxide ions in pure water. At 25°C, the concentrations of H_3O^+ and OH^- ions are both $1.0 \times 10^{-7}\ M$.

- In any aqueous solution at 25°C, the product of hydronium ion and hydroxide ion concentrations is 1.0×10^{-14}.

- Solutions in which the hydronium ion concentration is higher than the hydroxide ion concentration are *acidic;* those in which hydroxide concentration is higher are *basic;* and those in which the concentrations are equal are *neutral.*

Section 12.4

- A *strong acid* is one that undergoes complete ionization in water. When a strong acid dissolves, all of the molecules separate into the hydronium ion and the acid's conjugate base—no intact acid molecules remain in solution.

- An acid with just one ionizable hydrogen atom is called a *monoprotic acid.* An acid with two ionizable hydrogen atoms is called a *diprotic acid.*

- There are only seven strong acids, all but one of which are monoprotic; but the only diprotic strong acid, H_2SO_4, loses just one of its protons when dissolved in water.

- A *strong base* is an ionic hydroxide that dissociates completely in water. The list of strong bases is relatively short, consisting of the Group 1 hydroxides and the heaviest Group 2 hydroxides.

Section 12.5

- The pH scale is used to express the acidity of solutions in which the hydronium ion concentration can range over many orders of magnitude. pH values typically range from 1 to 14, with 7 being neutral, less than 7 being acidic, and greater than 7 being basic. A change of one pH unit corresponds to a 10-fold change in hydronium ion concentration.

- The pOH scale is analogous to the pH scale, but is calculated using the *hydroxide* ion concentration. For any aqueous solution at 25°C, the sum of pH and pOH is 14.

Section 12.6

- Most acids are weak rather than strong. A *weak acid* is one that ionizes only partially in water, leaving most of the acid molecules intact in solution.

- The chemical equation used to represent the ionization of a weak acid uses a double arrow to indicate that both the forward and the reverse reaction occur; and that not all of the reactants are converted to products.

- A *carboxylic acid* is one in which the ionizable hydrogen atom is part of a specific group of atoms known as a carboxyl group. A carboxyl group consists of a carbon atom bonded to two oxygen atoms and an ionizable hydrogen atom bonded to one of the oxygen atoms.

- The formulas of carboxylic acids may be written with their ionizable hydrogen atoms first, or with the carboxyl group at the end of the formula. For example, acetic acid can be written either as $HC_2H_3O_2$ or as CH_3COOH—both are correct.

- Most bases are weak rather than strong. A *weak base* is a molecular compound that ionizes only partly in water—producing hydroxide ion by accepting protons from water molecules. Many bases resemble ammonia (NH_3) in that they consist of a nitrogen atom surrounded by three groups.

- Equations used to represent the ionization of weak bases also employ double arrows to indicate that the ionization is not complete.

Section 12.7

- *Titration* involves the carefully measured addition of one reactant to another. Typically, in acid-base titration, a base of precisely known concentration is added from a buret to a sample of acid whose concentration is unknown. Using the volume and concentration of base required to neutralize the acid, we can calculate the concentration of the unknown acid—provided we know its volume.

- The point at which all of the acid has been neutralized is called the *equivalence point.* Most acid-base titrations require the use of an *indicator,* a dye that changes color at a certain pH, to show when the equivalence point has been reached.

Section 12.8

- A *buffer* is a solution that resists drastic pH change upon the addition of small amounts of strong acid or strong base. Buffers consist of comparable amounts of a weak acid and its conjugate base (or a weak base and its conjugate acid). Drastic *decrease* in the pH of a buffer (when strong acid is added) is prevented by the conjugate base, which reacts with added hydronium ions. Drastic *increase* in pH (when strong base is added) is prevented by the weak acid, which reacts with added hydroxide ions.

- Adding more strong acid than can be consumed by the conjugate base in a buffer—or adding more strong base than can be consumed by the weak acid in a buffer—will result in drastic pH change because the buffer's capacity has been exceeded.

- Buffers or buffered solutions are important to many industrial and biological processes. Many important processes in living systems can occur only within a very narrow range of pH values.

Key Terms

Amphoteric 432	Buffer 452	Equivalence point 450	pOH 442
Arrhenius acid 429	Carboxylic acid 446	Hydronium ion 430	Strong acid 434
Arrhenius base 429	Conjugate acid 430	Indicator 450	Strong base 435
Autoionization of water 432	Conjugate base 430	Ionizable hydrogen atom 427	Titration 450
Brønsted acid 429	Conjugate pair 430	Monoprotic acid 435	Weak acid 445
Brønsted base 429	Diprotic acid 435	pH 437	Weak base 448

Key Equations

12.1	at 25° C $[H_3O^+] \times [OH^-] = 1.0 \times 10^{-14}$	The product of hydronium and hydroxide ion concentrations is a constant. At 25°C, the product is 1.0×10^{-14}.
12.2	$pH = -\log[H_3O^+]$ or $pH = -\log[H^+]$	The pH of a solution is calculated as the negative log base 10 of hydronium ion concentration.
12.3	$[H_3O^+] = 10^{-pH}$	Hydronium ion concentration can be calculated from pH by taking the antilog of $-pH$.
12.4	$pOH = -\log[OH^-]$	The pOH of a solution is calculated as the negative log base 10 of hydroxide ion concentration.
12.5	$[OH^-] = 10^{-pOH}$	Hydroxide ion concentration can be calculated from pOH by taking the antilog of $-pOH$.
12.6	$pH + pOH = 14.00$	The sum of pH and pOH at 25°C is 14.00.

KEY SKILLS   Identifying Acids and Bases

It is important for you to be able to recognize acids and bases by looking at their chemical formulas. This spreadsheet gives you a process for determining if a compound is an acid or a base, and whether it is strong or weak.

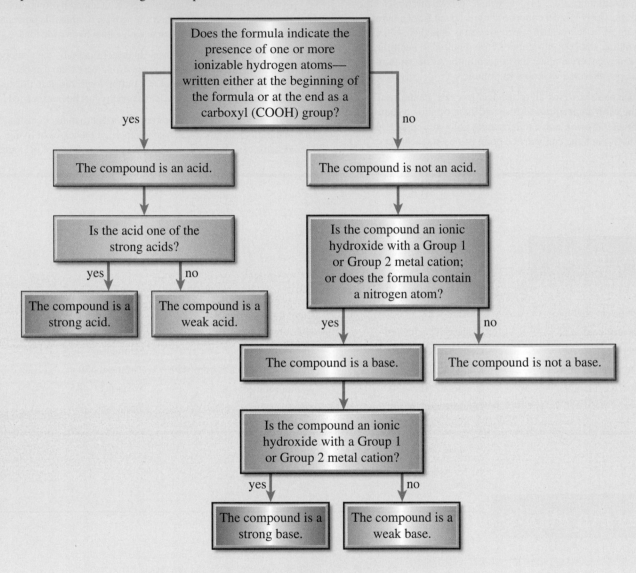

Key Skills Problems

12.1
Which of the following compounds are strong acids?
(Select all that apply.)
a) NaOH
b) CH_3COOH
c) H_2SO_4
d) $Ba(OH)_2$
e) HF

12.2
Which of the following compounds are strong bases?
(Select all that apply.)
a) NH_3
b) KOH
c) HNO_2
d) HNO_3
e) LiOH

12.3
Which of the following compounds are weak acids?
(Select all that apply.)
a) NaOH
b) NH_3
c) CH_3COOH
d) HNO_2
e) HF

12.4
Which of the following compounds are weak bases?
(Select all that apply.)
a) CH_3NH_2
b) NaCl
c) NH_3
d) HNO_2
e) $Ba(OH)_2$

Questions and Problems

SECTION 12.1: PROPERTIES OF ACIDS AND BASES

12.1 Identify each of the following statements as describing an acid, a base, or neither.
(a) Produces hydroxide ions in solution
(b) Produces hydronium ions in solution
(c) Tastes bitter
(d) Tastes sour
(e) Reacts with hydroxide ions to produce a neutral solution
(f) Reacts with hydronium ions to produce a neutral solution
(g) Reacts with some metals to produce H_2 gas
(h) Is good for unclogging drains
(i) Reacts with carbonate ions to produce water and CO_2 gas
(j) Feels slippery to the touch

12.2 List a few household products that contain acid.

12.3 List a few household products that contain base.

SECTION 12.2: DEFINITIONS OF ACIDS AND BASES

12.4 Define the term *Arrhenius acid*.

12.5 Define the term *Arrhenius base*.

12.6 Define the term *Brønsted acid*.

12.7 Define the term *Brønsted base*.

12.8 How do Arrhenius and Brønsted acids differ from one another?

12.9 Select the Arrhenius acids from the list.
(a) $HC_2H_3O_2$ (d) CH_3OH
(b) CH_4 (e) HCN
(c) NH_3

12.10 Select the Brønsted acids from the list.
(a) $HCOOH$ (d) CH_3NH_2
(b) C_2H_4 (e) HCN
(c) $LiOH$

12.11 Select the Arrhenius bases from the list.
(a) CH_3OH (d) $HCOOH$
(b) $Ca(OH)_2$ (e) HF
(c) NH_3

12.12 Select the Brønsted bases from the list.
(a) CH_3CH_2OH (d) CH_3COOH
(b) NH_3 (e) CH_3CH_3
(c) KOH

12.13 Identify the conjugate acid-base pairs in the following reactions.
(a) $HCN(aq) + H_2O(l) \rightleftharpoons H_3O^+(aq) + CN^-(aq)$
(b) $HCO_2H(aq) + H_2O(l) \rightleftharpoons$
$\quad\quad\quad\quad\quad H_3O^+(aq) + CO_2H^-(aq)$
(c) $CH_3NH_2(aq) + H_2O(l) \rightleftharpoons$
$\quad\quad\quad\quad\quad CH_3NH_3^+(aq) + OH^-(aq)$
(d) $F^-(aq) + H_2O(l) \rightleftharpoons HF(aq) + OH^-(aq)$

12.14 Identify the conjugate acid-base pairs in the following reactions.
(a) $(CH_3)_2NH(aq) + H_2O(l) \rightleftharpoons$
$\quad\quad\quad\quad\quad (CH_3)_2NH_2^+(aq) + OH^-(aq)$
(b) $CH_3CO_2H(aq) + H_2O(l) \rightleftharpoons$
$\quad\quad\quad\quad\quad CH_3CO_2^-(aq) + H_3O^+(aq)$
(c) $HCO_3^-(aq) + H_2O(l) \rightleftharpoons$
$\quad\quad\quad\quad\quad H_2CO_3(aq) + OH^-(aq)$
(d) $C_5H_5N(aq) + H_2O(l) \rightleftharpoons$
$\quad\quad\quad\quad\quad C_5H_5NH^+(aq) + OH^-(aq)$

12.15 Identify the conjugate acid of each of the following substances.
(a) NH_3 (c) HSO_3^-
(b) OH^- (d) CO_3^{2-}

12.16 Identify the conjugate base of each of the following substances.
(a) H_3O^+ (c) H_2O
(b) NH_4^+ (d) H_2CO_3

12.17 Which of the following substances can behave as a Brønsted base? For any that behave as Brønsted bases, write out the balanced chemical equation showing the reaction with water.
(a) $CH_3NH_3^+$ (d) H_2SO_3
(b) CH_3OH (e) CH_2CH_2
(c) OH^-

12.18 Which of the following substances can behave as a Brønsted acid? For any that behave as Brønsted acids, write out the balanced chemical equation showing the reaction with water.
(a) NH_3 (d) NH_4^+
(b) C_2H_6 (e) $HClO_4$
(c) H_3O^+

SECTION 12.3: WATER AS AN ACID; WATER AS A BASE

12.19 What is the concentration of H_3O^+ in pure water?

12.20 What is the concentration of OH^- in pure water?

12.21 Is it possible for an aqueous solution to have $[H_3O^+] = 0.10\ M$ and $[OH^-] = 0.10\ M$ at the same time? Explain.

12.22 Determine whether solutions described with the following concentrations are acidic, basic, or neutral.
(a) $[H_3O^+] = [OH^-]$
(b) $[OH^-] = 5.22 \times 10^{-9}\ M$
(c) $[H_3O^+] < [OH^-]$
(d) $[H_3O^+] = 5.22 \times 10^{-9}\ M$
(e) $[OH^-] > 1.00 \times 10^{-7}\ M$

12.23 Determine whether solutions described with the following concentrations are acidic, basic, or neutral.
(a) $[OH^-] = 9.13 \times 10^{-4}\ M$
(b) $[OH^-] > [H_3O^+]$
(c) $[H_3O^+] = 2.44 \times 10^{-8}\ M$
(d) $[H_3O^+] = 1.00 \times 10^{-7}\ M$
(e) $[OH^-] < 1.00 \times 10^{-7}\ M$

12.24 Determine the concentration of $[H_3O^+]$ in each solution, given $[OH^-]$. Identify the solution as acid, basic, or neutral.
(a) $[OH^-] = 3.88 \times 10^{-9} \, M$
(b) $[OH^-] = 2.45 \times 10^{-4} \, M$
(c) $[OH^-] = 7.94 \times 10^{-8} \, M$
(d) $[OH^-] = 1.00 \times 10^{-7} \, M$

12.25 Determine the concentration of $[H_3O^+]$ in each solution, given $[OH^-]$. Identify the solution as acid, basic, or neutral.
(a) $[OH^-] = 5.13 \times 10^{-8} \, M$
(b) $[OH^-] = 7.99 \times 10^{-12} \, M$
(c) $[OH^-] = 3.52 \times 10^{-2} \, M$
(d) $[OH^-] = 1.87 \times 10^{-6} \, M$

12.26 Determine the concentration of $[OH^-]$ in each solution, given $[H_3O^+]$. Identify the solution as acid, basic, or neutral.
(a) $[H_3O^+] = 8.11 \times 10^{-12} \, M$
(b) $[H_3O^+] = 7.28 \times 10^{-7} \, M$
(c) $[H_3O^+] = 4.37 \times 10^{-11} \, M$
(d) $[H_3O^+] = 6.49 \times 10^{-4} \, M$

12.27 Determine the concentration of $[OH^-]$ in each solution, given $[H_3O^+]$. Identify the solution as acid, basic, or neutral.
(a) $[H_3O^+] = 1.88 \times 10^{-3} \, M$
(b) $[H_3O^+] = 5.32 \times 10^{-12} \, M$
(c) $[H_3O^+] = 4.75 \times 10^{-2} \, M$
(d) $[H_3O^+] = 9.34 \times 10^{-13} \, M$

SECTION 12.4: STRONG ACIDS AND BASES

12.28 Define the term *strong acid*.

12.29 Define the term *strong base*.

12.30 Select the strong acids from the list.
(a) HBr   (d) HCOOH
(b) H_2SO_3 (e) CH_3OH
(c) CH_3CH_3 (f) HNO_3

12.31 Select the strong acids from the list.
(a) $HClO_4$ (d) $HC_2H_3O_2$
(b) HF   (e) NaOH
(c) NH_3 (f) HI

12.32 Select the strong bases from the list.
(a) C_2H_5OH (d) LiOH
(b) CH_3NH_2 (e) NH_4^+
(c) HF   (f) $Sr(OH)_2$

12.33 Select the strong bases from the list.
(a) HCN   (d) $Ba(OH)_2$
(b) CH_3COOH (e) HCO_3^-
(c) C_5H_5N (f) KOH

12.34 Determine the concentration of H_3O^+ in each of the following solutions:
(a) $[HNO_3] = 0.112 \, M$ (d) $[HCl] = 0.0686 \, M$
(b) $[HBr] = 0.0877 \, M$ (e) $[HClO_4] = 0.194 \, M$
(c) $[HI] = 0.284 \, M$

12.35 Determine the concentration of H_3O^+ in each of the following solutions:
(a) $[Ca(OH)_2] = 0.0500 \, M$
(b) $[NaOH] = 0.215 \, M$
(c) $[KOH] = 0.0436 \, M$

(d) $[LiOH] = 0.0755 \, M$
(e) $[Sr(OH)_2] = 0.422 \, M$

12.36 Determine the concentration of OH^- in each of the following solutions:
(a) $[Ba(OH)_2] = 0.0825 \, M$
(b) $[LiOH] = 0.316 \, M$
(c) $[KOH] = 0.278 \, M$
(d) $[Ca(OH)_2] = 0.0895 \, M$
(e) $[Sr(OH)_2] = 0.0747 \, M$

12.37 Determine the concentration of OH^- in each of the following solutions:
(a) $[HCl] = 0.417 \, M$
(b) $[HNO_3] = 0.0819 \, M$
(c) $[HBr] = 0.0565 \, M$
(d) $[HClO_4] = 0.788 \, M$
(e) $[HI] = 0.0741 \, M$

12.38 Determine the concentration of both H_3O^+ and OH^- in each of the following solutions:
(a) $[LiOH] = 0.100 \, M$
(b) $[HNO_3] = 0.100 \, M$
(c) $[Sr(OH)_2] = 0.100 \, M$
(d) $[HBr] = 0.100 \, M$

12.39 Determine the concentration of both H_3O^+ and OH^- in each of the following solutions:
(a) $[HI] = 0.0550 \, M$
(b) $[LiOH] = 0.0550 \, M$
(c) $[HClO_4] = 0.0550 \, M$
(d) $[Ba(OH)_2] = 0.0550 \, M$

SECTION 12.5: pH AND pOH SCALES

12.40 What is the difference between $[H_3O^+]$ in a solution that has a pH of 4.00 and $[H_3O^+]$ in a solution with a pH of 3.00? How does this compare to the $[H_3O^+]$ difference between a solution with a pH of 5.00 and one with a pH of 6.00?

12.41 Determine the pH of each solution, given $[H_3O^+]$.
(a) $[H_3O^+] = 2.33 \times 10^{-3} \, M$
(b) $[H_3O^+] = 8.13 \times 10^{-5} \, M$
(c) $[H_3O^+] = 9.43 \times 10^{-6} \, M$
(d) $[H_3O^+] = 1.73 \times 10^{-10} \, M$

12.42 Determine the pH of each solution, given $[OH^-]$.
(a) $[OH^-] = 7.21 \times 10^{-2} \, M$
(b) $[OH^-] = 4.87 \times 10^{-4} \, M$
(c) $[OH^-] = 9.47 \times 10^{-8} \, M$
(d) $[OH^-] = 3.82 \times 10^{-11} \, M$

12.43 Determine the pH of each of the following solutions:
(a) $[HBr] = 0.0821 \, M$ (c) $[HClO_4] = 0.204 \, M$
(b) $[HNO_3] = 0.103 \, M$ (d) $[HI] = 0.0613 \, M$

12.44 Determine the pH of each of the following solutions:
(a) $[HCl] = 0.118 \, M$
(b) $[HI] = 0.0616 \, M$
(c) $[HNO_3] = 0.00421 \, M$
(d) $[HCl] = 0.00908 \, M$

12.45 Determine the pH of each of the following solutions:
(a) $[LiOH] = 0.0855 \, M$
(b) $[KOH] = 0.0917 \, M$
(c) $[Ca(OH)_2] = 0.101 \, M$
(d) $[NaOH] = 0.0746 \, M$

12.46   Determine the pH of each of the following solutions:
(a) $[Sr(OH)_2] = 0.00287 \ M$
(b) $[LiOH] = 0.0522 \ M$
(c) $[KOH] = 0.0614 \ M$
(d) $[Ba(OH)_2] = 0.00386 \ M$

12.47   Determine the pOH of each of the following solutions:
(a) $[Ca(OH)_2] = 0.0211 \ M$
(b) $[LiOH] = 0.184 \ M$
(c) $[NaOH] = 0.0399 \ M$
(d) $[Ba(OH)_2] = 0.0866 \ M$

12.48   Determine the pOH of each of the following solutions:
(a) $[HNO_3] = 0.0712 \ M$ (c) $[HClO_4] = 0.109 \ M$
(b) $[HBr] = 0.00974 \ M$ (d) $[HI] = 0.0587 \ M$

12.49   Determine $[H_3O^+]$ in solutions with the following pH values.
(a) 1.35   (c) 6.83   (e) 13.69
(b) 3.78   (d) 9.44

12.50   Determine the $[OH^-]$ in solutions with the following pH values.
(a) 2.89   (c) 8.42   (e) 12.54
(b) 4.75   (d) 10.19

12.51   Determine the $[H_3O^+]$ in solutions with the following pOH values.
(a) 3.88   (c) 7.69   (e) 13.87
(b) 5.91   (d) 11.79

12.52   Determine $[OH^-]$ in solutions with the following pOH values.
(a) 2.11   (c) 6.83   (e) 12.68
(b) 4.22   (d) 10.72

SECTION 12.6: WEAK ACIDS AND BASES

12.53   Define the term *weak acid*.

12.54   Define the term *weak base*.

12.55   Using benchmark pH values, estimate the pH of a 0.0441 M HF solution.

12.56   Using benchmark pH values, estimate the pH of a 0.0755 M $HC_2H_3O_2$ solution.

12.57   Using benchmark pH values, estimate the pH of a 0.0890 M NH_3 solution.

12.58   Using benchmark pH values, estimate the pH of a 0.117 M C_5H_5N solution.

12.59   Place the following solutions in order of increasing $[H_3O^+]$.
(a) 0.10 M HF, 0.10 M NH_3, 0.10 M HNO_3
(b) 0.10 M HCN, 0.20 M HCN, 0.20 M $HClO_4$
(c) 0.10 M NH_3, 0.10 M NaOH, 0.10 M $Ca(OH)_2$

12.60   Place the following solutions in order of decreasing $[H_3O^+]$.
(a) 0.15 M $C_6H_5NH_2$, 0.15 M NaOH, 0.15 M $Ba(OH)_2$
(b) 0.10 M HCl, 0.10 M NH_3, 0.10 M NaOH
(c) 0.025 M CH_3NH_2, 0.025 M HClO,
        0.025 M $HClO_4$

12.61   Select the solution in each group with the lowest pH.
(a) 0.089-M solutions of HF, HNO_2, HNO_3
(b) 0.145-M solutions of NH_3, HCN, $HClO_4$
(c) 0.0779-M solutions of $HC_2H_3O_2$, CH_3NH_2, LiOH

12.62   Select the solution in each group with the lowest pH.
(a) 0.029-M solutions of HCO_2H, HI, NH_3
(b) 0.341-M solutions of H_3PO_4, CH_3NH_2, LiOH
(c) 0.096-M solutions of NH_3, LiOH, $Ba(OH)_2$

12.63   Select the solution in each group with the highest pH.
(a) 0.0955-M solutions of H_2SO_3, HBr, NH_3
(b) 0.0225-M solutions of H_2CO_3, CH_3NH_2, LiOH
(c) 0.0079-M solutions of NH_3, KOH, $Ca(OH)_2$

12.64   Select the solution in each group with the highest pH.
(a) 0.337-M solutions of HF, NH_3, HCl
(b) 0.115-M solutions of HNO_2, HNO_3, HI
(c) 0.214-M solutions of $Sr(OH)_2$, LiOH, NaOH

SECTION 12.7: ACID-BASE TITRATIONS

12.65   Describe what is meant by the term *titration*.

12.66   Are all titrations acid-base titrations? Explain.

12.67   Define the term *equivalence point*.

12.68   A 25.00-mL sample of an $HClO_4$ solution is titrated to the equivalence point using 0.114 M LiOH. If the titration requires 39.64 mL of LiOH, what was the concentration of $HClO_4$ in the original 25.00-mL sample?

12.69   A 125.00-mL sample of an HI solution is titrated to the equivalence point using 0.209 M KOH. If the titration requires 79.55 mL of KOH, what was the concentration of HI in the original 125.00-mL sample?

12.70   A 50.0-mL sample of an HNO_3 solution is titrated to the equivalence point using 0.107 M $Ca(OH)_2$. If the titration requires 65.0 mL of $Ca(OH)_2$, what was the concentration of HNO_3 in the original 50.0-mL sample?

12.71   A 75.0-mL sample of an H_2SO_4 solution is neutralized using 0.131 M NaOH. If the neutralization requires 142.2 mL of NaOH, what was the concentration of H_2SO_4 in the original 75.0-mL sample?

12.72   Determine the volume (in mL) of 0.304 M NaOH required to neutralize 235.5 mL of 0.100 M H_3PO_4.

12.73   Determine the volume (in mL) of 0.224 M $Ba(OH)_2$ required to neutralize 187.5 mL of 0.100 M HCl.

SECTION 12.8: BUFFERS

12.74   Does a buffer solution form if 0.50 mole of solid NaOH is added to 1.00 L of a 1.00 M HCN solution? Explain.

12.75   Does a buffer solution form if 0.050 mole of HCl gas is dissolved in 500.0 mL of 0.20 M NH_3 solution? Explain.

12.76   Which of the following solution pairs, when combined in equal volumes, would behave as a buffer?
(a) 0.200 M HF and 0.200 M NaF
(b) 0.150 M NaOH and 0.150 M NaCl
(c) 0.267 M HCN and 0.0010 M LiCN
(d) 0.145 M LiOH and 0.105 M HNO_3

12.77 Which of the following solution pairs, when combined in equal volumes, would behave as a buffer?
(a) 0.050 M KCl and 0.50 M KOH
(b) 0.075 M HCN and 0.0750 M HCl
(c) 0.115 M NH$_3$ and 0.145 M NH$_4$Br
(d) 0.100 M HC$_2$H$_3$O$_2$ and 0.080 M NaC$_2$H$_3$O$_2$

12.78 Which of the following compounds, when mixed with a solution of HCOOH, would form a buffer solution?
(a) HCOOLi   (c) CH$_3$COOH
(b) NaC$_2$H$_3$O$_2$ (d) HCO$_2$K

12.79 Which of the following compounds, when mixed with a solution of HCN, would form a buffer solution?
(a) CH$_3$NH$_2$ (c) NaCl
(b) NaCN   (d) NH$_4$Cl

12.80 Identify a compound that could be added to the following aqueous solutions to form a buffer.
(a) NH$_3$ (c) HNO$_2$
(b) HClO   (d) H$_2$CO$_3$

12.81 A buffer contains 0.0750 mole of HNO$_2$ and 0.0650 mole of LiNO$_2$.
(a) Determine the moles of strong acid that the buffer could absorb without a significant change in pH.
(b) Determine the moles of strong base that the buffer could absorb without a significant change in pH.

12.82 A 100.0-mL sample of a buffer is 0.050 M in HCOOH and 0.050 M in LiCOOH.
(a) Determine the volume of 0.10 M HCl that the buffer can absorb without a significant change in pH.
(b) Determine the volume of 0.10 M NaOH that the buffer can absorb without a significant change in pH.

(c) Determine the volume of 0.10 M Sr(OH)$_2$ that the buffer can absorb without a significant change in pH.

12.83 A 250.0-mL sample of a buffer is 0.250 M in HCN and 0.200 M in NaCN.
(a) Determine the volume of 0.200 M HNO$_3$ that the buffer can absorb without a significant change in pH.
(b) Determine the volume of 0.100 M LiOH that the buffer can absorb without a significant change in pH.
(c) Determine the moles of Ca(OH)$_2$ that the buffer can absorb without a significant change in pH.

12.84 Determine the pH that results when the following aqueous solutions are combined.
(a) 155 mL of 0.13 M KOH and 225 mL of 0.16 M HBr
(b) 155 mL of 0.13 M KOH and 101 mL of 0.16 M HBr
(c) 155 mL of 0.16 M KOH and 155 mL of 0.16 M HBr

12.85 Determine the pH that results when the following aqueous solutions are combined.
(a) 207 mL of 0.029 M Ca(OH)$_2$ and 375 mL of 0.021 M HNO$_3$
(b) 207 mL of 0.029 M Ca(OH)$_2$ and 715 mL of 0.021 M HNO$_3$
(c) 207 mL of 0.029 M Ca(OH)$_2$ and 207 mL of 0.029 M HNO$_3$

12.86 A recipe for making a vinegar–baking soda volcano (see Familiar Chemistry in Section 12.5) calls for 16.5 g sodium bicarbonate and 215 mL vinegar. Assuming the vinegar has a density of 1.02 g/mL and is 5.0 percent acetic acid by mass, determine the mass of carbon dioxide produced by the reaction and its volume at STP.

Answers to In-Chapter Materials

Answers to Practice Problems

12.1A (a) HClO$_4$, (b) HS$^-$, (c) HS$^-$, (d) HC$_2$O$_4^-$. **12.1B** SO$_3^{2-}$, H$_2$SO$_3$. **12.2A** (a) NH$_4^+$: acid, H$_2$O: base, NH$_3$: conjugate base, H$_3$O$^+$: conjugate acid, (b) CN$^-$: base, H$_2$O: acid, HCN: conjugate acid, OH$^-$: conjugate base. **12.2B** (a) HSO$_4^-$ + H$_2$O $\rightleftharpoons$ SO$_4^{2-}$ + H$_3$O$^+$, (b) HSO$_4^-$ + H$_2$O $\rightleftharpoons$ H$_2$SO$_4$ + OH$^-$. **12.3A** 2.0 × 10^{-11} M. **12.3B** 2.8 × 10^{-13} M. **12.4A** (a) [H$_3$O$^+$] = 0.118 M, [OH$^-$] = 8.5 × 10^{-14} M, (b) [H$_3$O$^+$] = 7.84 × 10^{-3} M, [OH$^-$] = 1.3 × 10^{-12} M, (c) [H$_3$O$^+$] = 9.33 × 10^{-2} M, [OH$^-$] = 1.1 × 10^{-13} M. **12.4B** (a) 8.3 × 10^{-7} M, (b) 2.7 × 10^{-6} M, (c) 2.0 × 10^{-3} M. **12.5A** (a) [OH$^-$] = 0.118 M, [H$_3$O$^+$] = 8.5 × 10^{-14} M, (b) [OH$^-$] = 7.84 × 10^{-3} M, [H$_3$O$^+$] = 1.3 × 10^{-12} M, (c) [OH$^-$] = 1.87 × 10^{-1} M, [H$_3$O$^+$] = 5.4 × 10^{-14} M. **12.5B** (a) 8.3 × 10^{-7} M, (b) 2.7 × 10^{-6} M, (c) 2.0 × 10^{-3} M. **12.6A** (a) 8.49, (b) 7.40, (c) 1.25. **12.6B** (a) 6.92, (b) 10.52, (c) 11.08. **12.7A** (a) 1.26 × 10^{-10} M, (b) 0.035 M,

(c) 9.8 × 10^{-8} M. **12.7B** (a) 7.8 × 10^{-3} M, (b) 2.6 × 10^{-12} M, (c) 1.3 × 10^{-7} M. **12.8A** (a) 11.24, (b) 2.14, (c) 5.07. **12.8B** (a) 6.45, (b) 6.00, (c) 3.00. **12.9A** (a) 9.5 × 10^{-14} M, (b) 7.2 × 10^{-6} M, (c) 1.0 × 10^{-7} M. **12.9B** (a) 5.5 × 10^{-12} M, (b) 2.0 × 10^{-4} M, (c) 2.5 × 10^{-2} M. **12.10A** >1.11. **12.10B** 0.13 M < 0.093 M < 0.050 M. **12.11A** pH < 13.17. **12.11B** 0.0664 M < 0.150 M < 0.393 M. **12.12A** 22.6 mL. **12.12B** 91.2 mL. **12.13A** b. **12.13B** b and c. The pair in (a) cannot be used to prepare a buffer because it is not a conjugate pair. **12.14A** 0.011 mol. **12.14B** 0.0090 mol.

Answers to Checkpoints

12.2.1 b, d. **12.2.2** c, d. **12.3.1** d. **12.3.2** a. **12.3.3** a. **12.3.4** e. **12.4.1** b. **12.4.2** c. **12.5.1** c. **12.5.2** a. **12.5.3** e. **12.5.4** b. **12.6.1** e. **12.6.2** a. **12.6.3** a. **12.6.4** b. **12.7.1** d. **12.7.2** a. **12.7.3** c. **12.7.4** a. **12.8.1** b, c. **12.8.2** b.

CHAPTER 13

Equilibrium

13.1 Reaction Rates

13.2 Chemical Equilibrium

13.3 Equilibrium Constants

- Calculating Equilibrium Constants
- Magnitude of the Equilibrium Constant

13.4 Factors That Affect Equilibrium

- Addition or Removal of a Substance
- Changes in Volume
- Changes in Temperature

Many biological processes, including the maintenance of body temperature in warm-blooded animals, resemble equilibrium processes—although strictly speaking they are "steady states," rather than true equilibria.

SDI Productions/E+/Getty Images

In This Chapter, You Will Learn

About the dynamic state known as chemical equilibrium, and about the factors that influence it.

Things To Review Before You Begin

- Balancing chemical equations [◄◄ Section 10.3]
- The gas laws [◄◄ Section 8.4]

The normal body temperature for a healthy human being is about 37°C. A healthy human body immersed in cold water—or out in very warm weather—will, except under extraordinary circumstances, remain at 37°C. Numerous processes, some chemical and some physical, contribute to the maintenance of normal body temperature. Some of these processes cause temperature to increase and some cause it to decrease. On balance, though, the effect of the many competing processes is to maintain the normal temperature. In a far simpler way, chemical reactions maintain their own kind of "normal," which we refer to as equilibrium. In this chapter, we describe chemical equilibrium and the factors that affect it.

13.1 Reaction Rates

The *rate of reaction* refers to how fast a reaction occurs. Many familiar reactions, such as the initial steps in vision and explosions, happen almost instantaneously, whereas others, such as the rusting of iron or the conversion of diamond to graphite, take place on a timescale of days or even millions of years. Figure 13.1 shows one process with a *high* rate of reaction (one that occurs rapidly), and one process with a *low* rate of reaction (one that occurs slowly). Often, the job of an industrial chemist is to work on increasing the *rate* of an important reaction rather than maximizing its yield or developing a new process.

There are several ways that chemists can increase the rate of a chemical reaction, including increasing the *temperature* at which the reaction takes place and increasing *reactant concentrations*. We can understand how and why such changes affect reaction rates using **collision theory.** Collision theory tells us that chemical reactions happen when the reactant particles (atoms, molecules, or ions) *collide*. Consider the gas-phase reaction of atomic chlorine (Cl) with nitrosyl chloride (NOCl) to form molecular chlorine (Cl_2) and nitrogen monoxide (NO):

$$Cl(g) + NOCl(g) \longrightarrow Cl_2(g) + NO(g)$$

For the Cl atoms and NOCl molecules to react, they must collide with each other. Moreover, they must collide with sufficient energy to break the N—O bonds in the NOCl molecules. Only then can the product molecules form. The reaction of Cl and NOCl is exothermic. Nevertheless, it is necessary to *add* energy to the reactants to initiate this reaction. The amount of energy necessary to *initiate* a reaction is called the **activation energy.** We can think of the activation energy as an energy *barrier* that prevents less energetic molecules from reacting. Figure 13.2 shows how

465

Figure 13.1 (a) The explosive decomposition of sodium azide in an automobile airbag could not serve its intended purpose if it did not happen, in effect, instantaneously. Reactions such as this have a very high rate of reaction. (b) The rusting of steel wool can occur over days, weeks, or months, depending on conditions. Note that the chemical equation used to represent the rusting of iron here is a simplified version. Water is also involved, and rust itself consists of several different iron(III) compounds.

(both) David A. Tietz/McGraw Hill

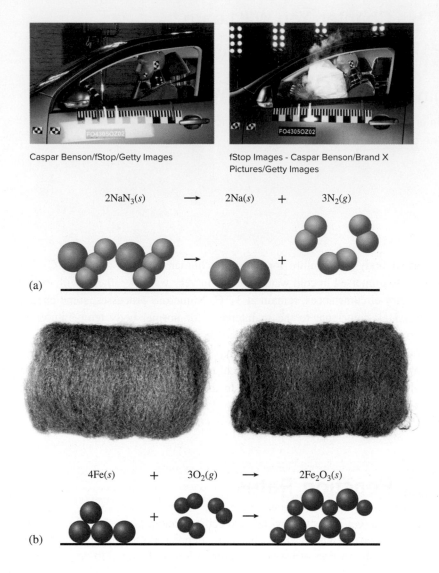

Caspar Benson/fStop/Getty Images

fStop Images - Caspar Benson/Brand X Pictures/Getty Images

$$2NaN_3(s) \longrightarrow 2Na(s) + 3N_2(g)$$

(a)

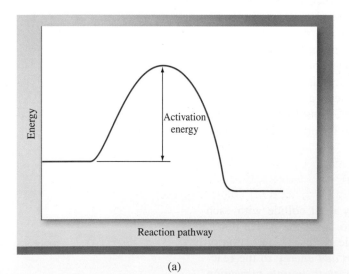 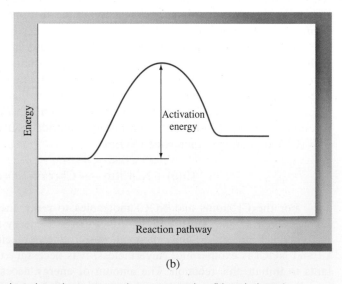

$$4Fe(s) + 3O_2(g) \longrightarrow 2Fe_2O_3(s)$$

(b)

(a)

Reaction pathway

Energy

Activation energy

(b)

Reaction pathway

Energy

Activation energy

Figure 13.2 Regardless of whether a reaction is (a) exothermic (products have less energy than reactants) or (b) endothermic (products have more energy than reactants), there is an activation energy barrier that must be overcome for the reaction to occur. The higher the activation energy, the lower the reaction rate.

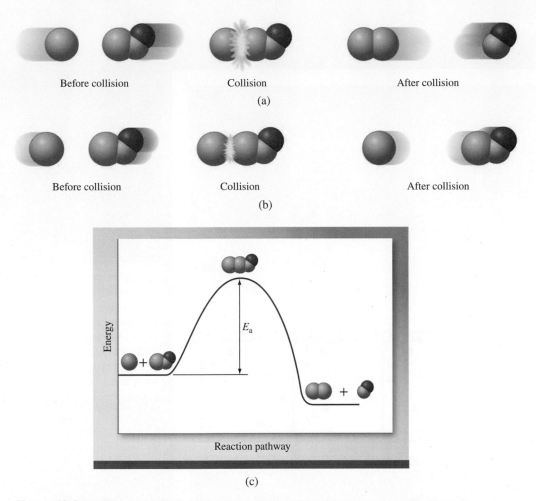

Figure 13.3 (a) When reactants collide with sufficient kinetic energy to overcome the activation energy of the reaction, products can form. (b) When slower-moving reactants collide with less than the necessary kinetic energy, they do not react. (c) The energy profile for the reaction of Cl with NOCl shows that although the reaction is exothermic, energy must be added to initiate it.

we can visualize the activation energy barrier in exothermic and endothermic reactions. Because the number of reactant molecules in an ordinary reaction is very large, the kinetic energies vary greatly. Normally, only a small fraction of the colliding molecules—the fastest moving ones—have sufficient kinetic energy to overcome the activation energy. Only these molecules can take part in the reaction. Figure 13.3 shows two possible collisions between Cl and NOCl, one that leads to product formation and one that does not.

According to collision theory, then, the rate of a reaction depends on the *number* of reactant collisions and on the *energy* of those collisions. The number of reactant collisions can be increased by increasing the concentrations of reactants. The more reactant molecules there are in a given volume, the more frequent the collisions between them will be. Moreover, with a greater number of collisions overall, there will be more collisions with sufficient energy to result in reaction. Another way to increase the frequency of collision is to increase the temperature at which the reaction takes place. At higher temperatures, molecules move faster [◄◄ Section 8.1]. Because they are moving faster, they encounter one another more frequently—resulting in more frequent collisions. In addition, increasing temperature—because it increases the speed and kinetic energy of the reactant molecules—causes the collisions to be more *energetic*. This makes a larger fraction of the collisions energetic enough to result in reaction. Figure 13.4 summarizes the effects of increased reactant concentrations and increased temperature on reaction rates.

Collision Theory

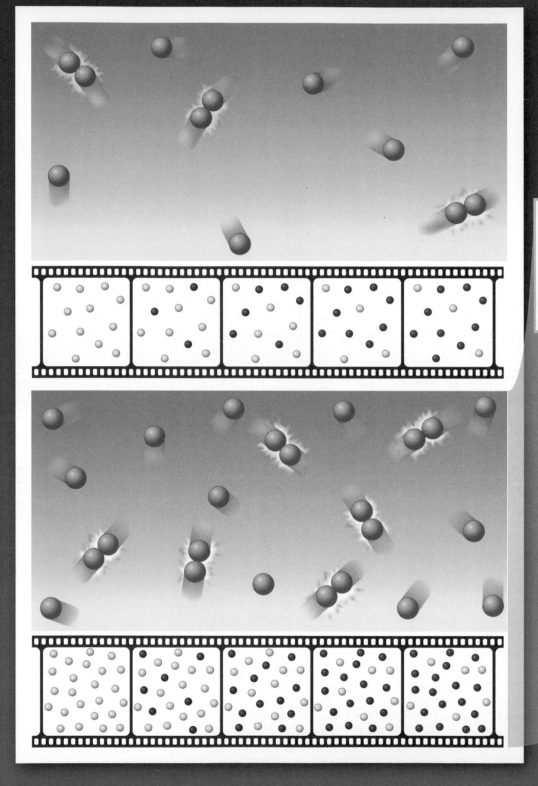

At higher concentration, reactant molecules collide more often, giving rise to a greater number of collisions that are sufficiently energetic to result in reaction—thereby increasing reaction rate.

Figure 13.4 Effects of Reactant Concentration and Temperature on Reaction Rate

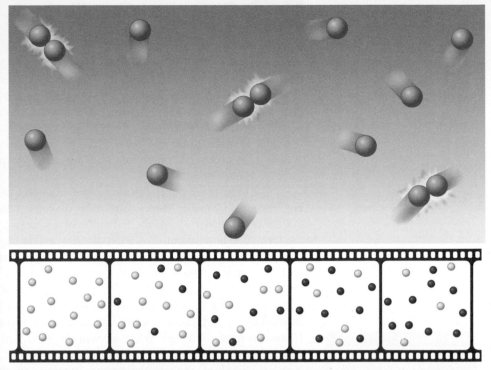

At higher temperature, reactant molecules are moving faster, causing more frequent collisions and greater energy at impact. Both factors increase the number of collisions that are sufficiently energetic to result in reaction.

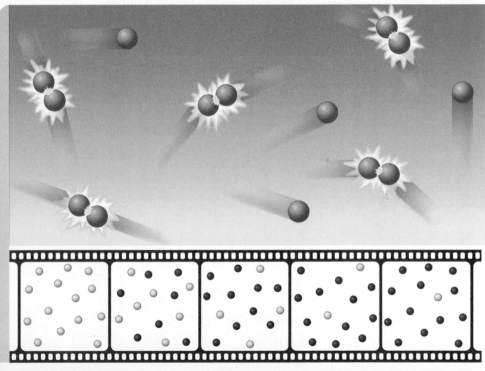

What's the point?

Two of the factors that influence the rate at which a chemical reaction occurs are *reactant concentration* and *temperature*. In general:

- Reaction rate increases as reactant concentration increases
- Reaction rate increases as temperature increases

13.2 Chemical Equilibrium

Until now, for the most part, we have treated chemical equations as processes that go to completion; that is, we start with only *reactants* and end up with only *products*. In reality, this is not what happens with most chemical reactions. Instead, if we start with only reactants, the typical reaction will proceed, causing reactant concentrations to decrease (as reactants are consumed) and product concentrations to increase (as products are produced). Eventually, though, the concentrations of both reactants and products will stop changing; and we will be left with a *mixture* of reactants and products. A system in which the concentrations of reactants and products remain constant is said to be *at equilibrium*.

We have already encountered several examples of *physical* processes that involve equilibrium—including the establishment of vapor pressure over a liquid in a closed system [◄◄ Section 7.4], and the formation of a saturated solution [◄◄ Section 9.1]. Using the formation of a saturated solution of silver iodide (AgI) as an example (Figure 13.5), let's examine the concept of equilibrium. The dissolution process can be represented with the chemical equation $AgI(s) \rightleftharpoons Ag^+(aq) + I^-(aq)$, where the double arrow [◄◄ Section 12.6] indicates that this is a **reversible process,** meaning that both the forward process and the reverse process can occur. In this case, the *forward* process is the dissolution of AgI and the *reverse* process is the recombination of aqueous Ag^+ and I^- ions to form solid AgI. When we add the solid AgI to the water, initially only the forward process can occur because there are no Ag^+ or I^- ions in solution. When some AgI has dissolved, though, the reverse process can also occur. Initially, the forward process occurs at a higher rate than the reverse process, simply because of the low number of ions in solution. After some time has passed, there are enough ions in solution that the reverse process can occur at the same rate as the forward process—and the concentration of dissolved AgI stops changing. A system in which both forward and reverse processes are occurring at the same rate is at **equilibrium.**

Now let's consider a *chemical* example of equilibrium, the decomposition of dinitrogen tetroxide (N_2O_4) to yield nitrogen dioxide (NO_2). This process, like most chemical reactions, is also reversible:

$$N_2O_4(g) \rightleftharpoons 2NO_2(g)$$

Student Note: A system being at *equilibrium* means that both forward and reverse reactions are occurring at the same *rate*. It does *not* mean that reactant and product concentrations are equal.

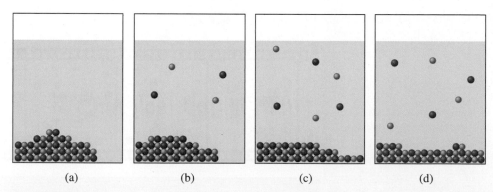

(a)   (b)   (c)   (d)

Figure 13.5 Preparation of a saturated solution of AgI: $AgI(s) \rightleftharpoons Ag^+ (aq) + I^-(aq)$. (a) Solid AgI is added to water. (b) Initially, only the forward process (dissolution of solid AgI) can occur, and AgI begins to dissolve. (c) When there are Ag^+ and I^- ions in solution, the reverse process (formation of solid AgI) can also occur. (d) Equilibrium has been achieved when both forward and reverse processes continue to occur at equilibrium; because they are occurring at the same rate, the concentration of dissolved AgI remains constant.

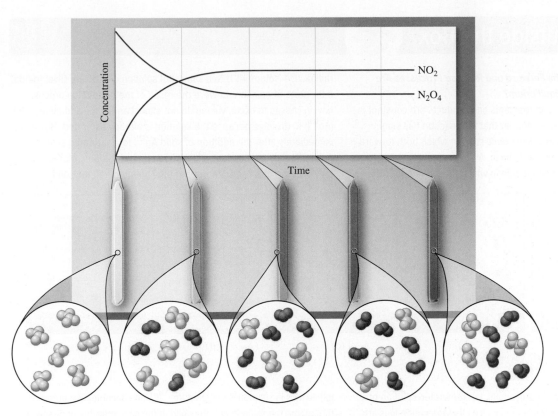

Figure 13.6 Reaction of colorless N_2O_4 to form brown NO_2. Initially, only N_2O_4 is present and only the forward reaction (decomposition of N_2O_4 to give NO_2) is occurring. As NO_2 forms, the reverse reaction (recombination of NO_2 to give N_2O_4) begins to occur. The brown color continues to intensify until the forward and reverse reactions are occurring at the same rate.

N_2O_4 is a colorless gas, whereas NO_2 is brown. (NO_2 is the cause of the brown appearance of some polluted air.) If we begin by placing a sample of pure N_2O_4 in an evacuated flask, the contents of the flask change from colorless to brown as the decomposition produces NO_2 (Figure 13.6). At first, the brown color intensifies as the concentration of NO_2 increases. Eventually, though, the intensity of the brown color stops increasing, indicating that the concentration of NO_2 has stopped changing. At this point, both the forward and reverse processes are occurring at the same rate, and the system has reached equilibrium. In the reaction shown in Figure 13.6, initially:

- N_2O_4 concentration is high, and the rate of the forward reaction is high.
- NO_2 concentration is zero, making the rate of the reverse reaction zero.

As the reaction proceeds:

- N_2O_4 concentration falls, decreasing the rate of the forward reaction.
- NO_2 concentration rises, increasing the rate of the reverse reaction.

Thinking Outside the Box

How Do We Know That the Forward and Reverse Processes Are Ongoing in a System at Equilibrium?

Because the concentrations of reactants and products are constant in a system at equilibrium, it may appear that the reaction has simply stopped. In fact, equilibrium is a *dynamic* state in which both forward and reverse processes continue to occur. An experiment that illustrates this involves adding solid silver iodide in which the iodide ion is the

radioactive isotope I-131 to a saturated solution of ordinary silver iodide. If the state of equilibrium were the result of the dissolution process having ceased to occur, we would not expect any of the additional solid ($Ag^{131}I$) to dissolve because the solution is already saturated. However, immediately after the addition of solid $Ag^{131}I$ to a saturated solution of AgI, radioactive iodide ions ($^{131}I^-$) appear in the solution. Moreover, they become distributed throughout the solution and the solid.

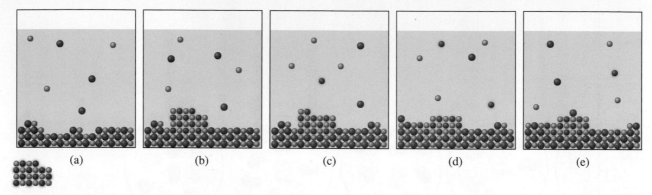

(a)   (b)   (c)   (d)   (e)

Evidence that equilibrium is dynamic. (a) A saturated solution of ordinary AgI, and a piece of solid AgI in which all of the iodide ions are the radioactive isotope ^{131}I. (b) The solid is added to the saturated solution. (c) $^{131}I^-$ ions immediately appear in the solution. (d) The number of $^{131}I^-$ ions in solution is not constant, although the *total* number of

iodide ions in solution *is* constant. (e) $^{131}I^-$ ions become distributed throughout the solution and the solid. If the concentration of dissolved AgI in the original saturated solution were constant because the reaction had stopped occurring, none of the $Ag^{131}I$ would have dissolved.

> **Student Note:** It is a common error to think that equilibrium means equal *concentrations* of reactants and products—it does not. Equilibrium refers to the state in which forward and reverse reactions are occurring at the same rate.

Some important things to remember about equilibrium are:

- Equilibrium is a *dynamic* state—both forward and reverse reactions continue to occur, although there is no net change in reactant and product concentrations over time.
- At equilibrium, the rates of the forward and reverse reactions are equal.

13.3 Equilibrium Constants

> **Student Note:** The *equilibrium expression* is the equation:
>
> equilibrium expression
> $$K = \frac{[NO_2]^2_{eq}}{[N_2O_4]_{eq}}$$
> equilibrium constant
>
> The equilibrium expression enables us to calculate the equilibrium constant, which is simply *K*.

In the mid-nineteenth century, Norwegian chemists Cato Guldberg and Peter Waage studied the equilibrium mixtures (mixtures in which reactant and product concentrations were constant) of a wide variety of chemical reactions. They made countless observations and found that at constant temperature, there is a constant relationship between the equilibrium concentrations of reactants and products. Specifically, the product of *product* concentrations, each raised to its coefficient in the balanced chemical equation, divided by the product of *reactant* concentrations, each raised to its coefficient in the balanced equation, was a constant. We call this constant the ***equilibrium constant, K.*** For the decomposition of N_2O_4 the ***equilibrium expression*** is derived from the balanced chemical equation as

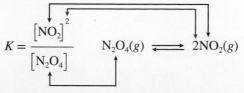

$$K = \frac{\left[NO_2\right]^2}{\left[N_2O_4\right]} \qquad N_2O_4(g) \rightleftharpoons 2NO_2(g)$$

In this equation the coefficient of N_2O_4 is 1, which generally is not written either as a coefficient or as an exponent.

Familiar Chemistry

Sweet Tea

If you like your iced tea especially sweet, you probably have figured out that stirring sugar into the tea while it is still hot enables you to dissolve more—making the tea sweeter than it would be if you were to add the sugar to tea already cold. The reason for this is that equilibrium constants are *constant* only at constant *temperature*. We can represent the dissolution of sugar in tea with the following chemical equation:

$$\text{sugar} + \text{tea} \rightleftharpoons \text{sweet tea}$$

This process has an equilibrium constant associated with it—whether or not we know its actual value. Remember, the magnitude of the equilibrium constant tells us the relative amounts of reactants and products at equilibrium. The K value for this process increases as temperature increases. In fact, it increases by approximately a factor of three when going from 0°C (roughly the temperature of iced tea) to 95°C (a common tea-steeping temperature). That means that *three times* as much sugar can be dissolved in the tea while it is still hot! If we dissolve as much sugar as possible while the tea is hot and *then* pour it over ice, we end up with much sweeter tea—in the form of a supersaturated solution [◄◄ Section 9.1]. And because supersaturated aqueous solutions of sugar are especially stable, the excess dissolved sugar will *not* crystallize out of solution when the solution is cooled significantly by the addition of ice.

Student Note: The magnitude of an equilibrium constant for an *endothermic* process *increases* with increasing temperature. The magnitude of an equilibrium constant for an *exothermic* process *decreases* with increasing temperature.

David A. Tietz/McGraw Hill

Calculating Equilibrium Constants

Table 13.1 lists the starting and equilibrium concentrations of N_2O_4 and NO_2 in a series of experiments carried out at 25°C. Using the equilibrium concentrations from each of the experiments in the table, we do indeed get a constant value for the ratio of $[NO_2]^2$ to $[N_2O_4]$—within the limits of experimental error. (The average value is 4.63×10^{-3}.) Therefore, the equilibrium constant (K) for this reaction at 25°C is 4.63×10^{-3}.

Equilibrium expressions can be written for any reaction for which we know the balanced chemical equation. Further, knowing the equilibrium expression for a reaction, we can use equilibrium concentrations to calculate the value of the corresponding equilibrium constant.

Sample Problem 13.2 shows how to use an equilibrium expression and equilibrium concentrations to calculate the value of K.

TABLE 13.1	Equilibrium Concentrations of N_2O_4 and NO_2 at 25°C		
	Equilibrium Concentrations (M)		
Experiment	**$[N_2O_4]$**	**$[NO_2]$**	**$\dfrac{[NO_2]^2}{[N_2O_4]}$**
1	0.643	0.0547	4.65×10^{-3}
2	0.448	0.0457	4.66×10^{-3}
3	0.491	0.0475	4.60×10^{-3}
4	0.594	0.0523	4.60×10^{-3}
5	0.0898	0.0204	4.63×10^{-3}

Sample Problems 13.1 and 13.2 let you practice writing equilibrium expressions from balanced chemical equations, and determining equilibrium constants from equilibrium expressions and equilibrium concentrations.

SAMPLE PROBLEM ⟨13.1⟩ Writing Equilibrium Expressions from Balanced Equations

Write equilibrium expressions for the following reactions:

(a) $N_2(g) + 3H_2(g) \rightleftharpoons 2NH_3(g)$

(b) $H_2(g) + I_2(g) \rightleftharpoons 2HI(g)$

(c) $Ag^+(aq) + 2NH_3(aq) \rightleftharpoons Ag(NH_3)_2^+(aq)$

(d) $2O_3(g) \rightleftharpoons 3O_2(g)$

(e) $Cd^{2+}(aq) + 4Br^-(aq) \rightleftharpoons CdBr_4^{2-}(aq)$

(f) $2NO(g) + O_2(g) \rightleftharpoons 2NO_2(g)$

Strategy Use the balanced equations to construct equilibrium expressions.

Setup The equilibrium expression for each reaction has the form of the concentrations of *products* over the concentrations of *reactants*, each raised to a power equal to its stoichiometric coefficient in the balanced chemical equation.

Solution

(a) $K = \dfrac{[NH_3]^2}{[N_2][H_2]^3}$ (b) $K = \dfrac{[HI]^2}{[H_2][I_2]}$ (c) $K = \dfrac{[Ag(NH_3)_2^+]}{[Ag^+][NH_3]^2}$ (d) $K = \dfrac{[O_2]^3}{[O_3]^2}$ (e) $K = \dfrac{[CdBr_4^{2-}]}{[Cd^{2+}][Br^-]^4}$ (f) $K = \dfrac{[NO_2]^2}{[NO]^2[O_2]}$

THINK ABOUT IT

With practice, writing equilibrium expressions becomes second nature. Without sufficient practice, it will seem inordinately difficult. It is important that you become proficient at this. It is very often the first step in solving equilibrium problems.

Practice Problem ⒶTTEMPT Write the equilibrium expression for each of the following reactions:

(a) $2N_2O(g) \rightleftharpoons 2N_2(g) + O_2(g)$

(b) $2NOBr(g) \rightleftharpoons 2NO(g) + Br_2(g)$

(c) $HF(aq) \rightleftharpoons H^+(aq) + F^-(aq)$

(d) $CO(g) + H_2O(g) \rightleftharpoons CO_2(g) + H_2(g)$

(e) $CH_4(g) + 2H_2S(g) \rightleftharpoons CS_2(g) + 4H_2(g)$

(f) $H_2C_2O_4(aq) \rightleftharpoons 2H^+(aq) + C_2O_4^{2-}(aq)$

Practice Problem ⒷUILD Write the equation for the equilibrium that corresponds to each of the following equilibrium expressions.

(a) $K = \dfrac{[HCl]^2}{[H_2][Cl_2]}$ (b) $K = \dfrac{[HF]}{[H^+][F^-]}$ (c) $K = \dfrac{[Cr(OH)_4^-]}{[Cr^{3+}][OH^-]^4}$ (d) $K = \dfrac{[H^+][ClO^-]}{[HClO]}$ (e) $K = \dfrac{[H^+][HSO_3^-]}{[H_2SO_3]}$ (f) $K = \dfrac{[NOBr]^2}{[NO]^2[Br_2]}$

Practice Problem Ⓒ ONCEPTUALIZE Fill in the missing information in the equilibrium expression and chemical equation that represent the same chemical reaction.

$$K = \dfrac{[CO_2]^?[H_2O]^?}{[C_3H_8][O_2]^?} \qquad C_3H_8(g) + \_?\_O_2(g) \rightleftharpoons \_?\_CO_2(g) + 4H_2O(g)$$

SAMPLE PROBLEM 13.2 Calculating Equilibrium Constant from Equilibrium Concentrations

Carbonyl chloride ($COCl_2$), also called phosgene, is a highly poisonous gas that was used on the battlefield in World War I. It is produced by the reaction of carbon monoxide with chlorine gas:

$$CO(g) + Cl_2(g) \rightleftharpoons COCl_2(g)$$

In an experiment conducted at 74°C, the equilibrium concentrations of the species involved in the reaction were as follows: $[CO] = 1.2 \times 10^{-2}$ M, $[Cl_2] = 0.054$ M, and $[COCl_2] = 0.14$ M. (a) Write the equilibrium expression, and (b) determine the value of the equilibrium constant for this reaction at 74°C.

Strategy Write the equilibrium expression and plug in the equilibrium concentrations of all three species to evaluate K.

Setup The equilibrium expression has the form of concentrations of products over concentrations of reactants, each raised to the appropriate power—in the case of this reaction, all the coefficients are 1, so all the powers will be 1.

Solution

1. $K = \dfrac{[COCl_2]}{[CO][Cl_2]}$

2. $K = \dfrac{(0.14)}{(1.2 \times 10^{-2})(0.054)} = 216$ or 2.2×10^2

K for this reaction at 74°C is 2.2×10^2.

THINK ABOUT IT

When putting the equilibrium concentrations into the equilibrium expression, we leave out the units. It is common practice to express equilibrium constants without units.

Practice Problem ATTEMPT In an analysis of the following reaction at 100°C,

$$Br_2(g) + Cl_2(g) \rightleftharpoons 2BrCl(g)$$

the equilibrium concentrations were found to be $[Br_2] = 2.3 \times 10^{-3}$ M, $[Cl_2] = 1.2 \times 10^{-2}$ M, and $[BrCl] = 1.4 \times 10^{-2}$ M. Write the equilibrium expression, and calculate the equilibrium constant for this reaction at 100°C.

Practice Problem BUILD In another analysis at 100°C involving the same reaction, the equilibrium concentrations of the reactants were found to be $[Br_2] = 4.1 \times 10^{-3}$ M and $[Cl_2] = 8.3 \times 10^{-3}$ M. Determine the value of $[BrCl]$.

Practice Problem CONCEPTUALIZE Consider the reaction 2A $\rightleftharpoons$ B. The diagram shown on the top right represents a system at equilibrium where A = ○ and B = ●. Which of the following diagrams [(i)–(iv)] also represents a system at equilibrium? Select all that apply.

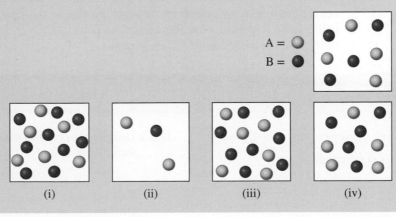

A = ○
B = ●

(i) (ii) (iii) (iv)

Magnitude of the Equilibrium Constant

One of the things the equilibrium constant tells us is the extent to which a reaction will proceed at a particular temperature if we combine stoichiometric amounts of reactants. To illustrate this, let's consider again the general reaction

$$A + B \rightleftharpoons C$$

If we combine stoichiometric amounts of reactants, 1 mol of A with 1 mol of B, three outcomes are possible:

1. The reaction will go essentially to completion, and the equilibrium mixture will consist predominantly of C (product).
2. The reaction will not occur to any significant degree, and the equilibrium mixture will consist predominantly of A and B (reactants).
3. The reaction will proceed to a significant degree, but will not go to completion, and the equilibrium mixture will contain comparable amounts of A, B, and C.

When the magnitude of K is very large, we expect the first outcome. The formation of the $Ag(NH_3)_2^+$ ion is an example of this possibility.

$$Ag^+(aq) + 2NH_3(aq) \rightleftharpoons Ag(NH_3)_2^+(aq) \qquad K = 1.5 \times 10^7 \text{ (at } 20°C)$$

Student Note: Stoichiometric amounts [◄◄ Section 10.3].

If we were to combine aqueous Ag^+ and aqueous NH_3 in a mole ratio of 1:2, the resulting equilibrium mixture would contain mostly $Ag(NH_3)_2^+$ with only very small amounts of reactants remaining. A reaction with a very large equilibrium constant is sometimes said to "lie to the right" or to "favor products."

When the magnitude of K is very small, we expect the second outcome. The chemical combination of nitrogen and oxygen gases to give nitrogen monoxide is an example of this possibility.

$$N_2(g) + O_2(g) \rightleftharpoons 2NO(g) \qquad K = 4.3 \times 10^{-25} \text{ (at } 25°C)$$

Nitrogen and oxygen gases do not react to any significant extent at room temperature. If we were to place a mixture of N_2 and O_2 in an evacuated flask and allow the system to reach equilibrium, the resulting mixture would be predominantly N_2 and O_2 with only a *very* small amount of NO. A reaction with a very small equilibrium constant is said to "lie to the left" or to "favor reactants."

The terms *very large* and *very small* are somewhat arbitrary when applied to equilibrium constants. Generally speaking, an equilibrium constant with a magnitude greater than about 1×10^2 can be considered large; one with a magnitude smaller than about 1×10^{-2} can be considered small.

An equilibrium constant that falls between 1×10^2 and 1×10^{-2} indicates that neither products nor reactants are strongly favored. In this case, a system at equilibrium will contain a *mixture* of reactants and products, the exact composition of which will depend on the stoichiometry of the reaction.

CHECKPOINT–SECTION 13.3   Equilibrium Constants

13.3.1 Select the correct equilibrium expression for the reaction

$$2CO(g) + O_2(g) \rightleftharpoons 2CO_2(g)$$

a) $K = \dfrac{2[CO_2]^2}{2[CO]^2[O_2]}$

b) $K = \dfrac{(2[CO_2])^2}{(2[CO])^2([O_2])}$

c) $K = \dfrac{2[CO_2]}{2[CO][O_2]}$

d) $K = \dfrac{2[CO_2]^2}{[2CO]^2[O_2]}$

e) $K = \dfrac{[CO_2]^2}{[CO]^2[O_2]}$

13.3.2 Determine the value of the equilibrium constant, K, for the reaction

$$A + 2B \rightleftharpoons C + D$$

if the equilibrium concentrations are [A] = 0.0115 *M*, [B] = 0.0253 *M*, [C] = 0.109 *M*, and [D] = 0.0110 *M*.

a) 163

b) 4.12

c) 2.06

d) 6.14×10^{-3}

e) 0.243

13.4 Factors That Affect Equilibrium

One of the interesting and useful features of chemical equilibria is that they can be manipulated in specific ways to maximize production of a desired substance. Consider, for example, the industrial production of ammonia from its constituent elements by the Haber process:

$$N_2(g) + 3H_2(g) \rightleftharpoons 2NH_3(g)$$

More than 100 million tons of ammonia is produced annually by this reaction, with most of the resulting ammonia being used for fertilizers to enhance crop production. It is in the best interest of industry, then, to maximize the yield of NH_3. In this section, we describe the various ways in which an equilibrium can be manipulated to accomplish this goal.

Le Châtelier's principle states that when a *stress* is applied to a system at equilibrium, the system will respond by *shifting* in the direction that minimizes the effect of the stress. In this context, "stress" refers to a disturbance of the system at equilibrium by any of the following means:

- The addition of a reactant or product
- The removal of a reactant or product
- A change in volume of the system, resulting in a change in concentration or partial pressure of the reactants and products
- A change in temperature

"Shifting" refers to the occurrence of either the forward or reverse reaction such that the effect of the stress is partially offset as the system reestablishes equilibrium. An equilibrium that shifts to the right is one in which more products are produced by the forward reaction. An equilibrium that shifts to the left is one in which more reactants are produced by the reverse reaction. Using Le Châtelier's principle, we can predict the direction in which an equilibrium will shift, given the specific stress that is applied.

Familiar Chemistry

Hemoglobin Production at High Altitude

Rapid ascent to a high altitude can cause altitude sickness. The symptoms of altitude sickness, including dizziness, headache, and nausea, are caused by hypoxia, an insufficient oxygen supply to body tissues. In severe cases, without prompt treatment, a victim may slip into a coma and die. And yet a person staying at a high altitude for weeks or months can recover gradually from altitude sickness, adjust to the low oxygen content in the atmosphere, and live and function normally.

The combination of dissolved oxygen with the hemoglobin (Hb) molecule, which carries oxygen through the blood, is a complex reaction, but for our purposes it can be represented by the following simplified equation:

$$Hb(aq) + O_2(aq) \rightleftharpoons HbO_2(aq)$$

where HbO_2 is oxyhemoglobin, the hemoglobin-oxygen complex that actually transports oxygen to tissues. The equilibrium expression for this process is

$$K = \frac{[HbO_2]}{[Hb][O_2]}$$

At an altitude of 3 km, the partial pressure of oxygen in the air is only about 0.14 atm, compared with 0.20 atm at sea level. A lowered partial pressure of oxygen in the air results in a lower dissolved oxygen concentration in the blood. According to Le Châtelier's principle, this decrease in oxygen concentration will shift the hemoglobin-oxyhemoglobin equilibrium from right to left. This change depletes the supply of oxyhemoglobin, causing hypoxia. Over time, a healthy body can cope with this problem by producing more hemoglobin molecules. As the concentration of Hb increases, the equilibrium gradually shifts back toward the right (toward the formation of oxyhemoglobin). It can take several weeks for the increase in hemoglobin production to meet the body's oxygen needs adequately. A return to full capacity may require several years to occur. Studies show that long-time residents of high-altitude areas have high hemoglobin levels in their blood—sometimes as much as 50 percent more than individuals living at sea level. The production of more hemoglobin and the resulting increased capacity of the blood to deliver oxygen to the body have made high-altitude training popular among some athletes.

Sam Edwards/age fotostock

Addition or Removal of a Substance

Again using the Haber process as an example,

$$N_2(g) + 3H_2(g) \rightleftharpoons 2NH_3(g)$$

consider a system at 700 K, in which the equilibrium concentrations are as follows:

$$[N_2] = 2.05 \, M \qquad [H_2] = 1.56 \, M \qquad [NH_3] = 1.52 \, M$$

Using these concentrations in the equilibrium expression, we can calculate the value of K for the reaction at this temperature as follows:

$$K = \frac{[NH_3]^2}{[N_2][H_2]^3} = \frac{(1.52)^2}{(2.05)(1.56)^3} = 0.297$$

If we were to apply stress to this system by adding more N_2, increasing its concentration from 2.05 M to 3.51 M, the system would no longer be at equilibrium. To see that this is true, use the new concentration of nitrogen in the equilibrium expression. The new calculated value (0.173) is no longer equal to the value of K (0.297).

$$\frac{[NH_3]^2}{[N_2][H_2]^3} = \frac{(1.52)^2}{(3.51)(1.56)^3} = 0.173 \neq K$$

For this system to reestablish equilibrium, the net reaction will have to shift in such a way that the equilibrium expression with the new concentrations plugged in is again equal to K, which is constant at a given temperature. Figure 13.7 shows how the concentrations of N_2, H_2, and NH_3 change when N_2 is added to the original equilibrium mixture.

Conversely, if we were to remove N_2 from the original equilibrium mixture, the lower concentration in the denominator of the equilibrium expression would result in a calculated number being greater than K. In this case the reaction will shift to the left. That is, the reverse reaction will take place, thereby increasing the concentrations of N_2 and H_2 and decreasing the concentration of NH_3 until the number we calculate using the equilibrium expression is once again equal to K.

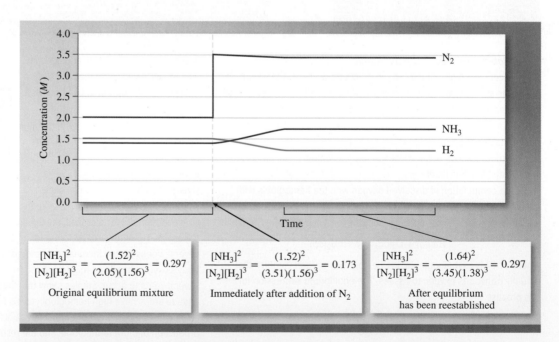

$$\frac{[NH_3]^2}{[N_2][H_2]^3} = \frac{(1.52)^2}{(2.05)(1.56)^3} = 0.297$$

Original equilibrium mixture

$$\frac{[NH_3]^2}{[N_2][H_2]^3} = \frac{(1.52)^2}{(3.51)(1.56)^3} = 0.173$$

Immediately after addition of N_2

$$\frac{[NH_3]^2}{[N_2][H_2]^3} = \frac{(1.64)^2}{(3.45)(1.38)^3} = 0.297$$

After equilibrium has been reestablished

Figure 13.7 Adding more of a reactant to a system at equilibrium causes the equilibrium position to shift toward product. The system responds to the addition of N_2 by consuming some of the added N_2 (and some of the other reactant, H_2) to produce more NH_3.

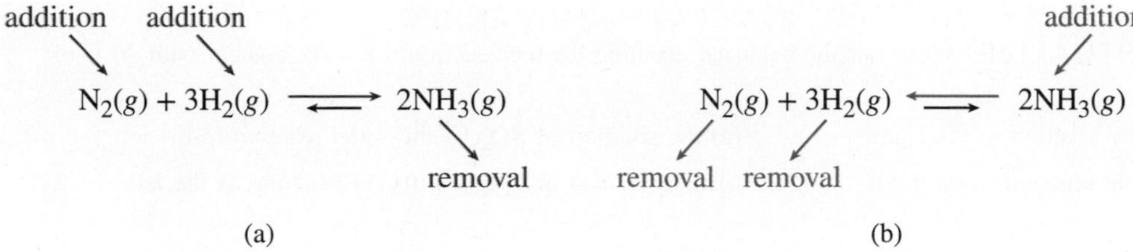

addition   addition   addition

$N_2(g) + 3H_2(g) \rightleftharpoons 2NH_3(g)$ $N_2(g) + 3H_2(g) \rightleftharpoons 2NH_3(g)$

removal   removal   removal

(a)   (b)

Figure 13.8 (a) Addition of a reactant or removal of a product will cause an equilibrium to shift to the right. (b) Addition of a product or removal of a reactant will cause an equilibrium to shift to the left.

The addition or removal of NH_3 will cause a shift in the equilibrium, too. The addition of NH_3 will cause a shift to the left; the removal of NH_3 will cause a shift to the right. Figure 13.8(a) shows the additions and removals that cause this equilibrium to shift to the right. Figure 13.8(b) shows those that cause it to shift to the left.

In essence, a system at equilibrium will respond to the addition of a species by consuming some of that species, and it will respond to the removal of a species by producing more of that species. It is important to remember that the addition or removal of a species from an equilibrium mixture does not change the value of the equilibrium constant, K.

Sample Problem 13.3 shows the effects of stress on a system at equilibrium.

SAMPLE PROBLEM 13.3 Determining Direction of Equilibrium Shift with Addition or Removal of a Substance

An industrially important reaction for the production of hydrogen gas is called the water-gas shift reaction.

$$CO(g) + H_2O(g) \rightleftharpoons H_2(g) + CO_2(g)$$

For each of the following scenarios, determine whether the equilibrium will shift to the right, shift to the left, or neither: (a) addition of $CO(g)$, (b) removal of $H_2O(g)$, (c) removal of $H_2(g)$, (d) addition of $CO_2(g)$.

Strategy Use Le Châtelier's principle to predict the direction of shift for each case.

Setup Begin by writing the expression for the equilibrium constant, K.

$$K = \frac{[H_2][CO_2]}{[CO][H_2O]}$$

Addition of a reactant or removal of a product will cause the reaction to shift to the right or toward more products. Addition of a product or removal of a reactant will cause the reaction to shift to the left or toward more reactants.

Solution

(a) Shift to the right   (b) Shift to the left   (c) Shift to the right   (d) Shift to the left

THINK ABOUT IT

In each case, analyze the effect the change will have on the values in the K expression. The value of K must remain constant, so any change in a concentration must be compensated for. For example, when CO is added, [CO] increases temporarily, making the value of the right side of the K expression shown smaller than it can be. The reaction will shift to reduce the concentration of CO (and H_2O) and increase the concentrations of H_2 and CO_2 to reestablish equilibrium.

Practice Problem ATTEMPT For each change indicated, determine whether the equilibrium

$$PCl_3(g) + Cl_2(g) \rightleftharpoons PCl_5(g)$$

will shift to the right, shift to the left, or neither: (a) addition of $PCl_3(g)$, (b) removal of $PCl_3(g)$, (c) removal of $PCl_5(g)$, and (d) removal of $Cl_2(g)$.

Practice Problem BUILD What can be (a) added to the equilibrium that will shift it to the left, (b) removed from the equilibrium that will shift it to the left, (c) added to the equilibrium that will shift it to the right?

$$Ag^+(aq) + 2NH_3(aq) \rightleftharpoons Ag(NH_3)_2^+(aq)$$

(Continued on next page)

Practice Problem Ⓒ**ONCEPTUALIZE** Write out the balanced equation for the reaction that is consistent with ALL of the following information:

(a) the addition of $H_2S(g)$ shifts the reaction to the right

(c) the addition of $SO_2(g)$ shifts the reaction to the left

(b) the removal of $H_2O(g)$ shifts the reaction to the right

(d) the removal of $O_2(g)$ shifts the reaction to the left

Changes in Volume

If we were to start with a gaseous system at equilibrium in a cylinder with a movable piston, we could change the volume of the system, thereby changing the concentrations of the reactants and products.

Consider again the equilibrium between N_2O_4 and NO_2:

$$N_2O_4(g) \rightleftharpoons 2NO_2(g)$$

At 25°C the equilibrium constant for this reaction is 4.63×10^{-3}. Suppose we have an equilibrium mixture of 0.643 M N_2O_4 and 0.0547 M NO_2 in a cylinder fitted with a movable piston. If we push down on the piston, the equilibrium will be disturbed and will shift in the direction that minimizes the effect of this disturbance. Consider what happens to the concentrations of both species if we decrease the volume of the cylinder by half. Both concentrations are initially *doubled:* $[N_2O_4] = 1.286$ M and $[NO_2] = 0.1094$ M. If we plug the new concentrations into the equilibrium expression, we get

$$\frac{[NO_2]^2_{eq}}{[N_2O_4]_{eq}} = \frac{(0.1094)^2}{1.286} = 9.31 \times 10^{-3}$$

which is not equal to K, so the system is no longer at equilibrium. Because the calculated number is greater than K, the equilibrium will have to shift to the left for equilibrium to be reestablished (Figure 13.9).

Figure 13.9 The effect of a volume decrease (pressure increase) on the $N_2O_4(g) \rightleftharpoons 2NO_2(g)$ equilibrium. When volume is decreased, the equilibrium is driven toward the side with the smallest number of moles of gas.

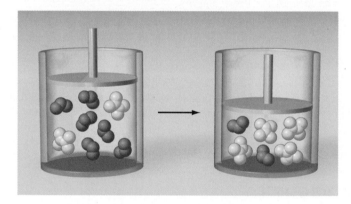

In general, a decrease in volume of a reaction vessel will cause a shift in the equilibrium in the direction that minimizes the total number of moles of gas. Conversely, an increase in volume will cause a shift in the direction that maximizes the total number of moles of gas.

Sample Problem 13.4 shows how to predict the equilibrium shift that will be caused by a volume change.

SAMPLE PROBLEM **13.4** **Determining Direction of Equilibrium Shift with Volume Change**

For each reaction, predict in what direction the equilibrium will shift when the volume of the reaction vessel is decreased.

(a) $PCl_5(g) \rightleftharpoons PCl_3(g) + Cl_2(g)$

(b) $3O_2(g) \rightleftharpoons 2O_3(g)$

(c) $H_2(g) + I_2(g) \rightleftharpoons 2HI(g)$

Strategy Determine which direction minimized the number of moles of gas in the reaction.

Setup We have (a) 1 mole of gas on the reactant side and 2 moles of gas on the product side, (b) 3 moles of gas on the reactant side and 2 moles of gas on the product side, and (c) 2 moles of gas on each side.

Solution

(a) Shift to the left

(b) Shift to the right

(c) No shift

THINK ABOUT IT

When there is no difference in the number of moles of gas, changing the volume of the reaction vessel will change the concentrations of reactant(s) and product(s)—but the system will remain at equilibrium.

Practice Problem ATTEMPT For each reaction, predict the direction of shift caused by increasing the volume of the reaction vessel.

(a) $2NOCl(g) \rightleftharpoons 2NO(g) + Cl_2(g)$

(b) $2O_3(g) \rightleftharpoons 3O_2(g)$

(c) $2H_2(g) + O_2(g) \rightleftharpoons 2H_2O(g)$

Practice Problem BUILD For the following equilibrium, give an example of a stress that will cause a shift to the right, a stress that will cause a shift to the left, and one that will cause no shift:

$$H_2(g) + F_2(g) \rightleftharpoons 2HF(g)$$

Practice Problem CONCEPTUALIZE Consider the reaction $A(g) + B(g) \rightleftharpoons AB(g)$. The diagram shown on the top right represents a system at equilibrium. Which of the following diagrams [(i)–(iii)] best represents the system when equilibrium has been reestablished following a volume increase of 50 percent?

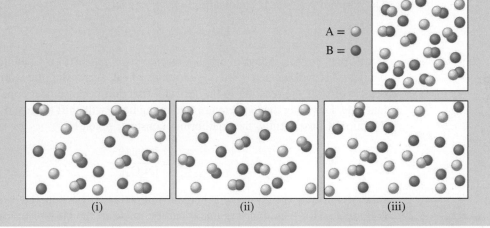

It is possible to change the total pressure of a system without changing its volume—by adding an inert gas such as helium to the reaction vessel. Because the total volume remains the same, the concentrations of reactant and product gases do not change. Therefore, the equilibrium is not disturbed, and no shift will occur.

Animation
Le Châtelier's Principle - Temperature Change

Changes in Temperature

A change in concentration or volume may alter the position of an equilibrium (i.e., the relative amounts of reactants and products), but it does *not* change the value of the

Animation
Le Châtelier's Principle - Volume Change

Figure 13.10 (a) N_2O_4-NO_2 equilibrium. (b) Because the reaction is endothermic, at higher temperature, the $N_2O_4(g) \rightleftharpoons 2NO_2(g)$ equilibrium shifts toward product, making the reaction mixture darker.

(a) Charles D. Winters/Timeframe Photography/McGraw Hill;
(b) Charles D. Winters/McGraw Hill

(a)

(b)

equilibrium constant. Only a change in temperature can alter the value of the equilibrium constant. To understand why, consider the following reaction:

$$N_2O_4(g) \rightleftharpoons 2NO_2(g) \qquad \text{heat of reaction} = 58.0 \text{ kJ}$$

The forward reaction is endothermic (absorbs heat):

$$\text{heat} + N_2O_4(g) \rightleftharpoons 2NO_2(g)$$

If we treat heat as though it were a reactant, we can use Le Châtelier's principle to predict what will happen if we add or remove heat. Increasing the temperature (adding heat) will shift the reaction in the forward direction because heat appears on the reactant side. Lowering the temperature (removing heat) will shift the reaction in the reverse direction. Consequently, the equilibrium constant, given by

$$K = \frac{[NO_2]^2}{[N_2O_4]}$$

increases when the system is heated and decreases when the system is cooled (Figure 13.10). A similar argument can be made in the case of an exothermic reaction, where heat can be considered to be a product. Increasing the temperature of an exothermic reaction causes the equilibrium constant to decrease, shifting the equilibrium toward reactants.

Another example of this phenomenon is the equilibrium between the following ions:

$$CoCl_4^{2-} + 6H_2O \rightleftharpoons Co(H_2O)_6^{2+} + 4Cl^- + \text{heat}$$

The reaction as written, the formation of $Co(H_2O)_6^{2+}$, is exothermic. Thus, the reverse reaction, the formation of $CoCl_4^{2-}$, is endothermic. On heating, the equilibrium shifts to the left and the solution turns blue. Cooling favors the exothermic reaction [the formation of $Co(H_2O)_6^{2+}$] and the solution turns pink (Figure 13.11).

In summary, a temperature increase favors an endothermic reaction, and a temperature decrease favors an exothermic process. Temperature affects the position of an equilibrium by changing the value of the equilibrium constant.

(a) (b) (c)

Figure 13.11 (a) An equilibrium mixture of $COCl_4^{2-}$ ions and $Co(H_2O)_6^{2+}$ ions appears violet. (b) Heating with a Bunsen burner favors the formation of $COCl_4^{2-}$, making the solution look more blue. (c) Cooling with an ice bath favors the formation of $Co(H_2O)_6^{2+}$, making the solution look more pink.

(all) Charles D. Winters/Timeframe Photography/McGraw Hill

CHECKPOINT–SECTION 13.4   Factors That Affect Equilibrium

13.4.1 Which of the following equilibria will shift to the right when H_2 is added? (Select all that apply.)

a) $2H_2 + O_2 \rightleftarrows 2H_2O$

b) $2HI \rightleftarrows H_2 + I_2$

c) $H_2 + CO_2 \rightleftarrows H_2O + CO$

d) $2NaHCO_3 \rightleftarrows Na_2CO_3 + H_2O + CO_2$

e) $2CO_2 + O_2 \rightleftarrows 2CO_2$

13.4.2 Which of the following equilibria will shift to the left when the temperature is increased? (Select all that apply.)

a) $S + H_2 \rightleftarrows H_2S$ (exothermic)

b) $C + H_2O \rightleftarrows CO + H_2$ (endothermic)

c) $H_2 + CO_2 \rightleftarrows H_2O + CO$ (endothermic)

d) $MgO + CO_2 \rightleftarrows MgCO_3$ (exothermic)

e) $2CO + O_2 \rightleftarrows 2CO_2$ (exothermic)

13.4.3 The diagram shows the gaseous reaction $2A \rightleftarrows A_2$ at equilibrium. How will the numbers of A and A_2 change if the volume of the container is increased at constant temperature?

a) A_2 will increase and A will decrease.

b) A will increase and A_2 will decrease.

c) A_2 and A will both decrease.

d) A_2 and A will both increase.

e) Neither A_2 nor A will change.

13.4.4 The diagrams show equilibrium mixtures of A_2, B_2, and AB at two different temperatures ($T_2 > T_1$). Is the reaction $A_2 + B_2 \rightleftarrows 2AB$ endothermic or exothermic?

a) Endothermic

b) Exothermic

c) Neither

d) There is not enough information to determine.

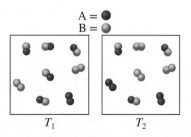

Profiles in Science

Student Note: In Michael Crichton's 1990 novel *Jurassic Park*, frog DNA was used to repair ancient dinosaur DNA to facilitate cloning of the extinct animals. In the story, the scientists believed the cloned animals could not reproduce because the population was designed to be entirely female. However, some of the dinosaurs became males—something known to occur in the frogs from which the DNA for repair had been taken—and the population of dinosaurs grew out of control.

Ruth Erica and Reinhold Benesch

In chemistry, and in biology, when a necessary ingredient for a process is in short supply, nature sometimes responds in a way that compensates for the shortage. For example, in an isolated population of certain types of fish or frogs consisting of a single sex, some of the individuals will change sexes, enabling the population to reproduce. Another example of nature responding to a shortage is the production of more red blood cells in low-oxygen, most often high-altitude, environments. The hemoglobin in red blood cells is responsible for delivering oxygen from the lungs to the rest of the body. Although biological processes such as increased production of hemoglobin in response to low O_2 are not strictly *equilibria* (they are known as *steady states* instead), many of the principles of equilibrium apply in our efforts to describe and understand them.

Ruth Erica Benesch (1925–2000) was evacuated from Nazi Germany in 1939 via *Kindertransport,* an organized rescue program that relocated children in the months leading up to the outbreak of World War II. She completed high school and then earned her bachelor's degree in Britain. Eventually, she and her husband, Reinhold Benesch (1919–1986), both earned Ph.D.s in biochemistry at Northwestern University in Illinois. They spent their careers collaborating to study hemoglobin and to elucidate its function. Together, they published more than 100 scientific papers on the topic and their work transformed scientists' understanding of the oxygen transport system.

Chapter Summary

Section 13.1

- *Reaction rate* refers to how fast a chemical reaction occurs. Reaction rate is influenced by several things, including reactant concentration and temperature. Usually, reaction rate increases as reactant concentration and/or temperature increases.

- The theory that explains how concentration and temperature affect reaction rate is *collision theory.* According to collision theory, reaction rate is related to the frequency with which reactant molecules collide, which is affected by concentration; and to the energy with which they collide, which is affected by temperature. Only collisions sufficiently energetic to overcome the *activation energy* barrier will result in reaction.

- Both endothermic and exothermic reactions have an activation energy that must be overcome for the reaction to proceed.

Section 13.2

- Nearly all physical and chemical processes are *reversible processes,* meaning that both forward and reverse reactions occur.

- A state of *equilibrium* exists when both the forward and reverse reactions are occurring at the same rate, resulting in no net change in reactant or product concentrations over time.

Section 13.3

- Increasing the volume of the container will shift a gaseous equilibrium in the direction of the greater number of gaseous moles. Decreasing the volume will shift the equilibrium in the direction of the smaller number of gaseous moles. In a reaction with no net change in the number of gaseous moles, changing volume will not cause a shift in the equilibrium position.

- The magnitude of the *equilibrium constant (K)* indicates whether products or reactants are favored at equilibrium. A large K value indicates that products are favored; a small K value indicates that reactants are favored. An equilibrium in which products are favored is said to "lie to the right." One in which reactants are favored is said to "lie to the left."

- The *equilibrium expression* can be derived from the balanced chemical equation, and can be used along with equilibrium concentrations to calculate the value of the equilibrium constant.

Section 13.4

- *Le Châtelier's principle* describes how stressing an equilibrium (by adding or removing a substance, or by changing the pressure of a gaseous reactant or product) will cause the equilibrium to shift in a direction that minimizes the effect of the stress. Adding a reactant or removing a product causes a shift to the right. Adding a product or removing a reactant causes a shift to the left.

- Increasing the volume of the container will shift a gaseous equilibrium in the direction of the greater number of gaseous moles. Decreasing the volume will shift the equilibrium in the direction of the smaller number of gaseous moles. In a reaction with no net change in the number of gaseous moles, changing volume will not cause a shift in the equilibrium position.

- Equilibrium constants are constant only at constant temperature. A temperature increase will cause the equilibrium of an endothermic process to shift toward products (to the right), and will cause the equilibrium of an exothermic process to shift toward reactants (to the left).

Key Terms

Activation energy 465

Collision theory 465

Equilibrium 470

Equilibrium constant (K) 472

Equilibrium expression 472

Le Châtelier's principle 477

Rate of reaction 465

Reversible process 470

Questions and Problems

SECTION 13.1: REACTION RATES

13.1 What factors impact the rate of a reaction?

13.2 Explain why increasing the concentration of a reactant will increase the rate of reaction.

13.3 Explain why increasing the temperature of a reaction will usually increase the rate of reaction.

13.4 Consider the reaction between CO and O_2 gas. Describe the difference in the rate of reaction that would be observed in each of the following:

$$2CO(g) + O_2(g) \rightleftharpoons 2CO_2(g)$$

(a) $[O_2] = 0.20\ M$ compared with $0.40\ M$, temperature constant

(b) Increasing temperature from 298 K to 398 K, initial reactant concentration constant

(c) Rate at time = 0 minutes compared with the rate observed 10 minutes later

SECTION 13.2: CHEMICAL EQUILIBRIUM

13.5 How is equilibrium related to reaction rates?

13.6 What is meant by the term *dynamic equilibrium*?

13.7 Give an example of dynamic equilibrium in your everyday life.

13.8 The first diagram represents a system at equilibrium for the reaction A $\rightleftharpoons$ B where A = 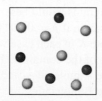 and B = ⬤. How many red spheres must be added to the second diagram for it also to represent a system at equilibrium?

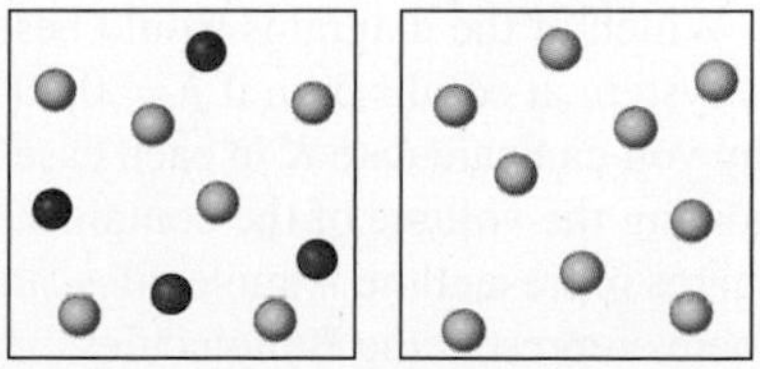

13.9 Which of the following statements is correct about a system at equilibrium?

(a) The concentrations of reactants are equal to the concentrations of products.

(b) The concentrations of reactants must be present in exact stoichiometric amounts.

(c) The rate of product formation is equal to the rate of reactant formation.

13.10 Sketch a graph of [A] and [B] (concentration on the y-axis) as a function of time (x-axis). Use the balanced reaction shown and start with [A] = 2 M and [B] = 0 M. Indicate on the graph where equilibrium has been reached.
$A(aq) \rightleftharpoons 2B(aq)$

SECTION 13.3: EQUILIBRIUM CONSTANTS

13.11 Write an equilibrium expression for each of the following reactions:

(a) $SO_2(g) + O_2(g) \rightleftharpoons SO_3(g)$

(b) $4NH_3(g) + 3O_2(g) \rightleftharpoons 2N_2(g) + 6H_2O(g)$

(c) $2H_2S(g) + SO_2(g) \rightleftharpoons 3S(g) + 2H_2O(g)$

(d) $CO(g) + 2H_2(g) \rightleftharpoons CH_3OH(g)$

13.12 Write an equilibrium expression for each of the following reactions:

(a) $2CO(g) + O_2(g) \rightleftharpoons 2CO_2(g)$

(b) $2N_2H_4(g) + N_2O_4(g) \rightleftharpoons 3N_2(g) + 4H_2O(g)$

(c) $2C_2H_2(g) + 5O_2(g) \rightleftharpoons 4CO_2(g) + 2H_2O(g)$

(d) $CS_2(g) + 3Cl_2(g) \rightleftharpoons CCl_4(g) + S_2Cl_2(g)$

13.13 Write an equilibrium expression for each of the following reactions:

(a) $2Cl_2(g) + O_2(g) \rightleftharpoons 2OCl_2(g)$

(b) $4NO_2(g) + O_2(g) \rightleftharpoons 2N_2O_5(g)$

(c) $2NO(g) + O_2(g) \rightleftharpoons 2NO_2(g)$

(d) $2H_2S(g) + 3O_2(g) \rightleftharpoons 2SO_2(g) + 2H_2O(g)$

13.14   Write an equilibrium expression for each of the following reactions:
(a) $CO(g) + Cl_2(g) \rightleftharpoons COCl(g) + Cl(g)$
(b) $2HI(g) \rightleftharpoons H_2(g) + I_2(g)$
(c) $2O_3(g) \rightleftharpoons 3O_2(g)$
(d) $2HBr(g) + I_2(g) \rightleftharpoons 2HI(g) + Br_2(g)$

13.15   Consider the reaction

$$2NO(g) + 2H_2(g) \rightleftharpoons N_2(g) + 2H_2O(g)$$

At a certain temperature, the equilibrium concentrations are $[NO] = 0.31\ M$, $[H_2] = 0.16\ M$, $[N_2] = 0.082\ M$, and $[H_2O] = 4.64\ M$. (a) Write the equilibrium expression for the reaction. (b) Determine the value of the equilibrium constant.

13.16   Determine the equilibrium constant for the reaction shown, given the following equilibrium concentrations: $[HI] = 0.200\ M$, $[H_2] = 0.028\ M$, and $[I_2] = 0.028\ M$.

$$2HI(g) \rightleftharpoons H_2(g) + I_2(g)$$

13.17   Determine the equilibrium constant for the reaction shown, given the following equilibrium concentrations: $[N_2] = 0.150\ M$, $[H_2] = 0.300\ M$, and $[NH_3] = 0.070\ M$.

$$N_2(g) + 3H_2(g) \rightleftharpoons 2NH_3(g)$$

13.18   Determine the equilibrium constant for the reaction shown, given the following equilibrium concentrations: $[A] = 0.0334\ M$, $[B] = 0.0225\ M$, $[C] = 0.00161\ M$, and $[D] = 0.0511\ M$.

$$2A(aq) + B(aq) \rightleftharpoons C(aq) + 3D(aq)$$

13.19   Determine the equilibrium constant for the reaction shown, given the following equilibrium concentrations: $[SO_2] = 0.0022\ M$, $[O_2] = 0.12\ M$, and $[SO_3] = 1.97\ M$.

$$2SO_2(g) + O_2(g) \rightleftharpoons 2SO_3(g)$$

13.20   Determine the equilibrium constant for the reaction shown, given the following equilibrium concentrations: $[NOCl] = 0.035\ M$, $[NO] = 0.0133\ M$, and $[Cl_2] = 0.0461\ M$.

$$2NOCl(g) \rightleftharpoons 2NO(g) + Cl_2(g)$$

13.21   Determine the equilibrium constant for the reaction shown, given the following equilibrium concentrations: $[CH_4] = 0.098\ M$, $[H_2S] = 0.75\ M$, $[H_2] = 1.08\ M$, and $[CS_2] = 1.54\ M$.

$$CH_4(g) + 2H_2S(g) \rightleftharpoons CS_2(g) + 4H_2(g)$$

13.22   Using the equilibrium constant, determine the equilibrium concentration of HF, given that the equilibrium concentrations of H_2 and F_2 are both $0.0422\ M$.

$$H_2(g) + F_2(g) \rightleftharpoons 2HF(g) \quad K = 2.35 \times 10^3$$

13.23   Using the equilibrium constant, determine the equilibrium concentration of B, given the following equilibrium concentrations: $[A] = 0.0212\ M$, $[C] = 0.0410\ M$, and $[D] = 0.0127\ M$.

$$A(aq) + 3B(aq) \rightleftharpoons$$
$$2C(aq) + D(aq) \quad K = 1.57 \times 10^{-2}$$

13.24   Determine the balanced chemical equation that corresponds to each equilibrium expression.

(a) $K = \dfrac{[NO]^4[H_2O]^6}{[NH_3]^4[O_2]^5}$

(b) $K = \dfrac{[SO_3]^2}{[SO_2]^2[O_2]}$

(c) $K = \dfrac{[IBr]^2}{[I_2][Br_2]}$

13.25   Determine the balanced chemical equation that corresponds to each equilibrium expression.

(a) $K = \dfrac{[H_2]^4[CS_2]}{[H_2S]^2[CH_4]}$

(b) $K = \dfrac{[H_2O]^2[Cl_2]^2}{[HCl]^4[O_2]}$

(c) $K = \dfrac{[NO_2]^2}{[N_2O_4]}$

13.26   Consider the following equilibrium process at 700°C:

$$2H_2(g) + S_2(g) \rightleftharpoons 2H_2S(g)$$

Analysis shows that there are 2.50 moles of H_2, 1.35×10^{-5} mole of S_2, and 8.70 moles of H_2S present in a 12.0-L flask at equilibrium. Calculate the equilibrium constant K for the reaction.

13.27   The equilibrium constant for the reaction

$$2SO_2(g) + O_2(g) \rightleftharpoons 2SO_3(g)$$

is 2.8×10^2 at a particular temperature. If $[SO_2] = 0.0124\ M$ and $[O_2] = 0.031\ M$, what is $[SO_3]$?

13.28   The equilibrium constant for the reaction

$$2H_2(g) + CO(g) \rightleftharpoons CH_3OH(g)$$

is 1.6×10^{-2} at a particular temperature. A 5.60-L flask at equilibrium contains 1.17×10^{-2} mol H_2 and 3.46×10^{-3} mol CH_3OH. Determine the concentration of CO in the flask.

13.29   The equilibrium constant for the reaction $A \rightleftharpoons B$ is $K = 10$ at a particular temperature. (1) Starting with only reactant A, which of the diagrams shown here best represents the system at equilibrium? (2) Which of the diagrams would best represent the system at equilibrium if $K = 0.10$? Explain why you can calculate K in each case without knowing the volume of the container. The grey spheres represent the A molecules, and the green spheres represent the B molecules.

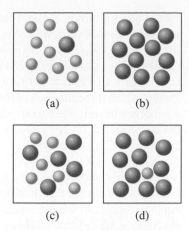

(a)   (b)

(c)   (d)

13.30 The following diagrams represent the equilibrium state for three different reactions of the type

$$A + X \rightleftharpoons AX \ (X = B, C, or \ D)$$

A + B ⇌ AB   A + C ⇌ AC   A + D ⇌ AD

(a) Which reaction has the largest equilibrium constant?

(b) Which reaction has the smallest equilibrium constant?

13.31 Consider the reaction A + B ⇌ 2C. The following top diagram represents a system at equilibrium where A = ⬤, B = ⬤, and C = ⬤. Which of the following diagrams [(a)–(d)] also represents a system at equilibrium? Select all that apply.

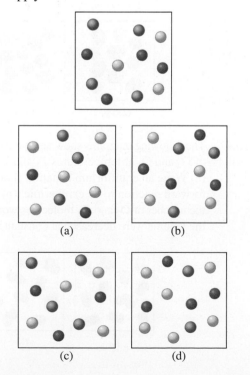

(a)   (b)

(c)   (d)

13.32 Consider the reaction 2D + 2E ⇌ F. The top diagram represents a system at equilibrium where D = ⬤, E = ⬤, and F = ⬤. Which of the other diagrams [(a)–(d)] also represents a system at equilibrium?

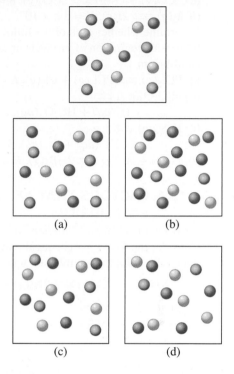

(a)   (b)

(c)   (d)

13.33 Consider the reaction 2X + Y ⇌ 2Z. The top diagram represents a system at equilibrium where X = ⬤, Y = ⬤, and Z = ⬤. Which of the other diagrams [(a)–(d)] also represents a system at equilibrium?

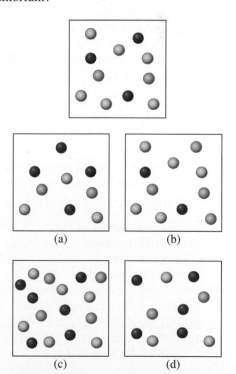

(a)   (b)

(c)   (d)

13.34 Determine whether each of the following reactions has a higher concentration of products or reactants at equilibrium.
(a) $2SO_2(g) + O_2(g) \rightleftharpoons 2SO_3(g)$, $K = 8.0 \times 10^{35}$
(b) $2HCl(g) \rightleftharpoons H_2(g) + Cl_2(g)$, $K = 4.2 \times 10^{-34}$
(c) $2CO(g) + O_2(g) \rightleftharpoons 2CO_2(g)$, $K = 2.24 \times 10^{22}$
(d) $I_2(g) \rightleftharpoons 2I(g)$, $K = 3.8 \times 10^{-5}$

13.35 Determine whether each of the following has a higher concentration of products or reactants at equilibrium.
(a) $PCl_5(g) \rightleftharpoons PCl_3(g) + Cl_2(g)$, $K = 12.6$
(b) $H_2C_2O_4(aq) \rightleftharpoons$
$$H^+(aq) + HC_2O_4^-(aq), K = 6.5 \times 10^{-2}$$
(c) $H_2S(aq) \rightleftharpoons$
$$2H^+(aq) + S^{2-}(aq), K = 9.5 \times 10^{-27}$$
(d) $H_2(g) + Br_2(g) \rightleftharpoons 2HBr(g)$, $K = 2.2 \times 10^6$

SECTION 13.4: FACTORS THAT AFFECT EQUILIBRIUM

13.36 Describe the term *Le Châtelier's principle*.

13.37 Consider the following chemical reaction. Predict which direction the equilibrium will shift when:

$$4NH_3(g) + 3O_2(g) \rightleftharpoons 2N_2(g) + 6H_2O(g)$$

(a) more NH_3 is added.
(b) more N_2 is added.
(c) water is removed.
(d) oxygen is removed.

13.38 Consider the following chemical reaction. Predict which direction the equilibrium will shift when:

$$2NO_2(g) \rightleftharpoons 2NO(g) + O_2(g)$$

(a) NO_2 is added.
(b) oxygen is added.
(c) NO is removed.
(d) NO_2 is removed.

13.39 Consider the following chemical reaction. Predict which direction the equilibrium will shift when:

$$CS_2(g) + 4H_2(g) \rightleftharpoons CH_4(g) + 2H_2S(g)$$

(a) hydrogen gas is added.
(b) more CH_4 is added.
(c) CS_2 is removed.
(d) H_2S is removed.

13.40 Consider the following chemical reaction. Predict which direction the equilibrium will shift when:

$$4HCl(g) + O_2(g) \rightleftharpoons 2Cl_2(g) + 2H_2O(g)$$

(a) HCl is removed.
(b) more O_2 is added.
(c) more chlorine gas is added.
(d) water is removed.

13.41 Predict which direction, if either, each of the following equilibria will shift when volume is reduced.
(a) $PCl_3(g) + 3NH_3(g) \rightleftharpoons P(NH_2)_3(g) + 3HCl(g)$
(b) $2NO(g) + Br_2(g) \rightleftharpoons 2NOBr(g)$
(c) $2HI(g) \rightleftharpoons H_2(g) + I_2(g)$

13.42 Predict which direction, if either, each of the following equilibria will shift when volume is increased.
(a) $2NO(g) + 2H_2(g) \rightleftharpoons N_2(g) + 2H_2O(g)$
(b) $2NClO_2(g) \rightleftharpoons Cl_2(g) + 2NO_2(g)$
(c) $2H_2(g) + CO(g) \rightleftharpoons CH_3OH(g)$

13.43 Predict which direction each of the following equilibria will shift when temperature is increased.
(a) $2H_2O(g) \rightleftharpoons 2H_2(g) + O_2(g)$ (endothermic)
(b) $H_2(g) + Cl_2(g) \rightleftharpoons 2HCl(g)$ (exothermic)
(c) $2C_2H_6(g) + 7O_2(g) \rightleftharpoons$
$$4CO_2(g) + 6H_2O(g)$$ (exothermic)

13.44 Predict which direction each of the following equilibria will shift when temperature is decreased.
(a) $N_2(g) + 3H_2(g) \rightleftharpoons 2NH_3(g)$ (exothermic)
(b) $CH_4(g) + H_2O(g) \rightleftharpoons$
$$CO(g) + 3H_2(g)$$ (endothermic)
(c) $NO(g) + O_3(g) \rightleftharpoons$
$$NO_2(g) + O_2(g)$$ (exothermic)

13.45 The following diagram represents a gas-phase equilibrium mixture for the exothermic reaction $AB \rightleftharpoons A + B$ at a certain temperature. Describe what would happen to the system after each of the following changes: (a) the temperature is decreased, (b) the volume is increased.

13.46 The following diagrams show the reaction $A + B \rightleftharpoons AB$ at two different temperatures. Is the forward reaction endothermic or exothermic?

600 K   800 K

13.47 The following diagrams show an equilibrium mixture of O_2 and O_3 at temperatures T_1 and T_2 ($T_2 > T_1$). (a) Write an equilibrium equation showing the forward reaction to be exothermic. (b) Predict how the number of O_2 and O_3 molecules would change if the volume were decreased at constant temperature.

T_1 T_2

Answers to In-Chapter Materials

Answers to Practice Problems

13.1A (a) $K = \dfrac{[N_2]^2[O_2]}{[N_2O]^2}$, (b) $K = \dfrac{[NO]^2[Br_2]}{[NOBr]^2}$, (c) $K = \dfrac{[H^+][F^-]}{[HF]}$,

(d) $K = \dfrac{[CO_2][H_2]}{[CO][H_2O]}$, (e) $K = \dfrac{[CS_2][H_2]^4}{[CH_4][H_2S]^2}$, (f) $K = \dfrac{[H^+]^2[C_2O_4^{2-}]}{[H_2C_2O_4]}$.

13.1B (a) $H_2(g) + Cl_2(g) \rightleftharpoons 2HCl(g)$,
(b) $H^+(aq) + F^-(aq) \rightleftharpoons HF(aq)$,
(c) $Cr^{3+}(aq)\ 4OH^-(aq) \rightleftharpoons Cr(OH)_4^-(aq)$,
(d) $HClO(aq) \rightleftharpoons H^+(aq) + ClO^-(aq)$,
(e) $H_2SO_3(aq) \rightleftharpoons H^+(aq)\ HSO_3^-(aq)$,
(f) $2NO(g) + Br_2(g) \rightleftharpoons 2NOBr(g)$.

13.2A $K = \dfrac{[BrCl]^2}{[Br_2][Cl_2]} = \dfrac{(1.4 \times 10^{-2})^2}{(2.3 \times 10^{-3})(1.2 \times 10^{-2})} = 7.1$

13.2B $1.6 \times 10^{-2}\,M$. **13.3A** (a) right, (b) left, (c) right, (d) left.

13.3B (a) $Ag(NH_3)_2^+$, (b) Ag^+ and/or NH_3, (c) Ag^+ and/or NH_3.

13.4A (a) right, (b) right, (c) left. **13.4B** Right: addition of H_2 or F_2 or removal of HF, Left: removal of H_2 and/or F_2 or addition of HF, Neither: volume change.

Answers to Checkpoints

13.3.1 e. **13.3.2** a. **13.4.1** a, c. **13.4.2** a, d, e. **13.4.3** b. **13.4.4** b.

CHAPTER 14

Organic Chemistry

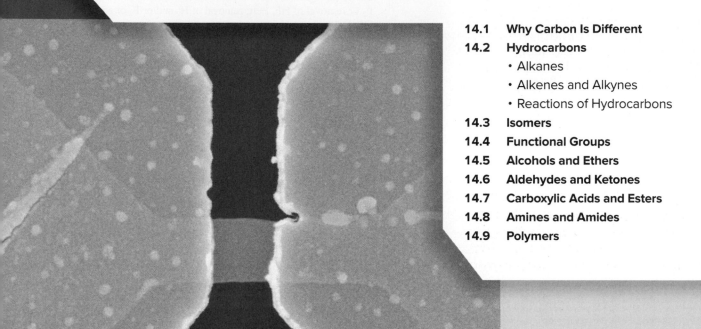

14.1 **Why Carbon Is Different**

14.2 **Hydrocarbons**
- Alkanes
- Alkenes and Alkynes
- Reactions of Hydrocarbons

14.3 **Isomers**

14.4 **Functional Groups**

14.5 **Alcohols and Ethers**

14.6 **Aldehydes and Ketones**

14.7 **Carboxylic Acids and Esters**

14.8 **Amines and Amides**

14.9 **Polymers**

An atom-thick sheet of carbon known as graphene connects two sides of a transistor. Graphene has shown tremendous promise in electronic applications because of its high strength, flexibility, and conductivity.

Andre Geim & Kostya Novoselov/Science Source

In This Chapter, You Will Learn

About the unique chemistry of substances that contain carbon.

Things To Review Before You Begin

- Lewis structures [◀◀ Section 6.1]
- Molecular shape [◀◀ Section 6.4]
- Chapter 6 Key Skills [◀◀ pages 236–237]

Organic chemistry is usually defined as the study of compounds that contain the element *carbon*. This definition is not entirely satisfactory, though, because it would include cyanides and metal carbonates, such as sodium cyanide (NaCN) and barium carbonate ($BaCO_3$)—which are considered to be *inorganic* compounds. A somewhat more useful definition is the study of compounds that contain carbon and hydrogen, although some organic compounds do not contain hydrogen, and many contain other elements, such as oxygen, sulfur, nitrogen, phosphorus, or halogens. Examples of organic compounds are:

CH_4	C_2H_5OH	$C_5H_7O_4COOH$	CH_3NH_2	CCl_4
Methane	Ethanol	Ascorbic acid	Methylamine	Carbon tetrachloride

Early in the study of organic chemistry there was thought to be some fundamental difference between compounds that came from living things, such as plants and animals, and those that came from nonliving things, such as rocks. Compounds obtained from plants or animals were called *organic,* whereas compounds obtained from nonliving sources were called *inorganic.* In fact, until early in the nineteenth century, scientists believed that only nature could produce organic compounds. In 1829, however, Friedrich Wöhler prepared urea, a well-known organic compound, by combining the inorganic substances lead cyanate and aqueous ammonia:

$$Pb(OCN)_2 + 2NH_3 + 2H_2O \xrightarrow{\text{heat}} 2NH_2CONH_2 + Pb(OH)_2$$
$$\text{Urea}$$

Wöhler's synthesis of urea dispelled the notion that organic compounds were fundamentally different from inorganic compounds—and that they could only be produced by nature. We now know that it is possible to synthesize a wide variety of organic compounds in the laboratory; in fact, many thousands of new organic compounds are produced in research laboratories each year. In this chapter, we learn about several types of organic compounds that are important biologically.

14.1 Why Carbon Is Different

Carbon's position in the periodic table (Group 14, Period 2) gives it some unique characteristics that enable it to form millions of different compounds.

- The electron configuration of carbon ($[He]2s^22p^2$) effectively prohibits ion formation. This and carbon's electronegativity [◀◀ Section 6.5], which is intermediate between that of metals and nonmetals, cause carbon to complete its octet by *sharing* electrons. In nearly all of its compounds, carbon forms four covalent bonds, which can be oriented in as many as four different directions.

Student Note: To form an ion that is isoelectronic [◀◀ Section 2.7] with a noble gas, a C atom would have to either gain or lose *four* electrons—something that is energetically impossible under ordinary conditions.

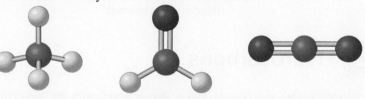

Methane   Formaldehyde   Carbon dioxide

- Carbon's small atomic radius allows the atoms to approach one another very closely, giving rise to short, *strong,* carbon-carbon bonds and *stable* carbon compounds.
- The short distances between carbon atoms also facilitate the formation of double and triple bonds.

These attributes enable carbon to form chains (straight, branched, and *cyclic*) containing single, double, and triple carbon-carbon bonds.

One of the most important cyclic organic compounds is benzene, C_6H_6, which can be represented with two different resonance structures [◄◄ Section 6.3]:

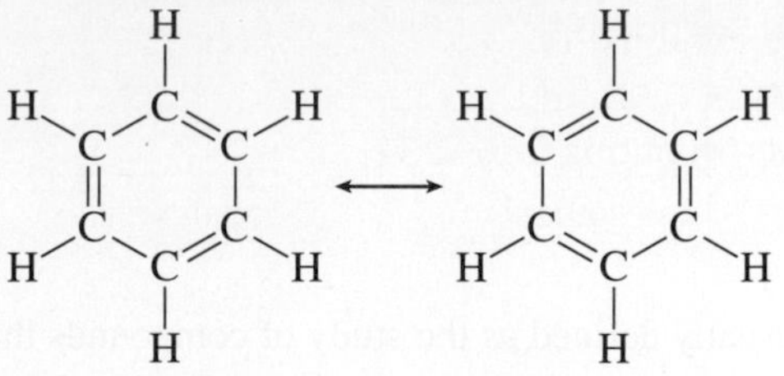

These two structures can also be represented with the single structure:

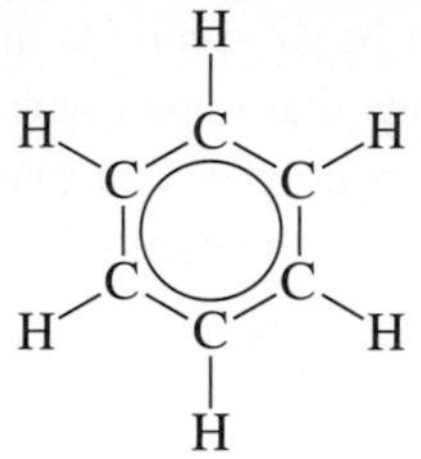

where the circle represents the alternating single and double bonds depicted in the two resonance structures. (This structure is more realistic in terms of its representation of the C—C bonds in benzene, which are all identical—rather than some being single and others being double.)

The benzene ring occurs commonly as part of other molecules, one of which (benzoic acid) we encountered in Chapter 12. Benzene itself and molecules that contain the benzene ring are called *aromatic* compounds. Aromatic compounds were so named originally because of their characteristic *aromas.* Only later were their structures determined and found to have the benzene ring in common. Organic molecules that do not contain the benzene ring are called *aliphatic* compounds.

> Benzoic acid is one of the weak acids listed in Table 12.6.

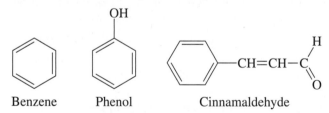

| Benzene | Phenol | Cinnamaldehyde |

Aromatic compounds

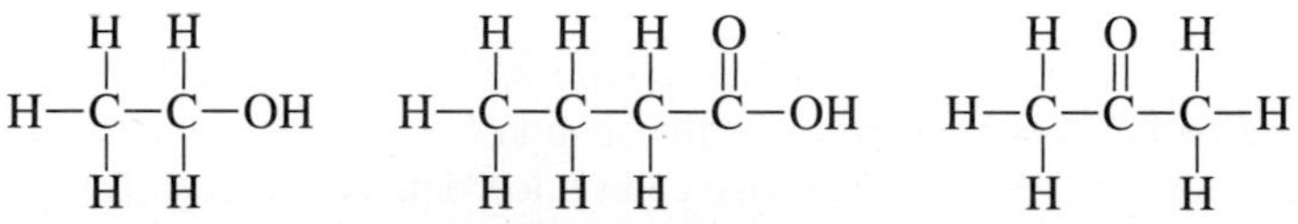

| Ethanol | Butyric acid | Acetone |

Aliphatic compounds

14.2 Hydrocarbons

The simplest organic compounds, those that contain only the elements carbon and hydrogen, are called **hydrocarbons.** Because of carbon's ability to form single, double, and triple bonds, hydrocarbons take many different forms, each with characteristic structural features and chemical properties. In this section, we look at three types of simple, aliphatic hydrocarbons, and some of the reactions they can undergo.

Alkanes

A hydrocarbon that contains only single carbon-carbon bonds is called an *alkane.* Four of the most common and familiar alkanes are *methane*—the principal component of natural gas; *ethane*—a minor component of natural gas; *propane*—the fuel used in gas grills; and *butane*—the fuel that is used in disposable lighters.

Student Note: Under ordinary conditions, both propane and butane are gases. They are both routinely compressed to *liquids* for use as fuels.

CH_4	C_2H_6	C_3H_8	C_4H_{10}
methane	ethane	propane	butane

Comstock Images/Alamy Stock Photo   mphillips007/iStockphoto/Getty Images   Ingram Publishing/SuperStock

The names of simple alkanes such as these depend on the number of carbon atoms they contain. Table 14.1 gives the molecular formulas, names, and structural formulas of the first 10 straight-chain alkanes. The prefix "*n*" stands for *normal,* and indicates that the carbons in a molecule are arranged in a straight, unbranched chain.

Note that an alkane that has *n* carbon atoms has $2n + 2$ hydrogen atoms. Simple alkanes have the general formula:

$$C_nH_{2n+2}$$

Alkanes are known as *saturated* hydrocarbons—meaning that they contain no multiple bonds and that they have the highest possible ratio of hydrogen to carbon atoms. You may have heard the term *saturated* used in the context of dietary fats. Saturated fats are often animal in origin and are typically solid at room temperature, whereas *unsaturated* fats are generally vegetable products and are commonly liquid at room temperature.

Alkenes and Alkynes

A hydrocarbon that contains a carbon-carbon double bond is called an *alkene,* and one that contains a carbon-carbon triple bond is called an *alkyne.* Simple alkenes have the general formula:

$$C_nH_{2n}$$

and simple alkynes have the general formula:

$$C_nH_{2n-2}$$

The most familiar alkene is *ethene,* also known commonly as *ethylene,* which is used commercially to accelerate the ripening of certain types of fruit, including bananas, mangoes, and papayas:

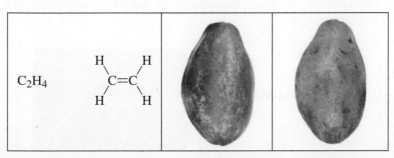

(both) David A. Tietz/McGraw Hill

TABLE 14.1	Formulas, Names, and Structures of Some Simple Alkanes	
Formula	Name	Structure
CH_4	Methane	H H—C—H H
C_2H_6	Ethane	H H H—C—C—H H H
C_3H_8	Propane	H H H H—C—C—C—H H H H
C_4H_{10}	Butane	H H H H H—C—C—C—C—H H H H H
C_5H_{12}	Pentane	H H H H H H—C—C—C—C—C—H H H H H H
C_6H_{14}	Hexane	H H H H H H H—C—C—C—C—C—C—H H H H H H H
C_7H_{16}	Heptane	H H H H H H H H—C—C—C—C—C—C—C—H H H H H H H H
C_8H_{18}	Octane	H H H H H H H H H—C—C—C—C—C—C—C—C—H H H H H H H H H
C_9H_{20}	Nonane	H H H H H H H H H H—C—C—C—C—C—C—C—C—C—H H H H H H H H H H
$C_{10}H_{22}$	Decane	H H H H H H H H H H H—C—C—C—C—C—C—C—C—C—C—H H H H H H H H H H H

The most familiar alkyne is *ethyne,* more commonly known as *acetylene,* which is used as fuel in welding torches:

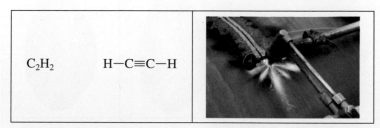

C_2H_2 H—C≡C—H

Alkenes and alkynes are **unsaturated** hydrocarbons. Unsaturated hydrocarbons have a lower hydrogen-to-carbon ratio than saturated hydrocarbons.

Sample Problem 14.1 lets you practice classifying hydrocarbons using their chemical formulas.

SAMPLE PROBLEM (14.1) Classifying Hydrocarbons Using Their Chemical Formulas

Determine whether each of the following hydrocarbons is an alkane, alkene, or alkyne: (a) C_4H_{10}, (b) C_5H_{10}, (c) C_8H_{14}, and (d) C_6H_{12}.

Strategy Use the carbon-to-hydrogen ratio to determine if the compound is an alkane, alkene, or alkyne.

Setup Alkenes have the general formula C_nH_{2n+2}, alkenes C_nH_{2n}, and alkynes C_nH_{2n-2}.

Solution (a) C_4H_{10} fits the formula, C_nH_{2n+2}, for alkanes. $n = 4$, $2n + 2 = 10$

(b) C_5H_{10} fits the formula, C_nH_{2n}, for alkenes. $n = 5$, $2n = 10$

(c) C_8H_{14} fits the formula, C_nH_{2n-2}, for alkynes. $n = 8$, $2n - 2 = 14$

(d) C_6H_{12} fits the formula, C_nH_{2n}, for alkenes. $n = 6$, $2n = 12$

THINK ABOUT IT

As the carbon:hydrogen ratio gets larger, the hydrocarbon becomes less saturated.

Practice Problem ATTEMPT Determine whether each of the following hydrocarbons could be an alkane, an alkene, or an alkyne: (a) C_6H_{10}, (b) C_5H_{12}, (c) C_8H_{16}, and (d) C_5H_8.

Practice Problem BUILD A given hydrocarbon contains 18 hydrogen atoms. Determine the number of carbons that the hydrocarbon would contain to make it an (a) alkane, (b) alkene, and (c) alkyne.

Practice Problem CONCEPTUALIZE Isoprene is a hydrocarbon that is emitted by many trees and plants, and found in human breath. It contains two C—C double bonds. Determine the general formula for a hydrocarbon that contains two double bonds.

Reactions of Hydrocarbons

We saw several examples in Chapter 10 of hydrocarbons combining with oxygen to produce carbon dioxide, water vapor, and heat—a reaction called *combustion*. Highly exothermic combustion reactions of hydrocarbons are what make fossil fuels so integral a part of our lives. Fossil fuels, including coal, natural gas, oil, and gasoline, consist almost entirely of carbon and hydrocarbons—and the burning of fossil fuels accounts for the vast majority of energy production in the United States. The chemical equation representing combustion of any hydrocarbon is

$$C_nH_m + \text{excess } O_2 \longrightarrow nCO_2 + \frac{m}{2}H_2O$$

- For propane (C_3H_8), where $n = 3$ and $m = 8$:

$$C_3H_8 + \text{excess } O_2 \longrightarrow 3CO_2 + 4H_2O$$

- For n-pentane (C_5H_{12}), where $n = 5$ and $m = 12$:

$$C_5H_{12} + \text{excess } O_2 \longrightarrow 5CO_2 + 6H_2O$$

- For benzene (C_6H_6), where $n = 6$ and $m = 6$:

$$C_6H_6 + \text{excess } O_2 \longrightarrow 6CO_2 + 3H_2O$$

- And for ethyne (acetylene, C_2H_2), where $n = 2$ and $m = 2$:

$$C_2H_2 + \text{excess } O_2 \longrightarrow 2CO_2 + H_2O$$

In addition to combustion, hydrocarbons can react with halogens, although the reactions of saturated hydrocarbons differ somewhat from those of unsaturated ones. Alkanes react with halogens to produce *substituted* alkanes, meaning that one or more hydrogen atoms has been *replaced* by a halogen atom. Unlike combustion, these *substitution* reactions require the *input* of energy, in the form of either heat or light. Examples include the chlorination of methane:

$$CH_4 + Cl_2 \longrightarrow CH_3Cl + HCl$$

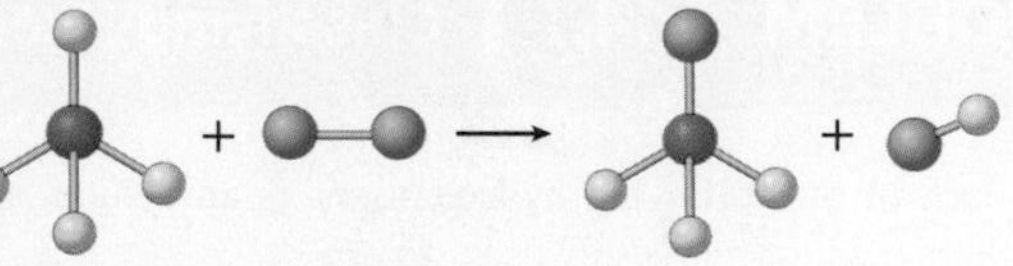

and the bromination of ethane:

In each case, and depending on conditions, it is possible for more than one hydrogen atom to be replaced by a halogen atom—which can result in a variety of products.

When an *unsaturated* hydrocarbon reacts with a halogen, the reaction is an *addition* rather than a substitution. For example, when ethene (ethylene, C_2H_4) reacts with Cl_2, the result is the addition of a chlorine atom to each carbon atom:

$$C_2H_4 + Cl_2 \longrightarrow C_2H_4Cl_2$$

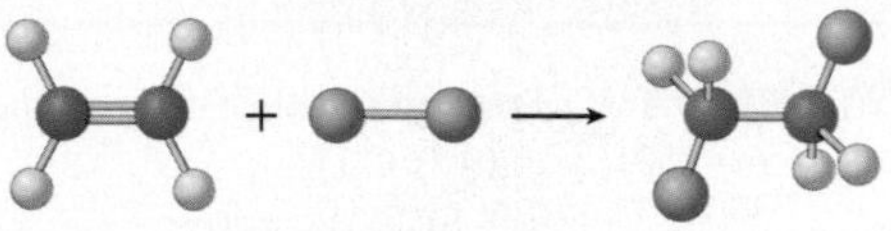

This is known as addition "across the double bond," and it converts an unsaturated hydrocarbon to a substituted, *saturated* hydrocarbon.

Unsaturated hydrocarbons can also react with hydrogen (H_2) in an addition reaction specifically known as **hydrogenation**. Hydrogen also adds across the double bond, converting an *unsaturated* hydrocarbon to a *saturated* one.

$$C_2H_4 + H_2 \longrightarrow C_2H_6$$

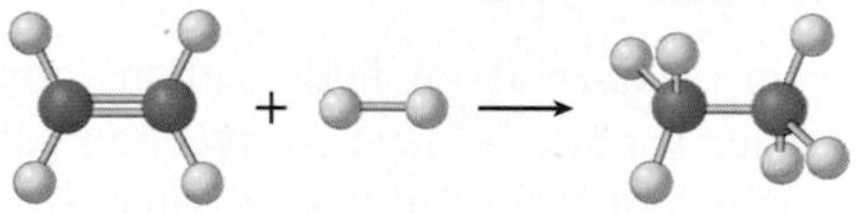

Hydrogenation of ethyne (acetylene, C_2H_2) to produce ethane requires two moles of H_2 for every mole of acetylene:

$$C_2H_2 + 2H_2 \longrightarrow C_2H_6$$

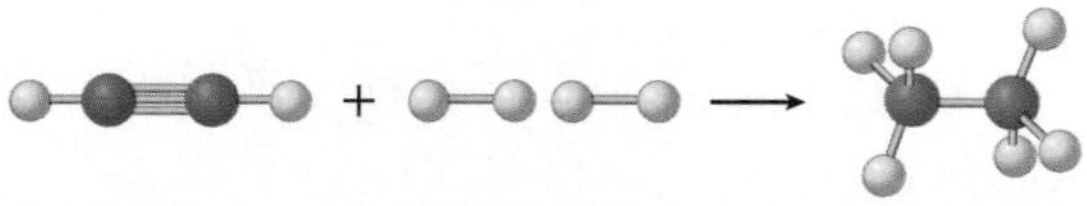

14.3 Isomers

Because of the versatility of carbon, and the number of different ways it can satisfy its four-bond requirement, the molecular formulas of most hydrocarbons can refer to multiple different compounds. For example, the formula C_4H_{10} may refer to *n*-butane:

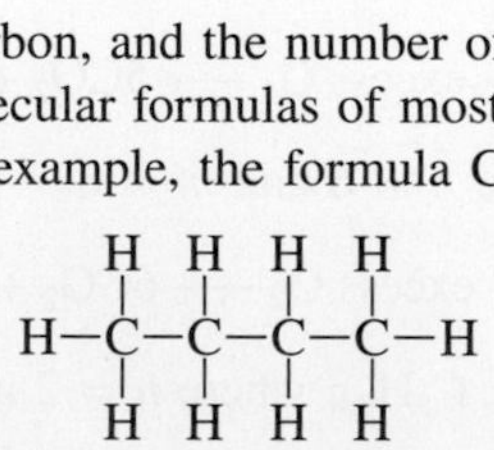

or it may refer to isobutane:

Each of these molecules has the same number of carbon atoms (4), the same number of hydrogen atoms (10), the same number of carbon-carbon bonds (3), and the same number of carbon-hydrogen bonds (10). However, these two compounds have distinct properties, including different melting points and boiling points—simply because the arrangements of atoms and bonds in the two molecules are different. Different structures of molecules with the same chemical formula are known as **structural isomers**.

As the number of carbon atoms in the formula of a hydrocarbon increases, so does the number of possible isomers. For example, as we have just seen, there are two isomers of C_4H_{10}. There are *three* isomers of C_5H_{12}; *five* isomers of C_6H_{14}; and *nine* isomers of C_7H_{16}. When we get to the alkane formula with 15 carbon atoms, $C_{15}H_{32}$, there are more than *four thousand* isomers!

In the case of an unsaturated hydrocarbon, different isomers can arise from different placements of the multiple bond. For example, an alkene with molecular formula C_4H_8 may be either 1-butene or 2-butene, depending on whether the double bond is between the first and second carbons or between the second and third carbons:

1-butene   2-butene

Again, although their molecular formulas are the same, because they are different isomers, these two compounds have distinct properties. Compounds such as 2-butene also exhibit another type of isomerism, in which the geometrical arrangement differs around the double bond.

cis-2-butene   *trans*-2-butene

The molecules *cis*-2-butene and *trans*-2-butene are known as **geometrical isomers**.

> **Student Note:** Structural isomers are also known as *constitutional isomers*.

Familiar Chemistry

Partially Hydrogenated Vegetable Oils

Most likely, you have seen and heard some of the bad press surrounding *trans fats*. Trans fats have been implicated in increasing blood levels of low-density lipoproteins, also known as "bad cholesterol," and are therefore believed by much of the medical community to increase the risk of heart disease. There are relatively few natural sources of trans fats, though, so where do they come from—and how do they get into the food supply?

Fats are used in the production and processing of an enormous percentage of the food available for purchase in the United States. There are two kinds of fats available for this purpose: animal and vegetable. Animal fats typically are *saturated*, meaning that the carbon chains that make up their molecules contain no carbon-carbon double bonds. Vegetable fats typically are *unsaturated*, meaning that their molecules contain at least *some* double

bonds. Although vegetable oils are believed to be a healthier option, they generally have shorter shelf lives and can produce inconsistent results in food texture and flavor.

A popular food-industry solution to this problem has been the partial hydrogenation of vegetable oils to render them solid, thus increasing their resemblance to saturated fats in terms of consistency, flavor, and shelf life. Hydrogenation is the addition of hydrogen to molecules that contain carbon-carbon double bonds, rendering the bonds *single*—and increasing the ratio of hydrogen to carbon. As the name suggests, *partial* hydrogenation adds hydrogen to some, but not all, of the carbon-carbon double bonds in an oil. However, the conditions required to make partial hydrogenation happen also cause many of the remaining double bonds in an oil to convert from the naturally occurring *cis* to the more stable *trans* isomer—thus producing *trans fats*.

The FDA has declared trans fats to be deleterious to human health and has announced its intention to have them phased out of processed foods entirely—a move that will purportedly save thousands of lives each year.

Sample Problem 14.2 lets you practice drawing isomers of hydrocarbons.

SAMPLE PROBLEM 14.2 Drawing Isomers of Hydrocarbons

Draw the isomers of (a) pentane and (b) pentene.

Strategy Determine the number of carbons in each substance from the prefix. The prefix *pent–* indicates five carbon atoms are present. The *–ane* ending indicates only carbon-carbon single bonds are present, whereas the *–ene* ending indicates there is a carbon-carbon double bond.

Setup It is helpful to only draw the carbon "skeleton" of the molecules first, leaving the hydrogen atoms off. Draw the long straight chain first and then take one carbon off the end and attach it to another carbon on the chain to create the next isomer. In the last step, add enough hydrogen atoms to each carbon to give it four bonds.

(a)

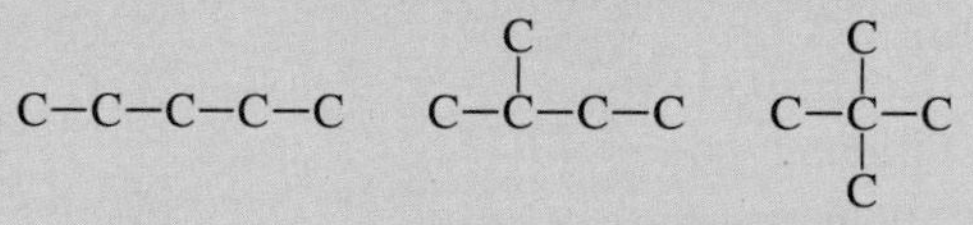

(b) In this molecule, the double bond can be moved as well.

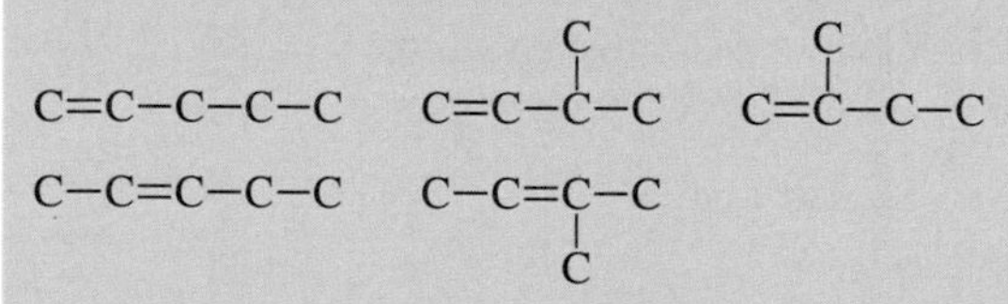

Solution

(a)

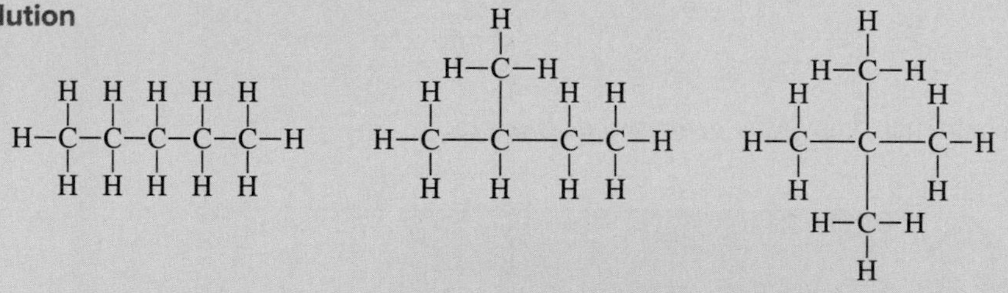

(b)

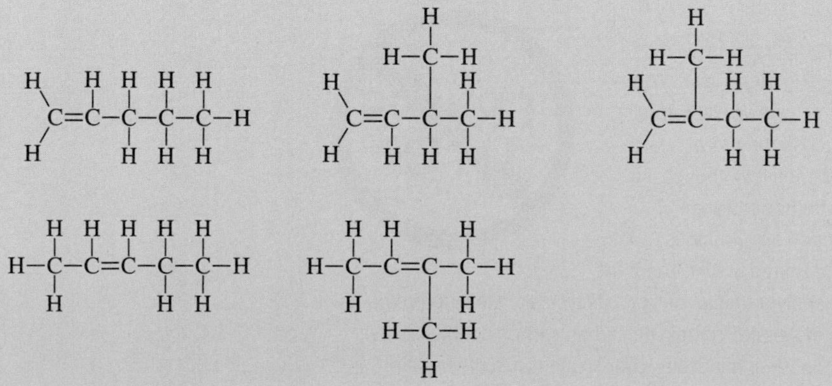

THINK ABOUT IT

Note that all of the following structures are the same, as you can simply "flip" them over and make them match the second structure shown in the setup for part (a).

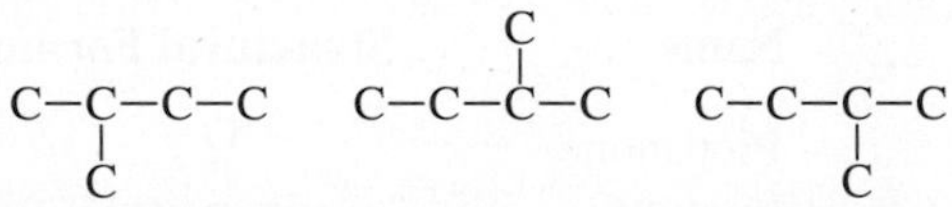

Practice Problem **A**TTEMPT Draw the isomers of (a) hexane and (b) hexyne.

Practice Problem **B**UILD Determine whether or not the alkene with the formula C_4H_8 can have any isomers in addition to 1-butene and 2-butene (shown above the Familiar Chemistry box). Draw any additional isomers.

Practice Problem **C**ONCEPTUALIZE Identify which of the following pairs of molecules are isomers of one another. Only the skeletons are shown; the hydrogen atoms have been left off to simplify the drawings.

(a) C—C—C—C—C—C—C

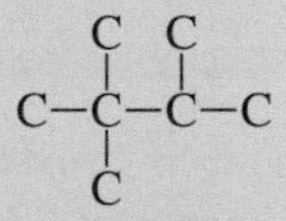

                         C C
                         | |
                      C—C—C—C
                           |
                           C

(b) 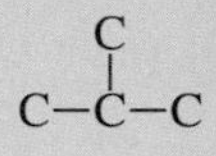 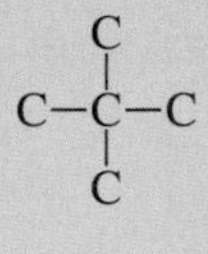 C—C—C (with a C branch above the middle C)

                         C
                         |
                     C—C—C
                         |
                         C

(c) C—C—C—C (with a C branch above the third C)

                         C
                         |
                     C—C=C—C

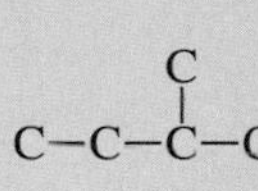

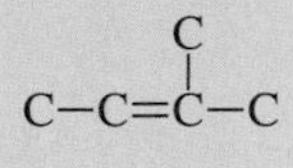

Thinking Outside the Box

Representing Organic Molecules with Bond-Line Structures

Bond-line structures are especially useful for representing complex organic molecules. A ***bond-line structure*** consists of straight lines that represent carbon-carbon bonds. The carbon atoms themselves (and the attached hydrogen atoms) are not shown, but you need to know that they are there.

The structural formulas and bond-line structures for several hydrocarbons are as follows:

Name	Structural Formula	Bond-Line Structure
Pentane	$CH_3CH_2CH_2CH_2CH_3$	 (bond-line zigzag)
Isopentane	$CH_3CHCH_2CH_3$ with CH_3 above	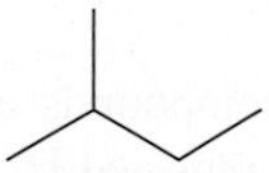 (bond-line branched)
Neopentane	CH_3CCH_3 with CH_3 above and CH_3 below	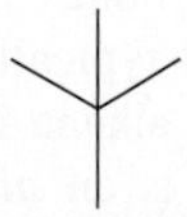 (bond-line branched)

The end of each straight line in a bond-line structure corresponds to a carbon atom (unless another atom of a different type is explicitly shown at the end of the line). Additionally, there are as many hydrogen atoms attached to each carbon atom as are necessary to give each carbon atom a total of four bonds.

When a molecule contains an element *other* than carbon or hydrogen, those atoms, called **heteroatoms,** are shown explicitly in the bond-line structure. Furthermore, while the hydrogens attached to carbon atoms typically are not shown, hydrogens attached to heteroatoms *are* shown, as illustrated by the molecule ethylamine in the following:

Name	Structural Formula	Bond-Line Structure
Propanone (acetone)	$CH_3\overset{\displaystyle O}{\overset{\|}{C}}CH_3$	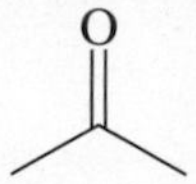
Ethylamine	$CH_3CH_2NH_2$	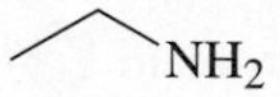
Tetrahydrofuran	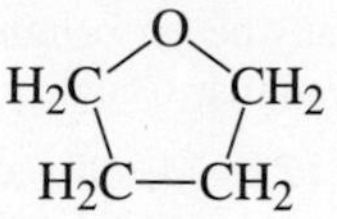 $H_2C\overset{\displaystyle O}{\diagup}\,CH_2$ $H_2C\!-\!CH_2$	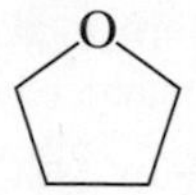

Although carbon and hydrogen atoms need not be shown in a bond-line structure, some of the C and H atoms *can* be shown for the purpose of emphasizing a particular part of a molecule. Often when we show a molecule, we choose to emphasize the *functional group(s),* which are largely responsible for the properties and reactivity of the compound.

Consider the following examples:

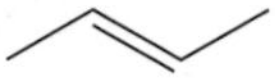

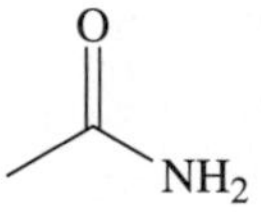

We can determine the molecular formulas as follows: Each line represents a bond. Double lines represent double bonds. We count one C atom at the end of each line unless another atom is shown there. Finally, we count the number of H atoms necessary to complete the octet of each C atom.

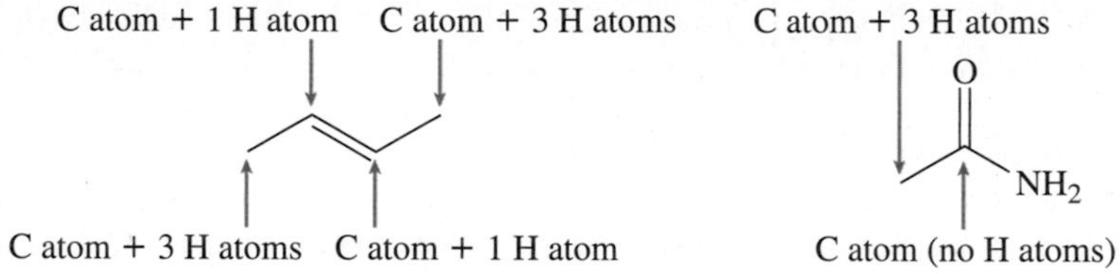

C atom + 1 H atom   C atom + 3 H atoms   C atom + 3 H atoms

C atom + 3 H atoms   C atom + 1 H atom   C atom (no H atoms)

The molecular formulas for these two compounds are C_4H_8 and C_2H_5NO, respectively.

14.4 Functional Groups

Many organic compounds are derivatives of alkanes in which one of the hydrogen atoms has been replaced by a group of atoms known as a **functional group.** A functional group determines many of the chemical properties of a compound because chemical reactions typically involve one or more of the atoms that make up the group. For example, any alkane in which one of the hydrogen atoms has been replaced by the —OH group is an **alcohol.** The best known alcohols are those derived from methane (methanol) and from ethane (ethanol), but alcohols as a class of compounds all exhibit similar chemical properties—because the reactions of alcohols involve the —OH group.

A class of organic compounds often is represented with a general formula that shows the atoms of the functional group explicitly and the remainder of the molecule

TABLE 14.2	Alkyl Groups	
Name	**Formula**	**Model**
Methyl	$-CH_3$	
Ethyl	$-CH_2CH_3$	
Propyl	$-CH_2CH_2CH_3$	
Isopropyl	$-CH(CH_3)_2$	
Butyl	$-CH_2CH_2CH_2CH_3$	
tert-butyl	$-C(CH_3)_3$	

using the letter R to represent one or more alkyl groups. An ***alkyl group*** is a portion of an organic molecule that resembles an *alkane*. In fact, an alkyl group is formed by removing one hydrogen atom from the corresponding alkane. The methyl group ($-CH_3$), for example, is formed by removing a hydrogen atom from methane (CH_4), the simplest alkane. Methyl groups are found in many organic molecules. Table 14.2 lists some of the simplest alkyl groups. Examples of eight of the most commonly encountered functional groups are listed and illustrated in Table 14.3.

14.5 Alcohols and Ethers

The structures of alcohols and ethers are really quite similar. The hydroxy and alkoxy functional groups differ only in one detail. The hydroxy group (alcohol) has a hydrogen atom bonded to the oxygen; whereas the alkoxy group (ether) has an *alkyl* group.

Student Note: In fact, both alcohols and ethers can be thought of as substituted *water* molecules—where one or both hydrogens has been replaced by an alkyl group.

$$R-O-H$$

$$R-O-R$$

Despite this similarity, the properties of alcohols and ethers are quite different. Alcohols are polar molecules, capable of hydrogen bonding, and most are liquids with relatively high boiling points. Ethers are significantly less polar and are not capable of hydrogen bonding; the simplest ether (dimethyl ether), although liquid at room temperature, boils at a relatively low 35°C. Table 14.4 lists some familiar alcohols.

Although alcohols are synthesized in the laboratory, the most familiar of the alcohols, ethanol, is a natural product of the fermentation of glucose, the sugar found in fruit and other plants.

The names of alcohols are based on the names of the corresponding alkanes, with the *–ane* ending replaced by *–ol*.

TABLE 14.3 General Formulas and Models of Select Functional Groups

Class of Compound	Functional Group Name	Structure	Model
Alcohol	hydroxy group	—OH	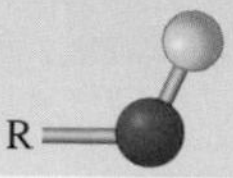
Aldehyde	carbonyl group	$\begin{matrix} O \\ \parallel \\ -C-H \end{matrix}$	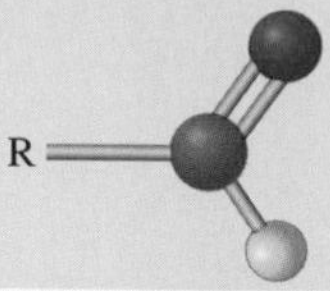
Carboxylic acid	carboxy group	$\begin{matrix} O \\ \parallel \\ -C-O-H \end{matrix}$	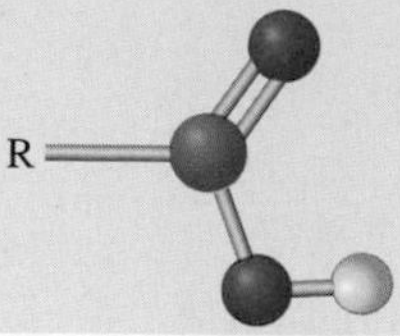
Amine	amino group	—NH$_2$	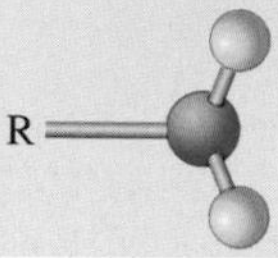
Ether	alkoxy group	—O—R	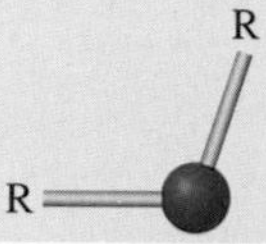
Ester	ester group	$\begin{matrix} O \\ \parallel \\ -C-O-R \end{matrix}$	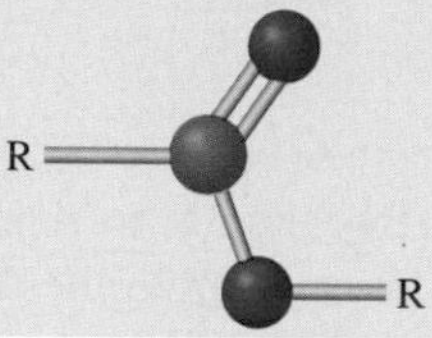
Ketone	carbonyl group	$\begin{matrix} O \\ \parallel \\ -C-R \end{matrix}$	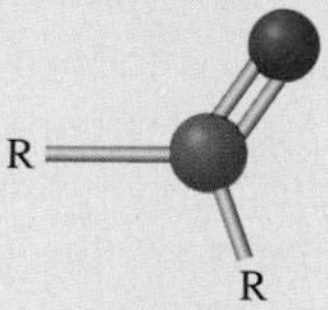
Amide	amide group	$\begin{matrix} O \\ \parallel \\ -C-N-R \\ H \end{matrix}$	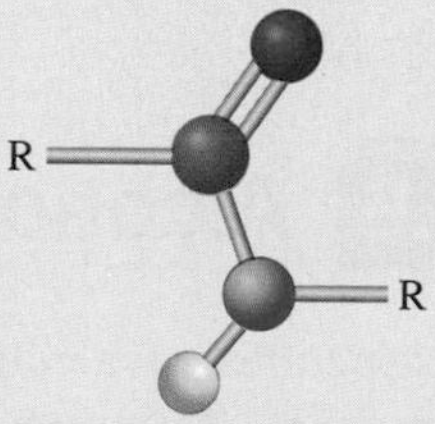

TABLE 14.4 Familiar Alcohols

CH$_3$OH	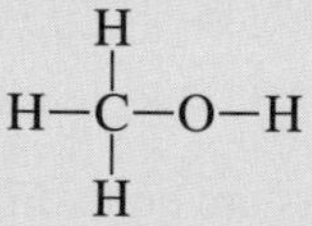 $\begin{matrix} H \\ \mid \\ H-C-O-H \\ \mid \\ H \end{matrix}$	Methanol, also known as *methyl alcohol* or "wood alcohol," is highly poisonous—causing blindness, coma, and potentially death.
C$_2$H$_5$OH	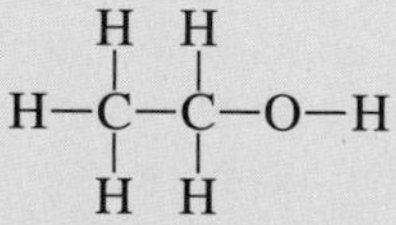 $\begin{matrix} H \quad H \\ \mid \quad \mid \\ H-C-C-O-H \\ \mid \quad \mid \\ H \quad H \end{matrix}$	Ethanol or *ethyl alcohol* is the alcohol in alcoholic beverages.
C$_3$H$_7$OH	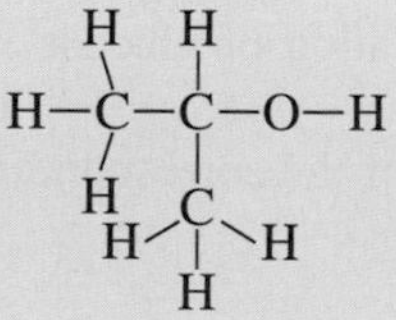 $\begin{matrix} H \quad H \\ \mid \quad \mid \\ H-C-C-O-H \\ \mid \quad \mid \\ H \quad C \\ \quad H \diagup \mid \diagdown H \\ \quad H \end{matrix}$	Isopropanol, also known as *isopropyl alcohol* or "rubbing alcohol," is used as a disinfectant.

TABLE 14.5	Familiar Ethers	
CH_3OCH_3	H—C—O—C—H (with H atoms on each carbon)	Dimethyl ether is used as an aerosol propellant.
$C_2H_5OC_2H_5$	H—C—C—O—C—C—H (with H atoms)	Diethyl ether was once commonly used as an anesthetic.
$C_6H_5OCH_3$	anisole structure (aromatic ring with OCH₃)	Anisole is an aromatic ether with a licorice-like scent. It is the essential oil in anise seed, which is used in baked goods and a variety of liqueurs—including ouzo, anisette, and absinthe.

Table 14.5 lists some of the more familiar ethers.

Many ethers are known by common names that simply specify the alkyl groups attached to the oxygen atom.

Dimethyl ether   Diethyl ether   Ethyl methyl ether

14.6 Aldehydes and Ketones

Aldehydes and ketones also share similar functional groups—both contain a **carbonyl group**—a carbon atom doubly bonded to an oxygen atom. The difference between them is a *hydrogen* atom bonded to the carbonyl carbon (aldehyde) versus an *alkyl* group (ketone). (Note that the two alkyl groups in a ketone need not be the same.)

$$\underset{|}{\overset{O}{R-C-H}}$$

$$\underset{|}{\overset{O}{R-C-R}}$$

The simplest aldehydes are formaldehyde, HCHO, and acetaldehyde, CH_3CHO. Aqueous solutions of formaldehyde have long been used to preserve biological specimens. It has an irritating, pungent odor and is toxic to humans and animals.

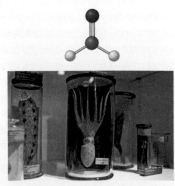

Jennifer Santolla/Alamy Stock Photo

Acetaldehyde is the metabolic product of alcohol consumption thought to be responsible for hangover symptoms. Ethyl alcohol is converted to acetaldehyde in the body by an enzyme known as ADH (alcohol dehydrogenase). The acetaldehyde is then converted to acetic acid (the relatively harmless, characteristic component of vinegar) by another enzyme, ALDH (aldehyde dehydrogenase). An effective but potentially miserable part of the treatment for alcohol abuse is to administer a drug called *disulfiram*—marketed under the name Antabuse. Disulfiram blocks the action of ALDH, preventing the conversion of acetaldehyde to acetic acid. The resulting buildup of acetaldehyde causes the patient to feel very sick almost immediately, making the next cocktail *far* less appealing. The action of disulfiram was discovered accidentally when Danish pharmaceutical researchers who were taking the drug as an experimental treatment for parasitic diseases became very ill every time they consumed even a small amount of alcohol.

The simplest ketone is acetone, CH_3COCH_3. Acetone is used as a solvent in the chemical industry and in consumer products such as nail polish remover.

Radius Images/Alamy Stock Photo

David A. Tietz/McGraw Hill

Profiles in Science

Percy Lavon Julian

The grandson of slaves, Percy Lavon Julian (1899–1975) was one of the first African Americans to earn a Ph.D. in chemistry—and to be inducted into the National Academy of Sciences. Julian received more than a hundred chemical patents and is probably best known for his development of the large-scale, cost-effective synthesis of desperately needed pharmaceutical compounds.

Prior to Julian's work, corticosteroids (steroids) were prohibitively expensive. They were effective in the treatment of a wide variety of ailments, including hormonal deficiencies and the crippling pain of rheumatoid arthritis. However, the source of these compounds was the spinal cords of animals—making it possible to produce them only in tiny quantities. Julian developed new methods to synthesize steroids that began with natural *sterols*, compounds that are plentiful in plants. Switching from animal to plant sources brought the cost of steroids down from roughly $700 per gram to less than 50 cents per gram, making the treatment of many chronic conditions affordable to millions.

Dr. Julian's considerable body of work was not limited to steroid synthesis. He also isolated a protein from soybeans that ultimately was used to develop Aer-O-Foam, a fire-fighting agent used by the U.S. military. And he developed a synthesis of physostigmine, a drug used to treat glaucoma, using the African calabar bean. His ingenious use of plant sources for organic synthesis resulted in numerous scientific advancements that have improved the human condition immeasurably.

Percy Lavon Julian

Walter Oleksy/Alamy Stock Photo

TABLE 14.6	Typical Aldehydes and Ketones, Their Structures, and Their Scents/Uses

Aldehyde	Formula	Scent and Uses
Benzaldehyde	C_6H_5CHO 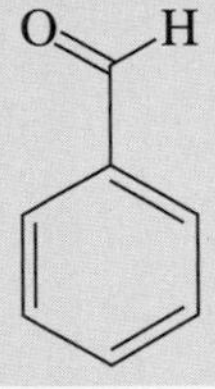	Almond scent—used as a flavorant in food and an odorant in cosmetics, lotions, and soaps.
Cinnamaldehyde	C_8H_7CHO 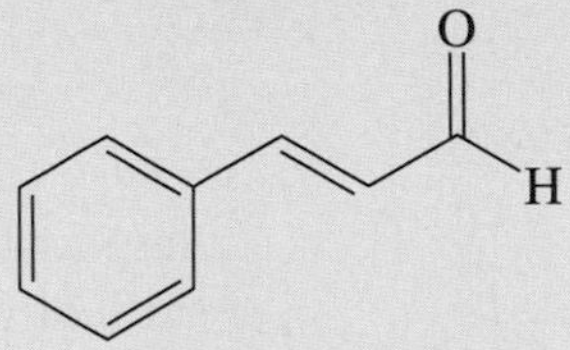	Cinnamon scent—used to flavor candy, gum, baked goods; used as an agricultural antifungal and insecticide; gaining favor as a dietary supplement; and being investigated for its potential anticancer properties.
Hexanal	$C_5H_{11}CHO$ 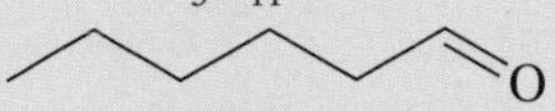	Grassy, new-mown-lawn scent—used in the flavor industry as a component of fruity flavors.

Ketone	Formula	Scent and Uses
Carvone	$C_{10}H_{14}O$ 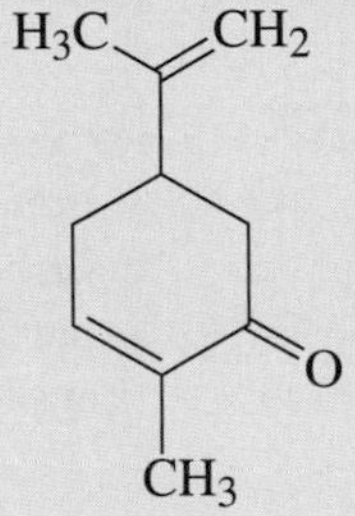	Two different forms of carvone occur naturally. One has a characteristic spearmint scent; the other has the scent of caraway. Both are used as flavorants.
Muscone	$C_{16}H_{30}O$ 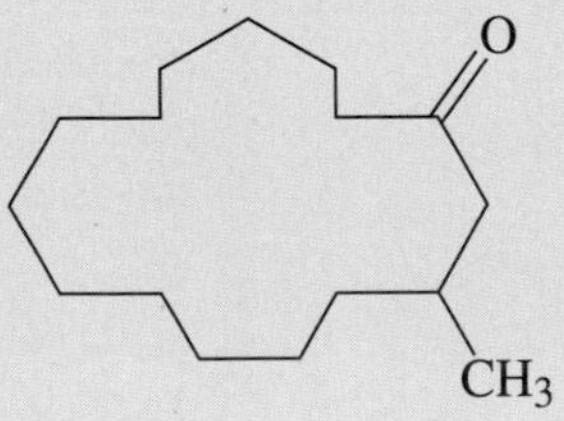	Musky scent, produced by a gland on the abdomen of the mature male musk deer, a species native to the Himalayas and nearly hunted to extinction to supply the fragrance industry.

Both aldehydes and ketones are ubiquitous in nature. Many of the compounds responsible for the characteristic aromas of fruits, herbs, and flowers are aldehydes and ketones. Because of their evocative scents, many aldehydes and ketones are used in the fragrance industry. Table 14.6 lists some aldehydes and ketones and their characteristic aromas and uses.

14.7 Carboxylic Acids and Esters

Again, we have two functional groups that differ only by the replacement of a hydrogen atom (carboxylic acid) with an alkyl group (ester).

TABLE 14.7	Typical Carboxylic Acids and Esters, Their Structures, and Their Origins

Carboxylic Acid	Formula	Origin
Formic acid	HCOOH 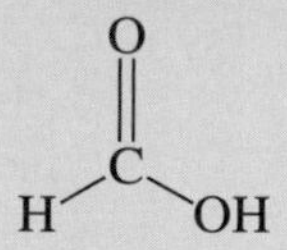 	The acid in many insect stings. (The name *formic* acid actually derives from the French word for "ant": *fourmi*.)
Acetic acid	CH_3COOH 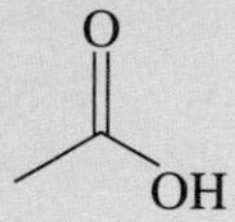 	Characteristic component of vinegar.
Butyric acid	C_3H_7COOH 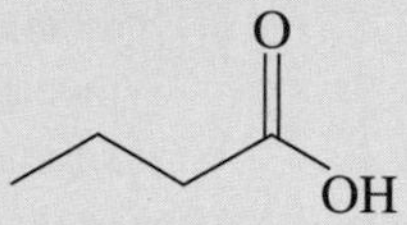 	Responsible for the unpleasant smell of vomit and rancid butter.

Ester	Formula	Origin
Butyl acetate	$C_4H_9OCOCH_3$ 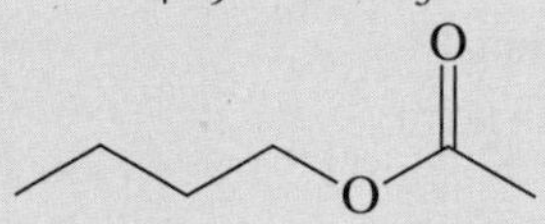 	Present in many fruits and honey. Used to flavor candies and other confections.
Benzyl acetate	$C_6H_5CH_2OCOCH_3$ 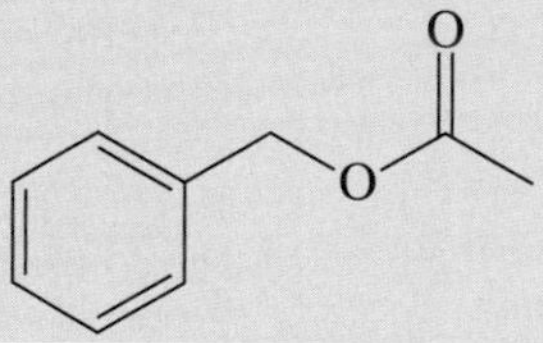 	Found in many flowers, including jasmine. Used in the perfume industry and to impart apple and pear flavors to foods.
Methyl salicylate	$C_6H_4OHOCOCH_3$ 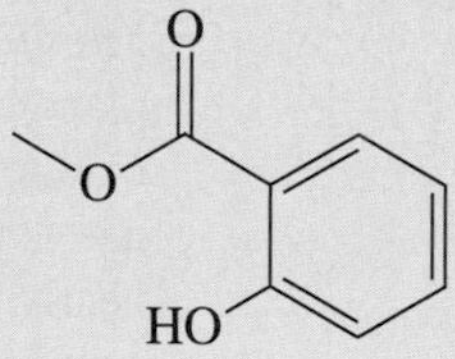 	Produced naturally by wintergreen plants. Used as a flavoring in food and as an odorant in liniments.

Many of the most familiar carboxylic acids have strong, typically unpleasant odors, whereas the corresponding esters, in which the hydrogen atom is replaced by an alkyl group, typically have pleasant odors—many are used in the flavor industry. Table 14.7 lists some carboxylic acids and esters along with their characteristic scents and origins.

14.8 Amines and Amides

Amine and amide functional groups are similar in that they both contain nitrogen atoms, but the chemical properties of amines and amides are very different. Amines are essentially ammonia (NH_3) in which one or more of the hydrogen atoms has been replaced by an alkyl group. Amines are classified according to how many hydrogen atoms have been replaced. A primary amine is one that has just one alkyl group. Examples include methylamine, a fishy-smelling product of putrefaction; and histamine, a biological molecule that operates as part of the immune system.

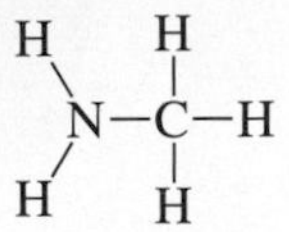

Methylamine

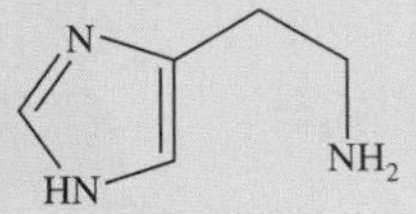

A secondary amine is one in which two of ammonia's hydrogen atoms have been replaced by other groups. An example of a secondary amine is dimethylamine, which is produced and used by the chemical industry but is also an insect pheromone.

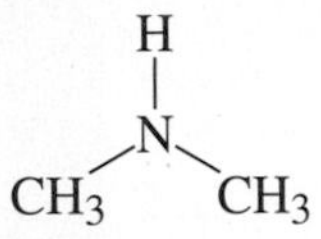

Dimethylamine

Tertiary amines are those with three alkyl groups (not hydrogen atoms) bonded to the nitrogen atom. An example of a tertiary amine is diphenhydramine, a common antihistamine.

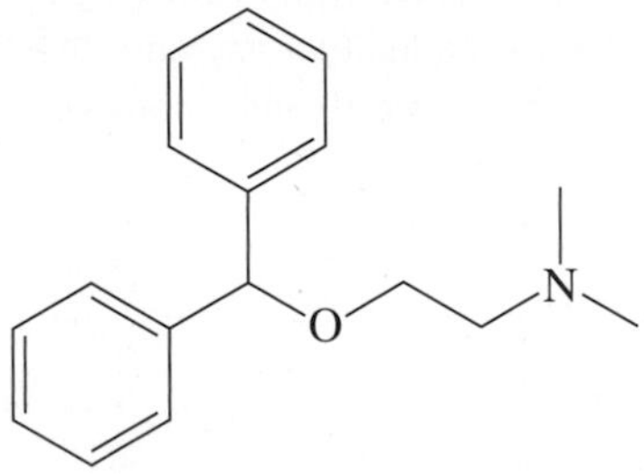

Diphenhydramine

Amides contain both the amino group and the carbonyl group. They are very common among pharmaceuticals, many of which are based on the molecule benzamide. Examples of substituted benzamides are the analgesic and anti-inflammatory drug ethenzamide, and Mocetinostat, a drug currently undergoing clinical trials for the treatment of several types of cancer.

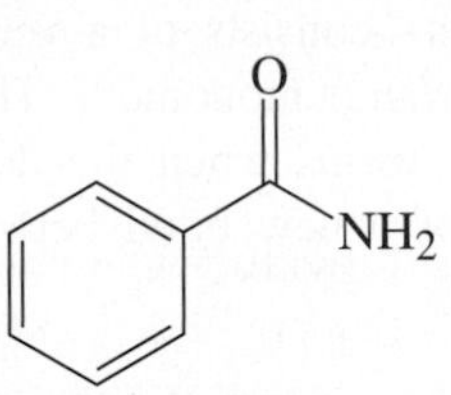

Benzamide

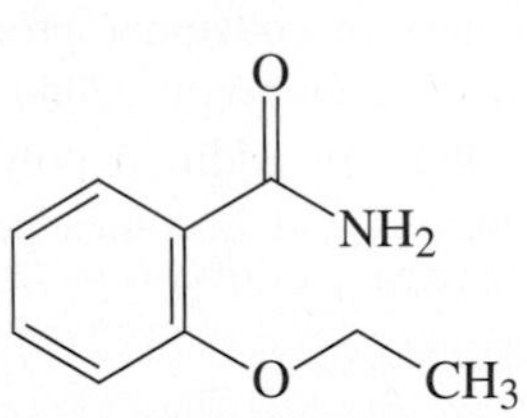

Ethenzamide

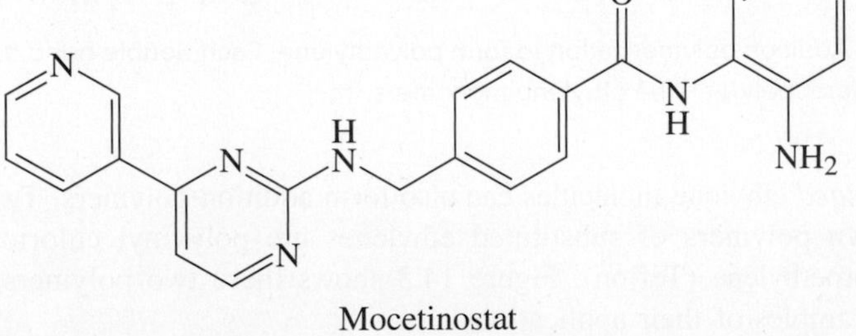

Mocetinostat

Student Note: A molecule may contain more than one type of amine group. Histamine, shown here, contains primary, secondary, and tertiary amine groups.

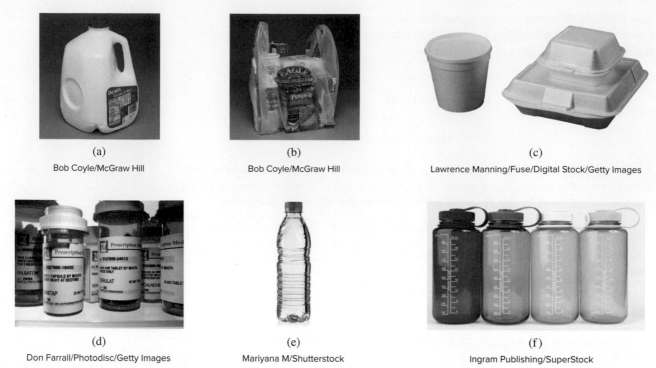

(a) Bob Coyle/McGraw Hill

(b) Bob Coyle/McGraw Hill

(c) Lawrence Manning/Fuse/Digital Stock/Getty Images

(d) Don Farrall/Photodisc/Getty Images

(e) Mariyana M/Shutterstock

(f) Ingram Publishing/SuperStock

Figure 14.1 Common polymers. (a) Milk is typically sold in jugs made of a particular type of polyethylene. (b) Grocery bags are made from another type of polyethylene. (c) Foam cups and takeout containers are made of polystyrene. (d) Prescription bottles are made of polypropylene. (e) Disposable water bottles are made of polyethylene terephthalate. (f) Some hard-sided, reusable water bottles are made of polycarbonate.

14.9 Polymers

Many of the materials we encounter commonly are substances known as *polymers*. Figure 14.1 shows examples of polymers that you probably encounter every day. A ***polymer*** consists of very *large* molecules, which form when a very large *number* of *small* molecules, known in this context as ***monomers,*** combine in a particular way.

The simplest type of polymer is one that is made entirely of one type of monomer, via a process known as ***addition polymerization.*** *Polyethylene,* for example—one of the most common polymers in consumer products—consists of a very large number of connected ethene (more commonly called *ethylene*) monomers. The ethene molecule contains a double bond. An addition polymer forms when the double bond in each ethylene molecule breaks and becomes part of a new bond between the monomers (Figure 14.2).

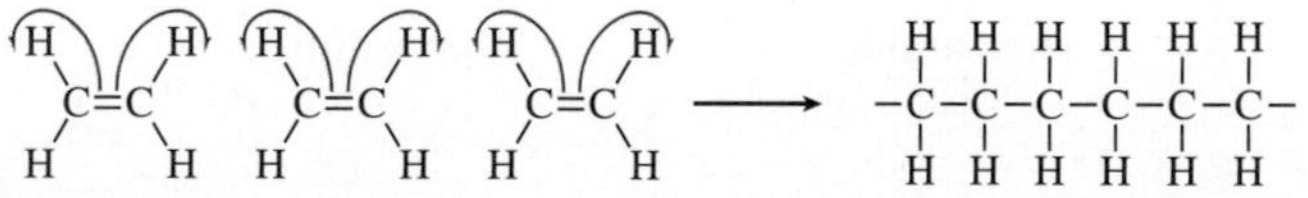

Figure 14.2 Addition polymerization to form polyethylene. Each double bond breaks and new single bonds form between the ethylene monomers.

Substituted ethylene molecules can also form addition polymers. Two of the most widely known polymers of substituted ethylenes are polyvinyl chloride (PVC) and polytetrafluoroethylene (Teflon). Figure 14.3 shows these two polymers, their monomers, and examples of their applications.

Table 14.8 shows some of the common addition polymers and their popular uses.

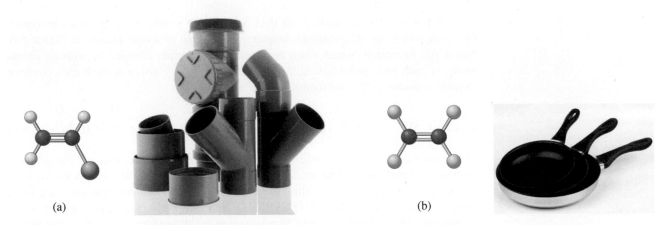

Figure 14.3 (a) Chloroethylene, commonly known as vinyl chloride, undergoes addition polymerization to form polyvinyl chloride. PVC is used in modern plumbing. It has also been used to make the traditional (and retro) "vinyl" records. (b) Tetrafluoroethylene undergoes addition polymerization to form polytetrafluoroethylene. Teflon is used to coat the interior surfaces of cookware, making it nonstick.

(a) IvonneW/Essentials/iStockphoto; (b) Gastrofotos/Gastromedia/Alamy Stock Photo

TABLE 14.8	Some Common Addition Polymers		
Polymer	**Structure**	**Monomer**	**Uses**
Polyethlene	$-(CH_2-CH_2)_n-$	$H_2C=CH_2$	Plastic bags and wraps, bottles, toys, Tyvek wrap
Polystyrene			Disposable cups and plates, insulation, packing materials
Polyvinyl chloride			Pipes and fittings, plastic wraps, records
Polyacrylonitrile			Fibers for carpeting and clothing (Orlon)
Polybutadiene			Synthetic rubber

A polymer that contains more than one type of monomer is called a ***copolymer.*** One copolymer we all encounter frequently is a type of nylon known as Nylon 6,6. Nylon 6,6 formation begins when hexamethylenediamine, a molecule with an *amino* group at each end, and adipic acid, which has a *carboxy* group at each end, combine to form a ***dimer***—the combination of two monomers.

$$H_2N-(CH_2)_6-NH_2 + HO-\overset{\overset{O}{\|}}{C}-(CH_2)_4-\overset{\overset{O}{\|}}{C}-OH \longrightarrow H_2N-(CH_2)_6-NH-\overset{\overset{O}{\|}}{C}-(CH_2)_4-\overset{\overset{O}{\|}}{C}-COOH$$

In this type of combination, when the monomers come together, some of the atoms in the two terminal groups (the amino group on hexamethylenediamine; and the carboxy group on the adipic acid) are eliminated. In this case, the atoms eliminated are an H from the amino group, and an H and an O from the carboxy group. Together, these atoms constitute a water molecule. Subsequent to the dimer formation, additional monomer units can add to the chain, resulting in the formation of the polymer; and one water molecule is eliminated each time another monomer is added to the chain. Overall, the process can be represented by:

$$H_2N-(CH_2)_6-NH_2 + HO-\overset{\overset{O}{\|}}{C}-(CH_2)_4-\overset{\overset{O}{\|}}{C}-OH \longrightarrow H_2N-(CH_2)_6-NH-\overset{\overset{O}{\|}}{C}-(CH_2)_4-\overset{\overset{O}{\|}}{C}-OH$$

$$+H_2O$$

This process, whereby the combination of monomers results in the elimination of water or some other small molecule, is an example of a ***condensation reaction.*** As we explain in Chapter 15, this type of condensation polymerization is very important in biological processes.

Chapter Summary

Section 14.1

- Organic chemistry is the study of carbon-containing substances. Although once thought to be exclusively produced by living things, organic compounds can be synthesized in the laboratory—thousands of new such substances are produced by scientists every year.

- Carbon's small size and moderate electronegativity make strong single and multiple bonds between carbon atoms possible.

- Aromatic compounds generally contain one or more benzene rings. Aliphatic compounds are generally organic compounds without benzene rings.

Section 14.2

- The simplest organic compounds, containing only carbon and hydrogen, are *hydrocarbons*. Hydrocarbons in which all C—C bonds are single bonds are called *alkanes* and are said to be *saturated*. An alkane has the general formula C_nH_{2n+2}.

- A hydrocarbon with a C—C double bond is called an *alkene;* and a hydrocarbon with a C—C triple bond is called an *alkyne.* Alkenes have the general formula C_nH_{2n}; and alkynes have the general formula C_nH_{2n-2}. Hydrocarbons that contain carbon-carbon multiple bonds are said to be *unsaturated.*

Section 14.3

- Compounds with the same chemical formula but different structures are called *structural isomers.* The greater the numbers of carbon atoms in a molecule, the greater the number of structural isomers are possible.

- Different arrangement of groups about a C—C double bond give rise to *cis* and *trans* *geometrical isomers.*

- Organic molecules can be represented with *bond-line structures,* in which lines represent C—C bonds and most atoms are not shown explicitly. Atoms other than carbon and hydrogen are called *heteroatoms.* Heteroatoms are shown explicitly in bond-line structures.

Section 14.4

- A *functional group* is a collection of atoms that impart specific properties to a molecule. One of the simplest organic functional groups is the —OH group, known as the *alcohol* group. An *alkyl group,* represented as —R, is simply an alkane—less one of its H atoms.

Section 14.5

- Alcohols and *ethers* can be thought of as substituted water molecules in which one and two H atoms, respectively, have been replaced by an alkyl group. The properties of alcohols and ethers differ because of their differing polarities and hydrogen-bonding capabilities.

Section 14.6

- Aldehydes and ketones contain the *carbonyl group,* which consists of a carbon-oxygen double bond. Aldehydes have a H atom bonded to the carbonyl carbon, whereas ketones have an R group.

- Aldehydes and ketones are common in nature, many being responsible for the flavors and aromas of fruits, flowers, and herbs.

Section 14.7

- Carboxylic acids and esters also contain the carbonyl group, along with an additional O atom. Carboxylic acids then contain another hydrogen atom and esters contain another R group.

- Carboxylic acids typically have unpleasant odors, whereas many esters are used as flavorants.

Section 14.8

- Amines and amides contain nitrogen and are derivatives of ammonia, NH_3. Amides also contain a carbonyl group.

- Both amines and amides are common in pharmaceuticals.

Section 14.9

- A *polymer* is a large molecule made up of repeating units of smaller molecules called *monomers.*

- Polymers may be natural or synthetic, and may form by *addition polymerization* or by a *condensation reaction.* In a condensation reaction, a small molecule (often water) is eliminated in each step.

- A *copolymer* is one consisting of more than one type of monomer. A *dimer* consists of just two monomer units.

- Polymers have found ubiquitous use in society, including for containers and packaging, clothing, cookware, and medical applications.

Key Terms

Addition polymerization 508	Bond-line structures 499	Geometrical isomers 497	Polymer 508
Alcohol 500	Carbonyl group 503	Heteroatom 500	Saturated 493
Alkane 493	Condensation reaction 510	Hydrocarbon 492	Structural isomer 497
Alkene 493	Copolymer 510	Hydrogenation 496	Unsaturated 495
Alkyl group 501	Dimer 510	Monomer 508	
Alkyne 493	Functional group 500		

Questions and Problems

SECTION 14.1: WHY CARBON IS DIFFERENT

14.1 Explain why carbon is able to form so many more compounds than any other element.

14.2 Explain the difference between aliphatic and aromatic hydrocarbons.

14.3 Draw an example of an aliphatic hydrocarbon.

14.4 Draw an example of an aromatic hydrocarbon.

SECTION 14.2: HYDROCARBONS

14.5 Identify the compounds that are hydrocarbons.
(a) $Ca(C_2H_3O_2)_2$ (c) $CH_3CH_2CH_3$
(b) CH_4O (d) CH_3COOH

14.6 Identify the compounds that are hydrocarbons.
(a) $LiHCO_3$ (c) NH_4CN
(b) CH_3CH_2OH (d) CH_3CHCH_2

14.7 Identify each hydrocarbon as an *alkane, alkene,* or *alkyne.*
(a) CH_3CHCH_2 (c) CH_3CCH
(b) $CH_3CH_2CH_2CH_3$ (d) $CH_2CHCH_2CH_2CH_3$

14.8 Identify each hydrocarbon as an *alkane, alkene,* or *alkyne.*
(a) C_4H_8
(b) $CH_2CHCH_2CH_2CH_3$
(c) C_5H_8
(d) $CHCCH_2CH_2CH_2CH_3$

14.9 Draw an alkane that contains
(a) four carbon atoms.
(b) six carbon atoms.
(c) eight carbon atoms.

14.10 Draw an alkene that contains one carbon-carbon double bond and
(a) five carbon atoms.
(b) seven carbon atoms.
(c) three carbon atoms.

14.11 Predict the product(s) of the reaction:
$$CH_3CH_2CH=CH_2 + H_2 \longrightarrow ?$$

14.12 Predict the product(s) of the reaction:
$$CH_3CH=CHCH_3 + Cl_2 \longrightarrow ?$$

14.13 Predict the product(s) of the reaction:
$$CH_3CCH + 2H_2 \longrightarrow ?$$

14.14 Predict the product(s) of the reaction:
$$CH_3CH_2CH_3 + Cl_2 \longrightarrow ?$$

SECTION 14.3: ISOMERS

14.15 Draw all of the isomers of an alkane that has 7 carbon atoms.

14.16 Draw four of the isomers of an alkene that has 7 carbon atoms.

14.17 Draw four of the isomers of an alkyne that has 7 carbon atoms.

14.18 Draw all possible isomers for the molecule C_4H_8.

14.19 Draw all possible isomers for the molecule C_3H_5Br.

14.20 Write the structural formula of an aldehyde that is a structural isomer of acetone.

14.21 Determine whether each pair of molecules are isomers of one another.

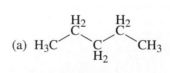

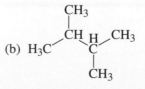

14.22 Determine whether each pair of molecules are isomers of one another.

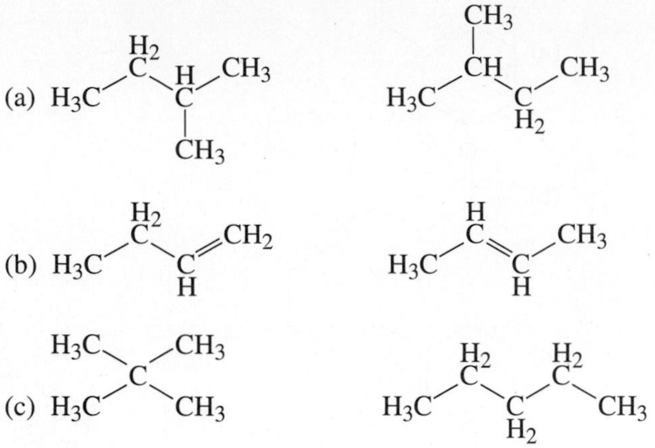

(a) H₃C—CH₂—CH—CH₃ with CH₃ branch, CH₃ ... | H₃C—CH—C—CH₃ with CH₃ groups

(b) H₃C—CH₂—C=CH₂ with H | H₃C—C=CH—CH₃ with H

(c) H₃C—C(CH₃)₂—CH₃ | H₃C—CH₂—CH₂—CH₂—CH₃

14.23 Given a structural formula or skeletal (line) structure, rewrite it in the other style.

(a) CH₃CH₂CHCH₂CH₂CH₂CH₃
 |
 OH

(b) skeletal structure with Br and CHO (H)

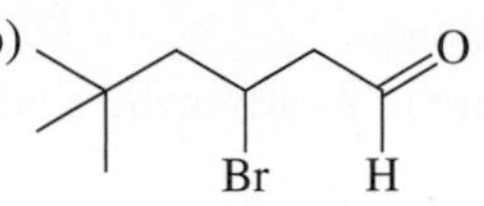

(c) CH(CH₃)₂ ring structure with O

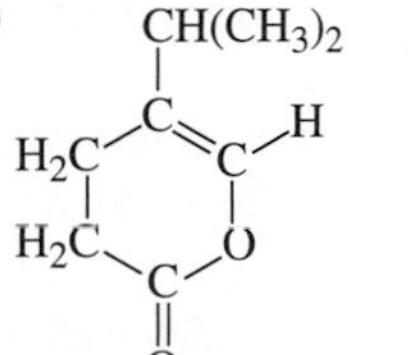

14.24 Given a structural formula or skeletal (line) structure, rewrite it in the other style.

(a) ClCH₂CH₂CH₂CH=CH₂

(b) line structure with N—H

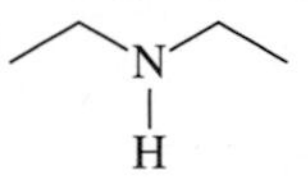

(c) line structure with two C=O and O

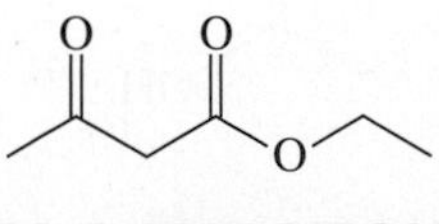

SECTION 14.4: FUNCTIONAL GROUPS

14.25 What are functional groups? Why is it logical and useful to classify organic compounds according to their functional groups?

14.26 Draw the Lewis structure for each of the following functional groups: alcohol, aldehyde, ketone, carboxylic acid, amine.

14.27 Name the classes to which the following compounds belong:
(a) C₄H₉OH   (c) C₆H₅COOH
(b) C₂H₅CHO   (d) CH₃NH₂

14.28 Classify each of the following molecules as alcohol, aldehyde, ketone, carboxylic acid, or amine:
(a) CH₃—CH₂—NH₂

(b) CH₃—CH₂—C⟨=O, H⟩

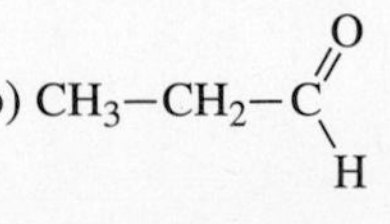

(c) CH₃—C(=O)—CH₂—CH₃

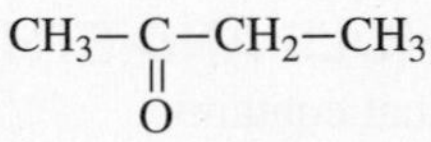

(d) H—C(=O)—OH

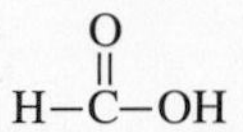

(e) CH₃—CH₂—CH₂—OH

14.29 Draw structures for molecules with the following formulas:
(a) CH₄O   (c) C₃H₆O₂
(b) C₂H₆O   (d) C₃H₈O

14.30 Identify the functional group(s) present in the following compound.

HO—CH₂—C(=O)—OH

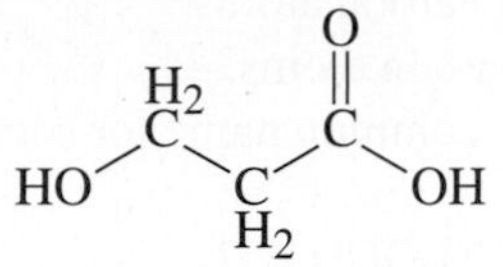

14.31 Identify the functional group(s) present in the following compound.

H₂N—C(=O)—CH₂—CH₂—C(=O)—CH₃

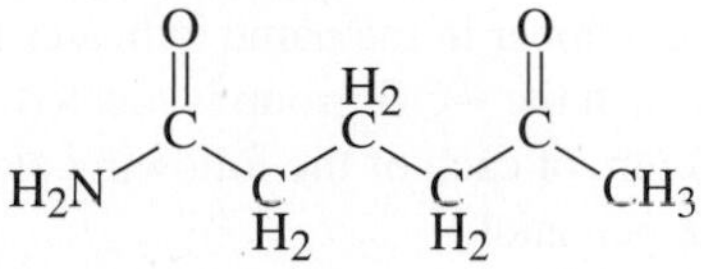

14.32 Classify the oxygen-containing groups in the plant hormone abscisic acid.

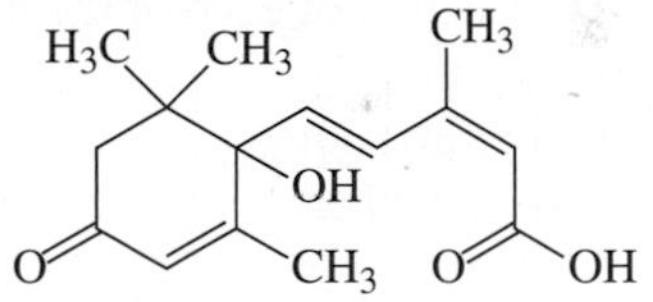

(abscisic acid structure with CH₃, OH, CH₃, O, OH groups)

14.33 Identify the functional groups in the antipsychotic drug haloperidol.

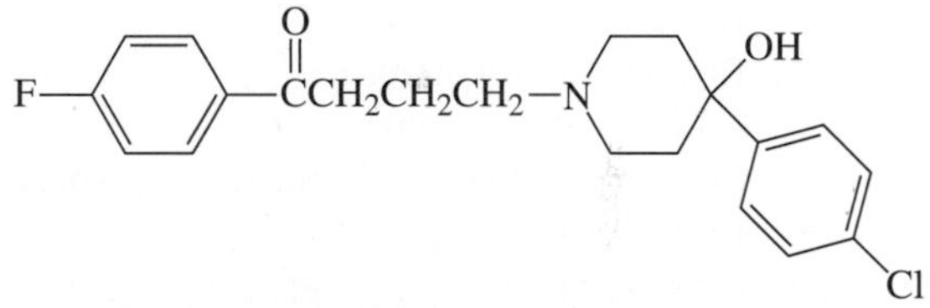

(haloperidol structure: F—phenyl—C(=O)CH₂CH₂CH₂—N piperidine ring with OH and phenyl—Cl)

14.34 PABA was the active UV-absorbing compound in earlier versions of sunblock creams. What functional groups are present in PABA?

14.35 Lidocaine (C₁₄H₂₂N₂O) is a widely used local anesthetic. Classify its nitrogen-containing functional groups.

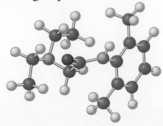

SECTION 14.5: ALCOHOLS AND ETHERS

14.36   Draw an alcohol that contains
(a) two carbon atoms.
(b) six carbon atoms.
(c) nine carbon atoms.

14.37   Draw an ether that contains
(a) four carbon atoms.
(b) five carbon atoms.
(c) seven carbon atoms.

14.38   Draw an alcohol and ether that contain
(a) three carbon atoms.
(b) eight carbon atoms.
(c) ten carbon atoms.

14.39   Give the common name for each of the following ethers:
(a) $CH_3CH_2OCH_2CH_3$
(b) $CH_3CH_2CH_2OCH_3$
(c) $CH_3OCH_2CH_2CH_2CH_3$

14.40   If the number in the name indicates the carbon on which the —OH group is located, draw the structure of each of the following alcohols:
(a) 2-pentanol
(b) 3-hexanol
(c) 2-propanol

14.41   Draw the structure of each of the following ethers:
(a) dipropyl ether
(b) methyl propyl ether
(c) ethyl pentyl ether

SECTION 14.6: ALDEHYDES AND KETONES

14.42   Draw an aldehyde and ketone that each contain
(a) three carbon atoms.
(b) four carbon atoms.
(c) five carbon atoms.

14.43   Draw an aldehyde with the given formula:
(a) C_2H_4O
(b) C_4H_8O
(c) $C_5H_{10}O$

14.44   Draw a ketone with the given formula:
(a) C_3H_6O
(b) C_4H_8O
(c) $C_5H_{10}O$

14.45   Determine the formula for each aldehyde structure:

(a)

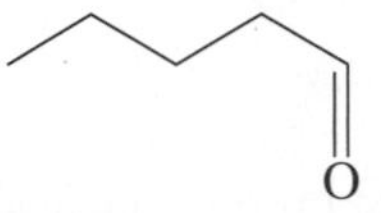

(b)

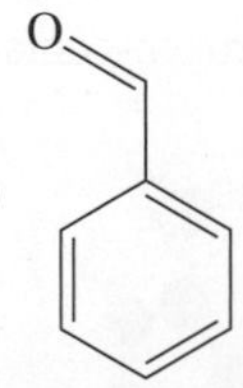

(c)

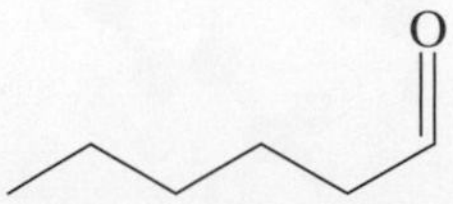

14.46   Determine the formula for each ketone structure:

(a)

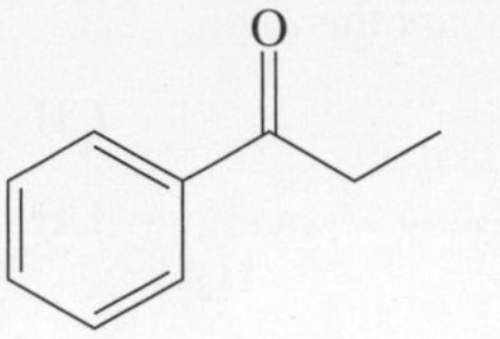

(b)

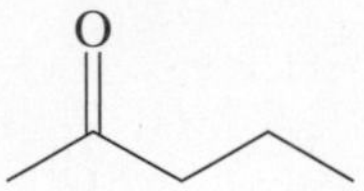

(c)

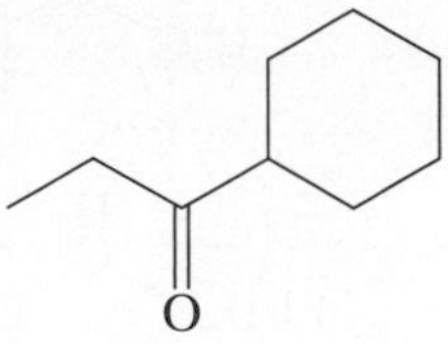

SECTION 14.7: CARBOXYLIC ACIDS AND ESTERS

14.47   Draw a carboxylic acid and an ester that each contain
(a) four carbon atoms.   (c) six carbon atoms.
(b) five carbon atoms.

14.48   Draw the structure of a carboxylic acid with the formula:
(a) $C_4H_8O_2$ (c) $C_6H_{12}O_2$
(b) $C_5H_{10}O_2$

14.49   Draw the structure of an ester with the formula:
(a) $C_4H_8O_2$ (c) $C_6H_{12}O_2$
(b) $C_5H_{10}O_2$

14.50   Determine the formula for each carboxylic acid structure:

(a)

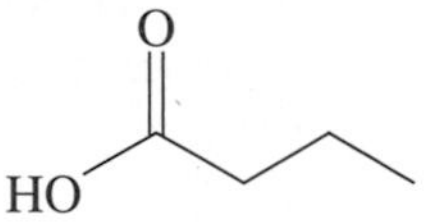

(b)

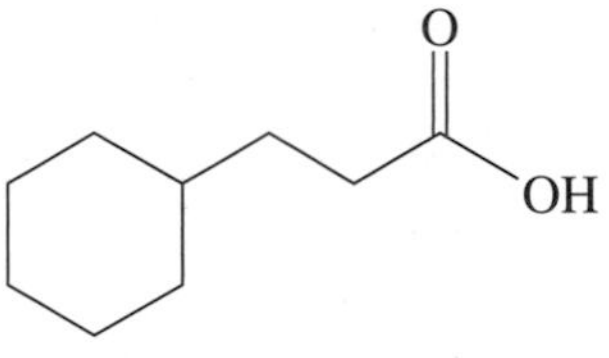

(c)

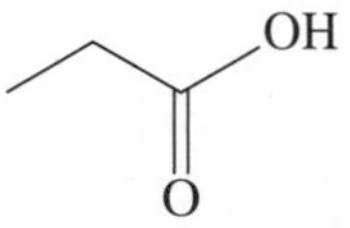

14.51   Determine the formula for each ester structure:

(a)

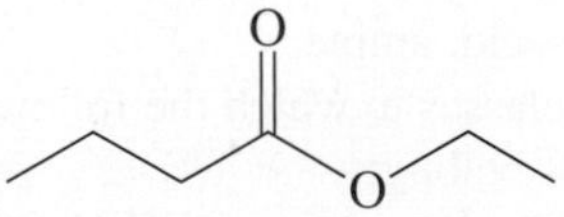

(b)

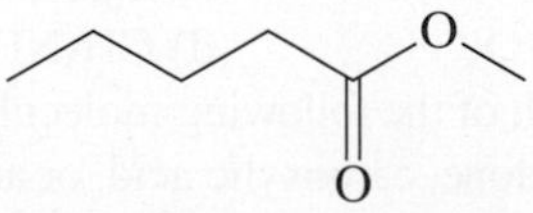

(c)

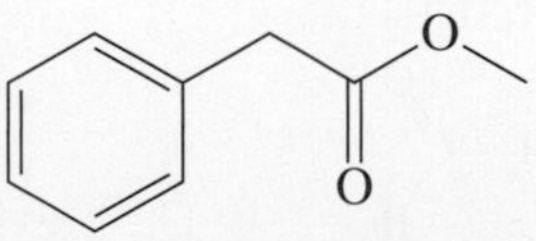

SECTION 14.8: AMINES AND AMIDES

14.52 Draw an amine and an amide that each contain
(a) three carbon atoms.   (c) five carbon atoms.
(b) four carbon atoms.

14.53 Draw the structure of an amine with the formula:
(a) C_3H_9N (c) $C_5H_{13}N$
(b) $C_4H_{11}N$

14.54 Draw the structure of an amide with the formula:
(a) C_3H_7NO (c) $C_5H_{11}NO$
(b) C_4H_9NO

14.55 Determine the formula from the amine structure:

(a)

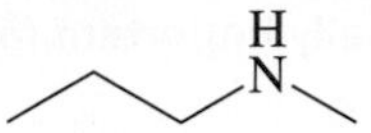

(b)

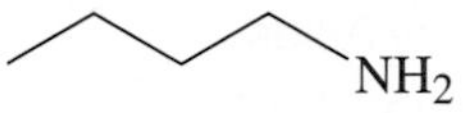

(c)

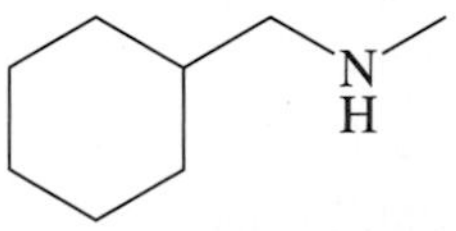

14.56 Determine the formula from the amide structure:

(a)

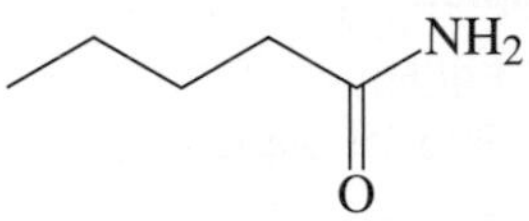

(b)

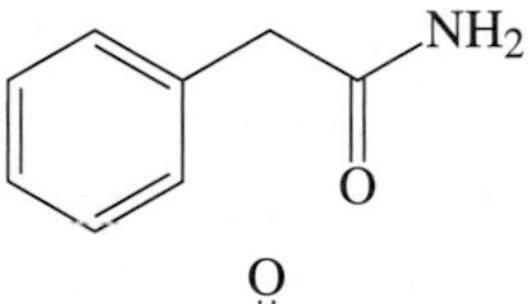

(c)

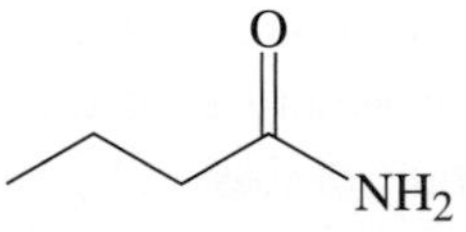

SECTION 14.9: POLYMERS

14.57 Define the following terms: *monomer, polymer, copolymer.*

14.58 Identify 10 familiar objects that contain synthetic organic polymers.

14.59 Calculate the molar mass of a particular polyethylene sample, $-(CH_2-CH_2)_n$, where $n = 4600$.

14.60 Vinyl chloride ($H_2C=CHCl$) undergoes copolymerization with 1,1-dichloroethylene ($H_2C=CCl_2$) to form a polymer commercially known as Saran®. Draw the structure of the polymer, showing the repeating monomer units.

14.61 Deduce plausible monomers for polymers with the following repeating units:

(a) $-(CH_2-CH_2)_n$

(b)

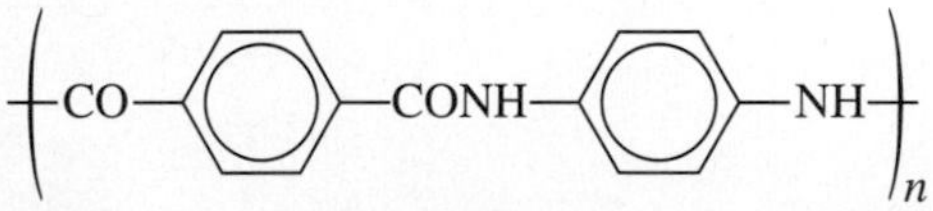

14.62 Deduce plausible monomers for polymers with the following repeating units:

(a) $-(CH_2-CH=CH-CH_2)_n$

(b) $-(CO-(CH_2)_6-NH)_n$

Answers to In-Chapter Materials

Answers to Practice Problems

14.1A (a) alkyne, (b) alkane, (c) alkene, (d) alkyne.
14.1B (a) 8, (b) 9, (c) 10.

14.2A (a)

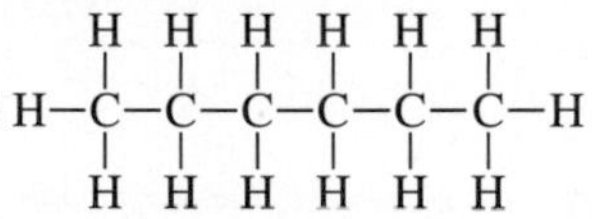

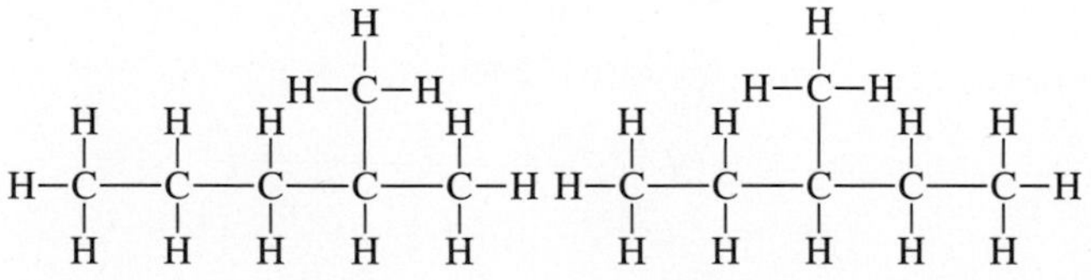

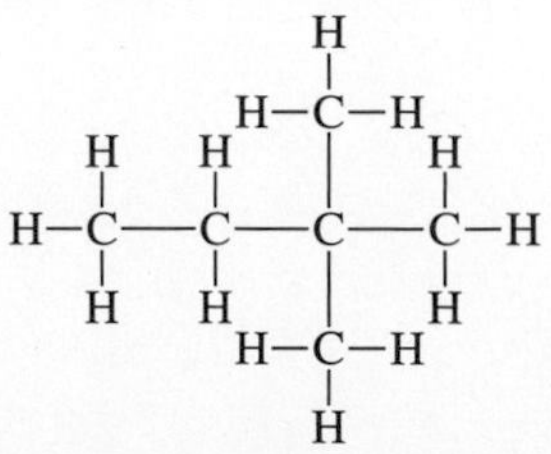

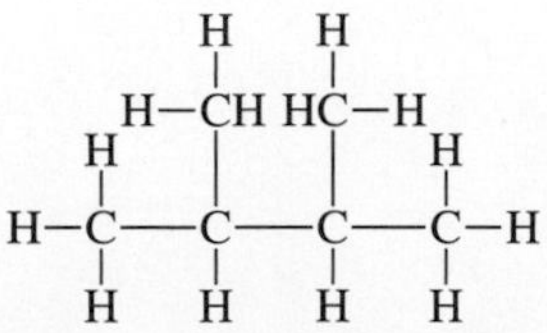

(b)

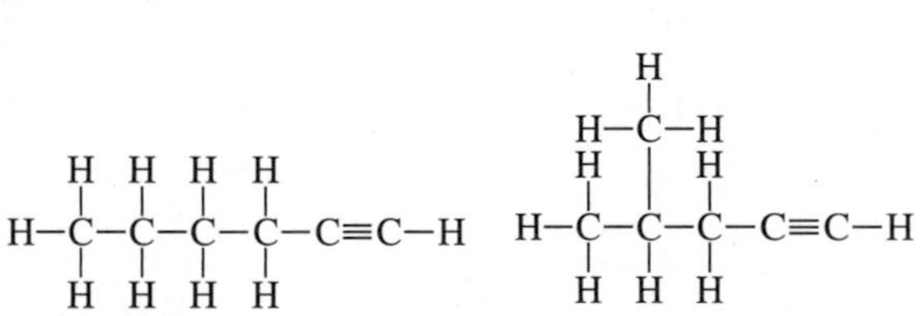

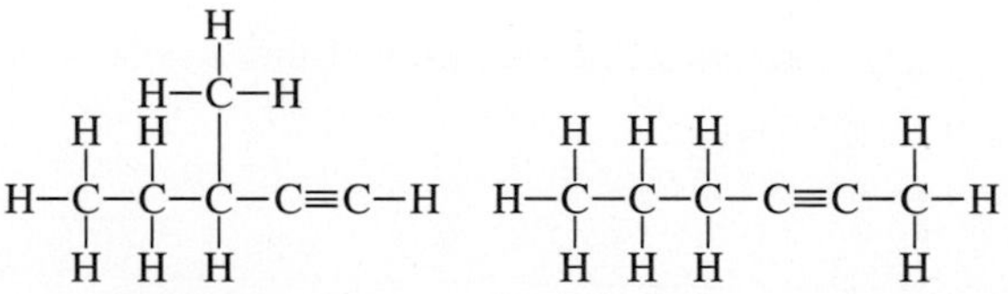

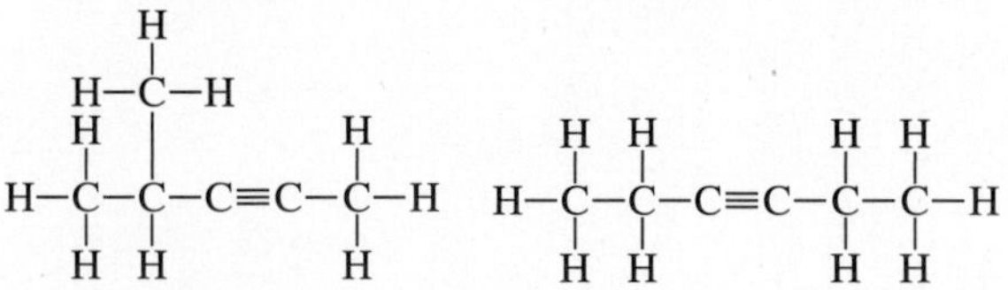

14.2B

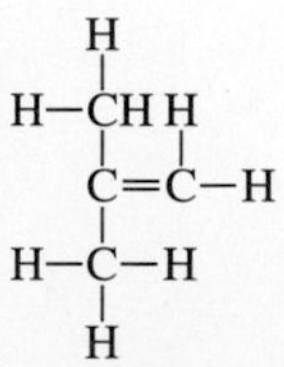

CHAPTER 15

Biochemistry

The brilliant colors displayed by chameleons are the product of complex biochemical processes.

hlansdown/iStock/Getty Images

15.1   Biologically Important Molecules
- Glycerol
- Fatty Acids
- Amino Acids
- Sugars
- Phosphates
- Organic Bases

15.2   Lipids
- Fats
- Phospholipids
- Steroids

15.3   Proteins
- Primary Structure
- Secondary Structure
- Tertiary Structure
- Quaternary Structure

15.4   Carbohydrates
- Monosaccharides
- Disaccharides
- Polysaccharides

15.5   Nucleic Acids

We have seen that the chemistry of many carbon-containing molecules constitutes a branch of chemistry known as organic chemistry. Within organic chemistry, there is another branch known as *biochemistry*. Biochemistry encompasses the study of specific chemical species, many of them organic, that play a variety of important roles in the chemistry of living systems.

15.1 Biologically Important Molecules

For us to understand basic biochemistry, it is necessary to recognize the importance of some specific molecules and molecule types. In this section, we introduce several new types of molecules and recall the structures of some we have already encountered.

Glycerol

Glycerol, $C_3H_8O_3$, contains a familiar functional group—the *hydroxy* group. The glycerol molecule is a three-carbon chain, in which each carbon atom is bonded to a hydroxy group:

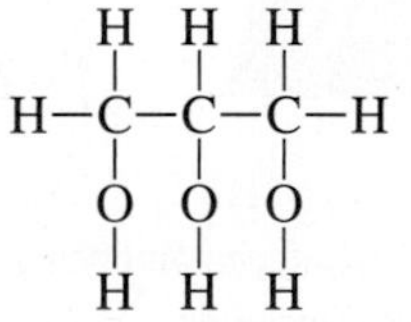

$$
\begin{array}{c}
\text{H} \quad \text{H} \quad \text{H} \\
| \quad\;\; | \quad\;\; | \\
\text{H}-\text{C}-\text{C}-\text{C}-\text{H} \\
| \quad\;\; | \quad\;\; | \\
\text{O} \quad \text{O} \quad \text{O} \\
| \quad\;\; | \quad\;\; | \\
\text{H} \quad \text{H} \quad \text{H}
\end{array}
$$

Fatty Acids

Fatty acids are carboxylic acids [◄◄ Section 14.7] in which the alkyl group (–R) is a relatively long hydrocarbon chain. The alkyl group in a fatty acid may be saturated, containing no carbon-carbon multiple bonds; or unsaturated, containing one or more carbon-carbon multiple bonds (usually double bonds, sometimes triple bonds). Table 15.1 gives the names, formulas, and structures of some common fatty acids.

Rewind: A carboxylic acid is a molecule with the carboxy functional group:

$$
\begin{array}{c}
\text{O} \\
\|\;\; \\
\text{R}-\text{C}-\text{O}-\text{H}
\end{array}
$$

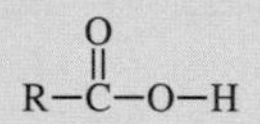

Amino Acids

Amino acids contain both the amino functional group and the carboxy functional group. They have the general structure:

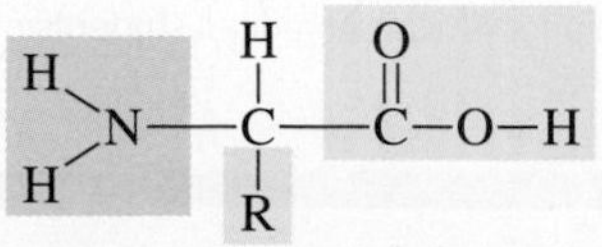

$$
\begin{array}{c}
\text{H} \quad\quad\quad \text{H} \quad\; \text{O} \\
\diagdown \quad\quad\quad | \quad\;\; \| \\
\;\;\;\;\;\text{N}-\text{C}-\text{C}-\text{O}-\text{H} \\
\diagup \quad\quad\quad | \\
\text{H} \quad\quad\quad \text{R}
\end{array}
$$

where the R group determines the identity of the amino acid.

TABLE 15.1	Common Fatty Acids	
Compound	**Formula**	**Structure**
Saturated		
Butyric acid	$C_4H_8O_2$	
Palmitic acid	$C_{16}H_{32}O_2$	
Stearic acid	$C_{18}H_{36}O_2$	
Unsaturated		
Linoleic acid	$C_{18}H_{32}O_2$	
Oleic acid	$C_{18}H_{34}O_2$	

Profiles in Science

Marie Maynard Daly

Marie Maynard Daly (1921–2003) was the first African American woman to earn a Ph.D. in chemistry. Her work included investigations of the effects of aging and disease on the biochemical processes involved in the digestion of food; the changes in protein metabolism during feeding and fasting conditions in mice; and the effects of cigarette smoke on the lungs. She was the recipient of numerous fellowships, including from the American Association for the Advancement of Science, the New York Academy of Sciences, and the American Cancer Society.

Daly was a professor at the Albert Einstein College of Medicine and was a member of the board of governors of the New York Academy of Sciences. She dedicated much of her professional life to increasing opportunities for minorities to study medicine. After she retired from teaching and research, she established a scholarship for African American students of chemistry and physics at Queens College in New York.

Marie Maynard Daly
Archive PL/Alamy Stock Photo

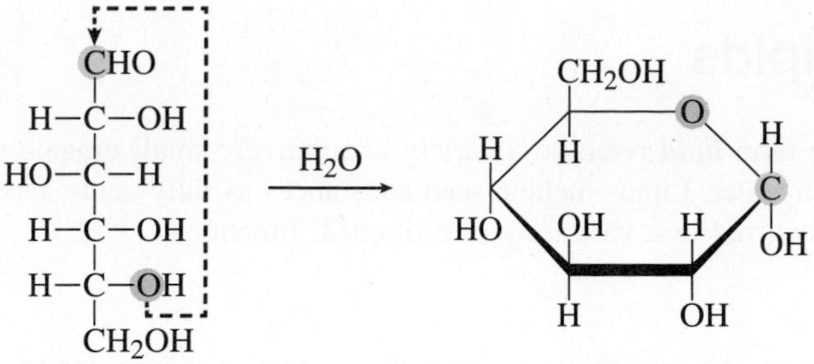

Figure 15.1 Glucose molecule. Sugar molecules, such as glucose, form cyclic structures when dissolved in water. Highlighting indicates the locations of the atoms between which the new covalent bond forms when the cyclic structure forms.

Sugars

Typically, the word *sugar* refers to table sugar. But in general, a *sugar* is a relatively small, water-soluble, typically *sweet* organic molecule—either a ketone or an aldehyde—consisting of carbon, hydrogen, and oxygen. A familiar example is *glucose,* sometimes referred to as "blood sugar," which we generally represent as:

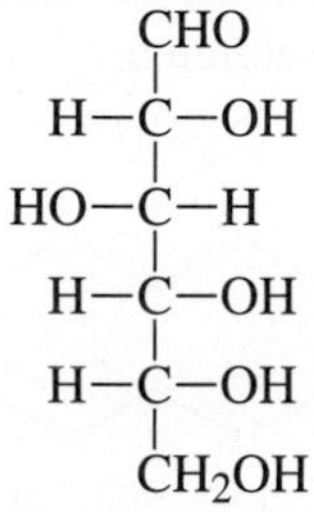

Like the other sugars presented in this chapter, though, glucose forms a cyclic molecule when dissolved in water, as shown in Figure 15.1:

Phosphates

Phosphoric acid, H_3PO_4, is a weak acid with the structure:

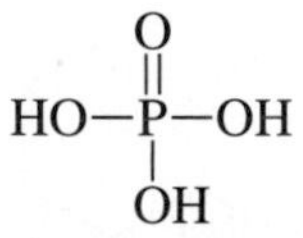

Organic phosphates are *esters* [◄◄ Section 14.7] of phosphoric acid—compounds in which one or more of the ionizable hydrogen atoms has been replaced by an R group:

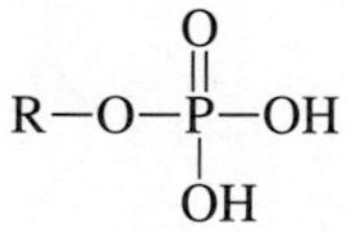

Organic Bases

Organic bases are molecules that contain carbon, hydrogen, and nitrogen atoms. These molecules exhibit Brønsted base behavior because a nitrogen atom has a pair of non-bonded electrons and can accept a proton. One such base is pyridine, C_5H_5N, which we have already encountered in the context of weak bases [◄◄ Section 12.6, Table 12.6].

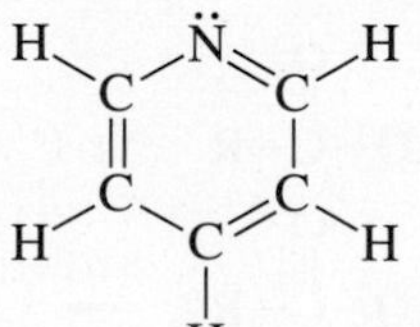

As we will see, the most important organic bases contain ring structures and multiple nitrogen atoms.

We now look at some of the molecules of life—including lipids, proteins, carbohydrates, and nucleic acids—and see how the *chemical* properties of these substances give rise to their *biological* properties.

15.2 Lipids

In general, the term *lipid* refers to a variety of relatively small organic molecules that are insoluble in water. Lipids include such substances as fatty acids, fats, and steroids, and they are involved in a vast array of biological functions.

Fats

We have all heard and used the term *fat* in the context of dietary intake. Moreover, we have been taught to distinguish *saturated* fats (purportedly unhealthy, animal fats) and *unsaturated* fats (purportedly healthier, vegetable fats). Chemically, a *fat* is a tri-ester of glycerol—meaning that each of the hydroxy-group hydrogen atoms in glycerol has been replaced by the alkyl group of a fatty acid (Figure 15.2):

Fats are also known as *triglycerides.* What determines whether a fat is saturated or unsaturated is whether or not the carbon chains it contains are saturated or unsaturated. For example, tristearin (also known as "stearin"), the fat derived from glycerol and three stearic acid molecules, is saturated:

Student Note: Tristearin is a by-product of the processing of beef, and is a major constituent of *tallow*—rendered beef fat.

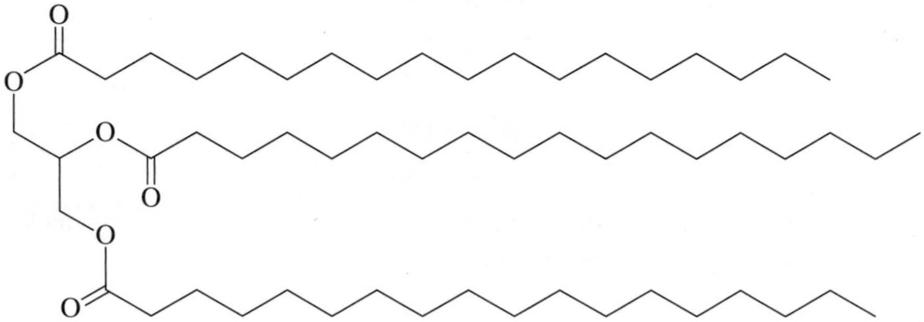

Triolein, the fat derived from glycerol and three oleic acid molecules, is unsaturated:

Student Note: Triolein is a component of olive oil.

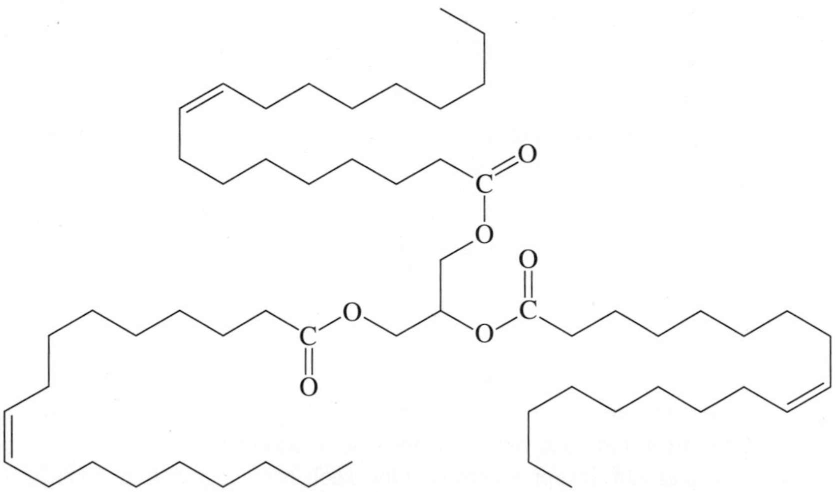

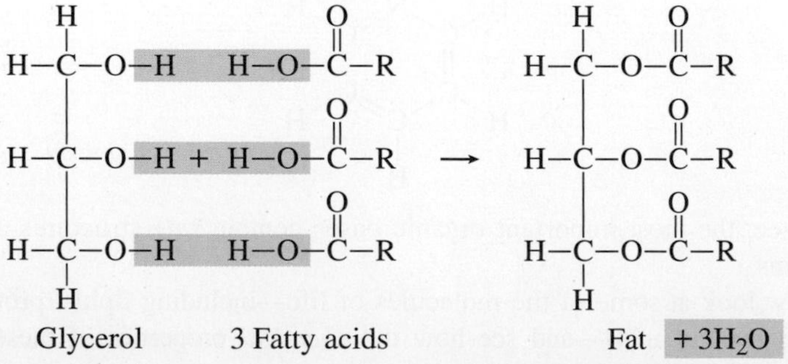

Figure 15.2 Combination of glycerol and three fatty acids to form a fat molecule. Note that a water molecule is eliminated for each addition of a fatty acid to the glycerol molecule. The atoms that are eliminated are highlighted.

Glycerol   3 Fatty acids   Fat $+3H_2O$

Unlike the examples of tristearin and triolein, most fats are *asymmetrical,* meaning that the fatty acids they contain are not all the same.

Phospholipids

Like fats, most phospholipids are derivatives of the glycerol molecule. Unlike fats, *phospholipids* are *di*glycerides with an organic phosphate group in place of one of the fatty acids:

$$
\begin{array}{c}
\underset{|}{\overset{H}{}} \quad \underset{}{\overset{O}{}} \\
H-C-O-C-R \\
| \qquad\quad O \\
\|\\
H-C-O-C-R \\
| \qquad\quad O \\
\|\\
H-C-O-P-R \\
| \qquad\quad | \\
H \qquad\quad O \\
| \\
H
\end{array}
$$

Nonpolar "tail"

Polar "head"

Having a phosphate group in place of one of the hydrocarbon chains gives phospholipids significantly different properties from those of fats. Unlike hydrocarbon chains, which are nonpolar and *hydrophobic,* a phosphate group is polar and *hydrophilic.* **Hydrophobic** literally means "water-fearing," whereas **hydrophilic** means "water-loving." Hydrophobic species are not water soluble. Fats, for example, consist of three long hydrocarbon chains that are nonpolar. Fats and oils (fats that are liquid at room temperature), as you know, do not mix with water. Phospholipids, on the other hand, have two distinct molecular regions: one nonpolar (the hydrocarbon chain) and one polar (the phosphate group). This causes phospholipids to arrange themselves in what is known as a **lipid bilayer.** When a sample of phospholipids is added to water, the hydrophobic "tails" (hydrocarbon chains) will avoid contact with the water by associating only with one another. This puts the hydrophilic "heads" (phosphate groups) in contact with the water—on the outside of the bilayer (Figure 15.3).

Lipid bilayers are important parts of cell structure, forming the very membranes that *contain* cells and that contain and protect many of their interior structures—including the nucleus.

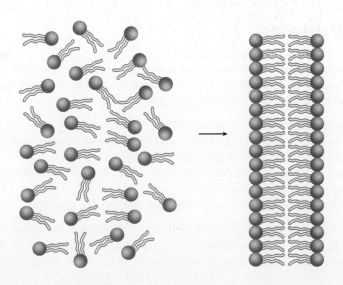

Figure 15.3 A collection of lipid molecules in water arranges itself into a bilayer. All of the hydrophobic tails (yellow) are together and away from the water. All of the hydrophilic heads (blue) are in contact with the water. Lipid bilayers are important structural features of living cells.

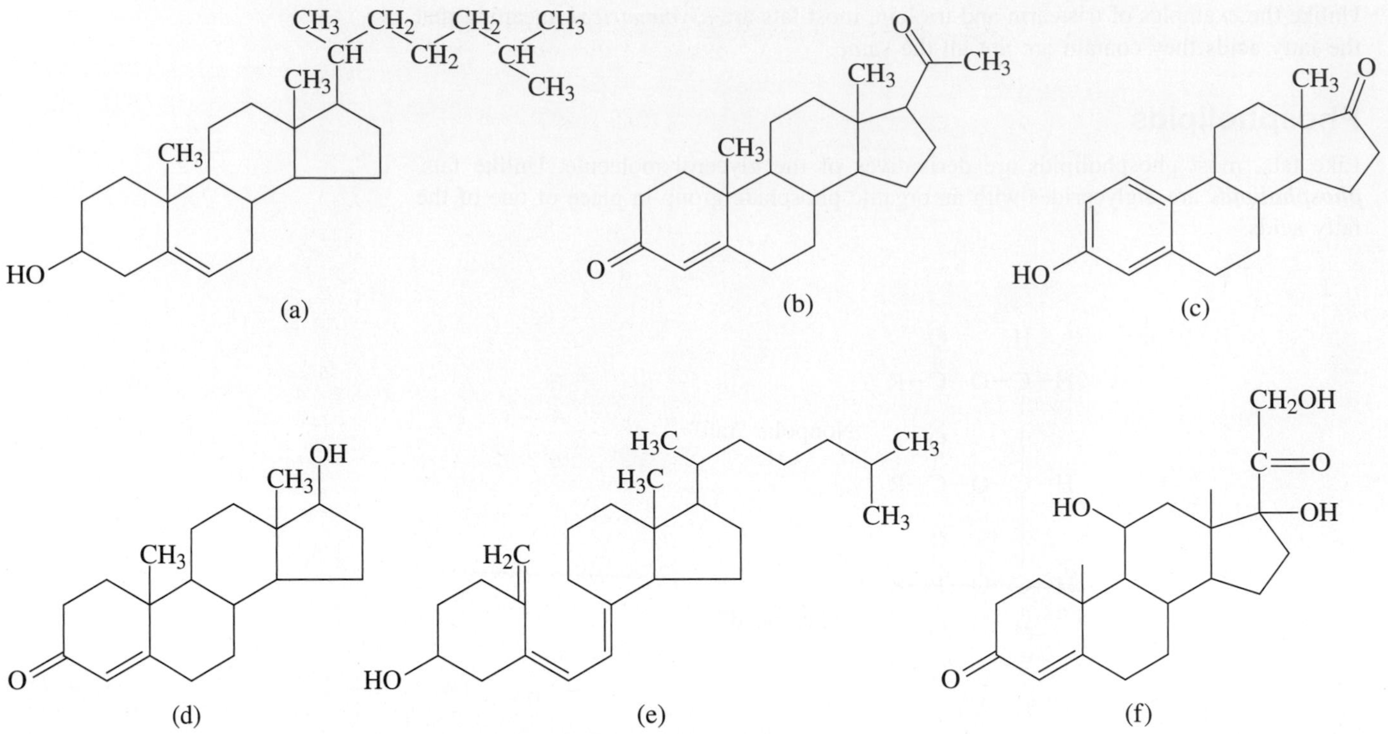

Figure 15.4 Common steroids.
(a) Cholesterol, (b) progesterone
(female sex hormone), (c) estrogen (female sex hormone),
(d) testosterone (male sex
hormone), (e) vitamin D_3,
(f) cortisol (stress hormone).

Steroids

Steroids are lipids that are derivatives of the molecule *sterol:*

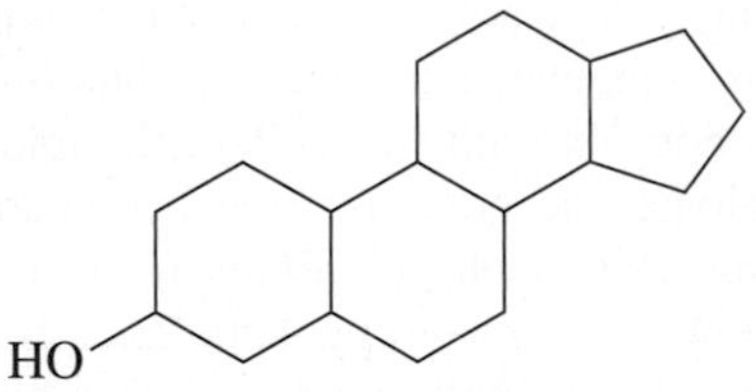

Probably the best known sterol derivative is cholesterol, which is a necessary component
of animal cell structure. It is used to build protective membranes in and around cells,
and to keep them fluid. It is also the building block necessary for production of other
steroids and steroid derivatives, including sex hormones, bile acids, and fat-soluble vita-
mins, such as vitamin D_3. Figure 15.4 shows the structures of several common steroids.

15.3 Proteins

Proteins play important roles in nearly all biological processes. The human body contains an estimated 100,000 different kinds of proteins, each of which has a specific
physiological function. These functions include:

- Structural substances, such as keratin (hair, nails, and skin)
- Hormones, such as oxytocin, insulin, and somatotropin (human growth hormone)
- Antibodies, such as immunoglobulin
- Enzymes, such as alcohol dehydrogenase, which breaks down ethanol; and
  sucrases, which break down sucrose
- Transport facilitators, such as hemoglobin (for oxygen transport)

A *protein* is a large biological polymer made up of a combination of any of the amino
acids shown in Figure 15.5. The identity of a protein depends on which of the 20 amino
acids it contains, and on the order in which those amino acids are assembled.

Name	Abbreviation	Structure
Alanine	Ala	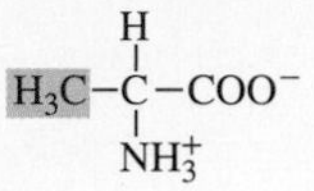 $H_3C-\overset{\overset{\displaystyle H}{\mid}}{\underset{\underset{\displaystyle NH_3^+}{\mid}}{C}}-COO^-$
Arginine	Arg	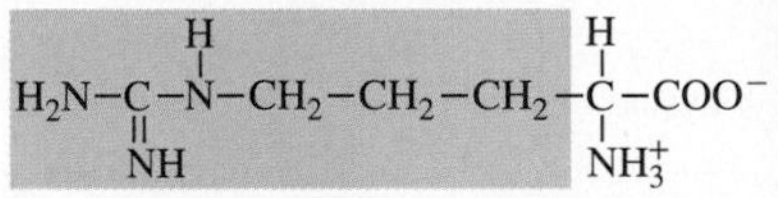 $H_2N-\overset{}{\underset{\underset{\displaystyle NH}{\parallel}}{C}}-N-CH_2-CH_2-CH_2-\overset{\overset{\displaystyle H}{\mid}}{\underset{\underset{\displaystyle NH_3^+}{\mid}}{C}}-COO^-$
Asparagine	Asn	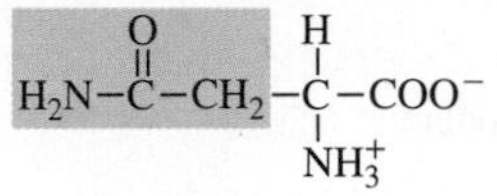 $H_2N-\overset{\overset{\displaystyle O}{\parallel}}{C}-CH_2-\overset{\overset{\displaystyle H}{\mid}}{\underset{\underset{\displaystyle NH_3^+}{\mid}}{C}}-COO^-$
Aspartic acid	Asp	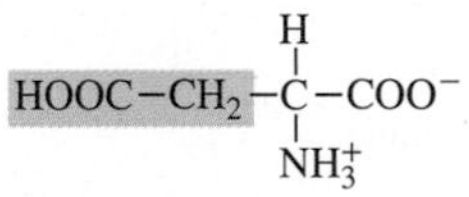 $HOOC-CH_2-\overset{\overset{\displaystyle H}{\mid}}{\underset{\underset{\displaystyle NH_3^+}{\mid}}{C}}-COO^-$
Cysteine	Cys	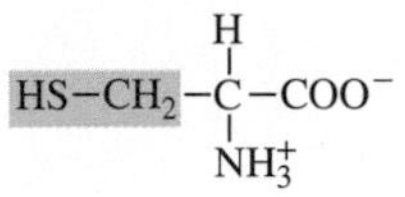 $HS-CH_2-\overset{\overset{\displaystyle H}{\mid}}{\underset{\underset{\displaystyle NH_3^+}{\mid}}{C}}-COO^-$
Glutamic acid	Glu	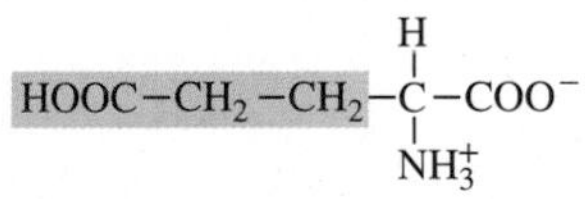 $HOOC-CH_2-CH_2-\overset{\overset{\displaystyle H}{\mid}}{\underset{\underset{\displaystyle NH_3^+}{\mid}}{C}}-COO^-$
Glutamine	Gln	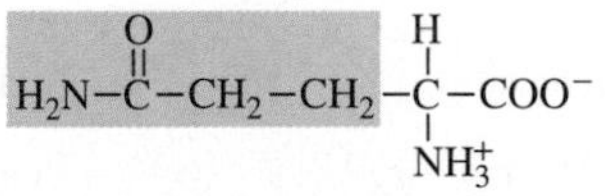 $H_2N-\overset{\overset{\displaystyle O}{\parallel}}{C}-CH_2-CH_2-\overset{\overset{\displaystyle H}{\mid}}{\underset{\underset{\displaystyle NH_3^+}{\mid}}{C}}-COO^-$
Glycine	Gly	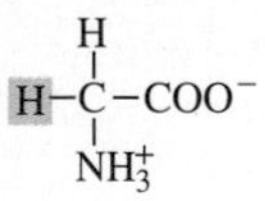 $H-\overset{\overset{\displaystyle H}{\mid}}{\underset{\underset{\displaystyle NH_3^+}{\mid}}{C}}-COO^-$
Histidine	His	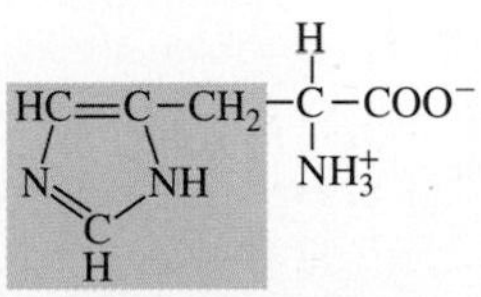 $HC=C-CH_2-\overset{\overset{\displaystyle H}{\mid}}{\underset{\underset{\displaystyle NH_3^+}{\mid}}{C}}-COO^-$
Isoleucine	Ile	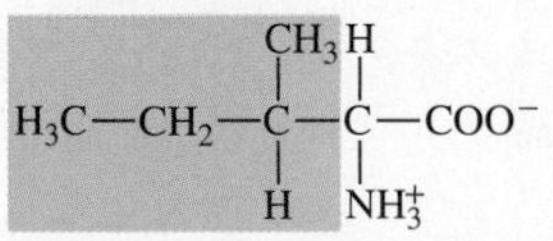 $H_3C-CH_2-\overset{\overset{\displaystyle CH_3}{\mid}}{\underset{\underset{\displaystyle H}{\mid}}{C}}-\overset{\overset{\displaystyle H}{\mid}}{\underset{\underset{\displaystyle NH_3^+}{\mid}}{C}}-COO^-$

Figure 15.5 The 20 amino acids essential to living organisms. The shaded area in each highlights the unique portion of each amino acid.

Name	Abbreviation	Structure
Leucine	Leu	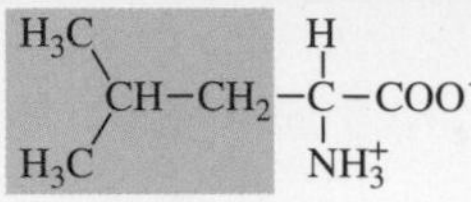
Lysine	Lys	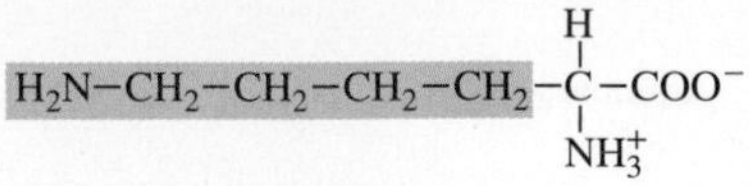
Methionine	Met	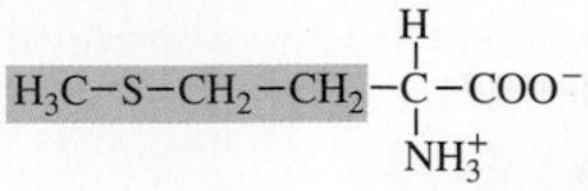
Phenylalanine	Phe	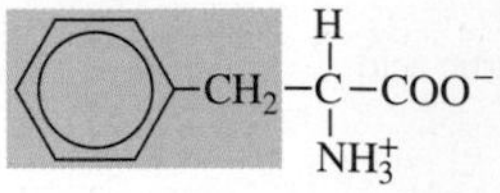
Proline	Pro	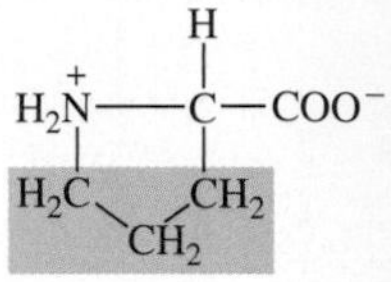
Serine	Ser	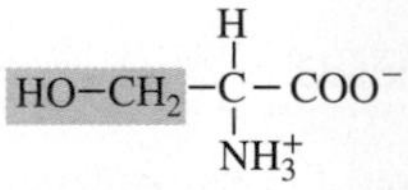
Threonine	Thr	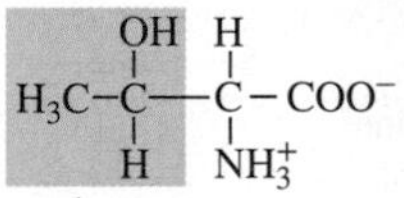
Tryptophan	Trp	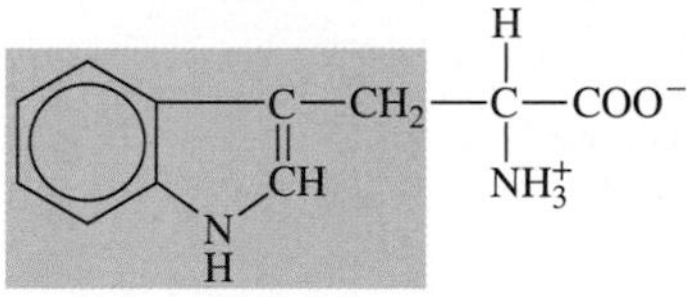
Tyrosine	Tyr	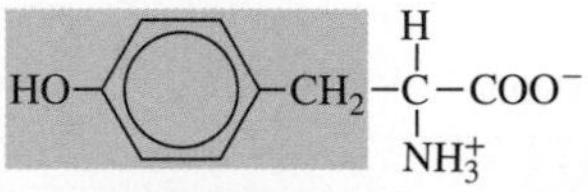
Valine	Val	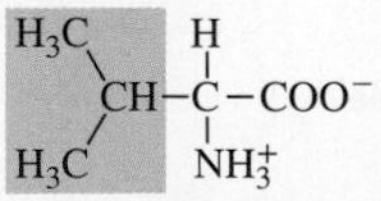

Figure 15.5 (*Continued*)

The bonds that form between amino acids when they undergo polymerization to form proteins are called *peptide bonds,* also known as *amide linkages.*

$$\underset{\substack{|\\H}}{\overset{\substack{H\\|}}{H-\overset{+}{N}}}-\underset{\substack{|\\R_1}}{\overset{\substack{H\\|}}{C}}-\overset{\substack{O\\\|}}{C}-O^- \;+\; \underset{\substack{|\\H}}{\overset{\substack{H\\|}}{H-\overset{+}{N}}}-\underset{\substack{|\\R_2}}{\overset{\substack{H\\|}}{C}}-\overset{\substack{O\\\|}}{C}-O^- \;\rightleftharpoons\; \underset{\substack{|\\H}}{\overset{\substack{H\\|}}{H-\overset{+}{N}}}-\underset{\substack{|\\R_1}}{\overset{\substack{H\\|}}{C}}-\overset{\substack{O\\\|}}{C}-\underset{\substack{|\\H}}{N}-\underset{\substack{|\\R_2}}{\overset{\substack{H\\|}}{C}}-\overset{\substack{O\\\|}}{C}-O^- \;+\; O-H$$

Peptide bond
(amide linkage)

The function of a protein depends on its structure, which depends on the attractive forces between the various groups it contains. A description of protein structure is done at three different levels—sometimes four.

Primary Structure

The *primary structure of a protein* refers to the sequence of amino acids that make up the protein chain. For example, consider the following partial sequence of amino acids in human hemoglobin:

The amino acids undergo condensation polymerization, eliminating one water molecule for each peptide bond formed. The primary structure of this part of hemoglobin is valine, histidine, leucine, threonine, proline, and glutamic acid.

Secondary Structure

Secondary structure (protein) refers to the shape the protein chain adopts as the result of hydrogen bonding between nearby groups. One of the possible secondary structures, and the one that this part of hemoglobin adopts, is a helix.

Student Note: The other possible secondary structure is a pleated sheet.

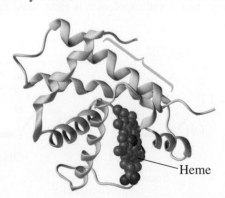

Tertiary Structure

Tertiary structure (protein) refers to the folding of the protein into a characteristic shape, which is stabilized by attractive forces between more distant groups.

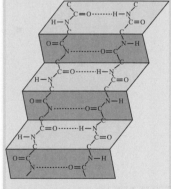

Note that within this tertiary structure, there is another molecule called heme. The heme molecule contains an iron atom and is responsible for the oxygen-carrying capacity of hemoglobin.

Quaternary Structure

Finally, in cases where the function of a protein requires that two or more folded protein chains be assembled into a single structure, **quaternary structure (protein)** refers to the shape resulting from that assembly. In the case of hemoglobin, four separate protein units, each containing a heme molecule, are assembled in a roughly spherical arrangement known as a *globular* protein.

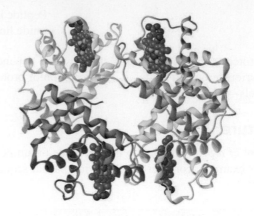

15.4 Carbohydrates

Student Note: Although most monosaccharides have this general formula, some do not. A very important exception to this formula is deoxyribose, which we discuss in Section 15.5.

The term **carbohydrate** refers to simple sugars with the general formula $C_n(H_2O)_m$, including such molecules as glucose—and the dimers and polymers that these sugars can form.

Monosaccharides

The term **monosaccharide** usually refers to sugar molecules in which the n and m of the general carbohydrate formula are equal. [This simplifies the general monosaccharide formula to $(CH_2O)_n$.] Some of the most biologically important monosaccharides are those in which $n = 5$ or 6; these are known respectively as *pentoses* and *hexoses*. Glucose, which we saw in Section 15.1, is a *hexose*. Figure 15.6 shows two other important monosaccharides, each in straight-chain form and cyclic form.

Disaccharides

A **disaccharide** is a dimer of monosaccharides. Dimers of monosaccharides form as the result of condensation reactions. The resulting linkage between the monomers is called a glycosidic linkage, and a water molecule is eliminated for each glycosidic linkage formed.

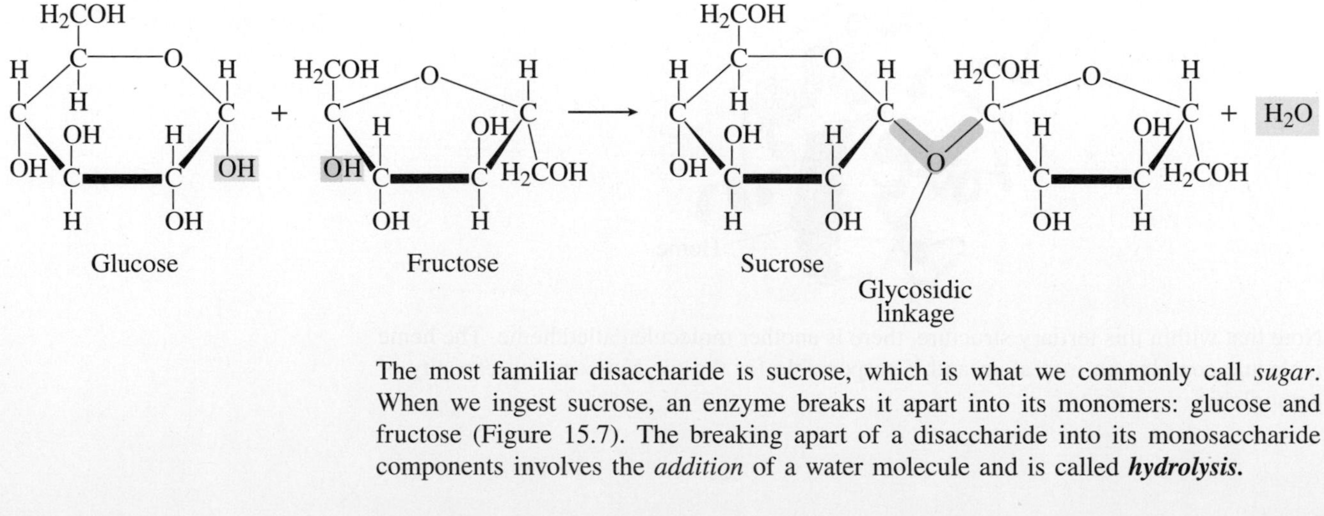

H₂COH

Glucose   +   Fructose   →   Sucrose   +   H₂O

Glycosidic linkage

The most familiar disaccharide is sucrose, which is what we commonly call *sugar*. When we ingest sucrose, an enzyme breaks it apart into its monomers: glucose and fructose (Figure 15.7). The breaking apart of a disaccharide into its monosaccharide components involves the *addition* of a water molecule and is called **hydrolysis.**

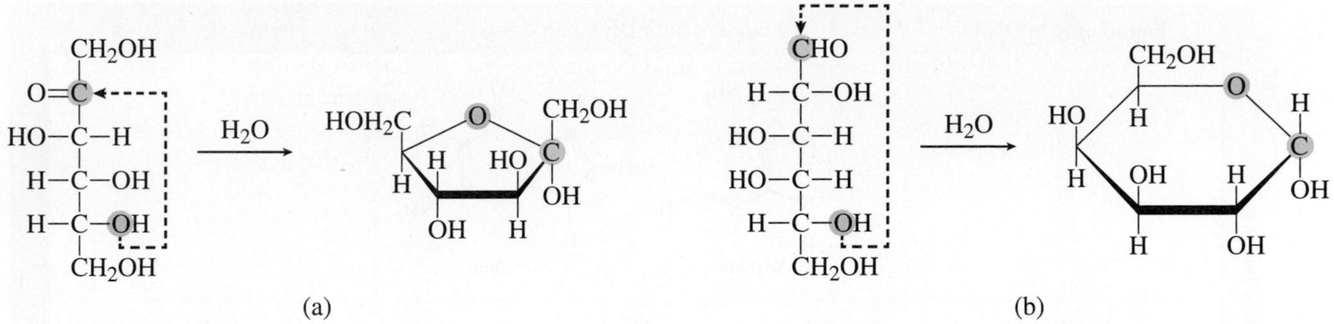

(a)

(b)

Figure 15.6 Important monosaccharides: (a) fructose, (b) galactose.

Polysaccharides

Polysaccharides are polymers of sugars such as glucose and fructose. Starch and cellulose are two polymers of glucose with slightly different linkages—and very different properties. In starch, glucose molecules are connected by what biochemists call α linkages. This enables animals, including humans, to digest the starch in such foods as corn, wheat, potatoes, and rice.

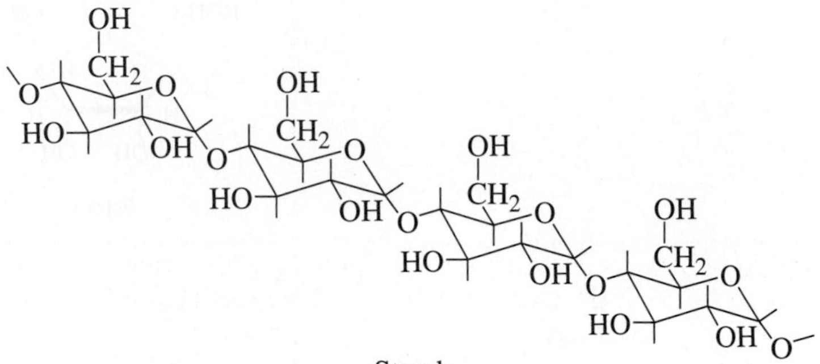

Starch

In cellulose, the glucose molecules are connected by β linkages. Digestion of cellulose requires enzymes that most animals do not have. Species that *do* digest cellulose, such as termites and ruminants (including cattle, sheep, and llamas), do so with the help of enzyme-producing symbiotic bacteria in the gut.

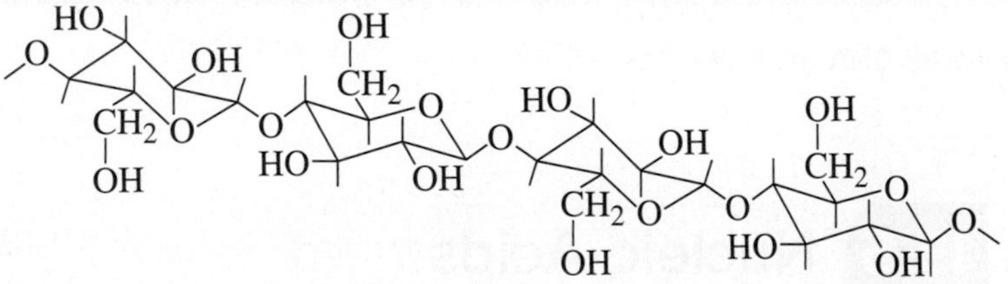

Cellulose

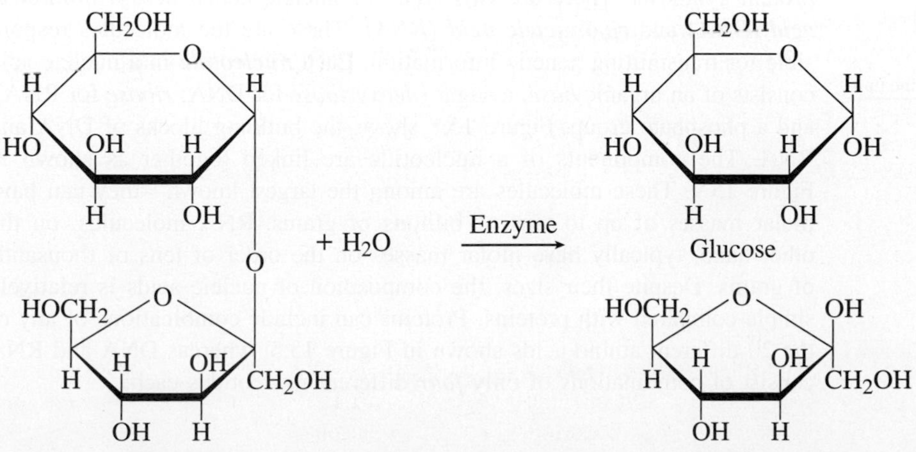

Sucrose

+ H₂O

Enzyme →

Glucose

Fructose

Figure 15.7 An enzyme called a sucrase catalyzes the hydrolysis of sucrose to yield glucose and fructose.

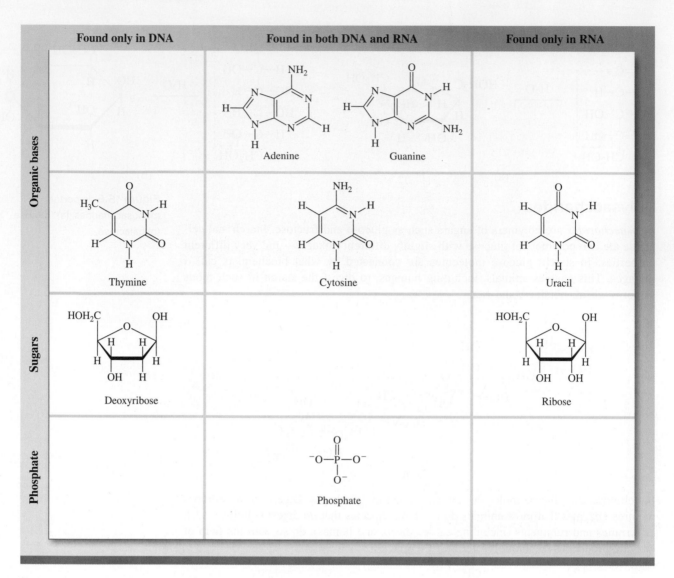

Figure 15.8 The components of the nucleic acids DNA and RNA.

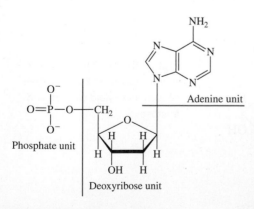

Figure 15.9 Structure of a nucleotide, one of the repeating units in DNA.

15.5 Nucleic Acids

Nucleic acids, which are polymers of *nucleotides,* play an important role in protein synthesis. There are two types of nucleic acids: *deoxyribonucleic acid (DNA)* and *ribonucleic acid (RNA).* These are the molecules responsible for transmitting genetic information. Each *nucleotide* in a nucleic acid consists of an organic *base,* a sugar (*deoxyribose* for DNA; *ribose* for RNA), and a phosphate group. Figure 15.8 shows the building blocks of DNA and RNA. The components of a nucleotide are linked together as shown in Figure 15.9. These molecules are among the largest known—they can have molar masses of up to tens of billions of grams. RNA molecules, on the other hand, typically have molar masses on the order of tens of thousands of grams. Despite their sizes, the composition of nucleic acids is relatively simple compared with proteins. Proteins can include combinations of any of the 20 different amino acids shown in Figure 15.5, whereas DNA and RNA consist of combinations of only *four* different nucleotides each.

Profiles in Science

Rosalind Franklin

Chemist and X-ray crystallographer Rosalind Franklin (1920–1958) earned her Ph.D. in physical chemistry at Cambridge University. Her research produced the groundbreaking crystallographic images of deoxyribonucleic acid (DNA) that were critical to the elucidation of its structure—the iconic double-helix. James Watson and Francis Crick, the names most often associated with the discovery of DNA's structure, were given Franklin's data without her knowledge by her colleague, Maurice Wilkins. Watson, Crick, and Wilkins were jointly awarded the 1962 Nobel Prize in Physiology or Medicine "for their discoveries concerning the molecular structure of nucleic acids and its significance for information transfer in living material." In truth, if not for Franklin's work, these three men may not have made their most famous discovery. Sadly, Franklin had died in 1958, at age 37. Her cause of death was ovarian cancer, which was almost certainly the result of repeated and prolonged radiation exposure in her research laboratory. The Nobel Prize has rarely been awarded

Rosalind Franklin
Jewish Chronicle Archive/Heritage
Image Partnership Ltd/Alamy Stock
Photo

posthumously and can only be shared by a maximum of three laureates. Nevertheless, this failure of the scientific community to recognize and laud Franklin's contributions during her lifetime have given rise to monikers such as the "forgotten heroine" and the "dark lady of DNA."

Chapter Summary

Section 15.1

- Biochemistry is the branch of chemistry concerned with the function of living systems. Biologically important molecules include *glycerol,* a molecule with three alcohol (–OH) groups; *fatty acids,* long-chain alkyl groups attached to a carboxy group; *amino acids,* molecules with an amine group, a carboxy group, and a unique group that includes carbon, hydrogen, and sometimes other elements such as N or S; *sugars,* water-soluble, typically sweet aldehydes or ketones; *organic phosphates,* esters of phosphoric acid (H_3PO_4); and *organic bases,* derivatives of ammonia (NH_3).

Section 15.2

- *Lipids* are small, water-insoluble organic molecules—including fatty acids, fats, and steroids. *Fats* are tri-esters of glycerol, in which each hydrogen atom from each alcohol group has been replaced by an alkyl group. Fats are also referred to as *triglycerides.* Saturated fats are those that contain no carbon-carbon double bonds in any of their alkyl groups. Unsaturated fats have carbon-carbon double bonds in one or more of their alkyl groups.

- Long hydrocarbon chains are *hydrophobic* (water-fearing) and are *not* water soluble. Phosphate groups, because of their

polarity, are *hydrophilic* (water-loving) and *are* water soluble. Lipids, which contain both long hydrocarbon chains and a phosphate group, form *lipid bilayers,* in which the hydrophilic ends of the molecules associate with water, and the hydrophobic chains associate only with one another.

- *Phospholipids* are diglycerides—derivatives of the glycerol molecule in which two of the hydrogen atoms from the alcohol groups have been replaced by hydrocarbon chains and one has been replaced by an organic phosphate group.

- *Steroids* are lipids derived from the *sterol* molecule. Cholesterol is a steroid that is necessary for the production of animal cells.

Section 15.3

- *Proteins* are biological polymers made up of amino acids. Proteins play numerous important roles in the function of living systems. The bonds between amino acids in a protein are known as *peptide bonds* or *amide linkages.*

- The structure of a protein is described in four different stages. *Primary structure* refers to the specific sequence of amino acids. *Secondary structure* refers to the shape adopted by the string of amino acids. Secondary structures are determined by the attractive forces between nearby amino acids, and include the *helix* and the *pleated sheet.* *Tertiary structure* refers to the way a protein folds into a characteristic shape, which is influenced by more distant amino acids. *Quaternary structure* refers to the shape that results when two or more folded protein chains combine to form a single structure.

reaction that separates a disaccharide into two monosaccharides is called *hydrolysis.* Hydrolysis of disaccharides is catalyzed by specialized proteins called enzymes. *Polysaccharides* are polymers of monosaccharides.

Section 15.4

• *Carbohydrates* are simple sugars, molecules made up of carbon, hydrogen, and oxygen with the general formula $C_n(H_2O)_m$. *Monosaccharides* are sugar molecules with the general formula $(CH_2O)_n$. *Disaccharides* are dimers of monosaccharides. The

Section 15.5

• A *nucleotide* is a molecule that consists of an organic base, a sugar, and a phosphate group. *Nucleic acids* are polymers of nucleotides. Nucleic acids include *deoxyribonucleic acid (DNA)* and *ribonucleic acid (RNA).*

Key Terms

Amide linkages 525

Amino acid 517

Carbohydrate 526

Deoxyribonucleic acid (DNA) 528

Disaccharide 526

Fat 520

Fatty acid 517

Glycerol 517

Hydrolysis 526

Hydrophilic 521

Hydrophobic 521

Lipid 520

Lipid bilayer 521

Monosaccharide 526

Nucleic acid 528

Nucleotide 528

Organic bases 519

Organic phosphates 519

Peptide bond 525

Phospholipid 521

Polysaccharide 527

Primary structure (protein) 525

Protein 522

Quaternary structure (protein) 526

Ribonucleic acid (RNA) 528

Secondary structure (protein) 525

Steroid 522

Sugar 519

Tertiary structure (protein) 525

Triglyceride 520

Questions and Problems

SECTION 15.1: BIOLOGICALLY IMPORTANT MOLECULES

15.1   Would you expect glycerol to be soluble or insoluble in water? Explain.

15.2   Identify the functional group that fatty acids contain. Select a fatty acid and sketch its structure.

15.3   What two functional groups do amino acids contain? Select an amino acid and sketch its structure.

15.4   What functional group(s) do sugars contain?

15.5   Give an example of an organic phosphate.

15.6   Give an example of an organic base and draw its structure.

SECTION 15.2: LIPIDS

15.7   Define the term *lipid.*

15.8   Name the three major types of lipids.

15.9   Give an example of each of the three major types of lipids.

15.10   Define the term *fat.*

15.11   Explain why fats are insoluble in water.

15.12   Identify the molecular components of a triglyceride.

15.13   Explain the difference between a saturated and an unsaturated fat.

15.14   Label each fat as either saturated or unsaturated.

(c)

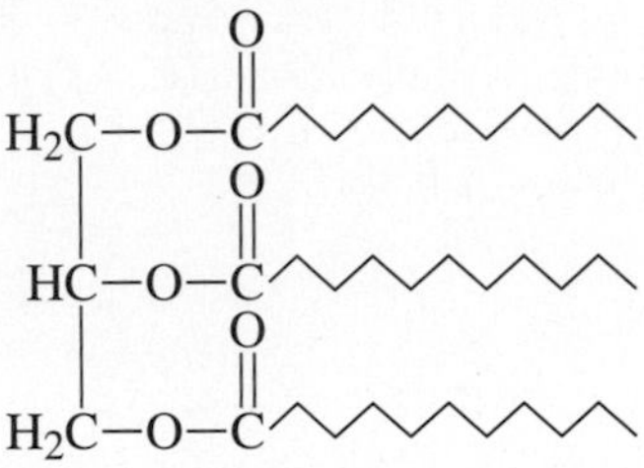

15.15 Draw the structure of the fat (triglyceride) that contains the fatty acid, butyric acid.

15.16 Define the term *phospholipid*. Explain how a phospholipid is different from a fat.

15.17 Define the terms *hydrophobic* and *hydrophilic*.

15.18 Describe a *lipid bilayer* and explain how it forms.

15.19 Define the term *diglyceride*.

15.20 Define the term *steroid*. Explain what all steroid structures have in common.

SECTION 15.3: PROTEINS

15.21 Explain the relationship between amino acids and proteins.

15.22 Define the term *peptide bond* and explain why peptide bonds are important.

15.23 Determine which of the following molecules are amino acids:

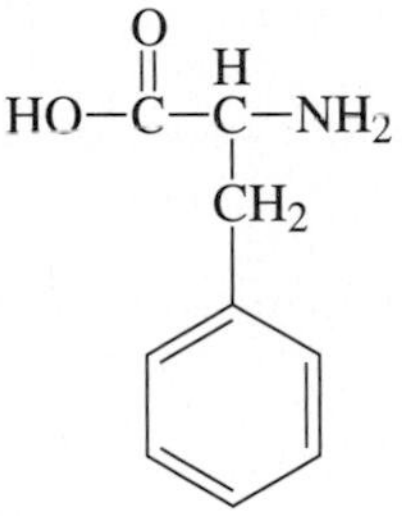

(a)

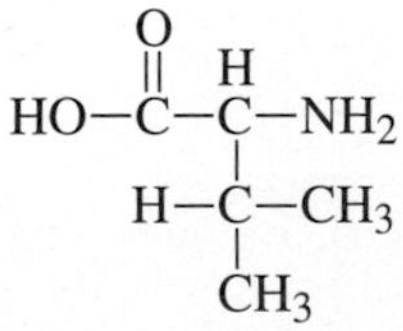

(b)

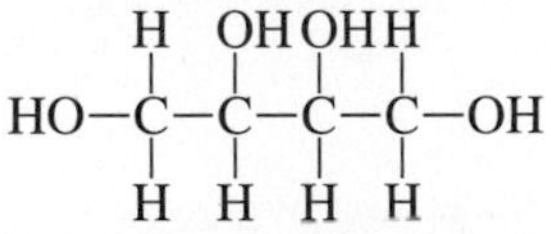

(c)

(d) 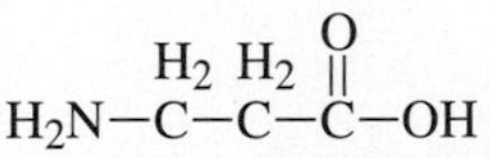 H₂N—C—C—C—OH

15.24 Determine which of the following molecules are amino acids:

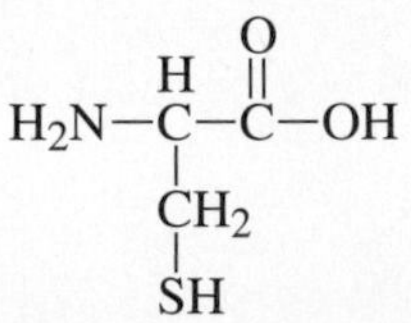

(a)

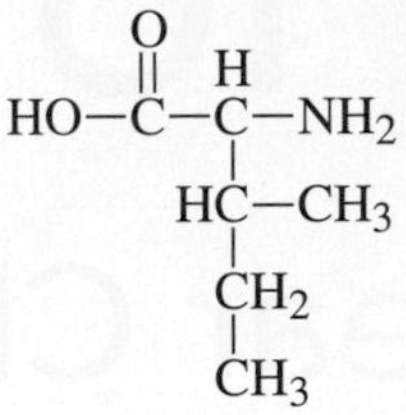

(b)

(c)

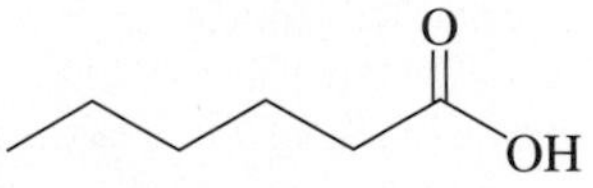

(d) 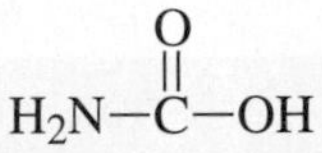 H₂N—C—OH

15.25 Show the product of the reaction of threonine and leucine when they form a peptide bond (Thr-Leu).

15.26 Explain how the structure of Leu-Thr is different from that of Thr-Leu.

15.27 Draw the structure for each polypeptide:
(a) Ser-Val-Lys
(b) Tyr-Gly-Ala
(c) His-Asp-Glu-Lys

15.28 Does the information Val-Ala-Gly-Leu-His-Ile-Asn-Pro-Ser describe a primary, secondary, tertiary, or quaternary structure?

15.29 Describe what is meant by the secondary structure of a protein and give two examples.

15.30 Explain the importance of a protein's tertiary structure.

15.31 Describe what is meant by quaternary structure.

SECTION 15.4: CARBOHYDRATES

15.32 Define the term *carbohydrate* and give the general chemical formula.

15.33 Give the general chemical formula for a monosaccharide. Give an example of a monosaccharide.

15.34 Show the reaction between fructose and galactose to form a dissacharide.

15.35 What term describes the bond formed when two monosaccharides combine to form a disaccharide?

15.36 Describe the hydrolysis of a disaccharide.

15.37 Define the term *polysaccharide* and give an example.

SECTION 15.5: NUCLEIC ACIDS

15.38 Define the term *nucleic acid*.

15.39 Explain how the components of DNA are different from the components of RNA.

CHAPTER 16

Nuclear Chemistry

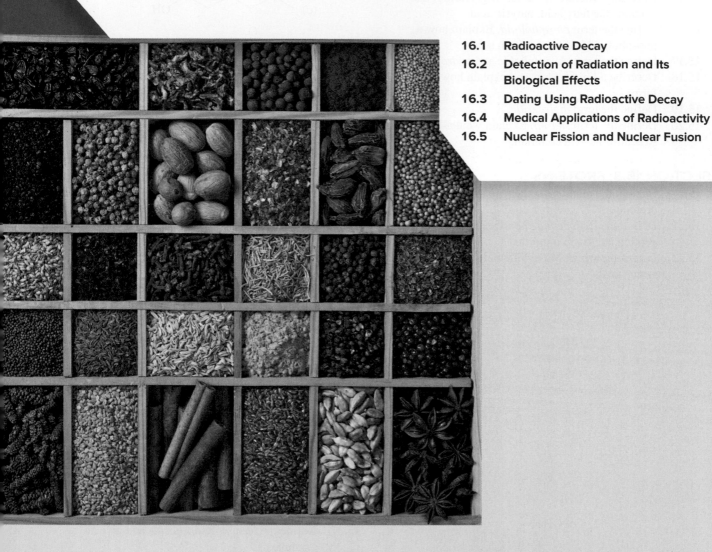

16.1 Radioactive Decay

16.2 Detection of Radiation and Its Biological Effects

16.3 Dating Using Radioactive Decay

16.4 Medical Applications of Radioactivity

16.5 Nuclear Fission and Nuclear Fusion

Irradiation of spices and seasonings, which often are produced and harvested in regions without adequate food-hygiene standards, ensures their safety for use in food.

Andrey Gorulko/iStockphoto/Getty Images

In This Chapter, You Will Learn

About several nuclear processes and their biological effects. You will also learn about the beneficial uses of some of these processes, including their use in determining the ages of fossils and rocks, medical procedures, food safety, and power production.

Things To Review Before You Begin

- Nuclear model of the atom [◄◄ Section 1.3]
- Reaction rates [◄◄ Section 13.1]

You learned in Chapter 1 that atomic nuclei are composed of the subatomic particles *protons* and *neutrons.* Some atomic nuclei are unstable and spontaneously emit particles and/or radiation to achieve stability—a process known as radioactive decay. Other nuclei can be manipulated to produce large amounts of energy. Many nuclear processes have proven useful in the treatment of cancer, the production of electricity, and the irradiation of certain foods to reduce the incidence of food-borne illness.

16.1 Radioactive Decay

With the exception of hydrogen ($_1^1$H), all nuclei contain protons and neutrons. Some nuclei are unstable and undergo *radioactive decay,* emitting particles [◄◄ Section 1.3] and/or electromagnetic radiation [◄◄ Section 2.1]. Radioactive nuclei can undergo a variety of decay types. Three types of radioactive decay are emission of ***beta particles;*** emission of ***gamma radiation;*** and emission of *alpha particles,* which we saw in Chapter 1. Spontaneous emission of particles or electromagnetic radiation is known as ***radioactivity.*** Figure 16.1 illustrates alpha, beta, and gamma emission by a radioactive substance.

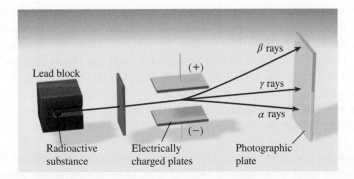

Figure 16.1 Three types of radiation from a radioactive substance. β radiation is actually a stream of electrons, which are negatively charged and therefore attracted to a positively charged plate. α radiation consists of positively charged, relatively massive particles. Each α particle consists of two protons and two neutrons—and is essentially a helium nucleus. Being positively charged, α particles are attracted to a negatively charged plate. γ radiation does not consist of particles, but is simply high-energy electromagnetic radiation.

TABLE 16.1	Comparison of Chemical Reactions and Nuclear Reactions	
Chemical Reactions	**Nuclear Reactions**	
1. Atoms are rearranged by the breaking and forming of chemical bonds.	1. Elements are converted to other elements (or isotopes).	
2. Only electrons in atomic or molecular orbitals are involved in the reaction.	2. Protons, neutrons, electrons, and other subatomic particles such as α particles may be involved.	
3. Reactions are accompanied by the absorption or release of relatively small amounts of energy.	3. Reactions are accompanied by the absorption or release of tremendous amounts of energy.	
4. Rates of reactions are influenced by temperature, pressure, concentration, and catalysts.	4. Rates of reactions normally are not affected by temperature, pressure, or catalysts.	

All elements having an atomic number greater than 83 are unstable and are therefore radioactive. Polonium-210 ($^{210}_{84}\text{Po}$), for example, decays spontaneously to Pb by emitting an α particle.

Another type of nuclear process, known as ***nuclear transmutation,*** results from the bombardment of nuclei by neutrons, protons, or other nuclei. An example of a nuclear transmutation is the conversion of atmospheric $^{14}_{7}\text{N}$ to $^{14}_{6}\text{C}$ and $^{1}_{1}\text{H}$, which results when the nitrogen isotope is bombarded by neutrons (from the Sun). In some cases, heavier elements are synthesized from lighter elements. This type of transmutation occurs naturally in outer space, but it can also be achieved artificially.

Radioactive decay and nuclear transmutation are *nuclear reactions,* which differ significantly from ordinary chemical reactions. Table 16.1 summarizes the differences.

To discuss nuclear reactions in any depth, we must understand how to write and balance nuclear equations. Writing a nuclear equation differs somewhat from writing equations for chemical reactions. In addition to writing the symbols for the various chemical elements, we must explicitly indicate the number of subatomic particles in *every* species involved in the reaction.

The symbols for subatomic particles are as follows:

$$\begin{array}{ccccc} ^{1}_{1}\text{H} \text{ or } ^{1}_{1}\text{p} & ^{1}_{0}\text{n} & ^{0}_{-1}e \text{ or } ^{0}_{-1}\beta & ^{0}_{+1}e \text{ or } ^{0}_{+1}\beta & ^{4}_{2}\alpha \text{ or } ^{4}_{2}\text{He} \\ \text{proton} & \text{neutron} & \text{electron} & \text{positron} & \alpha \text{ particle} \end{array}$$

In accordance with the notation introduced in Section 1.6, the superscript in each case denotes the mass number (the total number of neutrons and protons present) and the subscript is the atomic number (the number of protons). Thus, the "atomic number" of a proton is 1, because there is one proton present, and the "mass number" is also 1, because there is one proton but no neutrons present. On the other hand, the mass number of a neutron is 1, but its atomic number is zero, because there are no protons present. For the electron, the mass number is zero (there are neither protons nor neutrons present), but the atomic number is −1, because the electron possesses a unit negative charge.

The symbol $^{0}_{-1}e$ represents an electron in or from an atomic orbital. The symbol $^{0}_{-1}\beta$, on the other hand, represents an electron that, although physically identical to any other electron, comes from a nucleus (in a decay process in which a neutron is converted to a proton and an electron) and not from an atomic orbital. The ***positron*** has the same mass as the electron, but bears a charge of +1. The α particle has two protons and two neutrons, so its atomic number is 2 and its mass number is 4.

In balancing any nuclear equation, we must balance the total of all atomic numbers and the total of all mass numbers for the products and reactants. If we know the atomic numbers and mass numbers of all but one of the species in a nuclear equation, we can identify the *unknown* species by applying these rules, as shown in Sample Problem 16.1.

Student Note: An α particle is identical to a helium-4 nucleus and can be represented as either $^{4}_{2}\alpha$ or $^{4}_{2}\text{He}$.

SAMPLE PROBLEM 16.1 Balancing Nuclear Equations

Identify the missing species X in each of the following nuclear equations:

(a) $^{212}_{84}Po \longrightarrow ^{208}_{82}Pb + X$

(b) $^{90}_{38}Sr \longrightarrow X + ^{0}_{-1}\beta$

(c) $X \longrightarrow ^{18}_{8}O + ^{0}_{+1}\beta$

Strategy Determine the mass number for the unknown species, X, by summing the mass numbers on both sides of the equation:

$$\Sigma \text{ reactant mass numbers} = \Sigma \text{ product mass numbers}$$

Similarly, determine the atomic number for the unknown species:

$$\Sigma \text{ reactant atomic numbers} = \Sigma \text{ product atomic numbers}$$

Use the mass number and atomic number to determine the identity of the unknown species.

Setup (a) 212 = (208 + mass number of X); mass number of X = 4. 84 = (82 + atomic number of X); atomic number of X = 2.

(b) 90 = (mass number of X + 0); mass number of X = 90. 38 = [atomic number of X + (−1)]; atomic number of X = 39.

(c) Mass number of X = (18 + 0); mass number of X = 18. Atomic number of X = (8 + 1); atomic number of X = 9.

Solution (a) $X = ^{4}_{2}\alpha$: $^{212}_{84}Po \longrightarrow ^{208}_{82}Pb + ^{4}_{2}\alpha$

(b) $X = ^{90}_{39}Y$: $^{90}_{38}Sr \longrightarrow ^{90}_{39}Y + ^{0}_{-1}\beta$

(c) $X = ^{18}_{9}F$: $^{18}_{9}F \longrightarrow ^{18}_{8}O + ^{0}_{+1}\beta$

THINK ABOUT IT

The rules of summation that we apply to balance nuclear equations can be thought of as the *conservation of mass number* and the *conservation of atomic number.*

Practice Problem **A**TTEMPT Identify X in each of the following nuclear equations:

(a) $^{78}_{33}As \longrightarrow X + ^{0}_{-1}\beta$ (b) $^{1}_{1}H + ^{4}_{2}He \longrightarrow X$ (c) $^{258}_{100}Fm \longrightarrow ^{257}_{100}Fm + X$

Practice Problem **B**UILD Identify X in each of the following nuclear equations:

(a) $X + ^{0}_{-1}\beta \longrightarrow ^{244}_{94}Pu$ (b) $^{238}_{92}U \longrightarrow X + ^{4}_{2}He$ (c) $X \longrightarrow ^{14}_{7}N + ^{0}_{-1}e$

Practice Problem **C**ONCEPTUALIZE For each process, specify the identity of the product.

$\xleftarrow{\text{electron capture}} \quad ^{222}Rn \quad \xrightarrow{\text{alpha emission}} \qquad \xleftarrow{\text{electron capture}} \quad ^{132}Cs \quad \xrightarrow{\text{beta emission}}$

CHECKPOINT–SECTION 16.1 Radioactive Decay

16.1.1 Identify the species X in the following nuclear equation:

$$^{222}_{86}Rn \longrightarrow X + ^{4}_{2}\alpha$$

a) $^{226}_{84}Po$

b) $^{226}_{88}Ra$

c) $^{212}_{84}Po$

d) $^{218}_{84}Po$

e) $^{218}_{88}Ra$

16.1.2 Identify the species X in the following nuclear equation:

$$^{15}_{8}O \longrightarrow X + ^{0}_{-1}\beta$$

a) $^{15}_{9}F$

b) $^{14}_{8}O$

c) $^{16}_{9}F$

d) $^{15}_{7}N$

e) $^{14}_{7}N$

16.2 Detection of Radiation and Its Biological Effects

Radioactive isotopes are relatively easy to detect. Their presence, even in very small amounts, can be detected by photographic techniques—or by devices known as *counters*. Figure 16.2 is a diagram of a Geiger counter, an instrument used commonly in scientific work and medical laboratories to detect radiation.

The fundamental unit of radioactivity is the *curie* (Ci); 1 Ci corresponds to exactly 3.70×10^{10} nuclear disintegrations per second. This decay rate is equivalent to that of 1 g of radium. A *millicurie* (mCi) is one-thousandth of a curie. Thus, 10 mCi of a carbon-14 sample is the quantity that undergoes $(10 \times 10^{-3})(3.70 \times 10^{10}) = 3.70 \times 10^{8}$ disintegrations per second.

The intensity of radiation depends on the number of disintegrations as well as on the energy and type of radiation emitted. One common unit for the absorbed dose of radiation is the *rad* (radiation *a*bsorbed *d*ose), which is the amount of radiation that results in the absorption of 1×10^{-5} J per gram of irradiated material. The biological effect of radiation depends on the part of the body irradiated and the type of radiation. For this reason, the rad is often multiplied by a factor called the *RBE* (*r*elative *b*iological *e*ffectiveness). The product is called a *rem* (*r*oentgen *e*quivalent for *m*an):

$$\text{numbers of rems} = \text{number of rads} \times 1 \text{ RBE}$$

Of the three types of nuclear radiation, α particles usually have the least penetrating power. Beta particles are more penetrating than α particles, but less so than γ rays.

Gamma rays have very short wavelengths and high energies. Furthermore, because they carry no charge, they cannot be stopped by shielding materials as easily as α and β particles. If α- or β-emitters are ingested or inhaled, however, their damaging effects are greatly aggravated because the organs will be constantly subject to damaging radiation at close range. For example, strontium-90, a β-emitter, can replace calcium in bones, where it does the greatest damage.

Table 16.2 lists the average amounts of radiation an American receives every year. For short-term exposures to radiation, a dosage of 50 to 200 rems will cause a decrease in white blood cell counts and other complications, while a dosage of 500 rems or greater may result in death within weeks. Current safety standards permit nuclear

Figure 16.2 Schematic diagram of a Geiger counter. Radiation (α, β, or γ rays) entering through the window ionizes the argon gas to generate a small current flow between the electrodes. This current is amplified and is used to flash a light or operate a counter with a clicking sound.

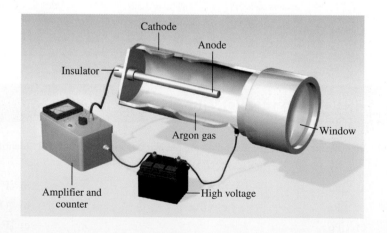

TABLE 16.2	Average Yearly Radiation Doses for Americans
Source	Dose (mrem/yr)*
Cosmic rays	20–50
Ground and surroundings	25
Human body†	26
Medical and dental X rays	50–75
Air travel	5
Fallout from weapons tests	5
Nuclear waste	2
Total	133–188

*1 mrem = millirem = 1×10^{-3} rem.
†The radioactivity in the body comes from food and air.

workers to be exposed to no more than 5 rems per year and specify a maximum of 0.5 rem of human-made radiation per year for the general public.

The chemical basis of radiation damage is that of ionizing radiation. Radiation (of either particles or γ rays) can remove electrons from atoms and molecules in its path, leading to the formation of ions and radicals. **Radicals** (also called *free radicals*) are molecular fragments having one or more unpaired electrons; they are usually short lived and highly reactive. When water is irradiated with γ rays, for example, the following reactions take place:

$$H_2O \xrightarrow{\text{radiation}} H_2O^+ + e^-$$

$$H_2O^+ + H_2O \longrightarrow H_3O^+ + \cdot OH$$
$$\text{Hydroxyl radical}$$

The electron (in the hydrated form) can subsequently react with water or with a hydrogen ion to form atomic hydrogen, and with oxygen to produce the superoxide ion (O_2^-) (a radical):

$$e^- + O_2 \longrightarrow \cdot O_2^-$$

In the tissues the superoxide ions and other free radicals attack cell membranes and a host of organic compounds, such as enzymes and DNA molecules. Organic compounds can themselves be directly ionized and destroyed by high-energy radiation.

Exposure to high-energy radiation can induce cancer in humans and other animals. Cancer is characterized by uncontrolled cellular growth. On the other hand, cancer cells can be destroyed by proper radiation treatment. In radiation therapy, a compromise is sought. The radiation to which the patient is exposed must be sufficient to destroy cancer cells without killing too many normal cells and, it is hoped, without inducing another form of cancer.

Radiation damage to living systems is generally classified as *somatic* or *genetic*. Somatic injuries are those that affect the organism during its own lifetime. Sunburn, skin rash, cancer, and cataracts are examples of somatic damage. Genetic damage means inheritable changes or gene mutations. For example, a person whose chromosomes have been damaged or altered by radiation may have deformed offspring.

Familiar Chemistry

Radioactivity in Tobacco

"SURGEON GENERAL'S WARNING: Smoking Is Hazardous to Your Health." Warning labels such as this appear on every package of cigarettes sold in the United States. The link between cigarette smoke and cancer has long been established. There is, however, another cancer-causing mechanism in smokers. The culprit in this case is a radioactive environmental pollutant present in the tobacco leaves from which cigarettes are made.

The soil in which tobacco is grown is heavily treated with phosphate fertilizers, which are rich in uranium and its decay products. Consider a particularly important step in the uranium-238 decay series:

$$^{226}_{88}Ra \longrightarrow ^{222}_{86}Rn + ^{4}_{2}\alpha$$

The product formed, radon-222, is an unreactive gas. (Radon is the only gaseous species in the uranium-238 decay series.) Radon-222 emanates from radium-226 and is present at high concentrations in soil gas and in the surface air layer under the vegetation canopy provided by the field of growing tobacco. In this layer some of the daughters of radon-222, such as polonium-218 and lead-214, become firmly attached to the surface and interior of tobacco leaves. As Figure 16.3 shows, the next few decay reactions leading to the formation of lead-210 proceed rapidly. Gradually, the concentration of radioactive lead-210 can build to quite a high level.

During the combustion of a cigarette, tiny insoluble smoke particles are inhaled and deposited in the respiratory tract of the smoker and are eventually transported and stored at sites in the liver, spleen, and bone marrow. Measurements indicate a high lead-210 content in these particles. The lead-210 content is not high enough to be hazardous *chemically* (it is insufficient to cause *lead poisoning*), but it is hazardous because it is radioactive. Because of its long half-life (20.4 years), lead-210 and its radioactive daughters—bismuth-210 and polonium-210—continue to build up in the body of a smoker over the years. Constant exposure of the organs and bone marrow to α- and β-particle radiation increases the probability the smoker will develop cancer. The overall impact on health is similar to that caused by indoor radon gas.

16.3 Dating Using Radioactive Decay

Animation
Half-Life

You may have heard the term "half-life" used to describe how long a drug remains in the system of a person who has taken it. In fact, with regard to radioactivity, the term *half-life* means something very specific: the **half-life ($t_{1/2}$)** of a radioactive substance is the amount of time necessary for the amount of the substance to be reduced, via nuclear decay, to one-*half* its original value. For example, the half-life of a radioactive isotope of carbon, carbon-14, is 5715 years. This means that if we start with a billion C-14 nuclei, after 5715 years we would be left with half a billion. If we started with just ten C-14 nuclei, after 5715 years we would be left with five. One of the distinguishing features of radioactive decay is the constancy of the half-life. Regardless of the starting amount, the same amount of time is required for it to be reduced by half. Further, the half-life of a radioactive isotope is independent of temperature, pressure, or any chemical reactions that may consume the radioactive isotope.

SAMPLE PROBLEM 16.2 Using Half-Life to Calculate Remaining Mass

Lead-210 has a half-life of 20.4 years. What mass of a 2.00-g sample of $^{210}_{82}Pb$ will remain after 61.2 years?

Strategy Remember that a half-life is the time it takes for half of the sample to decay. During each successive half-life, one-half of the remaining sample decays.

Setup Determine how many half-lives elapse in the given time frame.

$$61.2 \text{ years} \times \frac{1 \text{ half-life}}{20.4 \text{ years}} = 3 \text{ half-lives}$$

Solution

2.00 g $\times \frac{1}{2} = 1.00$ g remains after passing of first half-life

1.00 g $\times \frac{1}{2} = 0.500$ g remains after passing of second half-life

0.500 g $\times \frac{1}{2} = 0.250$ g remains after passing of third half-life

THINK ABOUT IT

It is common for students to think that there is no sample remaining after the passing of time equivalent to two half-lives. Do not fall into this trap; remember that only half of the remaining sample decays during each successive half-life.

Practice Problem ATTEMPT $^{74}_{36}\text{Kr}$ has a half-life of 11.5 minutes. What mass of a 250.0-mg sample will remain after 46.0 minutes?

Practice Problem BUILD $^{18}_{9}\text{F}$ has a half-life of 110.0 minutes. How long will it take for a 175-mg sample to be reduced to 10.9 mg?

Practice Problem CONCEPTUALIZE Isotope X decays to isotope Y with a half-life of 45 days. Which diagram most closely represents the sample of X after 105 days?

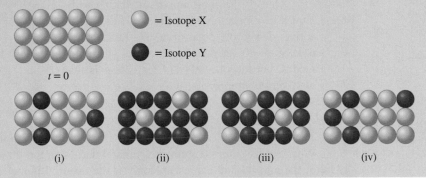

The half-lives of radioactive isotopes have been used as "atomic clocks" to determine the ages of certain objects. Some examples of dating by radioactive decay measurements are described here.

The carbon-14 isotope is produced when atmospheric nitrogen is bombarded by cosmic rays:

$$^{14}_{7}\text{N} + ^{1}_{0}\text{n} \longrightarrow ^{14}_{6}\text{C} + ^{1}_{1}\text{H}$$

The radioactive carbon-14 isotope decays according to the equation

$$^{14}_{6}\text{C} \longrightarrow ^{14}_{7}\text{N} + ^{0}_{-1}\beta$$

This reaction is the basis of radiocarbon or "carbon-14" dating. To determine the age of an object, we measure the *activity* (disintegrations per second) of ^{14}C and compare it to the activity of ^{14}C in living matter.

Figure 16.3 shows the natural radioactivity series of uranium-238, the most common isotope of uranium.

Because some of the intermediate products in the uranium decay series have very long half-lives (see Figure 16.3), this series is particularly suitable for estimating the ages of rocks found on Earth and of extraterrestrial objects. The half-life for the first step ($^{238}_{92}\text{U}$ to $^{234}_{90}\text{Th}$) is 4.51×10^9 years. This is about 20,000 times the second largest value (i.e., 2.47×10^5 years), which is the half-life for the fourth step ($^{234}_{92}\text{U}$ to $^{230}_{90}\text{Th}$). As a good approximation, therefore, we can assume that the half-life for the *overall* process (i.e., from $^{238}_{92}\text{U}$ to $^{206}_{82}\text{Pb}$) is equal to the half-life of the first step:

$$^{238}_{92}\text{U} \longrightarrow ^{206}_{82}\text{Pb} + 8^{4}_{2}\alpha + 6^{0}_{-1}\beta$$

In naturally occurring uranium minerals, we should and do find some lead-206 formed by radioactive decay. Assuming that no lead was present when the mineral was formed and that the mineral has not undergone chemical changes that would allow the lead-206 isotope to be separated from the parent uranium-238, it is possible to estimate the age of the rocks from the mass ratio of $^{206}_{82}\text{Pb}$ to $^{238}_{92}\text{U}$. According to the preceding nuclear

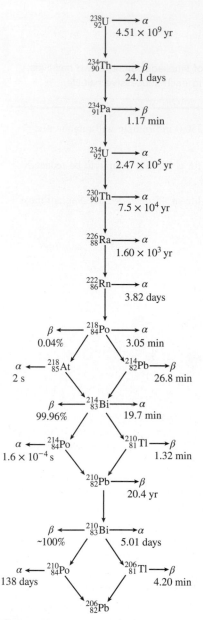

Figure 16.3 Decay series for uranium-238. (Times shown are half-lives.)

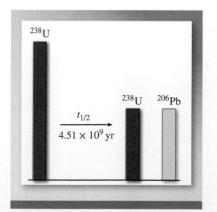

Figure 16.4 After one half-life, half of the original uranium-238 has been converted to lead-206.

equation, 1 mol (206 g) of lead is formed for every 1 mol (238 g) of uranium that undergoes complete decay. If only half a mole of uranium-238 has undergone decay, the mass ratio $^{206}Pb/^{238}U$ becomes

$$\frac{206 \text{ g}/2}{238 \text{ g}/2} = 0.866$$

and the process would have taken a half-life of 4.51×10^9 years to complete (Figure 16.4). Ratios lower than 0.866 mean that the rocks are less than 4.51×10^9 years old, and higher ratios suggest a greater age. Interestingly, studies based on the uranium series, as well as other decay series, put the age of the oldest rocks and, therefore, probably the age of Earth itself, at 4.5×10^9, or 4.5 billion, years.

One of the most important dating techniques in geochemistry is based on the radioactive decay of potassium-40. Radioactive potassium-40 decays by several different modes, but the one relevant for dating is that of *electron capture:*

> **Student Note:** In electron capture, an electron from outside the nucleus combines with a proton inside the nucleus to produce a neutron. The result is a nucleus with one *less* proton and one *more* neutron.
>
> $$_{-1}^{0}e + {}_{1}^{1}p \longrightarrow {}_{0}^{1}n$$

$$_{19}^{40}K + {}_{-1}^{0}e \longrightarrow {}_{18}^{40}Ar \qquad t_{1/2} = 1.2 \times 10^9 \text{ yr}$$

The accumulation of gaseous argon-40 is used to gauge the age of a specimen. When a potassium-40 atom in a mineral decays, argon-40 is trapped in the lattice of the mineral and can escape only if the material is melted. Melting, therefore, is the procedure for analyzing a mineral sample in the laboratory. The amount of argon-40 present can be conveniently measured with a mass spectrometer. Knowing the ratio of argon-40 to potassium-40 in the mineral and the half-life of decay makes it possible to establish the ages of rocks ranging from millions to billions of years old.

16.4 Medical Applications of Radioactivity

Radioactive isotopes are commonly used for diagnosis in medicine. Sodium-24 (a β-emitter with a half-life of 14.8 h) injected into the bloodstream as a salt solution can be monitored to trace the flow of blood and detect possible constrictions or obstructions in the circulatory system. Iodine-131 (a β-emitter with a half-life of 8 days) has been used to test the activity of the thyroid gland. A malfunctioning thyroid can be detected by giving the patient a drink of a solution containing a known amount of $Na^{131}I$ and measuring the radioactivity just above the thyroid to see if the iodine is absorbed at the normal rate. Another radioactive isotope of iodine, iodine-123 (a γ-ray emitter), is used to image the brain (Figure 16.5). In each of these cases, though, the amount of radioisotope used must be kept small to prevent the patient from suffering permanent damage from the high-energy radiation.

Technetium, the first artificially prepared element, is one of the most useful elements in nuclear medicine. Although technetium is a transition metal, all its isotopes are radioactive. In the laboratory it is prepared by the nuclear reactions

$$_{42}^{98}Mo + {}_{0}^{1}n \longrightarrow {}_{42}^{99}Mo$$

$$_{42}^{99}Mo \longrightarrow {}_{43}^{99m}Tc + {}_{-1}^{0}\beta$$

where the superscript "m" denotes that the technetium-99 isotope is produced in its *excited* nuclear state. This isotope has a half-life of about 6 hours, decaying by γ radiation to technetium-99 in its nuclear *ground* state. Thus, it is a valuable

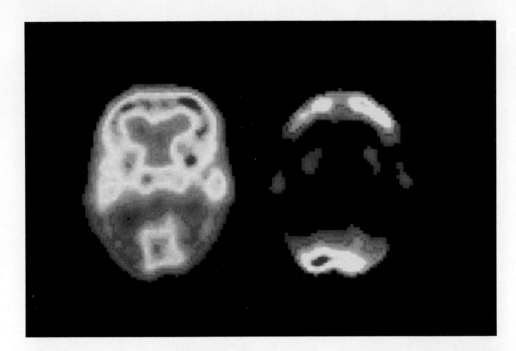

Figure 16.5 Image of a normal brain (left) and the brain of an Alzheimer's victim (right).
Science History Images/Alamy Stock Photo

diagnostic tool. The patient either drinks or is injected with a solution containing ^{99m}Tc. By detecting the γ rays emitted by ^{99m}Tc, doctors can obtain images of organs such as the heart, liver, and lungs.

Animation
Nuclear Medicine

Thinking Outside the Box

How Nuclear Chemistry Is Used to Treat Cancer

Brain tumors are some of the most difficult cancers to treat because the site of the malignant growth makes surgical excision difficult or impossible. Likewise, conventional radiation therapy using X rays or γ rays from outside the skull is usually not effective. An ingenious approach to this problem is *boron neutron capture therapy* (BNCT). This technique involves first administering a boron-10 compound that is selectively taken up by tumor cells and then applying a beam of low-energy neutrons to the tumor site. ^{10}B captures a neutron to produce ^{11}B, which disintegrates via the following nuclear reaction:

$$^{10}_{5}B + ^{1}_{0}n \longrightarrow ^{7}_{3}Li + ^{4}_{2}\alpha$$

The highly energetic particles produced by this reaction destroy the tumor cells in which the ^{10}B is concentrated. Because the particles are confined to just a few micrometers, they preferentially destroy tumor cells without damaging neighboring normal cells.

BNCT is a highly promising treatment and is an active area of research. One of the major goals of the research is to develop suitable compounds to deliver ^{10}B to the desired site. For such a compound to be effective, it must meet several criteria. It must have a high affinity for tumor cells, be able to pass through membrane barriers to reach the tumor site, and have minimal toxic effects on the human body.

This is one example of how *nuclear chemistry* is important in the treatment of cancer.

16.5 Nuclear Fission and Nuclear Fusion

Nuclear fission is the process in which a heavy nucleus (mass number > 200) divides to form smaller nuclei of intermediate mass and one or more neutrons. Because the heavy nucleus is less stable than its products, this process releases a large amount of energy.

The first nuclear fission reaction to be studied was that of uranium-235 bombarded with slow neutrons, whose speed is comparable to that of air molecules at room temperature. Under these conditions, uranium-235 undergoes fission, as shown in Figure 16.6. Actually, this reaction is very complex: more than 30 different elements have been found among the fission products. A representative reaction is

$$^{235}_{92}U + ^{1}_{0}n \longrightarrow ^{90}_{38}Sr + ^{143}_{54}Xe + 3^{1}_{0}n$$

In contrast to the nuclear fission process, *nuclear fusion,* the combining of small nuclei into larger ones, is largely exempt from the waste disposal problem.

Animation
Fission

Figure 16.6

Nuclear Fission and Fusion

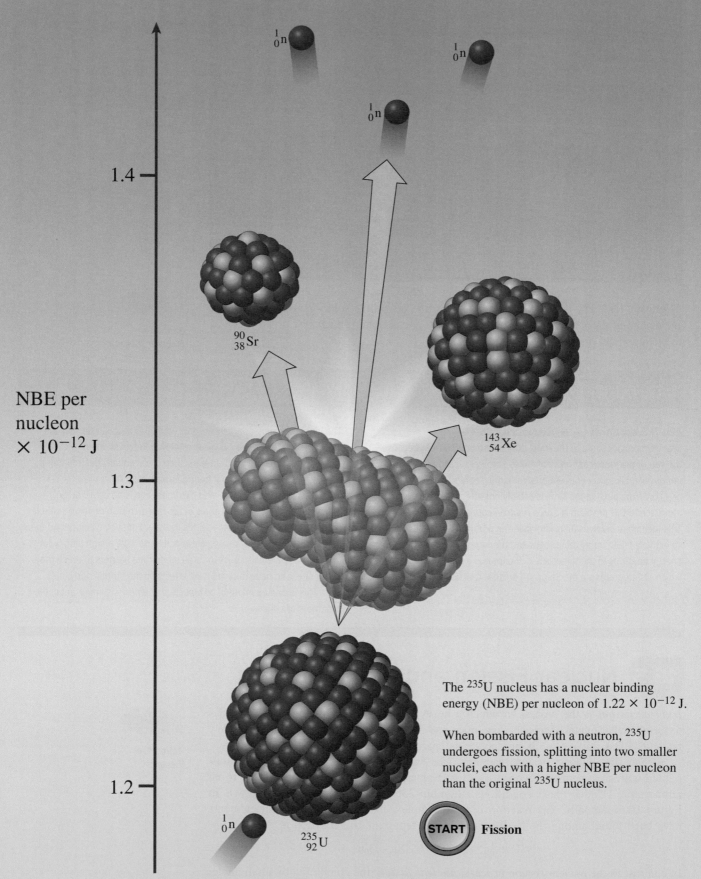

NBE per
nucleon
$\times 10^{-12}$ J

1.4 —

1.3 —

1.2 —

$^{1}_{0}n$

$^{1}_{0}n$

$^{1}_{0}n$

$^{90}_{38}Sr$

$^{143}_{54}Xe$

$^{235}_{92}U$

$^{1}_{0}n$

The ^{235}U nucleus has a nuclear binding
energy (NBE) per nucleon of 1.22×10^{-12} J.

When bombarded with a neutron, ^{235}U
undergoes fission, splitting into two smaller
nuclei, each with a higher NBE per nucleon
than the original ^{235}U nucleus.

START Fission

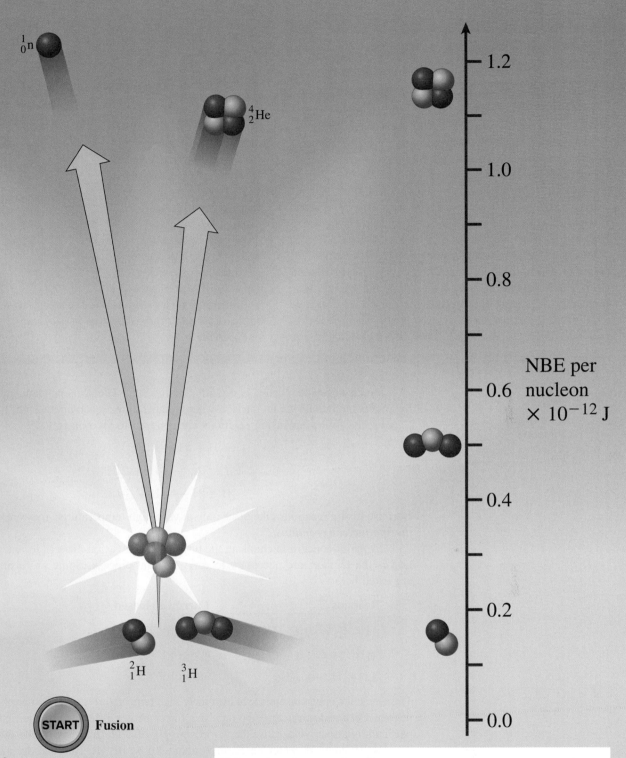

NBE per
nucleon
$\times 10^{-12}$ J

START Fusion

The ^{2}H and ^{3}H nuclei have the following nuclear binding energies per nucleon:

^{2}H: 0.185×10^{-12} J
^{3}H: 0.451×10^{-12} J

At very high temperatures, the ^{2}H and ^{3}H nuclei undergo fusion to produce a ^{4}He nucleus and a neutron. The ^{4}He nucleus has a significantly higher NBE per nucleon: 1.13×10^{-12} J

(See Visualizing Chemistry questions VC 16.1–VC 16.4 on page 547.)

What's the point?

Large nuclei, such as ^{235}U, can achieve greater nuclear stability by splitting into smaller nuclei with greater NBE per nucleon. Small nuclei achieve stability by undergoing fusion to produce a larger nucleus with a greater NBE per nucleon. Note that different scales are used to show the change in NBE per nucleon for the two processes. There is a much greater change in NBE per nucleon in the fusion process than in the fission process. As with chemical reactions, nuclear reactions are favored when the products are more stable than the reactants.

Profiles in Science

Lise Meitner

Lise Meitner was one of the most consequential physicists of the nineteenth century. Despite the systematic exclusion of women from most scientific fields of study at the time, and the prohibition against women attending institutions of higher learning, Meitner pursued a private education and ultimately earned a doctoral degree in physics at the University of Vienna. (She was only the second woman ever to have done so.)

Much of Meitner's work was done in Berlin, where she was the first woman to head the physics department at the Kaiser Wilhelm Institute for Chemistry at the University of Berlin. She collaborated and corresponded with many of the most famous scientists of her day, including Max Planck (Meitner is said to have been the first woman that Planck ever allowed to attend his lectures), Otto Hahn, Leó Szilárd, and James Chadwick. She and Hahn worked extensively together and amassed an extraordinary list of discoveries—including the discovery and characterization of nuclear fission.

Meitner's work, ultimately, was curtailed by the sociopolitical events associated with World War II. Being a Jew, she was forced to flee Germany in 1938—a fate that many believe contributed to her being excluded from consideration for the Nobel Prize. In 1945, Otto Hahn alone was awarded the 1944 Nobel Prize in Chemistry for the discovery of nuclear fission, a discovery that he and Meitner had made jointly.

Despite being passed over for this most coveted prize, Meitner received more than twenty prestigious awards in recognition of her contributions to science during her lifetime. After she died, meitnerium (element 109) was named for her—an honor bestowed on many fewer scientists than the Nobel Prize.

Lise Meitner
Photo Researchers/Science History Images/Alamy Stock Photo

Nuclear fusion occurs constantly in the Sun. The Sun is made up mostly of hydrogen and helium. In its interior, where temperatures reach about 15 million degrees Celsius, the following fusion reactions are believed to take place:

$$^1_1H + ^2_1H \longrightarrow ^3_2He$$

$$^3_2He + ^3_2He \longrightarrow ^4_2He + 2^1_1H$$

$$^1_1H + ^1_1H \longrightarrow ^2_1H + ^0_{+1}\beta$$

Because fusion reactions take place only at very high temperatures, they are often called **thermonuclear reactions.**

A major concern in choosing the proper nuclear fusion process for energy production is the temperature necessary to carry out the process. Some promising reactions are listed here:

Reaction	Energy Released
$^2_1H + ^2_1H \longrightarrow ^3_1H + ^1_1H$	6.3×10^{-13} J
$^2_1H + ^3_1H \longrightarrow ^3_2He + 2^1_0n$	2.8×10^{-12} J
$^6_3Li + ^2_1H \longrightarrow 2^4_2He$	3.6×10^{-12} J

These reactions must take place at extremely high temperatures, on the order of 100 million degrees Celsius, to overcome the repulsive forces between the nuclei. The first reaction is particularly attractive because the world's supply of deuterium is virtually inexhaustible. The total volume of water on Earth is about 1.5×10^{21} L. Because the natural abundance of deuterium is 0.015 percent, the total amount of deuterium present is roughly 4.5×10^{21} g, or 5.0×10^{15} tons. Although it is expensive to prepare deuterium, the cost is minimal compared to the value of the energy released by the reaction.

In contrast to the fission process, nuclear fusion looks like a very promising energy source, at least on paper. Although thermal pollution would be a problem, fusion has the following advantages: (1) the fuels are cheap and almost inexhaustible and (2) the process produces little radioactive waste. If a fusion machine were turned off, it would shut down completely and instantly, without any danger of a meltdown. Despite its apparent advantages, fusion is not something we are able to do reliably. Holding the necessary nuclei close enough to each other—and maintaining the necessary temperatures—are not things we are able to do as of yet.

Chapter Summary

Section 16.1

- Spontaneous emission of particles or radiation from unstable nuclei is known as *radioactivity.* Unstable nuclei can emit alpha particles, *beta particles,* and/or *gamma radiation.*

- *Nuclear transmutation* is the conversion of one nucleus to another. Nuclear reactions are balanced by summing the mass numbers and the atomic numbers of reactant and product species.

- A *positron* is a subatomic particle with the same mass as an electron, but with a positive charge.

Section 16.2

- Radioactivity can be measured using a Geiger counter.

- High-energy radiation damages living systems by causing ionization and the formation of *radicals,* or free radicals, which are chemical species with unpaired electrons.

Section 16.3

- The *half-life* of a radioactive decay process is the amount of time it takes for a quantity of substance to be reduced, via nuclear decay, by half.

- Uranium-238 is the parent of a natural radioactive decay series that can be used to determine the ages of rocks. Radiocarbon dating is done using carbon-14.

Section 16.4

- Radioactive isotopes are used in a wide variety of medical diagnostic procedures and treatments—including for many cancers.

Section 16.5

- *Nuclear fission* is the splitting apart of large nuclei to produce smaller nuclei and neutrons, along with large quantities of energy. Nuclear fission is the process used in nuclear power plants.

- *Nuclear fusion* is the combination of two small nuclei to produce one larger nucleus, which also produces large quantities of energy. Nuclear fusion is the source of power in the Sun.

- Nuclear fusion is sometimes referred to as a *thermonuclear reaction* because of the extraordinarily high temperatures required to achieve it.

Key Terms

Beta particles 533	Nuclear fission 541	Positron 534	Radioactivity 533
Gamma radiation 533	Nuclear fusion 541	Radicals 537	Thermonuclear reaction 544
Half-life ($t_{1/2}$) 538	Nuclear transmutation 534		

Questions and Problems

SECTION 16.1: RADIOACTIVE DECAY

16.1 How do nuclear reactions differ from ordinary chemical reactions?

16.2 What are the steps in balancing nuclear equations?

16.3 What is the difference between $_{-1}^{0}e$ and $_{-1}^{0}\beta$?

16.4 What is the difference between an electron and a positron?

16.5 Complete the following nuclear equations, and identify X in each case:

(a) $_{12}^{26}Mg + _{1}^{1}p \longrightarrow _{2}^{4}\alpha + X$

(b) $_{27}^{59}Co + _{1}^{2}H \longrightarrow _{27}^{60}Co + X$

(c) $_{92}^{235}U + _{0}^{1}n \longrightarrow _{36}^{94}Kr + _{56}^{139}Ba + 3X$

(d) $_{24}^{53}Cr + _{2}^{4}\alpha \longrightarrow _{0}^{1}n + X$

(e) $_{8}^{20}O \longrightarrow _{9}^{20}F + X$

16.6 Complete the following nuclear equations, and identify X in each case:

(a) $_{53}^{135}I \longrightarrow _{54}^{135}Xe + X$

(b) $_{19}^{40}K \longrightarrow _{-1}^{0}\beta + X$

(c) $_{27}^{59}Co + _{0}^{1}n \longrightarrow _{25}^{56}Mn + X$

(d) $_{92}^{235}U + _{0}^{1}n \longrightarrow _{40}^{99}Zr + _{52}^{135}Te + 2X$

16.7 Write the symbol for each of the following:
(a) alpha particle
(b) beta particle
(c) gamma ray
(d) positron

16.8 Write the symbol for the isotope of calcium that contains 28 neutrons.

16.9 Write the symbol for the isotope of zinc that contains 34 neutrons.

16.10 Identify X in each of the following nuclei:

(a) $_{39}^{90}X$

(b) $_{8}^{15}X$

(c) $_{38}^{90}X$

(d) $_{86}^{226}X$

16.11 Identify X in each of the following nuclei:

(a) $_{82}^{206}X$

(b) $_{84}^{210}X$

(c) $_{40}^{90}X$

(d) $_{83}^{209}X$

16.12 Determine the number of protons and the number of neutrons in each of the following nuclei:

(a) $_{94}^{236}Pu$

(b) $_{93}^{237}Np$

(c) $_{53}^{131}I$

16.13 Determine the number of protons and the number of neutrons in each of the following nuclei:

(a) $_{27}^{60}Co$

(b) $_{95}^{243}Am$

(c) $_{12}^{27}Mg$

16.14 Write a nuclear equation for the alpha decay of each nucleus:

(a) $_{83}^{211}Bi$

(b) $_{90}^{230}Th$

(c) $_{88}^{222}Ra$

16.15 Write a nuclear equation for the beta decay of each nucleus:

(a) $_{82}^{214}Pb$

(b) $_{90}^{234}Th$

(c) $_{89}^{227}Ac$

16.16 Write a nuclear equation for the positron decay of each nucleus:

(a) $_{11}^{22}Na$

(b) $_{6}^{11}C$

(c) $_{19}^{38}K$

16.17 Fill in the blanks in the following radioactive decay series:

(a) $^{232}Th \xrightarrow{\alpha} \underline{\quad} \xrightarrow{\beta} \underline{\quad} \xrightarrow{\beta} {}^{228}Th$

(b) $^{235}U \xrightarrow{\alpha} \underline{\quad} \xrightarrow{\beta} \underline{\quad} \xrightarrow{\alpha} {}^{227}Ac$

(c) $\underline{\quad} \xrightarrow{\alpha} {}^{233}Pa \xrightarrow{\beta} \underline{\quad} \xrightarrow{\alpha} \underline{\quad}$

16.18 The following equations are for nuclear reactions that are known to occur in the explosion of an atomic bomb. For each equation, identify X.

(a) $_{92}^{235}U + _{0}^{1}n \longrightarrow _{56}^{140}Ba + 3_{0}^{1}n + X$

(b) $_{92}^{235}U + _{0}^{1}n \longrightarrow _{55}^{144}Cs + _{37}^{90}Rb + 2X$

(c) $_{92}^{235}U + _{0}^{1}n \longrightarrow _{35}^{87}Br + 3_{0}^{1}n + X$

(d) $_{92}^{235}U + _{0}^{1}n \longrightarrow _{62}^{160}Sm + _{30}^{72}Zn + 4X$

16.19 Write complete nuclear equations for the following processes: (a) tritium (^{3}H) undergoes β decay, (b) ^{242}Pu undergoes α-particle emission, (c) ^{131}I undergoes β decay, (d) ^{251}Cf emits an α particle.

SECTION 16.2: DETECTION OF RADIATION AND ITS BIOLOGICAL EFFECTS

16.20 List the factors that affect the intensity of radiation from a radioactive element.

16.21 What are *rad* and *rem,* and how are they related?

16.22 Explain, with examples, the difference between somatic and genetic radiation damage.

16.23 Compare the extent of radiation damage done by α, β, and γ sources.

SECTION 16.3: DATING USING RADIOACTIVE DECAY

16.24 Define the term *half-life.*

16.25 What percentage of a radioactive isotope remains after 20.0 days if its half-life is 10.0 days?

16.26 The radioactive decay of Tl-206 to Pb-206 has a half-life of 4.20 min. Starting with 5.00×10^{22} atoms of Tl-206, calculate the number of such atoms left after 42.0 min.

16.27 A radioactive isotope of copper decays as follows:

$$^{64}Cu \longrightarrow {}^{64}Zn + {}_{-1}^{0}\beta \qquad t_{1/2} = 12.8 \text{ h}$$

Starting with 84.0 g of ^{64}Cu, calculate the quantity of ^{64}Zn produced after 25.6 h.

SECTION 16.4: MEDICAL APPLICATIONS OF RADIOACTIVITY

16.28 List three nuclei used in medicine and identify their specific use.

16.29 Describe boron neutron capture therapy.

SECTION 16.5: NUCLEAR FISSION AND NUCLEAR FUSION

16.30 Define the term *nuclear fission.*

16.31 Which isotopes can undergo nuclear fission?

16.32 Define the terms *nuclear fusion* and *thermonuclear reaction.*

16.33 Explain why heavy elements such as uranium undergo fission, whereas light elements such as hydrogen and lithium undergo fusion.

Visualizing Chemistry
Figure 16.6

VC 16.1 The fission of ^{235}U can result in a variety of products, including those shown in Figure 16.6. Which of the following equations does *not* represent another possible fission process?

(a) $_{0}^{1}n + {}_{92}^{235}U \longrightarrow {}_{52}^{137}Te + {}_{40}^{97}Zr + 2{}_{0}^{1}n$

(b) $_{0}^{1}n + {}_{92}^{235}U \longrightarrow {}_{56}^{142}Ba + {}_{36}^{91}Kr + 3{}_{0}^{1}n$

(c) $_{0}^{1}n + {}_{92}^{235}U \longrightarrow {}_{55}^{137}Cs + {}_{37}^{90}Rb + 3{}_{0}^{1}n$

VC 16.2 How many neutrons are produced in the fission reaction shown?

$$_{0}^{1}n + {}_{94}^{239}Pu \longrightarrow {}_{44}^{109}Ru + {}_{50}^{129}Sn + \underline{\quad} {}_{0}^{1}n$$

(a) 1

(b) 2

(c) 3

VC 16.3 The fission of ^{235}U shown in Figure 16.6 is represented by the equation

$$_{0}^{1}n + {}_{92}^{235}U \longrightarrow {}_{54}^{143}Xe + {}_{38}^{90}Sr + 3{}_{0}^{1}n$$

How does the combined mass of products compare to the combined mass of reactants for this process?

(a) The combined mass of products is smaller than the combined mass of reactants.

(b) The combined mass of products is larger than the combined mass of reactants.

(c) The combined mass of products is equal to the combined mass of reactants.

VC 16.4 The fusion of $_{1}^{2}H$ and $_{1}^{3}H$ shown in Figure 16.6 is represented by the equation

$$_{1}^{2}H + {}_{1}^{3}H \longrightarrow {}_{2}^{4}He + {}_{0}^{1}n$$

How does the combined mass of products compare to the combined mass of reactants for this process?

(a) The combined mass of products is smaller than the combined mass of reactants.

(b) The combined mass of products is larger than the combined mass of reactants.

(c) The combined mass of products is equal to the combined mass of reactants.

Answers to In-Chapter Materials

Answers to Practice Problems

16.1A (a) $_{34}^{78}Se$, (b) $_{3}^{5}Li$, (c) $_{0}^{1}n$. **16.1B** (a) $_{95}^{244}Am$, (b) $_{90}^{234}Th$, (c) $_{6}^{14}C$. **16.2A** 15.6 mg. **16.2B** 7 h 20 min.

Answers to Checkpoints

16.1.1 d. **16.1.2** a.

CHAPTER 17

Electrochemistry

17.1 Balancing Oxidation-Reduction Reactions Using the Half-Reaction Method

17.2 Batteries
- Dry Cells and Alkaline Batteries
- Lead Storage Batteries
- Lithium-Ion Batteries
- Fuel Cells

17.3 Corrosion

17.4 Electrolysis
- Electrolysis of Molten Sodium Chloride
- Electrolysis of Water

Potato clocks are popular as science demonstrations and experiments.

Science Photo Library/Alamy Stock Photo

We have learned about the importance of electrons in determining the properties of atoms, and about their role in chemical bonding. We also saw in Chapter 10 how some reactions (redox reactions) involve the transfer of electrons from one chemical species to another, and how we use oxidation numbers to keep track of the electrons that are transferred in a redox reaction. In this chapter, we examine redox reactions in more detail, and see how they can be useful for generating electricity, or detrimental—causing corrosion. Further, we explore how electricity can be used to drive redox reactions that would not happen on their own—and how that can be useful in a variety of ways.

17.1 Balancing Oxidation-Reduction Reactions Using the Half-Reaction Method

In Chapter 10 we briefly discussed oxidation-reduction or "redox" reactions, those in which electrons are transferred from one species to another. In this section we review how to identify a reaction as a redox reaction and look more closely at how such reactions are balanced.

A redox reaction is one in which there are *changes* in oxidation states, which we identify using the rules introduced in Chapter 10. The following are examples of redox reactions:

Student Note: Now is a good time to review how oxidation numbers are assigned [◄◄ Section 10.4].

$$2KClO_3(s) \longrightarrow 2KCl(s) + 3O_2(g)$$

$$CH_4(g) + 2O_2(g) \longrightarrow CO_2(g) + 2H_2O(l)$$

$$Sn(s) + Cu^{2+}(aq) \longrightarrow Cu(s) + Sn^{2+}(aq)$$

Equations for redox reactions, such as those shown here, can be balanced by inspection, the method of balancing introduced in Chapter 10 [◄◄ Section 10.3], but remember that redox equations must be balanced for mass (number of atoms) *and* for charge (number of electrons) [◄◄ Section 10.4]. In this section we introduce the *half-reaction method* to balance equations that cannot be balanced simply by inspection.

Consider the aqueous reaction of the iron(II) ion (Fe^{2+}) with the dichromate ion ($Cr_2O_7^{2-}$):

$$Fe^{2+} + Cr_2O_7^{2-} \longrightarrow Fe^{3+} + Cr^{3+}$$

Because there is no species containing oxygen on the product side of the equation, it would not be possible to balance this equation simply by adjusting the coefficients of reactants and products. However, there are two things about the reaction that make it possible to *add* species to the equation to balance it—without changing the chemical reaction it represents:

- The reaction takes place in aqueous solution, so we can add H_2O as needed to balance the equation.
- This particular reaction takes place in acidic solution, so we can add H^+ as needed to balance the equation. (Some reactions take place in basic solution, enabling us to add OH^- as needed for balancing. We will learn more about this shortly.)

After writing the unbalanced equation, we balance it stepwise as follows:

Student Note: Although the two half-reactions cannot happen independently, we can balance them separately and then add them back together to get the overall balanced equation.

1. Separate the unbalanced reaction into two ***half-reactions,*** one showing the oxidation and one showing the reduction. A half-reaction is an oxidation or a reduction that occurs as part of the overall redox reaction.

$$\begin{aligned} \textit{Oxidation:} \qquad & Fe^{2+} \longrightarrow Fe^{3+} \\ \textit{Reduction:} \qquad & Cr_2O_7^{2-} \longrightarrow Cr^{3+} \end{aligned}$$

Student Note: If it is not obvious that $Cr_2O_7^{2-} \longrightarrow Cr^{3+}$ is a reduction, use the method described in Section 10.4 to determine the oxidation number of Cr in the reactant and in the product. If the oxidation number is *reduced*, the process is a *reduction*.

2. Balance each of the half-reactions with regard to atoms other than O and H. In this case, no change is required for the oxidation half-reaction. We adjust the coefficient of the chromium(III) ion to balance the reduction half-reaction.

$$\begin{aligned} \textit{Oxidation:} \qquad & Fe^{2+} \longrightarrow Fe^{3+} \\ \textit{Reduction:} \qquad & Cr_2O_7^{2-} \longrightarrow 2Cr^{3+} \end{aligned}$$

3. Balance both half-reactions for O by adding H_2O. Again, the oxidation in this case requires no change, but we must add seven water molecules to the product side of the reduction.

$$\begin{aligned} \textit{Oxidation:} \qquad & Fe^{2+} \longrightarrow Fe^{3+} \\ \textit{Reduction:} \qquad & Cr_2O_7^{2-} \longrightarrow 2Cr^{3+} + 7H_2O \end{aligned}$$

4. Balance both half-reactions for H by adding H^+. Once again, the oxidation in this case requires no change, but we must add 14 hydrogen ions to the reactant side of the reduction.

$$\begin{aligned} \textit{Oxidation:} \qquad & Fe^{2+} \longrightarrow Fe^{3+} \\ \textit{Reduction:} \qquad & 14H^+ + Cr_2O_7^{2-} \longrightarrow 2Cr^{3+} + 7H_2O \end{aligned}$$

5. Balance both half-reactions for charge by adding electrons. To do this, we determine the total charge on each side and add electrons to make the total charges equal. In the case of the oxidation, there is a charge of $+2$ on the reactant side and a charge of $+3$ on the product side. Adding one electron to the product side makes the charges equal.

$$\begin{aligned} \textit{Oxidation:} \qquad & \underbrace{Fe^{2+}} \longrightarrow \underbrace{Fe^{3+} + e^-} \\ \textit{Total charge:} \qquad & \quad +2 \qquad\qquad +2 \end{aligned}$$

In the case of the reduction, there is a total charge of $[(14)(+1) + (1)(-2)] = +12$ on the reactant side and a total charge of $[(2)(+3)] = +6$ on the product side. Adding six electrons to the reactant side makes the charges equal.

$$\begin{aligned} \textit{Reduction:} \qquad & \underbrace{6e^- + 14H^+ + Cr_2O_7^{2-}} \longrightarrow \underbrace{2Cr^{3+} + 7H_2O} \\ \textit{Total charge:} \qquad & \qquad\qquad +6 \qquad\qquad\qquad\qquad +6 \end{aligned}$$

6. If the number of electrons in the balanced oxidation half-reaction is not the same as the number of electrons in the balanced reduction half-reaction, multiply one or both of the half-reactions by the number(s) required to make the number of electrons the same in both. In this case, with one electron in the oxidation and six in the reduction, multiplying the oxidation by 6 accomplishes this.

$$\textit{Oxidation:} \qquad 6(Fe^{2+} \longrightarrow Fe^{3+} + e^-)$$
$$6Fe^{2+} \longrightarrow 6Fe^{3+} + \boxed{6e^-}$$
$$\textit{Reduction:} \qquad \boxed{6e^-} + 14H^+ + Cr_2O_7^{2-} \longrightarrow 2Cr^{3+} + 7H_2O$$

7. Finally, add the balanced half-reactions back together and cancel the electrons, in addition to any other identical terms that appear on both sides.

$$6Fe^{2+} \longrightarrow 6Fe^{3+} + \cancel{6e^-}$$
$$\underline{\cancel{6e^-} + 14H^+ + Cr_2O_7^{2-} \longrightarrow 2Cr^{3+} + 7H_2O}$$
$$6Fe^{2+} + 14H^+ + Cr_2O_7^{2-} \longrightarrow 6Fe^{3+} + 2Cr^{3+} + 7H_2O$$

A final check shows that the resulting equation is balanced both for mass and for charge.

Some redox reactions occur in basic solution. When this is the case, balancing by the half-reaction method is done exactly as described for reactions in acidic solution, but it requires two additional steps:

8. For each H^+ ion in the final equation, add one OH^- ion to each side of the equation, combining the H^+ and OH^- ions to produce H_2O.
9. Make any additional cancellations made necessary by the new H_2O molecules.

Sample Problem 17.1 shows how to use the half-reaction method to balance a reaction that takes place in basic solution.

SAMPLE PROBLEM 17.1 Balancing Electrochemical Equations

Permanganate ion and iodide ion react in basic solution to produce manganese(IV) oxide and molecular iodine. Use the half-reaction method to balance the equation:

$$MnO_4^- + I^- \longrightarrow MnO_2 + I_2$$

Strategy The reaction takes place in basic solution, so apply steps 1 through 9 to balance for mass and for charge.

Setup Identify the oxidation and reduction half-reactions by assigning oxidation numbers.

$$MnO_4^- + I^- \longrightarrow MnO_2 + I_2$$
$$\scriptsize (+7)(-2) \quad (-1) \qquad (+4)(-2) \quad (0)$$

Solution

Step 1. Separate the unbalanced reaction into half-reactions.

$$\textit{Oxidation:} \qquad I^- \longrightarrow I_2$$
$$\textit{Reduction:} \qquad MnO_4^- \longrightarrow MnO_2$$

Step 2. Balance each half-reaction for mass, excluding O and H.

$$2I^- \longrightarrow I_2$$
$$MnO_4^- \longrightarrow MnO_2$$

Step 3. Balance both half-reactions for O by adding H_2O.

$$2I^- \longrightarrow I_2$$
$$MnO_4^- \longrightarrow MnO_2 + 2H_2O$$

Step 4. Balance both half-reactions for H by adding H^+.

$$2I^- \longrightarrow I_2$$
$$4H^+ + MnO_4^- \longrightarrow MnO_2 + 2H_2O$$

(Continued on next page)

Step 5. Balance the total charge of both half-reactions by adding electrons.

$$2I^- \longrightarrow I_2 + 2e^-$$

$$3e^- + 4H^+ + MnO_4^- \longrightarrow MnO_2 + 2H_2O$$

Step 6. Multiply the half-reactions to make the numbers of electrons the same in both.

$$3(2I^- \longrightarrow I_2 + 2e^-)$$

$$2(3e^- + 4H^+ + MnO_4^- \longrightarrow MnO_2 + 2H_2O)$$

Step 7. Add the half-reactions back together, canceling electrons.

$$6I^- \longrightarrow 3I_2 + \cancel{6e^-}$$

$$\cancel{6e^-} + 8H^+ + 2MnO_4^- \longrightarrow 2MnO_2 + 4H_2O$$

$$\overline{8H^+ + 2MnO_4^- + 6I^- \longrightarrow 2MnO_2 + 3I_2 + 4H_2O}$$

Step 8. For each H^+ ion in the final equation, add one OH^- ion to each side of the equation, combining the H^+ and OH^- ions to produce H_2O.

$$8H^+ + 2MnO_4^- + 6I^- \longrightarrow 2MnO_2 + 3I_2 + 4H_2O$$

$$+\ 8OH^- \hspace{6cm} +\ 8OH^-$$

$$\overline{\fbox{$8H_2O$} + 2MnO_4^- + 6I^- \longrightarrow 2MnO_2 + 3I_2 + \fbox{$4H_2O$} + 8OH^-}$$

Step 9. Carry out any cancellations made necessary by the additional H_2O molecules.

$$4H_2O + 2MnO_4^- + 6I^- \longrightarrow 2MnO_2 + 3I_2 + 8OH^-$$

THINK ABOUT IT

Verify that the final equation is balanced for mass and for charge. Remember that electrons cannot appear in the overall balanced equation.

Practice Problem **A**TTEMPT Use the half-reaction method to balance the following equation in basic solution:

$$CN^- + MnO_4^- \longrightarrow CNO^- + MnO_2$$

Practice Problem **B**UILD Use the half-reaction method to balance the following equation in acidic solution:

$$Fe^{2+} + MnO_4^- \longrightarrow Fe^{3+} + Mn^{2+}$$

Practice Problem **C**ONCEPTUALIZE In Chapter 10, you learned to balance chemical equations by changing coefficients only—it was not permissible to add species to the equation. Explain why it is all right to add water and hydronium (or hydroxide) to the equations in this chapter as part of the balancing process.

CHECKPOINT–SECTION 17.1   Balancing Oxidation-Reduction Reactions Using the Half-Reaction Method

17.1.1 Which of the following equations does *not* represent a redox reaction? (Select all that apply.)

a) $NH_3 + HCl \longrightarrow NH_4Cl$

b) $2H_2O_2 \longrightarrow 2H_2O + O_2$

c) $2O_3 \longrightarrow 3O_2$

d) $3NO_2 + H_2O \longrightarrow NO + 2HNO_3$

e) $LiCl \longrightarrow Li^+ + Cl^-$

17.1.2 MnO_4^- and $C_2O_4^{2-}$ react in basic solution to form MnO_2 and CO_3^{2-}. What are the coefficients of MnO_4^- and $C_2O_4^{2-}$ in the balanced equation?

a) 1 and 1

b) 2 and 1

c) 2 and 3

d) 2 and 6

e) 2 and 2

17.2 Batteries

When zinc metal is placed in a solution containing copper(II) ions, Zn is oxidized to Zn^{2+} ions, whereas Cu^{2+} ions are reduced to Cu [◀◀ Figure 10.6]:

$$Zn(s) + Cu^{2+}(aq) \longrightarrow Zn^{2+}(aq) + Cu(s)$$

The chemical species that is reduced in a redox reaction is called the **oxidizing agent,** because it is the "agent" that causes oxidation of the *other* species. Likewise, the species that is *oxidized* is called the **reducing agent.** In this case, the electrons are transferred directly from the reducing agent, Zn, to the oxidizing agent, Cu^{2+}, in solution. Although a transfer of electrons occurs, the result is not really useful. However, if we physically separate two half-reactions from each other, we can arrange it such that the electrons must travel through a wire to pass from the Zn atoms to the Cu^{2+} ions. As the reaction progresses, it generates a flow of electrons through the wire and thereby generates *electricity*.

> **Student Note:** A *galvanic* cell can also be called a *voltaic* cell. Both terms refer to a cell in which a spontaneous chemical reaction generates a flow of electrons.

The experimental apparatus for generating electricity through the use of a spontaneous reaction is called a **galvanic cell.** Figure 17.1 shows the essential components of a galvanic cell. A zinc bar is immersed in an aqueous $ZnSO_4$ solution in one container, and a copper bar is immersed in an aqueous $CuSO_4$ solution in another container. The cell operates on the principle that the oxidation of Zn to Zn^{2+} and the reduction of Cu^{2+} to Cu can be made to take place simultaneously in separate locations with the transfer of electrons between them occurring through an external wire. The zinc and copper bars are called **electrodes.** By definition, the **anode** in a galvanic cell is the electrode at which *oxidation* occurs and the **cathode** is the electrode at which *reduction* occurs. (Each combination of container, electrode, and solution is called a **half-cell.**) This particular arrangement of electrodes and electrolytes is called a Daniell cell.

> **Student Note:** This is the origin of the terms *cathode* and *anode*:
> • Cations move toward the cathode.
> • Anions move toward the anode.

The half-reactions for the galvanic cell shown in Figure 17.1 are

Oxidation: $\qquad\qquad Zn(s) \longrightarrow Zn^{2+}(aq) + 2e^-$

Reduction: $\qquad Cu^{2+}(aq) + 2e^- \longrightarrow Cu(s)$

To complete the electric circuit, and allow electrons to flow through the external wire, the solutions must be connected by a conducting medium through which the cations and anions can move from one half-cell to the other. This requirement is satisfied by a **salt bridge,** which, in its simplest form, is an inverted U tube containing an inert electrolyte solution, such as KCl or NH_4NO_3. The ions in the salt bridge must not react with the other ions in solution or with the electrodes (see Figure 17.1). During the course of the redox reaction, electrons flow through the external wire from the anode (Zn electrode) to the cathode (Cu electrode). In the solution, the cations (Zn^{2+}, Cu^{2+}, and K^+) move toward the cathode, while the anions (SO_4^{2-} and Cl^-) move toward the anode. Without the salt bridge connecting the two solutions, the buildup of positive charge in the anode compartment (due to the departure of electrons and the resulting formation of Zn^{2+}) and the buildup of negative charge in the cathode compartment (created by the arrival of electrons and the reduction of Cu^{2+} ions to Cu) would quickly prevent the cell from operating.

An electric current flows from the anode to the cathode because there is a difference in electrical potential energy between the electrodes. This flow of electric current is analogous to the flow of water down a waterfall, which occurs because there is a difference in the gravitational potential energy, or the flow of gas from a high-pressure region to a low-pressure region. Table 17.1 lists metals (and hydrogen) in order of their relative electrical potentials—an arrangement known as the *activity series*. The higher a metal appears on the list, the greater its potential to become oxidized. A metal will be oxidized by the aqueous ions of any species that appears below it. Consider again the example shown in Figure 17.1. Zinc appears higher in the activity series and so will be oxidized by aqueous ions of copper, which appears lower in the series.

Figure 17.1

Construction of a Galvanic Cell

We add a salt bridge, a tube containing a solution of a strong electrolyte—in this case Na_2SO_4. Having this solution in electrical contact with the two solutions in the beakers allows ions to migrate toward the electrodes, ensuring that the two compartments remain electrically neutral.

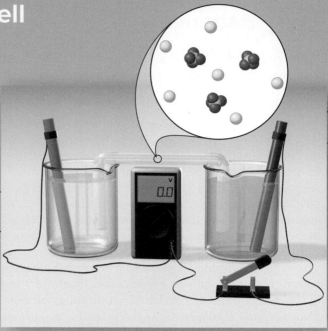

The two metal pieces are the electrodes in the galvanic cell. We connect the electrodes with a length of wire routed through a voltmeter and a switch so that we can complete the circuit when we have completed construction of the cell.

To make the reaction between zinc and copper more useful, we can construct a galvanic cell. In one beaker, we place a piece of zinc metal in a 1.00-*M* solution of Zn^{2+} ions. In the other, we place a piece of copper metal in a 1.00-*M* solution of Cu^{2+} ions.

START

As shown in Figure 10.6 on page 373, when zinc metal (Zn) is immersed in a solution containing copper ions (Cu^{2+}), the zinc is oxidized to zinc ions (Zn^{2+}), and copper ions are reduced to copper metal (Cu).

$$Zn(s) + Cu^{2+}(aq) \longrightarrow Zn^{2+}(aq) + Cu(s)$$

This is an oxidation-reduction reaction in which electrons flow spontaneously from zinc metal to the copper ions in solution. The lightening of the blue color indicates that the concentration of Cu^{2+} has decreased. Copper metal is deposited on the solid zinc surface. Some of the zinc metal has gone into solution as Zn^{2+} ions, which do not impart any color in the solution.

before   after

When we close the switch, we complete the circuit; the voltmeter indicates the initial potential of the cell: 1.10 V.

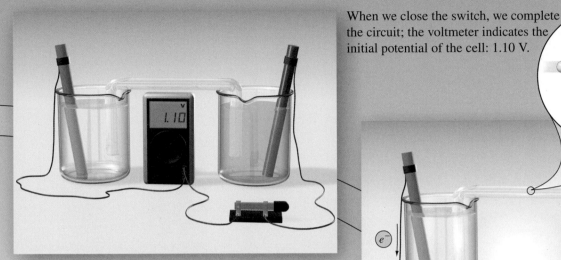

When we replace the voltmeter with a lightbulb, electrons flow from the zinc electrode (the anode) to the copper electrode (the cathode). The flow of electrons lights the bulb. Anions in the salt bridge migrate toward the anode; cations migrate toward the cathode.

As the reaction proceeds, zinc metal from the anode is oxidized, increasing the Zn^{2+} concentration in the beaker on the left; and additional copper metal is deposited on the cathode, decreasing the Cu^{2+} concentration in the beaker on the right. As the concentrations of both ions change, the potential of the cell decreases. After allowing the reaction to proceed for a time, we can reinsert the voltmeter to measure the decreased voltage.

What's the point?

Zinc is a stronger reducing agent than copper, so there is a natural tendency for electrons to flow from zinc metal to copper ions. We can harness this flow of electrons by forcing the half-reactions to occur in separate compartments. Electrons still flow from Zn to Cu^{2+}, but they must flow through the wire connecting the two electrodes. The potential of the cell decreases as the reaction proceeds, as the reading on the voltmeter shows.

(See Visualizing Chemistry questions VC 17.1–VC 17.4 on page 566.)

TABLE 17.1	Activity Series
Element	**Oxidation Half-Reaction**
Lithium	$Li \longrightarrow Li^+ + e^-$
Potassium	$K \longrightarrow K^+ + e^-$
Barium	$Ba \longrightarrow Ba^{2+} + 2e^-$
Calcium	$Ca \longrightarrow Ca^{2+} + 2e^-$
Sodium	$Na \longrightarrow Na^+ + e^-$
Magnesium	$Mg \longrightarrow Mg^{2+} + 2e^-$
Aluminum	$Al \longrightarrow Al^{3+} + 3e^-$
Manganese	$Mn \longrightarrow Mn^{2+} + 2e^-$
Zinc	$Zn \longrightarrow Zn^{2+} + 2e^-$
Chromium	$Cr \longrightarrow Cr^{3+} + 3e^-$
Iron	$Fe \longrightarrow Fe^{2+} + 2e^-$
Cadmium	$Cd \longrightarrow Cd^{2+} + 2e^-$
Cobalt	$Co \longrightarrow Co^{2+} + 2e^-$
Nickel	$Ni \longrightarrow Ni^{2+} + 2e^-$
Tin	$Sn \longrightarrow Sn^{2+} + 2e^-$
Lead	$Pb \longrightarrow Pb^{2+} + 2e^-$
Hydrogen	$H_2 \longrightarrow 2H^+ + 2e^-$
Copper	$Cu \longrightarrow Cu^{2+} + 2e^-$
Silver	$Ag \longrightarrow Ag^+ + e^-$
Mercury	$Hg \longrightarrow Hg^{2+} + 2e^-$
Platinum	$Pt \longrightarrow Pt^{2+} + 2e^-$
Gold	$Au \longrightarrow Au^{3+} + 3e^-$

Student Note: The volt is a derived SI unit: 1 V = 1 J/1 C.

Experimentally the difference in electrical potential between the anode and the cathode is measured by a voltmeter (Figure 17.2) and the reading (in volts) is called the ***cell potential (*E_{cell}*)***. The terms *cell potential, cell voltage, cell electromotive force,* and *cell emf* are used interchangeably and are all symbolized the same way with E_{cell}. The potential of a cell depends not only on the nature of the electrodes and the ions in solution, but also on the concentrations of the ions and the temperature at which the cell is operated.

Figure 17.2 The galvanic cell described in Figure 17.1. Note the U tube (salt bridge) connecting the two beakers. When the concentrations of Zn^{2+} and Cu^{2+} are 1 molar (1 *M*) at 25°C, the cell voltage is 1.10 V.
Ken Karp/McGraw Hill

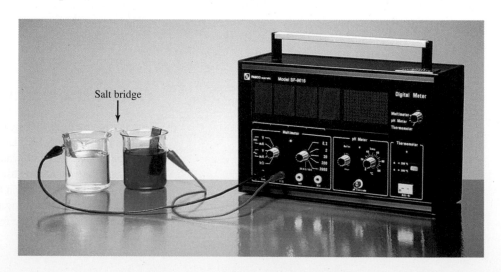

Salt bridge

The conventional notation for representing galvanic cells is the cell diagram. For the cell shown in Figure 17.1, if we assume that the concentrations of Zn^{2+} and Cu^{2+} ions are 1 M, the cell diagram is

$$Zn(s)\,|\,Zn^{2+}(1\ M)\,\|\,Cu^{2+}(1\ M)\,|\,Cu(s)$$

The single vertical line represents a phase boundary. For example, the zinc electrode is a solid and the Zn^{2+} ions are in solution. Thus, we draw a line between Zn and Zn^{2+} to show the phase boundary. The double vertical lines denote the salt bridge. By convention, the anode is written first, to the left of the double lines, and the other components appear in the order in which we would encounter them in moving from the anode to the cathode (from left to right in the cell diagram).

A **battery** is a galvanic cell, or a series of connected galvanic cells, that can be used as a portable, self-contained source of direct electric current.

SAMPLE PROBLEM (**17.2**) Writing Cell Notation from a Balanced Electrochemical Equation

Write the cell notation for the given balanced redox equation.

$$Al(s) + 3AgNO_3(aq) \longrightarrow Al(NO_3)_3(aq) + 3Ag(s)$$

Strategy Split the equation into the oxidation and reduction half-reactions, ignoring the coefficient and spectator ions. The reactant and product of each half-reaction is written in the correct order to place into the cell notation.

Setup $Al(s) \longrightarrow Al^{3+}(aq)$ (the nitrate ion, NO_3^-, is a spectator ion in this reaction) $Ag^+(aq) \longrightarrow Ag(s)$

Solution

$$Al(s)\,|\,Al^{3+}(aq)\,\|\,Ag^+(aq)\,|\,Ag(s)$$

> **THINK ABOUT IT**
>
> Remember that the oxidation half-reaction is written on the left side of the cell notation and the reduction half-reaction is on the right side.

Practice Problem **A**TTEMPT Write the cell notation for the given balanced redox equation.

$$Au(C_2H_3O_2)_3(aq) + Cr(s) \longrightarrow Cr(C_2H_3O_2)_3(aq) + Au(s)$$

Practice Problem **B**UILD Write the balanced chemical equation that represents the galvanic cell with the notation:
$Ni(s)\,|\,Ni^{2+}(aq)\,\|\,Cu^{2+}(aq)\,|\,Cu(s)$

Practice Problem **C**ONCEPTUALIZE Explain why the acetate ion from Practice Problem 17.2A does not appear in the cell notation for the reaction.

Dry Cells and Alkaline Batteries

The most common batteries, *dry cells* and *alkaline batteries,* are those used in flashlights, toys, and certain portable electronics such as CD players. The two are similar in appearance, but differ in the spontaneous chemical reaction responsible for producing a voltage. Although the reactions that take place in these batteries are somewhat complex, the reactions shown here approximate the overall processes.

A dry cell, so named because it has no fluid component, consists of a zinc container (the anode) in contact with manganese dioxide and an electrolyte (Figure 17.3). The electrolyte consists of ammonium chloride and zinc chloride in water, to which starch is added to thicken the solution to a paste so that it is less likely to leak. A

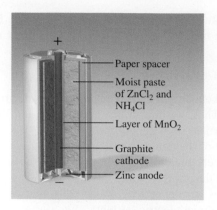

Figure 17.3 Interior view of the type of dry cell used in flashlights and other small devices.

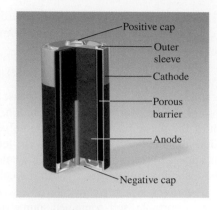

Figure 17.4 Interior view of an alkaline battery.

carbon rod, immersed in the electrolyte in the center of the cell, serves as the cathode. The cell reactions are:

Anode: $\qquad$ $Zn(s) \longrightarrow Zn^{2+}(aq) + 2e^-$

Cathode: $\quad 2NH_4^+(aq) + 2MnO_2(s) + 2e^- \longrightarrow Mn_2O_3(s) + 2NH_3(aq) + H_2O(l)$

Overall: $\quad Zn(s) + 2NH_4^+(aq) + 2MnO_2(s) \longrightarrow Zn^{2+}(aq) + Mn_2O_3(s) + 2NH_3(aq) + H_2O(l)$

The voltage produced by a dry cell is about 1.5 V.

An alkaline battery is also based on the reduction of manganese dioxide and the oxidation of zinc. However, the reactions take place in a *basic* medium, hence the name *alkaline* battery. The anode consists of powdered zinc suspended in a gel, which is in contact with a concentrated solution of KOH. The cathode is a mixture of manganese dioxide and graphite. The anode and cathode are separated by a porous barrier (Figure 17.4).

Anode: $\qquad$ $Zn(s) + 2OH^-(aq) \longrightarrow Zn(OH)_2(s) + 2e^-$

Cathode: $\quad 2MnO_2(s) + 2H_2O(l) + 2e^- \longrightarrow 2MnO(OH)(s) + 2OH^-(aq)$

Overall: $\quad Zn(s) + 2MnO_2(s) + 2H_2O(l) \longrightarrow Zn(OH)_2(s) + 2MnO(OH)(s)$

Alkaline batteries are more expensive than dry cells and offer superior performance and shelf life.

Lead Storage Batteries

The lead storage battery commonly used in automobiles consists of six identical cells joined together in series. Each cell has a lead anode and a cathode made of lead dioxide (PbO_2) packed on a metal plate (Figure 17.5). Both the cathode and the anode are immersed in an aqueous solution of sulfuric acid, which acts as the electrolyte. The cell reactions are

Anode: $\qquad$ $Pb(s) + SO_4^{2-}(aq) \longrightarrow PbSO_4(s) + 2e^-$

Cathode: $\quad PbO_2(s) + 4H^+(aq) + SO_4^{2-}(aq) + 2e^- \longrightarrow PbSO_4(s) + 2H_2O(l)$

Overall: $\quad Pb(s) + PbO_2(s) + 4H^+(aq) + 2SO_4^{2-}(aq) \longrightarrow 2PbSO_4(s) + 2H_2O(l)$

Under normal operating conditions, each cell produces 2 V. A total of 12 V from the six cells is used to power the ignition circuit of the automobile and its other electric systems. The lead storage battery can deliver large amounts of current for a short time, such as the time it takes to start the engine.

Unlike dry cells and alkaline batteries, the lead storage battery is rechargeable. Recharging the battery means reversing the normal electrochemical reaction by applying

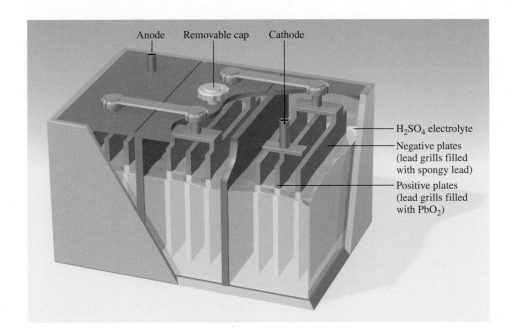

Anode   Removable cap   Cathode

— H_2SO_4 electrolyte
— Negative plates (lead grills filled with spongy lead)
— Positive plates (lead grills filled with PbO_2)

Figure 17.5 Interior view of a lead storage battery. Under normal operating conditions, the concentration of the sulfuric acid solution is about 38 percent by mass.

an external voltage at the cathode and the anode. (This kind of process is called *electrolysis,* which we discuss in Section 17.4.)

Lithium-Ion Batteries

Sometimes called "the battery of the future," lithium-ion batteries have several advantages over other battery types. The overall reaction that takes place in the lithium-ion battery is

$$Anode: \quad Li \longrightarrow Li^+ + e^-$$

$$Cathode: \quad \underline{Li^+ + CoO_2 + e^- \longrightarrow LiCoO_2}$$

$$Overall: \quad Li + CoO_2 \longrightarrow LiCoO_2$$

The cell produces 3.4 V, which is a relatively large voltage. Lithium is also the lightest metal—only 6.941 g of Li (its molar mass) are needed to produce 1 mole of electrons. Furthermore, a lithium-ion battery can be recharged hundreds of times. These qualities make lithium batteries suitable for use in portable devices such as cell phones, digital cameras, and laptop computers.

Fuel Cells

Fossil fuels are a major source of energy, but the conversion of fossil fuel into electric energy is a highly inefficient process. Consider the combustion of methane:

$$CH_4(g) + 2O_2(g) \longrightarrow CO_2(g) + 2H_2O(l) + energy$$

To generate electricity, heat produced by the reaction is first used to convert water to steam, which then drives a turbine, which then drives a generator. A significant fraction of the energy released in the form of heat is lost to the surroundings at each step (even the most efficient power plant converts only about 40 percent of the original chemical energy into electricity). Because combustion reactions are redox reactions, it is more desirable to carry them out directly by electrochemical means, thereby greatly increasing the efficiency of power production. This objective can be accomplished by a device known as a ***fuel cell,*** a galvanic cell that requires a continuous supply of reactants to keep functioning. Strictly speaking, a fuel cell is not a battery because it is not self-contained.

In its simplest form, a hydrogen-oxygen fuel cell consists of an electrolyte solution, such as a potassium hydroxide solution, and two inert electrodes. Hydrogen and

Figure 17.6 A hydrogen-oxygen fuel cell. The Ni and NiO embedded in the porous carbon electrodes are catalysts, which increase the rate of reaction.

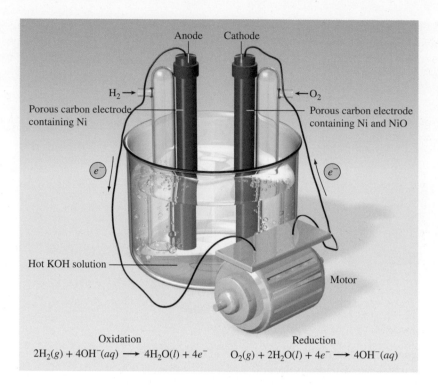

Anode   Cathode

H$_2$ →

Porous carbon electrode containing Ni

← O$_2$

Porous carbon electrode containing Ni and NiO

e^- e^-

Hot KOH solution

Motor

Oxidation

$2H_2(g) + 4OH^-(aq) \longrightarrow 4H_2O(l) + 4e^-$

Reduction

$O_2(g) + 2H_2O(l) + 4e^- \longrightarrow 4OH^-(aq)$

oxygen gases are bubbled through the anode and cathode compartments (Figure 17.6), where the following reactions take place:

Anode: $2H_2(g) + 4OH^-(aq) \longrightarrow 4H_2O(l) + 4e^-$

Cathode: $O_2(g) + 2H_2O(l) + 4e^- \longrightarrow 4OH^-(aq)$

Overall: $2H_2(g) + O_2(g) \longrightarrow 2H_2O(l)$

The H$_2$–O$_2$ fuel cell produces a voltage of 1.23 V.

In addition to the H$_2$–O$_2$ system, a number of other fuel cells have been developed. Among these is the propane-oxygen fuel cell. The corresponding half-cell reactions are

Anode: $C_3H_8(g) + 6H_2O(l) \longrightarrow 3CO_2(g) + 20H^+(aq) + 20e^-$

Cathode: $5O_2(g) + 20H^+(aq) + 20e^- \longrightarrow 10H_2O(l)$

Overall: $C_3H_8(g) + 5O_2(g) \longrightarrow 3CO_2(g) + 4H_2O(l)$

The overall reaction is identical to the burning of propane in oxygen.

Unlike batteries, fuel cells do not store chemical energy. Reactants must be constantly resupplied as they are consumed, and products must be removed as they form. However, properly designed fuel cells may be as much as 70 percent efficient, which is about twice as efficient as an internal combustion engine. In addition, fuel-cell generators are free of the noise, vibration, heat transfer, thermal pollution, and other problems normally associated with conventional power plants. Nevertheless, fuel cells are not yet in widespread use. One major problem is the expense of electrocatalysts that can function efficiently for long periods of time without contamination. One notable application of fuel cells is their use in space vehicles. Hydrogen-oxygen fuel cells provide electric power (and drinking water!) for space flight.

17.3 Corrosion

The term *corrosion* generally refers to the deterioration of a metal by an electrochemical process. There are many examples of corrosion, including rust on iron, tarnish on silver, and the green layer that forms on copper and brass. In this section we discuss the processes involved in corrosion and some of the measures taken to prevent it.

The formation of rust on iron requires oxygen and water. Although the reactions involved are quite complex and are not completely understood, the main steps are believed to be as follows: A region of the metal's surface serves as the anode, where the following oxidation occurs:

$$Fe(s) \longrightarrow Fe^{2+}(aq) + 2e^-$$

The electrons given up by iron reduce atmospheric oxygen to water at the cathode, which is another region of the same metal's surface:

$$O_2(g) + 4H^+(aq) + 4e^- \longrightarrow 2H_2O(l)$$

The overall redox reaction is

$$2Fe(s) + O_2(g) + 4H^+(aq) \longrightarrow 2Fe^{2+}(g) + 2H_2O(l)$$

Note that this reaction occurs in an *acidic* solution; the H^+ ions are supplied in part by the reaction of atmospheric carbon dioxide with water to form the weak acid, carbonic acid (H_2CO_3).

The Fe^{2+} ions formed at the anode are further oxidized by oxygen as follows:

$$4Fe^{2+}(aq) + O_2(g) + (4 + 2x)H_2O(l) \longrightarrow 2Fe_2O_3 \cdot xH_2O(s) + 8H^+(aq)$$

This hydrated form of iron(III) oxide is known as rust. The amount of water associated with the iron(III) oxide varies, so we represent the formula as $Fe_2O_3 \cdot xH_2O$.

Figure 17.7 shows the mechanism of rust formation. The electric circuit is completed by the migration of electrons and ions; this is why rusting occurs so rapidly in salt water. In cold climates, salts (NaCl or $CaCl_2$) spread on roadways to melt ice and snow are a major cause of rust formation on automobiles.

Other metals also undergo oxidation. Aluminum, for example, which is used to make airplanes, beverage cans, and aluminum foil, has a much greater tendency to oxidize than does iron.

Unlike the corrosion of iron, though, corrosion of aluminum produces an insoluble layer of protective coating (Al_2O_3) that prevents the underlying metal from additional corrosion.

Coinage metals such as copper and silver also corrode, but much more slowly than either iron or aluminum:

$$Cu(s) \longrightarrow Cu^{2+}(aq) + 2e^-$$

$$Ag(s) \longrightarrow Ag^+(aq) + e^-$$

In ordinary atmospheric exposure, copper forms a layer of copper carbonate ($CuCO_3$), a green substance referred to as *patina*, that protects the metal underneath from further

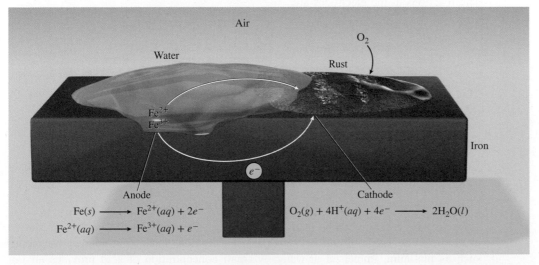

Figure 17.7 The electrochemical process involved in rust formation. The H^+ ions are supplied by H_2CO_3, which forms when CO_2 from air dissolves in water.

corrosion. Likewise, silverware that comes into contact with foodstuffs develops a layer of silver sulfide (Ag_2S).

A number of methods have been devised to protect metals from corrosion. Most of these methods are aimed at preventing rust formation. The most obvious approach is to coat the metal surface with paint to prevent exposure to the substances necessary for corrosion. If the paint is scratched or otherwise damaged, however, thus exposing even the smallest area of bare metal, rust will form under the paint layer.

A very popular method of preventing the corrosion of steel, which consists mostly of iron metal, is coating with zinc metal—a process known as ***galvanization.*** Zinc metal oxidizes more easily than steel and, in the process, becomes coated with a thin layer of zinc oxide. Like a layer of paint, the zinc oxide layer protects the underlying iron from exposure to oxygen and water. Unlike paint, however, a zinc coating provides protection even in the event that it is scratched or otherwise breached. A scratch in the protective zinc oxide layer exposes zinc metal, which will be oxidized more easily than the underlying iron. A new layer of zinc oxide forms, and the iron remains protected.

17.4 Electrolysis

In Section 17.2, we mentioned that lead storage batteries are rechargeable and that recharging means reversing the electrochemical processes by which the battery ordinarily operates through the application of an external voltage. This process, the use of electric energy to drive a nonspontaneous chemical reaction, is called ***electrolysis.*** An ***electrolytic cell*** is one used to carry out electrolysis. The same principles apply to the processes in both galvanic and electrolytic cells. In this section, we discuss two examples of electrolysis based on those principles. We then examine some of the quantitative aspects of electrolysis.

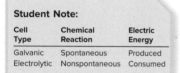

Student Note:

Cell Type	Chemical Reaction	Electric Energy
Galvanic	Spontaneous	Produced
Electrolytic	Nonspontaneous	Consumed

Electrolysis of Molten Sodium Chloride

In its molten (melted) state, sodium chloride, an ionic compound, can be electrolyzed to separate it into its constituent elements, sodium and chlorine. Figure 17.8(a) is a diagram of a Downs cell, which is used for the large-scale electrolysis of NaCl. In molten NaCl, the cations and anions are the Na^+ and Cl^- ions, respectively. Figure 17.8(b) is a simplified diagram showing the reactions that occur at the electrodes. The electrolytic cell contains a pair of electrodes connected to the battery. The battery serves to push electrons in the direction they would not flow spontaneously. The electrode toward which the electrons are pushed is the cathode, where reduction takes place. The electrode away from which electrons are drawn is the anode, where oxidation takes place. The reactions at the electrodes are

$$\begin{aligned} \textit{Anode (oxidation):} && 2Cl^-(l) &\longrightarrow Cl_2(g) + 2e^- \\ \textit{Cathode (reduction):} && 2Na^+(l) + 2e^- &\longrightarrow 2Na(l) \\ \hline \textit{Overall:} && 2Na^+(l) + 2Cl^-(l) &\longrightarrow 2Na(l) + Cl_2(g) \end{aligned}$$

This process is a major industrial source of pure sodium metal and chlorine gas. A relatively large voltage (~4 V) must be applied to drive this process.

Electrolysis of Water

Under ordinary atmospheric conditions (1 atm and 25°C), water will not spontaneously decompose to form hydrogen and oxygen gas:

$$2H_2O(l) \rightleftharpoons 2H_2(g) + O_2(g)$$

However, this reaction can be made to occur in an electrolytic cell like the one shown in Figure 17.9. This cell consists of a pair of electrodes made of a nonreactive metal, such as platinum, immersed in water. A small concentration of acid provides an ion concentration sufficient to conduct the necessary electricity.

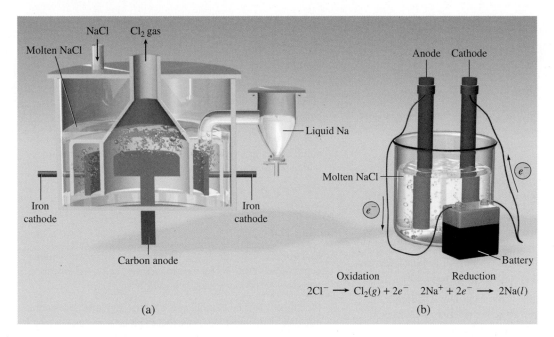

Figure 17.8 (a) A practical arrangement called a Downs cell for the electrolysis of molten NaCl (m.p. = 801°C). The sodium metal formed at the cathodes is in the liquid state. Because liquid sodium metal is lighter than molten NaCl, the sodium floats to the surface, as shown, and is collected. Chlorine gas forms at the anode and is collected at the top. (b) A simplified diagram showing the electrode reactions during the electrolysis of molten NaCl. The battery is needed to drive the nonspontaneous reaction.

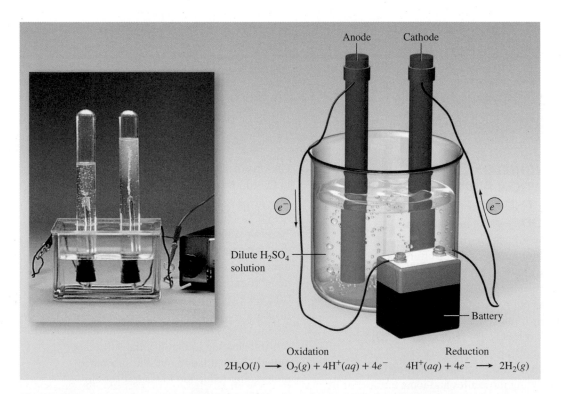

Figure 17.9 Apparatus for small-scale electrolysis of water. The volume of hydrogen gas generated at the cathode is twice that of oxygen gas generated at the anode.
Ken Karp/McGraw Hill

Chapter Summary

Section 17.1

- Redox reactions are those in which oxidation numbers change. *Half-reactions* are the separated oxidation and reduction reactions that make up the overall redox reaction.

- Redox equations can be balanced via the half-reaction method, which allows for the addition of H_2O to balance oxygen and H^+ to balance hydrogen. It also allows the addition of OH^- for reactions that take place in basic conditions.

Section 17.2

- A *battery* is a portable, self-contained source of electric energy consisting of a galvanic cell or a series of galvanic cells.

- A *fuel cell* is not really a battery, but is also used to supply electric energy via a spontaneous redox reaction. Reactants must be supplied continually for a fuel cell to operate.

Section 17.3

- *Corrosion* is the undesirable oxidation of metals.

- Corrosion can be prevented by coating the metal surface with paint; with a less easily oxidized metal, such as chromium; or with a more easily oxidized metal, such as zinc.

- The use of zinc to prevent oxidation of iron or steel is known as *galvanization.*

Section 17.4

- *Electrolysis* is the use of electric energy to drive a nonspontaneous redox reaction. An electrochemical cell used for this purpose is called an *electrolytic cell.*

- Electrolysis is used to recharge batteries, separate compounds into their constituent elements, and to process and purify metals.

Key Terms

Anode 553	Corrosion 560	Fuel cell 559	Half-reaction 550
Battery 557	Electrode 553	Galvanic cell 553	Oxidizing agent 553
Cathode 553	Electrolysis 562	Galvanization 562	Reducing agent 553
Cell potential (E_{cell}) 556	Electrolytic cell 562	Half-cell 553	Salt bridge 553

Questions and Problems

SECTION 17.1: BALANCING OXIDATION-REDUCTION REACTIONS USING THE HALF-REACTION METHOD

17.1 Arrange the following species in order of increasing oxidation number of the sulfur atom: (a) H_2S, (b) S_8, (c) H_2SO_4, (d) S^{2-}, (e) HS^-, (f) SO_2, (g) SO_3.

17.2 Phosphorus forms many oxoacids. Indicate the oxidation number of phosphorus in each of the following acids: (a) HPO_3, (b) H_3PO_2, (c) H_3PO_3, (d) H_3PO_4, (e) $H_4P_2O_7$, (f) $H_5P_3O_{10}$.

17.3 Give the oxidation numbers for the underlined atoms in the following molecules and ions: (a) $\underline{Cl}F$, (b) $\underline{I}F_7$, (c) $\underline{C}H_4$, (d) $\underline{C}_2H_2$, (e) $\underline{C}_2H_4$, (f) $K_2\underline{Cr}O_4$, (g) $K_2\underline{Cr}_2O_7$, (h) $K\underline{Mn}O_4$, (i) $NaH\underline{C}O_3$, (j) $\underline{Li}_2$, (k) $Na\underline{I}O_3$, (l) $K\underline{O}_2$, (m) $\underline{P}F_6^-$, (n) $K\underline{Au}Cl_4$.

17.4 Give the oxidation numbers for the underlined atoms in the following molecules and ions: (a) $\underline{Cs}_2O$, (b) $Ca\underline{I}_2$, (c) $\underline{Al}_2O_3$, (d) $H_3\underline{As}O_3$, (e) $\underline{Ti}O_2$, (f) $\underline{Mo}O_4^{2-}$, (g) $\underline{Pt}Cl_4^{2-}$, (h) $\underline{P}Cl_6^-$, (i) $\underline{Sn}F_2$, (j) $\underline{Cl}F_3$, (k) $\underline{Sb}F_6^-$.

17.5 Assign oxidation states to each element and use this information to identify the substance undergoing oxidation and the substance undergoing reduction in each reaction.
(a) $2AlCl_3(s) \longrightarrow 2Al(s) + 3Cl_2(g)$
(b) $Zn(s) + S(s) \longrightarrow ZnS(s)$

17.6 Assign oxidation states to each element and use this information to identify the substance undergoing oxidation and the substance undergoing reduction in each reaction.
(a) $Ni(s) + Cu(NO_3)_2(aq) \longrightarrow$
$$Ni(NO_3)_2(aq) + Cu(s)$$
(b) $2C_2H_2(g) + 5O_2(g) \longrightarrow 4CO_2(g) + 2H_2O(g)$

17.7 Identify the oxidizing agent and the reducing agent in each reaction.
(a) $2Sr + O_2 \longrightarrow 2SrO$
(b) $2Li + H_2 \longrightarrow 2LiH$
(c) $2Cs + Br_2 \longrightarrow 2CsBr$
(d) $3Mg + N_2 \longrightarrow Mg_3N_2$

17.8 Identify the oxidizing agent and the reducing agent in each reaction.
(a) $4Fe + 3O_2 \longrightarrow 2Fe_2O_3$
(b) $Cl_2 + 2NaBr \longrightarrow 2NaCl + Br_2$
(c) $Si + 2F_2 \longrightarrow SiF_4$
(d) $H_2 + Cl_2 \longrightarrow 2HCl$

17.9 Identify the oxidizing agent and the reducing agent in each reaction.
(a) $2H_2O_2 \longrightarrow 2H_2O + O_2$
(b) $Mg + 2AgNO_3 \longrightarrow Mg(NO_3)_2 + 2Ag$
(c) $NH_4NO_2 \longrightarrow N_2 + 2H_2O$
(d) $H_2 + Br_2 \longrightarrow 2HBr$

17.10 Identify the oxidizing agent and the reducing agent in each reaction.
(a) $P_4 + 10Cl_2 \longrightarrow 4PCl_5$
(b) $2NO \longrightarrow N_2 + O_2$
(c) $Cl_2 + 2KI \longrightarrow 2KCl + I_2$
(d) $3HNO_2 \longrightarrow HNO_3 + H_2O + 2NO$

17.11 Balance the following redox equations by the half-reaction method:
(a) $H_2O_2 + Fe^{2+} \longrightarrow Fe^{3+} + H_2O$
(in acidic solution)
(b) $Cu + HNO_3 \longrightarrow Cu^{2+} + NO + H_2O$
(in acidic solution)
(c) $CN^- + MnO_4^- \longrightarrow CNO^- + MnO_2$
(in basic solution)
(d) $Br_2 \longrightarrow BrO_3^- + Br^-$ (in basic solution)
(e) $S_2O_3^{2-} + I_2 \longrightarrow I^- + S_4O_6^{2-}$
(in acidic solution)

17.12 Balance the following redox equations by the half-reaction method:
(a) $Mn^{2+} + H_2O_2 \longrightarrow MnO_2 + H_2O$
(in basic solution)
(b) $Bi(OH)_3 + SnO_2^{2-} \longrightarrow SnO_3^{2-} + Bi$
(in basic solution)
(c) $Cr_2O_7^{2-} + C_2O_4^{2-} \longrightarrow Cr^{3+} + CO_2$
(in basic solution)
(d) $ClO_3^- + Cl^- \longrightarrow Cl_2 + ClO_2$
(in acidic solution)
(e) $Mn^{2+} + BiO_3^- \longrightarrow Bi^{3+} + MnO_4^-$
(in acidic solution)

17.13 Balance each oxidation-reduction equation in acidic solution using the half-reaction method.
(a) $S_2O_3^{2-}(aq) + Br_2(aq) \longrightarrow$
$S_4O_6^{2-}(aq) + Br^-(aq)$
(b) $Fe^{2+}(aq) + Cr_2O_7^{2-}(aq) \longrightarrow$
$Cr^{3+}(aq) + Fe^{3+}(aq)$

17.14 Balance each oxidation-reduction equation in acidic solution using the half-reaction method.
(a) $Zn(s) + NO_2^-(aq) \longrightarrow Zn^{2+}(aq) + N_2(g)$
(b) $Br_2(aq) + SO_2(aq) \longrightarrow Br^-(aq) + SO_4^{2-}(aq)$

17.15 Balance each oxidation-reduction equation in basic solution using the half-reaction method.
(a) $Cl_2O_7(g) + H_2O_2(aq) \longrightarrow ClO_2^-(aq) + O_2(g)$
(note that in H_2O_2, the oxygen has an oxidation state of -1)
(b) $Cr^{3+}(aq) + MnO_2(s) \longrightarrow$
$Mn^{2+}(aq) + CrO_4^{2-}(aq)$

17.16 Balance each oxidation-reduction equation in basic solution using the half-reaction method.
(a) $MnO_4^-(aq) + Br^-(aq) \longrightarrow$
$MnO_2(s) + BrO_3^-(aq)$
(b) $Al(s) + NO_3^-(aq) \longrightarrow$
$Al(OH)_4^-(aq) + NH_3(g)$

SECTION 17.2: BATTERIES
17.17 Which metal is the best reducing agent in each set?
(a) Li, Ca, Na
(b) Na, Zn, K
(c) Ni, Al, Sn

17.18 Which metal is the best reducing agent in each set?
(a) Au, Mg, Cr
(b) Fe, Mn, Pb
(c) Al, Cu, Zn

17.19 In each set, place the metals in order of increasing strength as a reducing agent.
(a) K, Al, Mg
(b) Ni, Pb, Ca
(c) K, Cr, Mn

17.20 In each set, place the metals in order of decreasing strength as a reducing agent.
(a) Li, Zn, Al
(b) Ag, Sn, Pb
(c) Na, Cu, Fe

17.21 Determine which of the following reactions will occur, complete the equation, and balance.
(a) $Mg(C_2H_3O_2)_2(aq) + Al(s) \longrightarrow$?
(b) $SnCl_2(aq) + Fe(s) \longrightarrow$?
(c) $Cu(s) + AgClO_4(aq) \longrightarrow$?

17.22 Determine which of the following reactions will occur, complete the equation, and balance.
(a) $Zn(s) + Pb(NO_3)_2(aq) \longrightarrow$?
(b) $Au(C_2H_3O_2)_3(aq) + Cr(s) \longrightarrow$?
(c) $Na(s) + CaCl_2(aq) \longrightarrow$?

17.23 Predict the outcome of the reactions represented by the following equations by using the activity series, and balance the equations.
(a) $Cu(s) + HCl(aq) \longrightarrow$
(b) $Au(s) + NaBr(aq) \longrightarrow$
(c) $Mg(s) + CuSO_4(aq) \longrightarrow$
(d) $Zn(s) + KBr(aq) \longrightarrow$

Visualizing Chemistry
Figure 17.1

VC 17.1 In the first scene of the animation, when a zinc bar is immersed in an aqueous copper sulfate solution, solid copper deposits on the bar. What reaction would take place if a *copper* bar were immersed in an aqueous *zinc* sulfate solution?
a) No reaction would take place.
b) Solid copper would still deposit on the bar.
c) Solid zinc would deposit on the bar.

VC 17.2 What causes the change in the potential of the galvanic cell in Figure 17.1 as the cell operates?
a) Changes in the sizes of the zinc and copper electrodes.
b) Changes in the concentrations of zinc and copper ions.
c) Changes in the volumes of solutions in the half-cells.

VC 17.3 Why does the color of the blue solution in the galvanic cell (Figure 17.1) fade as the cell operates?
a) Blue Cu^{2+} ions are replaced by colorless Zn^{2+} ions in solution.
b) Blue Cu^{2+} ions are removed from solution by reduction.
c) Blue Cu^{2+} ions are removed from solution by oxidation.

VC 17.4 What happens to the mass of the copper electrode in the galvanic cell in Figure 17.1 as the cell operates?
a) It increases.
b) It decreases.
c) It does not change.

17.24   Define the terms *anode* and *cathode*.
17.25   Sketch and label the parts of a galvanic cell.
17.26   Describe the basic features of a galvanic cell. Why are the two components of the cell separated from each other?
17.27   What is the function of a salt bridge? What kind of electrolyte should be used in a salt bridge?
17.28   What is a cell diagram? Write the cell diagram for a galvanic cell consisting of an Al electrode placed in a 1 *M* $Al(NO_3)_3$ solution and an Ag electrode placed in a 1 *M* $AgNO_3$ solution.

17.29   Sketch the galvanic cell that is described by the cell notation:
$$Fe(s)\,|\,Fe^{2+}(aq)\,\|\,Ag^+(aq)\,|\,Ag(s)$$

17.30   Sketch the galvanic cell that is described by the cell notation:
$$Ni(s)\,|\,Ni^{2+}(aq)\,\|\,Cu^{2+}(aq)\,|\,Cu(s)$$

17.31   Determine the cell notation for the following balanced redox equations.
(a) $2Al(s) + 3Sn(NO_3)_2(aq) \longrightarrow$
$$3Sn(s) + 2Al(NO_3)_3(aq)$$
(b) $Zn(s) + 2AgC_2H_3O_2(aq) \longrightarrow$
$$2Ag(s) + Zn(C_2H_3O_2)_2(aq)$$
(c) $2Cr(s) + 3Cu(ClO_4)_2(aq) \longrightarrow$
$$3Cu(s) + 2Cr(ClO_4)_3(aq)$$

17.32   Determine the cell notation for the following balanced redox equations.
(a) $2Na(s) + MnSO_4(aq) \longrightarrow$
$$Mn(s) + Na_2SO_4(aq)$$
(b) $3Mg(s) + 2CrBr_3(aq) \longrightarrow$
$$2Cr(s) + 3MgBr_2(aq)$$
(c) $3Ni(s) + 2Au(NO_3)_3(aq) \longrightarrow$
$$2Au(s) + 3Ni(NO_3)_2(aq)$$

SECTION 17.3: CORROSION

17.33   Many motorcycles have chrome-plated parts, such as mufflers, mirrors, and wheels. Explain how a layer of chromium can protect steel parts from corrosion.
17.34   "Galvanized iron" is steel that has been coated with zinc; "tin" cans are made of steel sheet coated with tin. Discuss the functions of these coatings and the electrochemistry of the corrosion reactions that occur if an electrolyte contacts the scratched surface of a galvanized iron sheet or a tin can.
17.35   Which of the following metals would be best to use in coating a lead object to keep it from corroding: Cu, Au, or Sn?
17.36   Which of the following metals would be best to use in coating an iron object to keep it from corroding: Ni, Mn, or Ag?

SECTION 17.4: ELECTROLYSIS

17.37   Describe how the process of electrolysis is different from the process that occurs in a galvanic cell.

Answers to In-Chapter Materials

Answers to Practice Problems

17.1A $2MnO_4^- + H_2O + 3CN^- \longrightarrow 2MnO_2 + 2OH^- + 3CNO^-$.
17.1B $MnO_4^- + 5Fe^{2+} + 8H^+ \longrightarrow Mn^{2+} + 5Fe^{3+} + 4H_2O$.
17.2A $Cr(s)\,|\,Cr^{3+}(aq)\,\|\,Au^{3+}(aq)\,|\,Au(s)$.
17.2B $Ni(s) + Cu^{2+}(aq) \longrightarrow Ni^{2+}(aq) + Cu(s)$.

Answers to Checkpoints

17.1.1 a, c, e. **17.1.2** c.

Appendix

Mathematical Operations

Scientific Notation

Chemists often deal with numbers that are either extremely large or extremely small. For example, in 1 g of the element hydrogen there are roughly

$$602,200,000,000,000,000,000,000$$

hydrogen atoms. Each hydrogen atom has a mass of only

$$0.00000000000000000000000166 \text{ g}$$

These numbers are cumbersome to handle, and it is easy to make mistakes when using them in arithmetic computations. Consider the following multiplication:

$$0.0000000056 \times 0.00000000048 = 0.000000000000000002688$$

It would be easy for us to miss one zero or add one more zero after the decimal point. Consequently, when working with very large and very small numbers, we use a system called *scientific notation*. Regardless of their magnitude, all numbers can be expressed in the form

$$N \times 10^n$$

where N is a number between 1 and 10 and n, the exponent, is a positive or negative integer (whole number). Any number expressed in this way is said to be written in scientific notation.

Suppose that we are given a certain number and asked to express it in scientific notation. Basically, this assignment calls for us to find n. We count the number of places that the decimal point must be moved to give the number N (which is between 1 and 10). If the decimal point has to be moved to the left, then n is a positive integer; if it has to be moved to the right, n is a negative integer. The following examples illustrate the use of scientific notation:

1. Express 568.762 in scientific notation:

$$568.762 = 5.68762 \times 10^2$$

   Note that the decimal point is moved to the left by two places and $n = 2$.
2. Express 0.00000772 in scientific notation:

$$0.00000772 = 7.72 \times 10^{-6}$$

   Here the decimal point is moved to the right by six places and $n = -6$.

   Keep in mind the following two points. First, $n = 0$ is used for numbers that are not expressed in scientific notation. For example, 74.6×10^0 ($n = 0$) is equivalent to 74.6. Second, the usual practice is to omit the superscript when $n = 1$. Thus the scientific notation for 74.6 is 7.46×10 and not 7.46×10^1.

   Next, we consider how scientific notation is handled in arithmetic operations.

Addition and Subtraction

To add or subtract using scientific notation, we first write each quantity—say N_1 and N_2—with the same exponent, n. Then we combine N_1 and N_2; the exponents remain the same. Consider the following examples:

$$(7.4 \times 10^3) + (2.1 \times 10^3) = 9.5 \times 10^3$$
$$(4.31 \times 10^4) + (3.9 \times 10^3) = (4.31 \times 10^4) + (0.39 \times 10^4)$$
$$= 4.70 \times 10^4$$
$$(2.22 \times 10^{-2}) - (4.10 \times 10^{-3}) = (2.22 \times 10^{-2}) - (0.41 \times 10^{-2})$$
$$= 1.81 \times 10^{-2}$$

Multiplication and Division

To multiply numbers expressed in scientific notation, we multiply N_1 and N_2 in the usual way, but *add* the exponents together. To divide using scientific notation, we divide N_1 and N_2 as usual and subtract the exponents. The following examples show how these operations are performed:

$$(8.0 \times 10^4) \times (5.0 \times 10^2) = (8.0 \times 5.0)(10^{4+2})$$
$$= 40 \times 10^6$$
$$= 4.0 \times 10^7$$
$$(4.0 \times 10^{-5}) \times (7.0 \times 10^3) = (4.0 \times 7.0)(10^{-5+3})$$
$$= 28 \times 10^{-2}$$
$$= 2.8 \times 10^{-1}$$

$$\frac{6.9 \times 10^7}{3.0 \times 10^{-5}} = \frac{6.9}{3.0} \times 10^{7-(-5)}$$
$$= 2.3 \times 10^{12}$$

$$\frac{8.5 \times 10^4}{5.0 \times 10^9} = \frac{8.5}{5.0} \times 10^{4-9}$$
$$= 1.7 \times 10^{-5}$$

Basic Trigonometry

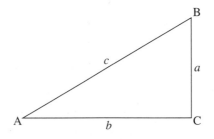

In the triangle shown, A, B, and C are angles (C = 90°) and a, b, and c are side lengths. To calculate unknown angles or sides, we use the following relationships:

$$a^2 + b^2 = c^2$$

$$\sin A = \frac{a}{c}$$

$$\cos A = \frac{b}{c}$$

$$\tan A = \frac{a}{b}$$

Logarithms

Common Logarithms

The concept of logarithms is an extension of the concept of exponents, which is discussed on page A-1. The common, or base-10, logarithm of any number is the power to which 10 must be raised to equal the number. The following examples illustrate this relationship:

Logarithm	Exponent
$\log 1 = 0$	$10^0 = 1$
$\log 10 = 1$	$10^1 = 10$
$\log 100 = 2$	$10^2 = 100$
$\log 10^{-1} = -1$	$10^{-1} = 0.1$
$\log 10^{-2} = -2$	$10^{-2} = 0.01$

In each case the logarithm of the number can be obtained by inspection.

Because the logarithms of numbers are exponents, they have the same properties as exponents. Thus, we have

Logarithm	Exponent
$\log AB = \log A + \log B$	$10^A \times 10^B = 10^{A+B}$
$\log \dfrac{A}{B} = \log A - \log B$	$\dfrac{10^A}{10^B} = 10^{A-B}$

Furthermore, $\log A^n = n \log A$.

Now suppose we want to find the common logarithm of 6.7×10^{-4}. On most electronic calculators, the number is entered first and then the log key is pressed. This operation gives us

$$\log 6.7 \times 10^{-4} = -3.17$$

Note that there are as many digits *after* the decimal point as there are significant figures in the original number. The original number has two significant figures, and the "17" in -3.17 tells us that the log has two significant figures. The "3" in -3.17 serves only to locate the decimal point in the number 6.7×10^{-4}. Other examples are

Number	Common Logarithm
62	1.79
0.872	-0.0595
1.0×10^{-7}	-7.00

Sometimes (as in the case of pH calculations) it is necessary to obtain the number whose logarithm is known. This procedure is known as taking the antilogarithm; it is simply the reverse of taking the logarithm of a number. Suppose in a certain calculation we have pH = 1.46 and are asked to calculate $[H^+]$. From the definition of pH (pH = $-\log [H^+]$) we can write

$$[H^+] = 10^{-1.46}$$

Many calculators have a key labeled $\log^{-1}$ or INV log to obtain antilogs. Other calculators have a 10^x or y^x key (where x corresponds to -1.46 in our example and y is 10 for base-10 logarithm). Therefore, we find that $[H^+] = 0.035$ M.

Natural Logarithms

Logarithms taken to the base e instead of 10 are known as natural logarithms (denoted by ln or $\log_e$); e is equal to 2.7183. The relationship between common logarithms and natural logarithms is as follows:

$$\log 10 = 1 \qquad\qquad 10^1 = 10$$
$$\ln 10 = 2.303 \qquad e^{2.303} = 10$$

Thus,

$$\ln x = 2.303 \log x$$

To find the natural logarithm of 2.27, say, we first enter the number on the electronic calculator and then press the ln key to get

$$\ln 2.27 = 0.820$$

If no ln key is provided, we can proceed as follows:

$$2.303 \log 2.27 = 2.303 \times 0.356$$
$$= 0.820$$

Sometimes we may be given the natural logarithm and asked to find the number it represents. For example,

$$\ln x = 59.7$$

On many calculators, we simply enter the number and press the e key:

$$e^{59.7} = 8 \times 10^{25}$$

Glossary

absolute temperature scale. A scale based on $-273°C$ (absolute zero) being the lowest point.

absolute zero. Theoretically the lowest obtainable temperature: $-273°C$ or 0 K.

acid. Substance that produces hydrogen ion in aqueous solution (Arrhenius acid), donates one or more protons (Brønsted acid).

activation energy (E_a). The minimum amount of energy to begin a chemical reaction.

actual yield. The amount of product actually obtained from a reaction.

addition polymerization. Process by which monomers combine to form polymers without the elimination of small molecules, such as water.

alcohol. A compound consisting of an alkyl group and the functional group OH.

alkali metal. An element from Group 1, with the exception of H (i.e., Li, Na, K, Rb, Cs, and Fr).

alkaline earth metal. An element from Group 2 (Be, Mg, Ca, Sr, Ba, and Ra).

alkane. Hydrocarbons having the general formula C_nH_{2n+2}, where $n = 1, 2, \ldots$

alkene. Describes a hydrocarbon that contains a carbon-carbon double bond.

alkyl group. A portion of a molecule that resembles an alkane.

alkyne. Describes a hydrocarbon that contains a carbon-carbon triple bond.

alpha (α) particle. A helium ion with a positive charge of +2.

amide linkages. The bonds that form between amino acids. Also known as peptide bonds.

amino acid. A compound that contains both an amino group and a carboxy group.

amorphous solid. A solid that lacks a regular three-dimensional arrangement of atoms.

amphoteric. Able either to donate or to accept a proton in a chemical reaction.

anion. An ion with a negative charge.

anode. The electrode at which oxidation occurs.

aqueous. Dissolved in water.

Arrhenius acid. Substance that produces hydrogen ions when dissolved in water.

Arrhenius base. Substance that produces hydroxide ions when dissolved in water.

atom. The basic unit of an element that can enter into chemical combination.

atomic ion. Atom that has lost or gained one or more electrons, giving it a positive or negative charge.

atomic mass (M). The mass of the atom given in atomic mass units (amu).

atomic mass unit (amu). A mass exactly equal to one-twelfth the mass of one carbon-12 atom.

atomic number (Z). The number of protons in the nucleus of each atom of an element.

atomic orbital. The wave function of an electron in an atom.

autoionization of water. Process by which one water molecule acts as an acid and another acts as a base to produce a hydrogen ion and a hydroxide ion.

average atomic mass. Weighted average mass of naturally-occurring isotopes of an element, given in atomic mass units.

Avogadro's law. The volume of a sample of gas (V) is directly proportional to the number of moles (n) in the sample at constant temperature and pressure: $V \propto n$.

Avogadro's number (N_A). The number of entities (atoms, molecules, ions, etc.) in a mole of substance: 6.0221413×10^{23}, usually rounded to 6.022×10^{23}.

barometer. An instrument used to measure atmospheric pressure.

base. A substance that produces hydroxide ion when dissolved in water.

battery. A portable, self-contained source of electric energy consisting of a galvanic cell or a series of galvanic cells.

beta particle. An electron.

binary ionic compounds. Ionic compound consisting of two elements.

Bohr atom. Model of the atom in which electrons occupy circular orbits around the nucleus.

boiling point. The temperature at which vapor pressure equals atmospheric pressure.

boiling-point elevation. The amount by which the boiling point of a solvent is increased by the presence of a solute.

boiling-point-elevation constant. Constant that describes by how much the boiling point of a specific solvent is increased by the presence of any solute.

bond pair. Pair of valence electrons shared between two atoms, constituting a covalent bond. Also known as a bonding pair.

bond-line structure. A simplified structure of an organic molecule in which each line represents a carbon-carbon bond and the atoms are not explicitly shown.

Boyle's law. The pressure of a fixed amount of gas at a constant temperature is inversely proportional to the volume of the gas: $P \propto 1/V$.

Brønsted acid. Substance that donates a proton in a chemical reaction.

Brønsted base. Substance that accepts a proton in a chemical reaction.

buffer. Solution containing comparable concentrations of a weak acid and its conjugate base (or a weak base and its conjugate acid) that prevents drastic pH change upon addition of small amounts of strong acid or base.

calorie (cal). Unit of energy originally defined as the amount of energy required to increase the temperature of 1 g water by 1 degree C.

carbohydrate. Sugar molecules with the general formula $C_n(H_2O)_m$.

carbonyl group. Functional group consisting of a carbon-oxygen double bond.

carboxylic acid. Organic acid containing the carboxyl group, −COOH.

cathode. The electrode at which reduction occurs.

cation. An ion with a positive charge.

cell potential (E_{cell}). The difference in electric potential between the cathode and the anode.

Celsius scale. Temperature scale based on the freezing point (0°C) and boiling point (100°C) of water.

central atom. Atom at the center of a molecule to which other atoms are bonded.

chalcogens. Elements in Group 16 (O, S, Se, Te, and Po).

Charles's law. The volume of a fixed amount of gas (V) maintained at constant pressure is directly proportional to absolute temperature (T): $V \propto T$.

chemical change. A process in which one or more substances are changed into one or more new substances.

chemical equation. Chemical symbols of reactants and products representing a chemical reaction.

chemical formula. Chemical symbols and numerical subscripts used to denote the composition of the substance.

chemical process. A process in which one or more substances are changed into one or more new substances.

chemical property. Any property of a substance that cannot be studied without converting the substance into some other substance.

chemical reaction. Process by which substances are changed into other substances.

chemistry. The study of matter and the changes it undergoes.

coefficients (or stoichiometric coefficients). Integers placed to the left of each species in a chemical equation for the purpose of balancing.

colligative properties. Properties that depend on the number of solute particles in solution but do not depend on the nature of the solute particles.

collision theory. The reaction rate is directly proportional to the number of molecular collisions per second.

combination reaction. Reaction in which two or more reactants combine to form a single product.

combined gas equation. The equation that relates the parameters (pressure, volume, and absolute temperature) of an ideal gas in one state to the parameters of the sample in another state.

combustion reaction. Reaction with oxygen to produce water and an oxide, most commonly carbon dioxide.

complete ionic equation. Chemical equation in which electrolytes are represented as separate ions in solution and nonelectrolytes are represented with their chemical formulas.

compound. A substance composed of atoms of two or more elements chemically united in fixed proportions.

concentrated. A solution containing relatively more solute.

condensation. The phase transition from gas to liquid.

condensation reaction. An elimination reaction in which two or more molecules become connected with the elimination of a small molecule, often water.

condensed phases. Solid and liquid phases.

conjugate acid. The species that results when a base has accepted a proton in a chemical reaction.

conjugate base. The species that remains when an acid has donated a proton in a chemical reaction.

conjugate pair. Two substances that differ only by the presence or absence of a proton.

conversion factor. A fraction in which the same quantity is expressed one way in the numerator and another way in the denominator.

copolymer. A polymer made of two or more different monomers.

core electrons. Electrons in completed shells.

corrosion. The undesirable oxidation of metals.

covalent bond. A shared pair of electrons.

covalent solid (or network solid). A crystalline solid in which the atoms are all held together by a network of covalent bonds.

crystalline solid. A solid that possesses rigid and long-range order; its atoms, molecules, or ions occupy specific positions.

Dalton's law of partial pressures. The total pressure exerted by a gas mixture is the sum of the partial pressures exerted by each component of the mixture.

decomposition reaction. Reaction in which a single substance reacts to form two or more simpler substances.

density. The ratio of mass to volume.

deoxyribonucleic acid (DNA). Biological polymer arranged in a double-helix shape, consisting of two long chains of nucleotides in which the sugar is deoxyribose.

deposition. The phase change from gas to solid.

diatomic molecule. A molecule that contains two atoms.

dilute. A solution containing relatively less solute.

dilution. The process of preparing a less concentrated solution from a more concentrated one.

dimensional analysis. The use of conversion factors in problem solving.

dimer. Molecule that is a combination of two monomers.

dipole. A separation of partial charges due to different electronegativities of bonded atoms.

dipole vector. A crossed arrow used to indicate the polarity of a bond. The head of the arrow points in the direction of the more electronegative atom.

dipole-dipole forces. Electrostatic attractive forces between polar molecules.

diprotic acid. Acid that has two ionizable hydrogen atoms (two protons).

disaccharide. Dimer of monosaccharides.

dispersion forces. Attractive forces between nonpolar molecules or atoms, caused by instantaneous dipoles and induced dipoles—both of which result from random motion of electrons.

dissociation. The process by which an ionic compound, upon dissolution, breaks apart into its constituent ions.

double bond. A multiple bond in which the atoms share two pairs of electrons.

double-displacement reaction. Reaction in which the cations in two soluble ionic compounds exchange anions.

driving force. Formation of a product that is distinct from the reactants. Products may include insoluble salts, water, gases, or oxidation-reduction products.

electrode. A piece of conducting metal in an electrochemical cell at which either oxidation or reduction takes place.

electrolysis. The use of electric energy to drive a nonspontaneous redox reaction.

electrolyte. A substance that dissolves in water to yield a solution that conducts electricity.

electrolytic cell. An electrochemical cell used for electrolysis.

electromagnetic spectrum. Consists of radio waves, microwave radiation, infrared radiation, visible light, ultraviolet radiation, X rays, and gamma rays.

electron. A negatively charged subatomic particle found outside the nucleus of all atoms.

electron configuration. The distribution of electrons in the atomic orbitals of an atom.

electron group. A group of electrons on a central atom. May be a single bond, a double bond, a triple bond, or a lone pair.

electron-group geometry. Three-dimensional arrangement of electron groups around a central atom.

electronegativity. The ability of an atom in a compound to draw electrons to itself.

element. A substance that cannot be separated into simpler substances by chemical means.

emission spectrum. The light emitted, either as a continuum or in discrete lines, by a substance in an excited electronic state.

empirical formula. The chemical formula that conveys with the smallest possible whole numbers the ratio of combination of elements in a compound.

endothermic process. A process that absorbs heat.

energy. The capacity to do work or transfer heat.

energy sublevel. Sublevel within a principal energy level designated by s, p, d, or f.

equilibrium. A state in which forward and reverse processes are occurring at the same rate.

equilibrium constant (K). A number equal to the ratio of the equilibrium concentrations of products to the equilibrium concentrations of reactants, with each concentration raised to the power of its stoichiometric coefficient.

equilibrium expression. The quotient of product concentrations and reactant concentrations, each raised to the power of its stoichiometric coefficient.

equivalence point. Point at which reactants in a titration have been combined in stoichiometric amounts.

exact number. Number containing no uncertainty because it is determined by counting or definition.

excess reactant. The reactant present in a greater amount than necessary to react with all the limiting reactant.

excited state. A state that is higher in energy than the ground state.

exothermic process. A process that gives off heat.

expanded octet. More than eight electrons around a central atom. This can happen with elements in the third period and beyond.

fat. Tri-ester of glycerol. Also known as triglycerides.

fatty acid. A carboxylic acid in which the alkyl group is a long hydrocarbon chain.

formula mass. The mass of a formula unit.

formula unit. Chemical formula representing the ratio of combination of elements in an ionic compound.

freezing. The phase transition from liquid to solid.

freezing-point depression. The amount by which the freezing point of a solvent is lowered by the presence of a solute.

freezing-point-depression constant. Constant that describes by how much the freezing point of a specific solvent is lowered by the presence of any solute.

frequency (ν). The number of waves that pass through a particular point in 1 s.

fuel cell. A voltaic cell in which reactants must be continually supplied.

functional group. The part of a molecule characterized by a special arrangement of atoms that is largely responsible for the chemical behavior of the parent molecule.

fusion. The phase transition from solid to liquid (melting). Also refers to the nuclear process by which two small nuclei combine to form one larger nucleus.

galvanic cell. An electrochemical cell in which a spontaneous chemical reaction generates a flow of electrons.

galvanization. The cathodic protection of iron or steel using zinc.

gamma radiation. High-energy radiation.

geometrical isomers. Molecules that contain the same atoms and bonds arranged differently in space.

glass. Commonly refers to an optically transparent fusion product of inorganic materials that has cooled to a rigid state without crystallizing.

glycerol. A simple three-carbon organic molecule containing three hydroxy groups.

ground state. The lowest energy state of an atom.

group. The elements in a vertical column of the periodic table.

half-cell. One compartment of an electrochemical cell containing an electrode immersed in a solution.

half-life ($t_{1/2}$). The time required for the reactant concentration to drop to half its original value.

half-reaction. The separated oxidation and reduction reactions that make up the overall redox reaction.

halogens. The elements in Group 17 (F, Cl, Br, I, and At).

heat of reaction. The amount of energy given off or absorbed during a chemical reaction.

heteroatom. Any atom in an organic molecule other than carbon or hydrogen.

heterogeneous mixture. A mixture in which the composition varies.

homogeneous mixture. A mixture in which the composition is uniform. Also called a solution.

Hund's rule. The most stable arrangement of electrons in orbitals of equal energy is the one in which the number of electrons with the same spin is maximized.

hydrocarbon. A compound containing only carbon and hydrogen.

hydrogen bonding. A special type of dipole-dipole interaction that occurs only in molecules that contain H bonded to a small, highly electronegative atom, such as N, O, or F.

hydrogenation. Addition of hydrogen to an unsaturated hydrocarbon to reduce all carbon-carbon multiple bonds to single bonds.

hydrolysis. Breaking apart of a disaccharide into its monosaccharide components by the addition of water.

hydronium ion. Hydrogen ion attached to a water molecule in aqueous solution.

hydrophilic. Water-loving.

hydrophobic. Water-fearing.

hypothesis. A tentative explanation for a set of observations.

ideal gas constant. A constant that relates the volume, pressure, temperature, and number of moles of an ideal gas.

ideal gas equation. An equation that describes the relationship among the four variables P, V, n, and T.

indicator. Soluble substance that changes colors in a specific pH range.

induced dipole. Temporary nonuniform distribution of electron density caused by proximity to a molecule with nonuniform electron distribution.

instantaneous dipole. A fleeting nonuniform distribution of electron density in a molecule without a permanent dipole.

intermolecular forces. The attractive forces that hold particles together in the condensed phases.

International System of Units (SI units). Internationally used system of scientific units based on metric measurements.

ionic bonding. An electrostatic attraction that holds oppositely charged ions together in an ionic compound.

ionizable hydrogen atom. Hydrogen atom that can separate from a molecule as H^+ in aqueous solution.

isoelectronic. Describes two or more species with identical electron configurations.

isotope. Atoms that have the same atomic number (Z) but different mass numbers (A).

joule (J). SI unit of energy, $1 \, kg \cdot m^2/s^2$.

kelvin (K). The SI base unit of temperature.

Kelvin scale. A temperature scale offset from the Celsius scale by 273. One kelvin (1 K) is equal in magnitude to one degree Celsius (1°C).

kinetic molecular theory. A theory that explains how the molecular nature of gases gives rise to their macroscopic properties.

law. A concise verbal or mathematical statement of a reliable relationship between phenomena.

law of conservation of mass. Law stating that the mass of all substances prior to a chemical reaction is equal to the mass of all substances after the reaction is complete.

law of constant composition. Different samples of a given compound always contain the same elements in the same mass ratio. Also known as the law of definite proportions.

law of definite proportions. Different samples of a given compound always contain the same elements in the same mass ratio. Also known as the law of constant composition.

law of multiple proportions. Different compounds made up of the same elements differ in the number of atoms of each kind that combine.

Le Châtelier's principle. When a stress is applied to a system at equilibrium, the system will respond by shifting in the direction that minimizes the effect of the stress.

Lewis dot symbol. An elemental symbol surrounded by dots, where each dot represents a valence electron.

Lewis structure. A representation of covalent bonding in which shared electron pairs are shown either as dashes or as pairs of dots between two atoms, and lone pairs are shown as pairs of dots on individual atoms.

limiting reactant. The reactant that is completely consumed and determines the amount of product formed.

line spectra. The emission or absorption of light only at discrete wavelengths.

lipid. Naturally occurring, small, water-insoluble organic molecules.

lipid bilayer. Thin polar membrane consisting of two layers of lipid molecules.

lone pair. A pair of valence electrons that are not involved in covalent bond formation.

manometer. A device used to measure the pressure of gases relative to atmospheric pressure.

mass. A measure of the amount of matter in an object or sample.

mass number (A). The number of neutrons and protons present in the nucleus of an atom of an element.

mass percent composition. The percent of the total mass contributed by each element in a compound.

matter. Anything that occupies space and has mass.

melting. The phase transition from solid to liquid.

melting point. The temperature at which solid and liquid phases are in equilibrium.

metal. Element with a tendency to lose electrons, located left of the zigzag line on the periodic table.

metallic bonding. Bonding in metallic elements in which all valence electrons are shared by all of the metal atoms.

metallic character. Collection of properties associated with metals, including electrical conductivity, shiny appearance, and malleability.

metalloid. Elements with properties intermediate between metals and nonmetals.

mixture. A combination of two or more substances in which the substances retain their distinct identities.

model. Theory used to explain a collection of scientific evidence.

molal concentration. The number of moles of solute dissolved in 1 kg (1000 g) of solvent. Also called *molality*.

molality (m). The number of moles of solute dissolved in 1 kg (1000 g) of solvent. Also called *molal concentration*.

molar concentration. The number of moles of solute per liter of solution. Also called *molarity*.

molar heat of fusion (ΔH_{fus}). The energy, usually expressed in kJ/mol, required to melt 1 mole of a solid.

molar heat of vaporization (ΔH_{vap}). The amount of heat required to vaporize a mole of substance at its boiling point.

molar mass ($\mathcal{M}$). The mass in grams of 1 mole of the substance.

molarity (M). The number of moles of solute per liter of solution. Also called *molar concentration*.

mole (mol). The amount of a substance that contains as many elementary entities (atoms, molecules, formula units, etc.) as there are atoms in exactly 0.012 kg (12 g) of carbon-12.

mole fraction (χ_i). The number of moles of a component divided by the total number of moles in a mixture.

molecular equation. Chemical equation in which all reactants and products are represented with their chemical formulas.

molecular formula. A chemical formula that gives the number of atoms of each element in a molecule.

molecular mass. The sum of the atomic masses (in amu) of the atoms that make up a molecule.

molecular shape. Spatial arrangement of atoms in a molecule or polyatomic ion.

molecule. A combination of two or more atoms in a specific arrangement held together by chemical bonds.

monomer. A small molecule that can be linked in large numbers to form a large molecule.

monoprotic acid. Acid with just one ionizable hydrogen atom (one *proton*).

monosaccharide. Sugar molecule with the general formula $(CH_2O)_n$.

multiple bond. A chemical bond in which two atoms share two or more pairs of electrons.

net ionic equation. Chemical equation from which spectator ions have been eliminated.

network solid (or covalent solid). A crystalline solid in which the atoms are all held together by a network of covalent bonds.

neutralization reaction. Reaction of acid and base to produce water and a salt.

neutron. An electrically neutral subatomic particle with a mass slightly greater than that of a proton.

newton (N). The SI unit of force.

noble gases. Elements in Group 18 (He, Ne, Ar, Kr, Xe, and Rn).

nomenclature. Systematic naming of chemical substances.

nonelectrolyte. A substance that dissolves in water to yield a solution that does not conduct electricity.

nonmetal. Element with a tendency to gain electrons, located in the upper right portion of the periodic table.

nonvolatile. Having no measurable vapor pressure.

nuclear fission. The splitting of a large nucleus into smaller nuclei and one or more neutrons.

nuclear fusion. The combination of two light nuclei to form one heavier nucleus.

nuclear transmutation. The conversion of one nucleus to another.

nucleic acid. Macromolecule formed by polymerization of nucleotides.

nucleons. Protons and neutrons.

nucleotide. A structural unit consisting of a sugar (ribose or deoxyribose) bonded to both a cyclic-amine base and a phosphate group.

nucleus. The central core of the atom that contains the protons and neutrons.

octet rule. Atoms will lose, gain, or share electrons to achieve a noble gas electron configuration.

orbit. Circular path around the nucleus in the Bohr atomic model.

organic bases. Organic molecules that contain nitrogen.

organic phosphates. Organic esters of phosphoric acid.

osmosis. The selective passage of solvent molecules through a porous membrane from a more dilute solution to a more concentrated one.

osmotic pressure (π). The pressure required to stop osmosis.

oxidation. Losing electrons.

oxidation number. The charge an atom in a compound would have if electrons were transferred completely. (Also called *oxidation state*.)

oxidation state. The charge an atom in a compound would have if electrons were transferred completely. (Also called *oxidation number*.)

oxidation-reduction reaction. Reaction in which electrons from one reactant are transferred to another.

oxidizing agent. A species that accepts electrons.

oxoacid. An acid consisting of one or more ionizable protons and an oxoanion.

oxoanion. A polyatomic anion that contains one or more oxygen atoms bonded to a central atom.

partial pressure (P_i). The pressure exerted by a component in a gas mixture.

pascal (Pa). The SI unit of pressure.

Pauli exclusion principle. No two electrons in an atom can have the same four quantum numbers.

peptide bond. The bond that forms between amino acids.

percent by mass. The ratio of the mass of an individual component to the total mass, multiplied by 100 percent.

percent by weight. The ratio of the weight of an individual component to the total weight, multiplied by 100 percent.

percent composition. The percent of the total mass contributed by each element in a compound.

percent yield. The ratio of actual yield to theoretical yield, multiplied by 100 percent.

period. A horizontal row of the periodic table.

periodic table. A chart in which elements having similar chemical and physical properties are grouped together.

pH. Logarithmic scale used to express degree of acidity.

phospholipid. Derivatives of glycerol in which two of the hydroxy hydrogens have been replaced by hydrocarbon chains and one has been replaced with a phosphate group.

photon. A particle of light.

physical change. A process in which the state of matter changes but the identity of the matter does not change.

physical property. A property that can be observed and measured without changing the identity of a substance.

pOH. Logarithmic scale used to express the basicity of a substance.

polar covalent bond. Bond in which electrons are unequally shared.

polyatomic ion. Electrically charged combination of two or more atoms that are covalently bonded to one another.

polyatomic molecules. Molecules containing more than two atoms.

polymer. Molecular compounds, either natural or synthetic, that are made up of many repeating units called monomers.

polysaccharide. Biological polymer consisting of sugars.

positron. A subatomic particle with the same mass as an electron, but with a positive charge.

precipitate. Insoluble ionic solid product of an aqueous reaction.

precipitation reaction. Reaction that produces an insoluble ionic solid product.

pressure. The force applied per unit area.

primary structure of a protein. Sequence of amino acids that make up a protein.

principal energy level. Electronic energy level within an atom designated with a non-zero integer value.

principal quantum number (n). Designates the size of the orbital.

product. Substance produced in a chemical reaction.

protein. A polymer of amino acids.

proton. A positively charged particle in the nucleus of an atom.

QM model. Model of the atom that treats electrons as waves, rather than as particles.

qualitative property. A property of a system that can be determined by general observation.

quantitative property. A property of a system that can be measured and expressed with a number.

quantized. Consisting of tiny, discrete entities.

quantum numbers. Numbers required to describe the arrangement of electrons in an atom.

quaternary structure. Overall shape adopted by two or more proteins, stabilized by attractive forces between groups of one protein and groups of another.

radical. Chemical species with one or more unpaired electrons.

radioactivity. The spontaneous emission of particles or radiation from unstable nuclei.

rate of reaction. The change in concentration of reactants or products per unit time.

reactant. Substance consumed in a chemical reaction.

redox reaction. Reaction in which electrons from one reactant are transferred to another.

reducing agent. A species that can donate electrons.

reduction. Gaining electrons.

resonance structures. Two or more equally valid Lewis structures for a single molecule that differ only in the positions of electrons.

reversible process. A process in which the products can react to form reactants.

ribonucleic acid (RNA). Biological polymer consisting of nucleotides in which the sugar is ribose.

salt. Ionic compound consisting of the cation from a base and the anion from an acid.

salt bridge. An inverted U tube containing an inert electrolyte solution, such as KCl or NH_4NO_3, that maintains electrical neutrality in an electrochemical cell.

saturated. Describes a hydrocarbon in which all carbon-carbon bonds are single bonds.

saturated solution. A solution that contains the maximum amount of a solute that will dissolve in a solvent at a specific temperature.

scientific method. A systematic approach to experimentation.

scientific notation. Method for expressing very small or very large numbers as a number between 1 and 10 times 10 raised to the appropriate power.

secondary structure. Shape adopted by a protein due to hydrogen bonding between nearby groups.

semipermeable membrane. A membrane that allows the passage of solvent molecules but blocks the passage of solute molecules.

significant figures. Meaningful digits in a measured or calculated value.

single-displacement reaction. Reaction in which one element replaces another in a compound.

skeletal structure. Simple arrangement of atoms in a molecule (or polyatomic ion) with shared pairs of electrons represented by dashes.

solubility The maximum amount of solute that will dissolve in a given quantity of solvent at a specific temperature.

solute. The dissolved substance in a solution.

solution. A homogeneous mixture consisting of a solvent and one or more solutes.

solvent. A substance in a solution that is present in the largest amount.

specific heat capacity or specific heat (s). The amount of heat required to raise the temperature of 1 g of a substance by 1°C.

spectator ion. Aqueous ions that do not participate in the chemical reaction.

spin. Intrinsic property of electrons with two possible values, $+\frac{1}{2}$ and $-\frac{1}{2}$.

standard temperature and pressure (STP). 0°C and 1 atm.

steroid. Lipid derivatives of the sterol molecule.

stoichiometric amounts. Quantities of reactants in the same relative amounts as those represented in the balanced chemical equation.

stoichiometric coefficients (or coefficients). Integers placed to the left of each species in a chemical equation for the purpose of balancing.

stoichiometry Numerical relationships between reactant and product amounts in a chemical reaction.

strong acid. Acid that ionizes completely in aqueous solution.

strong base. Metal hydroxide that dissociates completely in aqueous solution.

structural isomer. Molecules that have the same chemical formula but different arrangements of atoms.

sublimation. The phase change from solid to gas.

substance. Matter with a definite (constant) composition and distinct properties.

sugar. Small, typically sweet water-soluble ketone or aldehyde.

supersaturated solution. A solution that contains more dissolved solute than is present in a saturated solution.

surface tension. The amount of energy required to stretch or increase the surface of a liquid by a unit area.

terminal atoms. Atoms in a molecule or polyatomic ion that are bonded only to a central atom.

tertiary structure. Shape adopted by a protein due to attractive forces between distant groups.

theoretical yield. The maximum amount of product that can be obtained from a reaction.

theory. A unifying principle that explains a body of experimental observations and the laws that are based on them.

thermonuclear reaction. Generally refers to a fusion reaction that is set off by a fission reaction.

titration. Determination of one reactant concentration by careful, measured addition of another reactant of known concentration.

torr. Another name for the unit of pressure mmHg; named for the Italian scientist Torricelli—inventor of the mercury barometer.

triglyceride. Tri-ester of glycerol. Also known as fats.

triple bond. A multiple bond in which the atoms share three pairs of electrons.

type I compound. Ionic compound in which cation has only one possible charge.

type II compound. Ionic compound in which the cation may have more than one charge.

unsaturated. Describes a hydrocarbon that contains one or more carbon-carbon multiple bonds.

unsaturated solution. A solution that contains less solute than it has the capacity to dissolve.

valence electrons. The outermost electrons of an atom.

valence-shell electron-pair repulsion (VSEPR). A model that accounts for electron pairs in the valence shell of an atom repelling one another.

vapor pressure. Pressure exerted by molecules that have escaped from a condensed phase to the vapor phase.

vaporization. The phase change from liquid to gas at the boiling point.

viscosity. A measure of a fluid's resistance to flow.

volatile. Describes a substance that has a high vapor pressure.

wavelength (λ). The distance between identical points on successive waves.

weak acid. Acid that ionizes less than 100 percent in aqueous solution.

weak base. Base that does not ionize 100 percent in aqueous solution.

Answers

To Odd-Numbered Problems

Chapter 1

1.1 A theory (or model) is developed after a hypothesis has been tested extensively through experimentation. It is something that describes observations and is used to predict the outcomes of future experiments. **1.3** A hypothesis is an attempt to explain an observation and is testable. **1.5** Yes, an atom can be broken down into electrons, neutrons, and protons. If this is done, the "parts" do not have the same properties as the atom that we started with. **1.7** There would be 2 protons, 2 electrons, and 2 neutrons. **1.9** (a) False. A neutral atom always contains the same number of protons and electrons, but the number of neutrons can vary, depending on the isotope. (b) True. (c) True. (d) False. An atom is the smallest identifiable piece of an element that retains the properties of that element. **1.11** b and c **1.13** (a) Ca = calcium, C = carbon (b) B = boron, Br = bromine (c) correct (d) correct **1.15** (a) Pt = platinum, Pu = plutonium (b) Ni = nickel, N = nitrogen (c) correct (d) correct

1.17

Element Symbol	Element Name	Atomic Number	Mass Number	Number of Protons	Number of Neutrons	Number of Electrons
Si	**silicon**	14	**29**	14	15	14
Mg	magnesium	12	26	12	**14**	**12**
P	phosphorus	**15**	**31**	15	16	15
Zn	30	30	66	**30**	**36**	30
I	iodine	53	**127**	53	74	**53**

1.19 I = metals, II = metalloids, III = nonmetals; The main group elements are the first two columns on the left and the last six on the right. **1.21** Li, Ba, Cu, V **1.23** None of these elements are metalloids. **1.25** Ar & Kr **1.27** Ba **1.29** Br, Xe, Se, AC, P, N **1.31** Br **1.33** K

1.35

Symbol	Main Group Element	Transition Element	Metal	Nonmetal	Metalloid	Alkali Metal	Alkaline Earth Metal	Halogen	Noble Gas
Rb	X		X			X			
Be	✓		✓				✓		
Ag		✓	✓						
Zn		✓	✓						

1.37

Symbol	Main Group Element	Transition Element	Metal	Nonmetal	Metalloid	Alkali Metal	Alkaline Earth Metal	Halogen	Noble Gas
Cl	✓			✓				✓	
P	✓			✓					
Mg	✓		✓				✓		

1.39

Symbol	Main Group Element	Transition Element	Metal	Nonmetal	Metalloid	Alkali Metal	Alkaline Earth Metal	Halogen	Noble Gas
I	✓			✓				✓	
Ar	✓			✓					✓
K	✓		✓			✓			

1.41 (a) $^9_4$Be (b) $^{25}_{12}$Mg (c) $^{40}_{20}$Ca **1.43** (a) Your sketch should show 4 protons and 5 neutrons in the nucleus, and 4 electrons surrounding it. (b) Your sketch should show 2 protons and 2 neutrons in the nucleus, and 2 electrons surrounding it. (c) Your sketch should show 5 protons and 5 neutrons in the nucleus, and 5 electrons surrounding it.

1.45

Isotope Symbol	Element Name	Mass Number (A)	Neutrons ($n°$)	Protons (p^+)	Electrons (e^-)
^{109}Ag	silver	**109**	**62**	**47**	**47**
Si-28	silicon	28	**14**	**14**	**14**
Ar-40	**argon**	40	**22**	**18**	18

1.47 (b) It is the only answer where the protons and electrons are the same, but the neutrons differ. **1.49** (a) Ni-58 (b) K-39 (c) Fe-56 **1.51** None of the statements could be true, based on the average atomic mass shown for each element on the periodic table.
1.53 39.093 amu **1.55** 28.085 amu **1.57** 55.682 amu **1.59** B-10 abundance = 20.0%, B-11 abundance = 80.0% **1.61** 335.654 amu

Chapter 2

2.1 They are inversely proportional—the longer the wavelength, the shorter the frequency. **2.3** $E = \frac{hc}{\lambda}$, where h = Planck's constant 6.626×10^{-34} Js, c = speed of light 3.00×10^8 m/s, and λ = wavelength in meters. **2.5** red **2.7** blue **2.9** 400 nm > 550 nm > 700 nm **2.11** microwave < visible < gamma **2.13** blue > green > orange **2.15** A "packet" or particle of light energy. **2.17** The atom has absorbed energy and at least one electron has moved to a higher energy level than in the ground state. **2.19** b **2.21** a **2.23** a & b **2.25** 410 nm matches the $n = 6$ to $n = 2$ transition; 434 nm matches the $n = 5$ to $n = 2$ transition; 486 nm matches the $n = 4$ to $n = 2$ transition; 657 nm matches the $n = 3$ to $n = 2$ transition. **2.27** The volume where an electron is most likely to be found.

2.29 $3s$ =

,

$3p$ =

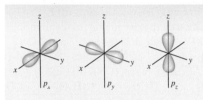

,

$3d$ =

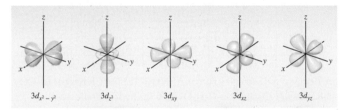

2.31 The $4p$ orbitals are larger, but have the same shape. **2.33** (a) $4s$, (b) they are equal in size, (c) $4p$ **2.35** (a) $3d$, (b) $1s$, (c) $2p_x$ **2.37** (a) $1s$, (b) $2p$, (c) $3d$ **2.39** (a) Yes, the fifth shell (level) contains p orbitals. (b) Yes, the fourth shell (level) contains an s orbital. (c) No, there are no f orbitals in the second shell (level). (d) No, there are no p orbitals in the first shell (level). **2.41** (a) sublevel, (b) orbital and sublevel, (c) single orbital, (d) sublevel **2.43** $1s$ = spherical orbital in the first level; $2p$ = dumbbell-shaped orbital in the second level; $4s$ = spherical orbital in the fourth level; $3d$ = cloverleaf-shaped orbital in the third level **2.45** (a) $4d > 4p > 4s$ (b) $4p > 3p > 2p$ (c) $3d > 2p > 1s$ **2.47** 2 electrons **2.49** (a) 1 (b) 5 (c) 3 (d) 7 **2.51** (a) 6 (b) 10 (c) 2 (d) 6 **2.53** (a) $1s^2 2s^2 2p^6 3s^2 3p^6 4s^1$ (b) $1s^2 2s^2 2p^6 3s^2 3p^6 4s^2 3d^{10} 4p^3$ (c) $1s^2 2s^2 2p^6 3s^2 3p^6 4s^2 3d^{10} 4p^4$ **2.55** (a) $1s^2 2s^1$ (b) $1s^2 2s^2 2p^6 3s^2 3p^2$ (c) $1s^2 2s^2 2p^6 3s^2$

2.57 (a) $1s$ (b) $1s$

2.59 (a) $1s$ (b) $1s$ (c) $1s$

2.61 (a) $[\mathrm{Ar}]4s^2 3d^6$ (b) $[\mathrm{Ar}]4s^2 3d^{10}$ (c) $[\mathrm{Ar}]4s^2 3d^8$ **2.63** (a) $[\mathrm{Kr}]5s^2 4d^{10}$ (b) $[\mathrm{Kr}]5s^2 4d^8$ (c) $[\mathrm{Ar}]4s^2 3d^3$ **2.65** (a) 10 core electrons, 5 valence electrons (b) 46 core electrons, 7 valence electrons (c) 18 core electrons, 2 valence electrons (d) 18 core electrons, 1 valence electron **2.67** I = s block, II = p block, III = d block **2.69** (a) $[\mathrm{Kr}]5s^2 4d^{10} 5p^3$, 5 (b) $[\mathrm{Xe}]6s^2$, 2 (c) $[\mathrm{Kr}]5s^2 4d^{10} 5p^2$, 4

2.71 (a) $[\mathrm{Kr}]5s^2 4d^{10} 5p^5$, $5s$ $5p$

(b) $[\mathrm{Ar}]4s^2 3d^{10} 4p^4$, $4s$ $4p$

(c) $[\mathrm{Ar}]4s^2 3d^{10} 4p^6$, $4s$ $4p$

(d) $[\mathrm{Kr}]5s^2$, $5s$

2.73 (a) 1 (b) 2 (c) 0 (d) 0 **2.75** (a) $[\mathrm{He}]2s^2 2p^5$, 7 valence electrons (b) $[\mathrm{Ne}]3s^2 3p^5$, 7 valence electrons (c) $[\mathrm{Ar}]4s^2 3d^{10} 4p^5$, 7 valence electrons. All three of these elements would be predicted to form 1− ions by gaining one electron to fill the valence shell. They are all found in Group 17, so the charge can be predicted from their location on the periodic table. **2.77** (a) tin, Sn (b) cesium, Cs (c) bromine, Br **2.79** (a) This element contains 14 electrons (add up superscripts), so can be identified as Si. $1s^2 2s^2 2p^6 3s^2 3p^2$ (b) This element contains 6 electrons and can be identified as C. $1s^2 2s^2 2p^2$ (c) This element contains 10 electrons and can be identified as Ne. $1s^2 2s^2 2p^6$ **2.81** Br, because it has the same number of valence electrons and is located in the same group on the periodic table. **2.83** (a) ·Mg· (b) :P̈· (c) :F̈: (d) :Är:

2.85 (a) :N̈· (b) :B̈r· (c) ·Ca· (d) ·Li **2.87** (a) :N̈· (b) :P̈· (c) :Äs· They have the same number of valence electrons (dots). They are in the same group on the periodic table. **2.89** Nonmetals gain electrons most easily.

2.91   **2.93**

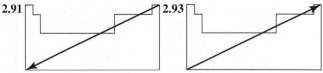

2.95 (a) S < Sr < Rb (b) Li < Mg < Ca (c) Br < Ca < K **2.97** (a) S (b) Si (c) K **2.99** (a) S (b) F (c) P **2.101** It is an atom that has lost or gained one or more electrons, leaving it with either a positive or negative charge. Atoms are neutral and become ions when they gain or lose electron(s). **2.103** An atom that has gained one or more electrons. It has a negative charge. **2.105** (a) Mg^{2+} (b) K^+ (c) P^{3-} (d) O^{2-} (e) I^- **2.107** (a) $[\mathrm{He}]2s^2 2p^6$ (b) $[\mathrm{Ar}]4s^2 3d^{10} 4p^6$

(c) $[Ar]4s^23d^{10}4p^6$ **2.109** (a) $[Ar]4s^23d^{10}4p^6$ (b) $[He]2s^22p^6$
(c) $[Ne]3s^23p^6$ **2.111** (a) $[Ne]3s^23p^6$ (b) $[Ne]3s^23p^6$ (c) $[Ne]3s^23p^6$
The ions of these elements all have the same electron configuration and the same number of electrons. They only differ by the number of protons (and neutrons) in their nuclei. **2.113** (a) Ca^{2+} (b) K^+
(c) $:\ddot{\text{F}}:^-$ (d) $:\ddot{\text{O}}:^{2-}$ (e) $:\ddot{\text{N}}:^{3-}$ **2.115** K^+ $:\ddot{\text{Cl}}:^-$

2.117 $[Ar]4s^23d^9$ (Please note that some elements defy our predicted electron configurations. Copper actually has a configuration of $[Ar]4s^13d^{10}$ but you are not expected to know this at this stage.)

4p	___ ___ ___	4p	___ ↑ ___
4s	↑↓	4s	↑

ground state   excited state
(atom absorbed energy and the 4s electron moved to an empty 4p orbital)

2.119 $[He]2s^1$, $2s$ ↑

$$E = \frac{hc}{\lambda} = \frac{6.626 \times 10^{-34} \text{ Js} \times 3.00 \times 10^8 \text{ m/s}}{6.70 \times 10^{-7} \text{ m}} = 2.97 \times 10^{-19} \text{ J}$$

2.121 $[He]2p^1$

Visualizing Chemistry
VC 2.1 b. **VC 2.2** a. **VC 2.3** c. **VC 2.4** b.

Chapter 3

3.1 (a) mixture (b) mixture (c) pure substance (d) pure substance
3.3 (a) mixture (b) mixture (c) pure substance (d) pure substance
3.5 (a) homogeneous (b) heterogeneous (c) heterogeneous
(d) heterogeneous **3.7** (a) physical (b) chemical (c) physical
(d) physical (e) chemical **3.9** (a) physical (b) physical (c) chemical
(d) chemical (e) physical **3.11** The attraction between opposite charges on two different ions constitutes an ionic bond. It is a very strong attraction. **3.13** There is no net charge on the compound. If it is an ionic compound, there are equal numbers of positive and negative charges.

3.15 $\cdot\text{Ca}\cdot + :\ddot{\text{O}}\cdot \longrightarrow Ca^{2+} :\ddot{\text{O}}:^{2-}$ **3.17** NaF

3.19 (a) K_2O (b) Li_2O (c) MgF_2 (d) Sr_3N_2
3.21

Ions	N^{3-}	Cl^-	O^{2-}
Fe^{2+}	Fe_3N_2	$FeCl_2$	FeO
Fe^{3+}	FeN	$FeCl_3$	Fe_2O_3
Zn^{2+}	Zn_3N_2	$ZnCl_2$	ZnO
Al^{3+}	AlN	$AlCl_3$	Al_2O_3
Sr^{2+}	Sr_3N_2	$SrCl_2$	SrO
NH_4^+	$(NH_4)_3N$	NH_4Cl	$(NH_4)_2O$

3.23 (a) +1 (b) −2 (c) −3 (d) −1 (e) +3 **3.25** (a) +3 (b) +2 (c) +2
(d) +3 **3.27** (a) sodium ion (b) magnesium ion (c) aluminum ion
(d) sulfide ion (e) fluoride ion **3.29** (a) titanium(II) ion (b) silver ion
(c) nickel(IV) ion (d) lead(II) ion (e) zinc ion **3.31** (a) protons = 11, electrons = 10 (b) protons = 12, electrons = 10 (c) protons = 13, electrons = 10 (d) protons = 16, electrons = 18 (e) protons = 9, electrons = 10 **3.33** (a) protons = 22, electrons = 20 (b) protons = 47, electrons = 46 (c) protons = 28, electrons = 24 (d) protons = 82, electrons = 80 (e) protons = 30, electrons = 28

3.35 (a) Na^+ (b) Mg^{2+} (c) Al^{3+} (d) $\left[:\ddot{\text{S}}:\right]^{2-}$ (e) $\left[:\ddot{\text{F}}:\right]^-$ **3.37** (a) rubidium chloride (b) sodium oxide (c) copper(II) chloride (d) nickel(IV)

chloride **3.39** (a) chromium(III) fluoride (b) silver iodide
(c) lithium sulfide (d) cobalt(II) oxide **3.41** (a) cesium nitride
(b) strontium phosphide (c) iron(III) phosphide (d) lead(IV)
nitride **3.43** (a) Sr_3N_2 (b) Li_3P (c) Al_2S_3 (d) BaO **3.45** (a) TiF_4
(b) Fe_2O_3 (c) CuO (d) NiS_2 **3.47** The smallest whole-number ratio between the elements present in the compound.
3.49 (a) CN (b) CH (c) CH_2 (d) P_2O_5 **3.51** nonmetals
3.53 $H\cdot + \cdot\ddot{\text{O}}\cdot + \cdot H \longrightarrow H-\ddot{\text{O}}-H$ **3.55** (i) mixture of two elements (ii) mixture of element (yellow) and compound
(iii) mixture of element (yellow only) and compound (iv) pure substance, element **3.57** (a) SiS_2 (b) SF_4 (c) $SeBr_6$ (d) PH_3
3.59 (a) carbon disulfide (b) sulfur hexafluoride (c) sulfur dioxide
(d) iodine pentachloride **3.61** (a) NF_3, nitrogen trifluoride
(b) PBr_5, phosphorus pentabromide (c) SCl_2, sulfur dichloride
3.63 Yes, some ionic compounds are composed of polyatomic ions which only contain nonmetals. The presence of ions indicates the compound is ionic, even if no metal ion is present.
3.65 (a) phosphite ion (b) nitrite ion (c) cyanide ion (d) hydroxide ion **3.67** (a) $LiClO_3$ (b) $BaSO_3$ (c) $Ca(C_2H_3O_2)_2$ (d) $Al(ClO_4)_3$
3.69 (a) NH_4HCO_3 (b) $Ca_3(PO_4)_2$ (c) $Al(NO_2)_3$ (d) $K_2Cr_2O_7$
3.71 (a) lead(II) bicarbonate (b) zinc nitrate (c) titanium(IV) chlorate (d) titanium(IV) sulfate **3.73** (b) 1 Ca^{2+} and 2 CN^-
(c) 2 Fe^{3+} and 3 SO_4^{2-} (d) 1 Sr^{2+} and 2 ClO_3^- (e) 3 NH_4^+ and 1 PO_3^{3-}
3.75 (a) Na_3PO_4 (b) $Al_2(SO_4)_3$ (c) $Mg(CN)_2$ (d) $CaCO_3$
3.77

Ions	$Cr_2O_7^{2-}$	HCO_3^-	$C_2H_3O_2^-$	CO_3^{2-}
Ni^{2+}	$NiCr_2O_7$	$Ni(HCO_3)_2$	$Ni(C_2H_3O_2)_2$	$NiCO_3$
Ti^{4+}	$Ti(Cr_2O_7)_2$	$Ti(HCO_3)_4$	$Ti(C_2H_3O_2)_4$	$Ti(CO_3)_2$
Ca^{2+}	$CaCr_2O_7$	$Ca(HCO_3)_2$	$Ca(C_2H_3O_2)_2$	$CaCO_3$
Cr^{3+}	$Cr_2(Cr_2O_7)_3$	$Cr(HCO_3)_3$	$Cr(C_2H_3O_2)_3$	$Cr_2(CO_3)_3$
Ag^+	$Ag_2Cr_2O_7$	$AgHCO_3$	$AgC_2H_3O_2$	Ag_2CO_3
Li^+	$Li_2Cr_2O_7$	$LiHCO_3$	$LiC_2H_3O_2$	Li_2CO_3

3.79 c & e **3.81** (a) NH_4Br (b) $Al(NO_3)_3$ (c) $Ca(ClO_3)_2$ (d) Li_2CO_3
3.83 $K_3C_6H_5O_7$ and $Na_3C_6H_5O_7$ **3.85** (a) hydroselenic acid
(b) hydrofluoric acid (c) hydroiodic acid **3.87** (a) chloric acid
(b) hydrothiocyanic acid (c) carbonic acid **3.89** (a) $HC_2H_3O_2$
(b) H_2CrO_4 (c) $HSCN$ **3.91** Ionic compounds are generally composed of a metal and nonmetal and are held together through the attraction of oppositely charged ions. Molecular compounds are composed of two or more nonmetals and are held together through the sharing of electrons in covalent bonds. **3.93** b & d
3.95 (a), (b), (e), (f). They are all composed of a metal and non-metallic element. **3.97** (a) compound (b) compound (c) compound
(d) element, molecular **3.99** (a) element, molecular (b) element, atomic (not molecular) (c) compound (d) compound

3.101 (a) (b) or (c) or (d) or
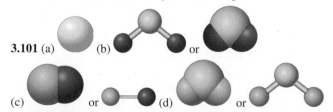

3.103 (a) Na (b) F_2 (c) CO_2 (d) NaCl (e) salt water (f) fizzy soft drink
Visualizing Chemistry
VC 3.1 c. **VC 3.2** a. **VC 3.3** b. **VC 2.4** b.

Chapter 4

4.1 nanometers (nm) **4.3** 1 nL < 1 mL < 1 kL **4.5** Kelvin and Celsius
4.7 temp in K = (temp in °C) + 273 K **4.9** (a) 100.°C (b) 0°C
(c) −273°C (d) 0°C (e) 177°C **4.11** (a) 519 K (b) 298 K (c) 1208 K
(d) 977 K (e) 412 K **4.13** 452°C (725 K) > 489 K > 193°C

$(466 \text{ K}) > 288°F (255 \text{ K}) > 212 \text{ K}$ **4.15** (a) 1.9×10^6 (b) 3.45×10^9
(c) 5.568×10^{-7} (d) 2.8×10^{-4} (e) 2.1×10^{10} **4.17** (a) 9,400,000,000
(b) 2,751,000 (c) 0.0000000094 (d) 0.000002751 (e) 48,000
4.19 It is not correct because scientific notation allows only one
digit to the left of the decimal point. The number should be written
as 4.55×10^5. **4.21** 1×10^5 (this assumes only one significant
figure) **4.23** (a) 18.5 mL (b) 18.49 mL (c) 10.9 mL (d) 10.92 mL
4.25 (a) 114.99 g (b) 115 g (c) 114.99 g (d) 114.986 g **4.27** (a) 3
(b) 4 (c) 1 for certain, but it is ambiguous in this format. The three
trailing zeros may or may not be significant. (d) 5 (e) 4 **4.29** (a) 5
(b) 5 for certain, but it is ambiguous in this format. The three trailing
zeros may or may not be significant. (c) 6 (d) 5 (e) 3 **4.31** (a) 2980
or 2.98×10^3 (b) 21.7 (c) 585 (d) 3.48×10^4 (e) 0.000111 or
1.11×10^{-4} **4.33** (a) 25.33 (b) 492.75 (c) 595.334 (d) 3696
(e) 3696 **4.35** (a) 1×10^2 (b) 2.06×10^4 (c) 2.4×10^3 (d) 5.7×10^{-2}
(e) 3.77×10^6 **4.37** (a) 4.44×10^{-4} (b) 5.7541×10^2 (c) 1.10×10^4
(d) 1.00×10^9 (e) 2.76×10^9 **4.39** g/mL, cm^3, m^2 as they are all
combinations of measurements.

4.41 (a) $\dfrac{1 \text{ g}}{1 \times 10^9 \text{ ng}}$ OR $\dfrac{1 \times 10^{-9} \text{ g}}{1 \text{ ng}}$ (b) $\dfrac{1 \text{ mg}}{1 \times 10^{-3} \text{ g}}$ OR $\dfrac{1 \times 10^3 \text{ mg}}{1 \text{ g}}$

(c) $\dfrac{1 \text{ L}}{1 \times 10^3 \text{ mL}}$ OR $\dfrac{1 \times 10^{-3} \text{ L}}{1 \text{ mL}}$ (d) $\dfrac{1 \text{ kL}}{1 \times 10^3 \text{ L}}$ OR $\dfrac{1 \times 10^{-3} \text{ kL}}{1 \text{ L}}$

(e) $\dfrac{1 \text{ m}}{1 \times 10^{-6} \text{ Mm}}$ OR $\dfrac{1 \times 10^6 \text{ m}}{1 \text{ mM}}$

4.43 Correct: c, d, e; Incorrect: a, it should be $\dfrac{1 \text{ kg}}{1000 \text{ g}}$;

b, it should be $\dfrac{1 \times 10^9 \text{ nL}}{1 \text{ L}}$

4.45 Both c and e are set up correctly.

(a) $2.8 \text{ mL} \times \dfrac{1 \text{ L}}{1 \times 10^3 \text{ mL}} = 2.8 \times 10^{-3} \text{ L}$

(b) $56 \text{ kg} \times \dfrac{1 \times 10^3 \text{ g}}{1 \text{ kg}} = 5.6 \times 10^4 \text{ g}$

(d) $1.35 \times 10^3 \text{ g} \times \dfrac{1 \times 10^9 \text{ ng}}{1 \text{ g}} = 1.35 \times 10^{12} \text{ g}$

4.47 (a) 9.651×10^{-6} g (b) 2.33×10^6 g (c) 0.499 g (d) 6.77×10^4 g
(e) 6.2×10^{-5} g **4.49** 13.6 g/mL **4.51** (a) 197 g (b) 317 mL
4.53 Correct: a and c

(b) $659 \text{ }\mu\text{L} \times \dfrac{1 \text{ L}}{1 \times 10^6 \text{ }\mu\text{L}} \times \dfrac{1 \times 10^9 \text{ nL}}{1 \text{ L}} = 6.59 \times 10^5 \text{ nL}$

(d) $9.42 \text{ ng} \times \dfrac{1 \text{ g}}{1 \times 10^9 \text{ ng}} \times \dfrac{1 \text{ kg}}{1 \times 10^3 \text{ g}} = 9.42 \times 10^{-12} \text{ kg}$

(e) $8.8 \text{ km} \times \dfrac{1 \times 10^3 \text{ m}}{1 \text{ km}} \times \dfrac{1 \times 10^9 \text{ nm}}{1 \text{ m}} = 8.8 \times 10^{12} \text{ nm}$

4.55 2.0×10^{-5} kg pollen/trip, 1.1×10^6 trips **4.57** 3.91×10^9 W
4.59 \$0.00297 **4.61** (a) larger, 4.92×10^7 mg (b) smaller,
7.542×10^6 mL (c) larger, 2.99×10^{20} nm (d) larger, 1.75×10^2 cg
(e) smaller, 0.322 dm **4.63** (a) 1.50×10^2 m (b) 1.04×10^3 km
(c) 1.08×10^5 g (d) 1.18×10^3 L **4.65** (a) 8.32×10^2 oz
(b) 1.08×10^4 oz (c) 7.82×10^4 m (d) 1.47×10^4 mL
4.67 (a) 9.78×10^6 mg/L (b) 92.9 km/hour (c) 0.593 m/s
(d) 1.76×10^{-3} g/L (e) 1.4×10^4 μg/cL **4.69** (a) 3.38×10^4 mm^3
(b) 2.89×10^{-23} m^3 (c) 7.36×10^{28} μm^3 (d) 2.49×10^5 cm^3
(e) 5.75×10^{-14} μm^3 **4.71** 1.3×10^8 kg ore, 93 ft square

Chapter 5

5.1 objects per mole $\dfrac{\text{objects}}{\text{mole of objects}}$ **5.3** $\dfrac{6.022 \times 10^{23} \text{ He atoms}}{\text{mole He atoms}}$

or $\dfrac{4.003 \text{ g He}}{\text{mole of He}}$ or $\dfrac{\text{mole of He}}{4.003 \text{ g He}}$ or $\dfrac{\text{mole He atoms}}{6.022 \times 10^{23} \text{ He atoms}}$

5.5 (a) 1.6×10^{24} atoms Br (b) 4.9×10^{24} atoms Mg (c) 3.0×10^{24}
atoms Ar **5.7** (a) 88.4 g Ne (b) 11.5 g Ne **5.9** (a) 52.7 g K
(b) 117 g K **5.11** (a) 3.00 mol Si (b) 0.845 mol Si (c) 0.251 mol
Ag (d) 6.34 mol Ag **5.13** a. **5.15** (a) 1.39×10^{-21} g Fe
(b) 5.03×10^{-22} g Ne (c) 9.74×10^{-22} g K (d) 2.18×10^{-21} g Sr

5.17

Moles of sample	Mass of sample	Atoms in sample
3.75 moles Ag	**405 g Ag**	**2.26×10^{24} Ag atoms**
1.62 moles Fe	90.3 g Fe	**9.74×10^{23} Fe atoms**
0.395 mole N	**5.54 g N**	2.38×10^{23} N atoms

5.19 0.644 g Mg **5.21** First, locate the mass of each element on the
periodic table. Next, multiply each element mass by the number of
atoms of the element present in the compound (subscripts in the
formula). Last, add together the masses of each element present in
the compound. **5.23** (a) 3.52×10^{23} molecules CO$_2$ (b) 3.57×10^{23}
molecules C$_2$Cl$_4$ (c) 3.64×10^{23} molecules SO$_2$ (d) 3.09×10^{23}
molecules SF$_4$ **5.25** 7.39×10^{21} SO$_3$ molecules, 2.22×10^{22} O
atoms **5.27** 1.623×10^4 g SO$_2$ **5.29** (a) 7.5×10^{-22} g H$_2$O (b) $5.7 \times$
10^{-21} g PCl$_3$ (c) 2.9×10^{-21} g LiNO$_3$ (d) 9.3×10^{-21} g Mg(ClO$_4$)$_2$
5.31 (a) 12.6 g H (b) 3.17 g H (c) 14.6 g H (d) 3.12 g H **5.33** If
there is one molecule of CH$_4$, it contains 4 atoms of H, just like
one car is made up of many components, like 4 tires. A mole is
just a larger number of CH$_4$ molecules where there will always be
4 times as many hydrogen atoms as CH$_4$ molecules.

5.35 (a) $\dfrac{12 \text{ moles O}}{1 \text{ mole Al}_2(\text{SO}_4)_3}$ (b) $\dfrac{5 \text{ moles O}}{1 \text{ mole N}_2\text{O}_5}$ (c) $\dfrac{8 \text{ moles O}}{1 \text{ mole Mg}_3(\text{PO}_4)_2}$

(d) $\dfrac{1 \text{ mole O}}{1 \text{ mole OCl}_2}$ **5.37** (a) $\dfrac{4 \text{ moles H}}{2 \text{ moles C}}$ (b) $\dfrac{6 \text{ moles H}}{2 \text{ moles C}}$ (c) $\dfrac{9 \text{ moles H}}{6 \text{ moles C}}$

(d) $\dfrac{2 \text{ moles H}}{1 \text{ mole C}}$ **5.39** (a) 2.14×10^{24} molecules C$_2$H$_6$, 7.10 moles C
(b) 1.07×10^{24} molecules C$_3$H$_8$, 5.34 moles C (c) 3.47×10^{24}
molecules H$_2$CO$_3$, 5.77 moles C (d) 1.27×10^{24} molecules C$_6$H$_{12}$O$_6$,
12.7 moles C **5.41** (a) 3.63×10^{24} formula units NaCN, 6.03 moles N
(b) 6.32×10^{23} formula units Ca(NO$_3$)$_2$, 2.10 moles N (c) 6.20×10^{24}
formula units (NH$_4$)$_2$SO$_4$, 20.6 moles N (d) 5.15×10^{24} formula
units Cr(CN)$_3$, 25.7 moles N **5.43** d. **5.45** (a) 190.52 g/mol
(b) 230.95 g/mol (c) 52.08 g/mol (d) 294.17 g/mol **5.47** (a) $1.13 \times$
10^{24} molecules AsCl$_3$, 3.40×10^{24} Cl atoms (b) 1.45×10^{23}
molecules AsCl$_3$, 4.36×10^{23} Cl atoms **5.49** (a) 2.58 g S (b) 1.73 g S
(c) 5.05 g S (d) 2.10 g S **5.51** (a) 0.0535 mol Mg$_3$P$_2$ (b) 0.0535 mol
Mg$_3$(PO$_3$)$_2$ (c) 0.161 mol Mg(CN)$_2$ (d) 0.161 mol MgS
5.53 (a) 1.70×10^{-3} g C$_2$H$_4$ (b) 4.81×10^{-3} g Ca(C$_2$H$_3$O$_2$)$_2$
(c) 8.98×10^{-3} g Li$_2$CO$_3$ (d) 6.83×10^{-3} g MgC$_2$O$_4$

5.55 (a) should be $\dfrac{32.00 \text{ g O}_2}{1 \text{ mole O}_2}$ (e) should be $\dfrac{1 \text{ mole O}_2}{32.00 \text{ g O}_2}$

5.57 (b) should be $\dfrac{8 \text{ moles H}}{1 \text{ mole (NH}_4)_2\text{O}}$ (c) should be $\dfrac{52.08 \text{ g (NH}_4)_2\text{O}}{1 \text{ mole (NH}_4)_2\text{O}}$

(e) should be $\dfrac{2 \text{ moles NH}_4^+}{1 \text{ mole (NH}_4)_2\text{O}}$

5.59

Moles of sample	Mass of sample	N atoms in sample
6.44 moles Al(NO$_2$)$_3$	**1.06×10^3 g Al(NO$_2$)$_3$**	**1.16×10^{25} atoms N**
0.0427 mole Mg$_3$N$_2$	4.31 g Mg$_3$N$_2$	**5.14×10^{22} atoms N**
0.198 mole (NH$_4$)$_2$CO$_3$	**19.0 g (NH$_4$)$_2$CO$_3$**	2.38×10^{23} atoms N

5.61 100% (within rounding error) **5.63** (a) 18.89% (b) 16.50%
(c) 25.94% (d) 5.522% **5.65** (a) 31.56% (b) 44.91% (c) 53.55%
(d) 65.31% **5.67** (a) 33.73% (b) 79.89% (c) 27.74% (d) 52.92%
5.69 C_5H_7N **5.71** $C_4H_5N_2O$ **5.73** N_2O **5.75** Cr_2S_3 **5.77** $Ca_3P_2O_8$
5.79 TiO_2 **5.81** N_2H_4 **5.83** C_6H_{12} **5.85** (a) 4.13×10^{23} Cl atoms
(b) 1.42×10^{24} Cl atoms (c) 7.62×10^{24} Cl atoms (d) 1.06×10^{24} Cl
atoms (e) 1.04×10^{23} Cl atoms **5.87** (a) 6.06×10^{21} H atoms
(b) 7.04×10^{25} H atoms (c) 8.94×10^{24} H atoms **5.89** (a) 26.6 g N_2O_5;
0.247 mol N_2O_5 (b) 12.8 g $(NH_4)_2O$; 0.247 mol $(NH_4)_2O$ (c) 35.0 g
$Al(NO_3)_3$; 0.164 mol $Al(NO_3)_3$ **5.91** (a) 1.44×10^{25} Na^+ (b) $8.72 \times$
10^{24} Na^+ (c) 3.29×10^{24} Na^+ **5.93** (a) 672 gold coins (b) 576 silver
coins (c) 6.78 lb (d) 12.0 lb silver coins (e) 1.25×10^{25} Au atoms,
3.04×10^{25} Ag atoms

Chapter 6

6.1 They are the outermost electrons and the ones that interact
with other atoms to form bonds. **6.3** Polyatomic ions have a charge
that must be accounted for when counting valence electrons.
6.5 (a) 32 (b) 20 (c) 26 (d) 20

6.7 (a) :Cl̈—Ö—Cl̈: (b) :Ï—N̈—Ï: (c) :Ï—C—Ï: (d) :Br̈—Br̈:

6.9 (a) $\begin{bmatrix} :\ddot{F}: \\ :\ddot{F}—P—\ddot{F}: \\ :\ddot{F}: \end{bmatrix}^+$ (b) $\begin{bmatrix} :\ddot{O}—C—\ddot{O}: \\ \| \\ :\ddot{O}: \end{bmatrix}^{2-}$ (c) $\begin{bmatrix} :\ddot{O}=\ddot{N}—\ddot{O}: \end{bmatrix}^-$

(d) $\begin{bmatrix} :\ddot{Cl}—C—\ddot{Cl}: \\ :\ddot{Cl}: \end{bmatrix}^-$ **6.11** (a) :C≡S: (b) :S̈=C=S̈:

(c) :Cl̈—P̈—Cl̈: (d) H—C≡N: **6.13** (a) :N̈=N=Ö:

(b) :Cl̈—C—F̈: or :Cl̈—C—Cl̈: are equivalent
         :F̈:   :F̈:

(c) :Ö=S̈—Ö: (d) :N≡N: **6.15** (a) $\begin{bmatrix} :\ddot{O}=\ddot{N}—\ddot{O}: \end{bmatrix}^-$

(b) $\begin{bmatrix} :\ddot{O}—S—\ddot{O}: \\ :\ddot{O}: \end{bmatrix}^{2-}$ (c) $\begin{bmatrix} :\ddot{O}—Si=\ddot{O}: \\ :\ddot{O}: \end{bmatrix}^{2-}$ (d) $\begin{bmatrix} :\ddot{O}: \\ :\ddot{O}—S—\ddot{O}: \\ :\ddot{O}: \end{bmatrix}^{2-}$

6.17 (a) Lone pair of electrons missing from P. (b) Lone pair of
electrons missing from O. (c) Carbon should be the central atom
bonded directly to each hydrogen. The carbon should also have a
double bond to the oxygen, which should have two lone pairs of
electrons. (d) There should only be a single bond between the C
and Cl, and 3 pairs of electrons shown on the Cl. (e) There are too
many electrons (4 extra) drawn. Remove the two lone pairs from C
and use one lone pair from each S to form double bonds between
the C and S atoms :S̈=C=S̈:. **6.19** H, N, O, F, Cl, Br, and I. By
sharing one electron from each atom as a bond, the octet (duet for
H) for both atoms is fulfilled. **6.21** Resonance structures are a
series of Lewis structures that together represent the bonding in a
molecule or ion. This is necessary when Lewis theory can't
describe the bonding in a molecule or ion with a single structure.
Many times this occurs when a double bond can be drawn in two
or more equivalent places within a structure.

6.23 (a) $\begin{bmatrix} :\ddot{O}—\ddot{Cl}—\ddot{O}: \end{bmatrix}^-$ (no resonance structures needed)

(b) $\begin{bmatrix} :\ddot{O}—Si=\ddot{O}: \\ :\ddot{O}: \end{bmatrix}^{2-} \longleftrightarrow \begin{bmatrix} :O=Si—\ddot{O}: \\ :\ddot{O}: \end{bmatrix}^{2-} \longleftrightarrow \begin{bmatrix} :\ddot{O}—Si—\ddot{O}: \\ \| \\ .\ddot{O}. \end{bmatrix}^{2-}$

(c) $\begin{bmatrix} :\ddot{O}—N=\ddot{O}: \\ :\ddot{O}: \end{bmatrix}^- \longleftrightarrow \begin{bmatrix} :O=N—\ddot{O}: \\ :\ddot{O}: \end{bmatrix}^- \longleftrightarrow \begin{bmatrix} :\ddot{O}—N—\ddot{O}: \\ \| \\ .\ddot{O}. \end{bmatrix}^-$

(d) $\begin{bmatrix} :\ddot{O}: \\ :\ddot{O}—S—\ddot{O}: \\ :\ddot{O}: \end{bmatrix}^{2-}$ (no resonance structures needed)

6.25 (a) $\begin{bmatrix} :\ddot{O}=P=\ddot{O}: \end{bmatrix}^+$ (no resonance structures needed)

(b) $\begin{bmatrix} :\ddot{O}—As=\ddot{O}: \\ :\ddot{O}: \end{bmatrix}^- \longleftrightarrow \begin{bmatrix} :O=As—\ddot{O}: \\ :\ddot{O}: \end{bmatrix}^- \longleftrightarrow \begin{bmatrix} :\ddot{O}—As—\ddot{O}: \\ \| \\ .\ddot{O}. \end{bmatrix}^-$

(c) :O=S—Ö: $\longleftrightarrow$:Ö—S=Ö: $\longleftrightarrow$:Ö—S—Ö:
      :Ö: :Ö: ‖
                                                    .Ö.

(d) :O=Ö—Ö: $\longleftrightarrow$:Ö—Ö=Ö:

6.27

	Electron groups	Electron-group geometry	Molecular shape	Bond angles
(a)	2	linear	linear	180°
(b)	3	trigonal planar	bent	~120°
(c)	4	tetrahedral	bent	~109.5°

6.29 (a) EG = trigonal planar, MS = trigonal planar
(b) EG = tetrahedral, MS = trigonal pyramidal (c) EG = tetrahedral,
MS = tetrahedral (d) EG = tetrahedral, MS = tetrahedral
6.31 (a) EG = tetrahedral, MS = tetrahedral (b) EG = trigonal
planar, MS = bent (c) EG = tetrahedral, MS = tetrahedral
(d) EG = trigonal planar, MS = trigonal planar
6.33 (a) EG = tetrahedral, MS = bent (b) EG = trigonal planar,
MS = trigonal planar (c) EG = tetrahedral, MS = tetrahedral
(d) EG = tetrahedral, MS = trigonal pyramidal
6.35 (a) 120° (b) ~109.5° (c) 109.5° (d) 109.5° **6.37** (a) 109.5°
(b) ~120° (c) 109.5° (d) 120° **6.39** (a) ~109.5° (b) 120° (c) 109.5°
(d) ~109.5° **6.41** Electronegativity is an atom's ability to pull
shared electrons (bonds) toward its nucleus. The lowest electronegativity is in the lower left corner of the periodic table and
increases as elements get closer to the upper right corner of the
periodic table. **6.43** A polar molecule contains one or more polar
bonds that do not "cancel" one another due to the shape of the
molecule. Polar molecules contain a dipole or overall charge
separation. **6.45** Yes, a molecule like CO_2 contains two polar
bonds, but is nonpolar. It is the arrangement of those bonds in
a linear fashion that causes the "cancelling" of the dipoles.

6.47 a, c, d **6.49** (a) As⟶Cl (b) F_2 has no dipole (c) N⟵I (d) C⟶Cl
6.51 a, b **6.53** a, b, c, d **6.55** Dispersion, dipole-dipole, and
hydrogen bonding. Polar molecules contain dipole-dipole forces,
whereas nonpolar molecules do not. If a polar molecule contains
an O—H, F—H, or N—H bond, it exhibits hydrogen bonding.
6.57 a, c, d **6.59** a, c **6.61** a, c **6.63** (a) dispersion (b) dispersion
(c) dipole-dipole (d) dispersion **6.65** (a) $SO_2 < NCl_3 < PCl_3$
(b) $CH_4 < C_2H_6 < C_3H_8$ (c) $CO_2 < Cl_2 < Xe$ **6.67** (a) $H_2O >$
$OF_2 > BCl_3$ (b) $NH_3 > SeO_2 > CO_2$ (c) $PF_3 > SF_2 > BeF_2$

6.69 Mg^{2+} $\left[:\ddot{O}:\right]^{2-}$ There are no shared electrons in the ionic compound. In order to fulfill the octet of each atom in CO, multiple pairs of electrons are shared as bonds. **6.71** In order for a substance to exhibit hydrogen bonding, it must be polar AND contain an O—H, N—H, or F—H bond. A molecule of CH_2F_2 is polar, but does not exhibit hydrogen bonding because it lacks a direct H—F bond (all the H and F atoms are bonded directly to the central C atom).

Chapter 7

7.1 Solid, liquid, and gas. Liquids and gases are compressible.
7.3 Liquid, gas, gas
7.5

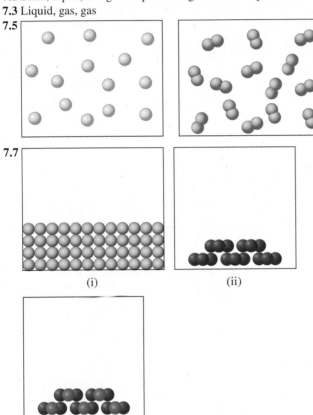

7.7

(i)   (ii)

(iii)

7.9 (a) molecular (b) molecular (c) molecular (d) atomic, metallic **7.11** (a) dispersion forces and metallic bonding (b) dispersion forces (c) dispersion forces (d) dispersion forces and ionic bonding **7.13** (a) dispersion forces, dipole-dipole forces, and hydrogen bonding (b) dispersion forces and ionic bonding (c) dispersion forces (d) dispersion forces **7.15** Ionic solids are held together by ionic bonds—the attraction between oppositely charged ions, whereas molecular solids are held together by much weaker dispersion, dipole-dipole, or hydrogen bonding intermolecular forces. The ionic solid could be NaCl, MgO, CaF_2, or one of many other ionic compounds. The molecular solid could be CO, CO_2, CH_4, N_2O, N_2, Br_2, or one of many other molecular substances. **7.17** Vapor pressure is the amount or pressure of the gaseous form that exists above a sample of a liquid or solid.

7.19

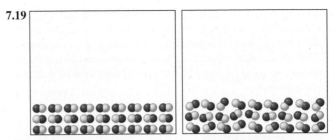

7.21 (a) $NH_3 < HCl < CH_4$ (b) $Kr < F_2 < Ne$ (c) $AsCl_3 < NCl_3 < BCl_3$ **7.23** (a) $BeBr_2$, both substances exhibit only dispersion forces, so the smaller one (smaller molar mass) has weaker attractions and therefore a higher vapor pressure. (b) BF_3, both substances exhibit only dispersion forces, so the smaller one (smaller molar mass) has weaker attractions and therefore a higher vapor pressure. (c) Cl_2 has weaker dispersion forces whereas HBr exhibits dipole-dipole forces. The weaker intermolecular forces in Cl_2 give it a higher vapor pressure. **7.25** Viscosity represents how easily a substance will flow. The stronger the intermolecular forces between particles, the more difficult for the particles to tumble past one another, and the higher the viscosity or "thicker" the substance. **7.27** The stronger the intermolecular forces in a sample of a substance, the higher the surface tension will be. **7.29** The stronger the intermolecular forces in a sample of a substance, the lower the vapor pressure will be. **7.31** (a) The stronger the intermolecular forces, the lower the vapor pressure. (b) The stronger the intermolecular forces, the higher the boiling point. (c) The stronger the intermolecular forces, the higher the viscosity. (d) The stronger the intermolecular forces, the higher the surface tension. **7.33** No, only the attractions between molecules (intermolecular attractions) are "broken" when a molecular substance boils. **7.35** (a) Acetone, because the —OH group in the isopropanol molecule exhibits stronger hydrogen bonding, giving it a lower vapor pressure. (b) $BeCl_2$, because it exhibits weaker dispersion forces than the dipole-dipole forces in $SeCl_2$. (c) HCl, because it has weaker dipole-dipole forces than the hydrogen bonding of HF. **7.37** (a) $CH_3F < CH_3Cl < CH_3OH$ (b) $O_2 < Xe < CaO$ (c) $Ar < CF_2H_2 < (CH_3)_2NH$

7.39

(a) H_3C [structure with CH_3] CH_3, because it has stronger dispersion forces than the other molecule.

(b) [structure: H—C—C—OH with H H above and H H below], because it exhibits stronger hydrogen bonding forces, where the other molecule only exhibits dispersion forces.

(c) [structure with NH_2] NH_2, because it exhibits stronger hydrogen bonding forces, where the other molecule only exhibits dipole-dipole forces.

7.41 6.18×10^3 J or 6.18 kJ **7.43** (a) 1.24×10^3 J or 1.24 kJ
(b) 3.38×10^3 J or 3.38 kJ (c) 5.29×10^2 J or 0.529 kJ **7.45** 36.9°C
7.47 0.303 $\frac{J}{g \cdot °C}$ **7.49** (a) endothermic (b) exothermic (c) endothermic
7.51 Both are kJ/mol. The molar heat of fusion must be used at the
melting point, and the molar heat of vaporization must be used at the
boiling point.

7.53

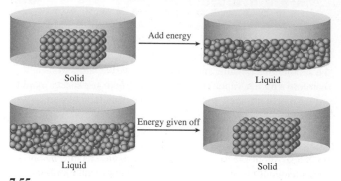

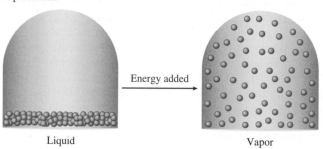

7.55
Vaporization

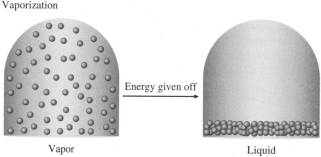

Vaporization

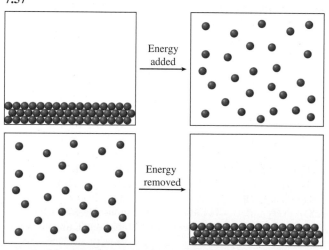

7.57

7.59 18.6 kJ **7.61** 7.06 kJ **7.63** 81.3 kJ

Chapter 8

8.1 Gases expand to fill their container and have a lot of empty
space separating each particle, whereas liquids and solids do not.
8.3 A gas consists primarily of empty space with gas particles
that are separated by large distances; Gas particles are in constant
random motion and change direction when they strike another gas
particle or container wall, without losing energy; Gas particles
don't interact with one another (attract or repel); The higher the
temperature of a gas, the higher its average kinetic energy, and the
faster the gas particles are moving. **8.5** Gases consist of a lot of
empty space, whereas liquids and solids do not. Typically, the
densities of gases are expressed in units of g/L. **8.7** (a) 0.625 atm,
(b) 2.18 atm, (c) 3.80 atm, (d) 0.979 atm **8.9** (a) 703 torr,
(b) 0.041 torr, (c) 1.56×10^3 torr, (d) 8.89×10^3 torr
8.11 (a) 22.5 in Hg, (b) 0.753 atm, (c) 7.63×10^4 Pa,
(d) 11.1 psi, (e) 572 torr, (f) 0.763 bar
8.13

psi	Pa	kPa	atm	mmHg	in Hg	torr	bar
35.5860	245,289	245.289	2.42081	1839.82	72.4308	1839.82	2.45289
17.3	1.19×10^5	119	1.18	895	35.2	895	1.19
10.5	7.25×10^4	72.5	0.716	544	21.4	544	0.725

8.15 4.90×10^5 Pa **8.17** $PV = nRT$, where P is pressure in atm,
V is volume in L, n represents moles of gas particles, R is the gas
constant 0.0821 L · atm/K · mol, and T is temperature in kelvins.
8.19 $V = \frac{nRT}{P}$ **8.21** $T = \frac{PV}{nR}$ **8.23** No, it only depends on the
number of gas particles present, independent of identity. **8.25** 5.23 L
8.27 (a) 38.2 L, (b) 49.3 L, (c) 31.4 L **8.29** (a) 49.5 L, (b) 40.3 L,
(c) 25.0 L **8.31** (a) 2.51 atm, (b) 5.02 atm, (c) 7.52 atm
8.33 (a) 2.75 atm, (b) 27.3 atm, (c) 14.4 atm **8.35** (a) 51.3 K,
(b) 649 K, (c) 1.62×10^3 K **8.37** (a) −25°C, (b) 222°C, (c) 718°C
8.39 (a) 1.79 mol, (b) 0.447 mol, (c) 0.841 mol **8.41** (a) 3.08 mol,
(b) 1.92 mol, (c) 3.33 mol **8.43** (a) C, (b) A = D **8.45** 44.0 g/mol
8.47 146 g/mol **8.49** 2.53 g/L **8.51** 133.623 g **8.53** a **8.55** 602 mmHg
or 0.792 atm
8.57 $\frac{V_1}{T_1} = \frac{V_2}{T_2}$ Charles's law relates the volume and temperature
of a gas sample, assuming constant pressure and number of moles
of gas in the sample. **8.59** 2.75 L **8.61** The combined gas law
becomes Boyle's law when temperature is held constant and
therefore can be removed from the equation: $\frac{P_1 V_1}{\cancel{T_1}} = \frac{P_2 V_2}{\cancel{T_2}}$.
Boyle's law is $P_1 V_1 = P_2 V_2$. The assumption is that the number
of moles and pressure stay constant. **8.63** As the external pressure
of the water lessens toward the surface, the lungs will continue
to expand with the constant number of moles of gas to match the
external pressure. **8.65** Partial pressure is used to describe the
pressure of one gas when it is part of a mixture of gases.
8.67 The hydrogen exerts a pressure of 0.8 atm and the oxygen
exerts a pressure of 1.2 atm.

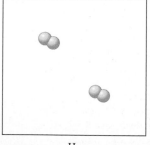

H_2

O_2

8.69 The pressure of Ne is 1.0 atm and the pressure of the gas mixture is 1.50 atm.

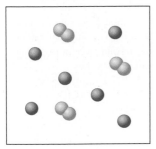

Ne & F_2

8.71 The mole fraction (χ) of H_2 is 2/5, or 0.40. The mole fraction of oxygen is 3/5, or 0.60. **8.73** The mole fraction of Ne is 6/9, or 0.67. The mole fraction of fluorine in the mixture is 3/9, or 0.33. **8.75** The second/larger box has a higher total pressure because it contains more than twice as many gas particles. The smaller box has a slightly higher partial pressure of helium with 5/9 as compared to the larger box with 11/20. **8.77** (a) 0.485 atm, (b) 0.692 atm, (c) 1.177 atm, (d) $\chi_{Ar} = 0.412$, $\chi_{N_2} = 0.588$ **8.79** 1.07 atm **8.81** 5.3×10^6 Tg CO_2 **8.83** $d_{dry\ air} = 1.29$ g/L, $d_{humid\ air} = 1.18$ g/L. Dry air has a higher density than humid air.

Chapter 9

9.1 A solution is a homogeneous mixture of two or more substances. No, there are gaseous and solid examples of solutions. **9.3** b, c, d **9.5** (a) Water is the solute and ethanol is the solvent. (b) Copper is the solute and gold is the solvent. (c) Water is the solute and isopropanol is the solvent. **9.7** The first image shows an unsaturated solution or supersaturated solution. Without knowing the process to form the solution, you can't visually distinguish between the two. The second image shows a saturated solution. **9.9** No, at some point we would be unable to dissolve any more NaCl in a given volume of water. **9.11** c **9.13** (a) soluble, (b) soluble, (c) soluble, (d) insoluble **9.15** (a) insoluble, (b) insoluble, (c) soluble, (d) soluble **9.17** a, b, c, d **9.19** (b) $Ni(OH)_2$, (c) $Mg(OH)_2$ **9.21** Mass percent can be used to describe the concentration of a solute in solution. It can be determined by dividing the mass of solute by the total mass of the solution (solute + solvent) and multiplying by 100 to obtain a percentage. **9.23** (a) 11.1% KF, (b) 11.8% CH_3CH_2OH, (c) 11.3% LiOH **9.25** (a) 8.1% NaF, (b) 31% K_3PO_4, (c) 6.3% CH_3OH **9.27** (a) 51.0 g, (b) 36.2 g, (c) 17.1 g **9.29** (a) 1.26×10^3 g, (b) 2.67×10^4 g, (c) 1.12×10^3 g **9.31** (a) 154 g, (b) 130 g, (c) 226 g **9.33** (a) 1.31 m, (b) 0.540 m, (c) 0.215 m **9.35** (a) 1.01 m, (b) 0.612 m, (c) 0.612 m **9.37** (a) 4.64 m, (b) 2.59 m, (c) 14.6 m **9.39** (a) 0.238 M, (b) 0.994 M, (c) 0.592 M **9.41** (a) 0.871 M, (b) 1.49 M, (c) 10.1 M **9.43** (a) 13.7 L, (b) 56.8 L, (c) 19.7 L **9.45** (a) 1.55 L, (b) 0.998 L, (c) 0.177 L **9.47** (a) 331 g, (b) 125 g, (c) 0.0696 g **9.49** 1. Choose 1.00 L of solution 2. Use density of solution to determine the mass of solution 3. Use moles of solute to determine mass of solute 4. Subtract mass of solute from mass of solution to find mass of solvent 5. Convert mass of solvent to units of kg 6. Divide moles of solute by kg of solvent **9.51** (a) 526 m, (b) 98.1% **9.53** (a) 15.6 M, (b) 69.7% **9.55** $1 \times 10^{-4}\%$ **9.57** $4 \times 10^{-4}\%$ and $1 \times 10^{-4}\%$ **9.59** Electrolytes form charged particles in solution and will conduct a current. Nonelectrolytes do not form charged particles in solution. **9.61** (a) electrolyte, (b) electrolyte, (c) electrolyte, (d) nonelectrolyte **9.63** (a) Has an incorrect ratio of sodium and chloride ions in the solution.

(b) Accurately represents the formation of the solution. (c) Has an incorrect ratio of aluminum and sulfate ions in the solution. (d) Accurately represents the formation of the solution. **9.65** Your drawing should show the correct ratio of *two* aluminum ions for every *three* sulfate ions. **9.67** (a) 0.10 M, (b) 0.20 M, (c) 0.30 M **9.69** 0.200 M **9.71** The more solute in a given solution, the more *concentrated* it is considered to be. A solution is said to be *dilute* if it has a relatively small amount of solute in a given volume of solution. **9.73** First, 0.340 gram of NaCl should be weighed out and placed in a 25.0-mL volumetric flask. Water should be added to a total volume of 25.0 mL and the solution mixed. **9.75** 26.2 g **9.77** A sample of 3.325 mL of the stock solution would be measured into a 25.0-mL volumetric flask. Water would be added to a total volume of 25.0 mL. **9.79** 1.61 mL **9.81** (a) 0.0800 M, 3.20×10^{-3} M, 1.28×10^{-4} M, 5.12×10^{-6} M (b) 0.0200 mol, 8.00×10^{-4} mol, 3.20×10^{-5} mol, 1.28×10^{-6} mol **9.83** Colligative properties are properties of a solution that only depend on the number of dissolved solute particles present, and not their identity. They include freezing-point depression, boiling-point elevation, and osmotic pressure. **9.85** No, because the addition of any solute to water will decrease its freezing point below 0.00°C. **9.87** The solution containing $CaCl_2$ has three solute particles for each dissolved formula unit, compared with two solute particles in the NaCl. Thus, the $CaCl_2$ solution will freeze at a lower temperature than the NaCl solution. **9.89** 3.9 m CH_3OH < 3.9 m $MgSO_4$ < 3.9 m $SrCl_2$ **9.91** −5.49°C **9.93** −9.09°C **9.95** 103.58°C **9.97** 103.25°C **9.99** i **9.101** 85.6 atm **9.103** 1.8 g, 0.023 mol **9.105** (a) 58%, (b) 18-carat, (c) 12.5%

Visualizing Chemistry
VC 9.1 b. **VC 9.2** c. **VC 9.3** c. **VC 9.4** a.

Chapter 10

10.1 color change, temperature change, fizzing/bubbling, light emission **10.3** The statue has a greenish hue, meaning the original copper coating (shiny and metallic) must have undergone a chemical reaction to turn green. **10.5** A chemical reaction is the process where the identity of one or more substances is changed. A chemical equation is the symbolic representation of the process. **10.7** (a) $N_2 + O_2 \longrightarrow NO_2$ (b) $N_2O_5 \longrightarrow N_2O_4 + O_2$ (c) $O_3 \longrightarrow O_2$ (d) $Cl_2 + NaI \longrightarrow I_2 + NaCl$ (e) $Mg + O_2 \longrightarrow MgO$ **10.9** (a) Sulfur reacts with oxygen to form sulfur dioxide. (b) Methane reacts with oxygen to form carbon dioxide and water. (c) Nitrogen reacts with hydrogen to form ammonia. (d) Tetraphosphorus decoxide reacts with water to form phosphoric acid. (e) Sulfur reacts with nitric acid to form sulfuric acid, nitrogen dioxide, and water. **10.11** Because atoms are not destroyed or created during a chemical reaction—they are merely bonded differently before and after reaction. The law of conservation of mass must be obeyed. **10.13** (a) $N_2 + 2O_2 \longrightarrow 2NO_2$ (b) $2N_2O_5 \longrightarrow 2N_2O_4 + O_2$ (c) $2O_3 \longrightarrow 3O_2$ (d) $Cl_2 + 2NaI \longrightarrow I_2 + 2NaCl$ (e) $2Mg + O_2 \longrightarrow 2MgO$ **10.15** (a) $S_8 + 8O_2 \longrightarrow 8SO_2$ (b) $CH_4 + 2O_2 + CO_2 + 2H_2O$ (c) $N_2 + 3H_2 \longrightarrow 2NH_3$ (d) $P_4O_{10} + 6H_2O \longrightarrow 4H_3PO_4$ (e) $S + 6HNO_3 \longrightarrow H_2SO_4 + 6NO_2 + 2H_2O$ **10.17** (a) $2C + O_2 \longrightarrow 2CO$ (b) $2CO + O_2 \longrightarrow 2CO_2$ (c) $H_2 + Br_2 \longrightarrow 2HBr$ (d) $2Ca + O_2 \longrightarrow 2CaO$ (e) $O_2 + 2Cl_2 \longrightarrow 2OCl_2$ **10.19** (a) $CH_4 + 4Br_2 \longrightarrow CBr_4 + 4HBr$ (b) $5N_2H_4 + 4HNO_3 \longrightarrow 7N_2 + 12H_2O$ (c) $2KNO_3 \longrightarrow 2KNO_2 + O_2$ (d) $NH_4NO_3 \longrightarrow N_2O + 2H_2O$ (e) $NH_4NO_2 \longrightarrow N_2 + 2H_2O$ **10.21** c. **10.23** The ionic equation shows exactly how all species exist in aqueous solution, whereas the molecular equation represents all substances in compound form. **10.25** If an ionic equation is written and all ions are spectator ions (and are "crossed out"), then this shows there is no reaction occurring. **10.27** (a) redox, (b) precipitation, (c) redox, (d) acid-base **10.29** (a) redox, (b) precipitation, (c) redox, (d) acid-base **10.31** (a) single displacement, (b) double displacement, (c) combination, (d) double displacement **10.33** (a) decomposition, (b) double displacement, (c) single displacement, (d) double displacement **10.35** The first beaker should show separate ions of H^+ and Cl^- in a 1:1 ratio. The second beaker should show Ca^{2+} and OH^- ions in a 1:2 ratio. When they are mixed, the final beaker should show only Ca^{2+} and Cl^- ions in a 1:2 ratio. The H^+ and OH^- ions have combined to form water. **10.37** (a) $Li_2S(aq)$ + $Mg(C_2H_3O_2)_2(aq) \longrightarrow MgS(s) + 2LiC_2H_3O_2(aq)$ (b) $(NH_4)_2SO_4(aq)$ + $BaBr_2(aq) \longrightarrow BaSO_4(s) + 2NH_4Br(aq)$ (c) No reaction (d) $H_3PO_4(aq) + 3LiOH(aq) \longrightarrow 3H_2O(l) + Li_3PO_4(aq)$ **10.39** (a) $K_2CO_3(aq) + 2HNO_3(aq) \longrightarrow H_2O(l) + CO_2(g)$ + $2KNO_3(aq)$ (b) $NH_4Cl(aq) + LiOH(aq) \longrightarrow NH_3(g) + H_2O(l)$ + $LiCl(aq)$ (c) $Na_2SO_3(aq) + 2HBr(aq) \longrightarrow H_2O(l) + SO_2(g)$ + $2NaBr(aq)$ (d) No reaction **10.41** (a) Molecular: $2AgNO_3(aq)$ + $Pb(s) \longrightarrow Pb(NO_3)_2(aq) + 2Ag(s)$ Complete ionic: $2Ag^+(aq)$ + $2NO_3^-(aq) + Pb(s) \longrightarrow Pb^{2+}(aq) + 2NO_3^-(aq) + 2Ag(s)$ Net ionic: $2Ag^+(aq) + Pb(s) \longrightarrow Pb^{2+}(aq) + 2Ag(s)$ (b) Molecular: $CaC_2(s) + 2H_2O(l) \longrightarrow C_2H_2(g) + Ca(OH)_2(aq)$ Complete ionic: $CaC_2(s) + 2H_2O(l) \longrightarrow C_2H_2(g) + Ca^{2+}(aq) + 2OH^-(aq)$ Net ionic: $CaC_2(s) + 2H_2O(l) \longrightarrow C_2H_2(g) + Ca^{2+}(aq) + 2OH^-(aq)$ (c) Molecular: $Ba(OH)_2(aq) + H_2SO_4(aq) \longrightarrow BaSO_4(s) + 2H_2O(l)$ Complete ionic: $Ba^{2+}(aq) + 2OH^-(aq) + 2H^+(aq) + SO_4^{2-}(aq) \longrightarrow$ $BaSO_4(s) + 2H_2O(l)$ Net ionic: $Ba^{2+}(aq) + 2OH^-(aq) + 2H^+(aq) +$ $SO_4^{2-}(aq) \longrightarrow BaSO_4(s) + 2H_2O(l)$ **10.43** (a) Ionic: $2Li^+(aq)$ + $S^{2-}(aq) + Mg^{2+}(aq) + 2C_2H_3O_2^-(aq) \longrightarrow MgS(s) + 2Li^+(aq) +$ $2C_2H_3O_2^-(aq)$ Net ionic: $S^{2-}(aq) + Mg^{2+}(aq) \longrightarrow MgS(s)$ (b) Ionic: $2NH_4^+(aq) + SO_4^{2-}(aq) + Ba^{2+}(aq) + 2Br^-(aq) \longrightarrow$ $BaSO_4(s) + 2NH_4^+(aq) + 2Br^-(aq)$ Net ionic: $SO_4^{2-}(aq) + Ba^{2+}(aq)$ $\longrightarrow BaSO_4(s)$ (c) No reaction (d) Ionic: $3H^+(aq) + PO_4^{3-}(aq) +$ $3Li^+(aq) + 3OH^-(aq) \longrightarrow 3H_2O(l) + 3Li^+(aq) + PO_4^{3-}(aq)$ Net ionic: $H^+(aq) + OH^-(aq) \longrightarrow H_2O(l)$ **10.45** (a) Ionic: $2K^+(aq)$ + $CO_3^{2-}(aq) + 2H^+(aq) + 2NO_3^-(aq) \longrightarrow H_2O(l) + CO_2(g) +$ $2K^+(aq) + 2NO_3^-(aq)$ Net ionic: $CO_3^{2-}(aq) + 2H^+(aq) \longrightarrow$ $H_2O(l) + CO_2(g)$ (b) Ionic: $NH_4^+(aq) + Cl^-(aq) + Li^+(aq) +$ $OH^-(aq) \longrightarrow NH_3(g) + H_2O(l) + Li^+(aq) + Cl^-(aq)$ Net ionic: $NH_4^+(aq) + OH^-(aq) \longrightarrow NH_3(g) + H_2O(l)$ (c) Ionic: $2Na^+(aq) +$ $SO_3^{2-}(aq) + 2H^+(aq) + 2Br^-(aq) \longrightarrow H_2O(l) + SO_2(g) +$ $2Na^+(aq) + 2Br^-(aq)$ Net ionic: $SO_3^{2-}(aq) + 2H^+(aq) \longrightarrow H_2O(l)$ + $SO_2(g)$ (d) No reaction **10.47** $Pb(NO_3)_2(aq)$ and $Na_2SO_4(aq)$ **10.49** (a) +4 (b) −4 (c) +4 (d) −3 (e) +4 (f) 0 (g) +4 (h) +6 (i) +3 **10.51** (a) $2K(s) + Cl_2(g) \longrightarrow 2KCl(s)$ (b) $2Al(s) + 3Cl_2(g) \longrightarrow$ $2AlCl_3(s)$ (c) $Sr(s) + Cl_2(g) \longrightarrow SrCl_2(s)$ **10.53** (a) Your sketch should show a solid lump of potassium reacting with gaseous molecules of Cl_2 above. The product should be a solid piece of alternating potassium and chloride ions in a 1:1 ratio. (b) Your sketch should show a solid lump of aluminum reacting with gaseous molecules of Cl_2 above. The product should be a solid piece of alternating aluminum and chloride ions in a 1:3 ratio. (c) Your sketch should show a solid lump of strontium reacting with gaseous molecules of Cl_2 above. The product should be a solid piece of alternating strontium and chloride ions in a 1:2 ratio. **10.55** (a) $2Na_2O(s) \longrightarrow 4Na(s) + O_2(g)$ (b) $MgS(s) \longrightarrow Mg(s) + S(s)$ (c) $2K_3N(s) \longrightarrow 6K(s) + N_2(g)$ **10.57** (a) Your sketch should show a solid piece of sodium oxide represented as alternating sodium and oxide ions (2:1) prior to reaction. After reaction, a solid piece of sodium should be shown with gaseous oxygen molecules above.

(b) Your sketch should show a solid piece of magnesium sulfide represented as alternating magnesium and sulfide ions (1:1) prior to reaction. After reaction, a solid piece of magnesium should be shown with a solid piece of sulfur. (c) Your sketch should show a solid piece of potassium nitride represented as alternating potassium and nitride ions (3:1) prior to reaction. After reaction, a solid piece of potassium should be shown with gaseous nitrogen molecules above. **10.59** (a) $3Zn(NO_3)_2(aq) + 2Al(s) \longrightarrow 2Al(NO_3)_3(aq) + 3Zn(s)$ (b) $FeCl_3(aq) + Al(s) \longrightarrow AlCl_3(aq) + Fe(s)$ (c) $3Pb(ClO_4)_2(aq) +$ $2Al(s) \longrightarrow 2Al(ClO_4)_3(aq) + 3Pb(s)$ **10.61** (a) Your sketch should show zinc and nitrate ions floating in solution with a solid piece of aluminum in the beaker. After reaction, aluminum and nitrate ions should be floating in solution with a solid piece of zinc metal in the beaker. (b) Your sketch should show iron(III) ions and chloride ions floating in solution with a solid piece of aluminum in the beaker. After reaction, aluminum and chloride ions should be floating in solution with a solid piece of iron metal in the beaker. (c) Your sketch should show lead(II) ions and perchlorate ions floating in solution with a solid piece of aluminum in the beaker. After reaction, aluminum and perchlorate ions should be floating in solution with a solid piece of lead metal in the beaker. **10.63** (a) H_2O (the hydrogen within water is being reduced) (b) HCl (the hydrogen within the HCl is being reduced) (c) O_2 is being reduced **10.65** A reaction or process is considered exothermic if it gives off heat to the surroundings. A reaction or process is considered endothermic if it absorbs heat from the surroundings. **10.67** (a) transfer of electrons (b) the formation of a precipitate (c) the formation of water molecules and CO_2 gas (d) the transfer of electrons **10.69** (a) decomposition, redox (b) double displacement, precipitation (c) double displacement, gas producing (d) combustion, redox

Chapter 11

11.1 The balanced reaction gives the exact ratio of "ingredients" needed and the amount of product(s) formed.

11.3 $\dfrac{4 \text{ mol } NH_3}{3 \text{ mol } O_2}, \dfrac{4 \text{ mol } NH_3}{2 \text{ mol } N_2}, \dfrac{4 \text{ mol } NH_3}{6 \text{ mol } H_2O}, \dfrac{2 \text{ mol } N_2}{3 \text{ mol } O_2}, \dfrac{6 \text{ mol } H_2O}{3 \text{ mol } O_2},$

$\dfrac{6 \text{ mol } H_2O}{2 \text{ mol } N_2}$, and the inverse of each (a) $\dfrac{2 \text{ mol } N_2}{4 \text{ mol } NH_3}$,

(b) $\dfrac{3 \text{ mol } O_2}{4 \text{ mol } NH_3}$, (c) $\dfrac{6 \text{ mol } H_2O}{2 \text{ mol } N_2}$ **11.5** 3.60 mol CO_2, 1.80 mol O_2

11.7 (a) 8.0 mol water, (b) 4.0 mol water, (c) 6.0 mol N_2, (d) 3.0 mol N_2 **11.9** (a) 6.66 mol CO_2, (b) 9.99 mol H_2O, (c) 8.65 mol O_2, (d) 2.48 mol C_2H_5OH, (e) 1.96 mol CO_2, (f) 1.18×10^{24} molecules CO_2 **11.11** (a) 3.45×10^{-3} g O_2, (b) 3.45 g O_2, (c) 3.45×10^3 g O_2 **11.13** (a) 908 g NO_2, (b) 0.165 g NO_2, (c) 6.62×10^{-7} g NO_2 **11.15** (a) 0.750 mol & 142 g $TiCl_4$, (b) 1.00 mol & 103 g MnO_3, (c) 0.750 mol & 114 g Cr_2O_3 **11.17** (a) 127 g HBO_2, (b) 52.1 g H_2O, (c) 139 g O_2 **11.19** 65.8 g Al_4C_3 **11.21** No. The limiting reactant is N_2 and limits the mass of NH_3 formed to 12.2 g. **11.23** (a) Cu_2O is the limiting reactant, 102 g of CuO forms. (b) Cu_2O is the limiting reactant, 13.3 g of CuO forms. (c) Cu_2O is the limiting reactant, 13.3 g of CuO forms. **11.25** (a) 0.321 kg $AlCl_3$ (b) 50.0 g $AlCl_3$ (c) 0.782 mol $AlCl_3$ **11.27** 64.6 g HNO_3, 92.1% **11.29** Ca_3P_2 is the limiting reactant, 30.1 g $Ca(OH)_2$ forms, 13.8 g H_2O remains **11.31** 0.361 g Cu **11.33** 0.0184 kg PbI_2 **11.35** 2.57×10^4 mL **11.37** 44.8 mL **11.39** 0.166 g Al **11.41** 21.2 mL **11.43** 12.8 L **11.45** (a) 27.0 L, (b) 36.9 L, (c) 9.23 L, (d) 23.4 L **11.47** (a) 1.21×10^5 mL, (b) 5.92×10^4 mL, (c) 1.18×10^5 mL, (d) 2.04×10^5 mL **11.49** 94.7 L **11.51** 1.96×10^5 mL

11.53 $\dfrac{1 \text{ mol } C_3H_8}{-2044 \text{ kJ}}, \dfrac{5 \text{ mol } O_2}{-2044 \text{ kJ}}, \dfrac{3 \text{ mol } CO_2}{-2044 \text{ kJ}}, \dfrac{4 \text{ mol } H_2O}{-2044 \text{ kJ}},$ and
the inverse of each **11.55** (a) 508 kJ, (b) 23.7 kJ, (c) 136 kJ,
(d) 23.7 kJ **11.57** (a) 1.50×10^4 kJ, (b) 334 kJ, (c) 93.5 kJ,
(d) 117 kJ **11.59** (a) 421 kJ, (b) 682 kJ, (c) 89.9 kJ, (d) 1.4×10^5 kJ
11.61 Your sketch should show $6NH_3$ molecules and one H_2
molecule remaining. **11.63** 1.01 g $NaHCO_3$, 0.269 L CO_2
11.65 4.21×10^5 kJ **11.67** 27.7 g CO **11.69** 7.03×10^{23} Cl_2
molecules **11.71** 0.0167 M Pb^{2+} **11.73** 17.7 L of $Mg(C_2H_3O_2)_2$
solution and 8.85 L of Li_3PO_4 solution **11.75** 0.103 L

Chapter 12

12.1 (a) base, (b) acid, (c) base, (d) acid, (e) acid, (f) base,
(g) acid, (h) base, (i) acid, (j) base **12.3** drain cleaner, ammonia/
window cleaner **12.5** A substance that produces OH^- ions in water.
12.7 A substance that accepts a proton (H^+). **12.9** a, e **12.11** b
12.13 (a) HCN/CN^- and H_2O/H_3O^+ (b) HCO_2H/CO_2H^- and H_2O/H_3O^+
(c) $CH_3NH_2/CH_3NH_3^+$ and H_2O/OH^- (d) F^-/HF and H_2O/OH^-
12.15 (a) NH_4^+, (b) H_2O, (c) H_2SO_3, (d) HCO_3^- **12.17** (c) $OH^-(aq) +$
$H_2O(l) \rightleftharpoons H_2O(l) + OH^-(aq)$ **12.19** 1.0×10^{-7} M at 25°C.
12.21 No, if the $[H_3O^+]$ goes above 1.0×10^{-7} M at 25°C,
then the $[OH^-]$ must be smaller than 1.0×10^{-7} M.
12.23 (a) basic, (b) basic, (c) basic, (d) neutral, (e) acidic
12.25 (a) 1.95×10^{-7} M, acidic; (b) 1.25×10^{-3} M, acidic;
(c) 2.84×10^{-13} M, basic; (d) 5.35×10^{-9} M, basic **12.27** (a) 5.32×10^{-12} M, acidic; (b) 1.88×10^{-3} M, basic; (c) 2.11×10^{-13} M,
acidic; (d) 1.07×10^{-2} M, basic **12.29** A base that dissociates
completely in water. **12.31** a, f **12.33** d, f **12.35** (a) 1.00×10^{-13} M,
(b) 4.65×10^{-14} M, (c) 2.29×10^{-13} M, (d) 1.32×10^{-13} M,
(e) 1.18×10^{-14} M **12.37** (a) 2.40×10^{-14} M, (b) 1.22×10^{-13} M,
(c) 1.77×10^{-13} M, (d) 1.27×10^{-14} M, (e) 1.35×10^{-13} M
12.39 (a) $[H_3O^+] = 0.0550$ M, $[OH^-] = 1.82 \times 10^{-13}$ M
(b) $[H_3O^+] = 1.82 \times 10^{-13}$ M, $[OH^-] = 0.0550$ M (c) $[H_3O^+] =$
0.0550 M, $[OH^-] = 1.82 \times 10^{-13}$ M (d) $[H_3O^+] = 9.09 \times 10^{-14}$ M,
$[OH^-] = 0.110$ M **12.41** (a) 2.633, (b) 4.090, (c) 5.025, (d) 9.762
12.43 (a) 1.086, (b) 0.987, (c) 0.690, (d) 1.213 **12.45** (a) 12.932,
(b) 12.962, (c) 13.305, (d) 12.873 **12.47** (a) 1.375, (b) 0.735,
(c) 1.399, (d) 0.761 **12.49** (a) 4.5×10^{-2} M, (b) 1.7×10^{-4} M,
(c) 1.5×10^{-7} M, (d) 3.6×10^{-10} M, (e) 2.0×10^{-14} M
12.51 (a) 7.6×10^{-11} M, (b) 8.1×10^{-9} M, (c) 4.9×10^{-7} M,
(d) 6.2×10^{-3} M, (e) 7.4×10^{-1} M **12.53** An acid that does not
dissociate to a large extent in solution. **12.55** If this were a strong
acid, the pH would be 1.356, so this weak acid would have to
have a pH above this value. **12.57** If this were a strong base, the
pH would be 12.949, so the pH must be below this value because
it is a weak base. **12.59** (a) 0.10 M NH_3 < 0.10 M HF < 0.10 M
HNO_3 (b) 0.10 M HCN < 0.20 M HCN < 0.20 M $HClO_4$ (c) 0.10
M $Ca(OH)_2$ < 0.10 M NaOH < 0.10 M NH_3 **12.61** (a) HNO_3,
(b) $HClO_4$, (c) $HC_2H_3O_2$ **12.63** (a) NH_3, (b) LiOH, (c) $Ca(OH)_2$
12.65 A titration is a process where an unknown concentration of an
acid is reacted with a known quantity of a known concentration of
base. The stoichiometry of the neutralization reaction can be used to
determine the concentration of the acid. **12.67** The equivalence point
in a titration is when there is exactly the same number of moles of
hydronium ion reacted with hydroxide—there is no limiting reactant.
12.69 0.133 M **12.71** 0.124 M **12.73** 41.9 mL **12.75** Yes. The added
HCl will convert half the moles of NH_3 to NH_4^+, which will then have
equal concentrations. The NH_3 and its conjugate base NH_4^+ in equal
concentrations are a buffer system. **12.77** c, d **12.79** b **12.81** (a) Up to
0.0650 mole, although the buffer would not work well as the amount
gets close to 0.0650 mole. (b) Up to 0.0750 mole, although the
buffer would not work well as the amount gets close to 0.0750 mole.

12.83 (a) up to 250 mL (b) up to 625 mL (c) up to 0.0313 mole
12.85 (a) 11.85 (b) 2.49 (c) 12.15; note that for part (c), if you
were to carry all of the digits in intermediate answers and not
round until the final result, you would get pH = 12.16.

Chapter 13

13.1 temperature and reactant concentrations **13.3** The rate depends
on the number of collisions between reactants and the energy of each
collision. The higher the temperature, the higher the number of
collisions and the higher the energy of each collision, making the
energy barrier to reaction (activation energy) more readily overcome.
13.5 Equilibrium occurs when the rate of the forward reaction is equal
to the rate of the reverse reaction. **13.7** During rush hour, the total
number of cars on the freeway stays nearly constant, but there are
always cars entering and leaving at about the same rate. **13.9** (c)

13.11 (a) $\dfrac{[SO_3]}{[SO_2][O_2]}$, (b) $\dfrac{[N_2]^2[H_2O]^6}{[NH_3]^4[O_2]^3}$, (c) $\dfrac{[S]^3[H_2O]^2}{[H_2S]^2[SO_2]}$,
(d) $\dfrac{[CH_3OH]}{[H_2]^2[CO]}$ **13.13** (a) $\dfrac{[OCl_2]^2}{[Cl_2]^2[O_2]}$, (b) $\dfrac{[N_2O_5]^2}{[NO_2]^4[O_2]}$,
(c) $\dfrac{[NO_2]^2}{[NO]^2[O_2]}$, (d) $\dfrac{[SO_2]^2[H_2O]^2}{[H_2S]^2[O_2]^3}$ **13.15** (a) $\dfrac{[N_2]^2[H_2O]^2}{[NO]^2[H_2]^2}$,
(b) 7.2×10^2 **13.17** 1.2 **13.19** 6.7×10^6 **13.21** 38 **13.23** 0.400
13.25 (a) $2H_2S + CH_4 \rightleftharpoons 4H_2 + CS_2$ (b) $4HCl + O_2 \rightleftharpoons$
$2H_2O + 2Cl_2$ (c) $N_2O_4 \rightleftharpoons 2NO_2$ **13.27** 3.7×10^{-2} **13.29** (1) a,
(2) d. The equilibrium constant is a simple ratio of products
over reactants since the stoichiometry is 1:1. **13.31** b, d **13.33** c
13.35 (a) products, (b) reactants, (c) reactants, (d) products
13.37 (a) to the right (toward products), (b) to the left (toward
reactants), (c) to the right (toward products), (d) to the left (toward
reactants) **13.39** (a) to the right (toward products), (b) to the left
(toward reactants), (c) to the left (toward reactants), (d) to the
right (toward products) **13.41** (a) neither, (b) to the right (toward
products), (c) neither **13.43** (a) to the right (toward products),
(b) to the left (toward reactants), (c) to the left (toward reactants)
13.45 (a) The equilibrium would shift to the right (toward
products) and make more A and B, (b) The equilibrium would
shift to the right (toward products) and make more A and B
13.47 (a) $2O_3 \rightleftharpoons 3O_2$ (b) A higher ratio of O_3 would be produced
to reduce the higher pressure caused by a decrease in volume.

Chapter 14

14.1 Carbon's electron configuration and electronegativity cause
it to form four covalent bonds by sharing electrons rather than
forming ions. Its small size allows it to form strong and/or multiple
bonds to other carbons.
14.3 One possibility:

$$\begin{array}{c} \ \ \ \ H\ \ \ H\ \ \ H \\ \ \ \ \ |\ \ \ \ \ |\ \ \ \ \ | \\ H-C-C-C-H \\ \ \ \ \ |\ \ \ \ \ |\ \ \ \ \ | \\ \ \ \ \ H\ \ \ H\ \ \ H \end{array}$$

14.5 c **14.7** (a) alkene, (b) alkane, (c) alkyne, (d) alkene

14.9 (a)
$$\begin{array}{c} \ \ \ \ H\ \ \ H\ \ \ H\ \ \ H \\ \ \ \ \ |\ \ \ \ \ |\ \ \ \ \ |\ \ \ \ \ | \\ H-C-C-C-C-H \\ \ \ \ \ |\ \ \ \ \ |\ \ \ \ \ |\ \ \ \ \ | \\ \ \ \ \ H\ \ \ H\ \ \ H\ \ \ H \end{array}$$

(b)
$$\begin{array}{c} \ \ \ \ H\ \ \ H\ \ \ H\ \ \ H\ \ \ H\ \ \ H \\ \ \ \ \ |\ \ \ \ \ |\ \ \ \ \ |\ \ \ \ \ |\ \ \ \ \ |\ \ \ \ \ | \\ H-C-C-C-C-C-C-H \\ \ \ \ \ |\ \ \ \ \ |\ \ \ \ \ |\ \ \ \ \ |\ \ \ \ \ |\ \ \ \ \ | \\ \ \ \ \ H\ \ \ H\ \ \ H\ \ \ H\ \ \ H\ \ \ H \end{array}$$

(c)
$$H-\overset{\overset{\displaystyle H}{|}}{\underset{\underset{\displaystyle H}{|}}{C}}-\overset{\overset{\displaystyle H}{|}}{\underset{\underset{\displaystyle H}{|}}{C}}-\overset{\overset{\displaystyle H}{|}}{\underset{\underset{\displaystyle H}{|}}{C}}-\overset{\overset{\displaystyle H}{|}}{\underset{\underset{\displaystyle H}{|}}{C}}-\overset{\overset{\displaystyle H}{|}}{\underset{\underset{\displaystyle H}{|}}{C}}-\overset{\overset{\displaystyle H}{|}}{\underset{\underset{\displaystyle H}{|}}{C}}-\overset{\overset{\displaystyle H}{|}}{\underset{\underset{\displaystyle H}{|}}{C}}-\overset{\overset{\displaystyle H}{|}}{\underset{\underset{\displaystyle H}{|}}{C}}-H$$

14.11 $CH_3CH_2CH_2CH_3$ **14.13** $CH_3CH_2CH_3$

14.15 $H_3C-CH_2-CH_2-CH_2-CH_2-CH_2-CH_3$

$H_3C-CH_2-CH_2-\overset{\overset{\displaystyle CH_3}{|}}{CH}-CH_2-CH_3$

$H_3C-CH_2-CH_2-CH_2-\overset{\overset{\displaystyle CH_3}{|}}{CH}-CH_3$

$H_3C-CH_2-\overset{\overset{\displaystyle CH_3}{|}}{\underset{\underset{\displaystyle CH_3}{|}}{C}}-CH_2-CH_3$ $H_3C-\overset{\overset{\displaystyle CH_3}{|}}{\underset{\underset{\displaystyle CH_3}{|}}{C}}-CH_2-CH_2-CH_3$

$H_3C-\overset{\overset{\displaystyle CH_3}{|}}{CH}-\overset{\overset{\displaystyle CH_3}{|}}{CH}-CH_2-CH_3$ $H_3C-\overset{\overset{\displaystyle CH_3}{|}}{\underset{\underset{\displaystyle CH_3}{|}}{C}}-\overset{\overset{\displaystyle CH_3}{|}}{CH}-CH_3$

$H_3C-CH_2-\overset{\overset{\displaystyle CH_3}{|}}{\underset{\underset{\displaystyle CH_2}{|}}{CH}}-CH_2-CH_3$ $H_3C-\overset{\overset{\displaystyle CH_3}{|}}{CH}-CH_2-\overset{\overset{\displaystyle CH_3}{|}}{CH}-CH_3$

14.17 Any four of the following:

$H_3C-CH_2-CH_2-CH_2-CH_2-C\equiv CH$

$H_3C-CH_2-\overset{\overset{\displaystyle CH_3}{|}}{CH}-C\equiv C-CH_3$

$H_3C-CH_2-CH_2-CH_2-C\equiv C-CH_3$

$H_3C-\overset{\overset{\displaystyle CH_3}{|}}{CH}-CH_2-C\equiv C-CH_3$

$H_3C-CH_2-CH_2-C\equiv C-CH_2-CH_3$

$H_3C-\overset{\overset{\displaystyle CH_3}{|}}{\underset{\underset{\displaystyle CH_3}{|}}{C}}-C\equiv C-CH_3$ $H_3C-CH_2-CH_2-\overset{\overset{\displaystyle CH_3}{|}}{CH}-C\equiv CH$

$H_3C-\overset{\overset{\displaystyle CH_3}{|}}{CH}-C\equiv C-CH_2-CH_3$

$H_3C-CH_2-\overset{\overset{\displaystyle CH_3}{|}}{CH}-CH_2-C\equiv CH$

$H_3C-\overset{\overset{\displaystyle CH_3}{|}}{CH}-CH_2-CH_2-C\equiv CH$

14.19 $H_2C=CH-CH_2-Br$

$\underset{\underset{\displaystyle H_3C}{}}{\overset{\overset{\displaystyle Br}{}}{C}}=\underset{\underset{\displaystyle H}{}}{\overset{\overset{\displaystyle H}{}}{C}}$ $\underset{\underset{\displaystyle H}{}}{\overset{\overset{\displaystyle Br}{}}{C}}=\underset{\underset{\displaystyle CH_3}{}}{\overset{\overset{\displaystyle H}{}}{C}}$

$\underset{\underset{\displaystyle H}{}}{\overset{\overset{\displaystyle Br}{}}{C}}=\underset{\underset{\displaystyle H}{}}{\overset{\overset{\displaystyle CH_3}{}}{C}}$ **14.21** (a) isomers, (b) isomers, (c) different chemical formula (not isomers)

14.23 (a)

(b) $(CH_3)_3CCH_2CHBrCH_2CHO$, (c) $C_8H_{12}O_2$

14.25 A functional group is a group of atoms bonded together in a specific way. This group of atoms behaves similarly to any other group of the same type, no matter what else is bonded to it.

14.27 (a) alcohol, (b) aldehyde, (c) carboxylic acid, (d) amine

14.29 (a) H_3C-OH, (b) $H_3C-O-CH_3$ H_3C-CH_2-OH

(c) $H_3C-CH_2-\overset{\overset{\displaystyle O}{||}}{C}-OH$ $\underset{\underset{\displaystyle H}{}}{\overset{\overset{\displaystyle H_3C}{}}{C}}=\underset{\underset{\displaystyle OH}{}}{\overset{\overset{\displaystyle OH}{}}{C}}$

(d) $H_3C-CH_2-CH_2-OH$ $H_3C-O-CH_2-CH_3$

14.31 amide and ketone **14.33** ketone, alcohol, amine

14.35 amine and amide

14.37 (a) $H_3C-CH_2-O-CH_2-CH_3$

(b) $H_3C-CH_2-O-CH_2-CH_2-CH_3$

(c) $H_3C-CH_2-O-CH_2-CH_2-CH_2-CH_2-CH_3$

14.39 (a) diethyl ether, (b) methyl propyl ether, (c) butyl methyl ether

14.41 (a) $H_3C-CH_2-CH_2-O-CH_2-CH_2-CH_3$

(b) $H_3C-O-CH_2-CH_2-CH_3$

(c) $H_3C-CH_2-O-CH_2-CH_2-CH_2-CH_2-CH_3$

14.43 (a) $H_3C-\overset{\overset{\displaystyle O}{||}}{C}-H$ (b) $H_3C-CH_2-CH_2-\overset{\overset{\displaystyle O}{||}}{C}-H$

(c) $H_3C-CH_2-CH_2-CH_2-\overset{\overset{\displaystyle O}{||}}{C}-H$

14.45 (a) $C_5H_{10}O$, (b) C_7H_6O, (c) $C_6H_{12}O$

14.47 (a) $H_3C-CH_2-CH_2-\overset{\overset{\displaystyle O}{||}}{C}-OH$

$H_3C-CH_2-\overset{\overset{\displaystyle O}{||}}{C}-O-CH_3$

(b) $H_3C-CH_2-CH_2-CH_2-\overset{\overset{\displaystyle O}{||}}{C}-OH$

$H_3C-CH_2-CH_2-\overset{\overset{\displaystyle O}{||}}{C}-O-CH_3$

(c) $H_3C-CH_2-CH_2-CH_2-CH_2-\overset{\overset{\displaystyle O}{||}}{C}-OH$

$H_3C-CH_2-CH_2-\overset{\overset{\displaystyle O}{||}}{C}-O-CH_2-CH_3$

14.49 (a) $H_3C-\overset{\overset{\displaystyle O}{||}}{C}-O-CH_2-CH_3$

(b) $H_3C-CH_2-\overset{\overset{\displaystyle O}{||}}{C}-O-CH_2-CH_3$

(c) $H_3C-CH_2-CH_2-\overset{\overset{\displaystyle O}{||}}{C}-O-CH_2-CH_3$

14.51 (a) $C_6H_{12}O_2$ or $CH_3CH_2CH_2CO_2CH_2CH_3$

(b) $C_6H_{12}O_2$ or $CH_3CH_2CH_2CH_2CO_2CH_3$

(c) $C_9H_{10}O_2$ or $C_6H_5CH_2CO_2CH_3$

14.53 (a) $H_3C-CH_2-CH_2-NH_2$

(b) $H_3C-CH_2-CH_2-CH_2-NH_2$

(c) $H_3C-CH_2-CH_2-CH_2-CH_2-NH_2$

14.55 (a) $C_4H_{11}N$ or $CH_3CH_2CH_2NHCH_3$
(b) $C_4H_{11}N$ or $CH_3CH_2CH_2CH_2NH_2$
(c) $C_8H_{17}N$ or $C_6H_{11}CH_2NHCH_3$
14.57 A *monomer* is a small molecule that can combine with itself or other small molecules to form a polymer.
A *polymer* is a very large molecule made up of many small molecules combined together.
A *copolymer* is a polymer that contains more than one type of monomer.
14.59 1.290×10^5 g/mol
14.61 (a) $CH_2{=}CH_2$

(b) 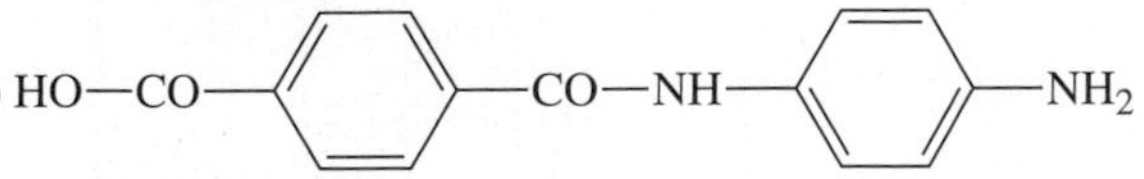 HO—CO—⬡—CO—NH—⬡—NH₂

Chapter 15

15.1 Soluble because it exhibits hydrogen bonding like water.
15.3 Amine and carboxylic acid groups. Alanine:

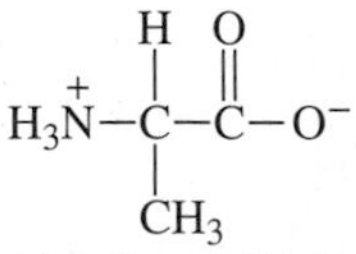

$H_3\overset{+}{N}$—$\overset{\overset{\displaystyle H}{|}}{\underset{\underset{\displaystyle CH_3}{|}}{C}}$—$\overset{\overset{\displaystyle O}{\|}}{C}$—$O^-$ **15.5** H_3C—O—$\overset{\overset{\displaystyle O}{\|}}{\underset{\underset{\displaystyle OH}{|}}{P}}$—OH

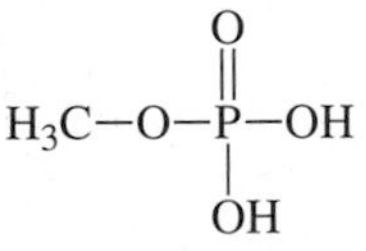

15.7 Generally, lipids are smaller organic molecules that are insoluble in water. Examples include fatty acids, fats, and steroids. **15.9** Fats, phospholipids, and steroids **15.11** Fats contain very long hydrocarbon chains which are hydrophobic and insoluble in water. **15.13** Saturated fats do not contain any carbon-carbon double bonds, whereas unsaturated fats contain one or more.

15.15

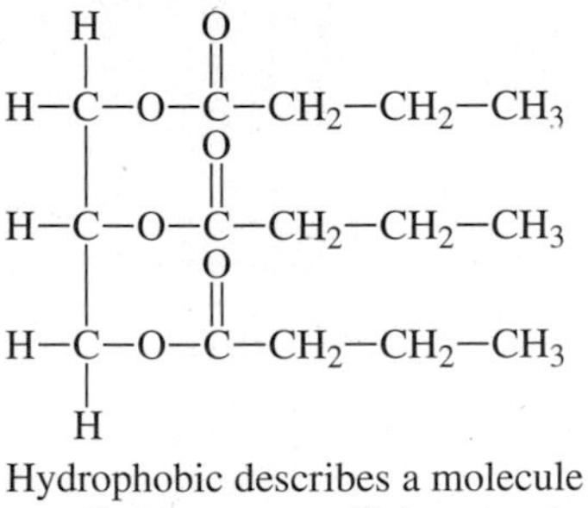

H—$\overset{\overset{\displaystyle H}{|}}{C}$—O—$\overset{\overset{\displaystyle O}{\|}}{C}$—$CH_2$—$CH_2$—$CH_3$

H—$\overset{\overset{\displaystyle }{|}}{C}$—O—$\overset{\overset{\displaystyle O}{\|}}{C}$—$CH_2$—$CH_2$—$CH_3$

H—$\overset{\overset{\displaystyle }{|}}{\underset{\underset{\displaystyle H}{|}}{C}}$—O—$\overset{\overset{\displaystyle O}{\|}}{C}$—$CH_2$—$CH_2$—$CH_3$

15.17 Hydrophobic describes a molecule or portion of a molecule that is repelled by water, which means it is not soluble in water. Hydrophilic describes a molecule or portion of a molecule that is attracted to water and will be soluble. **15.19** Molecules based on glycerol with one of the OH groups replaced by an organic phosphate group and two of the OH groups replaced by fatty acids. **15.21** Proteins are large biological polymers built from combinations of any of the amino acids. Essentially, amino acids are the building blocks of proteins. **15.23** a, b

15.25 $H_3C{-}\overset{\overset{\displaystyle OH}{|}}{\underset{\underset{\displaystyle H}{|}}{C}}{-}\overset{\overset{\displaystyle H}{|}}{\underset{\underset{\displaystyle NH_3^+}{|}}{C}}{-}\overset{\overset{\displaystyle O}{\|}}{C}{-}O^- + H{-}\overset{\overset{\displaystyle H}{|}}{\underset{\underset{\displaystyle H}{|}}{N^+}}{-}\overset{\overset{\displaystyle H}{|}}{\underset{\underset{\underset{\underset{H_3C \quad CH_3}{CH}}{CH_2}}{|}}{C}}{-}\overset{\overset{\displaystyle O}{\|}}{C}{-}O^- \longrightarrow$

$H_3C{-}\overset{\overset{\displaystyle OH}{|}}{\underset{\underset{\displaystyle H}{|}}{C}}{-}\overset{\overset{\displaystyle H}{|}}{\underset{\underset{\displaystyle NH_3^+}{|}}{C}}{-}\overset{\overset{\displaystyle O}{\|}}{C}{-}N{-}\overset{\overset{\displaystyle H}{|}}{\underset{\underset{\underset{\underset{H_3C \quad CH_3}{CH}}{CH_2}}{|}}{C}}{-}\overset{\overset{\displaystyle O}{\|}}{C}{-}O^-$

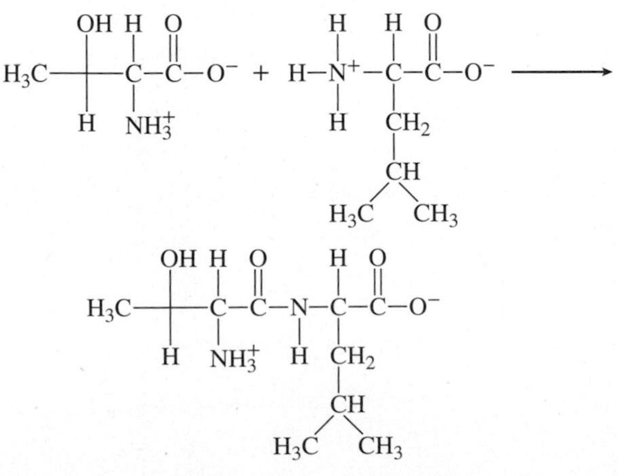

15.27 (a)

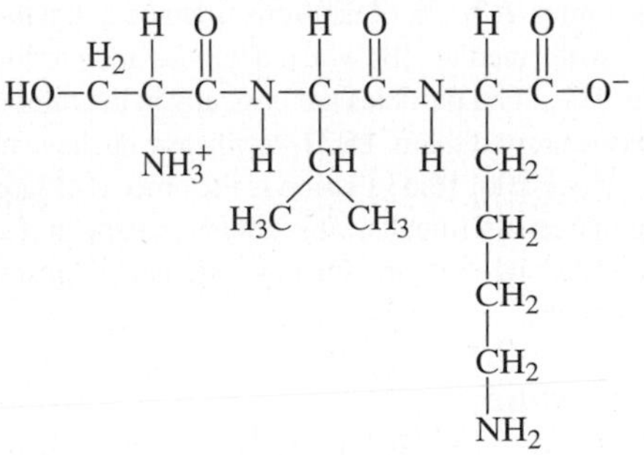

HO—$\overset{\overset{\displaystyle H_2}{|}}{C}$—$\overset{\overset{\displaystyle H}{|}}{\underset{\underset{\displaystyle NH_3^+}{|}}{C}}$—$\overset{\overset{\displaystyle O}{\|}}{C}$—N—$\overset{\overset{\displaystyle H}{|}}{\underset{\underset{H_3C \;\; CH_3}{\underset{\displaystyle CH}{|}}}{C}}$—$\overset{\overset{\displaystyle O}{\|}}{C}$—N—$\overset{\overset{\displaystyle H}{|}}{\underset{\underset{\underset{\underset{\underset{NH_2}{|}}{CH_2}}{CH_2}}{\underset{\displaystyle CH_2}{|}}}{C}}$—$\overset{\overset{\displaystyle O}{\|}}{C}$—$O^-$

(b)

HO—⬡—$\overset{\overset{\displaystyle H_2}{|}}{C}$—$\overset{\overset{\displaystyle H}{|}}{\underset{\underset{\displaystyle NH_3^+}{|}}{C}}$—$\overset{\overset{\displaystyle O}{\|}}{C}$—N—$\overset{\overset{\displaystyle H}{|}}{C}$—$\overset{\overset{\displaystyle O}{\|}}{C}$—N—$\overset{\overset{\displaystyle H}{|}}{\underset{\underset{\displaystyle CH_3}{|}}{C}}$—$\overset{\overset{\displaystyle O}{\|}}{C}$—$O^-$

(c)

$\underset{\underset{\displaystyle H}{\underset{|}{\overset{N{\diagdown}{\diagup}NH}{\underset{C}{}}}}}{HC}{=}\overset{\overset{\displaystyle H_2}{|}}{C}{-}\overset{\overset{\displaystyle H}{|}}{\underset{\underset{\displaystyle NH_3^+}{|}}{C}}{-}\overset{\overset{\displaystyle O}{\|}}{C}{-}\overset{\overset{\displaystyle H}{|}}{N}{-}\overset{\overset{\displaystyle H}{|}}{\underset{\underset{\underset{\displaystyle OH}{C=O}}{CH_2}}{C}}{-}\overset{\overset{\displaystyle O}{\|}}{C}{-}\overset{\overset{\displaystyle H}{|}}{N}{-}\overset{\overset{\displaystyle H}{|}}{\underset{\underset{\underset{\displaystyle OH}{C=O}}{CH_2}}{C}}{-}\overset{\overset{\displaystyle O}{\|}}{C}{-}\overset{\overset{\displaystyle H}{|}}{N}{-}\overset{\overset{\displaystyle H}{|}}{\underset{\underset{\underset{\underset{\displaystyle NH_2}{CH_2}}{CH_2}}{CH_2}}{C}}{-}\overset{\overset{\displaystyle O}{\|}}{C}{-}O^-$

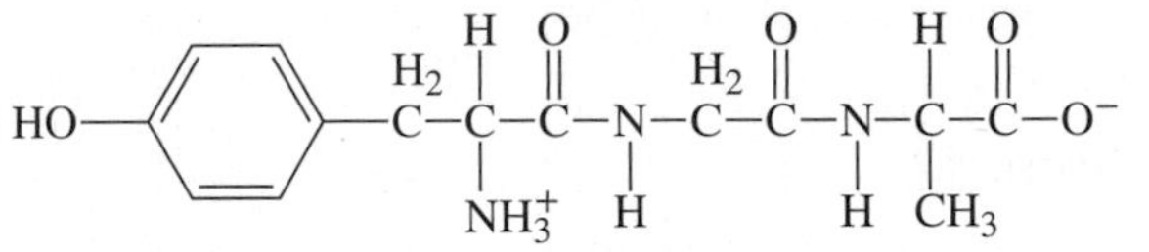

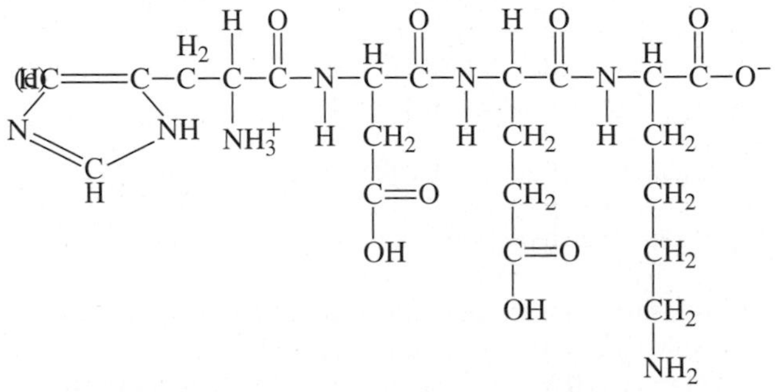

15.29 The shape of a protein chain that results from hydrogen bonding between nearby groups. Two examples include a helix and a pleated sheet. **15.31** The shape that results from the assembly of two or more folded protein chains. **15.33** $(CH_2O)_n$, fructose **15.35** Glycosidic linkage **15.37** Polysaccharides are polymers of sugars. Starch and cellulose are two examples. **15.39** DNA contains the sugar deoxyribose whereas RNA contains ribose. RNA is typically much smaller in size than DNA, though it is still a very large biopolymer.

Chapter 16

16.1 In nuclear reactions, elements are converted to other elements or isotopes, where subatomic particles, other than electrons, can be involved. Nuclear reactions are accompanied by the absorption or release of large amounts of energy. Nuclear reaction rates are not normally affected by temperature, pressure, or catalysts. **16.3** They both represent electrons, but the $_{-1}^{0}e$ comes from (or is in) an atomic orbital, and the $_{-1}^{0}\beta$ is from the decay of a neutron within the nucleus. **16.5** (a) $_{11}^{23}Na$, (b) $_{1}^{1}H$, (c) $_{0}^{1}n$, (d) $_{26}^{56}Fe$, (e) $_{-1}^{0}\beta$ **16.7** (a) $_{2}^{4}\alpha$ or $_{2}^{4}He$, (b) $_{-1}^{0}\beta$ or $_{-1}^{0}e$, (c) γ, (d) $_{+1}^{0}\beta$ or $_{+1}^{0}e$ **16.9** $_{30}^{64}Zn$ **16.11** (a) Pb, (b) Po, (c) Zr, (d) Bi **16.13** (a) 27p⁺ and 33 n°, (b) 95 p⁺ and 148 n°, (c) 12 p⁺ and 15 n° **16.15** (a) $_{82}^{214}Pb \longrightarrow {}_{-1}^{0}\beta + {}_{83}^{214}Bi$ (b) $_{90}^{234}Th \longrightarrow {}_{-1}^{0}\beta + {}_{91}^{234}Pa$ (c) $_{89}^{227}Ac \longrightarrow {}_{-1}^{0}\beta + {}_{90}^{227}Th$ **16.17** (a) $_{88}^{228}Ra$, $_{89}^{228}Ac$, (b) $_{90}^{231}Th$, $_{91}^{231}Pa$, (c) $_{93}^{237}Np$, $_{92}^{233}U$, $_{90}^{229}Th$ **16.19** (a) $_{1}^{3}H \longrightarrow {}_{-1}^{0}\beta + {}_{2}^{3}He$, (b) $_{94}^{242}Pu \longrightarrow {}_{2}^{4}\alpha + {}_{92}^{238}U$, (c) $_{53}^{131}I \longrightarrow {}_{-1}^{0}\beta + {}_{54}^{131}Xe$, (d) $_{98}^{251}Cf \longrightarrow {}_{2}^{4}\alpha + {}_{96}^{247}Cm$ **16.21** A rad is a radiation absorbed dose, a unit for describing the amount of radiation absorbed in units of 1×10^{-5} J/gram of material. Since the effect on an organism depends upon the type of radiation, the rad is multiplied by a factor called relative biological effectiveness to give the rem or roentgen equivalent for man. The rem is more meaningful in terms of biological impact. **16.23** Gamma (γ) rays penetrate the farthest and have the highest energy to cause damage. Beta particles (β) are the next most penetrating particle, and then alpha (α) particles. **16.25** 25% **16.27** 63.0 g Zn **16.29** $_{5}^{10}B$ is administered to the patient where it collects in the tumor. A beam of neutrons is aimed at the tumor, where $^{10}_{5}B$ is transformed to $^{11}_{5}B$, which degrades and produces an alpha particle. The alpha particles are basically in the tumor and destroy the tissue around them. **16.31** Any heavy nucleus, meaning a mass number above 200. **16.33** Fission is the process of large nuclei decomposing into smaller nuclei. This could not happen if small nuclei were the starting material. Fusion is the nuclear reaction between small nuclei.

Visualizing Chemistry
VC 16.1 c. **VC 16.2** b. **VC 16.3** a. **VC 16.4** a. The explanation of this is beyond the scope of this book.

Chapter 17

17.1 H_2S, S^{2-}, HS^- (−2) < S_8 (0) < SO_2 (+4) < H_2SO_4, SO_3 (+6)
17.3 (a) +1, (b) +7, (c) −4, (d) −1, (e) −2, (f) +6, (g) +6, (h) +7, (i) +4, (j) 0, (k) +5, (l) −1/2, (m) +5, (n) +3
17.5 (a) $2AlCl_3(s) \longrightarrow 2Al(s) + 3Cl_2(g)$
　　　　 +3 −1　　　 0　　　 0
The Al is gaining electrons and is therefore undergoing reduction. The Cl is losing electrons and is undergoing oxidation.
(b) $Zn(s) + S(s) \longrightarrow ZnS(s)$
　　 0　　 0　　　 +2 −2
The Zn is losing electrons and is therefore undergoing oxidation. The S is gaining electrons and undergoing reduction.
17.7 (a) OA = O_2, RA = Sr (b) OA = H_2, RA = Li (c) OA = Br_2, RA = Cs (d) OA = N_2, RA = Mg **17.9** (a) OA = O^- (in H_2O_2), RA = O^- (in H_2O_2) (b) OA = $AgNO_3$, RA = Mg (c) OA = NH_4NO_2, RA = NH_4NO_2 (d) OA = Br_2, RA = H_2 **17.11** (a) $2H^+(aq) + H_2O_2(aq) + 2Fe^{2+}(aq) \longrightarrow 2H_2O(l) + 2Fe^{3+}(aq)$
(b) $6H^+(aq) + 2HNO_3(aq) + 3Cu(s) \longrightarrow 4H_2O(l) + 3Cu^{2+}(aq) + 2NO(g)$ (c) $H_2O(l) + 2MnO_4^-(aq) + 3CN^-(aq) \longrightarrow 3CNO^-(aq) + 2MnO_2(s) + 2OH^-(aq)$ (d) $12OH^-(aq) + 6Br_2(aq) \longrightarrow 2BrO_3^-(aq) + 10Br^-(aq) + 6H_2O(l)$ (e) $2S_2O_3^{2-}(aq) + I_2(aq) \longrightarrow S_4O_6^{2-}(aq) + 2I^-(aq)$ **17.13** (a) $2S_2O_3^{2-}(aq) + Br_2(aq) \longrightarrow S_4O_6^{2-}(aq) + 2Br^-(aq)$ (b) $6Fe^{2+}(aq) + 14H^+(aq) + Cr_2O_7^{2-}(aq) \longrightarrow 6Fe^{3+}(aq) + 7H_2O(l) + 2Cr^{3+}(aq)$ **17.15** (a) $2OH^-(aq) + 4H_2O_2(aq) + Cl_2O_7(aq) \longrightarrow 2ClO_2^-(aq) + 4O_2(g) + 5H_2O(l)$ (b) $2Cr^{3+}(aq) + 3MnO_2(s) + 4OH^-(aq) \longrightarrow 2H_2O(l) + 2CrO_4^{2-}(aq) + 3Mn^{2+}(aq)$ **17.17** (a) Li, (b) K, (c) Al **17.19** (a) Al < Mg < K (b) Pb < Ni < Ca (c) Cr < Mn < K **17.21** (a) No reaction (b) $SnCl_2(aq) + Fe(s) \longrightarrow FeCl_2(aq) + Sn(s)$ (c) $Cu(s) + 2AgClO_4(aq) \longrightarrow 2Ag(s) + Cu(ClO_4)_2(aq)$ **17.23** (a) NR (b) NR (c) $Mg(s) + CuSO_4(aq) \longrightarrow Cu(s) + MgSO_4(aq)$, (d) NR **17.25** A = anode (where oxidation takes place); B = cathode (where reduction takes place); C = solution of metal ions that are forming from anode; D = solution of metal ions that are reduced at the cathode; E = salt bridge

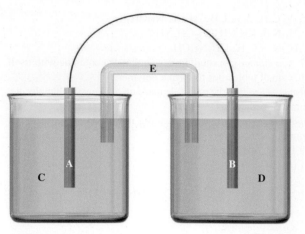

17.27 It keeps charge neutrality as ions are formed on one side of the cell and ions are reduced on the other. The strong electrolyte chosen cannot react with the solutions or electrodes on either side of the cell.

17.29

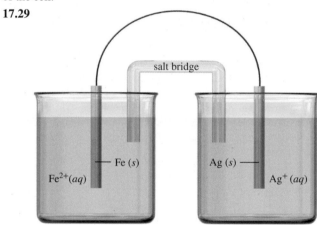

17.31 (a) $Al(s) | Al^{3+}(aq) \| Sn^{2+}(aq) | Sn(s)$
(b) $Zn(s) | Zn^{2+}(aq) \| Ag^+(aq) | Ag(s)$
(c) $Cr(s) | Cr^{3+}(aq) \| Cu^{2+}(aq) | Cu(s)$
17.33 Chromium is more reactive than iron (the major component of steel), so a coating will oxidize preferentially before iron will. **17.35** Gold would be the least reactive. **17.37** The process that occurs in a galvanic cell produces electrical current, whereas electrolysis requires the input of energy to cause reaction.

Visualizing Chemistry
VC 17.1 a. **VC 17.2** b. **VC 17.3** b. **VC 17.4** a.

Index

A

absolute temperature, 280, 285, 286, 288, 296, 411
absolute temperature scale, 130–131
absolute zero, 130, 296
acetaldehyde, 503–504
acetic acid, 370, 446, 453, 506
acetone, 492, 500
acetylene, 494, 495, 496
acetylsalicylic acid (aspirin), 402–403
acid(s), 107–109
　arrhenius, 429
　Brønsted, 429–434
　buffers and, 452–455
　commonly encountered, 444
　conjugate, 430–432
　defined, 429–432
　diprotic, 435
　familiar, 426, 444
　hydrochloric, 429
　identifying, 458
　key equations, 457
　monoprotic, 435
　neutralization, 408–409
　properties of, 427–428
　strong, 434–437
　volume of base to neutralize, 408–409
　weak, 445–449
acid-base indicator, 450
acid-base reactions, 383, 406
　gas-producing, 380–381, 383
　neutralization, 371–373, 383, 450–452
acid-base titrations, 450–452
acid reflux, 429
activation energy, 465–467
activity series, 553, 556
actual yield, 401–403
addition polymerization, 508–509
addition reactions, 496
ADH (alcohol dehydrogenase), 504
Aer-O-Foam, 504
air bags, 298, 411–412, 466
alanine, 523
alcohol, 500, 501–502
alcohol dehydrogenase (ADH), 504
aldehyde(s), 501, 503–505, 519
aldehyde dehydrogenase (ALDH), 504
ALDH (aldehyde dehydrogenase), 504
aliphatic compounds, 492
alkali metals, 14–15
alkaline batteries, 557–558
alkaline earth metals, 14–15
alkanes, 493, 496, 501
Alka-Seltzer, 404–405, 444

alkenes, 493–495
alkoxy group, 502
alkyl group, 501–503, 505–506, 517
alkynes, 493–495
allotropes, 95
alloys, 317
α (alpha) linkages, 527
alpha particles, 7, 533, 536
altitude
　and cooking, 260
　and hemoglobin production, 477
aluminum, 380
aluminum ion, 88, 89
amalgam, 380
amide group, 501, 506–507
amide linkages, 525
amines, 502, 506–507
amino acids, 517–518, 522–526
amino group, 502, 517
ammonia, 370, 371, 428, 440, 444
　amines from replacement in, 506–507
　ionization in water, 429, 430, 448–449
　production of, 284, 399–400, 477
　urea production from, 394–395, 491
　as weak base, 448–449
ammonium chloride, 101
ammonium ion, 429, 430
amorphous solid, 244
amphoteric species, 432
anemia, iron-deficiency, 21
aniline, 448
anions, 63, 83–91
　defined, 63, 83
　naming of, 89
　polyatomic, 101
　simple, 107
anisole, 503
anode, 553–557
Anson, George, 428
Antabuse, 504
anthrax, 209
aqueous reactions, 392, 406–410
　precipitation reactions in, 365–370
aqueous solubility, 319–320, 366–367
aqueous solutions, 316–322
　boiling point of, 341–342
　freezing point of, 340–341
　preparation of, 334–337
arginine, 523
argon-40, 540
aromatic compounds, 492
Arrhenius acids and bases, 429
ascorbic acid, 370, 409, 427
asparagine, 523
aspirin, 402–403
asymmetrical fats, 521

atmospheres (atm), 281–282
atmospheric pressure, 258–260, 280–282
atom(s), 5–10
　Bohr model of, 36–42
　central, 205, 206
　counting, by weighing, 169–170
　defined, 5
　electron configurations of, 48–57
　nuclear model of, 6–10
　properties of, and bonding, 109–111
　rearrangement in reactions, 355
　terminal, 204–205, 206, 207
atomic clocks, 539
atomic ion, 63. See also ion(s)
atomic (monatomic) ion, 86, 88
atomic mass, 19–22
atomic mass unit (amu), 19, 125, 172
atomic number, 10–13, 14, 18, 24, 25, 534–535
atomic orbitals, 42–57
atomic orbits, 37, 43
atomic size, 57–60
atomic solids, 246, 249
attraction, Coulombic, 91
Australia Center for Precision Optics (ACPO), 181
autoionization of water, 432–433
automobile air bags, 298, 411–412, 466
average atomic mass, 20, 22
Avogadro, Amedeo, 169, 298
Avogadro Project, 168, 181, 182
Avogadro's law, 298–300, 412
Avogadro's number, 169–170, 173–174, 175

B

baking, chemical reactions in, 355, 356
balanced chemical equations
　equilibrium expressions from, 474
　usage of, 392–425
balancing
　chemical equations, 360–365
　nuclear equations, 534–535
　oxidation-reduction reactions, 549–552
Ball, Alice, 321
ball-and-stick models, 94
Bardeen, John, 231
barometer, 283, 289
base(s) (chemical), 370. See also acid-base reactions; acid-base titrations
　Arrhenius, 429
　Brønsted, 429–434
　buffers and, 452–455

base(s) (chemical)—(*Cont.*)
　commonly encountered, 444
　conjugate, 430–432
　defined, 429–432
　familiar, 426
　identifying, 458
　key equations, 457
　properties of, 427–428
　strong, 435–437
　volume to neutralize acid, 408–409
　weak, 448–449
bases, organic, 519, 528
base units, 125–126
Bath, Patricia, 36
batteries, 553–560
　alkaline, 557–558
　dry cells, 557–558
　lead storage, 558–559
　lithium-ion, 559
bent shape, 211, 214–215
benzaldehyde, 505
Benzamide, 507
benzene, 492, 495
benzoic acid, 446
benzyl acetate, 506
beryllium, electronic configuration of, 49
β (beta) linkages, 527
beta particles, 533, 536
binary ionic compounds, 84–91
　defined, 85
　formulas for, 84–87
　naming of, 87–91
　type I, 85
　type II, 86–87, 89–90
binary molecular compounds, naming
　　of, 99–101
biochemistry, 516–531
　biologically important molecules in,
　　516–519
　carbohydrates in, 526–527
　lipids in, 520–522
　nucleic acids in, 528
　proteins in, 522–526
biologically important molecules, 516–519
bismuth-210, 538
bleaching, 209
blood alcohol level, 392
blood plasma
　osmotic pressure in, 343
　pH of, 453
body temperature, 131, 132–133, 464–465
Bohr, Niels, 36, 37
Bohr atom, 36–42
boiling point, 130–131, 258–260
　altitude and, 260
　energy and, 261–262
　of halogens, 229
　intermolecular forces and, 226–231,
　　258–260
　of noble gases, 230
　of water, 245, 259–260
boiling-point elevation, 341–342, 345
boiling-point-elevation constant, 341, 345

bond(s). *See also specific types*
　carbon, 491–492, 499–500
　covalent, 91–107, 201, 203, 207
　double, 207, 212
　electronegativity in, 219–220, 231
　hydrogen, 227–228, 244, 319–320
　ionic, 83–91, 221–222
　metallic, 246
　multiple, 207, 235
　polarity of, 221–223
　properties of atoms and, 109–111
　spectrum of, 219, 221
　triple, 207, 235
bond angles, 216–217
bond-line structures, 213, 499–500
bond pair, 203, 204
borax, 248
boron, electronic configuration of, 50
boron neutron capture therapy
　　(BNCT), 541
Bosch, Carl, 98, 284
Boyle, Robert, 293
Boyle's law, 293–295, 300
Brady, St. Elmo, 447
brain tumors, nuclear medicine for, 541
brass, 317, 560
breathalyzer tests, 392
Brønsted, Johannes, 429
Brønsted acids and bases, 429–434
buffers, 452–455
　capacity of, 455
　suitable components for, 454
butane, 360–361, 493, 494, 496–497
butyl acetate, 506
butyric acid, 362–363, 492, 506, 518

C

Cade, Mary, 332
Cade, Robert, 332
calcium chloride, 342
calcium phosphate, 101
calculators
　logarithmic functions on, 438–439
　scientific notation function on, 137–138
calorie (cal), 261, 269
cancer
　nuclear medicine for, 541
　radiation as cause of, 537
　radiation as therapy for, 534, 541
carbohydrates, 364, 526–527
carbon, 491–492
　allotropes of, 95
　atomic radius of, 491
　electron configuration of, 491
carbon-14, 538–540
carbon bonds, 491–492, 499–500
carbon dioxide
　as fire extinguisher, 251, 292
　solid form of, 251
carbonic acid, 380, 444
carbonyl group, 502, 503–505
carboxy group, 502, 510, 517

carboxyl group, 446
carboxylic acids, 446, 502, 505–506, 517
Carver, George Washington, 99
carvone, 505
cathode, 553–557
cations, 63, 83–91
　defined, 63, 83
　naming of, 88
　polyatomic, 101
cell electromotive force, 556
cell emf, 556
cell potential (E_{cell}), 556
cellulose, 527
cell voltage, 556
Celsius scale, 130–133, 159
centimeters, 126
central atom, 203, 204, 205, 206,
　　212–217
Cepheid variable, 127
cereal, iron-fortified, 21
Chadwick, James, 544
chalcogens, 14
chameleons, 516
Charles, Jacques, 296
Charles's law, 295–298, 300
chemical bonding. *See* bond(s)
chemical change, 80
chemical equations, 331
　balanced (*see* balanced chemical
　　equations)
　balancing, 360–365
　mass to mass conversions in, 395–397
　mole to mole conversions, 393–395
　representing chemical reactions with,
　　358–359
chemical equilibrium
　addition or removal of substance and,
　　478–480
　concept of, 470–472
　equilibrium constants in, 472–476
　equilibrium expressions in, 472–476
　temperature changes and, 481–483
　volume changes and, 480–481
chemical formula, 83, 359
　calculating mass percent composition
　　from, 182–183
　determining, from names, 105
　determining molar mass using, 175–177
　for ionic compounds, 83–87, 105
chemical process, 80
chemical property, 80, 81
chemical reaction(s)
　aqueous, 392, 406–410
　driving force of, 382
　energy and, 382
　fire as, 354
　gases in, 411–414
　and heat, 415–416
　limitations on yield in, 398–405, 418–419
　mass to mass conversions, 395–397
　mole to mole conversions, 393–395
　nuclear reactions *vs.*, 534
　percent yield of, 401–403, 417

rate of, 465–469
recognizing, 355–358
representing with equations, 358–359
types of, 365–382
chemical statement, 358
chemical symbol, 11–12, 25
chemical warfare, 284
chemistry
defined, 3
organic, 490–515
reasons for studying, 3
success in introductory class, 157–158
"chemist's dozen," 169–170
chloric acid, 435
chloride ion, 89, 406–407
chlorine, 92
chlorine dioxide, 209
chlorine ions, 64–65, 83
chlorine isotopes, 17
chloroethylene, 509
cholesterol, 497–498, 522
chromatography, 79
chromium(III) ion, 88, 89
chromium(II) ion, 88
cinnamaldehyde, 492, 505
citric acid, 444
cloves, 213
coefficients, 360
colligative properties, 340–344
collision theory, 465–469
color, as qualitative property, 80
combination reaction, 377, 379, 383
combined gas equation, 289–290, 307
combustion, 360–363, 378–379, 383, 495, 559
combustion analysis, 403–404
complete ionic equations, 368, 369–370, 372–373
compound(s), 77. See also specific types
acids as, 107–109
aliphatic, 492
aromatic, 492
elements vs., 112–113
gaseous, 278–279
inorganic, 491
ionic, 83–91, 113
mass-mole conversions of, 177–181
molar mass of, 175–176
molecular, 91–101, 113–114
naming of, 87–91, 99–101, 113–114, 116–117
organic, 491, 492
with polyatomic ions, 101–107
on product labels, 102–104
properties of atoms and, 109–111
solubility of, 319–320
type I, 85
type II, 86–87, 90–91
computer, woman employed as, 127
concentration, of reactants, 465, 470
concentration, of solutions, 322–330, 345
comparison of, 327–329
molal (molality), 327–329, 340–341

molar (molarity), 325–329, 334, 346–347, 406
condensation, 257–258, 259, 264–267
condensation reaction, 510
condensed phases, 243. See also liquids; solids
conjugate acid, 430–432
conjugate acid-base pair (conjugate pair), 430–432
conjugate base, 430–432
conservation of mass, law of, 356, 360, 378
constant temperature, 263
conversion factor, 148–151
defined, 148
dimensional analysis using, 155–157, 160–161
mole to mole conversions with, 393–395
copolymer, 510
copper, 317
copper corrosion, 560–562
copper electrodes, 553–557
copper sulfate, 106
core electrons, 55
Cori, Gerty, 231
corrosion, 80, 560–562
corticosteroids, 504
cortisol, 522
Coulombic attraction, 91
counters, radiation, 536
covalent bonding, 91–107
in ionic species, 101–107
in Lewis structures, 201, 203, 207
in molecules, 91–101
multiple, 207
in network solids, 246, 247–249
polar, 221
covalent solids, 247–249
crystalline solids, 244–250
cubic boron nitride, 248
cubic centimeter, 152
curie (Ci), 534
Curie, Marie Skłodowska, 10
cyclic organic compounds, 492

D

Dalton, John, 5, 93, 355
Dalton's law of partial pressures, 301–306
Daly, Marie Maynard, 518
Daniell cell, 553
dating, radioactive, 538–540
Davy, Humphry, 414
de Broglie, Louis, 43
decane, 494
decomposition reaction, 377, 378, 383
decontamination, 209
degree
Celsius, 130–132
Fahrenheit, 130–132

density, 80, 152–153, 159
calculation from mass, 153
of gases, 276, 279, 292
of solution, 328
dental amalgam, 380
dental pain, 380
deoxyribonucleic acid (DNA), 528
deoxyribose, 528
deposition, 265–267
derived units, 152–154
deuterium, 17, 544
diamond, 5, 95, 247–248, 359, 465
diatomic molecules, 94
diethyl ether, 503
dilute solutions, 322
dilution, 335–339
dimensional analysis, 155–157, 160–161
dimer, 510
dimethyl amine, 448
dimethylamine, 507
dimethyl ether, 503
dinitrogen tetroxide, 470–472
diphenhydramine, 507
dipole, 222–225
induced, 229
instantaneous, 229
dipole-dipole forces, 226–227, 232, 244
dipole vector, 222–224
disaccharides, 526–527
disinfection, 209
dispersion forces, 228–232, 244–246
dissociation, 330
distillation, 79
disulfiram, 504
DNA (deoxyribonucleic acid), 528
d orbitals, 43–47
double bond, 207, 212
double-displacement reaction, 366, 380
Douglas, Dwayne, 332
Downs cell, 562–563
driving force, 382
drug labels, 126
drug synthesis, 504
dry cells, 557–558
dry ice, 252
dynamic equilibrium, 257, 472

E

Earth
atmospheric pressure of, 280–281
crust of, 15–16
Einstein, Albert, 36
electric charge, 6
electrochemistry, 548–566
balancing equations in, 549–552
batteries in, 553–560
corrosion in, 560–562
electrolysis in, 558–559, 562–563
electrodes, 553–557
electrolysis, 558–559, 562–563
of molten sodium chloride, 562, 563
of water, 562–563

electrolytes, 330–333
electrolytic cell, 562
electromagnetic radiation, 33, 34
electromagnetic spectrum, 34
electron affinity, 62
electron capture, 540
electron configurations, 48–57
    of carbon, 491
    Hund's rule and, 49
    of ions, 63–65, 83
    Lewis dot symbols for, 56, 65–66
    Pauli exclusion principle and, 49
    periodic table and, 53–57, 68–69
    valence electrons in, 55–56, 68–69
electron density map, 221
electronegativity, 219–221, 231
electron gain
    in ionic bonding, 83–91, 221–223
    ions from, 63–66
    periodic trends in, 62
electron group, 212–218
electron-group geometry, 212–218
electron loss
    in ionic bonding, 83–91, 221–223
    ions from, 63–66
    periodic trends in, 59–61
electron(s), 6–9
    Bohr model and, 36–42
    bond pair of, 203, 204
    core, 55
    excited state, 37, 39
    ground state, 37, 68–69
    lone pair of, 203–205, 212
    number of, 18
    octet rule exceptions in, 208
    octet rule for, 201–202, 207, 208
    orbitals of, 42–48
    orbits of, 37, 43
    resonance structures and, 209–211
    sharing of (covalent bonding), 91–98,
        201–203
    spin of, 48–49
    transfer of (ionic bonding), 83–91
    valence, 55–56, 68–69, 201–202, 205
    wavelike properties of, 43
element(s), 5
    atomic number of, 10–13
    chemical symbol of, 12–13
    compounds vs., 112–113
    in Earth's crust, 15–16
    in human body, 12
    Lewis dot symbols for, 56
    periodic table of, 10–16
emission spectrum, 34
    and Bohr atom, 38–39
    and fireworks, 40
empirical formula, 96–97, 359
    calculating mass percent composition
        from, 182–183
    determining from mass percent
        composition, 185–188
    determining molecular formula from,
        188–190, 191

endothermic process, 263–264, 382, 415
energy
    and chemical reactions, 382, 465–469
    defined, 261
    and physical changes, 261–267
    SI unit of, 261
"energy crisis," 143–144
Energy Star (energy guide), 144
energy sublevel, 43
equilibrium, 464–489
    addition or removal of substance and,
        478–480
    concept of, 470–472
    defined, 470
    dynamic, 257, 472
    temperature changes and, 481–483
    volume changes and, 480–481
equilibrium constants, 472–476
    calculation of, 473–475
    magnitude of, 476
equilibrium expressions, 472–476
equivalence point, 450
erythrocytes, 343
ester(s), 505–506
ester group, 502
estimate, 140
estrogen, 522
ethane, 493, 494, 496
ethanol, 492, 500
ethene, 493, 496
Ethenzamide, 507
ethers, 501–503
ethylamine, 448, 500
ethylene, 493, 496, 508
ethylene glycol, viscosity of, 255
ethyne, 494, 495
eugenol, 213
evaporation, 257–258, 259, 264–267
exact number, 139, 140
excess reactant, 398–400
excited state, 37–39
exothermic process, 264, 382, 415
expanded octet, 208, 217–218

F

Fahrenheit, Daniel Gabriel, 131
Fahrenheit temperature scale, 130,
    131, 159
familiar acids, 426, 444
fat(s), 520–521
    asymmetrical, 521
    saturated, 493, 497–498, 517, 518, 520
    trans, 497–498
    unsaturated, 493, 497–498, 517,
        518, 520
fatty acids, 517, 520
fertilizers, 477
    fixed nitrogen in, 98, 284
    mass percent composition of, 187–188
    radioactivity from, 538
filtration, 79
fire, as chemical reaction, 354

fire extinguishers, 251, 292
fireworks, 2, 40
fission, nuclear, 541–544
fixed nitrogen, 98, 284
flavor
    acids and bases in, 426
    molecular shape and, 213
fluids, 278
fluorine, electronic configuration of, 50
food irradiation, 532, 533
f orbitals, 43–47
formaldehyde, 503
formic acid, 446, 506
formula(s). See chemical formula;
        empirical formula; molecular formula
formula mass, 175
formula units, 175–181
fossil fuels, 559
Franklin, Rosalind, 529
free radicals, 537
freezing, 263–264, 266–267
freezing point, 130–131
freezing-point depression, 340–341, 345
freezing-point-depression constant,
    340, 345
French Revolution, 397
frequency, 33
fructose, 526–527
fuel cells, 559–560
functional groups, 500–507
fusion (melting), 263–264, 266–267, 269
fusion, nuclear, 541–544

G

galactose, 527
gallium, 246–247, 264
galvanic cells, 553–557
galvanization, 562
gamma radiation, 533, 536
gas(es), 77, 243–244, 276–315
    in chemical reactions, 411–414
    compression of, 243, 278–279
    density of, 276, 279, 292
    kinetic molecular theory of, 279–280
    natural, 304
    noble, 14–15, 359
    partial pressures of, 301–306, 307
    phase changes of, 264–267
    pressure of, 280–285
    properties of, 277–280
    room temperature, 278
    as solvents, 317, 318
    standard state of, 288
    volume of, 243, 285–300
gaseous product, volume of, 411–412
gaseous reactant, required volume of,
    412–413
gaseous substances, 278–279
gas equations, 285–292
    combined, 289–290, 307
    ideal, 285–288, 307
    molar mass, 290–292, 307

gas laws, 293–300
    Avogadro's law, 298–300, 412
    Boyle's law, 293–295, 300
    Charles's law, 295–298, 300
gas mixtures, 279, 301–306
    mole fractions in, 302–305, 307, 308–309
    pressure in, 301–302
gas-producing reactions, 380–381, 383
gastric juice, 429, 439, 444
Gatorade, 330, 332, 409
Gay-Lussac, Joseph, 296
Gay-Lussac, Joseph Louis, 414
Geiger counter, 536
genetic radiation damage, 537
geochemistry, radioactive dating in, 538–540
geometrical isomers, 497
Germain, Sophie, 414
glass, 244–245
globular protein, 526
glucose, 404, 519, 526–527
glutamic acid, 523
glutamine, 523
glycerol, 255, 517, 520–521
glycine, 523
glycosidic linkage, 526
goiter, 184
gradations, 140–141
gram (g), 126
graphene, 490
graphite, 5, 95, 247, 359, 465
graphite cathode, 558
Greek prefixes, as metric multipliers, 127–130
ground state, 37, 68–69
groups, on periodic table, 14, 25
Guldberg, Cato, 472

H

Haber, Clara Immerwahr, 284
Haber, Fritz, 98, 284
Haber process, 98, 284, 477
Hahn, Otto, 544
half-cell, 553
half-life (t$_{1/2}$), 538–540
half-reaction method, 549–552
halogens, 14–15
    boiling points of, 229
    dispersion forces of, 229
    hydrocarbon reactions with, 496
    molar masses of, 229–230
heat of reaction, 415–416
Heliox, 317
helium, 14
    electron configuration of, 49
    emission spectrum of, 35
helix, 525
heme, 525–526
hemoglobin, 21, 477, 525–526
heptane, 494
heteroatoms, 500

heterogeneous mixture, 78–79
heteronuclear diatomic molecules, 94
hexanal, 505
hexane, 494
hexoses, 526
histamine, 507
histidine, 523
homogeneous mixture, 78, 279, 317–318. See also solutions
homonuclear diatomic molecules, 94
honey
    as supersaturated solutio, 318
    viscosity of, 255
hot packs, 319
Hubble, Edwin, 127, 231
Hund's rule, 49
hydrates, 106
hydrobromic acid, 435
hydrocarbons, 492–499
    bond-line structures of, 499–500
    isomers of, 496–500
    reactions of, 495–496
    saturated, 493, 496
    unsaturated, 495, 496, 497
hydrochloric acid, 370, 371, 380, 429, 434–435
hydrofluoric acid, 446
hydrogen
    emission spectrum of, 35, 37
    isotopes of, 17
hydrogenation, 496, 498
hydrogen atoms, ionizable, 427
hydrogen balloon, 414
hydrogen bonding, 227–228, 244, 319–320
hydrogen chloride, 107–108
hydrogen fluoride, 107–108
hydrogen-oxygen fuel cell, 559–560
hydrogen peroxide, mass percent composition of, 182–183
hydroiodic acid, 435
hydrolysis, 526–527
hydronium ions, 432–434
    in pH determination, 437–445, 457
    in strong acids and bases, 434–437
    in weak acids and bases, 446–449
hydrophilic species, 521
hydrophobic species, 521
hydroxide(s), 435
hydroxide ions, 370, 432–434, 442–444, 457
hydroxy group, 501, 502, 517
hypochlorous acid, 108
hypothesis, 4

I

ice, 243, 245–246, 263
ice melters, 342
ice storms, 242
ideal gas constant (R), 285
ideal gas equation, 285–288, 307, 411

indicator, acid-base, 450
induced dipole, 229
inorganic compounds, 491
instantaneous dipole, 229
instant hot packs, 319
intermediate products, 380
intermolecular forces, 226–234
    and boiling points, 226–231, 258–260
    dipole-dipole, 226–227, 232, 244
    dispersion, 228–232
    hydrogen bonding, 227–228
    and melting points, 252–254
    and solubility, 270–271
    in solutions, 319–320
International Bureau of Weights and Measures, 181
International prototype kilogram (IPK), 181
International System of Units (SI units), 125–130
International Units (IUs), 154
iodine-123, 540
iodine-131, 540
iodine, solid form of, 244–245, 252
iodine deficiency, 184
iodine solutions, 410
iodized salt, 184
ion(s), 63–66
    atomic (monatomic), 86, 88
    electrolytes and, 330–333
    electron configuration of, 63–65, 83
    isoelectronic with noble gases, 64, 201
    Lewis dot symbols for, 65–66, 83
    mass-mole conversions of, 180–181
    naming of, 87–91
    polyatomic, 86, 101–107, 203–205, 211–219, 374–376
    shapes of, 211–219
    spectator, 368–369, 371
ionic bonding, 83–91, 221–223
ionic compounds, 83–91
    binary, 84–91
    formulas for, 83–87, 105
    molar mass of, 192–193
    molecular compounds vs., 113
    naming of, 87–91, 104–105, 113–114, 117–118
    with polyatomic ions, 101–107
    solubility of, 319–321, 366
    type I, 85
    type II, 86–87
ionic equations, 368–370, 386–387
    for acid-base reactions, 372–373
ionic solids, 244, 249
ionizable hydrogen atoms, 427
ionization energy, 59
ionizing radiation, 537
iron, rusting of, 80, 355, 356, 560–561
iron-fortified cereal, 21
irradiated food, 532, 533
isobutane, 497
isoelectronic species, 64, 201, 491

isoeugenol, 213
isoleucine, 523
isomers, 496–500
isopentane, 499
isotopes, 17–19
IUs (International Units), 154

J

Jenner, Edward, 4
Jones, Amanda Theodosia, 298–299
Jones method, 298
joule (J), 261
Julian, Percy Lavon, 504

K

kelvin (K), 125, 130, 280
Kelvin scale, 130–133
ketones, 503–505, 519
kilocalorie (Cal), 261
kilogram (kg), 125–126, 168, 181
kilojoule (kJ), 261
kilometer (km), 126
Kimball, Oliver P., 184
kinetic energy, 257–258
kinetic molecular theory, 279–280
Kool-Aid®, 316

L

Lake Natron, 445
laser pointers, 35
laughing gas, 396
Lavoisier, Antoine, 378
Lavoisier, Marie-Anne Paulze, 397
Lavoisier's law, 397
law, 4
law of conservation of mass, 356, 360, 378
law of constant composition, 93
law of definite proportions, 93
law of multiple proportions, 93–94
lead-206, 539–540
lead-210, 538–540
lead ions, 407
lead poisoning, 538
lead storage batteries, 558–559
Leavitt, Henrietta Swan, 127
Le Châtelier's principle, 477
length, measurement of, 125–126
leucine, 524
Lewis dot symbols, 56, 65–66, 83
Lewis structures, 92, 201–208
    drawing, 201–206
    with less obvious skeletal
        structures, 206
    of molecules with central atom, 203
    with multiple bonds, 207
    of simple molecules, 201–202
    of simple polyatomic ions, 203–205
    two or more possible (resonance),
        209–211

light
    nature of, 33–36
    photons of, 36, 43
    ultraviolet, 34
    visible, 33–35
limiting reactant, 398–400, 418–419
Lind, James, 428
linear shape, 211, 215
line spectra, 35
linoleic acid, 363, 518
lipid bilayer, 521
lipids, 243–244, 520–522
lipoproteins, 497–498
liquids, 77, 243–244
    boiling point of, 258–260
    nonvolatile, 257
    phase changes of, 263–267
    physical properties of, 255–261
    as solvents, 317–318
    surface tension of, 255–257, 260
    vapor pressure of, 257–258, 259
    viscosity of, 255
    volatile, 257–258, 259
    volume of, 243
lithium, electronic configuration of, 49
lithium carbonate, 104
lithium-ion batteries, 559
logarithmic functions, on scientific
    calculator, 438–439
lone pair, 203–205, 212
low-density lipoproteins, 497–498
lysine, 524

M

Mackay, Helen, 21
magnesium ion, 88, 89
main-group elements, 14–15, 25, 65
    electron configurations for, 65
    Lewis dot symbols for, 65–66
malic acid, 444
manganese dioxide cathode, 558
manometer, 283
Marine, David, 184
Mars Climate Orbiter, 151
mass, 3, 125
    atomic, 19–22
    calculating density from, 153
    conservation of, 356, 360, 378
    formula, 175
    measurement of, 125–126
    molar (*See* molar mass)
    molecular, 175–181
    percent by, 322–325
    of product, determination of, 399–400
    reactant to product conversion,
        395–397
    solution calculations, 327–329
    weight *vs.,* 126
mass number, 8, 17–19
mass percent composition, 182–188
    calculation of, 182–183, 191
    defined, 182

determining empirical formula from,
        185–188
    of fertilizer, 187–188
mass spectrometry, 18, 540
matter, 3
    classification of, 77–82
    properties of, 80–82
    states of, 77 (*See also* gas(es);
        liquids; solids)
measured numbers, 139–147
    calculations with, 144–147
    derived units of, 152–154
    dimensional analysis in, 155–157,
        160–161
    gradations in, 140–141
    Greek prefixes for, 126–130
    International Units, 154
    metric multipliers of, 126–130, 148–151
    SI units of, 125–130
measurement, 125–133
    base units of, 125–126
    conversion of units, 130–133, 148–157
    pressure, 283
medical applications, of radioactivity,
        533, 540–541
Meitner, Lise, 544
meitnerium, 544
melting, 263–264, 266–267, 269
melting point, 80, 226–227
    of metals, 246
    and solids, 252–254
mercaptan, 304
mercury, as liquid, 246
metabolism, 364
metal(s), 14–15, 24, 59–61
    corrosion of, 561
    ionic bonding with nonmetals, 83
    melting points of, 246
    representation in equations, 359
metal cations, naming of, 89–90
metallic bonding, 246
metallic character, 59–61, 206
metallic crystalline solids, 246, 249
metalloids, 14–15, 24, 359
metal solutions (alloys), 317
meter (m), 125, 127
methane, 493, 494, 496, 559
methanol, 500
methionine, 524
methylamine, 448, 507
methyl group, 501
methyl salicylate, 506
metric multipliers, 126–130, 148–150
milk of magnesia, 372–373, 440
millicurie (mCi), 536
milligrams, 126
milliliter, 152
millimole, 335, 452
mixtures, 76, 77, 78–79
    gas, 279, 301–306
    heterogeneous, 78–79
    homogeneous, 78, 279, 317–318
    (*See also* solutions)

pure substances *vs.*, 81
separation of, 78–79
Mocetinostat, 507
model, 4
molality (m) (molal concentration),
    327–329, 334, 340–341, 345
molar heat of fusion, 263–264, 269
molar heat of vaporization, 265, 269
molarity (M) (molar concentration),
    325–329, 345, 346–347, 406
molar mass, 171–173
    calculating mass percent composition
        from, 182–183
    calculation of, 172, 175–176, 192–193
    defined, 172
    determining molecular formula from,
        188–190, 191
    determining using chemical formula,
        175–177
    and dispersion forces, 228–232
    of halogens, 229
    mass-mole conversions with, 172–174,
        177–181, 191, 192–193
    multiple conversions in single
        calculation, 179–181
    of noble gases, 230
molar mass gas equation, 290–292, 307
mole (mol), 169–170
    conversions with mass, 172–174,
        177–181, 191, 192–193
    ideal gas equation to calculate, 287
    multiple conversions in single
        calculation, 179–181
    reactant to product conversion,
        393–395
    solution calculations, 325–329, 345,
        346–347, 406
    volume and (Avogadro's law),
        298–300, 412
molecular art, 94–95
molecular compounds, 83–101
    acids as, 107–109
    binary, 99–101
    gaseous, 278–279
    ionic compounds *vs.*, 113
    naming of, 99–105, 113–114, 117–118
molecular equations, 366–368,
        369–370
molecular formula, 95–97, 100,
        188–190, 191
molecular mass, 175–181
molecular solids, 244–246, 249
molecular substances, 92
molecules, 91–98
    biologically important, 516–519
    counting, by weighing, 175–181
    defined, 92
    diatomic, 94
    forces between, 226–234
    gases as, 277–280
    Lewis structures of, 201–206
    models (art) of, 94–96
    polarity of, 223–225, 236–237

polyatomic, 94
    shape of, 200, 211–219, 236–237
mole fractions, 302–305, 307, 308–309
monatomic (atomic) ion, 86, 88
monomers, 508–510
monosaccharides, 526, 527
Muller, Derek, 182
multiple bond, 207
muriatic acid, 444
muscone, 505

N

naming. *See* nomenclature
nanometers, 34, 127, 128
nanoseconds, 126
naphthalene, solid form of, 251
Natron (lake), 445
natural gas, 304
NBE (nuclear binding energy), 542–543
neon, electronic configuration of, 51
neopentane, 499
net ionic equations, 368–370, 372–373,
        386–387
network solids, 247–249
neutralization reaction, 371–373, 383,
        408–409, 450–452
neutron(s), 8–9
    with protons, in mass number, 8, 17–19
    varying number of, 17–19
newton (N), 281
nitric acid, 435
nitric oxide, 208
nitride ion, 89
nitrogen
    electron configuration of, 50
    as solvent, 317
nitrogen dioxide, 393–394, 470–472
nitrogen fixation, 98, 284
nitrous acid, 446
nitrous oxide, 396
noble gas(es), 14–15, 359
    boiling points of, 230
    dispersion forces of, 230
    ions isoelectronic with, 64, 201
    molar masses of, 230
noble gas core, 51, 53–55
nomenclature, 87–88
    for acids, 107–109
    for ionic compounds, 87–91, 104–105,
        113–114, 117–118
    for ions and binary ionic compounds,
        87–91, 104–105, 113–114
    for molecular compounds, 99–105,
        113–114, 117–118
    for polyatomic ions, 104–105
    for product labels, 102–104
nonane, 494
nonbonding pair, 203, 204
nonelectrolytes, 330–331
nonmetal(s), 14–15, 24, 59–61
    ionic bonding with metals, 83
    representation in equations, 359

nonmetallic crystalline solids, 246, 249
nonvolatile liquids, 257
nuclear binding energy (NBE), 542–543
nuclear chemistry, 532–547. *See also*
        specific processes and reactions
nuclear equations, 534–535
nuclear fission, 541–544
nuclear fusion, 541–544
nuclear medicine, 540–541
nuclear model of atom, 6–10
nuclear reactions, chemical
        reactions *vs.*, 534
nuclear transmutation, 534
nucleic acids, 528
nucleons, 8
nucleotides, 528
nucleus, atomic, 7
numbers in chemistry. *See also specific*
        calculations and processes
    conversion of units, 129–130,
        132–133, 148–157
    scientific notation, 134–139
    significant figures, 139–148
    units of measurement, 127–133
nutmeg, 213
Nylon 6,6, 510

O

octane, 494
octet, expanded, 208, 217–218
octet rule, 201–202, 207, 208
    exceptions to, 208, 217
    resonance structures and, 209–211
oleic acid, 518, 520
orbital diagrams, 48–57
orbital(s), 42–57
orbit(s), 37, 43
organic bases, 519, 528
organic chemistry, 490–515
    defined, 491
    functional groups in, 500–507
    hydrocarbons in, 492–499
    polymers in, 508–510
    properties of carbon in, 491–492
organic chemistry,
    bond-line structures in, 499–500
organic compounds, 491, 492
organic phosphates, 519
osmosis, 343
osmotic pressure, 342–343
oxidation, 80, 374
oxidation numbers, 374–376
oxidation-reduction reactions, 373–381
    in batteries, 553–560
    in corrosion, 560–562
    in dental pain, 380
    half-reaction method for balancing,
        549–552
    oxidation numbers in, 374–376
    types of, 377–380, 383
oxidation state, 374
oxide ion, 89

oxidizing agent, 553
oxoacid, 108
oxoanions, 104, 107–108
oxygen, molar mass of, 176
oxygen generators, 371

P

palmitic acid, 518
paper chromatography, 79
partially hydrogenated vegetable oils, 497–498
partial pressures, 301–306, 307
particle emission, 533
parts per billion (ppb), 323–324
parts per million (ppm), 323–324
pascal (Pa), 281
patina, 561–562
Pauli exclusion principle, 49
Pauling, Linus, 231–232
pentane, 494, 495, 498, 499
pentene, 498
pentoses, 526
peptide bonds, 525
percent by mass (weight), 322–325, 345
percent composition, 182. *See also* mass percent composition
percent yield, 401–403, 417
perchloric acid, 108, 435
periodic table, 10–16
    atomic number on, 10–13, 14, 24–25
    chemical symbols on, 11–13, 25
    electron configurations and, 53–57, 68–69
    electronegativity values and, 219–220
    organization of, 14–16, 24–25
    trends on, 57–63
periods, on periodic table, 14, 24–25
pH, 437–445
    benchmark values of, 439
    of blood plasma, 453
    buffers and, 452–455
    defined, 437
    indicator of, 450
    key equations, 457
    in stomach, 440
    values of common fluids, 440
    of weak-acid solution, 447–448
phase changes, 261, 263–267
phenol, 492
phenolphthalein, 450
phenylalanine, 524
phlogiston theory, 378
pH meter, 439
phosphates, 519, 528, 538
phospholipids, 521
phosphoric acid, 395, 400, 444, 519
phosphorus pentachloride, 401–402
photoelectric effect, 41
photons, 36, 43
pH scale, 437–445
physical change, 80

physical process, 80
physical property, 80–81
physostigmine, 504
Planck, Max, 36, 544
plant sources, for drugs, 504
plasma
    osmotic pressure in, 343
    pH of, 453
pleated sheet of protein, 525
plum-pudding model, 7
pOH scale, 442–444, 457
polar covalent bond, 221
polarity, 221–225
    of bonds, 221–223
    of molecules, 223–225, 236–237
polonium-210, 534
polonium-218, 538
polyacrylonitrile, 509
polyatomic ions, 86, 101–107
    Lewis structures of, 203–205
    naming of, 104–105
    oxidation numbers in, 374–376
    shapes of, 211–219
polyatomic molecules, 94
polybutadiene, 509
polyethylene, 508–509
polymer(s), 508–510
polymerization, addition, 508–509
polysaccharides, 527
polystyrene, 509
polytetrafluoroethylene, 508–509
polyvinyl chloride (PVC), 508–509
p orbitals, 43–45
positron, 534
potassium-40, 540
potassium chloride, 32
potassium hydroxide, 400
potassium iodide, 184
potassium ion, 88, 89
potato clocks, 548
precipitate, 366
precipitation reactions, 365–370, 383, 406
pressure
    atmospheric, 258–260, 280–282
    defined, 281
    exerted by column of fluid, 289
    gas, 280–285
    measurement of, 283
    units of, 280–282
    volume and (Boyle's law), 293–295, 300
pressure cookers, 260
primary amine, 507
primary structure of protein, 525
principal energy level, 43
principal quantum number, 43–44
prism, 33, 34
probability density map, 43
product labels, 102–104
product of reaction, 358–359
    energy (heat) required for, 415–416
    gaseous, predicting volume of, 411–412

mass, determining, 399–400
    mass to mass conversion to, 395–397
    mole to mole conversions to, 393–395
progesterone, 522
proline, 524
propane, 493, 494, 495
propanone, 500
propanone (acetone), 492, 500
protein(s), 522–526
    functions of, 522
    primary structure of, 525
    quaternary structure of, 526
    secondary structure of, 525
    tertiary structure of, 525
proton(s), 8–9
    with neutrons, in mass number, 8, 17–19
    number of (atomic number), 10–13, 18–19, 24–25, 534–535
proton acceptors, 429
proton donors, 429
Proust, Joseph, 93
psi, 280–282
PVC (polyvinyl chloride), 508–509
pyridine, 448, 519

Q

Qingsongite, 248
QM (quantum mechanical) model, 42–57
qualitative properties, 80
quantitative properties, 80, 125
quantization, 36
quantum mechanics, 42–57
quantum number, 37, 43–44
quartz, 244, 245, 247
quaternary structure of protein, 526

R

rad (radiation absorbed dose), 536
radiating light, 35
radiation, 533–535
    average yearly doses for Americans, 536–537
    biological effects of, 536–538
    detection of, 536
    food treated with, 532, 533
    ionizing, 537
    units of, 536
radiation therapy, 533, 540
radicals, 537
radioactive decay, 533–535
    dating using, 538–540
    uranium series, 539–540
radioactive elements, 534
radioactive scans, 540–541
radioactive substances, 7
radioactivity
    defined, 533
    medical applications of, 533, 540–541
    in tobacco, 538
radium-226, 538
radon-222, 538

rate of reaction, 465–469
RBE (relative biological effectiveness), 536
reactant(s), 358–359
  aqueous, 406–410
  collision of, 465–469
  concentration of, 465, 470
  consumed, determining amount of, 413
  excess, 398–400
  gaseous, required volume of, 412–413
  limiting, 398–400, 418–419
  mass to mass conversions of, 395–397
  mole to mole conversion's from, 393–395
reactions. *See specific types*
red blood cells, 343
redox reactions. *See* oxidation-reduction reactions
reducing agent, 553
reduction, 374. *See also* oxidation-reduction reactions
reflecting light, 35
relative biological effectiveness (RBE), 536
rem (roentgen equivalent for man), 536–537
resonance structures, 209–211
reversible process, 470–472
ribonucleic acid (RNA), 528
ribose, 528
RNA (ribonucleic acid), 528
Robinson, Carol V., 250
roentgen equivalent for man (rem), 536–537
Rosenfeld, Arthur, 143–144
rounding, 138, 145–146
rounding error, 146, 158
rusting, 80, 355, 356, 466, 560–561
Rutherford, Ernest, 7–8

S

salt, 371. *See also* sodium chloride
  as electrolyte, 330–333
  iodized, 184
salt bridge, 553–557
saturated fats, 493, 497–498, 517, 518, 520
saturated hydrocarbons, 493, 496
saturated solutions, 318, 319
Sciencium (YouTube channel), 182
scientific calculator
  logarithmic functions on, 438–439
  scientific notation function on, 137–138
scientific method, 3–4
scientific notation, 134–139
  using calculator function for, 137–138
  for very large numbers, 135–136
  for very small numbers, 136–137
scurvy, 428
seawater, 76, 317
second (s), 125–126
secondary amine, 507

secondary structure of protein, 525
semipermeable membrane, 342–343
sequoias, height of, 124, 125–126
serial dilution, 338–339
serine, 524
shell (principal energy level), 43
shifting, in equilibrium, 477–483
significant figures, 139–148
  defined, 140
  determining number of, 142–143
  expressing calculated results with, 146–147
silicon, 378
silicon sphere, 168, 181
silver
  as solvent, 317
  tarnishing of, 560
silver chloride, 406–407
silver nitrate, 406–407
simple anion, 107
single-displacement reaction, 377, 378, 383
SI units, 125–130
skeletal structure, 202
smallpox vaccine, 4
smoking, radioactivity in, 538
snow, 242
snowflakes, 200
sodium-24, 540
sodium, electron configuration of, 51, 83
sodium acetate, 453–454
sodium azide, 411–412, 466
sodium bicarbonate, 444
sodium borate, 248
sodium chlorate, 371
sodium chloride, 32
  formation of, 83–84, 92, 111
  as ionic compound, 113
  molten, electrolysis of, 562, 563
  in seawater, 317
  solution in water, 330–333
sodium hydroxide, 370, 428, 429
sodium iodide, 184
sodium ions, 63–66, 83
sodium peroxide, 413
solids, 77, 243–255
  amorphous, 244
  atomic, 246, 249
  crystalline, 244–250
  dispersion forces in, 244–246
  fixed shape and volume of, 243
  formation in precipitation, 365–370
  ionic, 244–246, 249
  melting point of, 252–254
  molecular, 244–246, 249
  network (covalent), 247–249
  phase change of, 263–267
  physical properties of, 251–255
  preparation of solution from, 334–337
  as solvent, 317, 318
  types of, 244–250
  vapor pressure of, 251–252

solubility, 318
  aqueous, 319–320, 366–367
  intermolecular forces and, 270–271
solutes, 317
  determining amounts of, 324–325, 408–409
solutions, 316–353
  boiling-point elevation in, 341–342, 345
  colligative properties of, 340–344
  composition of, 317, 330–333
  concentrated, 322
  concentration of, 322–330, 345
  defined, 317
  density of, 328
  dilute, 322
  dilution of, 335–339
  electrolyte *vs.* nonelectrolyte, 330–333
  examples of, 317, 318
  freezing-point depression in, 340–341, 345
  general properties of, 317–319
  molality of, 327–329, 334, 340–341, 345
  molarity of, 325–329, 345, 346–347
  osmotic pressure of, 342–343
  percent by mass, 322–325
  preparation of, 334–339
  saturated, 318, 319
  stock, 335–339
  supersaturated, 318–319
  unsaturated, 318
solvent, 317–318
  universal, water as, 317, 432
somatic radiation damage, 537
s orbitals, 43–45
space-filling models, 94
spark density map, 43
specific gravity, 153
specific heat, 262, 269
specific heat capacity, 262, 269
spectator ions, 368–369, 371
spin, 48–49
sports drinks, 330, 332, 409
standard state of gas, 288
standard temperature and pressure (STP), 288
starch, 527
states of matter, 77. *See also* gas(es); liquids; solids
steady states, 464
stearic acid, 518, 520
stearin, 520
steel, 562
steroids, 504, 522
sterol, 522
stock solution, 335–339
stoichiometric amount, 394
stoichiometric coefficients, 360, 393
stoichiometry, 393
  of acid-base reactions, 371, 373
  of aqueous reactions, 406–410
  of mass to mass conversions, 395–397
  of metabolism, 364
  of mole to mole conversions, 393–395

stomach, pH balance in, 440
STP (standard temperature and pressure), 288
strong acid, 434–437
strong base, 435–437
strong electrolytes, 330
structural isomers, 497
subatomic particles, 6–10, 533–535
sublimation, 265–267
subshell (energy sublevel), 43
substance, 77, 81
substituted alkanes, 496
substituted ethylenes, 508
substitution reactions, 496
success, in chemistry class, 157–158
sucrase, 527
sucrose, 526, 527
sugars, 519, 526–527, 528
sulfuric acid, 435
sunlight, 33–35
supersaturated solutions, 318–319
surface tension, 255–257, 260
sweet tea, 473
Szilárd, Leó, 544

T

tartaric acid, 444
taste
    acid and bases in, 426
    molecular shape and, 213
technetium, 540–541
Teflon, 508–509
temperature
    absolute, 130, 132, 280, 285, 286, 288, 296, 411
    body, 131, 132, 464–465
    changes, and equilibrium, 481–483
    changes, energy and, 261–263
    gas, 285–292, 295–298
    measurement (scales) of, 125, 130–133, 159
    and reaction rate, 465
    and solubility, 318
    standard, 288
    volume and (Charles's law), 295–298, 300
terminal atoms, 203–205, 206
tertiary amine, 507
tertiary structure of protein, 525
testosterone, 522
tetrafluoroethylene, 509
tetrahedral shape, 211, 212–215
tetrahydrofuran, 500
Thenard, Louis Jacques, 414
theoretical yield, 401–403
theory, 4
thermometers, 131–132
thermonuclear reactions, 544
Thomson, J. J., 6–7
threonine, 524
thyroid gland, radioactive scans of, 540
time, measurement of, 125–126

titration, acid-base, 450–452
tobacco, radioactivity in, 538
torr, 281, 282, 283
trace concentrations, 323–324
trans fats, 497–498
transition elements, 14–15
transition metals, cations of, 85–86
tree heights, 124, 125–126
triglycerides, 520
trigonal planar shape, 211, 214–215
trigonal pyramidal shape, 211, 213–215
triolein, 520
triple bond, 207, 212
tristearin, 520
tritium, 17
tryptophan, 524
type I compound, 85
type II compound, 86–87, 89–90
tyrosine, 524

U

ultraviolet light, 34
unit conversion, 131–133, 148–157
units of measurement, 125–133
units of radiation, 536
universal solvent, water as, 317, 432
unsaturated fats, 493, 497–498, 517, 518, 520
unsaturated hydrocarbons, 495, 496, 497
unsaturated solutions, 318
uranium-235, 541–544
uranium-238, 538, 539–540
uranium, isotopes of, 17
uranium decay series, 539–540
urea, 394–395, 491

V

vacuum canning, 298
valence electrons, 55–56, 68–69, 201–205
valence-shell electron-pair repulsion (VSEPR) model, 212
valine, 524
vaporization, 257–258, 259, 264–267, 269
vapor pressure
    of liquids, 257–258, 259
    of solids, 251–252
Vertasium (YouTube channel), 182
2Vertasium (YouTube channel), 182
viscosity, 255
visible light, 33–35
vitamin C, 409
vitamin C (ascorbic acid), 370, 427
vitamin D_3, 522
vitamin supplements, 154
volatile liquids, 257–258, 259
voltaic (galvanic) cell, 553–557
voltmeter, 554–556
volume, 152–153
    of aqueous reactants, 406–410
    changes, and equilibrium, 480–481

of gas, 243, 285–300
of gaseous product, 411–412
of gaseous reactant, 412–413
of liquid, 243
moles and (Avogadro's law), 298–300, 412
pressure and (Boyle's law), 293–295, 300
of solid, 243
of solution, 325–329
temperature and (Charles's law), 295–298, 300

W

Waage, Peter, 472
water
    as acid or base, 432–434
    autoionization of, 432–433
    boiling point of, 245, 259–260
    as compound, 77, 113, 378
    electrolysis of, 562–563
    energy and temperature changes in, 261–263
    molecular mass of, 175
    as nonelectrolyte, 331
    osmosis of, 342–343
    physical states of, 77, 243, 245–246, 263–265, 278
    reactants in, 406–410
    solubility in, 319–321
    solution preparation in, 334–337
    surface tension of, 255–257
    as universal solvent, 317, 432
    vapor pressure of, 304
    viscosity of, 255
water drops, 255–257
wavelength, 33–35, 127–128
weak acids, 445–449
weak bases, 448–449
weak electrolytes, 330
weight, mass vs., 126
weighted average, 19
white light, 34, 35
Wöhler, Friedrich, 491

Y

yield
    actual, 401–403
    limitations on, 398–405, 418–419
    percent, 401–403, 417
    theoretical, 401–403
YouTube science channels, 182

Z

zero, absolute, 130, 296
zinc
    galvanization with, 562
    oxidation of, 373–374
zinc electrodes, 553–557